MW00844256

The IMA Volumes in Mathematics and its Applications

Volume 161

Institute for Mathematics and its Applications (IMA)

The Institute for Mathematics and its Applications (IMA) was established in 1982 as a result of a National Science Foundation competition. The mission of the IMA is to connect scientists, engineers, and mathematicians in order to address scientific and technological challenges in a collaborative, engaging environment, developing transformative, new mathematics and exploring its applications, while training the next generation of researchers and educators. To this end the IMA organizes a wide variety of programs, ranging from short intense workshops in areas of exceptional interest and opportunity to extensive thematic programs lasting nine months. The IMA Volumes are used to disseminate results of these programs to the broader scientific community.

The full list of IMA books can be found at the Web site of the Institute for Mathematics and its Applications:

http://www.ima.umn.edu/springer/volumes.html.

Presentation materials from the IMA talks are available at

http://www.ima.umn.edu/talks/.

Video library is at

http://www.ima.umn.edu/videos/.

Fadil Santosa, Director of the IMA

More information about this series at http://www.springer.com/series/811

Eric Carlen • Mokshay Madiman
Elisabeth M. Werner
Editors

Convexity and Concentration

Editors
Eric Carlen
Department of Mathematics
Rutgers University
Piscataway, NJ, USA

Mokshay Madiman
Department of Mathematical Sciences
University of Delaware
Newark, DE, USA

Elisabeth M. Werner
Department of Mathematics
Case Western Reserve University
Cleveland, OH, USA

ISSN 0940-6573 ISSN 2198-3224 (electronic)
The IMA Volumes in Mathematics and its Applications
ISBN 978-1-4939-7004-9 ISBN 978-1-4939-7005-6 (eBook)
DOI 10.1007/978-1-4939-7005-6

Library of Congress Control Number: 2016961208

Mathematics Subject Classification (2010): 46, 52, 58, 60

Printed on acid-free paper

This Springer imprint is published by Springer Nature
The registered company is Springer Science+Business Media LLC
The registered company address is: 233 Spring Street, New York, NY 10013, U.S.A.

Foreword

This volume is based on the research focus at the IMA during the Spring semester of 2015. The Annual Thematic Program covering this period was "Discrete Structures: Analysis and Applications." The program was organized by Sergey Bobkov, Jerrold Griggs, Penny Haxell, Michel Ledoux, Benny Sudakov, and Prasad Tetali. Many of the topics presented in this volume were discussed in the last five workshops that took place during the year. We thank the organizers of the workshops, the speakers, workshop participants, and visitors to the IMA who contributed to the scientific life at the institute and to the successful program. In particular, we thank volume editors Eric Carlen, Mokshay Madiman, and especially Elisabeth Werner, who also served as associate director of the IMA during the year. Finally, we are grateful to the National Science Foundation for its support of the IMA.

Minneapolis, MN, USA Fadil Santosa

Preface

The 2014–2015 Annual Thematic Program at the Institute for Mathematics and its Applications (IMA) was *Discrete Structures: Analysis and Applications*. The program was organized by Sergey Bobkov (University of Minnesota), Jerrold Griggs (University of South Carolina), Penny Haxell (University of Waterloo), Michel Ledoux (Paul Sabatier University of Toulouse), Benny Sudakov (University of California, Los Angeles), and Prasad Tetali (Georgia Institute of Technology).

Convexity and concentration phenomena were the focus during the spring semester of 2015, and this volume presents some of the research topics discussed during this period. We have particularly encouraged authors to write surveys of research problems, thus making state of the art results more conveniently and widely available. The volume addresses the themes of five workshops held during the spring semester of 2015:

- *Convexity and Optimization: Theory and Applications*, held February 23–27, 2015, at IMA and organized by Nina Balcan (Carnegie-Mellon University), Henrik Christensen (Georgia Institute of Technology), William Cook (University of Waterloo), Satoru Iwata (University of Tokyo), and Prasad Tetali (Georgia Institute of Technology)
- *The Power of Randomness in Computation*, held March 16–20, 2015, at Georgia Institute of Technology and organized by Dana Randall (Georgia Institute of Technology), Prasad Tetali (Georgia Institute of Technology), Santosh Vempala (Georgia Institute of Technology), and Eric Vigoda (Korea Advanced Institute of Science and Technology (KAIST))
- *Information Theory and Concentration Phenomena*, held April 13–17, 2015, at IMA and organized by Sergey Bobkov (University of Minnesota, Twin Cities), Michel Ledoux (Université de Toulouse III (Paul Sabatier)), and Joel Tropp (California Institute of Technology)
- *Analytic Tools in Probability and Applications*, held April 27–May 01, 2015, at IMA and organized by Sergey Bobkov (University of Minnesota, Twin Cities), Sergei Kislyakov (Russian Academy of Sciences), Michel Ledoux (Université de Toulouse III (Paul Sabatier)), and Andrei Zaitsev (Russian Academy of Sciences)

- *Graphical Models, Statistical Inference, and Algorithms (GRAMSIA)*, held May 18–22, 2015, at IMA and organized by David Gamarnik (Massachusetts Institute of Technology), Andrea Montanari (Stanford University), Devavrat Shah (Massachusetts Institute of Technology), Prasad Tetali (Georgia Institute of Technology), Rüdiger Urbanke (École Polytechnique Fédérale de Lausanne (EPFL)), and Martin Wainwright (University of California, Berkeley)

Discrete Structures: Analysis and Applications attracted intense interest from the mathematical science community at large. Each of the five workshops drew up to or more than 100 visitors. Aside from the workshops, an annual thematic year at the IMA provided an ideal environment for collaborative work. This program drew a mix of experts and junior researchers in various aspects of convex geometry and probability together with numerous people who apply these areas to other fields. This volume reflects many aspects of the semester, with chapters drawn from workshop talks, annual program seminars, and the research interests of many visitors.

The volume is organized into two parts. While the classification is of course arbitrary to some extent given the fluid boundaries between probability and analysis, **Part I: Probability and Concentration** contains those contributions that focus primarily on problems motivated by probability theory, while **Part II: Convexity and Concentration for Sets and Functions** contains those contributions that focus primarily on problems motivated by convex geometry and geometric analysis.

Acknowledgments No single volume could possibly cover all the active and important areas of research in convexity, probability, and related fields that were presented at the IMA, and we make no claim of comprehensiveness. However, we think that this volume presents a reasonable selection of interesting areas, written by leading experts who have surveyed the current state of knowledge and posed conjectures and open questions to stimulate further research. We thank the authors for their generous donation of time and expertise. Needless to say that without them, this volume would not have been possible.

We thank A. Beveridge, J. R. Griggs, L. Hogben, G. Musiker, and P. Tetali, the editors of the 2014–2015 fall semester volume *Recent Trends in Combinatorics*, for their advice and their many helpful comments.

We thank the IMA and their staff for wonderfully stimulating and productive long-term visits. We believe that the IMA is a critical national resource for mathematics. The *Discrete Structures: Analysis and Applications* program will have a lasting impact on research in convexity, probability, and related fields, and we hope this volume will enhance that impact. We are grateful for the opportunity to be part of it.

Piscataway, NJ, USA — Eric Carlen
Newark, DE, USA — Mokshay Madiman
Cleveland, OH, USA — Elisabeth M. Werner

Contents

Part I
Probability and Concentration

Interpolation of Probability Measures on Graphs

Erwan Hillion

Abstract These notes are a review of the author's works about interpolation of probability measures on graphs via optimal transportation methods. We give more detailed proofs and constructions in the particular case of an interpolation between two finitely supported probability measures on $\mathbb{Z}$, with a stochastic domination assumption. We also present other types of interpolations, in particular Léonard's entropic interpolations and discuss the relationships between these constructions.

1 Introduction

The main topic of these notes is the theory of optimal transportation on discrete metric spaces, and in particular on graphs. We recall in this introduction some basic facts about this theory. For additional information, the reader is referred to Villani's comprehensive textbooks [Vill03] and [Vill09] or to shorter lectures notes, for instance by [AG13] or [Sant15].

Let (X, d) be a metric space endowed with its Borel σ-algebra. We consider two probability measures μ_0, μ_1 and a parameter $p \geq 1$. The optimal transportation theory is the study of the Monge-Kantorovitch minimization problem

$$\inf_{\pi \in \Pi(\mu_0, \mu_1)} \mathcal{I}_p(\pi) := \inf_{\pi \in \Pi(\mu_0, \mu_1)} \int_{X \times X} d(x, y)^p d\pi(x, y), \tag{1}$$

where $\Pi(\mu_0, \mu_1)$ is the set of couplings between μ_0 and μ_1, i.e. the set of probability measures π on $X \times X$ with marginals μ_0 and μ_1.

Under mild assumptions which are always satisfied in these notes (it suffices, for instance, to assume that (X, d) is Polish, see [Vill09], Theorem 4.1.), the set of optimal couplings, i.e. the set of minimizers for the functional $\mathcal{I}_p(\pi)$, is non-empty. Moreover, the application W_p defined by

E. Hillion (✉)
Aix Marseille Univ, CNRS, Centrale Marseille, I2M, Marseille, France
e-mail: erwan.hillion@univ-amu.fr

E. Carlen et al. (eds.), *Convexity and Concentration*, The IMA Volumes in Mathematics and its Applications 161, DOI 10.1007/978-1-4939-7005-6_1

$$W_p(\mu_0, \mu_1) := \inf_{\pi \in \Pi(\mu_0, \mu_1)} \left(\int_{X \times X} d(x, y)^p d\pi(x, y) \right)^{1/p} \quad (2)$$

defines a distance on the set $\mathcal{P}_p(X)$ of probability measures on X having a finite p-th moment. Such distances are called Wasserstein distances.

We are interested in properties of the metric space $(\mathcal{P}_p(X), W_p)$. Recall that the length of a continuous curve $\gamma : [0, 1] \to X$ in a metric space (X, d) is given by

$$L(\gamma) := \sup_{0=t_0 \leq \cdots \leq t_N = 1} \sum_{i=0}^{N-1} d(\gamma(t_i), \gamma(t_{i+1})).$$

This induces a new distance $\tilde{d}$ on X given by

$$\tilde{d}(x, y) := \inf\{L(\gamma) \mid \gamma(0) = x, \gamma(1) = y\}. \quad (3)$$

If the distances d and $\tilde{d}$ coincide, the space (X, d) is said to be a length space. If furthermore the infimum is attained in (3) for some curve γ, then (X, d) is said to be a geodesic space and γ a geodesic curve. Compact length spaces are proven to be geodesic spaces (see [Stu06a], Lemma 2.3). Compact Riemannian manifolds and Euclidean spaces $\mathbb{R}^n$ are other important classes of geodesic spaces.

An important geometric property of Wasserstein spaces is the following :

Proposition 1 *If (X, d) is a geodesic space, then $(\mathcal{P}_p(X), W_p)$ is also a geodesic space.*

The study of Wasserstein W_p-geodesics, in particular for $p = 2$, has gained importance in the last decade, mainly because of the development of Sturm-Lott-Villani theory (see [Stu06a], [Stu06b] and [LV09]), which gives quite unexpected links between some geometric properties of a compact Riemannian manifold (M, g) and convexity properties of some functionals defined on $(\mathcal{P}_2(M), W_2)$. For instance, the Ricci curvature tensor Ric on M satisfies $\mathrm{Ric} \geq K$ if and only if every couple of measures $\mu_0, \mu_1 \in \mathcal{P}_2(M)$ can be joined by a W_2-geodesic $(\mu_t)_{t\in[0,1]}$ along which we have

$$H(\mu_t) \geq (1-t)H(\mu_0) + tH(\mu_1) + K\frac{t(1-t)}{2} W_2^2(\mu_0, \mu_1), \quad (4)$$

where the entropy functional $H(\mu)$ is defined by $H(\mu) := -\int_M \rho \log(\rho) dvol$ if $d\mu = \rho dvol$ is absolutely continuous w.r.t. the Riemannian volume measure, and $H(\mu) = -\infty$ elsewhere. (In the original papers by Sturm and Lott-Villani, $H(\mu)$ is defined as $+\int_M \rho \log(\rho) dvol$ and referred to as 'the Boltzmann functional', and equation (4) is thus stated with a different sign.)

Equation (4) is called K-geodesic concavity for the entropy on $\mathcal{P}_2(M)$. The purpose of Sturm-Lott-Villani theory is to use equation (4) to define the notion of a measured length space (i.e. a length space (X, d) with a reference measure ν), satisfying $\mathrm{Ric} \geq K$. This generalized notion of Ricci curvature bounds is consistent with the classical one in the Riemannian setting, and under the assumption $\mathrm{Ric} \geq K > 0$, one can recover geometric properties and functional inequalities holding on (X, d, ν).

The generalization of Sturm-Lott-Villani theory in the case where the underlying space (X, d) is discrete (and thus not a length space) has been the subject of several research works. Among them, papers by Erbar-Maas [EM12] and Léonard [Leo14] will be presented in these notes. These approaches are based on the same idea which can be loosely summed up as follows: given a couple of probability measures f_0, f_1 on a graph G, we first construct an interpolation $(f_t)_{t\in[0,1]}$ which shares some similarities with Wasserstein W_2-geodesics in length spaces. It is then possible to define a notion of 'Ricci curvature bounds' by considering the behaviour of the entropy functional along such interpolations.

A similar approach of discrete Ricci curvature is described in the paper [GRST14] by Gozlan et al. In this article, the authors study the behaviour of the entropy functional along particular interpolations in the space of probability measures on a graph. The notion of discrete Ricci curvature thus obtained is strong enough to imply interesting functional inequalities (for instance, a discrete HWI inequality, see Proposition 5.1. of [GRST14]). The interpolating curves constructed in [GRST14] are seen as mixtures of binomial families, which is also the case of the $W_{1,+}$-geodesics introduced in these notes in Section 4. Whether both interpolating families coincide or not is still an open question, for which there is a positive answer in particular cases (see Remark 2).

Other important works about discrete Ricci curvature which will not be discussed here are Ollivier's Ricci curvature, see [Oll09], Sturm-Bonciocat rough curvature bounds, see [SB09], and the recent Bochner-type approach by Klartag et al, see [KKRT15].

The main purpose of these notes is to present some of the results of the author's papers [Hill14a], [Hill14b], [Hill14c]. These articles are about the construction of an interpolating family $(f_t(x))_{t\in[0,1]}$ between two finitely supported probability measures f_0, f_1 on a graph G, and the study of the concavity properties of the entropy functional $H(t) := H(f_t)$ along this family. In these notes, we mainly focus on the simpler case when the underlying graph is $\mathbb{Z}$ and explain briefly how these constructions can be extended to the general case.

Section 2 is a study of a particular class of interpolations, known as thinning of measures. We define the thinning and give some of its properties, among them a result about the concavity of its entropy. We then give an overview of the paper [Hill14a], about the contraction of probability measures, which is a natural generalization of the thinning in the setting of graphs. In particular, we explain how to adapt the proof of the concavity of the entropy in this new framework.

In Section 3, we recall some more notions on optimal transportation theory in continuous spaces. As in the discrete case, we mainly focus on the one-dimensional setting. We recall the Benamou-Brenier formula and how the description of its solutions by the Hamilton-Jacobi equation can be used to obtain interesting properties about Wasserstein geodesics. In particular, we obtain a concavity of entropy result by applying Proposition 8, which is formally similar to the discrete Benamou-Brenier equation (11) used in Section 2. The methods used in this section, especially Proposition 8, will be extended to the discrete setting in the next section.

In Section 4, we explain how to construct interpolating family between two probability measures on a graph satisfying a generalized version of equation (11). This leads to the definition of $W_{1,+}$-geodesics (see Definition 13). In particular, the similarity between equation (30) and Proposition 8 show that $W_{1,+}$-geodesics share similarities with thinning (or contractions) of measures in the discrete setting and with Wasserstein geodesics in continuous spaces. This section is an overview of the papers [Hill14b] and [Hill14c].

In Section 5, we construct other types of interpolating curves in $\mathcal{P}(\mathbb{Z})$ along which concavity of entropy results hold. These curves, known as entropic interpolations, have been introduced by Léonard in a series of articles among which we can cite [Leo12], [Leo13a], [Leo13b]. We explain with heuristic arguments why $W_{1,+}$-geodesics can be seen as limits of entropic interpolation when a certain parameter is taken to 0.

In Section 6, we explain the proof of the Shepp-Olkin conjecture (see Theorem 7), which is based on the ideas introduced in the theory of $W_{1,+}$-geodesics. The section sums up the results detailed in the papers [HJ14] and [HJ16].

2 A First Example: Entropy and Thinning of Measures

In this section, we study a particular method of interpolation known as the thinning operation, and which is a natural way to interpolate a probability measure finitely supported on $\mathbb{Z}_+$ and the Dirac measure δ_0. Moreover, the entropy along the thinning of a measure is concave. We give a detailed proof of this result, which will serve as a template for other concavity of entropy results.

2.1 The Thinning Operation on $\mathbb{Z}_+$

Let f be a probability measure supported on $\{0, \ldots, N\}$ and X be a random variable distributed as f.

Definition 1 *The thinning of f is the family $(T_t f)_{t\in[0,1]}$ of probability measures on $\{0, \ldots, N\}$ defined by*

$$\forall k \in \{0, \ldots, N\}, \ (T_t f)(k) = \sum_{l\,:\,l\geq k} \mathrm{bin}_{l,t}(k) f(l), \tag{5}$$

where $\mathrm{bin}_{l,t}(k) := \binom{l}{k} t^k (1-t)^{l-k} 1_{k\in\{0,\ldots,l\}}$ *is the binomial measure.*

The map $t \mapsto T_t f$ can be seen as a curve in the space $\mathcal{P}(\mathbb{Z})$, interpolating between the Dirac measure $T_0 f = \delta_0$ and $T_1 f = f$. An equivalent point of view on the thinning is the following:

Proposition 2 *Let X be a random variable with probability mass function f. Let $(B_k)_{k\geq 1}$ be an i.i.d. family of Bernoulli variables of parameter t, independent of X. Then the random variable $T_t X := \sum_{i=1}^{X} B_i$ has probability mass function $T_t f$.*

The thinning operation has been introduced by Rényi in [Ren56], and can be seen as a discrete version of the scaling operation, which associates to a random variable X (on $\mathbb{R}$) the random variable tX, or equivalently associates to a density $(f(x))_{x\in\mathbb{R}}$ the density $f_t(x) := 1/tf(x/t)$, see the introduction of [HJK07] for further information. For instance, the thinning operation is used to state the following Poisson limit theorem, known as 'law of thin numbers' (see [HJK07]):

Theorem 1 *Let $f^{\star n}$ denote the n-th convolution of f, or equivalently the probability mass function of the independent sum $X_1 + \cdots + X_n$. Then $T_{1/n}(f^{\star n})$ converges pointwise to the Poisson distribution* $\mathrm{Poi}(\lambda)$, *where $\lambda := \mathbb{E}(f)$.*

2.2 *Concavity of the Entropy Along the Thinning*

We now state and prove a first concavity of entropy result.

Definition 2 *The entropy $H(f)$ of a finitely supported probabilty measure f on a discrete space E is defined by*

$$H(f) := -\sum_{x\in E} f(x)\log(f(x)),$$

where by convention $0\log(0) = 0$.

Theorem 2 *Let f be a probability measure supported on* $\{0, \ldots, N\}$. *The function $t \mapsto H(t) := H(T_t f)$ is concave on* $[0, 1]$.

Theorem 2 has been first proven by Johnson and Yu in [JY09]. Their proof is based on the decomposition $H(T_t f) = -D(t) - L(t)$, where $D(t)$ is the relative entropy of the measure f_t with respect to the Poisson measure $\mathcal{P}(\lambda_t)$ (with $\lambda_t = \mathbb{E}[T_t f] = t\mathbb{E}[f]$), and $L(t) := \mathbb{E}[\log(\mathcal{P}(X_t, \lambda_t))]$ where X_t as $T_t f$ as p.m.f. The convexity of $D(t)$ follows from the data-processing inequality, and the convexity of $L(t)$ is proven by computing directly $L''(t)$ and by using the formula

$$\frac{\partial}{\partial t} T_t f(k) = -\left(\frac{k+1}{t}(T_t f)(k+1) - \frac{k}{t}(T_t f)(k)\right). \tag{6}$$

We are giving a slightly different proof of Theorem 2, which does not need a decomposition of $H(T_t f)$, but relies on a formula for $\frac{\partial}{\partial t}T_t f(k)$ which is quite similar to (6).

Proof of Theorem 2 Using the equations $k\binom{n}{k} = n\binom{n-1}{k-1}$, $(n-k)\binom{n}{k} = n\binom{n-1}{k}$, we prove the following transport equation for the binomial distributions:

$$\frac{\partial}{\partial t}\mathrm{bin}_{n,t}(k) = -n\left(\mathrm{bin}_{n-1,t}(k) - \mathrm{bin}_{n-1,t}(k-1)\right), \tag{7}$$

where ∇ is the left derivative operator. Setting $f_t(k) := (T_t f)(k)$, it follows from equation (7) that we have the transport equation

$$\frac{\partial}{\partial t}f_t(k) = -\nabla g_t(k)\ , \text{where } g_t(k) := \sum_{l\geq k} l\,\mathrm{bin}_{l-1,t}(k)f(l). \tag{8}$$

There is a similar second-order transport equation:

$$\frac{\partial^2}{\partial t^2}f_t(k) = \nabla_2 h_t(k)\ , \text{where } h_t(k) := \sum_{l\geq k} l(l-1)\mathrm{bin}_{l-2,t}(k)f(l), \tag{9}$$

and where $\nabla_2 := \nabla \circ \nabla$ is the second left derivative operator.

Equations (8) and (9) allow us to express derivatives of $f_t(k)$ with respect to t as 'spatial derivatives' of other families of functions.

$$\begin{aligned}
-H''(t) &= \sum_k \left(\frac{\partial^2}{\partial t^2}f_t(k)\right)\log(f_t(k)) + \frac{1}{f_t(k)}\left(\frac{\partial}{\partial t}f_t(k)\right)^2 \\
&= \sum_k \nabla_2 h_t(k)\log(f_t(k)) + \frac{(\nabla_1 g_t(k))^2}{f_t(k)}.
\end{aligned}$$

Now, we notice that, for $k \geq 1$ and $l_1, l_2 \geq k+1$, we have

$$\begin{aligned}
l_1\binom{l_1-1}{k} l_2\binom{l_2-1}{k-1} &= \frac{l_1!l_2!}{(l_1-1-k)!k!(l_2-k)!(k-1)!} \\
&= l_1(l_1-1)\frac{(l_1-2)!}{((l_1-2)-(k-1))!(k-1)!}\,\frac{l_2!}{l!(l_2-k)!} \\
&= l_1(l_1-1)\binom{l_1-2}{k-1}\binom{l_2}{k},
\end{aligned}$$

which implies

$$l_1(l_1-1)\mathrm{bin}_{l_1-2,t}(k-1)\mathrm{bin}_{l_2,t}(k) = l_1\mathrm{bin}_{l_1-1,t}(k)l_2\mathrm{bin}_{l_2-1,t}(k-1). \tag{10}$$

With the usual convention that $\mathrm{bin}_{l,t}(k) = 0$ if $k \notin \{0, \ldots, l\}$, we notice that equation (10) is still true for any $k \geq 0$.

From equation (10) we deduce, by expanding the sums defining $f_t(k)$, $g_t(k)$ and $h_t(k)$, that for every $k \geq 0$,

$$f_t(k)h_t(k-1) = g_t(k)g_t(k-1). \tag{11}$$

Equation (11) allows us to write:

$$\begin{aligned}
\sum_k \nabla_2 h_t(k) \log(f_t(k)) &= \sum_k h_t(k) \left[\log(f_t(k)) - 2\log(f_t(k+1)) + \log(f_t(k+2))\right] \\
&= \sum_k h_t(k) \left[\log(f_t(k)) - \log\left(\frac{g_t(k+1)^2 g_t(k)^2}{h_t(k)^2}\right) + \log(f_t(k+2))\right] \\
&= \sum_k h_t(k) \left[\log\left(\frac{f_t(k)h_t(k)}{g_t(k)^2}\right) + \log\left(\frac{f_t(k+2)h_t(k)}{g_t(k+1)^2}\right)\right] \\
&\geq \sum_k h_t(k) \left[1 - \frac{g_t(k)^2}{f_t(k)h_t(k)} + 1 - \frac{g_t(k+1)^2}{f_t(k+2)h_t(k)}\right] \\
&= \sum_k 2h_t(k) - \frac{g_t(k)^2}{f_t(k)} - \frac{g_t(k+1)^2}{f_t(k+2)}.
\end{aligned}$$

The only inequality we have used is an elementary one: $\log(x) \geq 1 - 1/x$. On the other hand, we have:

$$\begin{aligned}
\sum_k \frac{(\nabla_1 g_t(k))^2}{f_t(k)} &= \sum_k \frac{g_t(k)^2}{f_t(k)} - 2\frac{g_t(k)g_t(k-1)}{f_t(k)} + \frac{g_t(k-1)^2}{f_t(k)} \\
&= \sum_k \frac{g_t(k)^2}{f_t(k)} - 2\frac{g_t(k)g_t(k+1)}{f_t(k+1)} + \frac{g_t(k+1)^2}{f_t(k+2)} \\
&= \sum_k -2h_t(k) + \frac{g_t(k)^2}{f_t(k)} + \frac{g_t(k+1)^2}{f_t(k+2)},
\end{aligned}$$

which finally proves that $-H''(t) \geq 0$. □

Remark 1 *The thinning operation can also be used to define an interpolation* $(f_t)_{t\in[0,1]}$ *between two probability measures* f_0 *and* f_1 *supported on* $\mathbb{Z}_+$, *by defining* f_t *as the convolution* $f_t := (T_{1-t}f_0) \star (T_t f_1)$. *Theorem 2, about the concavity of the entropy, is generalized to these interpolations under a technical assumption: if* f_0 *and* f_1 *are ultra-log-concave, which means that* $(k+1)f_i(k+1)^2 \geq (k+2)f_i(k)f_i(k+2)$ *for* $i = 0, 1$ *and* $k \in \mathbb{Z}_+$, *then the entropy* $H(f_t)$ *is a concave function of* t. *This is the main result of [JY09].*

2.3 Contraction of Measures on Graphs

There is a quite natural way to generalize the notion of thinning to the more general setting of a general connected, locally finite graph G. This has been done by the author in [Hill14a], and we recall here some definitions and theorems from this paper.

A curve on G of length n is any application $\gamma : \{0, \ldots, n\} \to G$. We will denote $L(\gamma) = n$. A geodesic between two vertices $x, y \in G$ is a curve which minimizes the length $L(\gamma)$ among the set of curves $\gamma : \{0, \ldots, n\} \to G$ with $\gamma(0) = x$ and $\gamma(n) = y$. The length of a geodesic path joining x to y id denoted $d_G(x, y)$, or $d(x, y)$ is there is no ambiguity, and this quantity defines a distance on the vertices of G, called the graph distance.

We will denote by $\Gamma(G)$ the set of geodesic paths on G, and $\Gamma_{x,y}$ the set of geodesic paths on G joining x to y. The cardinality of $\Gamma_{x,y}$ will be denoted by $|\Gamma_{x,y}|$.

Let $o \in G$ be a particular vertex, which will act as the vertex 0 in the thinning case. Let f be a finitely supported distribution on G.

Definition 3 *Let $\gamma \in \Gamma(G)$ be a geodesic path on G, of length n. The binomial family along γ is the family $(\mathrm{bin}_{\gamma,t})_{t\in[0,1]}$ of probability distributions supported on $\{\gamma(0), \ldots, \gamma(n)\}$, defined by $\mathrm{bin}_{\gamma,t}(\gamma(k)) := \mathrm{bin}_{n,t}(k)$.*

Definition 4 *The contraction of f on o is the family $(f_t)_{t\in[0,1]}$ of probability measures defined by*

$$f_t(x) := \sum_{z\in G} \left(\frac{1}{|\Gamma_{o,z}|} \sum_{\gamma\in\Gamma_{o,z}} \mathrm{bin}_{\gamma,t}(x) \right) f(z), \tag{12}$$

where, given a geodesic $\gamma \in \Gamma(G)$, the measure $\mathrm{bin}_{\gamma,t}$ is defined by

$$\forall j \in \{0, \ldots, l\}\ ,\ \mathrm{bin}_{\gamma,t}(\gamma(j)) := \mathrm{bin}_{l,t}(j),$$

where $l = L(\gamma)$, and by $\mathrm{bin}_{\gamma,t}(z) = 0$ if $z \neq \gamma(j)$.

Definition 5 *We orient the graph G as follows: given an edge $(xy) \in E(G)$, we set $x \to y$ if there exists a geodesic $\gamma \in \Gamma_{o,y}$ of length l with $\gamma(l-1) = x$. We then write that $x \in \mathcal{E}(y)$ and $y \in \mathcal{F}(x)$.*

Since G is connected, we have $\mathcal{E}(y) \neq \emptyset$ for every $y \neq o$. However, one may have $\mathcal{F}(x) = \emptyset$. Also notice that some edges may not be oriented with this definition, but they do not play any role in the construction of the contraction of f or in the study of the entropy of f_t.

The oriented graph $G, \to$ is itself naturally oriented as follows: we orient $(x_0y_0) \to (x_1y_1)$ if we have $x_0 \to y_0 = x_1 \to y_1$. The triple (x_0, x_1, y_1) is called an oriented triple. The set $T(G)$ of oriented triples on G can itself be oriented, the graph $(T(G), \to)$ being equal to $(E(E(G), \to), \to)$.

Definition 6 *The divergence of a function* $g : (E(G), \to) \to \mathbb{R}$ *is the function* $\nabla \cdot g : G \to \mathbb{R}$ *defined by*

$$\nabla \cdot g(x_1) := - \sum_{x_0 \in \mathcal{E}(x_1)} g(x_0 x_1) + \sum_{x_2 \in \mathcal{F}(x_1)} g(x_1 x_2).$$

The iterated divergence of $h : (T(G), \to) \to \mathbb{R}$ *is the function* $\nabla_2 \cdot h : G \to \mathbb{R}$ *defined by*

$$\nabla_2 \cdot h(x_2) := \sum_{x_1 \in \mathcal{E}(x_2)} \sum_{x_0 \in \mathcal{E}(x_1)} h(x_0 x_1 x_2) - 2 \sum_{x_1 \in \mathcal{E}(x_2)} \sum_{x_3 \in \mathcal{F}(x_2)} h(x_1 x_2 x_3)$$
$$+ \sum_{x_3 \in \mathcal{F}(x_2)} \sum_{x_4 \in \mathcal{E}(x_3)} h(x_2 x_3 x_4).$$

If we see $(T(G), \to)$ as $(E(E(G), \to), \to)$, then $\nabla_2 \cdot$ is simply $(\nabla \cdot) \circ (\nabla \cdot)$.

Definition 7 *Given a geodesic* $\gamma : \{0, \ldots, l\} \to G$, *we define the families of functions* $g_{t,\gamma} : (E(G), \to) \to \mathbb{R}$ *and* $h_{t,\gamma} : (T(G), \to) \to \mathbb{R}$ *by*

$$\forall j \in \{0, \ldots, l-1\}\ ,\ g_{t,\gamma}((\gamma(j)\gamma(j+1))) := l \operatorname{bin}_{l-1,t}(j)$$

and

$$\forall j \in \{0, \ldots, l-2\}, h_{t,\gamma}((\gamma(j)\gamma(j+1)\gamma(j+2))) := l(l-1) \operatorname{bin}_{l-2,t}(j),$$

the functions $g_{t,\gamma}$ *and* $h_{t,\gamma}$ *taking the value* 0 *elsewhere.*

The families $(g_{\gamma,t})$ and $(h_{\gamma,t})$ have been defined in order to have $\frac{\partial}{\partial t} \operatorname{bin}_{\gamma,t}(x) = -\nabla \cdot g_{\gamma,t}(x)$ and $\frac{\partial^2}{\partial t^2} \operatorname{bin}_{\gamma,t}(x) = \nabla_2 \cdot h_{\gamma,t}(x)$. From this fact we deduce easily:

Proposition 3 *Let* (f_t) *be a contraction family defined as in equation* (12). *We define the families of functions* $(g_t)_{t \in [0,1]}$ *and* $(h_t)_{t \in [0,1]}$, *respectively, on* $(E(G), \to)$ *and* $(T(G), \to)$ *by*

$$\forall (x_0 x_1) \in (E(G), \to)\ ,\ g_t(x_0 x_1) := \sum_{z \in G} \frac{1}{|\Gamma_{o,z}|} \sum_{\gamma \in \Gamma_{o,z}} g_{\gamma,t}(x_0 x_1) f(z), \tag{13}$$

$$\forall (x_0 x_1 x_2) \in (E(G), \to)\ ,\ h_t(x_0 x_1 x_2) := \sum_{z \in G} \frac{1}{|\Gamma_{o,z}|} \sum_{\gamma \in \Gamma_{o,z}} h_{\gamma,t}(x_0 x_1 x_2) f(z). \tag{14}$$

We then have the differential equations:

$$\frac{\partial}{\partial t} f_t(x) = -\nabla \cdot g_t(x)\ ,\ \frac{\partial^2}{\partial t^2} f_t(x) = \nabla_2 \cdot h_t(x). \tag{15}$$

The triple of functions (f_t, g_t, h_t) satisfies a generalized version of equation (11):

Proposition 4 *Let (f_t) be the contraction family of a measure f_1 to a point $o \in G$, and (g_t), (h_t) associated to f by equations* (13) *and* (14). *We then have:*

$$\forall (x_0 x_1 x_2) \in T(G)\ ,\ h_t(x_0 x_1 x_2) f_t(x_1) = g_t(x_0 x_1) g_t(x_1 x_2). \tag{16}$$

Equation (16) is then used to prove concavity of entropy results along contraction families on a graph. Explicit bounds on the second derivative $H''(t)$ can be found for particular graphs: the complete graph, the grid $\mathbb{Z}^n$, the cube $\{0, 1\}^n$ or trees. The reader is referred to [Hill14a] for detailed proofs and additional information.

Remark 2 *As there is only one coupling between a Dirac measure and another given probability measure on G, it is clear that the interpolation between δ_0 and f, as constructed in the paper [GRST14] (see in particular the beginning of Section 2), is identical to the contraction of f on o, as defined in Definition 4. In more general cases, the links between the interpolating families of [GRST14] and $W_{1,+}$-geodesics (as defined in Section 4) remain unclear.*

3 Optimal Transportation Theory

In this section, we recall some results about optimal transportation theory. We focus on equation (21), or equivalently on equation (22), which is a continuous, generalized version of equation (11) or equation (16) encountered in the previous section. This continuous equation, will be seen as a consequence of a Hamilton-Jacobi type equation (see equation (25)) which is satisfied by the velocity field associated to a Wasserstein geodesic.

The first paragraph is about the one-dimensional case and the second paragraph is about the general Riemannian setting. In both cases, the most important tool is the Benamou-Brenier formula, see equation (18) and equation (23), stated and proven in [BB99].

3.1 Optimal Transportation on the Real Line

We recall here some results from the continuous theory of optimal transportation, in the special case where the underlying metric space is the real line $\mathbb{R}$ with the usual Euclidian distance $d(x, y) = |x - y|$. In order to avoid technical difficulties that will not appear in the discrete setting, we will make the following additional assumption: the densities f_0 and f_1 with respect to the Lebesgue measure on $\mathbb{R}$ are supported on a compact interval K and are such that their respective cumulated distribution functions F_0 and F_1 are smooth bijections between K and $[0, 1]$.

We consider the Monge problem: we want to find

$$\inf_T \int_{\mathbb{R}} |x - T(x)|^p f_0(x) dx, \tag{17}$$

where the infimum is taken over the set of measurable maps $T : \mathbb{R} \to \mathbb{R}$ satisfying $T_* \mu_0 = \mu_1$. If there exists a solution T to the Monge problem, then the coupling $\pi = (Id \times T)_* \mu_0$ is solution to Monge-Kantorovitch problem (1).

It is possible to give an explicit expression for the optimal transport map (see [Vill03] for a proof):

Proposition 5 *The infimum in the Monge problem* (17) *is attained when* $T(x) = F_1^{-1} \circ F_0(x)$. *If* $p > 1$, *then this optimal* T *is unique.*

It is also possible to describe the Wasserstein geodesics:

Proposition 6 *Let* $T := F_1^{-1} \circ F_0$ *be the solution to Monge problem* (17). *We set, for* $t \in [0, 1]$, $T_t(x) := (1 - t)x + tT(x)$ *and* $\mu_t := (T_t)_* \mu_0$. *Then for any* $p \geq 1$, *the family* $(f_t)_{t \in [0,1]}$ *is a* W_p*-Wasserstein geodesic between* f_0 *and* f_1.

Proof We have:

$$\begin{aligned} W_p^p(\mu_s, \mu_t) &\leq \int_{\mathbb{R}} |x - T_t \circ T_s^{-1}(x)|^p d\mu_s \\ &= \int_{\mathbb{R}} |T_s(x) - T_t(x)|^p d\mu_0 \\ &= \int_{\mathbb{R}} |s - t|^p |T_0(x) - T_1(x)|^p d\mu_0 \\ &= |s - t|^p W_p^p(\mu_0, \mu_1). \end{aligned}$$

In particular, we have, for any $0 \leq s \leq t \leq 1$,

$$\begin{aligned} W_p(\mu_0, \mu_1) &\leq W_p(\mu_0, \mu_s) + W_p(\mu_s, \mu_t) + W_p(\mu_t, \mu_1) \\ &\leq ((s - 0) + (t - s) + (1 - t)) W_p(\mu_0, \mu_1) = W_p(\mu_0, \mu_1), \end{aligned}$$

so the previous inequalities are actually equalities, and in particular we have $W_p(\mu_s, \mu_t) = |t - s| W_p(\mu_0, \mu_1)$, which proves that $(\mu_t)_{t \in [0,1]}$ is a Wasserstein W_p-geodesic. □

In this one-dimensional framework, the Benamou-Brenier formula is written as follows (see [Sant15], Remark 9):

Theorem 3 *For* $p \geq 1$, *the Wasserstein distance* $W_p(f_0, f_1)$ *satisfies*

$$W_p^p(f_0, f_1) = \inf \int_{\mathbb{R}} \int_0^1 |v_t(x)|^p f_t(x) dx dt, \tag{18}$$

where the infimum is taken over the families $(f_t(x))_{x\in\mathbb{R},t\in[0,1]}$ *joining* f_0 *to* f_1 *and the velocity fields* $(v_t(x))_{x\in\mathbb{R},t\in[0,1]}$ *satisfying the continuity equation*

$$\frac{\partial}{\partial t}f_t(x) + \frac{\partial}{\partial x}(v_t(x)f_t(x)) = 0. \tag{19}$$

The velocity field $(v_t(x))$ associated to a Wasserstein geodesic by equation (19) satisfies interesting properties:

Proposition 7 *The Wasserstein geodesic* $(\mu_t)_{t\in[0,1]}$ *with* $d\mu_t(x) = f_t(x)dx$ *satisfies the continuity equation* (19) *with the velocity field* $(v_t(x))_{t\in[0,1],x\in\mathbb{R}}$ *defined by the equation*

$$\forall t \in [0,1]\,,\ \forall x \in \mathbb{R}\,,\ v_t(T_t(x)) = T(x) - x. \tag{20}$$

Moreover, the velocity field satisfies the differential equation

$$\frac{\partial}{\partial t}v_t(x) = -v_t(x)\frac{\partial}{\partial x}v_t(x). \tag{21}$$

Proof We first prove that equation (20) is unambiguous, i.e. that the mapping $x \mapsto T_t(x)$ is injective when t is fixed. The equation $T_t(x) = T_t(y)$ can be rewritten $(1-t)(x-y) + t(T(x) - T(y)) = 0$. This shows that $(x-y)(T(x) - T(y)) \leq 0$, with a strict inequality when $x \neq y$. But this is a contradiction with the fact that $T = F_0 \circ F_1^{-1}$ is increasing, so T_t is injective.

In order to prove equation (21), we differentiate both sides of equation (20) with respect to t. The differential of the right-hand side is clearly zero. To avoid ambiguities, we will use the notations $\partial_1 v(t,x) := \frac{\partial}{\partial t}v_t(x)$ and $\partial_2 v(t,x) := \frac{\partial}{\partial x}v_t(x)$. We then have:

$$\begin{aligned} 0 = \frac{\partial}{\partial t}v_t(T_t(x)) &= \left(\frac{\partial}{\partial t}T_t(x)\right)\partial_2 v(t, T_t(x)) + \partial_1 v(t, T_t(x)) \\ &= v(t, T_t(x))\partial_2 v(t, T_t(x)) + \partial_1 v(t, T_t(x)), \end{aligned}$$

which is equation (21) evaluated in t and $T_t(x)$. □

An equivalent form of Proposition 7 is stated as follows:

Proposition 8 *Let* $(f_t)_{t\in[0,1]}$ *be a Wasserstein geodesic on* $\mathbb{R}$. *We then have*

$$\frac{\partial^2}{\partial t^2}f_t(x) = \frac{\partial^2}{\partial x^2}\left(v_t(x)^2 f_t(x)\right). \tag{22}$$

Proof Equation (22) is simply obtained by differentiating the continuity equation (19) with respect to t and then by using equation (21). □

This description of Wasserstein geodesics can be used to prove a concavity of entropy property. Simple integration by parts allows us to prove the following (see [HJ16], Theorem 2.1. for a detailed proof):

Proposition 9 *Let $(f_t)_{t\in[0,1]}$ be a family of probability densities on $\mathbb{R}$. We suppose that there exist two families $(g_t(x))$ and $(h_t(x))$ such that*

$$\frac{\partial}{\partial t}f_t(x) = -\frac{\partial}{\partial x}g_t(x)\ , \ \frac{\partial^2}{\partial t^2}f_t(x) = \frac{\partial^2}{\partial x^2}h_t(x).$$

Then the entropy $H(t)$ of f_t satisfies

$$-H''(t) = \int_{\mathbb{R}} \left(h_t(x) - \frac{g_t(x)^2}{f_t(x)}\right)\frac{\partial^2}{\partial x^2}\left(\log(f_t(x))\right) + \frac{\left(\frac{\partial g_t(x)}{\partial x}f_t(x) - g_t(x)\frac{\partial f_t(x)}{\partial x}\right)^2}{f_t(x)^3}dx.$$

In the case where $(f_t)_{t\in[0,1]}$ is a Wasserstein geodesic, then $g_t(x) = v_t(x)f_t(x)$ and $h_t(x) = v_t(x)^2 f_t(x)$, and we find

$$-H''(t) = \int_{\mathbb{R}} \left(\frac{\partial v_t(x)}{\partial x}\right)^2 f_t(x)dx \geq 0.$$

3.2 Benamou-Brenier Theory in Higher Dimensions

The Benamou-Brenier formula stated in (18) in the one-dimensional setting can be stated in the Riemannian setting, at least for $p = 2$: more precisely, if μ_0, μ_1 are probability measures on a compact Riemannian manifold (M, g), with finite second moment, and with densities f_0, f_1 with respect to the Riemannian volume measure $dvol$, then

$$W_2(\mu, \mu_1)^2 = \inf \int_M \int_0^1 |v_t(x)|^2 f_t(x) dt dvol(x), \tag{23}$$

where the infimum is taken over the set of smooth families $(f_t(x))$ of probability densities joining f_0 to f_1, and where the velocity field $v_t : M \to TM$ satisfies the continuity equation

$$\frac{\partial}{\partial t}f_t(x) = -\nabla \cdot (v_t(x)f_t(x)), \tag{24}$$

where $\nabla\cdot$ is the divergence operator on the tangent bundle TM.

It can then be shown that this infimum is attained when (μ_t) is the Wasserstein W_2-geodesic between μ_0 and μ_1. Moreover, the associated velocity field $v_t(x)$ can be written under the form $v_t(x) = \nabla\Psi_t(x)$, where ∇ is the gradient operator and Ψ_t

is a convex function on M. Moreover, this function $\Psi_t(x)$ is solution to the Hamilton-Jacobi equation

$$\frac{\partial}{\partial t}\Psi_t(x) + \frac{1}{2}|\nabla\Psi_t(x)|^2 = 0. \tag{25}$$

The one-dimensional equation (21) can be seen as a consequence of this general Hamilton-Jacobi equation.

Hamilton-Jacobi equation is a powerful tool to prove concavity of the entropy results or functional inequalities holding on a Riemannian manifolds with Ricci curvature uniformly bounded from below, as done, for instance, in the 'heuristics section' of Otto-Villani's paper [OV00].

Among the several propositions for a definition of discrete curvature bounds, one of the most promising has been made by Erbar-Maas [EM12] and independently by Mielke [Miel13] and is based on a generalization of the Benamou-Brenier formula to the setting of discrete Markov chains:

Let $K : \mathcal{X} \times \mathcal{X}$ be an irreducible Markov Kernel on a finite space $\mathcal{X}$, admitting a reversible measure ν on $\mathcal{X}$. A density on $\mathcal{X}$ is a function $\rho : \mathcal{X} \to \mathbb{R}_+$ with $\sum_{x\in\mathcal{X}} \rho(x)\nu(x) = 1$.

The following definition by Erbar-Maas (see [EM12]) is directly inspired by Benamou-Brenier theory:

Definition 8 *We define a distance $\mathcal{W}$ on the set $\mathcal{P}(\mathcal{X})$ by*

$$\mathcal{W}(\rho_0, \rho_1)^2 := \inf_{\rho,\psi}\left\{\frac{1}{2}\int_0^1 \sum_{x,y\in\mathcal{X}} (\psi_t(x) - \psi_t(y))^2\, \hat{\rho}_t(x,y)K(x,y)\nu(x)dt\right\}, \tag{26}$$

where the infimum is taken on the set of regular families of densities $(\rho_t(x))_{t\in[0,1],x\in X}$ joining ρ_0 to ρ_1 and satisfying the transport equation

$$\frac{\partial}{\partial t}\rho_t(x) = -\sum_{y\in\mathcal{X}} (\psi_t(y) - \psi_t(x))\, \hat{\rho}_t(x,y)K(x,y),$$

where we define $\hat{\rho}_t(x,y) := \int_0^1 \rho(x)^{1-p}\rho(y)^p dp$.

As in classical Sturm-Lott-Villani theory, Erbar and Maas define a notion of Ricci curvature for the Markov kernel K by considering concavity properties of the entropy along generalized Wasserstein geodesics:

Definition 9 *The Markov kernel K satisfies* Ric $\geq \kappa$ *if for any geodesic $(\rho_t)_{t\in[0,1]}$ for the distance $\mathcal{W}$, we have*

$$H(t) \geq (1-t)H(0) + tH(1) - \kappa\frac{t(1-t)}{2}\mathcal{W}(\rho_0, \rho_1)^2,$$

where the entropy function H is defined by $H(t) = -\sum_{x\in\mathcal{X}} \rho_t(x)\log(\rho_t(x))\nu(x)$.

This definition of discrete Ricci curvature bounds is strong enough to recover several results which hold in the continuous setting, for instance a tensorization result ([EM12], Theorem 1.3) and functional inequalities satisfied on spaces with $Ric \geq \kappa > 0$, such as the modified logarithmic Sobolev inequality ([EM12], Theorem 1.5). Moreover, explicit bounds on the Ricci curvature can be computed in explicit fundamental examples, such as the complete graphs, the discrete hypercubes or circle graphs.

There are a lot of similarities and a lot of differences between the geodesics coming from the Erbar-Maas $\mathcal{W}$-distance and the $W_{1,+}$ interpolations presented below. In particular, both approaches are inspired by the continuous Benamou-Brenier theory. An important difference is that Erbar-Maas interpolations are based on a discrete version of the problem (23), whereas $W_{1,+}$-interpolations are based on the generalization of equation (21), which characterizes the solutions.

4 $W_{1,+}$-Geodesics on Graphs

In this section, we construct and study curves (f_t) in the space of finitely supported probability measures on a graph G along which equations similar to (11) and (16) hold. These curves will be called $W_{1,+}$-geodesics on G. We first focus on the case where $G = \mathbb{Z}$ and f_0 is stochastically dominated by f_1 (see Definition 10), before turning to the general case.

4.1 W_1-Geodesics on $\mathbb{Z}$

Before defining and constructing $W_{1,+}$-geodesics on $\mathbb{Z}$, we first recall in this paragraph some results about the geometry of the space $\mathcal{P}(\mathbb{Z})$ with the distance W_1. We consider the Monge-Kantorovitch problem

$$\inf_{\pi \in \Pi(f_0,f_1)} I_1(\pi) := \inf_{\pi \in \Pi(f_0,f_1)} \sum_{i,j} |i-j|\pi(i,j), \tag{27}$$

where f_0, f_1 are two finitely supported probability measures on $\mathbb{Z}$.

The minimization problem (27) has been extensively studied :

Proposition 10 *The set $\Pi_1(f_0,f_1)$ of minimizers for the problem (27) satisfy the following properties:*

- *$\Pi_1(f_0,f_1)$ is non-empty.*
- *$\Pi_1(f_0,f_1)$ is a convex subset of $\mathcal{P}(\mathbb{Z}\times\mathbb{Z})$.*
- *If $\Pi_1(f_0,f_1)$ has a unique element π, then either f_0 or f_1 is a Dirac measure.*
- *If neither f_0 nor f_1 is a Dirac measure, then $\Pi_1(f_0,f_1)$ has a non-empty interior.*

The proof of these facts is easy. The second and last points are proven by noticing that, if $\pi(i_1,j_1) \geq a > 0$ and $\pi(i_2,j_2) \geq a > 0$ and $(j_1 - i_1)(j_2 - i_2) \geq 2$, then the coupling $\tilde{\pi}$ defined by $\tilde{\pi}(i,j) = \pi(i,j) - a$ if $(i,j) = (i_1,j_1)$ or (i_2,j_2), $\tilde{\pi}(i,j) = \pi(i,j) + a$ if $(i,j) = (i_1,j_2)$ or (i_2,j_1) and $\tilde{\pi} = \pi$ elsewhere, satisfies $I_1(\tilde{\pi}) = I_1(\pi)$.

Definition 10 *Let f_0, f_1 be two probability measures on $\mathbb{Z}$. We say that f_0 is stochastically dominated by f_1, and we write $f_0 << f_1$, if we have $\forall k \in \mathbb{Z}$, $F_0(k) \geq F_1(k)$, where $F_i(k) := \sum_{l \leq k} f_i(l)$, $i = 0, 1$, is the cumulative distribution associated to f_i.*

Proposition 11 *We suppose that $f_0 << f_1$. Let π be in $\Pi_1(f_0,f_1)$. Then $\pi(i,j) > 0 \Rightarrow i \leq j$.*

Proof Let π be a coupling between f_0 and f_1. The stochastic domination assumption is equivalent to the following:

$$\forall k \in \mathbb{Z}, \sum_{i \leq k, j > k} \pi_{i,j} \geq \sum_{i > k, j \leq k} \pi_{i,j}. \tag{28}$$

Indeed, adding $\sum_{i,j \leq k} \pi_{i,j}$ to both sides of equation (28) gives $F_0(k) \geq F_1(k)$. Suppose that we have $\pi_{i_0 j_0} > 0$ for a couple $i_0 > j_0$. This means that the right-hand side of equation (28), with $k = j_0$, is non-zero, which implies that there exists a couple $i_1 \leq j_0 < j_1$ with $\pi_{i_1,j_1} > 0$. We now define $c(i_0,j_1) = c(i_1,j_0) := 1$, $c(i_0,j_0) = c(i_1,j_1) := -1$ and $c(i,j) := 0$ for other couples (i,j). If $0 < \varepsilon < \min(\pi_{i_0,j_0}, \pi_{i_1,j_1})$, then $\tilde{\pi} := \pi + \varepsilon c$ is a coupling between f_0 and f_1 and we have

$$\sum |j - i| \tilde{\pi}_{i,j} - \sum |j - i| \pi_{i,j} = 2\varepsilon(j_0 - \min(i_0,j_1)) < 0,$$

which means that π is not a W_1-optimal coupling. □

There is a converse to Proposition 11:

Proposition 12 *We suppose that there exists a coupling $\pi \in \Pi_1(f_0,f_1)$ such that for all $i,j \in \mathbb{Z}$, $\pi(i,j) > 0 \Rightarrow i \leq j$. Then $f_0 << f_1$.*

Proof It suffices to notice that the right-hand side of equation (28) is zero. □

We now prove that a W_1-geodesic between $f_0 << f_1$ is monotonic for the stochastic domination order:

Proposition 13 *Let $(f_t)_{t \in [0,1]}$ be a W_1-geodesic with $f_0 << f_1$. Then we have $f_s << f_t$ for any $0 \leq s \leq t \leq 1$. Moreover, we have $\sum_k kf_t(k) = tW_1(f_0,f_1) + \sum_k kf_0(k)$.*

Proof We fix $t \in [0, 1]$. Let $\pi_{0,t} \in \Pi_1(f_0,f_1)$, $\pi_{t,1} \in \Pi_1(f_t,f_1)$ be two optimal couplings. In particular, we have $\sum_k \pi_{0,t}(i, k) = f_0(i)$, $\sum_i \pi_{0,t}(i, k) = f_t(k)$ and $\sum_{i,k} |k-i| \pi_{0,t}(i, k) = W_1(f_0,f_t) = tW_1(f_0,f_1)$. We construct a coupling $\pi \in \Pi(f_0,f_1)$ by setting

$$\pi(i,j) := \sum_k \pi_{0,t}(i, k)\pi_{t,1}(k,j).$$

The triangle inequality $|i-j| \leq |i-k| + |k-j|$ easily implies

$$\sum_{i,j} |j-i|\pi(i,j) \leq W_1(f_0,f_t) + W_1(f_t,f_1) = W_1(f_0,f_1),$$

so we have $\pi \in \Pi_1(f_0,f_1)$. In particular, we have by Proposition 11 $\pi(i,j) > 0 \Rightarrow i \leq j$, and moreover there is equality in the triangle inequality: $j - i = |j-i| = |k-i| + |j-k|$ (thus $i \leq k \leq j$), whenever we have $\pi_{0,t}(i,k)\pi_{t,1}(k,j) > 0$, which implies by Proposition 12 that $f_0 << f_t << f_1$. The inequality $f_s << f_t$ for any $s \leq t$ comes from the fact that $(f_{\alpha t})_{\alpha \in [0,1]}$ is a W_1-geodesic between f_0 and f_t, and choosing $\alpha := s/t$ leads to $f_0 << f_s << f_t$.

In order to prove the second point, we write

$$\begin{aligned}\sum_k kf_t(k) &= \sum_k k \sum_i \pi_{0,t}(i,k)\\ &= \sum_i \sum_k (k-i+i)\pi_{0,t}(i,k)\\ &= \sum_{i,k} |k-i|\pi_{0,t}(i,k) + \sum_i i \sum_k \pi_{0,t}(i,k)\\ &= tW_1(f_0,f_1) + \sum_i if_0(i).\end{aligned}$$

□

The following result will be seen as a particular case of Proposition 15:

Proposition 14 *Let $(f_t)_{t\in[0,1]}$ be a W_1-geodesic on $\mathbb{Z}$ with $f_0 << f_1$. Then there exist two families of finitely supported functions $(g_t)_{t\in[0,1]}$, $(h_t)_{t\in[0,1]}$, such that:*

- $\frac{\partial}{\partial t}f_t(k) = -\nabla_1 g_t(k)$.
- $\frac{\partial}{\partial t}g_t(k) = -\nabla_1 h_t(k)$.
- $g_t(k) \geq 0$.

In this setting, the families (g_t) and (h_t) are defined by

$$g_t(k) := -\sum_{l\leq k} \frac{\partial}{\partial t} f_t(l)\ ,\ h_t(k) := -\sum_{l\leq k} \frac{\partial}{\partial t} g_t(l).$$

The fact that (f_t) is a W_1-geodesic between $f_0 << f_1$ is used as follows: we have $\sum_k kf_t(k) = W_1(f_0,f_1).t + \sum_k kf_0(k)$, so $\sum_k g_t(k)$ is constant, so $h_t(k)$ is finitely supported. The stochastic domination is also used to prove that $g_t(k) \geq 0$.

4.2 $W_{1,+}$-Geodesics on $\mathbb{Z}$

The proof of the concavity of the entropy along thinning of measures (Theorem 2) only uses the fact that

$$f_t(k)h_t(k-1) = g_t(k)g_t(k-1), \tag{29}$$

therefore a natural question is to find other families $(f_t(k))_{t\in[0,1],k\in\mathbb{Z}}$ of probability measures on $\mathbb{Z}$ satisfying the same equation (29), because along such families the concavity of the entropy would be already proven.

A very interesting fact is that the most natural continuous analogue of equation (29) is simply the equation $fh = g^2$, which is satisfied by Wasserstein geodesics on the real line. It thus seems natural to use equation (29) as a possible definition of Wasserstein interpolation on $\mathbb{Z}$.

We construct, with Theorem 5, an interpolation $(f_t)_{t\in[0,1]}$ satisfying equation (29), where the families $(g_t(k))$ and $(h_t(k))$ are defined from $(f_t(k))$ by Proposition 14. This interpolation is called $W_{1,+}$-geodesic between f_0 and f_1. We will later explain how to modify equation (29) in the general case to define $W_{1,+}$-geodesics between each couple f_0, f_1 of finitely supported probability measures on a graph.

Definition 11 *Let $f_0 << f_1$ be two finitely supported probability measures on $\mathbb{Z}$. A $W_{1,+}$-geodesic between f_0 and f_1 is a family $(f_t)_{t\in[0,1]}$ of probability measures on $\mathbb{Z}$ such that*

- *(f_t) is a W_1-geodesic.*
- *$f_t(k)h_t(k-1) = g_t(k)g_t(k-1)$.*
- *$g_t(k) > 0$ whenever we have $f_t(k) > 0$,*

where $(g_t)_{t\in[0,1]}$ and $(h_t)_{t\in[0,1]}$ are defined from $(f_t)_{t\in[0,1]}$ by Proposition 14.

The proof of Theorem 2 is still valid in this more general framework, which leads to the following:

Theorem 4 *Let $(f_t)_{t\in[0,1]}$ be a $W_{1,+}$-geodesic on $\mathbb{Z}$. The entropy $H(t)$ of f_t is then a concave function of t.*

In order to make Theorem 4 relevant, an important question is to show the existence of a $W_{1,+}$-geodesic with prescribed f_0 and f_1. This question is answered by the following:

Theorem 5 *Let $f_0 << f_1$ be two probability measures on $\mathbb{Z}$. There exists a unique $W_{1,+}$-geodesic $(f_t)_{t\in[0,1]}$ joining f_0 to f_1. Moreover, (f_t) can be written as a mixture of binomial families:*

$$f_t(k) = \sum_{i\le k\le j} \mathrm{bin}_{i,j,t}(k)\pi(i,j),$$

where the coupling π *is solution to the minimization problem*

$$\inf_{\pi \in \Pi_1(f_0, f_1)} \sum_{i \leq j} \pi(i,j) \log(\pi(i,j)(j-i)!) - \pi(i,j),$$

and where $\text{bin}_{i,j,t}(k) = \text{bin}_{j-i,t}(k-i)$.

The proof of Theorem 5 is quite similar to the proof of Theorem 6, of which a detailed proof is given. It is interesting to notice that the optimal coupling π can be written under the form $\pi(i,j) = \frac{a(i)b(j)}{(j-i)!}$ for a couple of functions $a, b : G \to \mathbb{R}$. A detailed proof of Theorem 5, stated in a more general form, can be found in [Hill14b]: see in particular Theorem 4.5 for the existence and Theorem 3.19 for the binomial mixture.

4.3 $W_{1,+}$-Geodesics on Graphs

In this paragraph we explain briefly how to define $W_{1,+}$-geodesics between a couple of finitely supported probability measures f_0, f_1 on a graph G. If we want an equation similar to (16) to make sense, we first need to define an orientation on G.

Definition 12 *We define the* W_1*-orientation on* G*, with respect to the couple* f_0, f_1*, in the following way: let* $(xy) \in E(G)$ *be any edge of* G*. If there exists an optimal coupling* $\pi \in \Pi_1(f_0, f_1)$*, a couple of vertices* $a, b \in G$ *with* $\pi(a,b) > 0$*, a geodesic path* $\gamma \in \Gamma(a,b)$ *and an integer* $l \in \{0, \ldots, d(a,b) - 1\}$ *such that* $x = \gamma(l)$ *and* $y = \gamma(l+1)$*, then we orient the edge* (xy) *by* $x \to y$.

It is proven in [Hill14b], Theorem 2.18, that the orientation $x \to y$ does not depend on π, a, b, γ, and so this orientation is well defined. As in the case of contraction of measures, some edges may not be oriented by this process, but they do not play any role in the construction of $W_{1,+}$-geodesics.

Having an orientation on G allows us to define the divergence and iterated divergence operators, respectively, on the oriented graphs $(E(G), \to)$ and $(T(G), \to)$. We now associate a family $(g_t)_{t \in [0,1]}$ to each W_1-geodesic as follows (see [Hill14b], Theorem 2.23 and Proposition 2.25):

Proposition 15 *Let* $(f_t)_{t \in [0,1]}$ *be a* W_1*-geodesic on* G*. We orient* G *with the* W_1*-orientation with respect to* f_0, f_1*. There exists a family* $(g_t)_{t \in [0,1]}$ *of functions defined on* $(E(G), \to)$ *such that* $\forall x, y \in (E(G), \to)$, $g_t(xy) > 0$ *and* $\frac{\partial}{\partial t} f_t(x) = -\nabla \cdot g_t(x)$.

Definition 13 *Let* G *be a graph,* W_1*-oriented with respect to* f_0, f_1*. A family* $(f_t)_{t \in [0,1]}$ *of probability measures on* G *is said to be a* $W_{1,+}$*-geodesic if:*

- $(f_t)_{t \in [0,1]}$ *is a* W_1*-geodesic.*
- *There exist two families* $(g_t)_{t \in [0,1]}$ *and* $(h_t)_{t \in [0,1]}$ *defined, respectively, on* $(E(G), \to)$ *and* $(T(G), \to)$ *such that*

$$\frac{\partial}{\partial t}f_t(x) = -\nabla \cdot g_t(x)\ , \ \frac{\partial^2}{\partial t^2}f_t(x) = \nabla_2 \cdot h_t(x).$$

- *For every* $(xy) \in (E(G), \rightarrow)$, *we have* $g_t(xy) > 0$.
- *The triple* (f_t, g_t, h_t) *satisfies the equation*

$$\forall (x_0x_1x_2) \in (T(G), \rightarrow)\ , \ f_t(x_1)h_t(x_0x_1x_2) = g_t(x_0x_1)g_t(x_1x_2). \tag{30}$$

One can then prove the existence and uniqueness of a $W_{1,+}$-geodesic $(f_t)_{t\in[0,1]}$ with prescribed f_0 and f_1, just as in the one-dimensional case. See the whole Section 4 of [Hill14b] for a rigorous construction of such geodesics.

5 About the Schrödinger-Léonard Theory

Proposition 9 can be used in setting different from Wasserstein geodesics. For instance, it can be used to study the entropy of a family $(f_t(k))_{t\in[0,1],k\in\mathbb{Z}}$ of probability measures satisfying a diffusion equation $\frac{\partial}{\partial t}f_t(k) = \nabla_2 f_t(k+1)$. Actually, we prove a stronger fact:

Proposition 16 *Let* $(u_t(x)), (v_t(x))$ *be two families of smooth positive functions on* $[0, 1] \times \mathbb{R}$ *satisfying*

$$\frac{\partial}{\partial t}u_t(x) = \frac{\partial^2}{\partial x^2}u_t(x)\ , \ \frac{\partial}{\partial t}v_t(x) = -\frac{\partial^2}{\partial x^2}v_t(x).$$

We set $f_t(x) := u_t(x)v_t(x)$. *Then the entropy* $H(t) := H(f_t)$ *is a concave function of* t.

Proof We can write $\frac{\partial}{\partial t}f_t(x) = -\frac{\partial}{\partial x}g_t(x)$ and $\frac{\partial^2}{\partial t^2}f_t(x) = \frac{\partial^2}{\partial x^2}h_t(x)$, with

$$g_t(x) := \frac{\partial u_t(x)}{\partial x}v_t(x) - u_t(x)\frac{\partial v_t(x)}{\partial x}$$

and

$$h_t(x) := \frac{\partial^2 u_t(x)}{\partial x^2}v_t(x) - 2\frac{\partial u_t(x)}{\partial x}\frac{\partial v_t(x)}{\partial x} + u_t(x)\frac{\partial^2 v_t(x)}{\partial x^2}.$$

We then apply Proposition 9, which, after simplifications, gives

$$-H''(t) = \int_{\mathbb{R}} \left(\frac{\partial^2}{\partial x^2}\log(u_t(x))\right)^2 v_t(x) + u_t(x)\left(\frac{\partial^2}{\partial x^2}\log(v_t(x))\right)^2 dx \geq 0. \quad \square$$

This proposition has a discrete analogue:

Proposition 17 *Let* $(u_t(k))$, $(v_t(k))$ *be two families of non-negative functions satisfying*

$$\frac{\partial}{\partial t}u_t(k) = \nabla_2 u_t(k+1)\,,\ \frac{\partial}{\partial t}v_t(k) = -\nabla_2 v_t(k+1)$$

Then the entropy $H(t)$ *of the family* $(f_t(k)) := (u_t(k)v_t(k))$ *is a concave function of* t.

Proof We can write $\frac{\partial f_t(k)}{\partial t} = -\nabla g_t(k)$ and $\frac{\partial^2 f_t(k)}{\partial t^2} = \nabla_2 h_t(k)$ with

$$\begin{aligned} g_t(k) &= u_t(k+1)v_t(k) - u_t(k)v_t(k+1)\,,\ h_t(k) \\ &= u_t(k+2)v_t(k) - 2u_t(k+1)v_t(k+1) + u_t(k)v_t(k+2). \end{aligned}$$

We then write

$$\begin{aligned} \sum_k \frac{\partial^2}{\partial t^2} f_t(k)\log(f_t(k)) &= \sum_k \nabla_2 h_t(k)\log(f_t(k)) \\ &= \sum_k h_t(k)\log\left(\frac{f_t(k)f_t(k+2)}{f_t(k+1)^2}\right) \\ &\sum_k (u_t(k+2)v_t(k) - 2u_t(k+1)v_t(k+1) \\ &\quad + u_t(k)v_t(k+2))\log\left(\frac{f_t(k)f_t(k+2)}{f_t(k+1)^2}\right). \end{aligned}$$

The elementary inequality $\log(x) \geq 1 - 1/x$ gives the bound

$$\begin{aligned} \log\left(\frac{f_t(k)f_t(k+2)}{f_t(k+1)^2}\right) &= \log\left(\frac{u_t(k)u_t(k+2)}{u_t(k+1)^2}\right) + \log\left(\frac{v_t(k)v_t(k+2)}{v_t(k+1)^2}\right) \\ &\geq 2 - \frac{u_t(k+1)^2}{u_t(k)u_t(k+2)} - \frac{v_t(k+1)^2}{v_t(k)v_t(k+2)}. \end{aligned}$$

Similarly, the inequality $-\log(x) = \log(1/x) \geq 1 - x$ gives

$$-\log\left(\frac{f_t(k)f_t(k+2)}{f_t(k+1)^2}\right) \geq 2 - \frac{u_t(k)u_t(k+2)}{u_t(k+1)^2} - \frac{v_t(k)v_t(k+2)}{v_t(k+1)^2}.$$

After simplifications, we obtain the inequality

$$\sum_k \frac{\partial^2}{\partial t^2} f_t(k)\log(f_t(k)) \geq \sum_k 2u_t(k+2)v_t(k) - \frac{u_t(k+1)^2}{u_t(k)}v_t(k) - \frac{v_t(k+1)^2}{v_t(k+2)}u_t(k+2)$$

$$+4u_t(k+1)v_t(k+1) - 2\frac{u_t(k)u_t(k+2)}{u_t(k+1)}v_t(k+1) - 2\frac{v_t(k)v_t(k+2)}{v_t(k+1)}u_t(k+1)$$

$$+2u_t(k)v_t(k+2) - \frac{u_t(k+1)^2}{u_t(k+2)}v_t(k+2) - \frac{v_t(k+1)^2}{v_t(k)}u_t(k).$$

On the other hand, we have

$$\begin{aligned}
\sum_k \frac{1}{f_t(k)}\left(\frac{\partial}{\partial t}f_t(k)\right)^2 &= \sum_k \frac{(g_t(k)-g_t(k-1))^2}{f_t(k)} \\
&= \sum_k \frac{g_t(k)^2}{f_t(k)} - 2\frac{g_t(k)g_t(k+1)}{f_t(k+1)} + \frac{g_t(k+1)^2}{f_t(k+2)} \\
&= \sum_k \frac{u_t(k+1)^2}{u_t(k)}v_t(k) - 2u_t(k+1)v_t(k+1) + \frac{v_t(k+1)^2}{v_t(k)}u_t(k) \\
&\quad -2u_t(k+2)v_t(k) + 2u_t(k+1)\frac{v_t(k)v_t(k+2)}{v_t(k+1)} \\
&\quad +2v_t(k+1)\frac{u_t(k)u_t(k+2)}{u_t(k+1)} - 2u_t(k)v_t(k+2) \\
&\quad +\frac{v_t(k+1)^2}{v_t(k+2)}u_t(k+2) - 2u_t(k+1)v_t(k+1) \\
&\quad +\frac{u_t(k+1)^2}{u_t(k+2)}v_t(k+2).
\end{aligned}$$

We then obtain immediately

$$-H''(t) = \sum_k \frac{\partial^2}{\partial t^2}f_t(k)\log(f_t(k) + \sum_k \frac{1}{f_t(k)}\left(\frac{\partial}{\partial t}f_t(k)\right)^2 \geq 0. \qquad \square$$

The equation $\frac{\partial}{\partial t^2}u_t(k) = \nabla_2 u_t(k+1)$ is a linear PDE often referred to as the heat equation on $\mathbb{Z}$. The second derivative operator $K = \nabla_2$ can be seen as a Markov semigroup generator: if u_0 is (the density of) a probability measure on $\mathbb{Z}$, then $(u_t)_{t\geq 0} := (e^{tK}u_0)_{t>0}$ is a family of probabilty measures on $\mathbb{Z}$ satisfying the heat equation. If $u_0 = \delta_0$ is the Dirac measure at 0, we denote $(P_t(k))_{k\in\mathbb{Z}}$ the measure $e^{tK}u_0$. If $u_0 = \delta_i$, then we have $(e^{tK}u_0)(k) = P_t(k-i)$.

Proposition 17 shows the concavity of the entropy functional for a class of families $(f_t) \in \mathcal{P}(\mathbb{Z})$. We now show that this class is large enough to contain interpolating families for any couple of prescribed initial and final measures f_0, f_1. We will actually prove a more explicit statement:

Theorem 6 *Let f_0, f_1 be two finitely supported probability measures on $\mathbb{Z}$. There exists a unique coupling $\pi \in \Pi(f_0, f_1)$ solution of the minimization problem*

$$\inf_{\pi' \in \Pi(f_0,f_1)} \sum_{i,j} \pi'(i,j) \log\left(\frac{\pi'(i,j)}{P_1(i-j)}\right). \tag{31}$$

Moreover, there are two functions $a, b : \mathbb{Z} \to \mathbb{R}$ such that $\pi(i,j) = a(i)b(j)P_1(i-j)$ and the family $(f_t(k))_{t\in[0,1],k\in\mathbb{Z}}$ defined by

$$f_t(k) := \sum_{i,j} \frac{P_t(k-i)P_{1-t}(k-j)}{P_1(i-j)} \pi(i,j) \tag{32}$$

satisfies $f_t(k) = u_t(k)v_t(k)$ with $\frac{\partial}{\partial t}u_t(k) = \nabla_2 u_t(k+1)$ and $\frac{\partial}{\partial t}v_t(k) = -\nabla_2 v_t(k+1)$. In particular, the entropy $H(t)$ of f_t is a concave function of t.

Sketch of proof We want to minimize the functional $I(\pi') := \sum_{i,j} \pi'(i,j) \log\left(\frac{\pi'(i,j)}{P_1(i-j)}\right)$ over the set $\Pi(f_0, f_1)$. The set $\Pi(f_0, f_1)$ can be seen as a convex subset of $\mathcal{P}(\mathbb{Z}\times\mathbb{Z})$. More precisely, this is the set of probability measures π' on $\{0,\ldots,N\}\times\{0,\ldots,N\}$ satisfying the constraints $\sum_j \pi'(i,j) = f_0(i)$ and $\sum_i \pi'(i,j) = f_1(j)$. If f_0 or f_1 are Dirac measures, then the set $\Pi(f_0, f_1)$ has a unique element, and the family (f_t) constructed as in equation (32) is similar to a thinning family. If neither f_0 nor f_1 are Dirac measures, then the interior of $\Pi(f_0, f_1)$ is non-empty.

The mapping $\pi' \mapsto I(\pi')$ is shown to be smooth and strictly convex on $\Pi(f_0, f_1)$, and thus attains its infimum at a unique coupling $\pi \in \Pi(f_0, f_1)$. Moreover, it is possible to show that this infimum cannot be attained on the boundary of $\Pi(f_0, f_1)$, which shows that π is a critical point for I. The coupling π is also a critical point for the functional

$$J(\pi') := I(\pi') - 1 = \sum_{i,j} \pi'(i,j) \log\left(P_1(i-j)\pi'(i,j)\right) - \pi'(i,j).$$

We have $\frac{\partial}{\partial \pi'(i,j)} J(\pi') = \log(\pi'(i,j)) - \log(P_1(i-j))$. Moreover, the constraints defining $\Pi(f_0, f_1)$ as a subset of $\mathcal{P}(\mathbb{Z}\times\mathbb{Z})$ show that there exist two functions $A, B : \mathbb{Z} \to \mathbb{R}$ such that

$$\forall i,j\,,\ \log(\pi'(i,j)) - \log(P_1(i-j)) = A(i) + B(j).$$

Taking $a(i) = \exp(A(i))$ and $b(j) = \exp(B(j))$ proves the first part of the theorem.

We now notice that $P_0(k-i)$ (resp. $P_1(k-j)$) is equal to zero unless $k = i$ (resp. $k = j$). This proves that the family $(f_t)_{t\in[0,1]}$ interpolates the measures f_0 and f_1. Moreover, we can write $f_t = u_t v_t$ with $u_t(k) := \sum_i P_t(k-i)a(i)$ and $v_t(k) :=$

$\sum_j P_{1-t}(j-k)b(j)$. The fact that $\frac{\partial}{\partial t}P_t(l) = \nabla_2 P_t(l+1)$ shows, by linearity, that $\frac{\partial}{\partial t}u_t(k) = \nabla_2 u_t(k+1)$ and similarly $\frac{\partial}{\partial t}v_t(k) = -\nabla_2 v_t(k+1)$. □

Such an interpolation between f_0 and f_1 is called an entropic interpolation. The continuous version of Theorem 6 is as follows:

Proposition 18 *Let $f_0(x)$, $f_1(x)$ be two smooth, compactly supported probability densities on $\mathbb{R}$. There exist two families of functions $(u_t(x))_{t\in[0,1],x\in\mathbb{R}}$, $(v_t(x))_{t\in[0,1],x\in\mathbb{R}}$ such that:*

- *$f_t(x) := u_t(x)v_t(x)$ interpolates the measures f_0 and f_1.*
- *$\frac{\partial}{\partial t}u_t(x) = \Delta u_t(x)$,*
- *$\frac{\partial}{\partial t}v_t(x) = -\Delta v_t(x)$,*

where Δ is the Laplacian on $\mathbb{R}$.

Moreover, there exists a coupling π between f_0 and f_1 which can be written $d\pi(x,y) = a(x)b(y)P_1(x-y)dxdy$, such that

$$f_t(z) = \int_{x,y\in\mathbb{R}\times\mathbb{R}} \frac{P_t(z-x)P_{1-t}(y-z)}{P_1(x-y)} d\pi(x,y),$$

and this coupling π is solution to the minimization problem

$$\inf_{\pi'\in\Pi(f_0,f_1)} \int_{\mathbb{R}\times\mathbb{R}} \log\left(\frac{\pi'(x-y)}{P_1(x-y)}\right) d\pi'(x-y),$$

where we set, for $t>0$, $P_t(z) := \frac{1}{\sqrt{t}}\exp\left(-\frac{z^2}{4t}\right)$.

The proof of Proposition 18 follows exactly the same lines as the proof of Theorem 6.

The constructions of entropic interpolations on $\mathbb{Z}$ and $\mathbb{R}$ have so far been quite similar. An important difference appears when we consider an additional parameter ϵ and study the same problem with equations

$$\frac{\partial}{\partial t}u_t(k) = \varepsilon\nabla_2 u_t(k+1)\ , \ \frac{\partial}{\partial t}v_t(k) = -\varepsilon\nabla_2 v_t(k+1), \tag{33}$$

and

$$\frac{\partial}{\partial t}u_t(x) = \varepsilon\Delta u_t(x)\ , \ \frac{\partial}{\partial t}v_t(x) = -\varepsilon\Delta v_t(x).$$

In that case, one can prove as in Theorem 6 the existence, when f_0 and f_1 are prescribed, of an interpolation $(f_t(k))_{t\in[0,1]} = (u_t(k)v_t(k))$, such that $u_t(k)$ and $v_t(k)$ satisfy (33). Moreover, $(f_t(k))$ can be written under the form

$$f_t(k) := \sum_{i,j} \frac{P_{\varepsilon t}(k-i)P_{\varepsilon(1-t)}(k-j)}{P_\varepsilon(i-j)} \pi_\varepsilon(i,j)$$

where the coupling π_ε is solution to the minimization problem

$$\inf_{\pi' \in \Pi(f_0, f_1)} I(\pi') = \inf_{\pi' \in \Pi(f_0, f_1)} \sum_{i,j} \pi'(i,j) \log\left(\frac{\pi'(i,j)}{P_\varepsilon(i-j)}\right).$$

A similar statement can be deduced in the continuous case from Proposition 18. In this setting, we consider families $(f_t(x))$ of the form

$$f_t(z) = \int_{x,y \in \mathbb{R}\times\mathbb{R}} \frac{P_{\varepsilon t}(z-x) P_{\nu\varepsilon(1-t)}(y-z)}{P_\varepsilon(x-y)} d\pi_\varepsilon(x,y),$$

where the coupling π_ε is solution to the minimization problem

$$\inf_{\pi' \in \Pi(f_0, f_1)} \int_{\mathbb{R}\times\mathbb{R}} \log\left(\frac{\pi'(x-y)}{P_\varepsilon(x-y)}\right) d\pi'(x-y),$$

The rigorous study of the behaviour of the minimization problems (5) when $\varepsilon \to 0$ is at the heart of the paper [Leo12] by Léonard. We explain here briefly why we must expect different behaviours in the discrete and continuous cases:

In the continuous case, we have $P_\varepsilon(z) = \frac{1}{\sqrt{\varepsilon}} \exp\left(-\varepsilon \frac{z^2}{4\varepsilon}\right)$, so the functional $I(\pi')$ we want to minimize can be written under the form

$$I(\pi') = \int_{\mathbb{R}\times\mathbb{R}} \left(-1/2 \log(\varepsilon) + \varepsilon(x-y)^2 + o(\varepsilon)\right) \pi'(x,y),$$

so we can expect, when $\varepsilon \to 0$, that the infimum of $I(\pi')$ is getting close to the solution of the minimization problem

$$\inf_{\pi' \in \Pi(f_0, f_1)} \int_{\mathbb{R}\times\mathbb{R}} (x-y)^2 d\pi'(x,y),$$

which is exactly the Monge-Kantorovitch problem for the quadratic cost.

In the discrete case, we no longer have $P_\varepsilon(z) = \frac{1}{\sqrt{\varepsilon}} \exp\left(-\varepsilon \frac{z^2}{4\varepsilon}\right)$, but $P_\varepsilon(l) = \frac{\varepsilon^l}{l!} + o(\varepsilon^l)$, so the minimization problem is written

$$I(\pi') = \sum_{i,j} \pi'(i,j) \left(-\log(\varepsilon)|i-j| + \log\left(|i-j|!\pi(i,j) + o(\varepsilon)\right)\right).$$

When $\varepsilon \to 0$, we have $-\log(\varepsilon) >> 1 >> \varepsilon$, so the minimizer π of I is expected to behave as follows:

- The coupling π minimizes the functional $I_0(\pi') := \sum_{i,j} |i-j|\pi(i,j)$, which means that π is a W_1-optimal coupling between f_0 and f_1.

- Among the minimzers of I_0, i.e. among the W_1-optimal couplings, π minimizes the functional

$$I_1(\pi') := \sum_{i,j} \pi'(i,j) \log(|i-j|!\pi'(i,j)).$$

Moreover, we can describe the behaviour of the interpolating curves $\frac{P_{\varepsilon t}(z-x)P_{\varepsilon(1-t)}(y-z)}{P_\varepsilon(x-y)}$ when $\varepsilon \to 0$. As before, the continuous and discrete cases are quite different:

- In the continuous case, we have

$$\frac{P_{\varepsilon t}(z-x)P_{\varepsilon(1-t)}(y-z)}{P_\varepsilon(y-x)} = \frac{1}{\sqrt{\varepsilon}} \exp\left(-\frac{\varepsilon}{4}\left(\frac{(x-z)^2}{t} + \frac{(y-z)^2}{1-t} - (x-y)^2\right)\right),$$

which, as a function of z, is a probability density more and more concentrated around $z = (1-t)x + ty$, so we can expect the family of measures $(\int_{\mathbb{R}\times\mathbb{R}} \frac{P_{\varepsilon t}(z-x)P_{\varepsilon(1-t)}(y-z)}{P_\varepsilon(x-y)} d\pi_\varepsilon(x,y)dz)_{t\in[0,1]}$ to converge in some sense (the rigorous statement use the Γ-convergence) to the family of measures $(\int_{\mathbb{R}\times\mathbb{R}} \delta_{(1-t)x+ty} d\pi(x,y))_{t\in[0,1]}$, which is exactly the W_2-Wasserstein geodesic between f_0 and f_1.
- In the discrete case, we have

$$\forall i \le k \le j \in \mathbb{Z}, \quad \frac{P_{\varepsilon t}(k-i)P_{\varepsilon(1-t)}(j-k)}{P_\varepsilon(j-i)} \to \mathrm{bin}_{i,j,t}(k),$$

so the entropic interpolation $\sum_{i,j} \frac{P_{\varepsilon t}(k-i)P_{\varepsilon(1-t)}(k-j)}{P_\varepsilon(i-j)} \pi_\varepsilon(i,j)$ is expected to converge, as $\varepsilon \to 0$, to the $W_{1,+}$-geodesic between f_0 and f_1.

In the general graph case, the question of the equivalence between $W_{1,+}$-geodesics and limit cases of entropic interpolations described as above remains unclear and future research works will focus on the comparison between these two constructions.

6 Shepp-Olkin Interpolations

We finish these notes by explaining how ideas coming from the optimal transportation theory on $\mathbb{Z}$ can be used to solve a problem about the entropy of sums of independent Bernoulli random variables.

Let $p_1(t), \ldots, p_n(t)$ be a family of affine functions with $p_i : [0,1] \to [0,1]$. The slope of $p_i(t)$ will be denoted by p_i'. Let $X_1(t), \ldots, X_n(t)$ be independent Bernoulli random variables of parameters $p_1(t), \ldots, p_n(t)$. We denote by $(f_t(k))$, or by $(B_t(k))$ the probability measure, supported on $\{0, \ldots, n\}$, defined by

$$f_t(k) = B_t(k) = \mathbb{P}\left(\sum_{i=1}^{n} X_i(t) = k\right).$$

The following is easily proven by induction on the number n of parameters:

Proposition 19 *When each function p_i is linear, i.e. when $p_i(t) = a_i \cdot t$ for some constant $a_i \in [0, 1]$, then (f_t) is the thinning family of the measure f_1.*

Theorem 2 then gives a concavity of entropy result in this particular case. This property actually holds in general:

Theorem 7 *The entropy $H(t)$ of f_t is a concave function of t.*

Theorem 7 has been proven by the author and Johnson in [HJ14] and [HJ16]. In the first paper, a restricted version of Theorem 7 is proven, in the particular case, called 'monotonic case', where all the slopes p_i' are non-negative. In this section, we explain the strategy of the proof in this case.

As Theorem 7 is a result about the concavity of the entropy along a family of probability measures on $\mathbb{Z}$, it is tempting to check if, by any chance, f_t is a $W_{1,+}$-geodesic on $\mathbb{Z}$. In that case, Theorem 4 would directly prove Theorem 7. The definition of $f_t(k)$ as a Bernoulli sum allows us to give quite explicit forms for the families $(g_t)_{t\in[0,1]}$ and $(h_t)_{t\in[0,1]}$:

Proposition 20 *We set:*

$$g_t(k) := \sum_i p_i' B_{i,t}(k) \ , \ h_t(k) := \sum_{i,j} p_i' p_j' B_{i,j,t}(k).$$

where, for $i \in \{1, \dots, n\}$, the family of measures $(B_{i,t}(k))$ is defined as $(B_t(k))$, but omitting the function $p_i(t)$, and where $B_{i,j,t}(k)$ is defined similarly. We then have:

$$\frac{\partial}{\partial t} f_t(k) = -\nabla_1 g_t(k) \ , \ \frac{\partial^2}{\partial t^2} f_t(k) = \nabla_2 h_t(k).$$

Apart from the thinning case, where $p_i(t) = a_i \cdot t$, one can check that the family (f_t) is not in general a $W_{1,+}$-geodesic. In the general case, the quantity $h_t(k)f_t(k+1) - g_t(k)g_t(k+1)$ is no longer equal to zero, but it is possible to prove the following upper bound:

$$(h_t(k)f_t(k+1) - g_t(k)g_t(k+1))^2 \leq (h_t(k)f_t(k) - g_t(k)^2)(h_t(k)f_t(k+2) - g_t(k+1)^2) \tag{34}$$

Inequality (34) can be seen as a weak form of the Benamou-Brenier equation (11).

As in Section 2, we have in the monotonic case:

$$
\begin{aligned}
-H''(t) = \sum_k h_t(k) \Bigg[& \log\left(\frac{f_t(k)h_t(k)}{g_t(k)^2}\right) - 2\log\left(\frac{f_t(k+1)h_t(k)}{g_t(k)g_t(k+1)}\right) \\
& + \log\left(\frac{f_t(k+2)h_t(k)}{g_t(k+1)^2}\right)\Bigg] && (35) \\
& + \sum_k \frac{g_t(k)^2}{f_t(k)} - 2\frac{g_t(k)g_t(k+1)}{f_t(k+1)} + \frac{g_t(k+1)^2}{f_t(k+2)}. && (36)
\end{aligned}
$$

The strategy to prove that $-H''(t) \geq 0$ consists in replacing the elementary inequality $\log(x) \geq 1 - 1/x$ by a second order inequality:

$$\forall x \leq 0, (1+x)\log(1+x) \geq x + x^2/2 \,, \forall x \geq 0, (1+x)\log(1+x) \leq x + x^2/2. \tag{37}$$

In the monotonic case, the following inequalities hold:

$$
\begin{aligned}
f_t(k)h_t(k) &\leq g_t(k)^2, \\
f_t(k+1)h_t(k) &\geq g_t(k)g_t(k+1), \\
f_t(k+2)h_t(k) &\leq g_t(k+1)^2,
\end{aligned}
$$

We use the first inequality as follows:

$$
\begin{aligned}
\sum_k h_t(k)\log\left(\frac{f_t(k)h_t(k)}{g_t(k)^2}\right) &= \sum_k \frac{g_t(k)^2}{f_t(k)}\left(1 + \frac{h_t(k)f_t(k) - g_t(k)^2}{g_t(k)^2}\right) \\
&\qquad \log\left(1 + \frac{h_t(k)f_t(k) - g_t(k)^2}{g_t(k)^2}\right) \\
&\geq h_t(k) - \frac{g_t(k)^2}{f_t(k)} + \frac{1}{2}\frac{(h_t(k)f_t(k) - g_t(k)^2)^2}{f_t(k)g_t(k)^2}.
\end{aligned}
$$

Similar inequalities can be obtained by using the second and third inequalities.

We can thus apply the second-order inequalities (37) to equation (35) to obtain the bound

$$
\begin{aligned}
-H''(t) \geq \sum_k & \frac{(h_t(k)f_t(k) - g_t(k)^2)^2}{2f_t(k)g_t(k)^2} - 2\frac{(h_t(k+1)f_t(k) - g_t(k)g_t(k+1))^2}{2f_t(k+1)g_t(k)g_t(k+1)} \\
& + \frac{(h_t(k)f_t(k+2) - g_t(k+1)^2)^2}{2f_t(k+2)g_t(k+1)^2}.
\end{aligned}
$$

We finally use the log-concavity property $f_t(k+1)^2 \leq f_t(k)f_t(k+2)$, which holds for any sum of independent Bernoulli variables, and inequality (34) to write:

$$-H''(t) \geq \sum_k \frac{1}{2}\left(\frac{h_t(k)f_t(k) - g_t(k)^2}{\sqrt{f_t(k)}g_t(k+1)} - \frac{h_t(k)f_t(k+2) - g_t(k+1)^2}{\sqrt{f_t(k+2)}g_t(k+1)}\right)^2 \geq 0,$$

which proves the concavity of the entropy.

References

[AG13] Ambrosio, L. ; Gigli, N. A user's guide to optimal transport. In *Modelling and optimisation of flows on networks* (pp. 1–155) (2013). Springer Berlin Heidelberg.

[BB99] Benamou, J-D ; Brenier, Y. A numerical method for the optimal time-continuous mass transport problem and related problems. In *Monge Ampere equation: applications to geometry and optimization (Deerfield Beach, FL, 1997)*, 1–11, Contemp. Math., 226, *Amer. Math. Soc., Providence, RI*, 1999.

[EM12] Erbar, M. ; Maas, J. Ricci curvature of finite Markov chains via convexity of the entropy. *Arch. Ration. Mech. Anal.* 206 (2012), no. 3, 997–1038.

[GRST14] Gozlan N., Roberto C., Samson P-M. and Tetali P. Displacement convexity of entropy and related inequalities on graphs. Probability Theory and Related Fields, 1–48, 2014.

[Hill14a] Hillion, E. Contraction of measures on graphs. *Potential Analysis*, 41(3) (2014), 679–698.

[Hill14b] Hillion, E. $W_{1,+}$-interpolation of probability measures on graphs. *Electron. Journ. Probab.*, 19 (2014)

[Hill14c] Hillion, E. Entropy along $W_{1,+}$-geodesics on graphs . arXiv preprint, arXiv:1406:5089.

[HJ14] Hillion, E ; Johnson, O. A proof of the Shepp–Olkin concavity conjecture. arXiv preprint, arXiv:1503.01570.

[HJ16] Hillion, E. ; Johsnon, O. Discrete versions of the transport equation and the Shepp–Olkin conjecture. *The Annals of Probability* 2016, Vol. 44, No. 1, 276–306

[HJK07] Harremoës, P. ; Johnson, O. ; Kontoyiannis, I. Thinning and the law of small numbers. In *Information Theory*, 2007. ISIT 2007. IEEE International Symposium on (pp. 1491–1495). IEEE.

[JY09] Johnson, O. ; Yu,Y. Concavity of entropy under thinning. In *Information Theory*, 2009. ISIT 2009. IEEE International Symposium on (pp. 144–148). IEEE.

[KKRT15] Klartag, B. ; Kozma, G. ; Ralli, P. ; Tetali, P. Discrete curvature and abelian groups. arXiv preprint arXiv:1501.00516.

[Leo12] Léonard, C. From the Schrödinger problem to the Monge–Kantorovich problem. *Journal of Functional Analysis*, 262(4) (2012), 1879–1920.

[Leo13a] Léonard, C. On the convexity of the entropy along entropic interpolations. arXiv preprint arXiv:1310.1274.

[Leo13b] Léonard, C. Lazy random walks and optimal transport on graphs. arXiv preprint, arXiv:1308.0226.

[Leo14] Léonard, C. A survey of the Schrödinger problem and some of its connections with optimal transport. *Discrete Contin. Dyn. Syst.* 34 (2014), no. 4, 1533–1574.

[LV09] Lott, J. and Villani, C. Ricci curvature for metric-measure spaces via optimal transport. *Ann. of Math.* 169 (2009), no. 3, 903–991.

[Miel13] Mielke, A. Geodesic convexity of the relative entropy in reversible Markov chains. *Calculus of Variations and Partial Differential Equations*, 2013, 48(1–2), 1–31.

[Oll09] Ollivier, Y. Ricci curvature of Markov chains on metric spaces. *J. Funct. Anal.* 256 (2009), no. 3, 810–864.

[OV00] Otto, F. ; Villani, C. Generalization of an inequality by Talagrand and links with the logarithmic Sobolev inequality. *Journal of Functional Analysis*, 2000, 173(2), 361–400.

[Ren56] Rényi, A. A characterization of Poisson processes. *Magyar Tud. Akad. Mat. Kutató Int. Közl.*, vol. 1, pp. 519–527, 1956.

[Sant15] Santambrogio, F. Optimal Transport for Applied Mathematicians: Calculus of Variations, PDEs, and Modeling (Vol. 87)(2015). Birkhäuser.

[Stu06a] Sturm, K-T. On the geometry of metric measure spaces. I. *Acta Math.* 196 (2006), no. 1, 65–131.

[Stu06b] Sturm, K-T. On the geometry of metric measure spaces. II. *Acta Math.* 196 (2006), no. 1, 133–177.

[SB09] Bonciocat, A-I ; Sturm, K-T. Mass transportation and rough curvature bounds for discrete spaces. *Journal of Functional Analysis*, 256 (2009), no. 9, 2944–2966.

[Vill03] Villani, C. Topics in optimal transportation. Graduate Studies in Mathematics, 58. *American Mathematical Society, Providence, RI*, 2003. ISBN: 0-8218-3312-X

[Vill09] Villani, C. Optimal transport. Old and new. Grundlehren der Mathematischen Wissenschaften [Fundamental Principles of Mathematical Sciences], 338. *Springer-Verlag, Berlin*, 2009. ISBN: 978-3-540-71049-3

Entropy and Thinning of Discrete Random Variables

Oliver Johnson

Abstract We describe five types of results concerning information and concentration of discrete random variables, and relationships between them, motivated by their counterparts in the continuous case. The results we consider are information theoretic approaches to Poisson approximation, the maximum entropy property of the Poisson distribution, discrete concentration (Poincaré and logarithmic Sobolev) inequalities, monotonicity of entropy and concavity of entropy in the Shepp–Olkin regime.

1 Results in Continuous Case

This paper gives a personal review of a number of results concerning the entropy and concentration properties of discrete random variables. For simplicity, we only consider independent random variables (though it is an extremely interesting open problem to extend many of the results to the dependent case). These results are generally motivated by their counterparts in the continuous case, which we will briefly review, using notation which holds only for Section 1.

For simplicity we restrict our attention in this section to random variables taking values in $\mathbb{R}$. For any probability density p, write $\lambda_p = \int_{-\infty}^{\infty} xp(x)dx$ for its mean and $\mathrm{Var}_p = \int_{-\infty}^{\infty}(x-\lambda_p)^2 p(x)dx$ for its variance. We write $h(p)$ for the differential entropy of p, and interchangeably write $h(X)$ for $X \sim p$. Similarly we write $D(p\|q)$ or $D(X\|Y)$ for relative entropy. We write $\phi_{\mu,\sigma^2}(x)$ for the density of Gaussian $Z_{\mu,\sigma^2} \sim N(\mu,\sigma^2)$.

Given a function f, we wish to measure its concentration properties with respect to probability density p; we write $\lambda_p(f) = \int_{-\infty}^{\infty} f(x)p(x)dx$ for the expectation of f with respect to p, write $\mathrm{Var}_p(f) = \int_{-\infty}^{\infty} p(x)(f(x)-\lambda_p(f))^2 dx$ for the variance and define

$$\mathrm{Ent}_p(f) = \int_{-\infty}^{\infty} p(x)f(x)\log f(x)dx - \lambda_p(f)\log\lambda_p(f). \tag{1}$$

O. Johnson (✉)
School of Mathematics, University of Bristol, University Walk, BS8 1TW Bristol, UK
e-mail: maotj@bristol.ac.uk

E. Carlen et al. (eds.), *Convexity and Concentration*, The IMA Volumes in Mathematics and its Applications 161, DOI 10.1007/978-1-4939-7005-6_2

We briefly summarize the five types of results we study in this paper, the discrete analogues of which are described in more detail in the five sections from Sections 3 to 7, respectively.

1. **[Normal approximation]** A paper by Linnik [70] was the first to attempt to use ideas from information theory to prove the Central Limit Theorem, with later contributions by Derriennic [37] and Shimizu [87]. This idea was developed considerably by Barron [14], building on results of Brown [25]. Barron considered independent and identically distributed X_i with mean 0 and variance σ^2. He proved [14] that relative entropy $D((X_1 + \ldots + X_n)/\sqrt{n} \| Z_{0,\sigma^2})$ converges to zero if and only if it ever becomes finite. Carlen and Soffer [29] proved similar results for non-identical variables under a Lindeberg-type condition.

   While none of these papers gave an explicit rate of convergence, later work [7, 11, 58] proved that relative entropy converges at an essentially optimal $O(1/n)$ rate, under the (too strong) assumption of finite Poincaré constant (see Definition 1.1 below). Typically (see [54] for a review), these papers did not manipulate entropy directly, but rather used properties of projection operators to control the behaviour of Fisher information on convolution. The use of Fisher information is based on the de Bruijn identity (see [14, 17, 88]), which relates entropy to Fisher information, using the fact that (see, for example, [88, Equation (5.1)] for fixed random variable X, the density p_t of $X + Z_{0,t}$ satisfies a heat equation of the form

$$\frac{\partial p_t}{\partial t}(z) = \frac{1}{2}\frac{\partial^2 p_t}{\partial z^2}(z) = \frac{1}{2}\frac{\partial}{\partial z}\left(p_t(z)\rho_t(z)\right), \tag{2}$$

   where $\rho_t(z) := \frac{\partial p_t}{\partial z}(z)/p_t(z)$ is the score function with respect to location parameter, in the Fisher sense.

   More recent work of Bobkov, Chistyakov and Götze [19, 20] used properties of characteristic functions to remove the assumption of finite Poincaré constant, and even extended this theory to include convergence to other stable laws [18, 21].

2. **[Maximum entropy]** In [84, Section 20], Shannon described maximum entropy results for continuous random variables in exponential family form. That is, given a test function ψ, positivity of the relative entropy shows that the entropy $h(p)$ is maximized for fixed values of $\int_{-\infty}^{\infty} p(x)\psi(x)dx$ by densities proportional to $\exp(-c\psi(x))$ for some value of c.

   This is a very useful result in many cases, and gives us well-known facts such as that entropy is maximized under a variance constraint by the Gaussian density. This motivates the information theoretic approaches to normal approximation described above, as well as the monotonicity of entropy result of [8] described later.

   However, many interesting densities cannot be expressed in exponential family form for a natural choice of ψ, and we would like to extend maximum entropy theory to cover these. As remarked in [40, p. 215], 'immediate candidates to be subjected to such analysis are, of course, stable laws'. For example, consider the Cauchy density, for which it is not at all natural to consider the expectation of $\psi(x) = \log(1+x^2)$ in order to define a maximum entropy class. In the discrete context of Section 4 we discuss the Poisson mass function, again believing that $\mathbb{E}\log X!$ is not a natural object of study.

3. **[Concentration inequalities]** Next we consider concentration inequalities; that is, relationships between senses in which a function is approximately constant. We focus on Poincaré and logarithmic Sobolev inequalities.

Definition 1.1 *We say that a probability density p has Poincaré constant* R_p, *if for all sufficiently well-behaved functions g,*

$$\mathrm{Var}_p(g) \leq R_p \int_{-\infty}^{\infty} p(x) g'(x)^2 dx. \tag{3}$$

Papers vary in the exact sense of 'well-behaved' in this definition, and for the sake of brevity we will not focus on this issue; we will simply suppose that both sides of (3) are well defined.

Example 1.2 *Chernoff [32] (see also [23]) used an expansion in Hermite polynomials (see [91]) to show that the Gaussian densities* ϕ_{μ,σ^2} *have Poincaré constant equal to* σ^2. *By considering* $g(t) = t$, *it is clear that for any random variable X with finite variance,* $R_p \geq \mathrm{Var}_p$. *Chernoff's result [32] played a central role in the information theoretic proofs of the Central Limit Theorem of Brown [25] and Barron [14].*

In a similar way, we define the log-Sobolev constant, in the form discussed in [10]:

Definition 1.3 *We say that a probability density p satisfies the logarithmic Sobolev inequality with constant C if for any function f with positive values:*

$$\mathrm{Ent}_p(f) \leq \frac{C}{2} \int_{-\infty}^{\infty} \frac{\Gamma_1(f,f)(x)}{f(x)} p(x) dx, \tag{4}$$

where we write $\Gamma_1(f,g) = f'g'$, *and view the RHS as a Fisher-type expression.*

This is an equivalent formulation of the log-Sobolev inequality first stated by Gross [42], who proved that the Gaussian density ϕ_{μ,σ^2} has log-Sobolev constant σ^2. In fact, Gross's Gaussian log-Sobolev inequality [42] follows from Shannon's Entropy Power Inequality [84, Theorem 15], as proved by the Stam–Blachman approach [17, 88].

However, we would like to find results that strengthen Chernoff and Gross's results, to provide inequalities of similar functional form, under assumptions that p belongs to particular classes of densities. Gross's Gaussian log-Sobolev inequality can be considerably generalized by the Bakry–Émery calculus [9] (see [6, 10, 43] for reviews of this theory). Assuming that the so-called

Bakry-Émery condition is satisfied, taking $U(t) = t^2$ in [9, Proposition 5] we deduce (see also [10, Proposition 4.8.1]):

Theorem 1.4 *If the Bakry-Émery condition holds with constant c, then the Poincaré inequality holds with constant* $1/c$.

Similarly [9, Theorem 1] (see also [10, Proposition 5.7.1]) gives that:

Theorem 1.5 *If the Bakry-Émery condition holds with constant c, then the logarithmic Sobolev inequality holds with constant* $1/c$.

If p is Gaussian with variance σ^2, since (see [43, Example 4.18]), the Bakry-Émery condition holds with constant $c = 1/\sigma^2$, we recover the original Poincaré inequality of Chernoff [32] and the log-Sobolev inequality of Gross [42]. Indeed, if the ratio $p/\phi_{\mu,\sigma^2}$ is a log-concave function, then this approach shows that the Poincare and log-Sobolev inequalities hold with constant σ^2.

4. **[Monotonicity of entropy]** While Brown [25] and Barron [14] exploited ideas of Stam [88] to deduce that entropy of $(X_1 + \ldots + X_n)/\sqrt{n}$ increases in the Central Limit Theorem regime along the 'powers of 2 subsequence' $n = 2^k$, it remained an open problem for many years to show that the entropy is monotonically increasing in all n. This conjecture may well date back to Shannon, and was mentioned by Lieb [68]. It was resolved by Artstein, Ball, Barthe and Naor [8], who proved a more general result using a variational characterization of Fisher information (introduced in [11]):

Theorem 1.6 ([8]) *Given independent continuous X_i with finite variance, for any positive α_i such that $\sum_{i=1}^{n+1} \alpha_i = 1$, writing $\alpha^{(j)} = 1 - \alpha_j$, then*

$$nh\left(\sum_{i=1}^{n+1} \sqrt{\alpha_i} X_i\right) \geq \sum_{j=1}^{n+1} \alpha^{(j)} h\left(\sum_{i \neq j} \sqrt{\alpha_i/\alpha^{(j)}} X_i\right). \tag{5}$$

This result is referred to as monotonicity because, choosing $\alpha_i = 1/(n+1)$ for independent and identically distributed (IID) X_i, (5) shows that $h\left(\sum_{i=1}^n X_i/\sqrt{n}\right)$ is monotonically increasing in n. Equivalently, relative entropy $D\left(\sum_{i=1}^n X_i/\sqrt{n} \middle\| Z\right)$ is monotonically decreasing in n. This means that the Central Limit Theorem can be viewed as an equivalent of the Second Law of Thermodynamics. An alternative proof of this monotonicity result is given by Tulino and Verdú [94], based on the MMSE characterization of mutual information of [44].

The form of (5) in the case $n = 2$ was previously a well-known result; indeed (see, for example, [36]) it is equivalent to Shannon's Entropy Power Inequality [84, Theorem 15]. However, Theorem 1.6 allows a stronger form of the Entropy Power Inequality to be proved, referring to 'leave-one-out' sums (see [8, Theorem 3]. This was generalized by Madiman and Barron [72], who considered the entropy power of sums of arbitrary subsets of the original variables. The more recent paper [73] reviews in detail relationships between the Entropy Power Inequality and results in convex geometry, including the Brunn-Minkowski inequality.

5. **[Concavity of entropy]** In the setting of probability measures taking values on the real line $\mathbb{R}$, there has been considerable interest in optimal transportation of measures; that is, given densities f_0 and f_1, to find a sequence of densities f_t smoothly interpolating between them in a way which minimizes some appropriate cost function. For example, using the natural quadratic cost function induces the Wasserstein distance W_2 between f_0 and f_1. This theory is extensively reviewed in the two books by Villani [95, 96], even for probability measures taking values on a Riemannian manifold. In this setting, work by authors including Lott, Sturm and Villani [71, 89, 90] relates the concavity of the entropy $h(f_t)$ as a function of t to the Ricci curvature of the underlying manifold.

   Key works in this setting include those by Benamou and Brenier [15, 16] (see also [30]), who gave a variational characterization of the Wasserstein distance between f_0 and f_1. Authors such as Caffarelli [27] and Cordero-Erausquin [33] used these ideas to give new proofs of results such as log-Sobolev and transport inequalities. Concavity of entropy also plays a key role in the field of information geometry (see [2, 3, 78]), where the Hessian of entropy induces a metric on the space of probability measures.

2 Technical Definitions in the Discrete Case

In the remainder of this paper, we describe results which can be seen as discrete analogues of the results of Section 1. In that section, a distinguished role was played by the set of Gaussian densities, a set which has attractive properties including closure under convolution and scaling, and an explicit value for their entropies. We argue that a similar role is played by the Poisson mass functions, a class which is preserved on convolution and thinning (see Definition 2.2 below), even though (see [1]) we only have entropy bounds.

We now make some definitions which will be useful throughout the remainder of the paper. From now, we will assume all random variables take values on the positive integers $\mathbb{Z}_+$. Given a probability mass function (distribution) P, we write $\lambda_P = \sum_{x=0}^{\infty} xP(x)$ for its mean and $\mathrm{Var}_P = \sum_{x=0}^{\infty}(x-\lambda_P)^2 P(x)$ for its variance. In particular, write $\Pi_\lambda(x) = e^{-\lambda}\lambda^x/x!$ for the mass function of a Poisson random variable with mean λ and $B_{n,p}(x) = \binom{n}{x}p^x(1-p)^{n-x}$ for the mass function of a Binomial $\mathrm{Bin}(n,p)$. We write $H(P)$ for the discrete entropy of a probability distribution P, and interchangeably write $h(X)$ for $X \sim P$. For any random variable X we write P_X, defined by $P_X(x) = \mathbb{P}(X = x)$, for its probability mass function.

Given a function f and mass function P, we write $\lambda_P(f) = \sum_{x=0}^{\infty} f(x)P(x)$ for the expectation of f with respect to P, write $\mathrm{Var}_P(f) = \sum_{x=0}^{\infty} P(x)(f(x)-\lambda_P(f))^2$ and define

$$\mathrm{Ent}_P(f) = \sum_{x=0}^{\infty} P(x) f(x) \log f(x) - \lambda_P(f) \log \lambda_P(f). \tag{6}$$

Note that if $f = Q/P$, where Q is a probability mass function, then

$$\mathrm{Ent}_P(f) = \sum_{x=0}^{\infty} Q(x) \log\left(\frac{Q(x)}{P(x)}\right) = D(Q\|P), \tag{7}$$

the relative entropy from Q to P. We define the operators Δ and Δ^* (which are adjoint with respect to counting measure) by $\Delta f(x) = f(x+1) - f(x)$ and $\Delta^* f(x) = f(x-1) - f(x)$, with the convention that $f(-1) = 0$.

We now make two further key definitions, in Condition 1 we define the ultra-log-concave (ULC) random variables (as introduced by Pemantle [77] and Liggett [69]) and in Definition 2.2 we define the thinning operation introduced by Rényi [79] for point processes.

Condition 1 (ULC) *For any λ, define the class of ultra-log-concave (ULC) mass functions P*

$$\mathbf{ULC}(\lambda) = \{P : \lambda_P = \lambda \textit{ and } P(v)/\Pi_\lambda(v) \textit{ is log-concave}\}.$$

Equivalently we require that $vP(v)^2 \geq (v+1)P(v+1)P(v-1)$, for all v.

The ULC property is preserved on convolution; that is, for $P \in \mathbf{ULC}(\lambda_P)$ and $Q \in \mathbf{ULC}(\lambda_Q)$ then $P \star Q \in \mathbf{ULC}(\lambda_P + \lambda_Q)$ (see [97, Theorem 1] or [69, Theorem 2]). The ULC class contains a variety of parametric families of random variables, including the Poisson and sums of independent Bernoulli variables (Poisson–Binomial), including the binomial distribution.

A weaker condition than ULC (Condition 1) is given in [57], which corresponds to Assumption A of [28]:

Definition 2.1 *Given a probability mass function P, write*

$$\mathcal{E}_P(x) := \frac{P(x)^2 - P(x-1)P(x+1)}{P(x)P(x+1)} = \frac{P(x)}{P(x+1)} - \frac{P(x-1)}{P(x)}. \tag{8}$$

Condition 2 (c log concavity) *If $\mathcal{E}_P(x) \geq c$ for all $x \in \mathbb{Z}_+$, we say that P is c-log-concave.*

[28, Section 3.2] showed that if P is ULC then it is c-log-concave, with $c = P(0)/P(1)$. In [57, Proposition 4.2] we show that DBE(c), a discrete form of the Bakry-Émery condition, is implied by c-log-concavity.

Another condition weaker than the ULC property (Condition 1) was discussed in [35], related to the definition of a size-biased distribution. For a mass function P, we define its size-biased version P^* by $P^*(x) = (x+1)P(x+1)/\lambda_P$ (note that some authors define this to be $xP(x)/\lambda_P$; the difference is simply an offset). Now

$P \in \mathbf{ULC}(\lambda_P)$ means that P^* is dominated by P in likelihood ratio ordering (write $P^* \leq_{lr} P$). This is a stronger assumption than that P^* is stochastically dominated by P (write $P^* \leq_{st} P$) – see [82] for a review of stochastic ordering results. As described in [34, 35, 75], such orderings naturally arise in many contexts where P is the mass function of a sum of negatively dependent random variables.

The other construction at the heart of this paper is the thinning operation introduced by Rényi [79]:

Definition 2.2 *Given probability mass function P, define the α-thinned version $T_\alpha P$ to be*

$$T_\alpha P(x) = \sum_{y=x}^{\infty} \binom{y}{x} \alpha^x (1-\alpha)^{y-x} P(y). \tag{9}$$

Equivalently, if $Y \sim P$, then $T_\alpha P$ is the distribution of a random variable $T_\alpha Y := \sum_{i=1}^{Y} B_i$, where $B_1, B_2 \ldots$ are IID Bernoulli(α) random variables, independent of Y.

A key result for Section 4 is the fact that (see [55, Proposition 3.7]) if $P \in \mathbf{ULC}(\lambda_P)$ then $T_\alpha P \in \mathbf{ULC}(\alpha\lambda_P)$. Further properties of this thinning operation are reviewed in [47], including the fact that T_α preserves several parametric families, such as the Poisson, binomial and negative binomial. In that paper, we argue that the thinning operation T_α is the discrete equivalent of scaling by $\sqrt{\alpha}$ and that the 'mean-preserving transform' $T_\alpha X + T_{1-\alpha} Y$ is equivalent to the 'variance-preserving transform' $\sqrt{\alpha} X + \sqrt{1-\alpha} Y$ in the continuous case.

3 Score Function and Poisson Approximation

A well-known phenomenon, often referred to as the 'law of small numbers', states that in a triangular array of 'small' random variables, the row sums converge in distribution to the Poisson probability distribution Π_λ. For example, consider the setting where binomial laws $\text{Bin}(n, \lambda/n) \to \Pi_\lambda$ as $n \to \infty$. This problem has been extensively studied, with a variety of strong bounds proved using techniques such as the Stein–Chen method (see, for example, [12] for a summary of this technique and its applications).

Following the pioneering work of Linnik [70] applying information theoretic ideas to normal approximation, it was natural to wonder whether Poisson approximation could be considered in a similar framework. An early and important contribution in this direction was made by Johnstone and MacGibbon [61], who made the following natural definition (also used in [62, 76]), which mirrors the Fisher-type expression of (4):

Definition 3.1 *Given a probability distribution P, define*

$$I(P) := \sum_{x=0}^{\infty} P(x) \left(\frac{P(x-1)}{P(x)} - 1 \right)^2. \tag{10}$$

Johnstone and MacGibbon showed that this quantity has a number of desirable features; first a Cramér-Rao lower bound $I(P) \geq 1/\mathrm{Var}_P$ (see [61, Lemma 1.2]), with equality if and only if P is Poisson, second a projection identity which implies the subadditivity result $I(P \star Q) \leq (I(P) + I(Q))/4$ (see [61, Proposition 2.2]). In theory, these two results should allow us to deduce Poisson convergence in the required manner. However, there is a serious problem. Expanding (10) we obtain

$$I(P) = \left(\sum_{x=0}^{\infty} \frac{P(x-1)^2}{P(x)} \right) - 1. \tag{11}$$

For example, if P is a Bernoulli(p) distribution, then the bracketed sum in (11) becomes $P(0)^2/P(1)+P(1)^2/P(2) = (1-p)^2/p+p^2/0 = \infty$. Indeed, for any mass function P with finite support $I(P) = \infty$. To counter this problem, a new definition was made by Kontoyiannis, Harremoës and Johnson in [65].

Definition 3.2 ([65]) *For probability mass function P with mean λ_P, define the scaled score function ρ_P and scaled Fisher information K by*

$$\rho_P(x) := \frac{(x+1)P(x+1)}{\lambda_P P(x)} - 1, \tag{12}$$

$$K(P) := \lambda_P \sum_{x=0}^{\infty} P(x)\rho_P(x)^2. \tag{13}$$

Note that this definition does not suffer from the problems of Definition 3.1 described above.

Example 3.3 *For P Bernoulli(p), the scaled score function can be evaluated as $\rho_P(0) = P(1)/\lambda_P P(0) - 1 = p/(1-p)$ and $\rho_P(1) = 2P(2)/\lambda_P P(1) - 1 = -1$, and the scaled Fisher information as $K(P) = p^2/(1-p)$.*

Further, K retains the desirable features of Definition 3.1; first positivity of the perfect square ensures that $K(P) \geq 0$ with equality if and only if P is Poisson and second a projection identity [65, Lemma, p. 471] ensures that the following subadditivity result holds:

Theorem 3.4 ([65], Proposition 3) *If $S = \sum_{i=1}^{n} X_i$ is the sum of n independent random variables X_i with means λ_i then writing $\lambda_S = \sum_{i=1}^{n} \lambda_i$:*

$$K(P_S) \leq \frac{1}{\lambda_S} \sum_{i=1}^{n} \lambda_i K(P_{X_i}), \tag{14}$$

As discussed in [65], this result implies Poisson convergence at rates of an optimal order, in a variety of senses including Hellinger distance and total variation. However, perhaps the most interesting consequence comes via the following modified logarithmic Sobolev inequality of Bobkov and Ledoux [22], proved via a tensorization argument:

Theorem 3.5 ([22], Corollary 4) *For any function f taking positive values:*

$$\mathrm{Ent}_{\Pi_\lambda}(f) \leq \lambda \sum_{x=0}^{\infty} \Pi_\lambda(x) \frac{(f(x+1) - f(x))^2}{f(x)}. \tag{15}$$

Observe that (see [65, Proposition 2]) taking $f = P/\Pi_\lambda$ for some probability mass function P, the RHS of (15) can be written as

$$\lambda \sum_{x=0}^{\infty} \Pi_\lambda(x) f(x) \left(\frac{f(x+1)}{f(x)} - 1 \right)^2 = \lambda \sum_{x=0}^{\infty} P(x) \left(\frac{(x+1)P(x+1)}{\lambda_P P(x)} - 1 \right)^2, \tag{16}$$

and so (15) becomes $D(P \| \Pi_\lambda) \leq K(P)$. Hence, combining Theorems 3.4 and 3.5, we deduce convergence in the strong sense of relative entropy in a general setting.

Example 3.6 (See [65, Example 1]) *Combining Example 3.3 and Theorem 3.5 we deduce that*

$$D(B_{n,\lambda/n} \| \Pi_\lambda) \leq \frac{\lambda^2}{n(n-\lambda)}, \tag{17}$$

so via Pinsker's inequality [66] we deduce that

$$\| B_{n,\lambda/n} - \Pi_\lambda \|_{TV} \leq (2 + \epsilon) \frac{\lambda}{n}, \tag{18}$$

giving the same order of convergence (if not the optimal constant) as results stated in [12].

We make the following brief remarks, the first of which will be useful in the proof of the Poisson maximum entropy property in Section 4:

Remark 3.7 *An equivalent formulation of the ULC property of Condition 1 is that $P \in \mathbf{ULC}(\lambda_P)$ if and only if the scaled score function $\rho_P(x)$ is a decreasing function of x.*

Remark 3.8 *In [13], definitions of score functions and subadditive inequalities are extended to the compound Poisson setting. The resulting compound Poisson approximation bounds are close to the strongest known results, due to authors such as Roos [81].*

Remark 3.9 *The fact that two competing definitions of the score function and Fisher information are possible (see Definition 3.1 and Definition 3.2) corresponds to the two definitions of the Stein operator in [67, Section 1.2].*

4 Poisson Maximum Entropy Property

We next prove a maximum entropy property for the Poisson distribution. The first results in this direction were proved by Shepp and Olkin [86] and by Mateev [74], who showed that entropy is maximized in the class of Poisson-binomial variables with mean λ by the binomial $\mathrm{Bin}(n, \lambda/n)$ distribution. Using a limiting argument, Harremoës [46] showed that the Poisson Π_λ maximizes the entropy in the closure of the set of Poisson-binomial variables with mean λ (though note that the Poisson itself does not lie in this set).

Johnson [55] gave the following argument, based on the idea of 'smart paths'. For some function $\Lambda(\alpha)$, surprisingly, it can be easiest to prove $\Lambda(1) \leq \Lambda(0)$ by proving the stronger result that $\Lambda'(\alpha) \leq 0$ for all $\alpha \in [0, 1]$. The key idea in this case is to introduce an interpolation between probability measures using the thinning operation of Definition 2.2, and to show that it satisfies a partial differential equation corresponding to the action of the $M/M/\infty$ queue (see also [31]).

Lemma 4.1 *Write $P_\alpha(z) = \mathbb{P}(T_\alpha X + T_{1-\alpha} Y = z)$, where $X \sim P_X$ with mean λ and $Y \sim \Pi_\lambda$ then (in a form reminiscent of* (2) *above):*

$$\frac{\partial}{\partial \alpha} P_\alpha(x) = \frac{\lambda}{\alpha} \Delta^* (P_\alpha(x) \rho_\alpha(x)), \tag{19}$$

where $\rho_\alpha := \rho_{P_\alpha}$ is the score in the sense of Definition 3.2 above.

The gradient form of the RHS of this partial derivative is key for us, and is reminiscent of the continuous probability models studied in [4]. Such representations will lie at the heart of the analysis of entropy in the Shepp–Olkin setting [50, 51, 86] in Section 7.

Theorem 4.2 *If $P \in \mathbf{ULC}(\lambda)$, then*

$$H(P) \leq H(\Pi_\lambda), \tag{20}$$

with equality if and only if $P \equiv \Pi_\lambda$.

Proof We consider the functional $\Lambda(\alpha) = -\sum_{x=0}^{\infty} P_\alpha(x) \log \Pi_\lambda(x)$ as a function of α. Using Lemma 4.1 and the adjoint property of Δ and Δ^* we deduce that

$$\Lambda'(\alpha) = -\sum_{x=0}^{\infty} \frac{\partial}{\partial \alpha} P_\alpha(x) \log \Pi_\lambda(x) = -\frac{\lambda}{\alpha} \sum_{x=0}^{\infty} \Delta^*(P_\alpha(x)\rho_\alpha(x)) \log \Pi_\lambda(x)$$

$$= \frac{\lambda}{\alpha} \sum_{x=0}^{\infty} P_\alpha(x)\rho_\alpha(x) \log\left(\frac{x+1}{\lambda}\right). \tag{21}$$

Now, the assumption that $P \in \mathbf{ULC}(\lambda)$ means that $P_\alpha \in \mathbf{ULC}(\lambda)$ so by Remark 3.7, the score function $\rho_\alpha(z)$ is decreasing in z. Hence (21) is the covariance of an increasing and a decreasing function, so by 'Chebyshev's other inequality' (see [63, Equation (1.7)]) it is negative.

In other words, $X \in \mathbf{ULC}(\lambda)$ makes $\Lambda'(\alpha) \leq 0$ so $\Lambda(\alpha)$ is a decreasing function in α. Since $P_0 = \Pi_\lambda$, and $P_1 = P$, we deduce that

$$H(X) \leq -\sum_{x=0}^{\infty} P(x) \log \Pi_\lambda(x) \tag{22}$$

$$= \Lambda(1) \leq \Lambda(0) = -\sum_{x=0}^{\infty} \Pi_\lambda(x) \log \Pi_\lambda(x) = H(\Pi_\lambda),$$

and the result is proved. Here (22) follows by the positivity of relative entropy. □

Remark 4.3 *These ideas were developed by Yu [103], who proved that for any n, λ the binomial mass function $B_{n,\lambda/n}$ maximizes entropy in the class of mass functions P such that $P/B_{n,\lambda/n}$ is log-concave (this class is referred to as the ULC(n) class by [77] and [69]).*

Remark 4.4 *A related extension was given in the compound Poisson case in [59], one assumption of which was removed by Yu [105]. Yu's result [105, Theorem 3] showed that (assuming it is log-concave) the compound Poisson distribution $CP(\lambda, Q)$ is maximum entropy among all distributions with 'claim number distribution' in* $\mathbf{ULC}(\lambda)$ *and given 'claim size distribution' Q. Yu [105, Theorem 2] also proved a corresponding result for compound binomial distributions.*

Remark 4.5 *The assumption in Theorem 4.2 that $P \in$* $\mathbf{ULC}(\lambda_P)$ *was weakened in recent work of Daly [34, Corollary 2.3], who proved the same result assuming that $P^* \leq_{st} P$.*

5 Poincaré and Log-Sobolev Inequalities

We give a definition which mimics Definition 1.1 in the discrete case:

Definition 5.1 *We say that a probability mass function P has Poincaré constant R_P, if for all sufficiently well-behaved functions g,*

$$\mathrm{Var}_P(g) \leq R_P \sum_{x=0}^{\infty} P(x) \left(\Delta g(x)\right)^2 . \tag{23}$$

Example 5.2 *Klaassen's paper [64] (see also [26]) showed that the Poisson random variable Π_λ has Poincaré constant λ. As in Example 1.2, taking $g(t) = t$, we deduce that $R_P \geq \mathrm{Var}_P$ for all P with finite variance. The result of Klaassen can also be proved by expanding in Poisson–Charlier polynomials (see [91]), to mimic the Hermite polynomial expansion of Example 1.2.*

However, we would like to generalize and extend Klaassen's work, in the spirit of Theorems 1.4 and 1.5 (originally due to Bakry and Émery [9]). Some results in this direction were given by Chafaï [31], however we briefly summarize two more recent approaches, using ideas described in two separate papers.

In [35, Theorem 1.1], using a construction based on Klaassen's original paper [64], it was proved that $R_P \leq \lambda_P$, assuming that $P^* \leq_{st} P$. Recall (see [82]) that this implies $P^* \leq_{lr} P$, which is a restatement of the ULC property (Condition 1). We deduce the following result [35, Corollary 2.4]:

Theorem 5.3 *For $P \in \mathbf{ULC}(\lambda)$:*

$$\mathrm{Var}_P \leq R_P \leq \lambda_P.$$

The second approach, appearing in [57], is based on a birth-and-death Markov chain construction introduced by Caputo, Dai Pra and Posta [28], generalizing the thinning-based interpolation of Lemma 4.1. In [57], ideas based on the Bakry-Émery calculus are used to prove the following two main results, taken from [57, Theorem 1.5] and [57, Theorem 1.3], which correspond to Theorems 1.4 and 1.5, respectively.

Theorem 5.4 (Poincaré inequality) *Any probability mass function P whose support is the whole of the positive integers $\mathbb{Z}_+$ and which satisfies the c-log-concavity condition (Condition 2) has Poincaré constant $1/c$.*

Note that (see the discussion at the end of [57]), Theorem 5.4 typically gives a weaker result than Theorem 5.3, under weaker conditions. Next we describe a modified form of log-Sobolev inequality proved in [57]:

Theorem 5.5 (New modified log-Sobolev inequality) *Fix probability mass function P whose support is the whole of the positive integers $\mathbb{Z}_+$ and which satisfies the c-log-concavity condition (Condition 2). For any function f with positive values:*

$$\mathrm{Ent}_P(f) \leq \frac{1}{c} \sum_{x=0}^{\infty} P(x) f(x+1) \left(\log \left(\frac{f(x+1)}{f(x)} \right) - 1 + \frac{f(x)}{f(x+1)} \right). \tag{24}$$

Remark 5.6 *As described in [57], Theorem 5.5 generalizes a modified log-Sobolev inequality of Wu [100], which holds in the case where $P = \Pi_\lambda$ (see [104] for a more direct proof of this result). Wu's result in turn strengthens Theorem 3.5, the* *modified logarithmic Sobolev inequality of Bobkov and Ledoux [22]. It can be seen directly that Theorem 5.5 tightens Theorem 3.5 using the bound* $\log u \leq u - 1$ *with* $u = f(x+1)/f(x)$.

Note that since Theorems 5.4 and 5.5 are proved under the c-log-concavity condition (Condition 2), they hold under the stronger assumption that $P \in \mathbf{ULC}(\lambda)$ (Condition 1), taking $c = P(0)/P(1)$.

Remark 5.7 *Note that (for reasons similar to those affecting the Johnstone–MacGibbon Fisher information quantity* $I(P)$ *of Definition 3.1), Theorems 5.4 and 5.5 are restricted to mass functions supported on the whole of* $\mathbb{Z}_+$. *In order to consider mass functions such as the Binomial* $B_{n,p}$, *it would be useful to remove this assumption.*

However, one possibility is that in this case, we may wish to modify the form of the derivative used in Definition 5.1. In the paper [52], for mass functions P supported on the finite set $\{0, 1, \ldots, n\}$, *a 'mixed derivative'*

$$\nabla_n f(x) = \left(1 - \frac{x}{n}\right)(f(x+1) - f(x)) + \frac{x}{n}(f(x) - f(x-1)) \tag{25}$$

was introduced. This interpolates between left and right derivatives, according to the position where the derivative is taken. In [52] it was shown that the Binomial $B_{n,p}$ *mass function satisfies a Poincaré inequality with respect to* ∇_n. *The proof was based on an expansion in Krawtchouk polynomials (see [91]), and exactly parallels the results of Chernoff (Example 1.2) and Klaassen (Example 5.2). It would be of interest to prove results that mimic Theorem 5.4 and 5.5 for the derivative* ∇_n *of (25).*

6 Monotonicity of Entropy

Next we describe results concerning the monotonicity of entropy on summation and thinning, in a regime described in [47] as the 'law of thin numbers'. The first interesting feature is that (unlike the Gaussian case of Theorem 1.6) monotonicity of entropy and of relative entropy is not equivalent. By convex ordering arguments, Yu [104] showed that:

Theorem 6.1 *Given a random variable X with probability mass function* P_X *and mean* λ, *we write* $D(X)$ *for* $D(P_X \| \Pi_\lambda)$. *For independent and identically distributed* X_i, *with mass function P and mean* λ:

1. *relative entropy* $D\left(\sum_{i=1}^n T_{1/n} X_i\right)$ *is monotone decreasing in n,*
2. *If* $P \in \mathbf{ULC}(\lambda)$, *the entropy* $H\left(\sum_{i=1}^n T_{1/n} X_i\right)$ *is monotone increasing in n.*

In fact, implicit in the work of Yu [104] is the following stronger theorem:

Theorem 6.2 *Given positive* α_i *such that* $\sum_{i-1}^{n+1} \alpha_i = 1$, *and writing* $\alpha^{(j)} = 1 - \alpha_j$, *then for any independent* X_i,

$$nD\left(\sum_{i=1}^{n+1} T_{\alpha_i} X_i\right) \leq \sum_{j=1}^{n+1} \alpha^{(j)} D\left(\sum_{i \neq j} T_{\alpha_i/\alpha^{(j)}} X_i\right).$$

In [60, Theorem 3.2], the following result was proved:

Theorem 6.3 *Given positive* α_i *such that* $\sum_{i=1}^{n+1} \alpha_i = 1$, *and writing* $\alpha^{(j)} = 1 - \alpha_j$, *then for any independent ULC* X_i,

$$nH\left(\sum_{i=1}^{n+1} T_{\alpha_i} X_i\right) \geq \sum_{j=1}^{n+1} \alpha^{(j)} H\left(\sum_{i \neq j} T_{\alpha_i/\alpha^{(j)}} X_i\right).$$

Comparison with (5) shows that this is a direct analogue of the result of Artstein et al. [8], replacing differential entropies h with discrete entropy H, and replacing scalings by $\sqrt{\alpha}$ by thinnings by α. Since the result of Artstein et al. is a strong one, and implies strengthened forms of the Entropy Power Inequality, this equivalence is good news for us. However, there are two serious drawbacks.

Remark 6.4 *First, the proof of Theorem 6.3, which is based on an analysis of 'free energy' terms similar to the* $\Lambda(\alpha)$ *of Section 4, does not lend itself to easy generalization or extension. It is possible that a different proof (perhaps using a variational characterization similar to that of [7, 11]) may lend more insight.*

Remark 6.5 *Second, Theorem 6.3 does not lead to a single universally accepted discrete Entropy Power Inequality in the way that Theorem 1.6 did. As discussed in [60], several natural reformulations of the Entropy Power Inequality fail, and while [60, Theorem 2.5] does prove a particular discrete Entropy Power Inequality, it is by no means the only possible result of this kind. Indeed, a growing literature continues to discuss the right formulation of this inequality.*

For example, [83, 99, 101, 102] consider replacing integer addition by binary addition, with the paper [102] introducing the so-called Mrs. Gerber's Lemma, extended to a more general group context by Anantharam and Jog [5, 53]. Harremoës and Vignat [48] (see also [83]) consider conditions under which the original functional form of the Entropy Power Inequality continues to hold. Other authors use ideas from additive combinatorics and sumset theory, with Haghighatshoar, Abbe and Telatar [45] adding an explicit error term and Wang, Woo and Madiman [98] giving a reformulation based on rearrangements. It appears that there is considerable scope for extensions and unifications of all this theory.

7 Entropy Concavity and the Shepp–Olkin Conjecture

In 1981, Shepp and Olkin [86] made a conjecture regarding the entropy of Poisson-binomial random variables. That is, for any vector $\mathbf{p} := (p_1, \ldots, p_m)$ where $0 \leq p_i \leq 1$, we can write $P_{\mathbf{p}}$ for the probability mass function of the random variable $S := X_1 + \ldots + X_m$, where X_i are independent Bernoulli(p_i) variables. Shepp and Olkin conjectured that the entropy of $H(P_{\mathbf{p}})$ is a concave function of the parameters. We can simplify this conjecture by considering the affine case, where each $p_i(t) = (1-t)p_i(0) + tp_i(1)$.

This result is plausible since Shepp and Olkin [86] (see also [74]) proved that for any m the entropy $H(B_{m,p})$ of a Bernoulli random variable is concave in p, and also claimed in their paper that it was true for the sum of m independent Bernoulli random variables when $m = 2, 3$. Since then, progress was limited. In [106, Theorem 2] it was proved that the entropy is concave when for each i either $p_i(0) = 0$ or $p_i(1) = 0$. (In fact, this follows from the case $n = 2$ of Theorem 6.3 above). Hillion [49] proved the case of the Shepp–Olkin conjecture where all $p_i(t)$ are either constant or equal to t.

However, in recent work of Hillion and Johnson [50, 51], the Shepp–Olkin conjecture was proved, in a result stated as [51, Theorem 1.2]:

Theorem 7.1 (Shepp-Olkin Theorem) *For any $m \geq 1$, function $\mathbf{p} \mapsto H(P_{\mathbf{p}})$ is concave.*

We briefly summarize the strategy of the proof, and the ideas involved. In [50], Theorem 7.1 was proved in the monotone case where each $p_i' := p_i'(t)$ has the same sign. In [51], this was extended to the general case; in fact, the monotone case of [50] is shown to be the worst case. In order to bound the second derivative of entropy $\frac{\partial^2}{\partial t^2} H(P_{\mathbf{p}(t)})$, we need to control the first and second derivatives of the probability mass function $P_{\mathbf{p}(t)}$, in the spirit of Lemma 4.1. Indeed, we write (see [51, Equations (8),(9)]) the derivative of the mass function in gradient form (see (2) and (19) for comparison):

$$\frac{\partial P_{\mathbf{p}(t)}}{\partial t}(k) = g(k-1) - g(k) \tag{26}$$

$$\frac{\partial^2 P_{\mathbf{p}(t)}}{\partial t^2}(k) = h(k-2) - 2h(k-1) + h(k) \tag{27}$$

for certain functions $g(k)$ and $h(k)$, about which we can be completely explicit. The key property (referred to as Condition 4 in [50]) is that for all k:

$$\begin{aligned} &h(k)\left(f(k+1)^2 - f(k)f(k+2)\right) \\ &\leq 2g(k)g(k+1)f(k+1) - g(k)^2 f(k+2) - g(k+1)^2 f(k), \end{aligned} \tag{28}$$

which allows us to bound the non-linear terms arising in the 2nd derivative $\frac{\partial^2}{\partial t^2} H(P_{\mathbf{p}(t)})$. The proof of (28) is based on a cubic inequality [50, Property(m)] satisfied by Poisson-binomial mass functions, which relates to an iterated log-concavity result of Brändén [24].

The argument in [50] was based around the idea of optimal transport of probability measures in the discrete setting, where it was shown that if all the p_i' have the same sign then the Shepp–Olkin path provides an optimal interpolation, in the sense of a discrete formula of Benamou–Brenier type [15, 16] (see also [38, 41] which also consider optimal transport of discrete measures, including analysis of discrete curvature in the sense of [28]).

8 Open Problems

We briefly list some open problems, corresponding to each research direction described in Sections 3 to 7. This is not a complete list of open problems in the area, but gives an indication of some possible future research directions.

1. **[Poisson approximation]** The information theoretic approach to Poisson approximation described in Section 3 holds only in the case of independent summands. In the original paper [65] bounds are given on the relative entropy, based on the data processing inequality, which hold for more general dependence structures. However, these data processing results do not give Poisson approximation bounds of optimal order. It is an interesting problem to generalize the projection identity [65, Lemma, P.471] to the dependent case (perhaps resulting in an inequality), under some appropriate model of (negative) dependence, allowing Poisson approximation results to be proved.

   Further, although (as described above) such information theoretic approximation results hold for Poisson and compound Poisson limit distributions, it would be of interest to generalize this theory to a wider class of discrete distributions.
2. **[Maximum entropy]** While not strictly a question to do with discrete random variables, as described in Section 1 it remains an interesting problem to find classes within which the stable laws are maximum entropy. As with the ULC class of Condition 1, or the fixed variance class within which the Gaussian maximizes entropy, we want such a class to be well-behaved on convolution and scaling. Preliminary work in this direction, including the fact that stable laws are not in general maximum entropy in their own domain of normal attraction (in the sense of [39]), is described in [56], and an alternative approach based on non-linear diffusion equations and fractional calculus is given by Toscani [92].
3. **[Concentration inequalities]** As described at the end of Section 5, it would be of interest to extend these results to the case of mass functions supported on a finite set $\{0, 1, \ldots, n\}$. We would like to introduce concavity conditions mirroring Conditions 1 or 2, under which results of the form Theorem 5.4 and 5.5 can be proved for the derivative ∇_n of (25). In addition, it would be of interest to see

whether a version of Theorem 5.5 holds under a stochastic ordering assumption $P^* \leq_{st} P$, in the spirit of [34].

4. **[Monotonicity]** As in Section 6, it is of interest to develop a new and more illuminating proof of Theorem 6.3, perhaps based on transport inequalities as in [8]. It is possible that such a proof will give some indication of the correct formulation of the discrete Entropy Power Inequality, which is itself a key open problem.
5. **[Entropy concavity]** As described in [51], it is natural to conjecture that the q-Rényi [80] and q-Tsallis [93] entropies are concave functions for q sufficiently small. Note that while they are monotone functions of each other, the q-Rényi and q-Tsallis entropies need not be concave for the same q. We quote the following generalized Shepp-Olkin conjecture from [51, Conjecture 4.2]:

   (a) There is a critical q_R^* such that the q-Rényi entropy of all Bernoulli sums is concave for $q \leq q_R^*$, and the entropy of some interpolation is convex for $q > q_R^*$.
   (b) There is a critical q_T^* such that the q-Tsallis entropy of all Bernoulli sums is concave for $q \leq q_T^*$, and the entropy of some interpolation is convex for $q > q_T^*$.

   Indeed we conjecture that $q_R^* = 2$ and $q_T^* = 3.65986\ldots$, the root of $2-4q+2^q = 0$. We remark that the form of discrete Entropy Power Inequality proposed by [98] (based on the theory of rearrangements) also holds for Rényi entropies.

   In addition, Shepp and Olkin [86] made another conjecture regarding Bernoulli sums, that the entropy $H(P_{\mathbf{p}})$ is a monotone increasing function in $\mathbf{p}$ if all $p_i \leq 1/2$, which remains open with essentially no published results.

Acknowledgements The author thanks the Institute for Mathematics and Its Applications for the invitation and funding to speak at the workshop '*Information Theory and Concentration Phenomena*' in Minneapolis in April 2015. In addition, he would like to thank the organizers and participants of this workshop for stimulating discussions. The author thanks Fraser Daly, Mokshay Madiman and an anonymous referee for helpful comments on earlier drafts of this paper.

References

1. J. A. Adell, A. Lekuona, and Y. Yu. Sharp bounds on the entropy of the Poisson law and related quantities. *IEEE Trans. Inform. Theory*, 56(5):2299–2306, May 2010.
2. S.-i. Amari, O. E. Barndorff-Nielsen, R. E. Kass, S. L. Lauritzen, and C. R. Rao. *Differential geometry in statistical inference.* Institute of Mathematical Statistics Lecture Notes—Monograph Series, 10. Institute of Mathematical Statistics, Hayward, CA, 1987.
3. S.-i. Amari and H. Nagaoka. *Methods of information geometry*, volume 191 of *Translations of Mathematical Monographs*. American Mathematical Society, Providence, RI, 2000.
4. L. Ambrosio, N. Gigli, and G. Savaré. *Gradient flows in metric spaces and in the space of probability measures.* Lectures in Mathematics ETH Zürich. Birkhäuser Verlag, Basel, second edition, 2008.

5. V. Anantharam. Counterexamples to a proposed Stam inequality on finite groups. *IEEE Trans. Inform. Theory*, 56(4):1825–1827, 2010.
6. C. Ané, S. Blachere, D. Chafaï, P. Fougeres, I. Gentil, F. Malrieu, C. Roberto, and G. Scheffer. Sur les inégalités de Sobolev logarithmiques. *Panoramas et Syntheses*, 10:217, 2000.
7. S. Artstein, K. M. Ball, F. Barthe, and A. Naor. On the rate of convergence in the entropic central limit theorem. *Probab. Theory Related Fields*, 129(3):381–390, 2004.
8. S. Artstein, K. M. Ball, F. Barthe, and A. Naor. Solution of Shannon's problem on the monotonicity of entropy. *J. Amer. Math. Soc.*, 17(4):975–982 (electronic), 2004.
9. D. Bakry and M. Émery. Diffusions hypercontractives. In *Séminaire de probabilités, XIX*, volume 1123 of *Lecture Notes in Math.*, pages 177–206. Springer, Berlin, 1985.
10. D. Bakry, I. Gentil, and M. Ledoux. *Analysis and geometry of Markov diffusion operators*, volume 348 of *Grundlehren der mathematischen Wissenschaften*. Springer, 2014.
11. K. Ball, F. Barthe, and A. Naor. Entropy jumps in the presence of a spectral gap. *Duke Math. J.*, 119(1):41–63, 2003.
12. A. Barbour, L. Holst, and S. Janson. *Poisson Approximation*. Clarendon Press, Oxford, 1992.
13. A. Barbour, O. T. Johnson, I. Kontoyiannis, and M. Madiman. Compound Poisson approximation via local information quantities. *Electronic Journal of Probability*, 15:1344–1369, 2010.
14. A. R. Barron. Entropy and the Central Limit Theorem. *Ann. Probab.*, 14(1):336–342, 1986.
15. J.-D. Benamou and Y. Brenier. A numerical method for the optimal time-continuous mass transport problem and related problems. In *Monge Ampère equation: applications to geometry and optimization (Deerfield Beach, FL, 1997)*, volume 226 of *Contemp. Math.*, pages 1–11. Amer. Math. Soc., Providence, RI, 1999.
16. J.-D. Benamou and Y. Brenier. A computational fluid mechanics solution to the Monge-Kantorovich mass transfer problem. *Numer. Math.*, 84(3):375–393, 2000.
17. N. M. Blachman. The convolution inequality for entropy powers. *IEEE Trans. Information Theory*, 11:267–271, 1965.
18. S. G. Bobkov, G. P. Chistyakov, and F. Götze. Convergence to stable laws in relative entropy. *Journal of Theoretical Probability*, 26(3):803–818, 2013.
19. S. G. Bobkov, G. P. Chistyakov, and F. Götze. Rate of convergence and Edgeworth-type expansion in the entropic central limit theorem. *Ann. Probab.*, 41(4):2479–2512, 2013.
20. S. G. Bobkov, G. P. Chistyakov, and F. Götze. Berry–Esseen bounds in the entropic central limit theorem. *Probability Theory and Related Fields*, 159(3–4):435–478, 2014.
21. S. G. Bobkov, G. P. Chistyakov, and F. Götze. Fisher information and convergence to stable laws. *Bernoulli*, 20(3):1620–1646, 2014.
22. S. G. Bobkov and M. Ledoux. On modified logarithmic Sobolev inequalities for Bernoulli and Poisson measures. *J. Funct. Anal.*, 156(2):347–365, 1998.
23. A. Borovkov and S. Utev. On an inequality and a related characterisation of the normal distribution. *Theory Probab. Appl.*, 28(2):219–228, 1984.
24. P. Brändén. Iterated sequences and the geometry of zeros. *J. Reine Angew. Math.*, 658:115–131, 2011.
25. L. D. Brown. A proof of the Central Limit Theorem motivated by the Cramér-Rao inequality. In G. Kallianpur, P. R. Krishnaiah, and J. K. Ghosh, editors, *Statistics and Probability: Essays in Honour of C.R. Rao*, pages 141–148. North-Holland, New York, 1982.
26. T. Cacoullos. On upper and lower bounds for the variance of a function of a random variable. *Ann. Probab.*, 10(3):799–809, 1982.
27. L. A. Caffarelli. Monotonicity properties of optimal transportation and the FKG and related inequalities. *Communications in Mathematical Physics*, 214(3):547–563, 2000.
28. P. Caputo, P. Dai Pra, and G. Posta. Convex entropy decay via the Bochner-Bakry-Emery approach. *Ann. Inst. Henri Poincaré Probab. Stat.*, 45(3):734–753, 2009.
29. E. Carlen and A. Soffer. Entropy production by block variable summation and Central Limit Theorems. *Comm. Math. Phys.*, 140(2):339–371, 1991.
30. E. A. Carlen and W. Gangbo. Constrained steepest descent in the 2-Wasserstein metric. *Ann. of Math. (2)*, 157(3):807–846, 2003.

31. D. Chafaï. Binomial-Poisson entropic inequalities and the M/M/∞ queue. *ESAIM Probability and Statistics*, 10:317–339, 2006.
32. H. Chernoff. A note on an inequality involving the normal distribution. *Ann. Probab.*, 9(3):533–535, 1981.
33. D. Cordero-Erausquin. Some applications of mass transport to Gaussian-type inequalities. *Arch. Ration. Mech. Anal.*, 161(3):257–269, 2002.
34. F. Daly. Negative dependence and stochastic orderings. *ESAIM: PS*, 20:45–65, 2016. https://doi.org/10.1051/ps/2016002.
35. F. Daly and O. T. Johnson. Bounds on the Poincaré constant under negative dependence. *Statistics and Probability Letters*, 83:511–518, 2013.
36. A. Dembo, T. M. Cover, and J. A. Thomas. Information theoretic inequalities. *IEEE Trans. Information Theory*, 37(6):1501–1518, 1991.
37. Y. Derriennic. Entropie, théorèmes limite et marches aléatoires. In H. Heyer, editor, *Probability Measures on Groups VIII, Oberwolfach*, number 1210 in Lecture Notes in Mathematics, pages 241–284, Berlin, 1985. Springer-Verlag. In French.
38. M. Erbar and J. Maas. Ricci curvature of finite Markov chains via convexity of the entropy. *Archive for Rational Mechanics and Analysis*, 206:997–1038, 2012.
39. B. V. Gnedenko and A. N. Kolmogorov. *Limit distributions for sums of independent random variables*. Addison-Wesley, Cambridge, Mass, 1954.
40. B. V. Gnedenko and V. Y. Korolev. *Random Summation: Limit Theorems and Applications*. CRC Press, Boca Raton, Florida, 1996.
41. N. Gozlan, C. Roberto, P.-M. Samson, and P. Tetali. Displacement convexity of entropy and related inequalities on graphs. *Probability Theory and Related Fields*, 160(1–2):47–94, 2014.
42. L. Gross. Logarithmic Sobolev inequalities. *Amer. J. Math.*, 97(4):1061–1083, 1975.
43. A. Guionnet and B. Zegarlinski. Lectures on logarithmic Sobolev inequalities. In *Séminaire de Probabilités, XXXVI*, volume 1801 of *Lecture Notes in Math.*, pages 1–134. Springer, Berlin, 2003.
44. D. Guo, S. Shamai, and S. Verdú. Mutual information and minimum mean-square error in Gaussian channels. *IEEE Trans. Inform. Theory*, 51(4):1261–1282, 2005.
45. S. Haghighatshoar, E. Abbe, and I. E. Telatar. A new entropy power inequality for integer-valued random variables. *IEEE Transactions on Information Theory*, 60(7):3787–3796, 2014.
46. P. Harremoës. Binomial and Poisson distributions as maximum entropy distributions. *IEEE Trans. Information Theory*, 47(5):2039–2041, 2001.
47. P. Harremoës, O. T. Johnson, and I. Kontoyiannis. Thinning, entropy and the law of thin numbers. *IEEE Trans. Inform. Theory*, 56(9):4228–4244, 2010.
48. P. Harremoës and C. Vignat. An Entropy Power Inequality for the binomial family. *JIPAM. J. Inequal. Pure Appl. Math.*, 4, 2003. Issue 5, Article 93; see also http://jipam.vu.edu.au/.
49. E. Hillion. Concavity of entropy along binomial convolutions. *Electron. Commun. Probab.*, 17(4):1–9, 2012.
50. E. Hillion and O. T. Johnson. Discrete versions of the transport equation and the Shepp-Olkin conjecture. *Annals of Probability*, 44(1):276–306, 2016.
51. E. Hillion and O. T. Johnson. A proof of the Shepp-Olkin entropy concavity conjecture. *Bernoulli (to appear)*, 2017. See also `arxiv:1503.01570`.
52. E. Hillion, O. T. Johnson, and Y. Yu. A natural derivative on $[0, n]$ and a binomial Poincaré inequality. *ESAIM Probability and Statistics*, 16:703–712, 2014.
53. V. Jog and V. Anantharam. The entropy power inequality and Mrs. Gerber's Lemma for groups of order 2^n. *IEEE Transactions on Information Theory*, 60(7):3773–3786, 2014.
54. O. T. Johnson. *Information theory and the Central Limit Theorem*. Imperial College Press, London, 2004.
55. O. T. Johnson. Log-concavity and the maximum entropy property of the Poisson distribution. *Stoch. Proc. Appl.*, 117(6):791–802, 2007.
56. O. T. Johnson. A de Bruijn identity for symmetric stable laws. In submission, see `arXiv:1310.2045`, 2013.

57. O. T. Johnson. A discrete log-Sobolev inequality under a Bakry-Émery type condition. In submission. Ann. L'Inst. Henri Poincaré Probab. Stat. http://imstat.org/aihp/accepted.html.
58. O. T. Johnson and A. R. Barron. Fisher information inequalities and the Central Limit Theorem. *Probability Theory and Related Fields*, 129(3):391–409, 2004.
59. O. T. Johnson, I. Kontoyiannis, and M. Madiman. Log-concavity, ultra-log-concavity, and a maximum entropy property of discrete compound Poisson measures. *Discrete Applied Mathematics*, 161:1232–1250, 2013.
60. O. T. Johnson and Y. Yu. Monotonicity, thinning and discrete versions of the Entropy Power Inequality. *IEEE Trans. Inform. Theory*, 56(11):5387–5395, 2010.
61. I. Johnstone and B. MacGibbon. Une mesure d'information caractérisant la loi de Poisson. In *Séminaire de Probabilités, XXI*, pages 563–573. Springer, Berlin, 1987.
62. A. Kagan. A discrete version of the Stam inequality and a characterization of the Poisson distribution. *J. Statist. Plann. Inference*, 92(1-2):7–12, 2001.
63. J. F. C. Kingman. Uses of exchangeability. *Ann. Probability*, 6(2):183–197, 1978.
64. C. Klaassen. On an inequality of Chernoff. *Ann. Probab.*, 13(3):966–974, 1985.
65. I. Kontoyiannis, P. Harremoës, and O. T. Johnson. Entropy and the law of small numbers. *IEEE Trans. Inform. Theory*, 51(2):466–472, 2005.
66. S. Kullback. A lower bound for discrimination information in terms of variation. *IEEE Trans. Information Theory*, 13:126–127, 1967.
67. C. Ley and Y. Swan. Stein's density approach for discrete distributions and information inequalities. See `arxiv:1211.3668`, 2012.
68. E. Lieb. Proof of an entropy conjecture of Wehrl. *Comm. Math. Phys.*, 62:35–41, 1978.
69. T. M. Liggett. Ultra logconcave sequences and negative dependence. *J. Combin. Theory Ser. A*, 79(2):315–325, 1997.
70. Y. Linnik. An information-theoretic proof of the Central Limit Theorem with the Lindeberg Condition. *Theory Probab. Appl.*, 4:288–299, 1959.
71. J. Lott and C. Villani. Ricci curvature for metric-measure spaces via optimal transport. *Ann. of Math. (2)*, 169(3):903–991, 2009.
72. M. Madiman and A. Barron. Generalized entropy power inequalities and monotonicity properties of information. *IEEE Trans. Inform. Theory*, 53(7):2317–2329, 2007.
73. M. Madiman, J. Melbourne, and P. Xu. Forward and reverse Entropy Power Inequalities in convex geometry. See: `arxiv:1604.04225`, 2016.
74. P. Mateev. The entropy of the multinomial distribution. *Teor. Verojatnost. i Primenen.*, 23(1):196–198, 1978.
75. N. Papadatos and V. Papathanasiou. Poisson approximation for a sum of dependent indicators: an alternative approach. *Adv. in Appl. Probab.*, 34(3):609–625, 2002.
76. V. Papathanasiou. Some characteristic properties of the Fisher information matrix via Cacoullos-type inequalities. *J. Multivariate Anal.*, 44(2):256–265, 1993.
77. R. Pemantle. Towards a theory of negative dependence. *J. Math. Phys.*, 41(3):1371–1390, 2000.
78. C. R. Rao. On the distance between two populations. *Sankhya*, 9:246–248, 1948.
79. A. Rényi. A characterization of Poisson processes. *Magyar Tud. Akad. Mat. Kutató Int. Közl.*, 1:519–527, 1956.
80. A. Rényi. On measures of entropy and information. In J. Neyman, editor, *Proceedings of the 4th Berkeley Conference on Mathematical Statistics and Probability*, pages 547–561, Berkeley, 1961. University of California Press.
81. B. Roos. Kerstan's method for compound Poisson approximation. *Ann. Probab.*, 31(4):1754–1771, 2003.
82. M. Shaked and J. G. Shanthikumar. *Stochastic orders*. Springer Series in Statistics. Springer, New York, 2007.
83. S. Shamai and A. Wyner. A binary analog to the entropy-power inequality. *IEEE Trans. Inform. Theory*, 36(6):1428–1430, Nov 1990.
84. C. E. Shannon. A mathematical theory of communication. *Bell System Tech. J.*, 27:379–423, 623–656, 1948.

85. N. Sharma, S. Das, and S. Muthukrishnan. Entropy power inequality for a family of discrete random variables. In *2011 IEEE International Symposium on Information Theory Proceedings (ISIT)*, pages 1945–1949. IEEE, 2011.
86. L. A. Shepp and I. Olkin. Entropy of the sum of independent Bernoulli random variables and of the multinomial distribution. In *Contributions to probability*, pages 201–206. Academic Press, New York, 1981.
87. R. Shimizu. On Fisher's amount of information for location family. In Patil, G. P. and Kotz, S. and Ord, J. K., editor, *A Modern Course on Statistical Distributions in Scientific Work, Volume 3*, pages 305–312. Reidel, 1975.
88. A. J. Stam. Some inequalities satisfied by the quantities of information of Fisher and Shannon. *Information and Control*, 2:101–112, 1959.
89. K.-T. Sturm. On the geometry of metric measure spaces. I. *Acta Math.*, 196(1):65–131, 2006.
90. K.-T. Sturm. On the geometry of metric measure spaces. II. *Acta Math.*, 196(1):133–177, 2006.
91. G. Szegő. *Orthogonal Polynomials*. American Mathematical Society, New York, revised edition, 1958.
92. G. Toscani. The fractional Fisher information and the central limit theorem for stable laws. *Ricerche di Matematica*, 65(1):71–91, 2016.
93. C. Tsallis. Possible generalization of Boltzmann-Gibbs statistics. *Journal of Statistical Physics*, 52:479–487, 1988.
94. A. Tulino and S. Verdú. Monotonic decrease of the non-Gaussianness of the sum of independent random variables: a simple proof. *IEEE Trans. Inform. Theory*, 52(9):4295–4297, 2006.
95. C. Villani. *Topics in optimal transportation*, volume 58 of *Graduate Studies in Mathematics*. American Mathematical Society, Providence, RI, 2003.
96. C. Villani. *Optimal transport: Old and New*, volume 338 of *Grundlehren der Mathematischen Wissenschaften*. Springer-Verlag, Berlin, 2009.
97. D. W. Walkup. Pólya sequences, binomial convolution and the union of random sets. *J. Appl. Probability*, 13(1):76–85, 1976.
98. L. Wang, J. O. Woo, and M. Madiman. A lower bound on the Rényi entropy of convolutions in the integers. In *2014 IEEE International Symposium on Information Theory (ISIT)*, pages 2829–2833. IEEE, 2014.
99. H. S. Witsenhausen. Some aspects of convexity useful in information theory. *IEEE Trans. Inform. Theory*, 26(3):265–271, 1980.
100. L. Wu. A new modified logarithmic Sobolev inequality for Poisson point processes and several applications. *Probab. Theory Related Fields*, 118(3):427–438, 2000.
101. A. D. Wyner. A theorem on the entropy of certain binary sequences and applications. II. *IEEE Trans. Information Theory*, 19(6):772–777, 1973.
102. A. D. Wyner and J. Ziv. A theorem on the entropy of certain binary sequences and applications. I. *IEEE Trans. Information Theory*, 19(6):769–772, 1973.
103. Y. Yu. On the maximum entropy properties of the binomial distribution. *IEEE Trans. Inform. Theory*, 54(7):3351–3353, July 2008.
104. Y. Yu. Monotonic convergence in an information-theoretic law of small numbers. *IEEE Trans. Inform. Theory*, 55(12):5412–5422, 2009.
105. Y. Yu. On the entropy of compound distributions on nonnegative integers. *IEEE Trans. Inform. Theory*, 55(8):3645–3650, 2009.
106. Y. Yu and O. T. Johnson. Concavity of entropy under thinning. In *Proceedings of the 2009 IEEE International Symposium on Information Theory, 28th June - 3rd July 2009, Seoul*, pages 144–148, 2009.

Concentration of Measure Principle and Entropy-Inequalities

Paul-Marie Samson

Abstract The concentration measure principle is presented in an abstract way to encompass and unify different concentration properties. We give a general overview of the links between concentration properties, transport-entropy inequalities, and logarithmic Sobolev inequalities for some specific transport costs. By giving few examples, we emphasize optimal weak transport costs as an efficient tool to establish new transport inequality and new concentration principles for discrete measures (the binomial law, the Poisson measure, the uniform law on the symmetric group).

Keywords Concentration of measure • Duality • Transport inequalities • Logarithmic-Sobolev inequalities

1991 *Mathematics Subject Classification.* 60E15, 32F32, 39B62, 26D10

1 Introduction

Of isoperimetric inspiration, the concentration of measure phenomenon has been pushed forward by V. Milman in the 70s in the study of the asymptotic geometry of Banach spaces and then in-depth studied by many authors including Gromov [GM83, Gro99], Talagrand [Tal95], Maurey [Mau91], Ledoux [Led97, BL97], Bobkov [Bob97, BL00]. This principle has applications in numerous fields of mathematics. The book by M. Ledoux [Led01] is devoted to this subject. It presents numerous examples and probabilistic, analytical, and geometrical technics related to this notion. We also refer to the monographs [BLM13, Mas07] for more applications of this principle in statistics and probability theory. We also warmly recommend the surveys [GL10, Goz15] by Gozlan and Léonard about transport-

P.-M. Samson (✉)
Laboratoire d'Analyse et de Mathématiques Appliquées (UMR 8050), UPEM, UPEC, CNRS, Université Paris-Est, F-77454 Marne-la-Vallée, France
e-mail: paul-marie.samson@u-pem.fr

E. Carlen et al. (eds.), *Convexity and Concentration*, The IMA Volumes in Mathematics and its Applications 161, DOI 10.1007/978-1-4939-7005-6_3

entropy inequalities. The main purpose of this paper is to complement these surveys in view of the recent developments.

In the present paper, the concentration of measure principle is formalized in an abstract way to encompass and unify different concentration properties investigated in the literature. The definition of this principle with enlargements of sets takes its origin from the papers by M. Talagrand [Tal95, Tal96b, Tal96a]. We propose a functional formulation of the concentration principle, rigorously introduced in [GRST14b]. We emphasize three types of cost functions that provide most of the enlargements of sets considered in the literature, the usual cost functions, the barycentric cost functions, and the universal cost functions.

The concentration properties associated to usual cost functions and its related functional inequalities have been widely studied these last years. Now, it is a challenge to develop new concentration inequalities that could capture precise dimensional concentration behavior for particular classes of functions, especially in discrete setting. In this document, we present some concentration results for discrete measures associated to the above weak transport costs. In the spirit of the early works by Talagrand [Tal95, Tal96b, Tal96a], we believe that entropy-functional inequalities associated to new weak transport costs could be adapted to understand some concentration challenging problems.

The third section of this paper put forward the transport-entropy inequality (also called transport inequality) associated to the above different cost functions, as a fundamental tool in the study of concentration properties in product spaces. This entropy-inequality is an alternative to the logarithmic Sobolev inequality and its variants, to establish concentration properties in product spaces. The main feature of these two inequalities is for each, a tensorization property, that provides concentration results in high dimension spaces. We will briefly recall the links between these two kinds of entropy-inequalities and the concentration of measure principle.

Section 4 is focussed on the concentration properties and transport inequalities related to the so-called barycentric costs. These transport costs, weaker than the usual one, are adapted to derive new transport inequalities for discrete measures (see Section 4.2). Indeed, let us recall that the Talagrand's transport inequality $\mathbf{T}_2$ is never satisfied by discrete measures. Recently, in the context of the study of curvature notion in discrete spaces, other transport inequalities have been proposed, mainly in the works by Erbar-Maas [Maa11, EM12]. However, due to the very abstract definition of the optimal transport costs, the associated concentration of measure phenomenon remains difficult to interpret.

Barycentric optimal transport costs can be expressed using optimal transport costs by considering the notion of convex order on probability measures (see Proposition 4.1). Moreover, most of the results with usual transport costs can be adapted for barycentric transport costs. First, barycentric transport inequalities are equivalent to logarithmic Sobolev inequalities restricted to a class of convex or concave functions (see Section 4.1). In an other direction, on the real line, as for usual transport costs, for any optimal barycentric transport cost, there exists an optimal coupling which is independent of the convex function involved in the barycentric cost (see Section 4.3). This independence property allows to characterize the probability measures satisfying a barycentric transport inequality on the real line. As a by-product of this characterization, the "convex" Poincaré inequality on the real line is equivalent to a barycentric transport inequality with some specific convex cost function (see Section 4.4).

Section 5 is devoted to examples of universal transport inequalities such as the so-called Csizár-Kullback-Pinsker inequality. The most emblematic universal transport inequality of this document is the Marton's transport inequality with its weak cost $\widetilde{\mathcal{T}}_2$ [Mar96b]. In the papers [Sam03, Sam07], the Marton's cost has been improved to reach optimal Bernstein bounds for suprema of empirical bounded independent processes (see Section 5.1). This method is an alternative to the so-called Herbst's method, first used by Ledoux to get deviation bounds for suprema of empirical processes [Led97]. In Section 5.2, we recall the results of [Sam00] and [Pau14] that extends Marton's inequality to any measure on a product space, with weak dependences of its marginals.

The last Section 5.3 concerns recent transport inequalities obtained for the uniform probability measure on the symmetric group. These inequalities are obtained from the Csizár-Kullback-Pinsker inequality or the Marton's inequality $\widetilde{\mathbf{T}}_2$, by using other tensorization arguments. The proofs are inspired by the work by Talagrand on the symmetric group [Tal95].

The works presented in this survey could be extended in different directions.

A first challenge is to define other cost functions that may capture new concentration's properties, as for the uniform measure on the symmetric group, for Gibbs measure or for other non product measures under dependence properties. The new cost functions presented in this survey are of particular interest in discrete setting (discrete cube, binomial law, Poisson measure) and we may use it in other discrete framework such as Poisson processes.

Another direction is to develop the multimarginal transport inequalities in discrete and continuous setting. We wonder whether this multimarginal approach allows to reach superconcentration-properties. In this field, the works by Dembo [Dem97] and Talagrand [Tal96a] are also a guideline.

These last years, the concept of curvature in discrete setting has emerged [Oll09, OV12, EM12, EMT15, Hil14] by analogy of the concept of lower bounded curvature in continuous setting in metric spaces [Vil09, AGS14, AGS15]. These notions could be revisited by relating the notion of curvatures to different cost functions, the paper [GRST14a] is a first attempt in that direction.

2 A General Concentration of Measure Principle

Let $(\mathcal{X}, d)$ be a Polish metric space (separable, complete), with Borel σ-field $\mathcal{B}(\mathcal{X})$. We denote by $\mathcal{P}(A)$ the set of probability measures on a subset $A \in \mathcal{B}(\mathcal{X})$, and for $q \geq 1$, we denote by $\mathcal{P}_q(A)$ the set of probability measures p on A such that $\int d(x_0, y)^q dp(y) < +\infty$ for a point x_0.

We assume that a notion of pseudo-distance from a point $x \in \mathcal{X}$ to a subset $A \subset \mathcal{X}$ is given, denoted by $c(x, A) \in [0, +\infty]$, and such that $c(x, A) \geq 0$ and $c(x, A) = 0$ if $x \in A$. The usual example is

$$c(x, A) = d(x, A) = \inf_{y \in A} d(x, y).$$

One could also choose $c(x, A) = \alpha(d(x, A))$, where $\alpha : \mathbb{R}^+ \to [0, +\infty]$ is such that $\alpha(0) = 0$.

For $r \geq 0$, *the enlargement of* A associated to this pseudo-distance is defined by

$$A_{r,c} = \{x \in \mathcal{X}, c(x, A) \leq r\}.$$

Definition 2.1 *Let $\beta : \mathbb{R}^+ \to [0, +\infty]$ be an increasing function with $\beta(0) = 0$. A probability measure μ satisfies a concentration principle with profile β and cost c, if there exist $a_1, a_2 > 0$ such that for all subsets $A \in \mathcal{B}(\mathcal{X})$ and for all $r \geq 0$,*

$$\mu(A)^{a_1} \mu(\mathcal{X} \setminus A_{r,c})^{a_2} \leq e^{-\beta(r)}.$$

If μ satisfies a concentration principle with profile β, then for all $A \in \mathcal{B}(\mathcal{X})$ with measure $\mu(A) \geq 1/2$, one has

$$\mu(\mathcal{X} \setminus A_{r,c}) \leq 2^{a_1/a_2} e^{-\beta(r)/a_2} = e^{-\tilde{\beta}(r)}, \qquad \forall r \geq 0.$$

This last property is the classical formulation of a concentration of measure principle. Actually, these two formulations are equivalent, up to constants, as soon as

$$A \subset \mathcal{X} \setminus (\mathcal{X} \setminus A_{r,c})_r, \qquad \forall r \geq 0, \ \forall A \in \mathcal{B}(\mathcal{X}),$$

(see Lemma 5.6 [GRST14b]). This inclusion depends on the kind of enlargement and is not satisfied for some enlargements.

In all this document, the pseudo-distance $c(x, A)$ is defined from a *cost function*.

Definition 2.2 *A cost function is a function*

$$c : \mathcal{X} \times \mathcal{P}(\mathcal{X}) \to [0, +\infty],$$

such that the function $p \mapsto c(x, p)$ is convex, and $c(x, \delta_x) = 0$ for all $x \in \mathcal{X}$ (δ_x denotes the Dirac measure at point x). Then we define the pseudo-distance c by

$$c(x, A) = \inf_{p \in \mathcal{P}(A)} c(x, p), \qquad x \in \mathcal{X}, \quad A \in \mathcal{B}(\mathcal{X}).$$

Here, for the sake of simplicity, the same notation c is used for the pseudo-distance and the cost function c.

In this document, we consider three kinds of cost functions with the following definitions.

The Usual Cost Functions

Usually, the transport cost is a function defined on $\mathcal{X} \times \mathcal{X}$ rather than $\mathcal{X} \times \mathcal{P}(\mathcal{X})$.

Definition 2.3 *A cost function $c : \mathcal{X} \times \mathcal{P}(\mathcal{X}) \to [0, +\infty]$ is called usual if there exists a measurable function $\omega : \mathcal{X} \times \mathcal{X} \to [0, +\infty]$ such that for all $x \in \mathcal{X}$ and all $p \in \mathcal{P}(\mathcal{X})$, $\omega(x, x) = 0$ and*

$$c(x, p) = \int \omega(x, y) dp(y).$$

In that case, $p \mapsto c(x, p)$ is an affine function. The pseudo-distance $\inf_{p \in \mathcal{P}(A)} c(x, p)$ is reached at Dirac measures, the extremal points of the convex set $\mathcal{P}(A)$, therefore

$$c(x, A) = \inf_{y \in A} \omega(x, y).$$

By the way, the pseudo-distance is exactly the classical one associated to the cost ω. The most studied cost functions ω are $\omega(x, y) = d(x, y)^q$, $q > 0$ or $\omega(x, y) = \alpha(d(x, y))$, with $\alpha : \mathbb{R}^+ \to \mathbb{R}^+$.

In the case $\omega(x, y) = d(x, y)$ one has $c(x, A) = d(x, A)$ and we simply denote by A_r the enlargement $A_{r,c}$.

The Universal Cost Functions

These cost functions have been introduced by Talagrand [Tal96b] and Marton [Mar96b] in order to solve different types of concentration's problems, for example the deviations of suprema of empirical processes, of the largest increasing subsequence, the bin-packing problem, etc. These transport costs are the main tools of the so-called convex hull method of [Tal96b, Tal96a] and in the papers [Mar97, Sam00, Sam03, Sam07, Pau14].

We denote by $\mathbb{1}_{x \neq y}$ *the Hamming distance* between two points x and y in $\mathcal{X}$, defined by $\mathbb{1}_{x \neq y} = \gamma(d(x, y))$, where $\gamma : \mathbb{R}^+ \to \mathbb{R}^+$ is the function defined by $\gamma(h) = 1$ if $h \neq 0$ and $\gamma(0) = 0$.

Definition 2.4 *Let $\alpha : \mathbb{R}^+ \to [0, +\infty]$ be a lower semi-continuous convex function and let μ_0 be a probability measure on $\mathcal{X}$. We define two classes of universal cost functions.*

- *A cost function $c : \mathcal{X} \times \mathcal{P}(\mathcal{X}) \to [0, +\infty]$ is called universal and associated to the function α, if*

$$c(x, p) = \alpha\left(\int \mathbb{1}_{x \neq y}\, dp(y)\right),$$

for all $x \in X$ and for all $p \in \mathcal{P}(\mathcal{X})$.

- *A cost function $c : \mathcal{X} \times \mathcal{P}(\mathcal{X}) \to [0, +\infty]$ is called universal, associated to the function α and to the measure μ_0, if*

$$c(x, p) = \int \alpha\left(\mathbb{1}_{x \neq y} \frac{dp}{d\mu_0}(y)\right) d\mu_0(y),$$

for all $(x, p) \in \mathcal{X} \times \mathcal{P}(\mathcal{X})$ such that p is absolutely continuous with respect to μ_0 on the set $X \setminus \{x\}$, and $c(x, p) = +\infty$ otherwise.

These universal cost functions are independent of the distance d on $\mathcal{X}$, and therefore of the geometry of the space $\mathcal{X}$.

The Barycentric Cost Functions

The so-called barycentric costs are defined on $\mathcal{X} = \mathbb{R}^n$ equipped with the Euclidean distance, $d(x, y) = |x - y|$, $x, y \in \mathbb{R}^n$. They have been introduced in the paper [GRST14b] to reach optimal concentration properties for discrete measures (see Section 4.2). As explained in [GRST14b], they are also related to the convex (τ)-property by Maurey [Mau91].

Definition 2.5 *A cost function $c : \mathcal{X} \times \mathcal{P}_1(\mathcal{X}) \to [0, +\infty]$ is called barycentric if there exists a lower semi-continuous convex function $\theta : \mathbb{R}^n \to [0, +\infty]$ such that for all $x \in \mathbb{R}^n$ and all $p \in \mathcal{P}_1(\mathbb{R}^n)$,*

$$c(x, p) = \theta\left(x - \int y\, dp(y)\right).$$

Let us observe that the concentration property associated to this cost function is weaker than the one associated to the usual cost function with $\omega(x, y) = \theta(x$ $y), x, y \in \mathcal{X}$, since by Jensen's inequality

$$c(x, p) \leq \int \theta(x - y) dp(y).$$

A functional formulation of the concentration principle of Definition 2.1 is presented in [GRST14b]. This second definition is associated to the following type of infimum-convolution operator, introduced in [Sam07, GRST14b]: for any measurable function $\varphi : \mathcal{X} \to \mathbb{R} \cup \{\infty\}$ bounded from below

$$R_c\varphi(x) = \inf_{p\in\mathcal{P}(\mathcal{X})} \left\{ \int \varphi dp + c(x,p) \right\}, \qquad x \in \mathcal{X}. \tag{1}$$

Since $c(x, \delta_x) = 0$, one has $R_c\varphi(x) \leq \varphi(x)$.

For a usual cost function, $c(x,p) = \int \omega(x,y)dp(y)$, since the function $p \mapsto \int \varphi dp + c(x,p)$ is affine, the operator $R_c\varphi$ is the classical infimum-convolution operator associated to the cost function ω,

$$R_c\varphi(x) = \inf_{y\in\mathcal{X}} \{\varphi(y) + \omega(x,y)\} = Q_\omega\varphi(x).$$

The functional formulation of the concentration principle is given by the following result.

Proposition 2.1 *Let $a_1, a_2 > 0$ and $\beta : \mathbb{R}^+ \to [0, +\infty]$ be a function. The following properties are equivalent.*

(i) For all $A \in \mathcal{B}(\mathcal{X})$, and all $r \geq 0$,

$$\mu(A)^{a_1}\mu(\mathcal{X} \setminus A_{r,c})^{a_2} \leq e^{-\beta(r)}.$$

(ii) For all measurable functions $\varphi : \mathcal{X} \to \mathbb{R} \cup \{\infty\}$ bounded from below,

$$\mu(\varphi \leq m)^{a_1}\mu(R_c\varphi > m + r)^{a_2} \leq e^{-\beta(r)} \qquad \forall m \in \mathbb{R}, \forall r \geq 0.$$

Proof Given $A \in \mathcal{B}(\mathcal{X})$, let i_A be the zero function on A and equal to $+\infty$ on $\mathcal{X} \setminus A$. By applying (ii) with the function $\varphi = i_A$ and with $m = 0$, we get (i) since $\{\varphi \leq 0\} = A$ and $R_c i_A(x) = c(x, A)$, $x \in \mathcal{X}$.

Conversely, given a function φ, we apply (1) with $A = \{\varphi \leq m\}$. Then, (2) follows from the fact that $\{R_c\varphi > m + r\} \subset (\mathcal{X} \setminus A_{r,c})$. Indeed, if $x \in A_{r,c}$, then for all $\varepsilon > 0$, there exists $p_\varepsilon \in \mathcal{P}(A)$ such that $c(x, p_\varepsilon) \leq r + \varepsilon$. Since

$$R_c\varphi(x) \leq \int \varphi dp_\varepsilon + c(x, p_\varepsilon) \leq m + r + \varepsilon,$$

when ε goes to 0, we get $x \in \{R_c\varphi \leq m+r\}$. As a consequence $A_{r,c} \subset \{R_c\varphi \leq m+r\}$ and $\mu(R_c\varphi > m + r) \leq \mu(\mathcal{X} \setminus A_{r,c})$. □

For a usual cost function of type $c(x,p) = \int \alpha(d(x,y))dp(y)$, $x \in \mathcal{X}$, $p \in \mathcal{P}(\mathcal{X})$, where $\alpha : \mathbb{R}^+ \to \mathbb{R}^+$ is one-to-one, the common way to write the concentration principle is to use the classical enlargement $A_t = \{x \in \mathcal{X}, d(x, A) \leq t\}$. Since $A_{r,c} = A_t$ for $r = \alpha(t)$, (i) can be rewritten as follows: for all $A \in \mathcal{B}(\mathcal{X})$ and all $t \geq 0$,

$$\mu(A)^{a_1}\mu(\mathcal{X} \setminus A_t)^{a_2} \leq e^{-\beta(\alpha(t))}.$$

In that case, going back to early P. Lévy's ideas, we may formalize the concentration property by using the class of 1-Lipschitz functions f (cf. [Led01]). As in the previous proof, by choosing $A = \{f \leq m\}$, $m \in \mathbb{R}$, we show that $A_t \subset \{f \leq m + t\}$, $t \geq 0$. This provides the following equivalent functional formulation : for all 1-Lipschitz functions $f : \mathcal{X} \to \mathbb{R}$,

$$\mu(f \leq m)^{a_1} \mu(f > m + t)^{a_2} \leq e^{-\beta(\alpha(t))}, \qquad \forall m \in \mathbb{R}, \ \forall t \geq 0. \tag{2}$$

Let us assume moreover that α is convex. The inequality (2) can be also derived from (ii) applied to the function $\varphi = \lambda f$, $\lambda > 0$, assuming first that f is bounded from below. Since f is 1-Lipschitz, $R_c\varphi$ is close to φ and its closeness is controlled by λ. More precisely, one has for any $x \in \mathcal{X}$

$$\begin{aligned} R_c\varphi(x) &= \inf_{y \in \mathcal{X}} \{\varphi(y) + \alpha(d(x, y))\} \\ &\geq \varphi(x) - \sup_{y \in \mathcal{X}} \{\lambda d(x, y) - \alpha(d(x, y))\} \geq \varphi(x) - \alpha^*(\lambda), \end{aligned}$$

with $\alpha^*(\lambda) = \sup_{v \geq 0}\{\lambda v - \alpha(v)\}$. Therefore, by replacing m by λm, (ii) provides: for all 1-Lipschitz function f, bounded from below, and for all $\lambda \geq 0$,

$$\mu(f \leq m)^{a_1} \mu(\lambda f > \lambda m + r + \alpha^*(\lambda))^{a_2} \leq e^{-\beta(r)}, \qquad \forall m \in \mathbb{R}, \ \forall r \geq 0.$$

Since $\alpha : \mathbb{R}^+ \to \mathbb{R}^+$ is one-to-one convex and $\alpha(0) = 0$, α is increasing and for all $t > 0$, $\partial\alpha(t) \subset (0, +\infty)$, where $\partial\alpha(t)$ denotes the subdifferential of α at point t. As a consequence, the last inequality implies property (2) for all 1-Lipschitz functions bounded from below by choosing $r = \alpha(t)$ and $\lambda \in \partial\alpha(t)$ such that $\lambda t = \alpha(t) + \alpha^*(\lambda)$. Then, by monotone convergence, the property (2) extends to all 1-Lipschitz functions.

Finally, the property (2) implies the classical concentration property for 1-Lipschitz functions f around their median m_f (see [Led01], Chapter 1). By applying (2) to f or to $-f$, and by choosing $m = m_f$ or $m = -m_f$, we get

$$\mu(|f - m_f| > t) \leq 2.2^{a_1/a_2} e^{-\beta(\alpha(t))/a_2}, \qquad \forall t \geq 0.$$

3 Transport-Entropy Inequalities

This section emphasizes the transport-entropy inequalities in the study of concentration of measure phenomenon on product spaces. As for the logarithmic Sobolev inequalities, the tensorization properties of the transport inequalities make it an effective tool to prove concentration properties in product spaces. The last part of this section briefly recalls the links between the transport-entropy inequalities and the logarithmic Sobolev inequalities.

Let us first recall the original links between isoperimetric properties and concentration of measure properties. Let μ be a measure on a metric space $(\mathcal{X}, d)$. For any Borel set A, the *surface measure* of A is defined by

$$\mu^+(\partial A) = \liminf_{t \to 0+} \frac{\mu(A_t) - \mu(A)}{t}.$$

The isoperimetric problem is to determine the smaller surface measure $\mu^+(\partial A)$ among all Borel set A of fixed measure $\mu(A)$. Namely, we want to find the largest function, denoted by $I_\mu : \mathbb{R}^+ \to \mathbb{R}^+$ such that for all $A \in \mathcal{B}(\mathcal{X})$,

$$\mu^+(\partial A) \geq I_\mu(\mu(A)). \tag{3}$$

The function I_μ is called *isoperimetric profile* of the measure μ.

If $I_\mu \geq v' \circ v^{-1}$, where $v : \mathbb{R} \to [0, \mu(X)]$ is an increasing smooth function, *the isoperimetric inequality* (3) provides a lower estimate of the measure of A_t (see Proposition 2.1 [Led01]): for all $t \geq 0$,

$$\mu(A_t) \geq v(v^{-1}(\mu(A)) + t). \tag{4}$$

Therefore, if μ is a probability measure, we get the following concentration property, for all $A \in \mathcal{B}(\mathcal{X})$ with $\mu(A) \geq 1/2$,

$$\mu(\mathcal{X} \setminus A_t) \leq 1 - v(v^{-1}(1/2) + t), \qquad \forall t \geq 0.$$

When $(X, d) = (\mathbb{R}^n, |\cdot|)$ is the Euclidean space and μ is the Lesbegue measure, or when $X = \mathbb{S}^n$ is the unit Euclidean sphere of $\mathbb{R}^{n+1}$ with its geodesic distance d and $\mu = \sigma^n$ is the uniform law on $\mathbb{S}^n$, the isoperimetric profile is given by $I_\mu = v' \circ v^{-1}$ where for all $r \geq 0$, $v(r)$ is the measure of a ball of radius r (see [Lév51]). As a consequence, if $\mu(A) = \mu(B) = v(r)$ where B is a ball of $\mathcal{X}$, then

$$v(v^{-1}(\mu(A)) + t) = v(r + t) = \mu(B_t).$$

Therefore the property (4) implies that the balls are extremal sets with the following meaning: for all Borel sets A and for all balls B with measure $\mu(B) = \mu(A)$,

$$\mu(A_t) \geq \mu(B_t),$$

for all $t \geq 0$. Actually, this property is equivalent to the isoperimetric inequality since it implies,

$$\mu^+(\partial A) = \liminf_{t \to 0+} \frac{\mu(A_t) - \mu(A)}{t} \geq \liminf_{t \to 0+} \frac{\mu(B_t) - \mu(B)}{t} = I_\mu(\mu(A)),$$

by using $\mu(A) = \mu(B)$.

In this way, the concentration profile of σ^n is given by the estimate of the measures of the spherical balls: for all $A \in \mathcal{B}(\mathbb{S}^n)$ with $\sigma^n(A) \geq 1/2$,

$$\sigma^n(\mathcal{X} \setminus A_t) \leq 1 - \sigma^n(B_t) = v(v^{-1}(1/2) + t) \leq e^{-(n-1)t^2/2}, \quad \forall t \geq 0, \tag{5}$$

where B is a half-sphere, $\sigma^n(B) = 1/2$. The proof of the estimate given by the last inequality is given after Corollary 2.2 in [Led01].

By volume expansion, since the uniform law of the sphere of $\mathbb{R}^{n+1}$ of radius $\sqrt{n}$ goes to the canonical Gaussian measure on $\mathbb{R}^{\mathbb{N}}$, we get the isoperimetric profile of the Gaussian measure (cf. [Led93, Led01]). Half hyperplanes are extremal sets for the standard Gaussian measure γ^n on $\mathbb{R}^n$ and one has

$$I_{\gamma^n}(s) = \varphi \circ \Phi^{-1}(s) \underset{0+}{\sim} s\sqrt{2 \log \frac{1}{s}},$$

with $\Phi(r) = \frac{1}{\sqrt{2\pi}} \int_{-\infty}^{r} e^{-u^2/2} du$, $r \in \mathbb{R}$, and $\varphi = \Phi'$, Φ is the cumulative distribution function of the standard Gaussian law on $\mathbb{R}$. It provides the following concentration property: for all $A \in \mathcal{B}(\mathbb{R}^n)$ such that $\gamma^n(A) \geq 1/2$,

$$\gamma^n(\mathbb{R}^n \setminus A_t) \leq 1 - \Phi(t) \leq e^{-t^2/2}, \quad t \geq 0. \tag{6}$$

On the discrete cube, $\mathcal{X} = \{0,1\}^n$, equipped with the uniform probability measure μ^n and the Hamming distance defined by

$$d(x, y) = \sum_{i=1}^{n} \mathbb{1}_{x_i \neq y_i}, \qquad x, y \in \{0,1\}^n,$$

the extremal sets A minimizing $\mu^n(A_t)$ for $\mu^n(A) \geq 1/2$ have been identified (see [Har66, WW77]). This provides the following concentration property: for any subset $A \subset \{0,1\}^n$ such that $\mu^n(A) \geq 1/2$,

$$\mu^n(\mathcal{X} \setminus A_t) \leq e^{-2t^2/n}, \quad t \geq 0. \tag{7}$$

In the last three basic examples, the sphere (5), the Gaussian space (6), and the discrete cube (7), we observe that in high dimension, $n >> 1$, dimension is a crucial parameter that quantifies the measure concentration phenomenon.

In high dimension, isoperimetric problems are often hard to establish and few of them are solved. Therefore, we need other methods to prove concentration properties. Moreover, it is well known that the concentration property does not tensorize properly. For that purpose, "entropic" methods are efficient alternative tools. They enable to enlarge considerably the class of examples of concentration properties on high dimensional spaces, thanks to the tensorization properties of the entropy.

For concentration, the main two useful entropic methods are the one associated to the logarithmic Sobolev inequality with the so-called Herbst's argument (cf. chapter 5. [Led01]), and the one based on transport inequalities with the so-called Marton's argument (cf. chapter 6. [Led01]). This paper mainly concerns this second one. For example, the concentration results (6) for the Gaussian measure on $\mathbb{R}^n$, or (7) for the uniform law on the discrete cube, are easy consequences of the tensorization property of the transport inequalities.

3.1 *Transport Inequalities and Concentration Properties*

Let $\Pi(\mu, \nu)$ denote the set of probability measures on the product space $\mathcal{X}\times\mathcal{X}$, with first marginal μ and second marginal ν. The probability space $\mathcal{P}(\mathcal{X})$ is endowed with the σ-field generated by the applications

$$\begin{array}{ccc} \mathcal{P}(\mathcal{X}) & \to & ([0,1], \mathcal{B}) \\ \nu & \mapsto & \nu(A), \end{array}$$

where A is any Borel set of $\mathcal{X}$ and $\mathcal{B}$ is the Borel σ-field on $[0, 1]$.

Since $\mathcal{X}$ is a Polish space, any measure $\pi \in \Pi(\mu, \nu)$ can be decomposed as follows:

$$d\pi(x, y) = d\mu(x)dp_x(y),$$

where $p : x \in \mathcal{X} \mapsto p_x \in \mathcal{P}(\mathcal{X})$ is a measurable map μ-almost-surely uniquely determined; p is a probability kernel satisfying

$$\mu p(A) = \int p_x(A)d\mu(x) = \nu(A), \qquad \forall A \in \mathcal{B}(\mathcal{X}).$$

The cost function c defines the following optimal transport cost T_c, introduced in [GRST14b].

Definition 3.1 *The optimal transport cost between probability measures μ and ν on $\mathcal{X}$, associated to $c : \mathcal{X} \times \mathcal{P}(\mathcal{X}) \to [0, +\infty]$, is the quantity*

$$T_c(\nu|\mu) = \inf_{\pi\in\Pi(\mu,\nu)} \int c(x, p_x)d\mu(x),$$

where, given $\pi \in \Pi(\mu, \nu)$, the kernel $p = (p_x)_{x\in X}$ is such that $d\pi(x, y) = d\mu(x)dp_x(y)$.

At first sight, $T_c(\nu|\mu)$ is not a symmetric quantity of μ and ν. As example, for a usual cost,

$$c(x, p_x) = \int \omega(x, y) dp_x(y), \qquad x \in \mathcal{X},$$

the optimal cost $T_c(\nu|\mu)$ corresponds to the usual optimal transport cost linked to the cost function ω,

$$\mathcal{T}_\omega(\mu, \nu) = \inf_{\pi \in \Pi(\mu,\nu)} \iint \omega(x, y) d\pi(x, y) = \inf_{\pi \in \Pi(\mu,\nu)} \int c(x, p_x) d\mu(x) = T_c(\nu|\mu).$$

If the function ω is symmetric, $\omega(x, y) = \omega(y, x)$ for all $x, y \in \mathcal{X}$, then $\mathcal{T}_\omega$ and therefore T_c is symmetric, $\mathcal{T}_\omega(\mu, \nu) = \mathcal{T}_\omega(\nu, \mu)$.

Let us present transport inequalities associated to the optimal transport costs $T_c(\nu|\mu)$. We emphasize a general version that exactly provides the concentration of measure property of Definition 2.1 by the Marton's argument.

Definition 3.2 *Let $a_1, a_2 > 0$ and let $\beta : \mathbb{R}^+ \to \mathbb{R}^+$ be a non-decreasing function. The probability measure $\mu \in \mathcal{P}(\mathcal{X})$ satisfies the transport inequality $\mathbf{T}_{c,\beta}(a_1, a_2)$ if*

$$\mathbf{T}_{c,\beta}(a_1, a_2) : \qquad \beta\left(T_c(\nu_1|\nu_2)\right) \le a_1 H(\nu_1|\mu) + a_2 H(\nu_2|\mu), \qquad \forall \nu_1, \nu_2 \in \mathcal{P}(\mathcal{X}),$$

where $H(\nu_1|\mu)$ is the relative entropy of ν_1 with respect to μ defined by

$$H(\nu_1|\mu) = \int \log \frac{d\nu_1}{d\mu} d\nu_1,$$

if ν_1 is absolutely continuous with respect to μ ($\nu_1 << \mu$), and $H(\nu_1|\mu) = +\infty$ otherwise.

In most cases, the inequality $\mathbf{T}_{c,\beta}(a_1, a_2)$ is called weak transport inequality, and for some particular costs $c : \mathcal{X} \times \times \mathcal{P}(\mathcal{X}) \to [0, +\infty]$, with inequality is called barycentric transport inequality (see Section 4), or universal transport inequality (see Section 5).

Generally, the transport inequalities $\mathbf{T}_{c,\beta}(0, a_2)$ or $\mathbf{T}_{c,\beta}(a_1, 0)$ do not make sense. Indeed, they should imply $\beta(T_c(\nu_1|\mu)) = 0, \forall \nu_1 \in \mathcal{P}(\mathcal{X})$, or even $\beta(T_c(\mu|\nu_2)) = 0, \forall \nu_2 \in \mathcal{P}(\mathcal{X})$, which are never satisfied, except in degenerated cases (for example, $c = 0$ or $\beta = 0$). However, with the convention $0.\infty = 0$, the transport inequality $\mathbf{T}_{c,\beta}(b, \infty)$ corresponds to the common transport inequality

$$\mathbf{T}^+_{c,\beta}(b) : \qquad \beta\left(T_c(\nu|\mu)\right) \le bH(\nu|\mu), \qquad \forall \nu \in \mathcal{P}(\mathcal{X}),$$

and $\mathbf{T}_{c,\beta}(\infty, b)$ corresponds to the common transport inequality

$$\mathbf{T}^-_{c,\beta}(b) : \qquad \beta\left(T_c(\mu|\nu)\right) \le bH(\nu|\mu), \qquad \forall \nu \in \mathcal{P}(\mathcal{X}).$$

These two inequalities are identical for symmetric optimal transport costs. When β is the identity, we simply denote by $\mathbf{T}_c(a_1, a_2)$, $\mathbf{T}_c^+(b)$ and $\mathbf{T}_c^-(b)$ the last transport inequalities.

Let us recall the Marton's argument. Given $A \in \mathcal{B}(\mathcal{X})$, if ν_1 is the renormalized restriction of μ to A and ν_2 the renormalized restriction of μ to $B = \mathcal{X} \setminus A_{r,c}$, $r \geq 0$,

$$\frac{d\nu_1}{d\mu} = \frac{\mathbb{1}_A}{\mu(A)} \quad \text{and} \quad \frac{d\nu_2}{d\mu} = \frac{\mathbb{1}_B}{\mu(B)},$$

then $H(\nu_1|\mu) = -\log \mu(A)$, $H(\nu_2|\mu) = -\log \mu(B)$, and $T_c(\nu_1|\nu_2) \geq r$ (since for all $x \in \mathcal{X} \setminus A_{r,c}$ and all $p \in \mathcal{P}(A)$, $c(x.p) \geq r$). Consequently, since β is non-decreasing, the transport inequality $\mathbf{T}_{c,\beta}(a_1, a_2)$ provides

$$\beta(r) \leq \log\left(\mu(A)^{-a_1}\right) + \log\left(\mu(B)^{-a_2}\right), \qquad \forall r \geq 0.$$

which is the concentration property of Definition 2.1.

For a better comprehension, let us illustrate Definition 3.2 by few examples of transport inequalities.

If c is the usual cost function $c(x,p) = \int d(x,y)^q p(dy)$, $q \geq 1$, then $T_c(\nu|\mu) = T_c(\mu|\nu)$ is associated to the Wasserstein distance W_q of order q,

$$T_c(\nu|\mu) = W_q^q(\mu, \nu) = \inf_{\pi \in \Pi(\mu,\nu)} \iint d(x,y)^q \pi(dx, dy).$$

The transport inequality $\mathbf{T}_2(b)$, first considered by Talagrand [Tal96c], and satisfied by the standard Gaussian measure $\mu = \gamma^n$ on $\mathbb{R}^n$ for $b = 2$, corresponds to the transport inequalities $\mathbf{T}_c^+(b)$ or $\mathbf{T}_c^-(b)$ with $q = 2$,

$$\mathbf{T}_2(b): \qquad W_2^2(\mu, \nu) \leq bH(\nu|\mu), \qquad \forall \nu \in \mathcal{P}(\mathcal{X}).$$

A special feature of the inequality $\mathbf{T}_2(b)$ is its equivalence to the family of transport inequalities $\mathbf{T}_c(b/t, b/(1-t))$, for $t \in (0,1)$. Indeed, if μ satisfies $\mathbf{T}_2(b)$, then by the triangular inequality for the Wasserstein metric,

$$W_2^2(\nu_1, \nu_2) \leq (W_2(\nu_1, \mu) + W_2(\mu, \nu_2))^2 \leq b\left(\sqrt{H(\nu_1|\mu)} + \sqrt{H(\nu_2|\mu)}\right)^2.$$

From the identity $\left(\sqrt{u} + \sqrt{v}\right)^2 = \inf_{t \in (0,1)} \left\{\frac{u}{t} + \frac{v}{1-t}\right\}$, we get that for all $t \in (0,1)$, μ satisfies $T_c(b/t, b/(1-t))$,

$$W_2^2(\nu_1, \nu_2) \leq \frac{b}{t} H(\nu_1|\mu) + \frac{b}{1-t} H(\nu_2|\mu), \quad \forall \nu_1, \nu_2 \in \mathcal{P}(\mathcal{X}).$$

Conversely if μ verifies $T_c(b/t, b/(1-t))$ for all $t \in (0,1)$, then by choosing $\nu_2 = \mu$ and then when t goes to 1, we recover the transport inequality $\mathbf{T}_2(b)$.

More generally, assume that c is a usual cost of type $c(x,p) = \int \alpha(d(x,y))p(dy)$, where $\alpha : \mathbb{R}^+ \to \mathbb{R}$ is a convex function. In that case we note $\mathcal{T}_\alpha(\mu,\nu) = T_c(\nu|\mu) = T_c(\mu|\nu)$. If moreover α is increasing, $\alpha(0) = \alpha'(0) = 0$ and α satisfies the following Δ_2*-condition*, [RR91]: there exists a positive constant C such that

$$\alpha(2h) \leq C\alpha(h), \qquad \forall h \geq 0, \tag{8}$$

then, we may use the following change of metric Lemma given in [GRS13].

Lemma 3.1 *With the above conditions, setting* $p_\alpha = \sup_{h>0} \frac{h\alpha'_+(h)}{\alpha(h)}$, *the function* $h \mapsto \alpha^{1/p_\alpha}(h)$ *is sub-additive, namely*

$$\alpha^{1/p_\alpha}(h+k) \leq \alpha^{1/p_\alpha}(h) + \alpha^{1/p_\alpha}(k), \qquad \forall h,k \in \mathbb{R}^+.$$

As a consequence, $d_\alpha(x,y) = \alpha^{1/p_\alpha}(d(x,y)), x,y \in \mathcal{X}$ *is a distance on* $\mathcal{X}$.

This lemma together with the triangular inequality gives for all $\nu_1,\nu_2 \in \mathcal{P}(\mathcal{X})$,

$$\begin{aligned}\mathcal{T}_\alpha(\nu_1,\nu_2) &= W_{p_\alpha}^{p_\alpha}(\nu_1,\nu_2)\\ &\leq \left(W_{p_\alpha}(\nu_1,\mu) + W_{p_\alpha}(\mu,\nu_2)\right)^{p_\alpha} = \left(\mathcal{T}_\alpha^{1/p_\alpha}(\nu_1,\mu) + \mathcal{T}_\alpha^{1/p_\alpha}(\mu,\nu_2)\right)^{p_\alpha},\end{aligned}$$

where the Wasserstein distance W_{p_α} is understood with respect to the distance d_α of Lemma 3.1 Then, observing that $p_\alpha > 1$ when the function α is not linear, and using the identity

$$\left(u^{1/p} + v^{1/p}\right)^p = \inf_{t\in(0,1)} \left\{\frac{u}{t^{p-1}} + \frac{v}{(1-t)^{p-1}}\right\}, \qquad p > 1,$$

we get that μ satisfies the usual transport inequality

$$\mathcal{T}_\alpha(\mu,\nu) \leq H(\nu|\mu), \qquad \forall \nu \in \mathcal{P}(\mathcal{X}),$$

if and only if μ satisfies the following transport inequalities: for all $t \in (0,1)$,

$$\mathcal{T}_\alpha(\nu_1,\nu_2) \leq \frac{H(\nu_1|\mu)}{t^{p_\alpha-1}} + \frac{H(\nu_2|\mu)}{(1-t)^{p_\alpha-1}}, \qquad \forall \nu_1,\nu_2 \in \mathcal{P}(\mathcal{X}). \tag{9}$$

Here is another example of transport inequality. When $c(x,p) = 2\int \mathbb{1}_{x\neq y}dp(y)$, the universal optimal transport cost $T_c(\nu|\mu)$ is in fact the total variation distance between the measures μ and ν

$$T_c(\nu|\mu) = \|\mu - \nu\|_{TV} = 2\sup_{A\subset\mathcal{X}} |\mu(A) - \nu(A)|.$$

The Csizár-Kullback-Pinsker inequality [Pin64, Csi67, Kul67] that holds for any (reference) probability measure μ,

$$\|\mu - \nu\|_{TV}^2 \leq 2H(\nu|\mu), \qquad \forall \nu \in \mathcal{P}(\mathcal{X}).$$

corresponds to the transport inequalities $\mathbf{T}_{c,\beta}^+(b)$ or $\mathbf{T}_{c,\beta}^-(b)$ with $\beta(r) = r^2/2, r \geq 0$. Here again, this inequality is equivalent to $\mathbf{T}_{c,\beta}(b/t, b/(1-t))$ for all $t \in (0,1)$. This inequality and its improvements are known for their numerous applications in probability, in analysis, and in information theory (cf. [Vil09], page 636).

As a last example, let us consider the universal cost function

$$c(x,p) = \left(\int \mathbb{1}_{x \neq y} dp(y)\right)^2, \qquad x \in \mathcal{X}, \quad p \in \mathcal{P}(\mathcal{X}).$$

Then, the transport inequalities $\mathbf{T}_c^+(2)$ and $\mathbf{T}_c^-(2)$ correspond to the weak transport inequalities introduced by Marton [Mar96b]. As for the Csizár-Kullback-Pinsker inequality, $\mathbf{T}_c^+(2)$ and $\mathbf{T}_c^-(2)$ hold for any (reference) probability measure μ. In that case, $\mathbf{T}_c^+(b)$ and $\mathbf{T}_c^-(b)$ are equivalent to the family of transport inequalities $\mathbf{T}_c(b/t, b/(1-t))$ for $t \in (0,1)$, since the weak-transport cost $\mathcal{T}_c$, also denoted by $\widetilde{\mathcal{T}}_2$, satisfies the following triangular inequality [Mar97],

$$\sqrt{\widetilde{\mathcal{T}}_2(\nu_1|\nu_2)} \leq \sqrt{\widetilde{\mathcal{T}}_2(\nu_1|\mu)} + \sqrt{\widetilde{\mathcal{T}}_2(\mu|\nu_2)}, \qquad \forall \mu, \nu_1, \nu_2 \in \mathcal{P}(\mathcal{X}). \tag{10}$$

3.2 Functional Formulation of Transport Inequality, the Dual Kantorovich Theorem

The dual functional formulation of usual transport inequalities has been obtained by Bobkov and Götze [BG99] and then expanded in the paper [GL10].

This dual form is based on the duality between the relative entropy and the log-Laplace transform. Namely, for any continuous bounded function $g : \mathcal{X} \to \mathbb{R}$,

$$\log \int e^g d\mu = \sup_{\nu \in \mathcal{P}(\mathcal{X})} \left\{ \int g d\mu - H(\nu|\mu) \right\}. \tag{11}$$

A simple proof of this identity is given in [GL10] and one more general in [GRS11a].

The second argument is the dual Kantorovich Theorem. This theorem is well known for usual lower semi-continuous cost functions $\omega : \mathcal{X} \times \mathcal{X} \to (-\infty, +\infty]$ (cf. [Vil09])

$$\mathcal{T}_\omega(\nu_1, \nu_2) = \sup_{\varphi \in C_b(\mathcal{X})} \left\{ \int Q_\omega \varphi \, d\nu_2 - \int \varphi d\nu_1 \right\}, \qquad \nu_1, \nu_2 \in \mathcal{P}(\mathcal{X}), \tag{12}$$

where $\mathcal{C}_b(\mathcal{X})$ is the set of continuous bounded functions on $\mathcal{X}$ and

$$Q_\omega\varphi(y) = \inf_{x\in\mathcal{X}} \{\varphi(x) + \omega(x,y)\}, \qquad y \in \mathcal{X}.$$

In the paper [GRST14b], as the function $p \in \mathcal{P}(\mathcal{X}) \to c(x,p)$ is convex, this result is extended to weak transport costs T_c, under weak regularity additional assumptions on the cost $c : \mathcal{X} \times \mathcal{P}(\mathcal{X}) \to [0, +\infty]$ (see Theorem 3.5, [GRST14b]). Overall, the result is the following,

$$T_c(\nu_1|\nu_2) = \sup_{\varphi\in\mathcal{C}_b(\mathcal{X})} \left\{ \int R_c\varphi\, d\nu_2 - \int \varphi d\nu_1 \right\}, \qquad \nu_1, \nu_2 \in \mathcal{P}(\mathcal{X}), \tag{13}$$

where $R_c\varphi$ is the infimum-convolution operator (1) previously defined,

$$R_c\varphi(x) = \inf_{p\in\mathcal{P}(\mathcal{X})} \left\{ \int \varphi dp + c(x,p) \right\}.$$

To be precise, we should slightly modify the sets $\mathcal{C}_b(\mathcal{X})$ and $\mathcal{P}(\mathcal{X})$, depending on the type of involved cost function c (see [GRST14b]).

The two duality identities (11) and (13) provide the following functional formulation of the transport-entropy inequality $\mathbf{T}_{c,\beta}(a_1, a_2)$.

Proposition 3.1 *Let $\mu \in \mathcal{P}(\mathcal{X})$ and $\beta : \mathbb{R}+ \to [0, +\infty]$ be a lower semi-continuous convex function such that $\beta(0) = 0$. The following statements are equivalent.*

(1) The probability measure μ satisfies $\mathbf{T}_{c,\beta}(a_1, a_2)$.
(2) For all functions $\varphi \in \mathcal{C}_b(\mathcal{X})$ and for all $\lambda \geq 0$,

$$\left(\int e^{\frac{\lambda R_c\varphi}{a_2}} d\mu\right)^{a_2} \left(\int e^{-\frac{\lambda\varphi}{a_1}} d\mu\right)^{a_1} \leq e^{\beta^*(\lambda)},$$

with $\beta^*(\lambda) = \sup_{t\geq 0} \{\lambda t - \beta(t)\}$.

Point (2) generalizes the infimum-convolution viewpoint of transport inequalities introduced by Maurey [Mau91], the so-called (τ)-property.

Idea of the proof (1) $\Rightarrow$ (2) One has $\beta(t) = \sup_{\lambda\geq 0} \{\lambda t - \beta^*(\lambda)\}$, $\forall t \geq 0$. If μ satisfies $\mathbf{T}_{c,\beta}(a_1, a_2)$, then the general dual Kantorovich identity (13) implies that for all $\varphi \in \mathcal{C}_b(\mathcal{X})$ and all $\lambda \geq 0$,

$$\lambda \left(\int R_c\varphi\, d\nu_2 - \int \varphi d\nu_1 \right) - \beta^*(\lambda) \leq a_2 H(\nu_2|\mu) + a_1 H(\nu_1|\mu), \qquad \forall \nu_1, \nu_2 \in \mathcal{P}(\mathcal{X}).$$

Point (2) follows by reordering the terms of this inequality, by optimizing over all probability measures ν_1 and ν_2, and then by applying the dual formula (11), with the function $g = \lambda R_c\varphi/a_2$ and with the function $g = -\lambda\varphi/a_1$. □

By density of the set of the bounded continuous functions in $L^1(\mu)$, and then by monotone convergence, (2) also holds for all measurable functions $\varphi : \mathcal{X} \to (-\infty, +\infty]$ bounded from below.

Given $A \in \mathcal{B}(\mathcal{X})$, let us consider the function i_A equal to 0 on A, and to $+\infty$ on its complement. Applying (2) to the function $\varphi = i_A$, since $R_c i_A(x) = c(x, A)$, $x \in \mathcal{X}$, the transport inequality $\mathbf{T}_{c,\beta}(a_1, a_2)$ provides the Talagrand's formulation of concentration properties (cf. [Tal95, Tal96a, Tal96b]):

$$\int e^{\frac{\lambda c(x,A)}{a_2}}\, d\mu \le \frac{e^{\beta^*(\lambda)/a_2}}{\mu(A)^{a_1/a_2}}, \qquad \forall \lambda \ge 0, \quad \forall A \in \mathcal{B}(\mathcal{X}).$$

When the function β is the identity, one has $\beta^* = i_{(-\infty,1]}$ and Proposition 3.1 is written as follows.

Proposition 3.2 ([GRST14b]) *The following statements are equivalent.*

(1) The probability μ satisfies $\mathbf{T}_c(a_1, a_2)$.
(2) For all function $\varphi \in \mathcal{C}_b(\mathcal{X})$,

$$\left(\int e^{\frac{R_c\varphi}{a_2}}\, d\mu\right)^{a_2} \left(\int e^{-\frac{\varphi}{a_1}}\, d\mu\right)^{a_1} \le 1.$$

3.3 Tensorization: Characterization by Dimension-Free Concentration Properties

The transport entropy inequalities tensorize, and this provides concentration results in high dimension.

Proposition 3.3 *Let $\mathcal{X}_1$ and $\mathcal{X}_2$ be Polish spaces. Let $\beta_1 : \mathbb{R}^+ \to \mathbb{R}^+$, $\beta_2 : \mathbb{R}^+ \to \mathbb{R}^+$ be convex functions and let $\beta : \mathbb{R}^+ \to \mathbb{R}^+$ be defined by*

$$\beta(t) = \beta_1 \Box \beta_2(t) = \inf\{\beta_1(t_1) + \beta_2(t_2), t = t_1 + t_2\}, \qquad t \ge 0.$$

If $\mu_1 \in \mathcal{P}(\mathcal{X}_1)$ and $\mu_2 \in \mathcal{P}(X_2)$ satisfy, respectively, the transport inequalities $\mathbf{T}_{c_1,\beta_1}(a_1, a_2)$ and $\mathbf{T}_{c_2,\beta_2}(a_1, a_2)$, then $\mu_1 \otimes \mu_2 \in \mathcal{P}(\mathcal{X}_1 \times \mathcal{X}_2)$ satisfies the transport inequality $\mathbf{T}_{c,\beta}(a_1, a_2)$ with for all $x = (x_1, x_2) \in \mathcal{X}_1 \times \mathcal{X}_2$, and all $p \in \mathcal{P}(\mathcal{X}_1 \times \mathcal{X}_2)$ with marginals $p_1 \in \mathcal{P}(\mathcal{X}_1)$ and $p_2 \in \mathcal{P}(\mathcal{X}_2)$

$$c(x, p) = c_1 \oplus c_2(x, p) = c_1(x_1, p_1) + c_2(x_2, p_2).$$

The proof of this proposition exactly follows the one of Theorem 4.11 [GRST14b] for which $\beta_1(t) = \beta_2(t) = \beta(t) = t$, $t \geq 0$. We could also follow the tensorization proof given in [Sam07] on the dual functional form of such transport inequalities.

This tensorization property is a consequence of the tensorization properties of the relative entropy and of the optimal transport cost: for any measure $\nu \in \mathcal{P}(\mathcal{X}_1 \times \mathcal{X}_2)$ with decomposition $d\nu(x_1, x_2) = d\nu_1(x_1) d\nu_2^{x_1}(x_2)$, one has

$$H(\nu|\mu) = H(\nu_1|\mu_1) + \int H(\nu_2^{x_1}|\mu_2) d\nu_1(x_1),$$

and for any other measure $\nu' \in \mathcal{P}(\mathcal{X}_1 \times \mathcal{X}_2)$ with decomposition $d\nu'(x_1', x_2') = d\nu_1(x_1') d\nu_2^{x_1'}(x_2')$, for all $\varepsilon \geq 0$, there exists $\pi_1^\varepsilon \in \Pi(\nu_1, \nu_1')$ such that

$$T_c(\nu|\nu') \leq T_{c_1}(\nu_1|\nu_1') + \iint T_{c_2}(\nu_2^{x_1}|\nu_2'^{x_1'}) \, d\pi_1^\varepsilon(x_1, x_1') + \varepsilon,$$

where $c = c_1 \oplus c_2$. The error term ε can be chosen equal to 0 when $\mathcal{X}_1$ and $\mathcal{X}_2$ are compact spaces.

Therefore, if $\mu \in \mathcal{P}(\mathcal{X})$ satisfies the transport inequality $\mathbf{T}_{c,\beta}(a_1, a_2)$, then $\mu^n = \mu \otimes \cdots \otimes \mu \in \mathcal{P}(\mathcal{X}^n)$ satisfies $\mathbf{T}_{c^n, \beta^{\square n}}(a_1, a_2)$, with for all $p \in \mathcal{P}(\mathcal{X}^n)$, with marginals $p_i \in \mathcal{P}(\mathcal{X})$, $i \in \{1, \ldots, n\}$,

$$c^n(x, p) = c^{\oplus n}(x, p) := c(x_1, p_1) + \cdots + c(x_n, p_n), \qquad x = (x_1, \ldots, x_n) \in \mathcal{X}^n,$$

and since β is convex,

$$\beta^{\square n}(t) := \beta \square \cdots \square \beta(t) = n\beta(t/n), \qquad t \geq 0.$$

From the transport inequality $\mathbf{T}_{c^n, \beta^{\square n}}(a_1, a_2)$, we get that μ^n satisfies the following concentration property, for all $A \in \mathcal{B}(\mathcal{X}^n)$,

$$\mu^n(A)^{a_1} \mu^n(\mathcal{X} \setminus A_{r,c^n})^{a_2} \leq e^{-n\beta(r/n)}, \qquad \forall r \geq 0.$$

The concentration profile in the right-hand side is independent of n if and only if β is linear. In that case, we say that μ satisfies a *dimension-free concentration property*.

Definition 3.3 *A measure $\mu \in \mathcal{P}(\mathcal{X})$ satisfies a dimension-free concentration property associated to a cost function $c : \mathcal{X} \times \mathcal{P}(\mathcal{X})$, if for all $n \geq 1$, and for all $A \in \mathcal{B}(\mathcal{X}^n)$,*

$$\mu^n(A)^{a_1} \mu^n(\mathcal{X} \setminus A_{r,c^n})^{a_2} \leq e^{-r}, \qquad \forall r \geq 0.$$

Actually, Gozlan has proved that for usual enlargements associated to the cost function $\omega(x, y) = \alpha(d(x, y))$, this dimension-free concentration property is equivalent to a transport inequality (see [Goz09]). Its proof is based on large deviation technics. In the paper [GRST14b], a simpler approach, starting from the dual formulation of transport inequalities, allows to extend Gozlan's result to any transport inequality $\mathbf{T}_c(a_1, a_2)$.

Proposition 3.4 ([GRST14b]) *The following statements are equivalent.*

(i) μ *satisfies* $\mathbf{T}_c(a_1, a_2)$*: for all functions* $\psi \in \mathcal{C}_b(\mathcal{X})$*,*

$$\left(\int e^{\frac{R_c\psi}{a_2}}\,d\mu\right)^{a_2}\left(\int e^{-\frac{\psi}{a_1}}\,d\mu\right)^{a_1} \leq 1.$$

(ii) For all integers $n \geq 1$ *and for all functions* $\varphi \in \mathcal{C}_b(\mathcal{X}^n)$*,*

$$\mu^n(\varphi \leq m)^{a_1}\,\mu^n(R_{c^n}\varphi > m + r)^{a_2} \leq e^{-r}, \qquad \forall m \in \mathbb{R}, \quad \forall r \geq 0. \tag{14}$$

Idea of the proof As already explained, $(i) \Rightarrow (ii)$ is a consequence of the tensorization properties of the inequality $\mathbf{T}_c(a_1, a_2)$.

In order to get $(ii) \Rightarrow (i)$, we estimate the product of exponential moments of $R_{c^{\oplus n}}\varphi$ and $-\varphi$ using the tail distribution estimates given by (ii). More precisely, (ii) provides: for all $\varepsilon > 0$,

$$\left(\int e^{\frac{R_{c^n}\varphi}{(1+\varepsilon)a_2}}\,d\mu^n\right)^{a_2}\left(\int e^{-\frac{\varphi}{(1-\varepsilon)a_1}}\,d\mu^n\right)^{a_1} \leq K(\varepsilon, a_1, a_2),$$

where $K(\varepsilon, a_1, a_2)$ is a constant independent of n. We want to "tighten" this inequality by replacing this constant by 1. For that purpose, let us choose $\varphi(x) = \psi(x_1) + \cdots + \psi(x_n)$, $x = (x_1, \ldots, x_n) \in \mathcal{X}^n$, for which

$$R_{c^n}\varphi(x) = R_c\psi(x_1) + \cdots + R_c\psi(x_n).$$

By independence, the last inequality can be rewritten as follows:

$$\left(\int e^{\frac{R_c\psi}{(1+\varepsilon)a_2}}\,d\mu\right)^{a_2}\left(\int e^{-\frac{\psi}{(1-\varepsilon)a_1}}\,d\mu\right)^{a_1} \leq K(\varepsilon, a_1, a_2)^{1/n}.$$

The result follows from this inequality as n goes to $+\infty$ and then ε goes to 0. □

3.4 Connections with Logarithmic Sobolev Inequalities

In this section, we assume that the closed balls of the Polish metric space $(\mathcal{X}, d)$ are compact. In this part, the transport costs are associated to usual cost functions on a metric space $(\mathcal{X}, d)$:

$$c(x,p) = \int \alpha(d(x,y))dp(y), \qquad x \in \mathcal{X}, p \in \mathcal{P}(\mathcal{X}),$$

where $\alpha : \mathbb{R}^+ \to \mathbb{R}^+$ is a convex function such that $\alpha(0) = \alpha'(0) = 0$ satisfying the Δ_2-condition (8). In this case, we note $\mathcal{T}_\alpha(\mu, \nu) = T_c(\mu|\nu)$ and $\mathbf{T}_\alpha(b)$ the transport inequality $\mathbf{T}_c^+(b)$ that coincides with $\mathbf{T}_c^-(b)$.

For any locally Lipschitz function $f : \mathcal{X} \to \mathbb{R}$, the gradient norms of f at a non-isolated point $x \in \mathcal{X}$ are defined by

$$|\nabla^+ f|(x) = \limsup_{y \to x} \frac{[f(y) - f(x)]_+}{d(x,y)}, \quad \text{or} \quad |\nabla^- f|(x) = \limsup_{y \to x} \frac{[f(y) - f(x)]_-}{d(x,y)},$$

and $|\nabla^+ f|(x) = |\nabla^- f|(x) = 0$ if x is an isolated point. If $\mathcal{X}$ is a Riemannian manifold and f is smooth, $|\nabla^+ f|(x)$ and $|\nabla^- f|(x)$ are the norm of $\nabla f(x)$ in the tangent space $T_x\mathcal{X}$ at point x.

Definition 3.4 *A measure $\mu \in \mathcal{P}(\mathcal{X})$ satisfies the modified logarithmic Sobolev inequality* $\mathbf{LogSob}_\alpha^+(b)$*, $b \geq 0$, associated to the cost α, if for any locally Lipschitz function $f : \mathcal{X} \to \mathbb{R}$, one has*

$$\mathbf{LogSob}_\alpha^+(b) : \qquad \mathrm{Ent}_\mu(e^f) \leq b \int \alpha^*(|\nabla^+ f|) e^f d\mu,$$

where $\alpha^*(h) = \sup_{t \geq 0}\{ht - \alpha(t)\}$ and for any function $g : \mathcal{X} \to \mathbb{R}^+$,

$$\mathrm{Ent}_\mu(g) = \int g \log g \, d\mu - \int g \, d\mu \log \int g d\mu.$$

In the same way, we define the logarithmic Sobolev inequality $\mathbf{LogSob}_\alpha^-(b)$ by replacing $|\nabla^+ f|$ by $|\nabla^- f|$. If $\mathcal{X}$ is a Riemannian manifold, we simply note $\mathbf{LogSob}_\alpha(b)$. When α is quadratic, $\alpha(t) = t^2$, $t \geq 0$, $\alpha^*(h) = h^2/4$, $h \geq 0$, the logarithmic Sobolev inequalities are denoted by $\mathbf{LogSob}_2^+(b)$ or $\mathbf{LogSob}_2^-(b)$.

As a first result, the well-known Otto-Villani Theorem asserts that the Talagrand's transport inequality is a consequence of the logarithmic Sobolev inequality.

Theorem 3.1 ([OV00]) *Let $\mathcal{X}$ be a Riemannian manifold. If $\mu \in \mathcal{P}_2(\mathcal{X})$ satisfies the logarithmic Sobolev inequality* $\mathbf{LogSob}_2(b)$*, then μ satisfies the Talagrand's transport inequality* $\mathbf{T}_2(b)$.

In the heuristic part of [OV00], Otto and Villani give the idea of their proof of this result by interpreting the Wasserstein space $(\mathcal{P}_2(X), W_2)$ as a Riemannian manifold and by considering the gradient flow of the relative entropy $\nu \mapsto H(\nu|\mu)$. Bobkov, Gentil, and Ledoux [BGL01] give another proof based on the Hopf-Lax formula for the solutions of the Hamilton-Jacobi equation. More precisely, on a Riemannian manifold $\mathcal{X}$, the infimum-convolution operator

$$v(x,t) = Q_t f(x) = \inf_{y \in \mathcal{X}} \left\{ f(y) + \frac{1}{2t} d(x,y)^2 \right\}, \qquad x \in \mathcal{X}, \quad t > 0,$$

is a semi-group, solution of the Hamilton-Jacobi equation

$$\frac{\partial v}{\partial t} = -\frac{1}{2}|\nabla v|^2, \qquad \text{avec} \quad v(x,0) = f(x), \ \forall x \in \mathcal{X}.$$

A counterexample, showing that Otto-Villani's Theorem cannot be reversed in full generality, has been given by Cattiaux and Guillin [CG06] (see also [Goz07]).

Then Otto-Villani's result has been complemented by Gozlan et al. in a series of papers [GRS11b, GRS13, GRS14]. Following the Hamilton-Jacobi approach by Bobkov-Gentil-Ledoux, the modified logarithmic Sobolev inequality $\mathbf{LogSob}_\alpha^-(b)$ is characterized in terms of hypercontractivity property of the operator $Q_t f$ defined by

$$Q_t f(x) = \inf_{y \in \mathcal{X}} \left\{ f(y) + t\alpha\left(\frac{d(x,y)}{t}\right) \right\}, \qquad x \in \mathcal{X},$$

for all bounded function $f : \mathcal{X} \to \mathbb{R}$.

Theorem 3.2 ([GRS14]) *Assume that α satisfies the Δ_2-condition* (8). *Then the exponents $r_\alpha \leq p_\alpha$ defined by*

$$r_\alpha = \inf_{x>0} \frac{x\alpha'_-(x)}{\alpha(x)} \geq 1 \qquad \textit{and} \qquad 1 < p_\alpha = \sup_{x>0} \frac{x\alpha'_+(x)}{\alpha(x)}$$

are both finite. Moreover, the measure μ satisfies $\mathbf{LogSob}_\alpha^-(b)$ if and only if for all $t > 0$, for all $t_o \leq b(p_\alpha - 1)$ and for all bounded continuous functions $f : X \to \mathbb{R}$,

$$\left\| e^{Q_t f} \right\|_{k(t)} \leq \left\| e^f \right\|_{k(0)},$$

with

$$k(t) = \begin{cases} \left(1 + \frac{b^{-1}(t-t_o)}{p_\alpha - 1}\right)^{p_\alpha - 1} \mathbf{1}_{t \leq t_o} + \left(1 + \frac{b^{-1}(t-t_o)}{r_\alpha - 1}\right)^{r_\alpha - 1} \mathbf{1}_{t > t_o} & \textit{if } r_\alpha > 1 \\ \min\left(1; \left(1 + \frac{b^{-1}(t-t_o)}{p_\alpha - 1}\right)^{p_\alpha - 1}\right) & \textit{if } r_\alpha = 1 \end{cases},$$

where $\|g\|_k = \left(\int |g|^k d\mu\right)^{1/k}$ for $k \neq 0$ and $\|g\|_0 = \exp\left(\int \log g \, d\mu\right)$.

By choosing $t_o = b(p_\alpha - 1)$ and after some easy computations this theorem implies the following Otto-Villani Theorem, extended to any metric space and for any cost function α satisfying the Δ_2-condition.

Theorem 3.3 ([GRS14]) *Suppose that α verifies the Δ_2-condition* (8). *If μ verifies* $\mathbf{LogSob}_\alpha^-(b)$, *then it verifies* $\mathbf{T}_\alpha(B)$, *with*

$$B = \max\left(((p_\alpha - 1)b)^{r_\alpha - 1}; ((p_\alpha - 1)b)^{p_\alpha - 1}\right),$$

where the numbers r_α, p_α are defined in Theorem 3.2.

This result exactly recovers Otto-Villani Theorem 3.1 since $p_\alpha = r_\alpha = 2$ for $\alpha(h) = h^2, h \geq 0$.

Popular functions α appearing as cost functions in the literature are the functions $\alpha = \alpha_{p_1,p_2}$, with $p_1 \geq 2$ and $p_2 \geq 1$ defined by

$$\alpha_{p_1,p_2}(h) = \begin{cases} h^{p_1} & \text{if } 0 \leq h \leq 1, \\ \frac{p_1}{p_2} h^{p_2} + 1 - \frac{p_1}{p_2} & \text{if } h \geq 1. \end{cases}$$

Any such function satisfies the Δ_2-condition with $r_\alpha = \min(p_1, p_2)$ and $p_\alpha = \max(p_1, p_2)$.

As examples, the best known measures on $\mathbb{R}^n$ satisfying the logarithmic Sobolev inequality $\mathbf{LogSob}_{\alpha_{2,p}}(b)$ for some $b > 0$ are the standard Gaussian measure for $p = 2$ [Gro75], the exponential measure for $p = 1$ [BL97], and more generally the probability measures $d\mu_p = e^{-|t|^p}/Z_p dt$, for $p \geq 1$ (see [GGM07, BR08, Goz07]). For these measures, Theorem 3.2 provides the related transport inequalities obtained in different papers [Tal91, BK08, GGM05].

To end the comparisons between logarithmic Sobolev inequalities and transport inequalities, let us recall the reversed Otto-Villani's Theorem obtained in the papers [GRS11b, GRS13, GRS14]. It characterizes the transport inequalities in terms of modified logarithmic Sobolev inequalities restricted to a class of $K - \alpha$-convex functions. By definition, a function $f : \mathcal{X} \to \mathbb{R}$ is *$K - \alpha$-convex* if there exists a function $h : \mathcal{X} \to \mathbb{R}$ such that

$$f(x) = \sup_{y \in \mathcal{X}} \{h(y) - K\alpha(d(x, y))\} = P_\alpha^K h(x), \qquad \forall x \in \mathcal{X}.$$

On the Euclidean space, if $\alpha(h) = h^2, h \geq 0$, then a smooth $K - \alpha$-convex function is exactly a function with Hessian bounded from below.

Let us summarize the results of Theorem 1.12 [GRS11b], Theorem 5.1 [GRS13] when $(\mathcal{X}, d)$ is a *geodesic space*: for any $x, y \in \mathcal{X}$, there exists a path $(\gamma_t)_{t \in [0,1]}$ in $\mathcal{X}$, such that $\gamma_0 = x$, $\gamma_1 = y$ and $d(\gamma_s, \gamma_t) = |t - s| d(x, y)$, for all $s, t \in [0, 1]$.

Theorem 3.4 ([GRS11b, GRS13]) *Let $(\mathcal{X}, d)$ be a geodesic space and $\alpha : \mathbb{R}^+ \to \mathbb{R}^+$ be a convex function satisfying the Δ_2-condition* (8) *and such that $\alpha(0) = \alpha'(0) = 0$. The following properties are equivalent.*

(1) There exists $C_1 > 0$, such that μ satisfies the transport inequality $\mathbf{T}_\alpha(C_1)$.
(2) There exist $C_2 > 0$ and $\lambda > 0$, such that μ satisfies the following (τ)-log-Sobolev inequality: for all locally Lipschitz functions $f : \mathcal{X} \to \mathbb{R}$,

$$(\tau) - \mathbf{LogSob}_\alpha(C_2, \lambda) \qquad \mathrm{Ent}_\mu(e^f) \leq C_2 \int (f - Q_\alpha^\lambda f) e^f \, d\mu,$$

where $Q_\alpha^\lambda f(x) := \inf_{y \in \mathcal{X}} \{f(y) + \lambda\alpha(d(x,y))\}$, $x \in \mathcal{X}$.

(3) *There exist* $C_3 > 0$ *and* $\lambda > 0$,*such that* μ *satisfies the following restricted modified logarithmic Sobolev inequality: for all* $K - \alpha$*-convex functions* $f : \mathcal{X} \to \mathbb{R}$, *with* $0 \leq K < \lambda$

$$\mathbf{r} - \mathbf{LogSob}_\alpha(C_3, \lambda) \qquad \mathrm{Ent}_\mu(e^f) \leq C_3 \int \alpha^*(|\nabla^+ f|) e^f \, d\mu.$$

The logarithmic Sobolev inequality at point (2) is called $(\tau) - \mathbf{LogSob}_\alpha(C_2, \lambda)$, as a reference to the (τ)-property by Maurey [Mau91] for which the infimum-convolution operator also occurs.

A main application of this characterization is the following perturbation result.

Corollary 3.1 *(Theorem 1.9 [GRS11b]) Let* $(\mathcal{X}, d)$ *be a geodesic space and* $\alpha : \mathbb{R}^+ \to \mathbb{R}^+$ *be a convex function satisfying the* Δ_2*-condition* (8) *and such that* $\alpha(0) = \alpha'(0) = 0$. *Let* $\mu \in \mathcal{P}(\mathcal{X})$ *and* $\tilde{\mu} \in \mathcal{P}(\mathcal{X})$ *with density* e^ϕ *with respect to* μ, $\phi : \mathcal{X} \to \mathbb{R}$. *If* μ *satisfies* $\mathbf{T}_\alpha(C)$, *then* $\tilde{\mu}$ *satisfies* $\mathbf{T}_\alpha\left(8Ce^{\mathrm{Osc}\,\phi}\right)$ *with* $\mathrm{Osc}\,\phi = \sup \phi - \inf \phi$.

This type of perturbation's result has been established by Holley and Stroock [HS87] for usual logarithmic Sobolev inequalities. This corollary follows by applying their arguments to logarithmic Sobolev inequalities restricted to a class of functions.

4 Some Results Around "Barycentric" Costs

In all this part, c is a barycentric cost function,

$$c(x, p) = \theta\left(x - \int y \, dp(y)\right), \qquad x \in \mathbb{R}^n, \, p \in \mathcal{P}_1(\mathbb{R}^n),$$

where $\theta : \mathbb{R}^n \to \mathbb{R}^+$ is a convex function. Most of the results of this section extend to lower semi-continuous convex functions $\theta : \mathbb{R}^n \to [0, +\infty]$.

In that case, the optimal transport cost $T_c(\nu|\mu)$ between μ and ν in $\mathcal{P}(\mathbb{R}^n)$ is denoted by $\overline{\mathcal{T}}_\theta(\nu|\mu)$. The following specific Kantorovich dual expression of $\overline{\mathcal{T}}_\theta(\nu|\mu)$ has been obtained in [GRST14b] (see Theorem 2.11)

$$\overline{\mathcal{T}}_\theta(\nu|\mu) = \sup\left\{\int Q\varphi \, d\mu - \int \varphi \, d\nu \,;\, \varphi \text{ convex, Lipschitz, bounded from below}\right\}. \tag{15}$$

In the supremum, $Q\varphi$ is the usual infimum-convolution operator,

$$Q\varphi(x) = \inf_{y\in\mathbb{R}^n} \{\varphi(y) + \theta(x-y)\}, \qquad x \in \mathbb{R}^n.$$

Therefore, by a restriction to convex functions, the operator $Q\varphi$ replaces the operator $R_c\varphi$ in the dual formula (13).

Let γ and ν be two probability measures on $\mathbb{R}^n$. By definition, the measure γ is *dominated by* ν *in the convex order*, and we note $\gamma \preceq \nu$, if for all convex functions $f : \mathbb{R}^n \to \mathbb{R}$,

$$\int f d\gamma \leq \int f d\nu.$$

The following Strassen's Theorem provides an alternative definition of the convex order.

Theorem 4.1 ([Str65]) *Let γ and ν be two probability measures on $\mathbb{R}^n$, then $\gamma \preceq \nu$ if and only if there exists a martingale (X, Y) for which X has law γ and Y has law ν. Namely, if π is the law of the couple (X, Y), with decomposition $d\pi(x, y) = d\gamma(x)dp_x(y)$, where p is a Markov Kernel such that $\gamma p = \nu$, then one has for γ-almost every x,*

$$\int y\, dp_x(y) = x.$$

Idea of the proof A simple proof follows from the dual Kantorovich expression of the optimal barycentric cost:

$$\begin{aligned}\overline{\mathcal{T}}_1(\nu|\gamma) &= \inf_{\pi\in\Pi(\mu,\nu)} \int \left| x - \int y\, dp_x(y) \right| d\gamma(x) \\ &= \sup \left\{ \int f\, d\gamma - \int f\, d\nu\, ; f \text{ convex, 1-lipschitz, lower bounded} \right\},\end{aligned}$$

(see Proposition 3.2. [GRST14b]). Therefore, if $d\pi^*(x, y) = d\gamma(x)dp_x^*(y)$ is the law of (X, Y), with marginals γ and ν, then

$$0 \leq \overline{\mathcal{T}}_1(\nu|\gamma) \leq \int \left| x - \int y\, dp_x^*(y) \right| d\gamma(x) = 0.$$

It follows that $\overline{\mathcal{T}}_1(\nu|\gamma) = 0$ and the dual expression of $\overline{\mathcal{T}}_1(\nu|\gamma)$ gives

$$\int f d\gamma \leq \int f d\nu,$$

for every convex, 1-Lipschitz, lower bounded function f. This inequality extends to any lower bounded convex function and then to any convex function by monotone convergence, which means that $\nu \preceq \gamma$. One way to prove this is to use the fact that if f is convex lower-bounded then the classical infimum convolution operator $Q_t f(x) := \inf_{y\in\mathbb{R}^n}\left\{f(y) + \frac{1}{t}|x-y|\right\}$ is convex $1/t$-Lipschitz and $Q_t f(x)$ is increasing to $f(x)$ as t goes to 0 for all $x \in \mathbb{R}^n$. □

Let $\mu \in \mathcal{P}(\mathbb{R}^n)$ be such that for any $\gamma \in \mathcal{P}(\mathbb{R}^n)$, there exists an optimal transport map $S^* : \mathbb{R}^n \to \mathbb{R}^n$ such that $S^*\#\mu = \gamma$ and

$$\mathcal{T}_\theta(\gamma,\mu) = \inf_{\pi\in\Pi(\mu,\nu)} \int \theta(x-y)d\pi(x,y) = \int \theta(x - S^*(x))d\mu(x).$$

This assumption is satisfied, for example, when μ is absolutely continuous with respect to the Lebesgue measure and θ is smooth and strictly convex (see, e.g., [Vil09], Theorem 9.4, Theorem 10.28). Then the optimal barycentric transport cost $\overline{\mathcal{T}}_\theta$ can be expressed in terms of the usual transport cost $\mathcal{T}_\theta$ as follows.

Proposition 4.1 *Under the above conditions on the probability measure* $\mu \in \mathcal{P}(\mathbb{R}^n)$*, for any probability measure* ν *such that* $\overline{\mathcal{T}}_\theta(\nu|\mu) < \infty$,

$$\overline{\mathcal{T}}_\theta(\nu|\mu) = \inf_{\gamma\in\mathcal{P}(\mathbb{R}^n),\gamma\preceq\nu} \mathcal{T}_\theta(\gamma,\mu).$$

Proof Let $\gamma \preceq \nu$. From the previous Strassen's Theorem, there exists a kernel p^* such that $\gamma p^* = \nu$ and $x = \int y\,dp^*_x(y)$ γ-almost surely. Furthermore, by hypotheses, there exists a transport map $S^* : \mathbb{R}^n \to \mathbb{R}^n$ such that $S^*\#\mu = \gamma$ and

$$\mathcal{T}_\theta(\gamma,\mu) = \int \theta(x - S^*(x))d\mu(x).$$

Let us consider the kernel defined by $p_x(dy) = p^*_{S^*(x)}(dy)$. We may simply check that $\mu p = \nu$ and for μ-almost every x, $\int y\,dp_x(y) = \int y\,dp^*_{S^*(x)}(y) = S^*(x)$. Therefore, $\mathcal{T}_\theta(\gamma,\mu) = \int \theta\left(x - \int y\,dp_x(y)\right) d\mu(x) \geq \overline{\mathcal{T}}_\theta(\nu|\mu)$, and by optimizing over all probability measures γ, with $\gamma \preceq \nu$, we get

$$\inf_{\gamma\in\mathcal{P}(\mathbb{R}^n),\gamma\preceq\nu} \mathcal{T}_\theta(\gamma,\mu) \geq \overline{\mathcal{T}}_\theta(\nu|\mu).$$

To prove the reverse inequality, let us consider a kernel p such that $\mu p = \nu$ and

$$\int \theta\left(x - \int y\,dp_x(y)\right) d\mu(x) < \infty.$$

Let $S : \mathbb{R}^n \to \mathbb{R}^n$ be the measurable map defined by $S(x) = \displaystyle\int y\,dp_x(y)$, for μ-almost every x. Let γ be the push forward measure of μ by the map S, $\gamma = S\#\mu$. Then, one has $\gamma \preceq \nu$, since by Jensen's inequality, for all convex functions $f : \mathbb{R}^n \to \mathbb{R}$,

$$\int f\, d\gamma = \int f\left(\int y\, dp_x(y)\right) d\mu(x) \leq \iint f(y)\, dp_x(y) d\mu(x) = \int f\, d\nu;$$

and moreover

$$\int \theta\left(x - \int y\, dp_x(y)\right) d\mu(x) = \int (x - S(x))\, d\mu(x) \geq \mathcal{T}_\theta(\gamma, \mu) \geq \inf_{\gamma \in \mathcal{P}(\mathbb{R}^n), \gamma \preceq \nu} \mathcal{T}_\theta(\gamma, \mu).$$

We get the expected inequality by optimizing this inequality over all kernels p such that $\mu p = \nu$. □

4.1 Barycentric Transport Inequality and Logarithmic Sobolev Inequalities

As for the usual transport costs, connections have been established between barycentric transport inequalities and logarithmic Sobolev inequalities restricted to a class of functions (see [GRST14b, AS15]). To simplify, in this section we only consider the case $\theta(h) = \|h\|^2$ where $\|\cdot\|$ is a fixed norm on $\mathbb{R}^n$. In that case we note $\overline{\mathcal{T}}_\theta = \overline{\mathcal{T}}_2$.

The next results have been obtained by Gozlan et al. [GRST14b] thanks to the Kantorovich dual expression (15) of $\overline{\mathcal{T}}_2$, and by applying the technics linked to the Hamilton-Jacobi equation satisfied by the semi-group $Q_t\varphi$,

$$Q_t\varphi(x) = \inf_{y \in \mathbb{R}^n} \left\{ \varphi(y) + \frac{1}{t}\|x - y\|^2 \right\}, x \in \mathbb{R}^n.$$

From the non-symmetry of the optimal transport cost $\overline{\mathcal{T}}_2(\nu|\mu)$, they establish two different results one corresponding to the transport inequality $\mathbf{T}_c^+(C)$ and the other associated to $\mathbf{T}_c^-(C)$.

Theorem 4.2 see Theorem 8.15 [GRST14b] *Let $\mu \in \mathcal{P}_1(\mathbb{R}^n)$. The following properties are equivalent.*

(1) There exists $C_1 > 0$ such that μ satisfies

$$\overline{\mathcal{T}}_2(\nu|\mu) \leq C_1 H(\nu|\mu), \qquad \forall \nu \in \mathcal{P}_1(\mathbb{R}^n).$$

(2) There exists $C_2 > 0$ such that for all convex Lipschitz functions $\varphi : \mathbb{R}^n \to \mathbb{R}$ bounded from below,

$$\int e^{-\varphi/C_2} d\mu \leq e^{-\int Q_1\varphi/C_2\, d\mu}.$$

(3) *There exist* $C_3 > 0$ *and* $\lambda > 0$, *such that for all concave Lipschitz functions* $\psi : \mathbb{R}^n \to \mathbb{R}$, *bounded from below and* $\lambda \|\cdot\|^2$*-convex,*

$$\mathrm{Ent}_\mu(e^\psi) \leq C_3 \int \|\nabla \psi\|_*^2 e^\psi \, d\mu,$$

where $\|\cdot\|_*$ *is the dual norm* $\|\cdot\|$ *on* $\mathbb{R}^n$.

Recall that if $\|\cdot\| = |\cdot|$ is the Euclidean norm, a function ψ is $\lambda\|\cdot\|^2$-convex if and only if its Hessian is bounded from below by $-2\lambda I$ (in the sense of quadratic forms).

Let us observe that in Theorem 4.2, point (1) is equivalent to point (2) with the same constant $C_1 = C_2$.

Theorem 4.3 see Theorem 8.8 [GRST14b] *Let* $\mu \in \mathcal{P}_1(\mathbb{R}^n)$. *The following properties are equivalent.*

(1) *There exists* $C_1 > 0$ *such that* μ *satisfies*

$$\overline{\mathcal{T}}_2(\mu|\nu) \leq C_1 H(\nu|\mu), \qquad \forall \nu \in \mathcal{P}_1(\mathbb{R}^n).$$

(2) *There exists* $C_2 > 0$ *such that for all Lipschitz convex functions* $\varphi : \mathbb{R}^n \to \mathbb{R}$, *bounded from below,*

$$\int e^{Q_1\varphi/C_2} d\mu < e^{\int \varphi/C_2 \, d\mu}.$$

(3) *There exists* $C_3 > 0$ *such that for all Lipschitz convex functions* $\varphi : \mathbb{R}^n \to \mathbb{R}$, *bounded from below,*

$$\mathrm{Ent}_\mu(e^\varphi) \leq C_3 \int \|\nabla \varphi\|_*^2 e^\varphi \, d\mu,$$

where $\|\cdot\|_*$ *is the dual norm of* $\|\cdot\|$ *on* $\mathbb{R}^n$.

(4) *There exists* $C_4 > 0$ *such that for all Lipschitz convex functions* $\varphi : \mathbb{R}^n \to \mathbb{R}$, *bounded from below,*

$$\left\| e^{Q_t\varphi} \right\|_{a+t/(2C_4),(\mu)} \leq \|e^\varphi\|_{a,(\mu)}, \qquad \forall t > 0,$$

where for all $h : \mathbb{R}^n \to \mathbb{R}$, $p \geq 0$, $\|h\|_{p,(\mu)} = \left(\int |h|^p d\mu\right)^{1/p}$.

Let us observe that in this theorem, point (1) is equivalent to point (2) with the same constant $C_1 = C_2$, and point (3) is equivalent to point (4) with the same constant $C_3 = C_4$. The other links between the constants in the two last theorems are given in [GRST14b].

The logarithmic Sobolev inequality restricted to the class of convex functions of point (3) has been investigated in [Ada05, AS15], where sufficient conditions on the probability measure μ are given for such an inequality to hold.

4.2 Barycentric Transport Inequalities for the Binomial Law and the Poisson Measure

Discrete probability measures do not generally satisfy the Talagrand's transport inequality $\mathbf{T}_2$. To be convinced, it suffices to consider μ_ρ, a convex combination of two Dirac measures at two distinct points a and b, $\mu_\rho = \rho\delta_a + (1-\rho)\delta_b$, $\rho \in (0,1)$. The measure μ_ρ satisfies $\mathbf{T}_2(C)$, $C > 0$ if and only if for all $q \in [0,1]$

$$W_2^2(\mu_\rho, \mu_q) = \mathcal{T}_2(\mu_\rho, \mu_q) = |a-b|^2|q-\rho| \leq CH(\mu_q|\mu_\rho)$$
$$= q\log\frac{q}{\rho} + (1-q)\log\frac{1-q}{1-\rho}.$$

We get a contradiction as q goes to ρ by observing that $H(\mu_q|\mu_\rho) = o(|q-\rho|)$.
However the measure μ_ρ satisfies the transport inequality

$$W_1^2(\mu_\rho, \nu) \leq \frac{d(a,b)^2}{2}H(\nu,\mu_\rho), \qquad \forall \nu << \mu_\rho.$$

To get other transport inequalities, one strategy is to replace the usual transport cost by a barycentric cost. To simplify, let μ_ρ be the Bernoulli measure of parameter $\rho \in (0,1)$ ($a = 0$ and $b = 1$). In [Sam03] and [GRST14b] (see Proposition 7.1), the barycentric transport inequality $\mathbf{T}_c(1/(1-t), 1/t)$, $t \in (0,1)$ with cost

$$c(x,p) = \theta_{\rho,t}\left(x - \int y\, dp(y)\right), \qquad x \in \mathbb{R}, \quad p \in \mathcal{P}(\mathbb{R}),$$

is established for the Bernoulli measure μ_ρ, with an optimal cost function $\theta_{\rho,t}$: one has

$$\overline{\mathcal{T}}_{\theta_{\rho,t}}(\nu_1|\nu_2) \leq \frac{1}{1-t}H(\nu_1|\mu_\rho) + \frac{1}{t}H(\nu_2|\mu_\rho), \qquad \forall \nu_1, \nu_2 << \mu_\rho.$$

By tensorization, it provides a barycentric transport inequality for the product measure $\mu_\rho^n = \mu_\rho \otimes \cdots \otimes \mu_\rho$ associated to the cost

$$c^n(x,p) = c(x_1,p_1) + \cdots + c(x_n,p_n), \qquad x = (x_1,\ldots,x_n) \in \mathbb{R}^n, p \in \mathcal{P}_1(\mathbb{R}^n).$$

By projection, and observing that by convexity

$$c^n(x,p) \geq n\, c\left(\frac{\sum_i^n x_i}{n}, \frac{\sum_i^n p_i}{n}\right),$$

a barycentric transport inequality follows for the binomial law $\mu_{n,\rho}$ with parameters n and ρ (see [GRST14b], Corollary 7.7),

$$\overline{\mathcal{T}}_{\theta_{\rho,t,n}}(\nu_1|\nu_2) \leq \frac{1}{1-t}H(\nu_1|\mu_{n,\rho}) + \frac{1}{t}H(\nu_2|\mu_{n,\rho}), \qquad \forall \nu_1, \nu_2 << \mu_{n,\rho},$$

where $\theta_{\rho,t,n}(h) = n\theta_{\rho,t}(h/n)$, $h \in \mathbb{R}$. Finally, by the weak convergence of the measure μ_{n,ρ_n} towards the Poisson measure p_λ with parameter $\lambda > 0$ when $\rho_n = \lambda/n$, an optimal barycentric transport inequality is obtained for the Poisson measure

$$\overline{\mathcal{T}}_{c_{\lambda,t}}(\nu_1|\nu_2) \leq \frac{1}{1-t}H(\nu_1|p_\lambda) + \frac{1}{t}H(\nu_2|p_\lambda), \qquad \forall \nu_1, \nu_2 << p_\lambda,$$

with $c_{\lambda,t}(h) = \lim_{n\to\infty} n\theta_{\rho_n,t}(h/n)$, $h \in \mathbb{R}$. One specific feature of the cost function $c_{\lambda,t}$ is to be zero for $h \geq 0$. For more details, we refer to Proposition 7.11 [GRST14b].

One other famous strategy to establish transport inequalities in discrete setting is coming from the notion of curvature on discrete spaces introduced by Maas [Maa11]. Transport inequalities for invariant reversible measures of Markov chains are obtained from curvature type conditions (see also [EM12, EM14, EMT15]). The optimal transport cost is defined by an abstract Benamou-Brenier type formula which is not associated to a transport cost function. This optimal cost is not comparable to a barycentric cost.

In an other direction, transport inequalities for Poisson processes of different types are proposed by Ma et al. in [MSWW11].

4.3 *Optimal Transport Coupling for Barycentric Costs on* $\mathbb{R}$

This part concerns the construction of an optimal coupling π^* that optimizes the optimal barycentric cost on the real line (in dimension one).

The cost function $\theta : \mathbb{R} \to \mathbb{R}$ is assumed to be even (we believe that this condition can be removed). For any probability measure μ, we denote by F_μ its cumulative distribution function, $F_\mu(x) = \mu(-\infty, x]$, and by F_μ^{-1} its general inverse

$$F_\mu^{-1}(u) = \inf\{x \in \mathbb{R}, F_\mu(x) \geq u\}, \quad u \in [0,1].$$

Let μ and γ be two probability measures on $\mathbb{R}$. Assume that μ has no atoms, then it is well known that for all convex cost functions θ, the usual optimal transport cost $T_\theta(\mu, \gamma)$ is reached for the optimal deterministic coupling measure

$$d\pi^*(x,y) = d\mu(x)d\delta_{S^*(x)}(y),$$

where S^* is the monotone transport map defined by $S^*(x) = F_\gamma^{-1} \circ F_\mu(x)$, $x \in \mathbb{R}$. In other words, there exists a monotone transport map S^*, independent of θ and such that

$$\mathcal{T}_\theta(\mu,\gamma) = \int \theta(x - S^*(x))\, d\mu(x).$$

We want the same kind of result of independence of the function θ for an optimal coupling of the barycentric cost $\overline{\mathcal{T}}_\theta(\nu|\mu)$. For that purpose, we will use the following preliminary result of [GRS$^+$15] (Theorem 1.3).

Theorem 4.4 *Let $\mu,\nu \in \mathcal{P}_1(\mathbb{R})$. There exists $\hat{\gamma} \in \mathcal{P}_1(\mathbb{R})$ such that $\hat{\gamma} \preceq \nu$ and for any even convex function θ, one has*

$$\overline{\mathcal{T}}_\theta(\nu|\mu) = \mathcal{T}_\theta(\hat{\gamma},\mu).$$

This result is still available when μ has atoms and it seems that the even condition on θ can be removed.

Based on the facts set out above, if μ has no atoms,

$$\overline{\mathcal{T}}_\theta(\nu|\mu) = \mathcal{T}_\theta(\hat{\gamma},\mu) = \int \theta(x - S^*(x))\, d\mu(x), \quad \text{with } S^* = F_{\hat{\gamma}}^{-1} \circ F_\mu,$$

and since $\hat{\gamma} \preceq \nu$, according to Strassen's Theorem 4.1, there exists a kernel p^* such that $\hat{\gamma}p^* = \nu$ and $\hat{\gamma}$-almost surely $\int y\, dp_x^*(y) = x$. Since $\hat{\gamma}$ is independent of θ, le kernel p^* is also independent of θ. Moreover, since $S^*\#\mu = \hat{\gamma}$, we get for μ-almost every x,

$$\int y\, dp^*_{S^*(x)}(y) = S^*(x),$$

and it finally gives

$$\overline{\mathcal{T}}_\theta(\nu|\mu) = \int \theta\left(x - \int y\, dp^*_{S^*(x)}(y)\right) d\mu(x).$$

This shows that if μ has no atoms, then the optimal barycentric cost $\overline{\mathcal{T}}_\theta(\nu|\mu)$ is reached for the optimal coupling

$$\pi^*(dx,dy) = \mu(dx)p^*_{S^*(x)}(dy),$$

which is independent of the even convex function θ.

4.4 *Characterization of Probability Measures Satisfying a Barycentric Transport Inequality on* $\mathbb{R}$

We know how to characterize the probability measures on $\mathbb{R}$ satisfying different functional inequalities as the Poincaré inequality [Muc72], the logarithmic Sobolev inequality [BG99, BR03] (see also chapter 6, [ABC+00]), the usual transport inequalities [Goz12]. In each of these cases, the characterization can be given by criteria of Hardy type, on the tails of distribution and on the densities of the involved measures. This section concerns the characterization of the barycentric transport inequalities. The approach is the one introduced by Gozlan [Goz12].

Let τ be the exponential law on $\mathbb{R}$, with density $e^{-|x|}/2$. For any $\mu \in \mathcal{P}(\mathbb{R})$, let us note U_μ the unique left-continuous monotone transport map from the measure τ to the measure μ, $U_\mu = F_\mu^{-1} \circ F_\tau$, namely,

$$U_\mu(x) = \begin{cases} F_\mu^{-1}\left(1 - \frac{1}{2}e^{-|x|}\right) & \text{if } x \geq 0, \\ F_\mu^{-1}\left(\frac{1}{2}e^{-|x|}\right) & \text{if } x \leq 0. \end{cases}$$

Here is one of the main results of [Goz12]: a probability measure μ satisfies the transport inequality $\mathbf{T}_2(C)$ with $C > 0$ if and only if it satisfies the Poincaré inequality and the following condition: there exists $b \geq 0$ such that

$$\sup_{x \in \mathbb{R}} \big(U_\mu(x+u) - U_\mu(x)\big) \leq b\sqrt{1+u}, \qquad \forall u \geq 0,$$

that enforces a particular behavior of the tails of distribution of the measure μ.

In [GRS+15], an analogous result is obtained for the barycentric transport inequality with costs $\overline{\mathcal{T}}_\theta$. Let us note $\theta^a(t) = \theta(at)$, $t \in \mathbb{R}$, for any $a > 0$.

Theorem 4.5 (Theorem 1.2, [GRS+15]) *Let θ be an even convex function such that $\theta(t) = t^2$ for all $|t| \leq t_0$, $t_0 > 0$. A probability measure μ satisfies the barycentric transport cost inequalities*

$$\overline{\mathbf{T}}^-_{\theta^a} : \qquad \overline{\mathcal{T}}_{\theta^a}(\mu|\nu) \leq H(\nu|\mu), \qquad \forall \nu \in \mathcal{P}_1(\mathbb{R}),$$

and

$$\overline{\mathbf{T}}^+_{\theta^a} : \qquad \overline{\mathcal{T}}_{\theta^a}(\nu|\mu) \leq H(\nu|\mu), \qquad \forall \nu \in \mathcal{P}_1(\mathbb{R}),$$

for some positive constant a if and only if there exists $b \geq 0$ such that

$$\sup_x \big(U_\mu(x+u) - U_\mu(x)\big) \leq b\,\theta^{-1}\left(u + t_o^2\right) \qquad \forall u \geq 0. \tag{16}$$

Probability measures satisfying a barycentric transport inequality do not necessarily verify a Poincaré inequality since there support is not necessarily connected (for example, the Bernoulli and the binomial laws as explained in Section 4.2). However, the condition (16) of the above theorem implies for $u = 1$: there exists $h > 0$ such that

$$\sup_{x\in\mathbb{R}} \big(U_\mu(x+1) - U_\mu(x)\big) \leq h. \tag{17}$$

Bobkov and Götze [BG99] have proved that this condition is equivalent to the existence of a Poincaré inequality restricted to convex functions satisfied by the measure μ. Therefore, probability measures satisfying (16) necessarily satisfy a so-called convex Poincaré inequality. More precisely, the following result has been established by Feldheim et al. [FMNW15], and also independently by Gozlan et al. [GRS+15], as an intermediate key result of their proof of Theorem 4.5.

Theorem 4.6 *Let μ be a probability measure on $\mathbb{R}$. The condition* (17) *is equivalent to each of the following properties.*

(a) There exists $C > 0$ such that for all convex functions f on $\mathbb{R}$,

$$\mathrm{Var}_\mu(f) \leq C \int_{\mathbb{R}} f'^2 \, d\mu.$$

(b) There exist $a, t_0 > 0$ such that

$$\overline{\mathbf{T}}^{-}_{\theta_1^a} : \qquad \overline{\mathcal{T}}_{\theta_1^a}(\mu|\nu) \leq H(\nu|\mu), \qquad \forall \nu \in \mathcal{P}_1(\mathbb{R}),$$

and

$$\overline{\mathbf{T}}^{+}_{\theta_1^a} : \qquad \overline{\mathcal{T}}_{\theta_1^a}(\nu|\mu) \leq H(\nu|\mu), \qquad \forall \nu \in \mathcal{P}_1(\mathbb{R}),$$

where the function θ_1 is defined by $\theta_1(t) = \begin{cases} t^2 & \text{if } |t| \leq t_0, \\ 2|t|t_0 - t_0^2 & \text{if } |t| > t_0. \end{cases}$

5 Universal Transport Inequalities

We call universal any transport inequality that holds for any (reference) probability measure μ on $\mathcal{X}$.

The most popular and commonly used universal transport inequality, mentioned in Section 3.1, is the Csizár-Kullback-Pinsker inequality [Csi67, Kul67, Pin64],

$$\frac{1}{2}\|\mu - \nu\|_{TV}^2 \leq H(\nu|\mu), \qquad \forall \mu, \nu \in \mathcal{P}(\mathcal{X}), \tag{18}$$

where $\|\mu - \nu\|_{TV}$ is the total variation distance between μ and ν,

$$\|\mu - \nu\|_{TV} = 2 \inf_{\pi \in \Pi(\mu,\nu)} \iint \mathbb{1}_{x \neq y} \, d\pi(x, y).$$

The functional dual formulation of the Csizár-Kullback-Pinsker inequality is the following exponential inequality, for any function $f : \mathcal{X} \to \mathbb{R}$ such that $\sup_{x,y \in \mathcal{X}} |f(x) - f(y)| \leq c$,

$$\int e^{tf} \, d\mu \leq e^{t \int f \, d\mu + t^2 c^2/8}, \qquad t \geq 0.$$

This inequality, commonly used, gives the Hoeffding inequality by applying Markov's inequality,

$$\mu \left(f \geq \int f \, d\mu + t \right) \leq e^{-2t^2/c^2} \qquad t \geq 0.$$

The Csizár-Kullback-Pinsker inequality (18) can be improved in different ways. We may change the function of the total variation distance on the left-hand side (see [FHT03, Gil10]), or we may replace the total variation distance by a comparable optimal weak transport cost of Definition 2.4. More precisely, given a convex function $\alpha : \mathbb{R}^+ \to [0, +\infty]$, and $\mu, \nu_1, \nu_2 \in \mathcal{P}(\mathcal{X})$, we note

$$\widetilde{\mathcal{T}}_\alpha(\nu_1|\nu_2) = \inf_{\pi \in \Pi(\nu_2,\nu_1)} \left\{ \int c(x, p_x) d\nu_2(x), d\pi(x, y) = d\nu_2(x) dp_x(y) \right\},$$

when $c(x, p) = \alpha \left(\int \mathbb{1}_{x \neq y} dp(y) \right)$, $x \in \mathcal{X}, p \in \mathcal{P}(\mathcal{X})$, and

$$\widehat{\mathcal{T}}_\alpha(\nu_1|\nu_2) = \inf_{\pi \in \Pi(\nu_2,\nu_1)} \left\{ \int c(x, p_x) d\nu_2(x), d\pi(x, y) = d\nu_2(x) dp_x(y) \right\},$$

when

$$c(x, p) = \int \alpha \left(\mathbb{1}_{x \neq y} \frac{dp}{d\mu}(y) \right) d\mu(y),$$

if (x, p) is such that p is absolutely continuous with respect to μ on $\mathcal{X} \setminus \{x\}$, and $c(x, p) = +\infty$ otherwise. In [Sam07], Theorem 1.1 and 1.2 give the following variants of the Csizár-Kullback-Pinsker inequality.

Theorem 5.1 ([Sam07, GRST14b]) *Let $\mathcal{X}$ be a compact metric space, $\mu \in \mathcal{P}(\mathcal{X})$ and $t \in (0, 1)$.*

(a) For any probability measures ν_1 and ν_2 on $\mathcal{X}$, one has

$$\widetilde{\mathcal{T}}_{\alpha_t}(\nu_1|\nu_2) \leq \frac{1}{1-t} H(\nu_1|\mu) + \frac{1}{t} H(\nu_2|\mu),$$

where the convex cost function $\alpha_t : \mathbb{R}^+ \to [0, +\infty]$ *is defined by*

$$\alpha_t(u) = \frac{t(1-u)\log(1-u) - (1-tu)\log(1-tu)}{t(1-t)}, \qquad 0 \le u \le 1,$$

and $\alpha_t(u) = +\infty$ *if* $u > 1$.
As t goes to 0, it implies

$$\widetilde{\mathcal{T}}_{\alpha_0}(\nu_1|\mu) \le H(\nu_1|\mu),$$

with $\alpha_0(u) = (1-u)\log(1-u) + u$ *if* $0 \le u \le 1$, *and* $\alpha_0(u) = +\infty$ *if* $u > 1$.
As t goes to 1, it implies

$$\widetilde{\mathcal{T}}_{\alpha_1}(\mu|\nu_2) \le H(\nu_2|\mu),$$

with $\alpha_1(u) = -\log(1-u) - u$, *if* $0 \le u < 1$ *and* $\alpha_0(u) = +\infty$ *if* $u \ge 1$.

(b) For any probability measures ν_1 *and* ν_2 *on* $\mathcal{X}$, *one has*

$$\widehat{\mathcal{T}}_{\beta_t}(\nu_1|\nu_2) \le \frac{1}{1-t}H(\nu_1|\mu) + \frac{1}{t}H(\nu_2|\mu),$$

where the convex cost function $\beta_t : \mathbb{R}^+ \to [0, +\infty]$ *is defined by*

$$\beta_t(u) := \sup_{s \in \mathbb{R}^+} \left\{ su - \beta_t^*(s) \right\}, \qquad u \in \mathbb{R}^+,$$

with

$$\beta_t^*(s) = \frac{te^{(1-t)s} + (1-t)e^{-ts} - 1}{t(1-t)}, \qquad s \in \mathbb{R}^+.$$

When t goes to 0 this implies

$$\widehat{\mathcal{T}}_{\beta_0}(\nu_1|\mu) \le H(\nu_1|\mu), \tag{19}$$

with $\beta_0(u) = (1+u)\log(1+u) - u$, $u \ge 0$, *and when t goes to 1, it implies*

$$\dot{\mathcal{T}}_{\beta_1}(\mu|\nu_2) \le H(\nu_2|\mu),$$

with $\beta_1(u) = (1-u)\log(1-u) + u$, *if* $u \le 1$ *and* $\beta_1(u) = +\infty$ *if* $u > 1$.

By using the estimate , $\alpha_t(u) \ge u^2/2 = \alpha(u)$, for all $t \in [0, 1]$, $u \ge 0$, the transport inequalities of point (a) provide the Marton's transport inequalities [Mar96b] associated to the quadratic cost function α:

$$\widetilde{\mathcal{T}}_2(\nu_1|\mu) \le 2H(\nu_1|\mu), \qquad \widetilde{\mathcal{T}}_2(\mu|\nu_2) \le 2H(\nu_2|\mu), \tag{20}$$

or even for every $t \in (0, 1)$,

$$\frac{1}{2}\widetilde{\mathcal{T}}_2(\nu_1|\nu_2) \leq \frac{1}{1-t}H(\nu_1|\mu) + \frac{1}{t}H(\nu_2|\mu).$$

By optimizing in t, this inequality can be rewritten

$$\frac{1}{2}\widetilde{\mathcal{T}}_2(\nu_1|\nu_2) \leq \left(\sqrt{H(\nu_1|\mu)} + \sqrt{H(\nu_2|\mu)}\right)^2.$$

As explained in Section 3.3, these transport inequalities can be tensorised on product spaces, and provide concentration results for product measures, or even weakly dependent measures. The concentration principle related to this kind of n-dimensional costs has been introduced by Talagrand [Tal96a, Tal96c], especially to prove deviations inequalities for suprema of empirical processes of Bernstein type.

The optimal deviation results for suprema of empirical processes that follow from Theorem 5.1 are briefly recalled in Section 5.1 below. Then in Section 5.2, we summarize results obtained in a weak dependence framework concerning the Marton's transport inequalities (20) in [Sam00] and [Pau14]. Finally, in Section 5.3, we suggest a different way to tensorize the Csizár-Kullback-Pinsker inequality or the Marton's inequality. It provides new weak transport inequalities for the uniform law on the symmetric group (see [Sam16]). These results are guided by the concentration results by Talagrand [Tal95].

5.1 *Bernstein's Type of Deviation Inequalities for Suprema of Independent Empirical Processes*

The first Bernstein's type of deviation inequalities for suprema of independent empirical processes have been obtained by Talagrand [Tal96b, Tal96a] with the so-called convex hull method. These inequalities are of particular interest in statistics [Mas00b, Mas07].

Ledoux [Led97] has proposed an "entropic" method that allows to simply recover the results by Talagrand. This approach is based on the tensorization property of the entropy and the so-called Herbst's argument. Then, it has been widely developed, mainly to reach optimal deviation bounds for the suprema of independent empirical processes [Mas00a, BLM00, Rio01, BLM03, Rio02, Bou03, KR05, Rio12, Rio13, BLM13].

In the continuation of the works by Talagrand, Marton [Mar96a, Mar96b], Dembo [Dem97] and Maurey [Mau91], the transport-entropy method has been developed in [Sam07] as an alternative approach to achieve the best constants in the deviation inequalities of suprema of empirical processes.

Another approach has been proposed by Panchenko, based on symmetrization technics [Pan01, Pan02, Pan03]. Finally, to complete the picture, Stein's methods have been pushed forward by Chatterjee to reach similar concentration properties to the one by Talagrand [Cha05, Cha07, CD10, Pau14]. The main interest of this last method is that it extends to dependence cases, under Dobrushin type of conditions.

Let us present some concentration results for suprema of empirical processes that follow from the transport inequalities of Theorem 5.1 after tensorization. Let $\mathcal{F}$ be a countable set and let $(X_{1,t})_{t\in\mathcal{F}}, \ldots, (X_{n,t})_{t\in\mathcal{F}}$ be n independent processes. We are interested by the deviations of the random variable

$$Z = \sup_{t\in\mathcal{F}} \sum_{i=1}^{n} X_{i,t}.$$

Let us note

$$V = \sup_{t\in\mathcal{F}} \sum_{i=1}^{n} \mathbb{E}\left[[X_{i,t} - X'_{i,t}]_+^2 \,\middle|\, X_{i,t} \right],$$

where $(X'_{i,t})_{t\in\mathcal{F}}$ is an independent copy of $X_i = (X_{i,t})_{t\in\mathcal{F}}$ and $\mathbb{E}[\cdot\,|X_{i,t}]$ denotes the conditional expectation, given $X_{i,t}$. In the following theorem, for all $t \in \mathcal{F}$ and all $i \in \{1,\ldots,n\}$, $M_{i,t}$ and $m_{i,t}$ are numerical constants limiting the random variables $X_{i,t}$.

Theorem 5.2 *(a) Assume that $X_{i,t} \le M_{i,t}$, and $\mathbb{E}\left[(M_{i,t} - X_{i,t})^2\right] \le 1$, for all i and all t, then for all $u \ge 0$,*

$$\mathbb{P}(Z \ge \mathbb{E}[Z] + u) \le \exp\left[-\frac{u}{2\left(1+\varepsilon\left(\frac{u}{\mathbb{E}[V]}\right)\right)} \log\left(1 + \frac{u}{\mathbb{E}[V]}\right)\right]$$
$$\le \exp\left[-\frac{u^2}{2\mathbb{E}[V] + 2u}\right],$$

with $\varepsilon(u) = \frac{\beta_0(u)}{(1+u)\log(1+u)}$ and $\beta_0(u) := (1+u)\log(1+u) - u$.

(b) Assume that $m_{i,t} \le X_{i,t} \le M_{i,t}$, with $M_{i,t} - m_{i,t} = 1$ for all i and all t, then for all $u \ge 0$,

$$\mathbb{P}(Z \le \mathbb{E}[Z] - u) \le \exp\left[-\mathbb{E}[V]\beta_0\left(\frac{u}{\mathbb{E}[V]}\right)\right] \le \exp\left[-\frac{u^2}{2\mathbb{E}[V] + \frac{2}{3}u}\right],$$

with $\beta_0(u) = (1+u)\log(1+u) - u$.

The optimality of these results is discussed in [Sam07]. Recall that by usual symmetrization's technics ([LT91], Lemma 6.3 and Theorem 4.12), the variance term $\mathbb{E}[V]$ can be estimated as follows:

$$\mathbb{E}[V] \leq \sup_{t\in\mathcal{F}} \sum_{i=1}^{n} \mathrm{Var}(X_{i,t}) + 16\,\mathbb{E}\left[\sup_{t\in\mathcal{F}} \left|\sum_{i=1}^{n} (X_{i,t} - \mathbb{E}[X_{i,t}])\right|\right].$$

In [Ada08], by using Hoffman-Jørgensen's inequality and some other results by Talagrand, Adamczak extends the concentration properties to suprema of unbounded random variables, by truncation arguments of the random variables.

Idea of the proof We only present one elementary proof of (b) to show the links between the transport inequality with cost $\widehat{\mathcal{T}}_{\beta_t}$ and the deviations of a function around its mean.

Let μ_i denote the law of the process $X_i = (X_{i,t})_{t\in\mathcal{F}}$. Point (b) simply follows from the dual form of the tensorized transport inequality (19): $\mathcal{X} = \mathbb{R}^{\mathcal{F}}$, for any function $g : \mathcal{X}^n \to \mathbb{R}$,

$$\int e^{-g} d\mu \leq \exp\left(-\int \hat{Q}g\, d\mu\right), \tag{21}$$

where $\mu = \mu_1 \otimes \cdots \otimes \mu_n$ and for all $x = (x_1, \ldots, x_n) \in \mathcal{X}^n$

$$\widehat{Q}g(x) = \inf_{p\in\mathcal{P}(\mathcal{X}^n)} \left\{\int g(y)\, dp(y) + \sum_{i=1}^{n} \int \beta_0\left(\mathbb{1}_{x_i\neq y_i} \frac{dp_i}{d\mu_i}(y_i)\right) d\mu_i(y_i)\right\},$$

the probability measures p_i are the marginals of p. Let us choose $g = \lambda f$ with $\lambda \geq 0$ and

$$f(x) = \sup_{t\in\mathcal{F}} \sum_{i=1}^{n} x_{i,t}, \qquad x = (x_1, \ldots, x_n) \in \mathcal{X}^n.$$

To simplify, we assume that for all $x \in \mathcal{X}^n$, the supremum is reached at a single point $\tau(x) \in \mathcal{F}$:

$$\sup_{t\in\mathcal{F}} \sum_{i=1}^{n} x_{i,t} = \sum_{i=1}^{n} x_{i,\tau(x)}.$$

Then for all $x, y \in \mathcal{X}^n$

$$f(y) \geq f(x) + \sum_{i=1}^{n} (y_{i,\tau(x)} - x_{i,\tau(x)}) = f(x) + \sum_{i=1}^{n} (y_{i,\tau(x)} - x_{i,\tau(x)}) \mathbb{1}_{x_i\neq y_i}.$$

as a consequence, for all x,

$$
\begin{aligned}
&\widehat{Q}g(x)\\
&\geq \lambda f(x) - \sup_p \left\{ \int \sum_{i=1}^n \lambda (x_{i,\tau(x)} - y_{i,\tau(x)}) \mathbb{1}_{x_i \neq y_i} dp(y) \right.\\
&\quad \left. - \sum_{i=1}^n \int \beta_0 \left(\mathbb{1}_{x_i \neq y_i} \frac{dp_i}{d\mu_i}(y_i) \right) d\mu_i(y_i) \right\}\\
&= \lambda f(x) - \sum_{i=1}^n \sup_{p_i \in \mathcal{P}(\mathcal{X})} \left\{ \int \lambda (x_{i,\tau(x)} - y_{i,\tau(x)}) \mathbb{1}_{x_i \neq y_i} \frac{dp_i}{d\mu_i}(y_i) d\mu_i(y_i) \right.\\
&\quad \left. - \int \beta_0 \left(\mathbb{1}_{x_i \neq y_i} \frac{dp_i}{d\mu_i}(y_i) \right) d\mu_i(y_i) \right\}\\
&\geq \lambda f(x) - \sum_{i=1}^n \int \sup_{h \geq 0} \left\{ \lambda (x_{i,\tau(x)} - y_{i,\tau(x)}) h - \beta_0(h) \right\} d\mu_i(y_i)\\
&= \lambda f(x) - \sum_{i=1}^n \int \beta_0^* \left(\lambda [x_{i,\tau(x)} - y_{i,\tau(x)}]_+ \right) d\mu_i(y_i)\\
&\geq \lambda f(x) - \beta_0^*(\lambda) \sum_{i=1}^n \int [x_{i,\tau(x)} - y_{i,\tau(x)}]_+^2 d\mu_i(y_i)\\
&\geq \lambda f(x) - \beta_0^*(\lambda) \sup_{t \in \mathcal{F}} \sum_{i=1}^n \int [x_{i,t} - y_{i,t}]_+^2 d\mu_i(y_i),
\end{aligned}
$$

where $\beta_0^*(s) = e^s - s - 1$, $s \geq 0$. The second last inequality is a consequence of the fact that $[x_{i,\tau(x)} - y_{i,\tau(x)}]_+ \leq M_{i,t} - m_{i,t} \leq 1$ and $\beta_0^*(\lambda u) \leq u^2 \beta_0^*(\lambda)$ for $0 \leq u \leq 1$. By inserting the previous estimate of $\widehat{Q}g(x)$ in the transport inequality (21), we get for all $\lambda \geq 0$,

$$
\mathbb{E}\left[e^{-\lambda Z}\right] \leq e^{-\lambda \mathbb{E}[Z] + \mathbb{E}[V] \beta_0^*(\lambda)}.
$$

The deviation inequality of (b) directly follows by the Markov inequality, optimizing over all $\lambda \geq 0$. □

5.2 *Marton's Transport Inequality for Weakly Dependent Random Variables*

The paper [Sam00] presents a tensorization scheme of the Marton's inequality (20) when μ is a probability measure on the product space $\mathcal{X}^n$, whose marginals are weakly dependent on each other, more precisely if μ is the law the n first random variables $X_1, \ldots, X_n$ of a Φ-mixing process. This tensorization scheme is based on the construction of couplings similar to the one of [Mar03].

To simplify, one may assume that $\mathcal{X}$ is a finite set. Given a sequence of random variables $X_1, \ldots, X_n$ with values in $\mathcal{X}$, for $1 \leq i < j \leq n$, let us note

$$\mathcal{L}(X_j^n | X_1^{i-1} = x_1^{i-1}, X_i = x_i)$$

the law of $(X_j, \ldots, X_n)$ knowing that $X_1 = x_1, \ldots, X_{i-1} = x_{i-1}, X_i = x_i$, and let $\Gamma = (\gamma_{i,j})_{1 \leq i,j \leq n}$ be the upper triangular matrix defined by

$$\gamma_{ij}^2 = \sup_{x_1^{i-1}, x_i, y_i} \left\| \mathcal{L}(X_j^n | X_1^{i-1} = x_1^{i-1}, X_i = x_i) - \mathcal{L}(X_j^n | X_1^{i-1} = x_1^{i-1}, X_i = y_i) \right\|_{TV},$$

for $i < j$ and $\gamma_{ii} = 1$.

Theorem 5.3 ([Sam00]) *According to the previous notations, for all probability measures μ and ν on $\mathcal{X}^n$, one has*

$$\widetilde{\mathcal{T}}_2(\nu|\mu) \leq 2\|\Gamma\|^2 H(\nu|\mu) \quad \textit{and} \quad \widetilde{\mathcal{T}}_2(\mu|\nu) \leq 2\|\Gamma\|^2 H(\nu|\mu),$$

where $\|\Gamma\|$ denotes the operator norm of the matrix Γ from $(\mathbb{R}^n, |\cdot|)$ to $(\mathbb{R}^n, |\cdot|)$.

Note that since the Marton's cost $\widetilde{\mathcal{T}}_2$ in dimension n also satisfies the triangular inequality (10), the two weak transport inequalities of this theorem are equivalent to the following family of transport inequalities, for all $t \in (0, 1)$, for all $\mu, \nu_1, \nu_2 \in \mathcal{P}(\mathcal{X}^n)$,

$$\frac{1}{2\|\Gamma\|^2} \widetilde{\mathcal{T}}_2(\nu_1|\nu_2) \leq \frac{1}{1-t} H(\nu_1|\mu) + \frac{1}{t} H(\nu_2|\mu),$$

or equivalently, applying Theorem 3.1 for all functions $g : \mathcal{X}^n \to \mathbb{R}$ bounded from below,

$$\left(\int e^{t\tilde{Q}g} d\mu \right)^{1/t} \left(\int e^{-(1-t)g} d\mu \right)^{1/(1-t)} \leq 1,$$

where

$$\tilde{Q}g(x) = \inf_{p \in \mathcal{P}(\mathcal{X}^n)} \left\{ \int g dp + \frac{1}{2\|\Gamma\|^2} \sum_{i=1}^n \left(\int \mathbb{1}_{x_i \neq y_i} dp(y) \right)^2 \right\}.$$

In particular, applying this inequality to the function $g = i_A$, with $A \subset \mathcal{X}^n$, we get the Talagrand's concentration result extended to any measure $\mu \in \mathcal{P}(\mathcal{X}^n)$: for all subsets $A \subset X^n$, for all $t \in (0, 1)$,

$$\int e^{\frac{t}{2\|\Gamma\|^2} D_T^2(x,A)} d\mu(x) \leq \frac{1}{\mu(A)^{t/(1-t)}}, \tag{22}$$

where $D_T(x,A)$ is the Talagrand's convex distance defined by

$$\begin{aligned}
D_T^2(x,A) &= \left(\sup_{\alpha\in\mathbb{R}^n, |\alpha|\leq 1} \inf_{y\in A} \sum_{i=1}^n \alpha_i \mathbb{1}_{x_i\neq y_i} \right)^2 \\
&= \left(\sup_{\alpha\in\mathbb{R}^n, |\alpha|\leq 1} \inf_{p\in\mathcal{P}(A)} \sum_{i=1}^n \alpha_i \int \mathbb{1}_{x_i\neq y_i} dp(y) \right)^2 \\
&= \left(\inf_{p\in\mathcal{P}(A)} \sup_{\alpha\in\mathbb{R}^n, |\alpha|\leq 1} \sum_{i=1}^n \alpha_i \int \mathbb{1}_{x_i\neq y_i} dp(y) \right)^2 \\
&= \inf_{p\in\mathcal{P}(A)} \sum_{i=1}^n \left(\int \mathbb{1}_{x_i\neq y_i} dp(y) \right)^2 = \tilde{Q}i_A(x).
\end{aligned}$$

The second equality follows from the linearity of the expression in p and from the fact that Dirac measures are the extremal points of the convex set $\mathcal{P}(A)$. The third equality is a consequence of Sion's minimax Theorem [Sio58, Kom88] since the expression is linear in p and α, and therefore convex in p and concave in α.

When μ is the law of some independent random variables, the property (22) exactly recovers Talagrand's concentration results [Tal95] since $\|\Gamma\| = \|\mathrm{Id}\| = 1$.

Theorem 5.3 complements the results by Marton for contracting Markov chains [Mar96b, Mar97]. More generally, assume that the sequence $(X_k)_{k\geq 1}$ is a Doeblin recurrent Markov chain with kernel K; in other words, there exists a probability measure m, an integer r and a real $\rho \in (0,1)$ such that for all $x \in \mathcal{X}$ and all subsets $A \subset \mathcal{X}$

$$K^r(x,A) \geq \rho m(A).$$

Then the coefficient $\|\Gamma\|$ is bounded independently of n,

$$\|\Gamma\| \leq \frac{\sqrt{2}}{1-\rho^{1/2r}}.$$

In any case, if $(\Phi_k)_{k\geq 1}$ represents the sequence of Φ-mixing coefficients of the sequence of random variables $X_1, \ldots, X_n$ (see [Sam00], [Dou94]), then one has

$$\|\Gamma\| \leq \sum_{k=1}^n \sqrt{\Phi_k}.$$

After this result, many authors have obtained transport inequalities under weak different dependence assumptions, for example under Dobrushin type conditions [DGW04, Mar04, Mar10, Wu06, Kon12, Pau12, WW14, Wan14, Pau14].

Among these results, we want to emphasize a result by Paulin [Pau14] that exactly concerns the Talagrand's concentration property (22) obtained by using Stein's methods, following Chatterjee's approach [Cha07].

For $1 \leq i \leq n$, let us note X_{-i} the random vector defined by

$$X_{-i} = (X_1, \ldots, X_{i-1}, X_{i+1}, \ldots, X_n).$$

The Dobrushin's interdependence matrix $D = (d_{ij})$ is a matrix of non-negative entries such that for every $i \in \{1, \ldots, n\}$, for every $x, y \in \mathcal{X}^n$,

$$\|\mathcal{L}(X_i|X_{-i} = x_{-i}) - \mathcal{L}(X_i|X_{-i} = y_{-i})\|_{TV} \leq \sum_{j,j\neq i} d_{ij}\mathbb{1}_{x_j\neq y_j}.$$

Theorem 5.4 (Theorem 3.3, [Pau14]) *Let* $\|D\|_1 = \max_{1\leq j\leq n}\sum_{i=1}^n d_{ij}$ *and* $\|D\|_\infty = \max_{1\leq i\leq n}\sum_{j=1}^n d_{ij}$. *If* $\|D\|_1 < 1$ *and* $\|D\|_\infty \leq 1$, *then for all subsets* $A \subset \mathcal{X}^n$, *one has*

$$\int e^{\frac{1-\|D\|_1}{26.1} D_T^2(x,A)} d\mu(x) \leq \frac{1}{\mu(A)}.$$

Examples of applications of this concentration result are presented in [Pau14] (the stochastic travelling salesman problem, Steiner trees).

5.3 Transport Inequalities for the Uniform Law on the Symmetric Group

In this section we present transport inequalities for the uniform law on the symmetric group S_n, that provide concentration results obtained by Maurey [Mau79] and Talagrand [Tal95].

Let μ be the uniform law on S_n, $\mu(\sigma) = \frac{1}{n!}, \sigma \in S_n$.

Theorem 5.5 ([Mau79]) *Let d be the Hamming distance on the symmetric group, for all* $\sigma, \tau \in S_n$,

$$d(\sigma, \tau) = \sum_{i=1}^n \mathbb{1}_{\sigma(i)\neq\tau(i)}.$$

Then for any subset $A \subset S_n$ *such that* $\mu(A) \geq 1/2$, *and for all* $t \geq 0$, *one has*

$$\mu(A_t) \geq 1 - 2e^{-\frac{t^2}{64n}},$$

where $A_t = \{y \in S_n, d(x, A) \leq t\}$.

This result has been generalized by Milman and Schechtman, to some groups whose distance is invariant by translation [MS86]. Talagrand has obtained another concentration property, stronger in terms of the dependence in n, obtained by the so-called convex-hull method. Here is the Talagrand's property with slightly modified notations. This implies the one of the previous theorem up to constant.

Theorem 5.6 ([Tal95]) *For any subset $A \subset S_n$,*

$$\int_{S_n} e^{f(A,\sigma)/16} d\mu(\sigma) \leq \frac{1}{\mu(A)},$$

where the quantity $f(A, \sigma)$ measures the distance from σ to A as follows:

$$f(A, \sigma) = \inf_{p \in \mathcal{P}(A)} \sum_{i=1}^{n} \left(\int \mathbb{1}_{\sigma(i) \neq \tau(i)} dp(\tau) \right)^2.$$

This result has been first generalized to product of symmetric groups by McDiarmid [McD02], and then further by Luczak and McDiarmid, to cover more general permutation groups which act suitably "locally" [LM03].

Theorems 5.5 and 5.6 are consequences of the following transport inequalities.

Theorem 5.7 ([Sam16]) *Let μ be the uniform law on the symmetric group.*

(a) For all probability measures ν_1 and ν_2 on S_n,

$$\frac{1}{2(n-1)} W_1^2(\nu_2, \nu_1) \leq \left(\sqrt{H(\nu_1|\mu)} + \sqrt{H(\nu_2|\mu)} \right)^2,$$

where $W_1(\nu_1, \nu_2) = \inf_{\pi \in \Pi(\nu_2, \nu_1)} \iint d(\sigma, \tau) d\pi(\sigma, \tau)$.

(b) For all probability measures ν_1 and ν_2 on S_n,

$$\frac{1}{8} \mathcal{T}_2(\nu_1|\nu_2) \leq \left(\sqrt{H(\nu_1|\mu)} + \sqrt{H(\nu_2|\mu)} \right)^2,$$

where $\mathcal{T}_2(\nu_1|\nu_2) = \inf_{\pi \in \Pi(\nu_2, \nu_1)} \int \sum_{i=1}^{n} \left(\int \mathbb{1}_{\sigma(i) \neq \tau(i)} dp_\sigma(\tau) \right)^2 d\nu_2(\sigma)$, with $\pi(\sigma, \tau) = \nu_2(\sigma) p_\sigma(\tau)$.

The proof of (b), inspired from the Talagrand's results, is given in the preprint [Sam16]. We present a simpler proof of (a) of the same nature at the end of this section. In fact, the dual formulation of the transport inequality of (a) is more popular: for all 1-Lipschitz functions $f : S_n \to \mathbb{R}$ (with respect to the Hamming distance d),

$$\int e^{f} d\mu \leq e^{\int f\,d\mu + (n-1)t^2/2}, \qquad \forall t \geq 0.$$

This exponential inequality is a consequence of Hoeffding inequalities for bounded martingales. It is widely commented and a proof is given in the paper [BHT06].

Point (b) implies the following useful concentration property.

Theorem 5.8 ([Sam16]) *Let μ be the uniform law on the symmetric group S_n. Let $g : S_n \to \mathbb{R}$ and $\alpha_k : S_n \to \mathbb{R}^+$, $k \in \{1, \ldots, n\}$ be functions such that for all $\tau, \sigma \in S_n$,*

$$g(\tau) - g(\sigma) \leq \sum_{k=1}^{n} \alpha_k(\tau) \mathbb{1}_{\tau(k) \neq \sigma(k)}.$$

Then, for all $t \geq 0$, one has

$$\mu\left(g \leq \int g\,d\mu - t\right) \leq \exp\left(-\frac{t^2}{8\int |\alpha|^2 d\mu}\right),$$

and

$$\mu\left(g \geq \int g\,d\mu + t\right) \leq \exp\left(-\frac{t^2}{8\,\sup_{\sigma \in S^n} |\alpha(\sigma)|^2}\right),$$

with $|\alpha(\sigma)|^2 = \sum_{k=1}^{n} \alpha_k^2(\sigma)$, $\sigma \in S_n$.

By applying this result to the particular function $g(\sigma) = \varphi(x_\sigma)$ where $\varphi : [0,1]^n \to \mathbb{R}$ is a Lipschitz convex function and given $(x_1, \ldots, x_n) \in [0,1]^n$, $x_\sigma = (x_{\sigma(1)}, \ldots, x_{\sigma(n)})$, we recover the deviation inequality by Adamczak, Chafaï, and Wolff [ACW14] (Theorem 3.1) obtained from Theorem 5.6 by Talagrand. This concentration property plays a key role in their approach, to study the convergence of the empirical spectral measure of random matrices with exchangeable entries, when the size of these matrices is increasing.

Proof of point (a) in Theorem 5.7 Since the distance W_1 satisfies a triangular inequality, it suffices to prove that for all probability measures ν_1 on S_n,

$$\frac{1}{2(n-1)} W_1^2(\nu_1, \mu) \leq H(\nu_1|\mu).$$

According to Proposition 3.1, the dual formulation of this inequality is the following, for all function φ on S_n and all $\lambda \geq 0$,

$$\int e^{\lambda Q\varphi} d\mu \leq e^{\int \lambda\varphi\,d\mu + (n-1)\lambda^2/2}, \tag{23}$$

with

$$Q\varphi(\sigma) = \inf_{p\in\mathcal{P}(S_n)} \left\{ \int \varphi dp + \int d(\sigma,\tau)\, dp(\tau) \right\}$$
$$= \inf_{p\in\mathcal{P}(S_n)} \left\{ \int \varphi dp + \sum_{k=1}^{n} \int \mathbb{1}_{\sigma(k)\neq\tau(k)}\, dp(\tau) \right\}.$$

We will prove the inequality (23) by induction on n.

When $n = 2$, S_n is the two point space

$$Q\varphi(\sigma) = \inf_{p\in\mathcal{P}(S_n)} \left\{ \int \varphi dp + 2\int \mathbb{1}_{\sigma\neq\tau}\, dp(\tau) \right\}.$$

The inequality (23) corresponds exactly to the dual form of the Csizár-Kullback-Pinsker inequality given by Proposition 3.1: for any probability measure ν on a separable metric space $\mathcal{X}$, for any measurable function $f : \mathcal{X} \to \mathbb{R}$,

$$\int e^{\lambda Rf} d\nu \leq e^{\lambda \int f\, d\nu + \lambda^2/2}, \qquad \forall \lambda \geq 0, \tag{24}$$

with $Rf(x) = \inf_{p\in\mathcal{P}(\mathcal{X})} \left\{ \int f dp + 2\int \mathbb{1}_{x\neq y} dp(y) \right\}, x \in \mathcal{X}$.

The induction step is also a consequence of the dual form (24) of the Csizár-Kullback-Pinsker inequality. Let $(H_i)_{1\leq i\leq n}$ be the partition of S_n defined by,

$$H_i = \{\sigma \in S_n, \sigma(i) = n\}$$

If p is a probability measure on S_n, it admits a unique decomposition defined by

$$p = \sum_{i=1}^{n} \hat{p}(i) p_i, \qquad \text{with} \quad p_i \in \mathcal{P}(H_i) \qquad \text{and} \quad \hat{p}(i) = p(H_i).$$

Thus, we define a probability measure $\hat{p}$ on $\{1,\ldots,n\}$. In particular, for the uniform law μ on S_n, one has

$$\mu = \frac{1}{n}\sum_{i=1}^{n} \mu_i,$$

where μ_i is the uniform law on H_i, $\mu_i(\sigma) = \frac{1}{(n-1)!}$, for any $\sigma \in H_i$. Therefore, one has

$$\int e^{\lambda Q\varphi} d\mu = \frac{1}{n}\sum_{i=1}^{n} \int e^{\lambda Q\varphi(\sigma)} d\mu_i(\sigma).$$

For any function $f : H_i \to \mathbb{R}$, let us note

$$Q_{H_i} f(\sigma) = \inf_{p \in \mathcal{P}(H_i)} \left\{ \int f\, dp + \sum_{k \neq i} \int \mathbb{1}_{\sigma(k) \neq \tau(k)} dp(\tau) \right\}.$$

We denote by τ_{ij} the transposition that exchanges the indices i and j. The application from H_i to H_n defined by $\tau \mapsto \tau\tau_{in}$ is one to one, and therefore by a change of index in the sum, we get

$$Q_{H_i} f(\sigma) = \inf_{p \in \mathcal{P}(H_i)} \left\{ \int f(\tau)\, dp(\tau) + \sum_{k \neq n} \int \mathbb{1}_{\sigma\tau_{in}(k) \neq \tau\tau_{in}(k)} dp(\tau) \right\}$$

$$= \inf_{q \in \mathcal{P}(H_n)} \left\{ \int f(\tau\tau_{in})\, dq(\tau) + \sum_{k \neq n} \int \mathbb{1}_{\sigma\tau_{in}(k) \neq \tau(k)} dq(\tau) \right\} = Q_{H_n} f^{\tau_{in}}(\sigma\tau_{in}).$$

where $f^{\tau_{in}}(\tau) = f(\tau\tau_{in})$ for all $\tau \in H_n$. Consequently, by induction, one has for all function $f : H_i \to \mathbb{R}$, for all $\lambda \geq 0$,

$$\int e^{\lambda Q_{H_i} f} d\mu_i = \int e^{\lambda Q_{H_n} f^{\tau_{in}}(\sigma\tau_{in})} d\mu_i(\sigma) = \int e^{\lambda Q_{H_n} f^{\tau_{in}}} d\mu_n$$

$$\leq \exp\left[\lambda \int f^{\tau_{in}} d\mu_n + (n-2)\frac{\lambda^2}{2}\right] = \exp\left[\lambda \int f d\mu_i + (n-2)\frac{\lambda^2}{2}\right].$$

Then the proof relies on the following Lemma.

Lemma 5.1 *For any function $\varphi : H_i \to \mathbb{R}$ and any $\sigma \in H_i$, one has*

$$Q\varphi(\sigma) \leq \inf_{\hat{p} \in \mathcal{P}(\{1,\ldots,n\})} \left\{ \sum_{l=1}^{n} Q_{H_l} \varphi^{\tau_{il}} \hat{p}(l) + 2 \sum_{l=1}^{n} \mathbb{1}_{l \neq i} \hat{p}(l) \right\}.$$

The proof of this lemma is by decomposition of the probability measures p on the H_j's, we get that if $\sigma \in H_i$ then

$$Q\varphi(\sigma) = \inf_{\hat{p} \in \mathcal{P}(\{1,\ldots,n\})} \inf_{p_1 \in \mathcal{P}(H_1),\ldots,p_n \in \mathcal{P}(H_n)}$$

$$\left\{ \sum_{l=1}^{n} \left[\int \varphi\, dp_l + \sum_{k \notin \{l,i\}} \int \mathbb{1}_{\sigma(k) \neq \tau\tau_{il}(k)} dp_l(\tau) \right] \hat{p}(l) + 2 \sum_{l=1}^{n} \mathbb{1}_{l \neq i} \hat{p}(l) \right\}.$$

The proof of (a) continues by applying consecutively this lemma, the Hölder inequality, and the induction hypotheses, this gives

$$\int e^{\lambda Q\varphi(\sigma)} d\mu_i(\sigma) \leq \inf_{\hat{p}\in\mathcal{P}(\{1,\ldots,n\})} \left\{ \prod_{l-1}^{n} \left(\int e^{\lambda Q_{H_i}\varphi^{\tau_{il}}} d\mu_i \right)^{\hat{p}(l)} e^{2\lambda \sum_{l=1}^{n} \mathbb{1}_{l\neq i}\hat{p}(l)} \right\}$$

$$\leq \exp\left[\inf_{\hat{p}\in\mathcal{P}(\{1,\ldots,n\})} \left\{ \lambda \sum_{l=1}^{n} \left(\int \varphi^{\tau_{il}} d\mu_i \right) \hat{p}(l) + (n-2)\frac{\lambda^2}{2} + 2\lambda \sum_{l=1}^{n} \mathbb{1}_{l\neq i}\hat{p}(l) \right\} \right]$$

$$= \exp\left[\lambda \inf_{\hat{p}\in\mathcal{P}(\{1,\ldots,n\})} \left\{ \sum_{l=1}^{n} \hat{\varphi}(l)\hat{p}(l) + 2\sum_{l=1}^{n} \mathbb{1}_{l\neq i}\hat{p}(l) \right\} + (n-2)\frac{\lambda^2}{2} \right],$$

where $\hat{\varphi}(l) = \int \varphi^{\tau_{il}} d\mu_i = \int \varphi d\mu_l$. Let us consider again the above infimum-convolution $R\hat{\varphi}$ defined on the space $\mathcal{X} = \{1,\ldots,n\}$, one has

$$R\hat{\varphi}(i) = \inf_{\hat{p}\in\mathcal{P}(\{1,\ldots,n\})} \left\{ \sum_{l=1}^{n} \hat{\varphi}(l)\hat{p}(l) + 2\sum_{l=1}^{n} \mathbb{1}_{l\neq i}\hat{p}(l) \right\}.$$

As a consequence, by applying (24) with the uniform law ν on $\{1,\ldots,n\}$, the previous inequality gives

$$\int e^{\lambda Q\varphi} d\mu = \frac{1}{n}\sum_{i=1}^{n} \int e^{\lambda Q\varphi(\sigma)} d\mu_i(\sigma) \leq \left(\frac{1}{n}\sum_{i=1}^{n} e^{\lambda R\hat{\varphi}(i)} \right) e^{(n-2)\lambda^2/2}$$

$$\leq \exp\left[\frac{\lambda}{n}\sum_{i=1}^{n} \hat{\varphi}(i) + \frac{\lambda^2}{2} + (n-2)\frac{\lambda^2}{2} \right] = \exp\left[\lambda \int \varphi\, d\mu + (n-1)\frac{\lambda^2}{2} \right].$$

□

Acknowledgements The author would like to thank Nathael Gozlan and an anonymous referee for their careful reading of the manuscript.
This work was supported by the grants ANR 2011 BS01 007 01, ANR 10 LABX-58.

References

[ABC+00] C. Ané, S. Blachère, D. Chafaï, P. Fougères, I. Gentil, F. Malrieu, C. Roberto, and G. Scheffer. *Sur les inégalités de Sobolev logarithmiques*, volume 10 of *Panoramas et Synthèses [Panoramas and Syntheses]*. Société Mathématique de France, Paris, 2000. With a preface by Dominique Bakry and Michel Ledoux.

[ACW14] R. Adamczak, D. Chafaï, and P. Wolff. Circular law for random matrices with exchangeable entries. *ArXiv e-prints*, February 2014.

[Ada05] R. Adamczak. Logarithmic Sobolev inequalities and concentration of measure for convex functions and polynomial chaoses. *Bull. Pol. Acad. Sci. Math.*, 53(2):221–238, 2005.

[Ada08] R. Adamczak. A tail inequality for suprema of unbounded empirical processes with applications to Markov chains. *Electron. J. Probab.*, 13:no. 34, 1000–1034, 2008.

[AGS14] L. Ambrosio, N. Gigli, and G. Savaré. Metric measure spaces with Riemannian Ricci curvature bounded from below. *Duke Math. J.*, 163(7):1405–1490, 2014.

[AGS15] L. Ambrosio, N. Gigli, and G. Savaré. Bakry-Émery curvature-dimension condition and Riemannian Ricci curvature bounds. *Ann. Probab.*, 43(1):339–404, 2015.

[AS15] R. Adamczak and M. Strzelecki. Modified log-Sobolev inequalities for convex functions on the real line. Sufficient conditions. *ArXiv e-prints*, May 2015.

[BG99] S.G. Bobkov and F. Götze. Exponential integrability and transportation cost related to logarithmic Sobolev inequalities. *J. Funct. Anal.*, 163(1):1–28, 1999.

[BGL01] S.G. Bobkov, I. Gentil, and M. Ledoux. Hypercontractivity of Hamilton-Jacobi equations. *J. Math. Pures Appl. (9)*, 80(7):669–696, 2001.

[BHT06] S. G. Bobkov, C. Houdré, and P. Tetali. The subgaussian constant and concentration inequalities. *Israel J. Math.*, 156:255–283, 2006.

[BK08] F. Barthe and A. V. Kolesnikov. Mass transport and variants of the logarithmic Sobolev inequality. *J. Geom. Anal.*, 18(4):921–979, 2008.

[BL97] S. G. Bobkov and M. Ledoux. Poincaré's inequalities and Talagrand's concentration phenomenon for the exponential distribution. *Probab. Theory Related Fields*, 107(3):383–400, 1997.

[BL00] S. G. Bobkov and M. Ledoux. From Brunn-Minkowski to Brascamp-Lieb and to logarithmic Sobolev inequalities. *Geom. Funct. Anal.*, 10(5):1028–1052, 2000.

[BLM00] S. Boucheron, G. Lugosi, and P. Massart. A sharp concentration inequality with applications. *Random Structures Algorithms*, 16(3):277–292, 2000.

[BLM03] S. Boucheron, G. Lugosi, and P. Massart. Concentration inequalities using the entropy method. *Ann. Probab.*, 31(3):1583–1614, 2003.

[BLM13] S. Boucheron, G. Lugosi, and P. Massart. *Concentration inequalities*. Oxford University Press, Oxford, 2013. A nonasymptotic theory of independence, With a foreword by Michel Ledoux.

[Bob97] S. G. Bobkov. An isoperimetric inequality on the discrete cube, and an elementary proof of the isoperimetric inequality in Gauss space. *Ann. Probab.*, 25(1):206–214, 1997.

[Bou03] O. Bousquet. Concentration inequalities for sub-additive functions using the entropy method. In *Stochastic inequalities and applications*, volume 56 of *Progr. Probab.*, pages 213–247. Birkhäuser, Basel, 2003.

[BR03] F. Barthe and C. Roberto. Sobolev inequalities for probability measures on the real line. *Studia Math.*, 159(3):481–497, 2003.

[BR08] F. Barthe and C. Roberto. Modified logarithmic Sobolev inequalities on $\mathbb{R}$. *Potential Anal.*, 29(2):167–193, 2008.

[CD10] S. Chatterjee and P.S. Dey. Applications of Stein's method for concentration inequalities. *Ann. Probab.*, 38(6):2443–2485, 2010.

[CG06] P. Cattiaux and A. Guillin. On quadratic transportation cost inequalities. *J. Math. Pures Appl. (9)*, 86(4):341–361, 2006.

[Cha05] S. Chatterjee. Concentration inequalities with exchangeable pairs (Ph.D. thesis). *ArXiv Mathematics e-prints*, July 2005.

[Cha07] S. Chatterjee. Stein's method for concentration inequalities. *Probab. Theory Related Fields*, 138(1–2):305–321, 2007.

[Csi67] I. Csiszár. Information-type measures of difference of probability distributions and indirect observations. *Studia Sci. Math. Hungar.*, 2:299–318, 1967.

[Dem97] A. Dembo. Information inequalities and concentration of measure. *Ann. Probab.*, 25(2):927–939, 1997.

[DGW04] H. Djellout, A. Guillin, and L. Wu. Transportation cost-information inequalities and applications to random dynamical systems and diffusions. *Ann. Probab.*, 32(3B):2702–2732, 2004.

[Dou94] P. Doukhan. *Mixing*, volume 85 of *Lecture Notes in Statistics*. Springer-Verlag, New York, 1994. Properties and examples.

[EM12] M. Erbar and J. Maas. Ricci curvature of finite Markov chains via convexity of the entropy. *Arch. Ration. Mech. Anal.*, 206(3):997–1038, 2012.

[EM14] M. Erbar and J. Maas. Gradient flow structures for discrete porous medium equations. *Discrete Contin. Dyn. Syst.*, 34(4):1355–1374, 2014.

[EMT15] M. Erbar, J. Maas, and P. Tetali. Discrete Ricci curvature bounds for Bernoulli-Laplace and random transposition models. *Ann. Fac. Sci. Toulouse Math. (6)*, 24(4):781–800, 2015.

[FHT03] A. A. Fedotov, P. Harremoës, and F. Topsøe. Refinements of Pinsker's inequality. *IEEE Trans. Inform. Theory*, 49(6):1491–1498, 2003.

[FMNW15] N. Feldheim, A. Marsiglietti, P. Nayar, and J. Wang. A note on the convex infimum convolution inequality. *ArXiv e-prints*, May 2015.

[GGM05] I. Gentil, A. Guillin, and L. Miclo. Modified logarithmic Sobolev inequalities and transportation inequalities. *Probab. Theory Related Fields*, 133(3):409–436, 2005.

[GGM07] I. Gentil, A. Guillin, and L. Miclo. Modified logarithmic Sobolev inequalities in null curvature. *Rev. Mat. Iberoam.*, 23(1):235–258, 2007.

[Gil10] G. L. Gilardoni. On Pinsker's and Vajda's type inequalities for Csiszár's f-divergences. *IEEE Trans. Inform. Theory*, 56(11):5377–5386, 2010.

[GL10] N. Gozlan and C. Léonard. Transport inequalities. A survey. *Markov Process. Related Fields*, 16(4):635–736, 2010.

[GM83] M. Gromov and V. D. Milman. A topological application of the isoperimetric inequality. *Amer. J. Math.*, 105(4):843–854, 1983.

[Goz07] N. Gozlan. Characterization of Talagrand's like transportation-cost inequalities on the real line. *J. Funct. Anal.*, 250(2):400–425, 2007.

[Goz09] N. Gozlan. A characterization of dimension free concentration in terms of transportation inequalities. *Ann. Probab.*, 37(6):2480–2498, 2009.

[Goz12] N. Gozlan. Transport-entropy inequalities on the line. *Electron. J. Probab.*, 17:no. 49, 18, 2012.

[Goz15] Gozlan, N. Transport inequalities and concentration of measure*. *ESAIM: Proc.*, 51:1–23, 2015.

[Gro75] L. Gross. Logarithmic Sobolev inequalities. *Amer. J. Math.*, 97(4):1061–1083, 1975.

[Gro99] M. Gromov. *Metric structures for Riemannian and non-Riemannian spaces*, volume 152 of *Progress in Mathematics*. Birkhäuser Boston, Inc., Boston, MA, 1999.

[GRS11a] N. Gozlan, C. Roberto, and P.-M. Samson. From concentration to logarithmic Sobolev and Poincaré inequalities. *J. Funct. Anal.*, 260(5):1491–1522, 2011.

[GRS11b] N. Gozlan, C. Roberto, and P.-M. Samson. A new characterization of Talagrand's transport-entropy inequalities and applications. *Ann. Probab.*, 39(3):857–880, 2011.

[GRS13] N. Gozlan, C. Roberto, and P.-M. Samson. Characterization of Talagrand's transport-entropy inequalities in metric spaces. *Ann. Probab.*, 41(5):3112–3139, 2013.

[GRS14] N. Gozlan, C. Roberto, and P.-M. Samson. Hamilton Jacobi equations on metric spaces and transport entropy inequalities. *Rev. Mat. Iberoam.*, 30(1):133–163, 2014.

[GRS+15] N. Gozlan, C. Roberto, P.-M. Samson, Y. Shu, and P. Tetali. Characterization of a class of weak transport-entropy inequalities on the line. *ArXiv e-prints*, September 2015.

[GRST14a] N. Gozlan, C. Roberto, P.-M. Samson, and P. Tetali. Displacement convexity of entropy and related inequalities on graphs. *Probab. Theory Related Fields*, 160(1–2):47–94, 2014.

[GRST14b] N. Gozlan, C. Roberto, P.-M. Samson, and P. Tetali. Kantorovich duality for general transport costs and applications. *ArXiv e-prints*, December 2014.

[Har66] L. H. Harper. Optimal numberings and isoperimetric problems on graphs. *J. Combinatorial Theory*, 1:385–393, 1966.

[Hil14] E. Hillion. Contraction of measures on graphs. *Potential Anal.*, 41(3):679–698, 2014.

[HS87] R. Holley and D. Stroock. Logarithmic Sobolev inequalities and stochastic Ising models. *J. Statist. Phys.*, 46(5–6):1159–1194, 1987.

[Kom88] H. Komiya. Elementary proof for Sion's minimax theorem. *Kodai Math. J.*, 11(1):5–7, 1988.

[Kon12] A. Kontorovich. Obtaining measure concentration from Markov contraction. *Markov Process. Related Fields*, 18(4):613–638, 2012.

[KR05] T. Klein and E. Rio. Concentration around the mean for maxima of empirical processes. *Ann. Probab.*, 33(3):1060–1077, 2005.

[Kul67] S. Kullback. Lower bound for discrimination information in terms of variation. *IEEE Trans. Information Theory*, 4:126–127, 1967.

[Led93] M. Ledoux. Inégalités isopérimétriques en analyse et probabilités. *Astérisque*, (216):Exp. No. 773, 5, 343–375, 1993. Séminaire Bourbaki, Vol. 1992/93.

[Led97] M. Ledoux. On Talagrand's deviation inequalities for product measures. *ESAIM Probab. Statist.*, 1:63–87 (electronic), 1997.

[Led01] M. Ledoux. *The concentration of measure phenomenon*, volume 89 of *Mathematical Surveys and Monographs*. American Mathematical Society, Providence, RI, 2001.

[Lév51] P Lévy. *Problèmes concrets d'analyse fonctionnelle. Avec un complément sur les fonctionnelles analytiques par F. Pellegrino*. Gauthier-Villars, Paris, 1951. 2d ed.

[LM03] M. J. Luczak and C. McDiarmid. Concentration for locally acting permutations. *Discrete Math.*, 265(1–3):159–171, 2003.

[LT91] M. Ledoux and M. Talagrand. *Probability in Banach spaces*, volume 23 of *Ergebnisse der Mathematik und ihrer Grenzgebiete (3) [Results in Mathematics and Related Areas (3)]*. Springer-Verlag, Berlin, 1991. Isoperimetry and processes.

[Maa11] J. Maas. Gradient flows of the entropy for finite Markov chains. *J. Funct. Anal.*, 261(8):2250–2292, 2011.

[Mar96a] K. Marton. Bounding $\overline{d}$-distance by informational divergence: a method to prove measure concentration. *Ann. Probab.*, 24(2):857–866, 1996.

[Mar96b] K. Marton. A measure concentration inequality for contracting Markov chains. *Geom. Funct. Anal.*, 6(3):556–571, 1996.

[Mar97] K. Marton. Erratum to: "A measure concentration inequality for contracting Markov chains" [Geom. Funct. Anal. **6** (1996), no. 3, 556–571; MR1392329 (97g:60082)]. *Geom. Funct. Anal.*, 7(3):609–613, 1997.

[Mar03] K. Marton. Measure concentration and strong mixing. *Studia Sci. Math. Hungar.*, 40(1–2):95–113, 2003.

[Mar04] K. Marton. Measure concentration for Euclidean distance in the case of dependent random variables. *Ann. Probab.*, 32(3B):2526–2544, 2004.

[Mar10] K. Marton. Correction: Measure concentration for Euclidean distance in the case of dependent random variables [mr2078549]. *Ann. Probab.*, 38(1):439–442, 2010.

[Mas00a] P. Massart. About the constants in Talagrand's concentration inequalities for empirical processes. *Ann. Probab.*, 28(2):863–884, 2000.

[Mas00b] P. Massart. Some applications of concentration inequalities to statistics. *Ann. Fac. Sci. Toulouse Math. (6)*, 9(2):245–303, 2000. Probability theory.

[Mas07] P. Massart. *Concentration inequalities and model selection*, volume 1896 of *Lecture Notes in Mathematics*. Springer, Berlin, 2007. Lectures from the 33rd Summer School on Probability Theory held in Saint-Flour, July 6–23, 2003, With a foreword by Jean Picard.

[Mau79] B. Maurey. Construction de suites symétriques. *C. R. Acad. Sci. Paris Sér. A-B*, 288(14):A679–A681, 1979.

[Mau91] B. Maurey. Some deviation inequalities. *Geom. Funct. Anal.*, 1(2):188–197, 1991.

[McD02] Colin McDiarmid. Concentration for independent permutations. *Combin. Probab. Comput.*, 11(2):163–178, 2002.

[MS86] V. D. Milman and G. Schechtman. *Asymptotic theory of finite-dimensional normed spaces*, volume 1200 of *Lecture Notes in Mathematics*. Springer-Verlag, Berlin, 1986. With an appendix by M. Gromov.

[MSWW11] Y. Ma, S. Shen, X. Wang, and L. Wu. Transportation inequalities: from Poisson to Gibbs measures. *Bernoulli*, 17(1):155–169, 2011.

[Muc72] B. Muckenhoupt. Hardy's inequality with weights. *Studia Math.*, 44:31–38, 1972. Collection of articles honoring the completion by Antoni Zygmund of 50 years of scientific activity, I.

[Oll09] Y. Ollivier. Ricci curvature of Markov chains on metric spaces. *J. Funct. Anal.*, 256(3):810–864, 2009.

[OV00] F. Otto and C. Villani. Generalization of an inequality by Talagrand and links with the logarithmic Sobolev inequality. *J. Funct. Anal.*, 173(2):361–400, 2000.
[OV12] Y. Ollivier and C. Villani. A curved Brunn-Minkowski inequality on the discrete hypercube, or: what is the Ricci curvature of the discrete hypercube? *SIAM J. Discrete Math.*, 26(3):983–996, 2012.
[Pan01] D. Panchenko. A note on Talagrand's concentration inequality. *Electron. Comm. Probab.*, 6:55–65 (electronic), 2001.
[Pan02] D. Panchenko. Some extensions of an inequality of Vapnik and Chervonenkis. *Electron. Comm. Probab.*, 7:55–65 (electronic), 2002.
[Pan03] D. Panchenko. Symmetrization approach to concentration inequalities for empirical processes. *Ann. Probab.*, 31(4):2068–2081, 2003.
[Pau12] D. Paulin. Concentration inequalities for Markov chains by Marton couplings and spectral methods. *ArXiv e-prints*, December 2012.
[Pau14] D. Paulin. The convex distance inequality for dependent random variables, with applications to the stochastic travelling salesman and other problems. *Electron. J. Probab.*, 19:no. 68, 34, 2014.
[Pin64] M. S. Pinsker. *Information and information stability of random variables and processes*. Translated and edited by Amiel Feinstein. Holden-Day Inc., San Francisco, Calif., 1964.
[Rio01] E. Rio. Inégalités de concentration pour les processus empiriques de classes de parties. *Probab. Theory Related Fields*, 119(2):163–175, 2001.
[Rio02] E. Rio. Une inégalité de Bennett pour les maxima de processus empiriques. *Ann. Inst. H. Poincaré Probab. Statist.*, 38(6):1053–1057, 2002. En l'honneur de J. Bretagnolle, D. Dacunha-Castelle, I. Ibragimov.
[Rio12] E. Rio. Sur la fonction de taux dans les inégalités de Talagrand pour les processus empiriques. *C. R. Math. Acad. Sci. Paris*, 350(5–6):303–305, 2012.
[Rio13] E. Rio. On McDiarmid's concentration inequality. *Electron. Commun. Probab.*, 18:no. 44, 11, 2013.
[RR91] M. M. Rao and Z. D. Ren. *Theory of Orlicz spaces*. Marcel Dekker Inc., 1991.
[Sam00] P.-M. Samson. Concentration of measure inequalities for Markov chains and Φ-mixing processes. *Ann. Probab.*, 28(1):416–461, 2000.
[Sam03] P.-M. Samson. Concentration inequalities for convex functions on product spaces. In *Stochastic inequalities and applications*, volume 56 of *Progr. Probab.*, pages 33–52. Birkhäuser, Basel, 2003.
[Sam07] P.-M. Samson. Infimum-convolution description of concentration properties of product probability measures, with applications. *Ann. Inst. H. Poincaré Probab. Statist.*, 43(3):321–338, 2007.
[Sam16] P.-M. Samson. Transport-entropy inequalities on locally acting groups of permutations. working paper or preprint, September 2016.
[Sio58] M. Sion. On general minimax theorems. *Pacific J. Math.*, 8:171–176, 1958.
[Str65] V. Strassen. The existence of probability measures with given marginals. *Ann. Math. Statist.*, 36:423–439, 1965.
[Tal91] M. Talagrand. A new isoperimetric inequality and the concentration of measure phenomenon. In *Geometric aspects of functional analysis (1989–90)*, volume 1469 of *Lecture Notes in Math.*, pages 94–124. Springer, Berlin, 1991.
[Tal95] M. Talagrand. Concentration of measure and isoperimetric inequalities in product spaces. *Inst. Hautes Études Sci. Publ. Math.*, (81):73–205, 1995.
[Tal96a] M. Talagrand. New concentration inequalities in product spaces. *Invent. Math.*, 126(3):505–563, 1996.
[Tal96b] M. Talagrand. A new look at independence. *Ann. Probab.*, 24(1):1–34, 1996.
[Tal96c] M. Talagrand. Transportation cost for Gaussian and other product measures. *Geom. Funct. Anal.*, 6(3):587–600, 1996.
[Vil09] C. Villani. *Optimal transport: old and new*, volume 338 of *Grundlehren der Mathematischen Wissenschaften [Fundamental Principles of Mathematical Sciences]*. Springer-Verlag, Berlin, 2009.

[Wan14] N.-Y. Wang. Concentration inequalities for Gibbs sampling under d_{l_2}-metric. *Electron. Commun. Probab.*, 19:no. 63, 11, 2014.

[Wu06] L. Wu. Poincaré and transportation inequalities for Gibbs measures under the Dobrushin uniqueness condition. *Ann. Probab.*, 34(5):1960–1989, 2006.

[WW77] D. L. Wang and P. Wang. Extremal configurations on a discrete torus and a generalization of the generalized Macaulay theorem. *SIAM. J. Appl. Math.*, 33(1):55–59, 1977.

[WW14] N.-Y. Wang and L. Wu. Convergence rate and concentration inequalities for Gibbs sampling in high dimension. *Bernoulli*, 20(4):1698–1716, 2014.

Structured Random Matrices

Ramon van Handel

Abstract Random matrix theory is a well-developed area of probability theory that has numerous connections with other areas of mathematics and its applications. Much of the literature in this area is concerned with matrices that possess many exact or approximate symmetries, such as matrices with i.i.d. entries, for which precise analytic results and limit theorems are available. Much less well understood are matrices that are endowed with an arbitrary structure, such as sparse Wigner matrices or matrices whose entries possess a given variance pattern. The challenge in investigating such structured random matrices is to understand how the given structure of the matrix is reflected in its spectral properties. This chapter reviews a number of recent results, methods, and open problems in this direction, with a particular emphasis on sharp spectral norm inequalities for Gaussian random matrices.

1 Introduction

The study of random matrices has a long history in probability, statistics, and mathematical physics, and continues to be a source of many interesting old and new mathematical problems [2, 25]. Recent years have seen impressive advances in this area, particularly in the understanding of universality phenomena that are exhibited by the spectra of classical random matrix models [8, 26]. At the same time, random matrices have proved to be of major importance in contemporary applied mathematics, see, for example, [28, 32] and the references therein.

Much of classical random matrix theory is concerned with highly symmetric models of random matrices. For example, the simplest random matrix model, the *Wigner matrix*, is a symmetric matrix whose entries above the diagonal are independent and identically distributed. If the entries are chosen to be Gaussian (and the diagonal entries are chosen to have the appropriate variance), this model is additionally invariant under orthogonal transformations. Such strong symmetry

R. van Handel (✉)
Princeton University, Sherrerd Hall 227, Princeton, NJ 08544, USA
e-mail: rvan@princeton.edu

E. Carlen et al. (eds.), *Convexity and Concentration*, The IMA Volumes in Mathematics and its Applications 161, DOI 10.1007/978-1-4939-7005-6_4

properties make it possible to obtain extremely precise analytic results on the asymptotic properties of macroscopic and microscopic spectral statistics of these matrices, and give rise to deep connections with classical analysis, representation theory, combinatorics, and various other areas of mathematics [2, 25].

Much less is understood, however, once we depart from such highly symmetric settings and introduce nontrivial structure into the random matrix model. Such models are the topic of this chapter. To illustrate what we mean by "structure," let us describe some typical examples that will be investigated in the sequel.

- A *sparse Wigner matrix* is a matrix with a given (deterministic) sparsity pattern, whose nonzero entries above the diagonal are i.i.d. centered random variables. Such models have interesting applications in combinatorics and computer science (see, for example, [1]), and specific examples such as random band matrices are of significant interest in mathematical physics (cf. [22]). The "structure" of the matrix is determined by its sparsity pattern. We would like to know how the given sparsity pattern is reflected in the spectral properties of the matrix.
- Let $x_1, \ldots, x_s$ be deterministic vectors. Matrices of the form

$$X = \sum_{k=1}^{s} g_k x_k x_k^*,$$

where $g_1, \ldots, g_s$ are i.i.d. standard Gaussian random variables, arise in functional analysis (see, for example, [20]). The "structure" of the matrix is determined by the positions of the vectors $x_1, \ldots, x_s$. We would like to know how the given positions are reflected in the spectral properties of the matrix.
- Let $X_1, \ldots, X_n$ be i.i.d. random vectors with covariance matrix Σ. Consider

$$Z = \frac{1}{n} \sum_{k=1}^{n} X_k X_k^*,$$

the *sample covariance matrix* [32, 10]. If we think of $X_1, \ldots, X_n$ are observed data from an underlying distribution, we can think of Z as an unbiased estimator of the covariance matrix $\Sigma = \mathbf{E}Z$. The "structure" of the matrix is determined by the covariance matrix Σ. We would like to know how the given covariance matrix is reflected in the spectral properties of Z (and particularly in $\|Z - \Sigma\|$).

While these models possess distinct features, we will refer to such models collectively as *structured random matrices*. We emphasize two important features of such models. First, the symmetry properties that characterize classical random matrix models are manifestly absent in the structured setting. Second, it is evident in the above models that it does not make much sense to investigate their asymptotic properties (that is, probabilistic limit theorems): as the structure is defined for the given matrix only, there is no natural way to take the size of these matrices to infinity.

Due to these observations, the study of structured random matrices has a significantly different flavor than most of classical random matrix theory. In the absence of asymptotic theory, our main interest is to obtain nonasymptotic *inequalities* that identify what structural parameters control the macroscopic properties of the underlying random matrix. In this sense, the study of structured random matrices is very much in the spirit of probability in Banach spaces [12], which is heavily reflected in the type of results that have been obtained in this area. In particular, the aspect of structured random matrices that is most well understood is the behavior of matrix norms, and particularly the spectral norm, of such matrices. The investigation of the latter will be the focus of the remainder of this chapter.

In view of the above discussion, it should come as no surprise that some of the earliest general results on structured random matrices appeared in the functional analysis literature [27, 13, 11], but further progress has long remained relatively limited. More recently, the study of structured random matrices has received renewed attention due to the needs of applied mathematics, cf. [28] and the references therein. However, significant new progress was made in the past few years. On the one hand, surprisingly sharp inequalities were recently obtained for certain random matrix models, particularly in the case of independent entries, that yield nearly optimal bounds and go significantly beyond earlier results. On the other hand, very simple new proofs have been discovered for some (previously) deep classical results that shed new light on the underlying mechanisms and that point the way to further progress in this direction. The opportunity therefore seems ripe for an elementary presentation of the results in this area. The present chapter represents the author's attempt at presenting some of these ideas in a cohesive manner.

Due to the limited capacity of space and time, it is certainly impossible to provide an encyclopedic presentation of the topic of this chapter, and some choices had to be made. In particular, the following focus is adopted throughout this chapter:

- The emphasis throughout is on spectral norm inequalities for *Gaussian* random matrices. The reason for this is twofold. On the one hand, much of the difficulty of capturing the structure of random matrices arises already in the Gaussian setting, so that this provides a particularly clean and rich playground for investigating such problems. On the other hand, Gaussian results extend readily to much more general distributions, as will be discussed further in section 4.4.
- For simplicity of presentation, no attempt was made to optimize the universal constants that appear in most of our inequalities, even though many of these inequalities can in fact be obtained with surprisingly sharp (even optimal) constants. The original references can be consulted for more precise statements.
- The presentation is by no means exhaustive, and many variations on and extensions of the presented material have been omitted. None of the results in this chapter are original, though I have done my best to streamline the presentation. On the other hand, I have tried to make the chapter as self-contained as possible, and most results are presented with complete proofs.

The remainder of this chapter is organized as follows. The preliminary section 2 sets the stage by discussing the basic methods that will be used throughout this chapter to bound spectral norms of random matrices. Section 3 is devoted to a family of powerful but suboptimal inequalities, the noncommutative Khintchine inequalities, that are applicable to the most general class of structured random matrices that we will encounter. In section 4, we specialize to structured random matrices with independent entries (such as sparse Wigner matrices) and derive nearly optimal bounds. We also discuss a few fundamental open problems in this setting. We conclude this chapter in the short section 5 by investigating sample covariance matrices.

2 How to Bound Matrix Norms

As was discussed in the introduction, the investigation of random matrices with arbitrary structure has by its nature a nonasymptotic flavor: we aim to obtain probabilistic inequalities (upper and lower bounds) on spectral properties of the matrices in question that capture faithfully the underlying structure. At present, this program is largely developed in the setting of spectral norms of random matrices, which will be our focus throughout this chapter. For completeness, we define:

Definition 2.1 *The* spectral norm $\|X\|$ *is the largest singular value of the matrix* X.

For convenience, we generally work with symmetric random matrices $X = X^*$. There is no loss of generality in doing so, as will be explained below.

Before we can obtain any meaningful bounds, we must first discuss some basic approaches for bounding the spectral norms of random matrices. The most important methods that are used for this purpose are collected in this section.

2.1 The Moment Method

Let X be an $n \times n$ symmetric random matrix. The first difficulty one encounters in bounding the spectral norm $\|X\|$ is that the map $X \mapsto \|X\|$ is highly nonlinear. It is therefore not obvious how to efficiently relate the distribution of $\|X\|$ to the distribution of the entries X_{ij}. One of the most effective approaches to simplifying this relationship is obtained by applying the following elementary observation.

Lemma 2.2 *Let* X *be an* $n \times n$ *symmetric matrix. Then*

$$\|X\| \asymp \mathrm{Tr}[X^{2p}]^{1/2p} \quad \textit{for } p \asymp \log n.$$

The beauty of this observation is that unlike $\|X\|$, which is a very complicated function of the entries of X, the quantity $\mathrm{Tr}[X^{2p}]$ is a *polynomial* in the matrix entries. This means that $\mathbf{E}[\mathrm{Tr}[X^{2p}]]$, the $2p$-th moment of the matrix X, can be evaluated explicitly and subjected to further analysis. As Lemma 2.2 implies that

$$\mathbf{E}[\|X\|^{2p}]^{1/2p} \asymp \mathbf{E}[\mathrm{Tr}[X^{2p}]]^{1/2p} \quad \text{for } p \asymp \log n,$$

this provides a direct route to controlling the spectral norm of a random matrix. Various incarnations of this idea are referred to as the *moment method.*

Lemma 2.2 actually has nothing to do with matrices. Given $x \in \mathbb{R}^n$, everyone knows that $\|x\|_p \to \|x\|_\infty$ as $p \to \infty$, so that $\|x\|_p \approx \|x\|_\infty$ when p is large. How large should p be for this to be the case? The following lemma provides the answer.

Lemma 2.3 *If $p \asymp \log n$, then $\|x\|_p \asymp \|x\|_\infty$ for all $x \in \mathbb{R}^n$.*

Proof It is trivial that

$$\max_{i \le n} |x_i|^p \le \sum_{i \le n} |x_i|^p \le n \max_{i \le n} |x_i|^p.$$

Thus $\|x\|_\infty \le \|x\|_p \le n^{1/p}\|x\|_\infty$, and $n^{1/p} = e^{(\log n)/p} \asymp 1$ when $\log n \asymp p$. □

The proof of Lemma 2.2 follows readily by applying Lemma 2.3 to the spectrum.

Proof (Proof of Lemma 2.2) Let $\lambda = (\lambda_1, \dots, \lambda_n)$ be the eigenvalues of X. Then $\|X\| = \|\lambda\|_\infty$ and $\mathrm{Tr}[X^{2p}]^{1/2p} = \|\lambda\|_{2p}$. The result follows from Lemma 2.3. □

The moment method will be used frequently throughout this chapter as the first step in bounding the spectral norm of random matrices. However, the moment method is just as useful in the vector setting. As a warmup exercise, let us use this approach to bound the maximum of i.i.d. Gaussian random variables (which can be viewed as a vector analogue of bounding the maximum eigenvalue of a random matrix). If $g \sim N(0, I)$ is the standard Gaussian vector in $\mathbb{R}^n$, Lemma 2.3 implies

$$[\mathbf{E}\|g\|_\infty^p]^{1/p} \asymp [\mathbf{E}\|g\|_p^p]^{1/p} \asymp [\mathbf{E}g_1^p]^{1/p} \quad \text{for } p \asymp \log n.$$

Thus the problem of bounding the maximum of n i.i.d. Gaussian random variables is reduced by the moment method to computing the $\log n$-th moment of a single Gaussian random variable. We will bound the latter in section 3.1 in preparation for proving the analogous bound for random matrices. For our present purposes, let us simply note the outcome of this computation $[\mathbf{E}g_1^p]^{1/p} \lesssim \sqrt{p}$ (Lemma 3.1), so that

$$\mathbf{E}\|g\|_\infty \le [\mathbf{E}\|g\|_\infty^{\log n}]^{1/\log n} \lesssim \sqrt{\log n}.$$

This bound is in fact sharp (up to the universal constant).

Remark 2.4 Lemma 2.2 implies immediately that

$$\mathbf{E}\|X\| \asymp \mathbf{E}[\mathrm{Tr}[X^{2p}]^{1/2p}] \quad \text{for } p \asymp \log n.$$

Unfortunately, while this bound is sharp by construction, it is essentially useless: the expectation of $\mathrm{Tr}[X^{2p}]^{1/2p}$ is in principle just as difficult to compute as that of $\|X\|$

itself. The utility of the moment method stems from the fact that we can explicitly compute the expectation of $\mathrm{Tr}[X^{2p}]$, a polynomial in the matrix entries. This suggests that the moment method is well suited in principle only for obtaining sharp bounds on the pth moment of the spectral norm

$$\mathbf{E}[\|X\|^{2p}]^{1/2p} \asymp \mathbf{E}[\mathrm{Tr}[X^{2p}]]^{1/2p} \quad \text{for } p \asymp \log n,$$

and not on the first moment $\mathbf{E}\|X\|$ of the spectral norm. Of course, as $\mathbf{E}\|X\| \leq [\mathbf{E}\|X\|^{2p}]^{1/2p}$ by Jensen's inequality, this yields an *upper* bound on the first moment of the spectral norm. We will see in the sequel that this upper bound is often, but not always, sharp. We can expect the moment method to yield a sharp bound on $\mathbf{E}\|X\|$ when the fluctuations of $\|X\|$ are of a smaller order than its mean; this was the case, for example, in the computation of $\mathbf{E}\|g\|_\infty$ above. On the other hand, the moment method is inherently dimension-dependent (as one must choose $p \sim \log n$), so that it is generally not well suited for obtaining dimension-free bounds.

We have formulated Lemma 2.2 for symmetric matrices. A completely analogous approach can be applied to non-symmetric matrices. In this case, we use that

$$\|X\|^2 = \|X^*X\| \asymp \mathrm{Tr}[(X^*X)^p]^{1/p} \quad \text{for } p \asymp \log n,$$

which follows directly from Lemma 2.2. However, this non-symmetric form is often somewhat inconvenient in the proofs of random matrix bounds, or at least requires additional bookkeeping. Instead, we recall a classical trick that allows us to directly obtain results for non-symmetric matrices from the analogous symmetric results. If X is any $n \times m$ rectangular matrix, then it is readily verified that $\|\tilde{X}\| = \|X\|$, where $\tilde{X}$ denotes the $(n+m) \times (n+m)$ symmetric matrix defined by

$$\tilde{X} = \begin{bmatrix} 0 & X \\ X^* & 0 \end{bmatrix}.$$

Therefore, to obtain a bound on the norm $\|X\|$ of a non-symmetric random matrix, it suffices to apply the corresponding result for symmetric random matrices to the doubled matrix $\tilde{X}$. For this reason, it is not really necessary to treat non-symmetric matrices separately, and we will conveniently restrict our attention to symmetric matrices throughout this chapter without any loss of generality.

Remark 2.5 A variant on the moment method is to use the bounds

$$e^{t\lambda_{\max}(X)} \leq \mathrm{Tr}[e^{tX}] \leq n e^{t\lambda_{\max}(X)},$$

which gives rise to the so-called matrix concentration inequalities. This approach has become popular in recent years (particularly in the applied mathematics literature) as it provides easy proofs for a number of useful inequalities. Matrix concentration bounds are often stated in terms of tail probabilities $\mathbf{P}[\lambda_{\max}(X) > t]$, and therefore appear at first sight to provide more information than expected norm bounds. This is not the case, however: the resulting tail bounds are highly suboptimal, and much sharper inequalities can be obtained by combining expected norm bounds with concentration inequalities [5] or chaining tail bounds [7]. As in the case of classical concentration inequalities, the moment method essentially subsumes the matrix concentration approach and is often more powerful. We therefore do not discuss this approach further, but refer to [28] for a systematic development.

2.2 The Random Process Method

While the moment method introduced in the previous section is very powerful, it has a number of drawbacks. First, while the matrix moments $\mathbf{E}[\mathrm{Tr}[X^{2p}]]$ can typically be computed explicitly, extracting useful information from the resulting expressions is a nontrivial matter that can result in difficult combinatorial problems. Moreover, as discussed in Remark 2.4, in certain cases the moment method *cannot* yield sharp bounds on the expected spectral norm $\mathbf{E}\|X\|$. Finally, the moment method can only yield information on the spectral norm of the matrix; if other operator norms are of interest, this approach is powerless. In this section, we develop an entirely different method that provides a fruitful approach for addressing these issues.

The present method is based on the following trivial fact.

Lemma 2.6 *Let X be an $n \times n$ symmetric matrix. Then*

$$\|X\| = \sup_{v \in B} |\langle v, Xv \rangle|,$$

where B denotes the Euclidean unit ball in $\mathbb{R}^n$.

When X is a symmetric random matrix, we can view $v \mapsto \langle v, Xv \rangle$ as a *random process* that is indexed by the Euclidean unit ball. Thus controlling the expected spectral norm of X is none other than a special instance of the general probabilistic problem of controlling the expected supremum of a random process. There exist a powerful methods for this purpose (see, e.g., [24]) that could potentially be applied in the present setting to generate insight on the structure of random matrices.

Already the simplest possible approach to bounding the suprema of random processes, the ε-net method, has proved to be very useful in the study of basic random matrix models. The idea behind this approach is to approximate the supremum over the unit ball B by the maximum over a finite discretization B_ε of the unit ball, which reduces the problem to computing the maximum of a finite number of random variables (as we did, for example, in the previous section when we computed $\|g\|_\infty$). Let us briefly sketch how this approach works in the following basic example. Let X be the $n \times n$ symmetric random matrix with i.i.d. standard Gaussian entries above the diagonal. Such a matrix is called a *Wigner matrix*.

Then for every vector $v \in B$, the random variable $\langle v, Xv \rangle$ is Gaussian with variance at most 2. Now let B_ε be a finite subset of the unit ball B in $\mathbb{R}^n$ such that every point in B is within distance at most ε from a point in B_ε. Such a set is called an *ε-net*, and should be viewed as a uniform discretization of the unit ball B at the scale ε. Then we can bound, for small ε,[1]

$$\mathbf{E}\|X\| = \mathbf{E}\sup_{v \in B} |\langle v, Xv \rangle| \lesssim \mathbf{E}\sup_{v \in B_\varepsilon} |\langle v, Xv \rangle| \lesssim \sqrt{\log |B_\varepsilon|},$$

where we used that the expected maximum of k Gaussian random variables with variance $\lesssim 1$ is bounded by $\lesssim \sqrt{\log k}$ (we proved this in the previous section using the moment method: note that independence was not needed for the upper bound.) A classical argument shows that the smallest ε-net in B has cardinality of order ε^{-n}, so the above argument yields a bound of order $\mathbf{E}\|X\| \lesssim \sqrt{n}$ for Wigner matrices. It turns out that this bound is in fact sharp in the present setting: Wigner matrices satisfy $\mathbf{E}\|X\| \asymp \sqrt{n}$ (we will prove this more carefully in section 3.2 below).

Variants of the above argument have proved to be very useful in random matrix theory, and we refer to [32] for a systematic development. However, ε-net arguments are usually applied to highly symmetric situations, such as is the case for Wigner matrices (all entries are identically distributed). The problem with the ε-net method is that it is sharp essentially only in this situation: this method cannot incorporate nontrivial structure. To illustrate this, consider the following typical structured example. Fix a certain sparsity pattern of the matrix X at the outset (that is, choose a subset of the entries that will be forced to zero), and choose the remaining entries to be independent standard Gaussians. In this case, a "good" discretization of the problem cannot simply distribute points uniformly over the unit ball B, but rather must take into account the geometry of the given sparsity pattern. Unfortunately, it is entirely unclear how this is to be accomplished in general. For this reason, ε-net methods have proved to be of limited use for *structured* random matrices, and they will play essentially no role in the remainder of this chapter.

Remark 2.7 Deep results from the theory of Gaussian processes [24] guarantee that the expected supremum of any Gaussian process and of many other random processes can be captured sharply by a sophisticated multiscale counterpart of the ε-net method called the generic chaining. Therefore, in principle, it should be possible to capture precisely the norm of structured random matrices if one is able to construct a near-optimal multiscale net. Unfortunately, the general theory only guarantees the existence of such a net, and provides essentially no mechanism to construct one in any given situation. From this perspective, structured random matrices provide a particularly interesting case study of inhomogeneous random processes whose investigation could shed new light on these more general mechanisms (this perspective provided strong motivation for this author's interest in

[1]The first inequality follows by noting that for every $v \in B$, choosing $\tilde{v} \in B_\varepsilon$ such that $\|v - \tilde{v}\| \leq \varepsilon$, we have $|\langle v, Xv \rangle| = |\langle \tilde{v}, X\tilde{v} \rangle + \langle v - \tilde{v}, X(v + \tilde{v}) \rangle| \leq |\langle \tilde{v}, X\tilde{v} \rangle| + 2\varepsilon\|X\|$.

random matrices). At present, however, progress along these lines remains in a very primitive state. Note that even the most trivial of examples from the random matrix perspective, such as the case where X is a diagonal matrix with i.i.d. Gaussian entries on the diagonal, require already a delicate multiscale net to obtain sharp results; see, e.g., [30].

As direct control of the random processes that arise from structured random matrices is largely intractable, a different approach is needed. To this end, the key idea that we will exploit is the use of *comparison theorems* to bound the expected supremum of one random process by that of another random process. The basic idea is to design a suitable comparison process that dominates the random process of Lemma 2.6 but that is easier to control. For this approach to be successful, the comparison process must capture the structure of the original process while at the same time being amenable to some form of explicit computation. In principle there is no reason to expect that this is ever possible. Nonetheless, we will repeatedly apply different variations on this approach to obtain the best known bounds on structured random matrices. Comparison methods are a recurring theme throughout this chapter, and we postpone further discussion to the following sections.

Let us note that the random process method is easily extended also to non-symmetric matrices: if X is an $n \times m$ rectangular matrix, we have

$$\|X\| = \sup_{v,w \in B} \langle v, Xw \rangle.$$

Alternatively, we can use the same symmetrization trick as was illustrated in the previous section to reduce to the symmetric case. For this reason, we will restrict attention to symmetric matrices in the sequel. Let us also note, however, that unlike the moment method, the present approach extends readily to other operator norms by replacing the Euclidean unit ball B by the unit ball for other norms. In this sense, the random process method is substantially more general than the moment method, which is restricted to the spectral norm. However, the spectral norm is often the most interesting norm in practice in applications of random matrix theory.

2.3 *Roots and Poles*

The moment method and random process method discussed in the previous sections have proved to be by far the most useful approaches to bounding the spectral norms of random matrices, and all results in this chapter will be based on one or both of these methods. We want to briefly mention a third approach, however, that has recently proved to be useful. It is well known from linear algebra that the eigenvalues of a symmetric matrix X are the roots of the characteristic polynomial

$$\chi(t) = \det(tI - X),$$

or, equivalently, the poles of the Stieltjes transform

$$s(t) := \mathrm{Tr}[(tI - X)^{-1}] = \frac{d}{dt} \log \chi(t).$$

One could therefore attempt to bound the extreme eigenvalues of X (and therefore the spectral norm $\|X\|$) by controlling the location of the largest root (pole) of the characteristic polynomial (Stieltjes transform) of X, with high probability.

The Stieltjes transform method plays a major role in random matrix theory [2], as it provides perhaps the simplest route to proving limit theorems for the spectral distributions of random matrices. It is possible along these lines to prove asymptotic results on the extreme eigenvalues, see [3] for example. However, as the Stieltjes transform is highly nonlinear, it seems to be very difficult to use this approach to address nonasymptotic questions for structured random matrices where explicit limit information is meaningless. The characteristic polynomial appears at first sight to be more promising, as this is a polynomial in the matrix entries: one can therefore hope to compute $\mathbf{E}\chi$ exactly. This simplicity is deceptive, however, as there is no reason to expect that $\mathrm{maxroot}(\mathbf{E}\chi)$ has any relation to the quantity $\mathbf{E}\,\mathrm{maxroot}(\chi)$ that we are interested in. It was therefore long believed that such an approach does not provide any useful tool in random matrix theory. Nonetheless, a deterministic version of this idea plays the crucial role in the recent breakthrough resolution of the Kadison-Singer conjecture [15], so that it is conceivable that such an approach could prove to be fruitful in problems of random matrix theory (cf. [23] where related ideas were applied to Stieltjes transforms in a random matrix problem). To date, however, these methods have not been successfully applied to the problems investigated in this chapter, and they will make no further appearance in the sequel.

3 Khintchine-Type Inequalities

The main aim of this section is to introduce a very general method for bounding the spectral norm of structured random matrices. The basic idea, due to Lust-Piquard [13], is to prove an analog of the classical Khintchine inequality for scalar random variables in the noncommutative setting. This *noncommutative Khintchine inequality* allows us to bound the moments of structured random matrices, which immediately results in a bound on the spectral norm by Lemma 2.2.

The advantage of the noncommutative Khintchine inequality is that it can be applied in a remarkably general setting: it does not even require independence of the matrix entries. The downside of this inequality is that it almost always gives rise to bounds on the spectral norm that are suboptimal by a multiplicative factor that is logarithmic in the dimension (cf. section 4.2). We will discuss the origin of this suboptimality and some potential methods for reducing it in the general setting of this section. Much sharper bounds will be obtained in section 4 under the additional restriction that the matrix entries are independent.

For simplicity, we will restrict our attention to matrices with Gaussian entries, though extensions to other distributions are easily obtained (for example, see [14]).

3.1 The Noncommutative Khintchine Inequality

In this section, we will consider the following very general setting. Let X be an $n \times n$ symmetric random matrix with zero mean. The only assumption we make on the distribution is that the entries on and above the diagonal (that is, those entries that are not fixed by symmetry) are centered and jointly Gaussian. In particular, these entries can possess an arbitrary covariance matrix, and are assumed to be neither identically distributed nor independent. Our aim is to bound the spectral norm $\|X\|$ in terms of the given covariance structure of the matrix.

It proves to be convenient to reformulate our random matrix model somewhat. Let $A_1, \ldots, A_s$ be *nonrandom* $n \times n$ symmetric matrices, and let $g_1, \ldots, g_s$ be independent standard Gaussian variables. Then we define the matrix X as

$$X = \sum_{k=1}^{s} g_k A_k.$$

Clearly X is a symmetric matrix with jointly Gaussian entries. Conversely, the reader will convince herself after a moment's reflection that any symmetric matrix with centered and jointly Gaussian entries can be written in the above form for some choice of $s \leq n(n+1)/2$ and $A_1, \ldots, A_s$. There is therefore no loss of generality in considering the present formulation (we will reformulate our ultimate bounds in a way that does not depend on the choice of the coefficient matrices A_k).

Our intention is to apply the moment method. To this end, we must obtain bounds on the moments $\mathbf{E}[\mathrm{Tr}[X^{2p}]]$ of the matrix X. It is instructive to begin by considering the simplest possible case where the dimension $n = 1$. In this case, X is simply a scalar Gaussian random variable with zero mean and variance $\sum_k A_k^2$, and the problem in this case reduces to bounding the moments of a scalar Gaussian variable.

Lemma 3.1 *Let* $g \sim N(0, 1)$. *Then* $\mathbf{E}[g^{2p}]^{1/2p} \leq \sqrt{2p-1}$.

Proof We use the following fundamental *gaussian integration by parts* property:

$$\mathbf{E}[gf(g)] = \mathbf{E}[f'(g)].$$

To prove it, simply note that integration by parts yields

$$\int_{-\infty}^{\infty} xf(x)\, \frac{e^{-x^2/2}}{\sqrt{2\pi}}\, dx = \int_{-\infty}^{\infty} \frac{df(x)}{dx}\, \frac{e^{-x^2/2}}{\sqrt{2\pi}}\, dx$$

for smooth functions f with compact support, and the conclusion is readily extended by approximation to any C^1 function for which the formula makes sense.

We now apply the integration by parts formula to $f(x) = x^{2p-1}$ as follows:

$$\mathbf{E}[g^{2p}] = \mathbf{E}[g \cdot g^{2p-1}] = (2p-1)\mathbf{E}[g^{2p-2}] \le (2p-1)\mathbf{E}[g^{2p}]^{1-1/p},$$

where the last inequality is by Jensen. Rearranging yields the conclusion. □

Applying Lemma 3.1 yields immediately that

$$\mathbf{E}[X^{2p}]^{1/2p} \le \sqrt{2p-1}\left[\sum_{k=1}^{s} A_k^2\right]^{1/2} \qquad \text{when } n = 1.$$

It was realized by Lust-Piquard [13] that the analogous inequality holds in any dimension n (the correct dependence of the bound on p was obtained later, cf. [17]).

Theorem 3.2 (Noncommutative Khintchine inequality) *In the present setting*

$$\mathbf{E}[\mathrm{Tr}[X^{2p}]]^{1/2p} \le \sqrt{2p-1}\,\mathrm{Tr}\left[\left(\sum_{k=1}^{s} A_k^2\right)^p\right]^{1/2p}.$$

By combining this bound with Lemma 2.2, we immediately obtain the following conclusion regarding the spectral norm of the matrix X.

Corollary 3.3 *In the setting of this section,*

$$\mathbf{E}\|X\| \lesssim \sqrt{\log n}\left\|\sum_{k=1}^{s} A_k^2\right\|^{1/2}.$$

This bound is expressed directly in terms of the coefficient matrices A_k that determine the structure of X, and has proved to be extremely useful in applications of random matrix theory in functional analysis and applied mathematics. To what extent this bound is sharp will be discussed in the next section.

Remark 3.4 Recall that our bounds apply to any symmetric matrix X with centered and jointly Gaussian entries. Our bounds should therefore not depend on the choice of representation in terms of the coefficient matrices A_k, which is not unique. It is easily verified that this is the case. Indeed, it suffices to note that

$$\mathbf{E}X^2 = \sum_{k=1}^{s} A_k^2,$$

so that we can express the conclusion of Theorem 3.2 and Corollary 3.3 as

$$\mathbf{E}[\mathrm{Tr}[X^{2p}]]^{1/2p} \lesssim \sqrt{p}\,\mathrm{Tr}[(\mathbf{E}X^2)^p]^{1/2p}, \qquad \mathbf{E}\|X\| \lesssim \sqrt{\log n}\,\|\mathbf{E}X^2\|^{1/2}$$

without reference to the coefficient matrices A_k. We note that the quantity $\|\mathbf{E}X^2\|$ has a natural interpretation: it measures the size of the matrix X "on average" (as the expectation in this quantity is *inside* the spectral norm).

We now turn to the proof of Theorem 3.2. We begin by noting that the proof follows immediately from Lemma 3.1 not just when $n = 1$, but also in any dimension n under the additional assumption that the matrices $A_1, \dots, A_s$ commute. Indeed, in this case we can work without loss of generality in a basis in which all the matrices A_k are simultaneously diagonal, and the result follows by applying Lemma 3.1 to every diagonal entry of X. The crucial idea behind the proof of Theorem 3.2 is that *the commutative case is in fact the worst case situation*! This idea will appear very explicitly in the proof: we will simply repeat the proof of Lemma 3.1, and the result will follow by showing that we can permute the order of the matrices A_k at the pivotal point in the proof. (The simple proof given here follows [29].)

Proof (Proof of Theorem 3.2) As in the proof of Lemma 3.1, we obtain

$$\begin{aligned}\mathbf{E}[\mathrm{Tr}[X^{2p}]] &= \mathbf{E}[\mathrm{Tr}[X \cdot X^{2p-1}]] \\ &= \sum_{k=1}^{s} \mathbf{E}[g_k \mathrm{Tr}[A_k X^{2p-1}]] \\ &= \sum_{\ell=0}^{2p-2} \sum_{k=1}^{s} \mathbf{E}[\mathrm{Tr}[A_k X^{\ell} A_k X^{2p-2-\ell}]]\end{aligned}$$

using Gaussian integration by parts. The crucial step in the proof is the observation that permuting A_k and X^ℓ inside the trace can only increase the bound.

Lemma 3.5 $\mathrm{Tr}[A_k X^\ell A_k X^{2p-2-\ell}] \le \mathrm{Tr}[A_k^2 X^{2p-2}]$.

Proof Let us write X in terms of its eigendecomposition $X = \sum_{i=1}^n \lambda_i v_i v_i^*$, where λ_i and v_i denote the eigenvalues and eigenvectors of X. Then we can write

$$\mathrm{Tr}[A_k X^\ell A_k X^{2p-2-\ell}] = \sum_{i,j=1}^{n} \lambda_i^\ell \lambda_j^{2p-2-\ell} |\langle v_i, A_k v_j\rangle|^2 \le \sum_{i,j=1}^{n} |\lambda_i|^\ell |\lambda_j|^{2p-2-\ell} |\langle v_i, A_k v_j\rangle|^2.$$

But note that the right-hand side is a convex function of ℓ, so that its maximum in the interval $[0, 2p-2]$ is attained either at $\ell = 0$ or $\ell = 2p-2$. This yields

$$\mathrm{Tr}[A_k X^\ell A_k X^{2p-2-\ell}] \le \sum_{i,j=1}^{n} |\lambda_j|^{2p-2} |\langle v_i, A_k v_j\rangle|^2 = \mathrm{Tr}[A_k^2 X^{2p-2}],$$

and the proof is complete. □

We now complete the proof of the noncommutative Khintchine inequality. Substituting Lemma 3.5 into the previous inequality yields

$$\begin{aligned}\mathbf{E}[\mathrm{Tr}[X^{2p}]] &\leq (2p-1)\sum_{k=1}^{s}\mathbf{E}[\mathrm{Tr}[A_k^2X^{2p-2}]] \\ &\leq (2p-1)\,\mathrm{Tr}\left[\left(\sum_{k=1}^{s}A_k^2\right)^p\right]^{1/p}\mathbf{E}[\mathrm{Tr}[X^{2p}]]^{1-1/p},\end{aligned}$$

where we used Hölder's inequality $\mathrm{Tr}[YZ] \leq \mathrm{Tr}[|Y|^p]^{1/p}\mathrm{Tr}[|Z|^{p/(p-1)}]^{1-1/p}$ in the last step. Rearranging this expression yields the desired conclusion. □

Remark 3.6 The proof of Corollary 3.3 given here, using the moment method, is exceedingly simple. However, by its nature, it can only bound the spectral norm of the matrix, and would be useless if we wanted to bound other operator norms. It is worth noting that an alternative proof of Corollary 3.3 was developed by Rudelson, using deep random process machinery described in Remark 2.7, for the special case where the matrices A_k are all of rank one (see [24, Prop. 16.7.4] for an exposition of this proof). The advantage of this approach is that it extends to some other operator norms, which proves to be useful in Banach space theory. It is remarkable, however, that no random process proof of Corollary 3.3 is known to date in the general setting.

3.2 How Sharp Are Khintchine Inequalities?

Corollary 3.3 provides a very convenient bound on the spectral norm $\|X\|$: it is expressed directly in terms of the coefficients A_k that define the structure of the matrix X. However, is this structure captured *correctly*? To understand the degree to which Corollary 3.3 is sharp, let us augment it with a lower bound.

Lemma 3.7 *Let $X = \sum_{k=1}^{s} g_k A_k$ as in the previous section. Then*

$$\left\|\sum_{k=1}^{s}A_k^2\right\|^{1/2} \lesssim \mathbf{E}\|X\| \lesssim \sqrt{\log n}\left\|\sum_{k=1}^{s}A_k^2\right\|^{1/2}.$$

That is, the noncommutative Khintchine bound is sharp up to a logarithmic factor.

Proof The upper bound in Corollary 3.3, and it remains to prove the lower bound. A slightly simpler bound is immediate by Jensen's inequality: we have

$$\mathbf{E}\|X\|^2 \geq \|\mathbf{E}X^2\| = \left\|\sum_{k=1}^{s}A_k^2\right\|.$$

It therefore remains to show that $(\mathbf{E}\|X\|)^2 \gtrsim \mathbf{E}\|X\|^2$, or, equivalently, that $\mathrm{Var}\|X\| \lesssim (\mathbf{E}\|X\|)^2$. To bound the fluctuations of the spectral norm, we recall an important property of Gaussian random variables (see, for example, [16]).

Lemma 3.8 (Gaussian concentration) *Let g be a standard Gaussian vector in $\mathbb{R}^n$, let $f : \mathbb{R}^n \to \mathbb{R}$ be smooth, and let $p \geq 1$. Then*

$$[\mathbf{E}(f(g) - \mathbf{E}f(g))^p]^{1/p} \lesssim \sqrt{p}\,[\mathbf{E}\|\nabla f(g)\|^p]^{1/p}.$$

Proof Let g' be an independent copy of g, and define $g(\varphi) = g\sin\varphi + g'\cos\varphi$. Then

$$f(g) - f(g') = \int_0^{\pi/2} \frac{d}{d\varphi} f(g(\varphi))\, d\varphi = \int_0^{\pi/2} \langle g'(\varphi), \nabla f(g(\varphi))\rangle\, d\varphi,$$

where $g'(\varphi) = \frac{d}{d\varphi} g(\varphi)$. Applying Jensen's inequality twice gives

$$\mathbf{E}(f(g) - \mathbf{E}f(g))^p \leq \mathbf{E}(f(g) - f(g'))^p \leq \frac{2}{\pi}\int_0^{\pi/2} \mathbf{E}(\tfrac{\pi}{2}\langle g'(\varphi), \nabla f(g(\varphi))\rangle)^p\, d\varphi.$$

Now note that $(g(\varphi), g'(\varphi)) \stackrel{d}{=} (g, g')$ for every φ. We can therefore apply Lemma 3.1 conditionally on $g(\varphi)$ to estimate for every φ

$$[\mathbf{E}\langle g'(\varphi), \nabla f(g(\varphi))\rangle^p]^{1/p} \lesssim \sqrt{p}\,\mathbf{E}[\|\nabla f(g(\varphi))\|^p]^{1/p} = \sqrt{p}\,\mathbf{E}[\|\nabla f(g)\|^p]^{1/p},$$

and substituting into the above expression completes the proof. □

We apply Lemma 3.8 to the function $f(x) = \|\sum_{k=1}^s x_k A_k\|$. Note that

$$\begin{aligned}|f(x) - f(x')| &\leq \left\|\sum_{k=1}^s (x_k - x_k')A_k\right\| = \sup_{v\in B}\left|\sum_{k=1}^s (x_k - x_k')\langle v, A_k v\rangle\right| \\ &\leq \|x - x'\| \sup_{v\in B}\left[\sum_{k=1}^s \langle v, A_k v\rangle^2\right]^{1/2} =: \sigma_*\|x - x'\|.\end{aligned}$$

Thus f is σ_*-Lipschitz, so $\|\nabla f\| \leq \sigma_*$, and Lemma 3.8 yields $\mathrm{Var}\|X\| \lesssim \sigma_*^2$. But as

$$\sigma_* = \sqrt{\frac{\pi}{2}}\sup_{v\in B}\mathbf{E}\left|\sum_{k=1}^s g_k\langle v, A_k v\rangle\right| \leq \sqrt{\frac{\pi}{2}}\mathbf{E}\|X\|,$$

we have $\mathrm{Var}\|X\| \lesssim (\mathbf{E}\|X\|)^2$, and the proof is complete. □

Lemma 3.7 shows that the structural quantity $\sigma := \|\sum_{k=1}^s A_k^2\|^{1/2} = \|\mathbf{E}X^2\|^{1/2}$ that appears in the noncommutative Khintchine inequality is very natural: the expected spectral norm $\mathbf{E}\|X\|$ is controlled by σ up to a logarithmic factor in the dimension. It is not at all clear, *a priori*, whether the upper or lower bound in Lemma 3.7 is sharp. It turns out that either the upper bound or the lower bound may be sharp in different situations. Let us illustrate this in two extreme examples.

Example 3.9 (Diagonal matrix) Consider the case where X is a diagonal matrix

$$X = \begin{bmatrix} g_1 & & & \\ & g_2 & & \\ & & \ddots & \\ & & & g_n \end{bmatrix}$$

with i.i.d. standard Gaussian entries on the diagonal. In this case,

$$\mathbf{E}\|X\| = \mathbf{E}\|g\|_\infty \asymp \sqrt{\log n}.$$

On the other hand, we clearly have

$$\sigma = \|\mathbf{E}X^2\|^{1/2} = 1,$$

so the upper bound in Lemma 3.7 is sharp. This shows that the logarithmic factor in the noncommutative Khintchine inequality *cannot* be removed.

Example 3.10 (Wigner matrix) Let X be a symmetric matrix

$$X = \begin{bmatrix} g_{11} & g_{12} & \cdots & g_{1n} \\ g_{12} & g_{22} & & g_{2n} \\ \vdots & & \ddots & \vdots \\ g_{1n} & g_{2n} & \cdots & g_{nn} \end{bmatrix}$$

with i.i.d. standard Gaussian entries on and above the diagonal. In this case

$$\sigma = \|\mathbf{E}X^2\|^{1/2} = \sqrt{n}.$$

Thus Lemma 3.7 yields the bounds

$$\sqrt{n} \lesssim \mathbf{E}\|X\| \lesssim \sqrt{n \log n}.$$

Which bound is sharp? A hint can be obtained from what is perhaps the most classical result in random matrix theory: the empirical spectral distribution of the matrix $n^{-1/2}X$ (that is, the random probability measure on $\mathbb{R}$ that places a point mass on every eigenvalue of $n^{-1/2}X$) converges weakly to the Wigner semicircle distribution $\frac{1}{2\pi}\sqrt{(4-x^2)_+}\,dx$ [2, 25]. Therefore, when the dimension n is large, the eigenvalues of X are approximately distributed according to the following density:

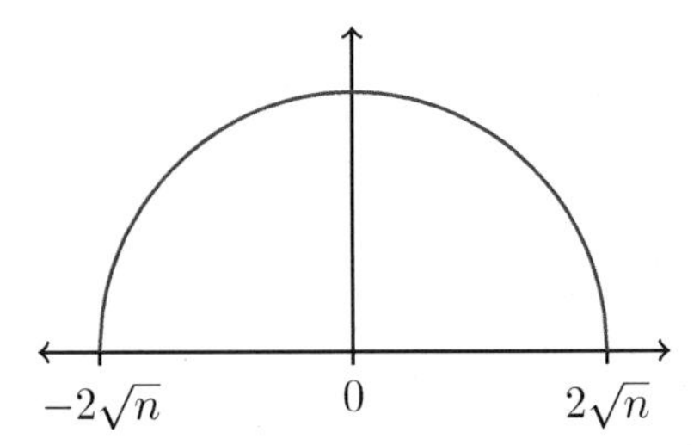

This picture strongly suggests that the spectrum of X is supported at least approximately in the interval $[-2\sqrt{n}, 2\sqrt{n}]$, which implies that $\|X\| \asymp \sqrt{n}$.

Lemma 3.11 *For the Wigner matrix of Example 3.10,* $\mathbf{E}\|X\| \asymp \sqrt{n}$.

Thus we see that in the present example it is the *lower* bound in Lemma 3.7 that is sharp, while the upper bound obtained from the noncommutative Khintchine inequality fails to capture correctly the structure of the problem.

We already sketched a proof of Lemma 3.11 using ε-nets in section 2.2. We take the opportunity now to present another proof, due to Chevet [6] and Gordon [9], that provides a first illustration of the *comparison methods* that will play an important role in the rest of this chapter. To this end, we first prove a classical comparison theorem for Gaussian processes due to Slepian and Fernique (see, e.g., [5]).

Lemma 3.12 (Slepian-Fernique inequality) *Let $Y \sim N(0, \Sigma^Y)$ and $Z \sim N(0, \Sigma^Z)$ be centered Gaussian vectors in $\mathbb{R}^n$. Suppose that*

$$\mathbf{E}(Y_i - Y_j)^2 \le \mathbf{E}(Z_i - Z_j)^2 \quad \textit{for all } 1 \le i, j \le n.$$

Then

$$\mathbf{E}\max_{i \le n} Y_i \le \mathbf{E}\max_{i \le n} Z_i.$$

Proof Let g, g' be independent standard Gaussian vectors. We can assume that $Y = (\Sigma^Y)^{1/2} g$ and $Z = (\Sigma^Z)^{1/2} g'$. Let $Y(t) = \sqrt{t}Z + \sqrt{1-t}Y$ for $t \in [0, 1]$. Then

$$\begin{aligned}
\frac{d}{dt}\mathbf{E}[f(Y(t))] &= \frac{1}{2}\mathbf{E}\left[\left\langle \nabla f(Y(t)), \frac{Z}{\sqrt{t}} - \frac{Y}{\sqrt{1-t}} \right\rangle\right] \\
&= \frac{1}{2}\mathbf{E}\left[\frac{1}{\sqrt{t}}\left\langle (\Sigma^Z)^{1/2}\nabla f(Y(t)), g' \right\rangle - \frac{1}{\sqrt{1-t}}\left\langle (\Sigma^Y)^{1/2}\nabla f(Y(t)), g \right\rangle\right] \\
&= \frac{1}{2}\sum_{i,j=1}^{n} (\Sigma^Z_{ij} - \Sigma^Y_{ij})\,\mathbf{E}\left[\frac{\partial^2 f}{\partial x_i \partial x_j}(Y(t))\right],
\end{aligned}$$

where we used Gaussian integration by parts in the last step. We would really like to apply this identity with $f(x) = \max_i x_i$: if we can show that $\frac{d}{dt}\mathbf{E}[\max_i Y_i(t)] \geq 0$, that would imply $\mathbf{E}[\max_i Z_i] = \mathbf{E}[\max_i Y_i(1)] \geq \mathbf{E}[\max_i Y_i(0)] = \mathbf{E}[\max_i Y_i]$ as desired. The problem is that the function $x \mapsto \max_i x_i$ is not sufficiently smooth: it does not possess second derivatives. We therefore work with a smooth approximation.

Previously, we used $\|x\|_p$ as a smooth approximation of $\|x\|_\infty$. Unfortunately, it turns out that Slepian-Fernique does *not* hold when $\max_i Y_i$ and $\max_i Z_i$ are replaced by $\|Y\|_\infty$ and $\|Z\|_\infty$, so this cannot work. We must therefore choose instead a *one-sided* approximation. In analogy with Remark 2.5, we choose

$$f_\beta(x) = \frac{1}{\beta}\log\left(\sum_{i=1}^{n} e^{\beta x_i}\right).$$

Clearly $\max_i x_i \leq f_\beta(x) \leq \max_i x_i + \beta^{-1}\log n$, so $f_\beta(x) \to \max_i x_i$ as $\beta \to \infty$. Also

$$\frac{\partial f_\beta}{\partial x_i}(x) = \frac{e^{\beta x_i}}{\sum_j e^{\beta x_j}} =: p_i(x), \qquad \frac{\partial^2 f_\beta}{\partial x_i \partial x_j}(x) = \beta\{\delta_{ij}p_i(x) - p_i(x)p_j(x)\},$$

where we note that $p_i(x) \geq 0$ and $\sum_i p_i(x) = 1$. The reader should check that

$$\frac{d}{dt}\mathbf{E}[f_\beta(Y(t))] = \frac{\beta}{4}\sum_{i\neq j}\{\mathbf{E}(Z_i - Z_j)^2 - \mathbf{E}(Y_i - Y_j)^2\}\,\mathbf{E}[p_i(Y(t))p_j(Y(t))],$$

which follows by rearranging the terms in the above expressions. The right-hand side is nonnegative by assumption, and thus the proof is easily completed. □

We can now prove Lemma 3.11.

Proof (Proof of Lemma 3.11) That $\mathbf{E}\|X\| \gtrsim \sqrt{n}$ follows from Lemma 3.7, so it remains to prove $\mathbf{E}\|X\| \lesssim \sqrt{n}$. To this end, define $X_v := \langle v, Xv\rangle$ and $Y_v = 2\langle v, g\rangle$, where g is a standard Gaussian vector. Then we can estimate

$$\mathbf{E}(X_v - X_w)^2 \leq 2\sum_{i,j=1}^{n}(v_iv_j - w_iw_j)^2 \leq 4\|v - w\|^2 = \mathbf{E}(Y_v - Y_w)^2$$

when $\|v\| = \|w\| = 1$, where we used $1-\langle v, w\rangle^2 \leq 2(1-\langle v, w\rangle)$ when $|\langle v, w\rangle| \leq 1$. It follows form the Slepian-Fernique lemma that we have

$$\mathbf{E}\lambda_{\max}(X) = \mathbf{E}\sup_{\|v\|=1}\langle v, Xv\rangle \leq 2\,\mathbf{E}\sup_{\|v\|=1}\langle v, g\rangle = 2\,\mathbf{E}\|g\| \leq 2\sqrt{n}.$$

But as X and $-X$ have the same distribution, so do the random variables $\lambda_{\max}(X)$ and $-\lambda_{\min}(X) = \lambda_{\max}(-X)$. We can therefore estimate

$$\begin{aligned}\mathbf{E}\|X\| &= \mathbf{E}(\lambda_{\max}(X) \vee -\lambda_{\min}(X)) \le \mathbf{E}\lambda_{\max}(X) + 2\,\mathbf{E}|\lambda_{\max}(X) - \mathbf{E}\lambda_{\max}(X)| \\ &= 2\sqrt{n} + O(1),\end{aligned}$$

where we used that $\mathrm{Var}(\lambda_{\max}(X)) = O(1)$ by Lemma 3.8. □

We have seen above two extreme examples: diagonal matrices and Wigner matrices. In the diagonal case, the noncommutative Khintchine inequality is sharp, while the lower bound in Lemma 3.7 is suboptimal. On the other hand, for Wigner matrices, the noncommutative Khintchine inequality is suboptimal, while the lower bound in Lemma 3.7 is sharp. We therefore see that while the structural parameter $\sigma = \|\mathbf{E}X^2\|^{1/2}$ that appears in the noncommutative Khintchine inequality always crudely controls the spectral norm up to a logarithmic factor in the dimension, it fails to capture correctly the structure of the problem and cannot in general yield sharp bounds. The aim of the rest of this chapter is to develop a deeper understanding of the norms of structured random matrices that goes beyond Lemma 3.7.

3.3 A Second-Order Khintchine Inequality

Having established that the noncommutative Khintchine inequality falls short of capturing the full structure of our random matrix model, we naturally aim to understand where things went wrong. The culprit is easy to identify. The main idea behind the proof of the noncommutative Khintchine inequality is that the case where the matrices A_k commute is the worst possible, as is made precise by Lemma 3.5. However, when the matrices A_k do not commute, the behavior of the spectral norm can be *strictly better* than is predicted by the noncommutative Khintchine inequality. The crucial shortcoming of the noncommutative Khintchine inequality is that it provides no mechanism to capture the effect of noncommutativity.

Remark 3.13 This intuition is clearly visible in the examples of the previous section: the diagonal example corresponds to choosing coefficient matrices A_k of the form $e_ie_i^*$ for $1 \le i \le n$, while to obtain a Wigner matrix we add additional coefficient matrices A_k of the form $e_ie_j^* + e_je_i^*$ for $1 \le i < j \le n$ (here $e_1, \ldots, e_n$ denotes the standard basis in $\mathbb{R}^n$). Clearly the matrices A_k commute in the diagonal example, in which case noncommutative Khintchine is sharp, but they do not commute for the Wigner matrix, in which case noncommutative Khintchine is suboptimal.

The present insight suggests that a good bound on the spectral norm of random matrices of the form $X = \sum_{k=1}^{s} g_kA_k$ should somehow take into account the algebraic structure of the coefficient matrices A_k. Unfortunately, it is not at all clear how this is to be accomplished. In this section we develop an interesting result in this spirit due to Tropp [29]. While this result is still very far from being sharp, the proof contains some interesting ideas, and provides at present the only known approach to improve on the noncommutative Khintchine inequality in the most general setting.

The intuition behind the result of Tropp is that the commutation inequality

$$\mathbf{E}[\mathrm{Tr}[A_k X^\ell A_k X^{2p-2-\ell}]] \le \mathbf{E}[\mathrm{Tr}[A_k^2 X^{2p-2}]]$$

of Lemma 3.5, which captures the idea that the commutative case is the worst case, should incur significant loss when the matrices A_k do not commute. Therefore, rather than apply this inequality directly, we should try to go to second order by integrating again by parts. For example, for the term $\ell = 1$, we could write

$$\begin{aligned}\mathbf{E}[\mathrm{Tr}[A_k X A_k X^{2p-3}]] &= \sum_{l=1}^{s} \mathbf{E}[g_l \mathrm{Tr}[A_k A_l A_k X^{2p-3}]] \\ &= \sum_{l=1}^{s} \sum_{m=0}^{2p-4} \mathbf{E}[\mathrm{Tr}[A_k A_l A_k X^m A_l X^{2p-4-m}]].\end{aligned}$$

If we could again permute the order of A_l and X^m on the right-hand side, we would obtain control of these terms not by the structural parameter

$$\sigma = \left\| \sum_{k=1}^{s} A_k^2 \right\|^{1/2}$$

that appears in the noncommutative Khintchine inequality, but rather by the second-order "noncommutative" structural parameter

$$\left\| \sum_{k,l=1}^{s} A_k A_l A_k A_l \right\|^{1/4}.$$

Of course, when the matrices A_k commute, the latter parameter is equal to σ and we recover the noncommutative Khintchine inequality; but when the matrices A_k do not commute, it can be the case that this parameter is much smaller than σ. This back-of-the-envelope computation suggests that we might indeed hope to capture noncommutativity to some extent through the present approach.

In essence, this is precisely how we will proceed. However, there is a technical issue: the convexity that was exploited in the proof of Lemma 3.5 is no longer present in the second-order terms. We therefore cannot naively exchange A_l and X^m as suggested above, and the parameter $\| \sum_{k,l=1}^{s} A_k A_l A_k A_l \|^{1/4}$ is in fact too small to yield any meaningful bound (as is illustrated by a counterexample in [29]). The key idea in [29] is that a classical complex analysis argument [18, Appendix IX.4] can be exploited to force convexity, at the expense of a larger second-order term.

Theorem 3.14 (Tropp) *Let $X = \sum_{k=1}^{s} g_k A_k$ as in the previous section. Define*

$$\sigma := \left\| \sum_{k=1}^{s} A_k^2 \right\|^{1/2}, \qquad \tilde{\sigma} := \sup_{U_1,U_2,U_3} \left\| \sum_{k,l=1}^{s} A_k U_1 A_l U_2 A_k U_3 A_l \right\|^{1/4},$$

where the supremum is taken over all triples U_1, U_2, U_3 of commuting unitary matrices.[2] *Then we have a second-order noncommutative Khintchine inequality*

$$\mathbf{E}\|X\| \lesssim \sigma \log^{1/4} n + \tilde{\sigma} \log^{1/2} n.$$

Due to the (necessary) presence of the unitaries, the second-order parameter $\tilde{\sigma}$ is not so easy to compute. It is verified in [29] that $\tilde{\sigma} \le \sigma$ (so that Theorem 3.14 is no worse than the noncommutative Khintchine inequality), and that $\tilde{\sigma} = \sigma$ when the matrices A_k commute. On the other hand, an explicit computation in [29] shows that if X is a Wigner matrix as in Example 3.10, we have $\sigma \asymp \sqrt{n}$ and $\tilde{\sigma} \asymp n^{1/4}$. Thus Theorem 3.14 yields in this case $\mathbf{E}\|X\| \lesssim \sqrt{n}(\log n)^{1/4}$, which is strictly better than the noncommutative Khintchine bound $\mathbf{E}\|X\| \lesssim \sqrt{n}(\log n)^{1/2}$ but falls short of the sharp bound $\mathbf{E}\|X\| \asymp \sqrt{n}$. We therefore see that Theorem 3.14 does indeed improve, albeit ever so slightly, on the noncommutative Khintchine bound. The real interest of Theorem 3.14 is however the very general setting in which it holds, and that it does capture explicitly the noncommutativity of the coefficient matrices A_k. In section 4, we will see that much sharper bounds can be obtained if we specialize to random matrices with independent entries. While this is perhaps the most interesting setting in practice, it will require us to depart from the much more general setting provided by the Khintchine-type inequalities that we have seen so far.

The remainder of this section is devoted to the proof of Theorem 3.14. The proof follows essentially along the lines already indicated: we follow the proof of the noncommutative Khintchine inequality and integrate by parts a second time. The new idea in the proof is to understand how to appropriately extend Lemma 3.5.

Proof (Proof of Theorem 3.14) We begin as in the proof of Theorem 3.2 by writing

$$\mathbf{E}[\mathrm{Tr}[X^{2p}]] = \sum_{\ell=0}^{2p-2} \sum_{k=1}^{s} \mathbf{E}[\mathrm{Tr}[A_k X^{\ell} A_k X^{2p-2-\ell}]].$$

Let us investigate each of the terms inside the first sum.

Case $\ell = 0, 2p-2$. In this case there is little to do: we can estimate

$$\sum_{k=1}^{s} \mathbf{E}[\mathrm{Tr}[A_k^2 X^{2p-2}]] \le \mathrm{Tr}\left[\left(\sum_{k=1}^{s} A_k^2\right)^{p}\right]^{1/p} \mathbf{E}[\mathrm{Tr}[X^{2p}]]^{1-1/p}$$

precisely as in the proof of Theorem 3.2.

[2]For reasons that will become evident in the proof, it is essential to consider (complex) unitary matrices U_1, U_2, U_3, despite that all the matrices A_k and X are assumed to be real.

Case $\ell = 1, 2p-3$. This is the first point at which something interesting happens. Integrating by parts a second time as was discussed before Theorem 3.14, we obtain

$$\sum_{k=1}^{s} \mathbf{E}[\mathrm{Tr}[A_k X A_k X^{2p-3}]] = \sum_{m=0}^{2p-4} \sum_{k,l=1}^{s} \mathbf{E}[\mathrm{Tr}[A_k A_l A_k X^m A_l X^{2p-4-m}]].$$

The challenge we now face is to prove the appropriate analogue of Lemma 3.5.

Lemma 3.15 *There exist unitary matrices U_1, U_2 (dependent on X and m) such that*

$$\sum_{k,l=1}^{s} \mathrm{Tr}[A_k A_l A_k X^m A_l X^{2p-4-m}] \le \left| \sum_{k,l=1}^{s} \mathrm{Tr}[A_k A_l A_k U_1 A_l U_2 X^{2p-4}] \right|.$$

Remark 3.16 Let us start the proof as in Lemma 3.5 and see where things go wrong. In terms of the eigendecomposition $X = \sum_{i=1}^{n} \lambda_i v_i v_i^*$, we can write

$$\sum_{k,l=1}^{s} \mathrm{Tr}[A_k A_l A_k X^m A_l X^{2p-4-m}] = \sum_{k,l=1}^{s} \sum_{i,j=1}^{n} \lambda_i^m \lambda_j^{2p-4-m} \langle v_j, A_k A_l A_k v_i \rangle \langle v_i, A_l v_j \rangle.$$

Unfortunately, unlike in the analogous expression in the proof of Lemma 3.5, the coefficients $\langle v_j, A_k A_l A_k v_i \rangle \langle v_i, A_l v_j \rangle$ can have arbitrary sign. Therefore, we cannot easily force convexity of the above expression as a function of m as we did in Lemma 3.5: if we replace the terms in the sum by their absolute values, we will no longer be able to interpret the resulting expression as a linear algebraic object (a trace).

However, the above expression is still an *analytic* function in the complex plane $\mathbb{C}$. The idea that we will exploit is that analytic functions have some hidden convexity built in, as we recall here without proof (cf. [18, p. 33]).

Lemma 3.17 (Hadamard three line lemma) *If $\varphi : \mathbb{C} \to \mathbb{C}$ is analytic, the function $t \mapsto \sup_{s \in \mathbb{R}} \log |\varphi(t + is)|$ is convex on the real line (provided it is finite).*

Proof (Proof of Lemma 3.15) We can assume that X is nonsingular; otherwise we may replace X by $X + \varepsilon$ and let $\varepsilon \downarrow 0$ at the end of the proof. Write $X = V|X|$ according to its polar decomposition, and note that as X is self-adjoint, $V = \mathrm{sign}(X)$ commutes with X and therefore $X^m = V^m |X|^m$. Define

$$\varphi(z) := \sum_{k,l=1}^{s} \mathrm{Tr}[A_k A_l A_k V^m |X|^{(2p-4)z} A_l V^{2p-4-m} |X|^{(2p-4)(1-z)}].$$

As X is nonsingular, φ is analytic and $\varphi(t + is)$ is a periodic function of s for every t. By the three line lemma, $\sup_{s \in \mathbb{R}} |\varphi(t + is)|$ attains its maximum for $t \in [0, 1]$ at either $t = 0$ or $t = 1$. Moreover, the supremum itself is attained at some $s \in \mathbb{R}$ by periodicity. We have therefore shown that there exists $s \in \mathbb{R}$ such that

$$\left|\sum_{k,l=1}^{s}\mathrm{Tr}[A_kA_lA_kX^mA_lX^{2p-4-m}]\right| = \left|\varphi\left(\frac{m}{2p-4}\right)\right| \leq |\varphi(is)| \vee |\varphi(1+is)|.$$

But, for example,

$$|\varphi(is)| = \left|\sum_{k,l=1}^{s}\mathrm{Tr}[A_kA_lA_kV^m|X|^{is(2p-4)}A_lV^{2p-4-m}|X|^{-is(2p-4)}X^{2p-4}]\right|,$$

so if this term is the larger we can set $U_1 = V^m|X|^{is(2p-4)}$ and $U_2 = V^{2p-4-m}|X|^{-is(2p-4)}$ to obtain the statement of the lemma (clearly U_1 and U_2 are unitary). If the term $|\varphi(1+is)|$ is larger, the claim follows in precisely the identical manner. □

Putting together the above bounds, we obtain

$$\sum_{k=1}^{s}\mathbf{E}[\mathrm{Tr}[A_kXA_kX^{2p-3}]]$$
$$\leq (2p-3)\,\mathbf{E}\left[\sup_{U_1,U_2}\left|\sum_{k,l=1}^{s}\mathrm{Tr}[A_kA_lA_kU_1A_lU_2X^{2p-4}]\right|\right]$$
$$\leq (2p-3)\sup_{U}\mathrm{Tr}\left[\left|\sum_{k,l=1}^{s}A_kA_lA_kUA_l\right|^{p/2}\right]^{2/p}\mathbf{E}[\mathrm{Tr}[X^{2p}]]^{1-2/p}.$$

This term will evidently yield a term of order $\tilde{\sigma}$ when $p \sim \log n$.

Case $2 \leq \ell \leq 2p-4$. These terms are dealt with much in the same way as in the previous case, except the computation is a bit more tedious. As we have come this far, we might as well complete the argument. We begin by noting that

$$\sum_{k=1}^{s}\mathbf{E}[\mathrm{Tr}[A_kX^{\ell}A_kX^{2p-2-\ell}]] \leq \sum_{k=1}^{s}\mathbf{E}[\mathrm{Tr}[A_kX^2A_kX^{2p-4}]]$$

for every $2 \leq \ell \leq 2p-4$. This follows by convexity precisely in the same way as in Lemma 3.5, and we omit the (identical) proof. To proceed, we integrate by parts:

$$\sum_{k=1}^{s}\mathbf{E}[\mathrm{Tr}[A_kX^2A_kX^{2p-4}]] = \sum_{k,l=1}^{s}\mathbf{E}[g_l\mathrm{Tr}[A_kA_lXA_kX^{2p-4}]]$$
$$= \sum_{k,l=1}^{s}\mathbf{E}[\mathrm{Tr}[A_kA_l^2A_kX^{2p-4}]] + \sum_{m=0}^{2p-5}\sum_{k,l=1}^{s}\mathbf{E}[\mathrm{Tr}[A_kA_lXA_kX^mA_lX^{2p-5-m}]].$$

We deal separately with the two types of terms.

Lemma 3.18 *There exist unitary matrices U_1, U_2, U_3 such that*

$$\sum_{k,l=1}^{s} \mathrm{Tr}[A_k A_l X A_k X^m A_l X^{2p-5-m}] \le \left| \sum_{k,l=1}^{s} \mathrm{Tr}[A_k A_l U_1 A_k U_2 A_l U_3 X^{2p-4}] \right|.$$

Proof Let $X = V|X|$ be the polar decomposition of X, and define

$$\varphi(y,z) := \sum_{k,l=1}^{s} \mathbf{E}[\mathrm{Tr}[A_k A_l V|X|^{(2p-4)y} A_k V^m |X|^{(2p-4)z} A_l V^{2p-5-m} |X|^{(2p-4)(1-y-z)}]].$$

Now apply the three line lemma to φ twice: to $\varphi(\cdot, z)$ with z fixed, then to $\varphi(y, \cdot)$ with y fixed. The omitted details are almost identical to the proof of Lemma 3.15. □

Lemma 3.19 *We have for $p \ge 2$*

$$\sum_{k,l=1}^{s} \mathrm{Tr}[A_k A_l^2 A_k X^{2p-4}] \le \mathrm{Tr}\left[\left(\sum_{k=1}^{s} A_k^2\right)^p\right]^{2/p} \mathrm{Tr}[X^{2p}]^{1-2/p}.$$

Proof We argue essentially as in Lemma 3.5. Define $H = \sum_{l=1}^{s} A_l^2$ and let

$$\varphi(z) := \sum_{k=1}^{s} \mathrm{Tr}[A_k H^{(p-1)z} A_k |X|^{(2p-2)(1-z)}],$$

so that the quantity we would like to bound is $\varphi(1/(p-1))$. By expressing $\varphi(z)$ in terms of the spectral decompositions $X = \sum_{i=1}^{n} \lambda_i v_i v_i^*$ and $H = \sum_{i=1}^{n} \mu_i w_i w_i^*$, we can verify by explicit computation that $z \mapsto \log \varphi(z)$ is convex on $z \in [0, 1]$. Therefore

$$\varphi(1/(p-1)) \le \varphi(1)^{1/(p-1)} \varphi(0)^{(p-2)/(p-1)} = \mathrm{Tr}[H^p]^{1/(p-1)} \mathrm{Tr}[H X^{2p-2}]^{(p-2)/(p-1)}.$$

But $\mathrm{Tr}[H|X|^{2p-2}] \le \mathrm{Tr}[H^p]^{1/p} \mathrm{Tr}[X^{2p}]^{1-1/p}$ by Hölder's inequality, and the conclusion follows readily by substituting this into the above expression. □

Putting together the above bounds and using Hölder's inequality yields

$$\begin{aligned}\sum_{k=1}^{s} \mathbf{E}[\mathrm{Tr}[A_k X^{\ell} A_k X^{2p-2-\ell}]] &\le \mathrm{Tr}\left[\left(\sum_{k=1}^{s} A_k^2\right)^p\right]^{2/p} \mathbf{E}[\mathrm{Tr}[X^{2p}]]^{1-2/p} \\ &\quad + (2p-4) \sup_{U_1,U_2} \mathrm{Tr}\left[\left|\sum_{k,l=1}^{s} A_k A_l U_1 A_k U_2 A_l\right|^{p/2}\right]^{2/p} \mathbf{E}[\mathrm{Tr}[X^{2p}]]^{1-2/p}.\end{aligned}$$

Conclusion Let $p \asymp \log n$. Collecting the above bounds yields

$$\mathbf{E}[\mathrm{Tr}[X^{2p}]] \lesssim \sigma^2 \mathbf{E}[\mathrm{Tr}[X^{2p}]]^{1-1/p} + p\,(\sigma^4 + p\,\tilde{\sigma}^4)\,\mathbf{E}[\mathrm{Tr}[X^{2p}]]^{1-2/p},$$

where we used Lemma 2.2 to simplify the constants. Rearranging gives

$$\mathbf{E}[\mathrm{Tr}[X^{2p}]]^{2/p} \lesssim \sigma^2 \mathbf{E}[\mathrm{Tr}[X^{2p}]]^{1/p} + p\,(\sigma^4 + p\,\tilde{\sigma}^4),$$

which is a simple quadratic inequality for $\mathbf{E}[\mathrm{Tr}[X^{2p}]]^{1/p}$. Solve this inequality using the quadratic formula and apply again Lemma 2.2 to conclude the proof. □

4 Matrices with Independent Entries

The Khintchine-type inequalities developed in the previous section have the advantage that they can be applied in a remarkably general setting: they not only allow an arbitrary variance pattern of the entries, but even an arbitrary dependence structure between the entries. This makes such bounds useful in a wide variety of situations. Unfortunately, we have also seen that Khintchine-type inequalities yield suboptimal bounds already in the simplest examples: the mechanism behind the proofs of these inequalities is too crude to fully capture the structure of the underlying random matrices at this level of generality. In order to gain a deeper understanding, we must impose some additional structure on the matrices under consideration.

In this section, we specialize to what is perhaps the most important case of the random matrices investigated in the previous section: we consider symmetric random matrices with *independent* entries. More precisely, in most of this section, we will study the following basic model. Let g_{ij} be independent standard Gaussian random variables and let $b_{ij} \geq 0$ be given scalars for $i \geq j$. We consider the $n \times n$ symmetric random matrix X whose entries are given by $X_{ij} = b_{ij} g_{ij}$, that is,

$$X = \begin{bmatrix} b_{11}g_{11} & b_{12}g_{12} & \cdots & b_{1n}g_{1n} \\ b_{12}g_{12} & b_{22}g_{22} & & b_{2n}g_{2n} \\ \vdots & & \ddots & \vdots \\ b_{1n}g_{1n} & b_{2n}g_{2n} & \cdots & b_{nn}g_{nn} \end{bmatrix}.$$

In other words, X is the symmetric random matrix whose entries above the diagonal are independent Gaussian variables $X_{ij} \sim N(0, b_{ij}^2)$, where the structure of the matrix is controlled by the given variance pattern $\{b_{ij}\}$. As the matrix is symmetric, we will write for simplicity $g_{ji} = g_{ij}$ and $b_{ji} = b_{ij}$ in the sequel.

The present model differs from the model of the previous section only to the extent that we imposed the additional independence assumption on the entries. In particular, the noncommutative Khintchine inequality reduces in this setting to

$$\mathbf{E}\|X\| \lesssim \max_{i \le n} \sqrt{\sum_{j=1}^{n} b_{ij}^2} \sqrt{\log n},$$

while Theorem 3.14 yields (after some tedious computation)

$$\mathbf{E}\|X\| \lesssim \max_{i \le n} \sqrt{\sum_{j=1}^{n} b_{ij}^2} (\log n)^{1/4} + \max_{i \le n} \left(\sum_{j=1}^{n} b_{ij}^4 \right)^{1/4} \sqrt{\log n}.$$

Unfortunately, we have already seen that neither of these results is sharp even for Wigner matrices (where $b_{ij} = 1$ for all i, j). The aim of this section is to develop much sharper inequalities for matrices with independent entries that capture *optimally* in many cases the underlying structure. The independence assumption will be crucially exploited to control the structure of these matrices, and it is an interesting open problem to understand to what extent the mechanisms developed in this section persist in the presence of dependence between the entries (cf. section 4.3).

4.1 Latała's Inequality and Beyond

The earliest nontrivial result on the spectral norm Gaussian random matrices with independent entries is the following inequality due to Latała [11].

Theorem 4.1 (Latała) *In the setting of this section, we have*

$$\mathbf{E}\|X\| \lesssim \max_{i \le n} \sqrt{\sum_{j=1}^{n} b_{ij}^2} + \left(\sum_{i,j=1}^{n} b_{ij}^4 \right)^{1/4}.$$

Latała's inequality yields a sharp bound $\mathbf{E}\|X\| \lesssim \sqrt{n}$ for Wigner matrices, but is already suboptimal for the diagonal matrix of Example 3.9 where the resulting bound $\mathbf{E}\|X\| \lesssim n^{1/4}$ is very far from the correct answer $\mathbf{E}\|X\| \asymp \sqrt{\log n}$. In this sense, we see that Theorem 4.1 fails to correctly capture the structure of the underlying matrix. Latała's inequality is therefore not too useful for *structured* random matrices; it has however been widely applied together with a simple symmetrization argument [11, Theorem 2] to show that the sharp bound $\mathbf{E}\|X\| \asymp \sqrt{n}$ remains valid for Wigner matrices with general (non-Gaussian) distribution of the entries.

In this section, we develop a nearly sharp improvement of Latała's inequality that can yield optimal results for many structured random matrices.

Theorem 4.2 ([31]) *In the setting of this section, we have*

$$\mathbf{E}\|X\| \lesssim \max_{i\le n} \sqrt{\sum_{j=1}^{n} b_{ij}^2} + \max_{i\le n} \left(\sum_{j=1}^{n} b_{ij}^4\right)^{1/4} \sqrt{\log i}.$$

Let us first verify that Latała's inequality does indeed follow.

Proof (Proof of Theorem 4.1) As the matrix norm $\|X\|$ is unchanged if we permute the rows and columns of X, we may assume without loss of generality that $\sum_{j=1}^{n} b_{ij}^4$ is decreasing in i (this choice minimizes the upper bound in Theorem 4.2). Now recall the following elementary fact: if $x_1 \ge x_2 \ge \cdots \ge x_n \ge 0$, then $x_k \le \frac{1}{k}\sum_{i=1}^{n} x_i$ for every k. In the present case, this observation and Theorem 4.2 imply

$$\mathbf{E}\|X\| \lesssim \max_{i\le n} \sqrt{\sum_{j=1}^{n} b_{ij}^2} + \left(\sum_{i,j=1}^{n} b_{ij}^4\right)^{1/4} \max_{1\le i<\infty} \frac{\sqrt{\log i}}{i^4},$$

which concludes the proof of Theorem 4.1. □

The inequality of Theorem 4.2 is somewhat reminiscent of the bound obtained in the present setting from Theorem 3.14, with a crucial difference: there is no logarithmic factor in front of the first term. As we already proved in Lemma 3.7 that

$$\mathbf{E}\|X\| \gtrsim \max_{i\le n} \sqrt{\sum_{j=1}^{n} b_{ij}^2},$$

we see that Theorem 4.2 provides an *optimal* bound whenever the first term dominates, which is the case for a wide range of structured random matrices. To get a feeling for the sharpness of Theorem 4.2, let us consider an illuminating example.

Example 4.3 (Block matrices) Let $1 \le k \le n$ and suppose for simplicity that n is divisible by k. We consider the $n \times n$ symmetric block-diagonal matrix X of the form

$$X = \begin{bmatrix} \mathbf{X}_1 & & & \\ & \mathbf{X}_2 & & \\ & & \ddots & \\ & & & \mathbf{X}_{n/k} \end{bmatrix},$$

where $\mathbf{X}_1, \ldots, \mathbf{X}_{n/k}$ are independent $k \times k$ Wigner matrices. This model interpolates between the diagonal matrix of Example 3.9 (the case $k = 1$) and the Wigner matrix of Example 3.10 (the case $k = n$). Note that $\|X\| = \max_i \|\mathbf{X}_i\|$, so we can compute

$$\mathbf{E}\|X\| \lesssim \mathbf{E}[\|\mathbf{X}_1\|^{\log n}]^{1/\log n} \le \mathbf{E}\|\mathbf{X}_1\| + \mathbf{E}[(\|\mathbf{X}_1\| - \mathbf{E}\|\mathbf{X}_1\|)^{\log n}]^{1/\log n} \lesssim \sqrt{k} + \sqrt{\log n}$$

using Lemmas 2.3, 3.11, and 3.8, respectively. On the other hand, Lemma 3.7 implies that $\mathbf{E}\|X\| \gtrsim \sqrt{k}$, while we can trivially estimate $\mathbf{E}\|X\| \geq \mathbf{E}\max_i X_{ii} \asymp \sqrt{\log n}$. Averaging these two lower bounds, we have evidently shown that

$$\mathbf{E}\|X\| \asymp \sqrt{k} + \sqrt{\log n}.$$

This explicit computation provides a simple but very useful benchmark example for testing inequalities for structured random matrices.

In the present case, applying Theorem 4.2 to this example yields

$$\mathbf{E}\|X\| \lesssim \sqrt{k} + k^{1/4}\sqrt{\log n}.$$

Therefore, in the present example, Theorem 4.2 fails to be sharp only when k is in the range $1 \ll k \ll (\log n)^2$. This suboptimal parameter range will be completely eliminated by the sharp bound to be proved in section 4.2 below. But the bound of Theorem 4.2 is already sharp in the vast majority of cases, and is of significant interest in its own right for reasons that will be discussed in detail in section 4.3.

An important feature of the inequalities of this section should be emphasized: unlike all bounds we have encountered so far, the present bounds are *dimension-free*. As was discussed in Remark 2.4, one cannot expect to obtain sharp dimension-free bounds using the moment method, and it therefore comes as no surprise that the bounds of the present section will therefore be obtained by the random process method. The original proof of Latała proceeds by a difficult and very delicate explicit construction of a multiscale net in the spirit of Remark 2.7. We will follow here a much simpler approach that was developed in [31] to prove Theorem 4.2.

The basic idea behind our approach was already encountered in the proof of Lemma 3.11 to bound the norm of a Wigner matrix (where $b_{ij} = 1$ for all i, j): we seek a Gaussian process Y_v that dominates the process $X_v := \langle v, Xv\rangle$ whose supremum coincides with the spectral norm. The present setting is significantly more challenging, however. To see the difficulty, let us try to adapt directly the proof of Lemma 3.11 to the present structured setting: we readily compute

$$\mathbf{E}(X_v - X_w)^2 \leq 2\sum_{i,j=1}^{n} b_{ij}^2(v_iv_j - w_iw_j)^2 \leq 4\max_{i,j\leq n} b_{ij}^2\,\|v-w\|^2.$$

We can therefore dominate X_v by the Gaussian process $Y_v = 2\max_{i,j} b_{ij}\,\langle v, g\rangle$. Proceeding as in the proof of Lemma 3.11, this yields the following upper bound:

$$\mathbf{E}\|X\| \lesssim \max_{i,j\leq n} b_{ij}\sqrt{n}.$$

This bound is sharp for Wigner matrices (in this case the present proof reduces to that of Lemma 3.11), but is woefully inadequate in any structured example. The problem with the above bound is that it always crudely estimates the behavior of the increments $\mathbf{E}[(X_v - X_w)]^{1/2}$ by a *Euclidean* norm $\|v-w\|$, regardless of the structure of the underlying matrix. However, the geometry defined by $\mathbf{E}[(X_v - X_w)]^{1/2}$ depends strongly on the structure of the matrix, and is typically highly non-Euclidean. For example, in the diagonal matrix of Example 3.9, we have $\mathbf{E}[(X_v - X_w)]^{1/2} = \|v^2 - w^2\|$ where $(v^2)_i := v_i^2$. As v^2 is in the simplex whenever $v \in B$, we see that the underlying geometry in this case is that of an ℓ_1-norm and not of an ℓ_2-norm. In more general examples, however, it is far from clear what is the correct geometry.

The key challenge we face is to design a comparison process that is easy to bound, but whose geometry nonetheless captures faithfully the structure of the underlying matrix. To develop some intuition for how this might be accomplished, let us consider in first instance instead of the increments $\mathbf{E}[(X_v - X_w)^2]^{1/2}$ only the standard deviation $\mathbf{E}[X_v^2]^{1/2}$ of the process $X_v = \langle v, Xv \rangle$. We easily compute

$$\mathbf{E}X_v^2 = 2\sum_{i \neq j} v_i^2 b_{ij}^2 v_j^2 + \sum_{i=1}^n b_{ii}^2 v_i^4 \leq 2\sum_{i=1}^n x_i(v)^2,$$

where we defined the nonlinear map $x : \mathbb{R}^n \to \mathbb{R}^n$ as

$$x_i(v) := v_i \sqrt{\sum_{j=1}^n b_{ij}^2 v_j^2}.$$

This computation suggests that we might attempt to dominate the process X_v by the process $Y_v = \langle x(v), g \rangle$, whose increments $\mathbf{E}[(Y_v - Y_w)^2]^{1/2} = \|x(v) - x(w)\|$ capture the non-Euclidean nature of the underlying geometry through the nonlinear map x. The reader may readily verify, for example, that the latter process captures automatically the correct geometry of our two extreme examples of Wigner and diagonal matrices.

Unfortunately, the above choice of comparison process Y_v is too optimistic: while we have chosen this process so that $\mathbf{E}X_v^2 \lesssim \mathbf{E}Y_v^2$ by construction, the Slepian-Fernique inequality requires the stronger bound $\mathbf{E}(X_v - X_w)^2 \lesssim \mathbf{E}(Y_v - Y_w)^2$. It turns out that the latter inequality does not always hold [31]. However, the inequality *nearly* holds, which is the key observation behind the proof of Theorem 4.2.

Lemma 4.4 *For every $v, w \in \mathbb{R}^n$*

$$\mathbf{E}(\langle v, Xv \rangle - \langle w, Xw \rangle)^2 \leq 4\|x(v) - x(w)\|^2 - \sum_{i,j=1}^n (v_i^2 - w_i^2) b_{ij}^2 (v_j^2 - w_j^2).$$

Proof We simply compute both sides and compare. Define for simplicity the seminorm $\|\cdot\|_i$ as $\|v\|_i^2 := \sum_{j=1}^n b_{ij}^2 v_j^2$, so that $x_i(v) = v_i \|v\|_i$. First, we note that

$$\begin{aligned}\mathbf{E}(\langle v, Xv \rangle - \langle w, Xw \rangle)^2 &= \mathbf{E}\langle v + w, X(v - w) \rangle^2 \\ &= \sum_{i=1}^n (v_i - w_i)^2 \|v + w\|_i^2 + \sum_{i,j=1}^n (v_i^2 - w_i^2) b_{ij}^2 (v_j^2 - w_j^2).\end{aligned}$$

On the other hand, as $2(x_i(v)-x_i(w)) = (v_i+w_i)(\|v\|_i - \|w\|_i) + (v_i - w_i)(\|v\|_i + \|w\|_i)$,

$$4\|x(v)-x(w)\|^2 = \sum_{i=1}^n (v_i+w_i)^2(\|v\|_i - \|w\|_i)^2 + \sum_{i=1}^n (v_i-w_i)^2(\|v\|_i + \|w\|_i)^2$$
$$+ 2\sum_{i,j=1}^n (v_i^2 - w_i^2) b_{ij}^2 (v_j^2 - w_j^2).$$

The result follows readily from the triangle inequality $\|v+w\|_i \le \|v\|_i + \|w\|_i$. □

We can now complete the proof of Theorem 4.2.

Proof (Proof of Theorem 4.2) Define the Gaussian processes

$$X_v = \langle v, Xv\rangle, \qquad Y_v = 2\langle x(v), g\rangle + \langle v^2, Y\rangle,$$

where $g \sim N(0, I)$ is a standard Gaussian vector in $\mathbb{R}^n$, $(v^2)_i := v_i^2$, and $Y \sim N(0, B^-)$ is a centered Gaussian vector that is independent of g and whose covariance matrix B^- is the negative part of the matrix of variances $B = (b_{ij}^2)$ (if B has eigendecomposition $B = \sum_i \lambda_i v_i v_i^*$, the negative part B^- is defined as $B^- = \sum_i \max(-\lambda_i, 0) v_i v_i^*$). As $-B \preceq B^-$ by construction, it is readily seen that Lemma 4.4 implies

$$\mathbf{E}(X_v - X_w)^2 \le 4\|x(v)-x(w)\|^2 + \langle v^2 - w^2, B^-(v^2-w^2)\rangle = \mathbf{E}(Y_v - Y_w)^2.$$

We can therefore argue by the Slepian-Fernique inequality that

$$\mathbf{E}\|X\| \lesssim \mathbf{E}\sup_{v\in B} Y_v \le 2\,\mathbf{E}\sup_{v\in B}\langle x(v), g\rangle + \mathbf{E}\max_{i\le n} Y_i$$

as in the proof of Lemma 3.11. It remains to bound each term on the right.

Let us begin with the second term. Using the moment method as in section 2.1, one obtains the dimension-dependent bound $\mathbf{E}\max_i Y_i \lesssim \max_i \mathrm{Var}(Y_i)^{1/2}\sqrt{\log n}$. This bound is sharp when all the variances $\mathrm{Var}(Y_i) = B_{ii}^-$ are of the same order, but can be suboptimal when many of the variances are small. Instead, we will use a sharp *dimension-free* bound on the maximum of Gaussian random variables.

Lemma 4.5 (Subgaussian maxima) *Suppose that $g_1, \ldots, g_n$ satisfy $\mathbf{E}[|g_i|^k]^{1/k} \lesssim \sqrt{k}$ for all k, i, and let $\sigma_1, \ldots, \sigma_n \ge 0$. Then we have*

$$\mathbf{E}\max_{i\le n}|\sigma_i g_i| \lesssim \max_{i\le n}\sigma_i\sqrt{\log(i+1)}.$$

Proof By a union bound and Markov's inequality

$$\mathbf{P}\left[\max_{i\le n}|\sigma_i g_i|\ge t\right]\le\sum_{i=1}^n\mathbf{P}[|\sigma_i g_i|\ge t]\le\sum_{i=1}^n\left(\frac{\sigma_i\sqrt{2\log(i+1)}}{t}\right)^{2\log(i+1)}.$$

But we can estimate

$$\sum_{i=1}^\infty s^{-2\log(i+1)}=\sum_{i=1}^\infty(i+1)^{-2}(i+1)^{-2\log s+2}\le 2^{-2\log s+2}\sum_{i=1}^\infty(i+1)^{-2}\lesssim s^{-2\log 2}$$

as long as $\log s>1$. Setting $s=t/\max_i\sigma_i\sqrt{2\log(i+1)}$, we obtain

$$\begin{aligned}\mathbf{E}\max_{i\le n}|\sigma_i g_i|&=\max_i\sigma_i\sqrt{2\log(i+1)}\int_0^\infty\mathbf{P}\left[\max_{i\le n}|\sigma_i g_i|\ge s\max_i\sigma_i\sqrt{2\log(i+1)}\right]ds\\&\lesssim\max_i\sigma_i\sqrt{2\log(i+1)}\left(e+\int_e^\infty s^{-2\log 2}ds\right)\lesssim\max_i\sigma_i\sqrt{\log(i+1)},\end{aligned}$$

which completes the proof. □

Remark 4.6 Lemma 4.5 does not require the variables g_i to be either independent or Gaussian. However, if $g_1,\ldots,g_n$ are independent standard Gaussian variables (which satisfy $\mathbf{E}[|g_i|^k]^{1/k}\lesssim\sqrt{k}$ by Lemma 3.1) and if $\sigma_1\ge\sigma_2\ge\cdots\ge\sigma_n>0$ (which is the ordering that optimizes the bound of Lemma 4.5), then

$$\mathbf{E}\max_{i\le n}|\sigma_i g_i|\asymp\max_{i\le n}\sigma_i\sqrt{\log(i+1)},$$

cf. [31]. This shows that Lemma 4.5 captures precisely the dimension-free behavior of the maximum of independent centered Gaussian variables.

To estimate the second term in our bound on $\mathbf{E}\|X\|$, note that $(B^-)^2\preceq B^2$ implies

$$\mathrm{Var}(Y_i)^2=(B^-_{ii})^2\le(B^-)^2_{ii}\le(B^2)_{ii}=\sum_{j=1}^n(B_{ij})^2=\sum_{j=1}^n b_{ij}^4.$$

Applying Lemma 4.5 with $g_i=Y_i/\mathrm{Var}(Y_i)^{1/2}$ yields the bound

$$\mathbf{E}\max_{i\le n}Y_i\lesssim\max_{i\le n}\left(\sum_{j=1}^n b_{ij}^4\right)^{1/4}\sqrt{\log(i+1)}.$$

Now let us estimate the first term in our bound on $\mathbf{E}\|X\|$. Note that

$$\sup_{v\in B}\langle x(v),g\rangle=\sup_{v\in B}\sum_{j=1}^n g_j v_j\sqrt{\sum_{i=1}^n v_i^2 b_{ij}^2}\le\sup_{v\in B}\sqrt{\sum_{i,j=1}^n v_i^2 b_{ij}^2 g_j^2}=\max_{i\le n}\sqrt{\sum_{j=1}^n b_{ij}^2 g_j^2},$$

where we used Cauchy-Schwarz and the fact that v^2 is in the ℓ_1-ball whenever v is in the ℓ_2-ball. We can therefore estimate, using Lemma 3.8 and Lemma 4.5,

$$\begin{aligned}\mathbf{E}\sup_{v\in B}\langle x(v), g\rangle &\le \max_{i\le n}\mathbf{E}\sqrt{\sum_{j=1}^n b_{ij}^2 g_j^2} + \mathbf{E}\max_{i\le n}\left|\sqrt{\sum_{j=1}^n b_{ij}^2 g_j^2} - \mathbf{E}\sqrt{\sum_{j=1}^n b_{ij}^2 g_j^2}\right| \\ &\lesssim \max_{i\le n}\sqrt{\sum_{j=1}^n b_{ij}^2} + \max_{i,j\le n} b_{ij}\sqrt{\log(i+1)}.\end{aligned}$$

Putting everything together gives

$$\mathbf{E}\|X\| \lesssim \max_{i\le n}\sqrt{\sum_{j=1}^n b_{ij}^2} + \max_{i,j\le n} b_{ij}\sqrt{\log(i+1)} + \max_{i\le n}\left(\sum_{j=1}^n b_{ij}^4\right)^{1/4}\sqrt{\log(i+1)}.$$

It is not difficult to simplify this (at the expense of a larger universal constant) to obtain the bound in the statement of Theorem 4.2. □

4.2 A Sharp Dimension-Dependent Bound

The approach developed in the previous section yields optimal results for many structured random matrices with independent entries. The crucial improvement of Theorem 4.2 over the noncommutative Khintchine inequality is that no logarithmic factor appears in the first term. Therefore, when this term dominates, Theorem 4.2 is sharp by Lemma 3.7. However, the second term in Theorem 4.2 is not quite sharp, as is illustrated in Example 4.3. While Theorem 4.2 captures much of the geometry of the underlying model, there remains some residual inefficiency in the proof.

In this section, we will develop an improved version of Theorem 4.2 that is essentially sharp (in a sense that will be made precise below). Unfortunately, it is not known at present how such a bound can be obtained using the random process method, and we revert back to the moment method in the proof. The price we pay for this is that we lose the dimension-free nature of Theorem 4.2.

Theorem 4.7 ([4]) *In the setting of this section, we have*

$$\mathbf{E}\|X\| \lesssim \max_{i\le n}\sqrt{\sum_{j=1}^n b_{ij}^2} + \max_{i,j\le n} b_{ij}\sqrt{\log n}.$$

To understand why this result is sharp, let us recall (Remark 2.4) that the moment method necessarily bounds not the quantity $\mathbf{E}\|X\|$, but rather the larger quantity $\mathbf{E}[\|X\|^{\log n}]^{1/\log n}$. The latter quantity is now however completely understood.

Corollary 4.8 *In the setting of this section, we have*

$$\mathbf{E}[\|X\|^{\log n}]^{1/\log n} \asymp \max_{i\le n} \sqrt{\sum_{j=1}^{n} b_{ij}^2} + \max_{i,j\le n} b_{ij}\sqrt{\log n}.$$

Proof The upper bound follows from the proof of Theorem 4.7. The first term on the right is a lower bound by Lemma 3.7. On the other hand, if $b_{kl} = \max_{i,j} b_{ij}$, then $\mathbf{E}[\|X\|^{\log n}]^{1/\log n} \ge \mathbf{E}[|X_{kl}|^{\log n}]^{1/\log n} \gtrsim b_{kl}\sqrt{\log n}$ as $\mathbf{E}[|X_{kl}|^p]^{1/p} \asymp b_{kl}\sqrt{p}$. □

The above result shows that Theorem 4.7 is in fact the *optimal* result that could be obtained by the moment method. This result moreover yields optimal bounds even for $\mathbf{E}\|X\|$ in almost all situations of practical interest, as it is true under mild assumptions that $\mathbf{E}\|X\| \asymp \mathbf{E}[\|X\|^{\log n}]^{1/\log n}$ (as will be discussed in section 4.3). Nonetheless, this is not always the case, and will fail in particular for matrices whose variances are distributed over many different scales; in the latter case, the *dimension-free* bound of Theorem 4.2 can give rise to much sharper results. Both Theorems 4.2 and 4.7 therefore remain of significant independent interest. Taken together, these results strongly support a fundamental conjecture, to be discussed in the next section, that would provide the ultimate understanding of the magnitude of the spectral norm of the random matrix model considered in this chapter.

The proof of Theorem 4.7 is completely different in nature than that of Theorem 4.2. Rather than prove Theorem 4.7 in the general case, we will restrict attention in the rest of this section to the special case of *sparse Wigner matrices*. The proof of Theorem 4.7 in the general case is actually no more difficult than in this special case, but the ideas and intuition behind the proof are particularly transparent when restricted to sparse Wigner matrices (which was how the authors of [4] arrived at the proof). Once this special case has been understood, the reader can extend the proof to the general setting as an exercise, or refer to the general proof given in [4].

Example 4.9 (Sparse Wigner matrices) Informally, a sparse Wigner matrix is a symmetric random matrix with a given sparsity pattern, whose nonzero entries are independent standard Gaussian variables. It is convenient to fix the sparsity pattern of the matrix by specifying a given undirected graph $G = ([n], E)$ on n vertices, whose adjacency matrix we denote as $B = (b_{ij})_{1\le i,j\le n}$. The corresponding sparse Wigner matrix X is the symmetric random matrix whose entries are given by $X_{ij} = b_{ij}g_{ij}$, where g_{ij} are independent standard Gaussian variables (up to symmetry $g_{ji} = g_{ij}$). Clearly our previous Examples 3.9, 3.10, and 4.3 are all special cases of this model.

For a sparse Wigner matrix, the first term in Theorem 4.7 is precisely the maximal degree $k = \deg(G)$ of the graph G, so that Theorem 4.7 reduces to

$$\mathbf{E}\|X\| \lesssim \sqrt{k} + \sqrt{\log n}.$$

We will see in section 4.3 that this bound is sharp for sparse Wigner matrices.

The remainder of this section is devoted to the proof of Theorem 4.7 in the setting of Example 4.9 (we fix the notation introduced in this example in the sequel). To understand the idea behind the proof, let us start by naively writing out the central quantity that appears in moment method (Lemma 2.2): we evidently have

$$\begin{aligned}\mathbf{E}[\mathrm{Tr}[X^{2p}]] &= \sum_{i_1,\ldots,i_{2p}=1}^{n} \mathbf{E}[X_{i_1i_2}X_{i_2i_3}\cdots X_{i_{2p-1}i_{2p}}X_{i_{2p}i_1}] \\ &= \sum_{i_1,\ldots,i_{2p}=1}^{n} b_{i_1i_2}b_{i_2i_3}\cdots b_{i_{2p}i_1}\,\mathbf{E}[g_{i_1i_2}g_{i_2i_3}\cdots g_{i_{2p}i_1}].\end{aligned}$$

It is useful to think of $\gamma = (i_1,\ldots,i_{2p})$ geometrically as a *cycle* $i_1 \to i_2 \to \cdots \to i_{2p} \to i_1$ of length $2p$. The quantity $b_{i_1i_2}b_{i_2i_3}\cdots b_{i_{2p}i_1}$ is equal to one precisely when γ defines a cycle in the graph G, and is zero otherwise. We can therefore write

$$\mathbf{E}[\mathrm{Tr}[X^{2p}]] = \sum_{\text{cycle } \gamma \text{ in } G \text{ of length } 2p} c(\gamma),$$

where we defined the constant $c(\gamma) := \mathbf{E}[g_{i_1i_2}g_{i_2i_3}\cdots g_{i_{2p}i_1}]$.

It turns out that $c(\gamma)$ does not really depend on the position of the cycle γ in the graph G. While we will not require a precise formula for $c(\gamma)$ in the proof, it is instructive to write down what it looks like. For any cycle γ in G, denote by $m_\ell(\gamma)$ the number of distinct edges in G that are visited by γ precisely ℓ times, and denote by $m(\gamma) = \sum_{\ell\ge1} m_\ell(\gamma)$ the total number of distinct edges visited by γ. Then

$$c(\gamma) = \prod_{\ell=1}^{\infty} \mathbf{E}[g^\ell]^{m_\ell(\gamma)},$$

where $g \sim N(0,1)$ is a standard Gaussian variable and we have used the independence of the entries. From this formula, we read off two important facts (which are the only ones that will actually be used in the proof):

- If any edge in G is visited by γ an odd number of times, then $c(\gamma) = 0$ (as the odd moments of g vanish). Thus the only cycles that matter are *even* cycles, that is, cycles in which every distinct edge is visited an even number of times.
- $c(\gamma)$ depends on γ only through the numbers $m_\ell(\gamma)$. Therefore, to compute $c(\gamma)$, we only need to know the *shape* $\mathbf{s}(\gamma)$ of the cycle γ.

The shape $\mathbf{s}(\gamma)$ is obtained from γ by relabeling its vertices in order of appearance; for example, the shape of the cycle $7 \to 3 \to 9 \to 7 \to 3 \to 9 \to 7$ is given by $1 \to 2 \to 3 \to 1 \to 2 \to 3 \to 1$. The shape $\mathbf{s}(\gamma)$ captures the topological properties of γ (such as the numbers $m_\ell(\gamma) = m_\ell(\mathbf{s}(\gamma))$) without keeping track of the manner in which γ is embedded in G. This is illustrated in the following figure:

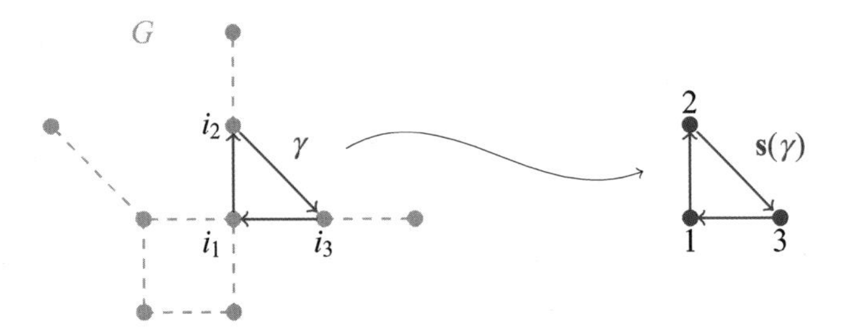

Putting together the above observations, we obtain the useful formula

$$\mathbf{E}[\mathrm{Tr}[X^{2p}]] = \sum_{\text{shape } \mathbf{s} \text{ of even cycle of length } 2p} c(\mathbf{s}) \times \#\{\text{embeddings of } \mathbf{s} \text{ in } G\}.$$

So far, we have done nothing but bookkeeping. To use the above bound, however, we must get down to work and count the number of shapes of even cycles that can appear in the given graph G. The problem we face is that the latter proves to be a difficult combinatorial problem, which is apparently completely intractable when presented with any given graph G that may possess an arbitrary structure (this is already highly nontrivial even in a complete graph when p is large!) To squeeze anything useful out of this bound, it is essential that we find a shortcut.

The solution to our problem proves to be incredibly simple. Recall that G is a given graph of degree $\deg(G) = k$. Of all graphs of degree k, which one will admit the most possible shapes? Obviously the graph that admits the most shapes is the one where every potential edge between two vertices is present; therefore, the graph of degree k that possesses the most shapes is the *complete graph on k vertices*. From the random matrix point of view, the latter corresponds to a Wigner matrix of dimension $k \times k$. This simple idea suggests that rather than directly estimating the quantity $\mathbf{E}[\mathrm{Tr}[X^{2p}]]$ by combinatorial means, we should aim to prove a *comparison principle* between the moments of the $n \times n$ sparse matrix X and the moments of a $k \times k$ Wigner matrix Y, which we already know how to bound by Lemma 3.11. Note that such a comparison principle is of a completely different nature than the Slepian-Fernique method used previously: here we are comparing two matrices of *different dimension*. The intuitive idea is that a large sparse matrix can be "compressed" into a much lower dimensional dense matrix without decreasing its norm.

The alert reader will note that there is a problem with the above intuition. While the complete graph on k points admits more shapes than the original graph G, there are less potential ways in which each shape can be embedded in the complete graph as the latter possesses less vertices than the original graph. We can compensate for this deficiency by slightly increasing the dimension of the complete graph.

Lemma 4.10 (Dimension compression) *Let X be the $n \times n$ sparse Wigner matrix (Example 4.9) defined by a graph $G = ([n], E)$ of maximal degree* $\deg(G) = k$*, and let Y_r be an $r \times r$ Wigner matrix (Example 3.10). Then, for every $p \geq 1$,*

$$\mathbf{E}[\mathrm{Tr}[X^{2p}]] \leq \frac{n}{k+p}\,\mathbf{E}[\mathrm{Tr}[Y_{k+p}^{2p}]].$$

Proof Let $\mathbf{s}$ be the shape of an even cycle of length $2p$, and let K_r be the complete graph on $r > p$ points. Denote by $m(\mathbf{s})$ the number of distinct vertices in $\mathbf{s}$, and note that $m(\mathbf{s}) \leq p+1$ as every distinct edge in $\mathbf{s}$ must appear at least twice. Thus

$$\#\{\text{embeddings of } \mathbf{s} \text{ in } K_r\} = r(r-1)\cdots(r-m(\mathbf{s})+1),$$

as any assignment of vertices of K_r to the distinct vertices of $\mathbf{s}$ defines a valid embedding of $\mathbf{s}$ in the complete graph. On the other hand, to count the number of embeddings of $\mathbf{s}$ in G, note that we have as many as n choices for the first vertex, while each subsequent vertex can be chosen in at most k ways (as $\deg(G) = k$). Thus

$$\#\{\text{embeddings of } \mathbf{s} \text{ in } G\} \leq nk^{m(\mathbf{s})-1}.$$

Therefore, if we choose $r = k+p$, we have $r - m(\mathbf{s}) + 1 \geq r - p \geq k$, so that

$$\#\{\text{embeddings of } \mathbf{s} \text{ in } G\} \leq \frac{n}{r}\,\#\{\text{embeddings of } \mathbf{s} \text{ in } K_r\}.$$

The proof now follows from the combinatorial expression for $\mathbf{E}[\mathrm{Tr}[X^{2p}]]$. □

With Lemma 4.10 in hand, it is now straightforward to complete the proof of Theorem 4.7 for the sparse Wigner matrix model of Example 4.9.

Proof (Proof of Theorem 4.7 in the setting of Example 4.9) We begin by noting that

$$\mathbf{E}\|X\| \leq \mathbf{E}[\|X\|^{2p}]^{1/2p} \leq n^{1/2p}\,\mathbf{E}[\|Y_{k+p}\|^{2p}]^{1/2p}$$

by Lemma 4.10, where we used $\|X\|^{2p} \leq \mathrm{Tr}[X^{2p}]$ and $\mathrm{Tr}[Y_r^{2p}] \leq r\|Y_r\|^{2p}$. Thus

$$\begin{aligned}
\mathbf{E}\|X\| &\lesssim \mathbf{E}[\|Y_{k+\lfloor \log n \rfloor}\|^{2\log n}]^{1/2\log n} \\
&\leq \mathbf{E}\|Y_{k+\lfloor \log n \rfloor}\| + \mathbf{E}[(\|Y_{k+\lfloor \log n \rfloor}\| - \mathbf{E}\|Y_{k+\lfloor \log n \rfloor}\|)^{2\log n}]^{1/2\log n} \\
&\lesssim \sqrt{k+\log n} + \sqrt{\log n},
\end{aligned}$$

where in the last inequality we used Lemma 3.11 to bound the first term and Lemma 3.8 to bound the second term. Thus $\mathbf{E}\|X\| \lesssim \sqrt{k} + \sqrt{\log n}$, completing the proof. □

4.3 *Three Conjectures*

We have obtained in the previous sections two remarkably sharp bounds on the spectral norm of random matrices with independent centered Gaussian entries: the slightly suboptimal dimension-free bound of Theorem 4.2 for $\mathbf{E}\|X\|$, and the sharp dimension-dependent bound of Theorem 4.7 for $\mathbf{E}[\|X\|^{\log n}]^{1/\log n}$. As we will shortly argue, the latter bound is also sharp for $\mathbf{E}\|X\|$ in almost all situations of practical interest. Nonetheless, we cannot claim to have a complete understanding of the mechanisms that control the spectral norm of Gaussian random matrices unless we can obtain a sharp dimension-free bound on $\mathbf{E}\|X\|$. While this problem remains open, the above results strongly suggest what such a sharp bound should look like.

To gain some initial intuition, let us complement the sharp lower bound of Corollary 4.8 for $\mathbf{E}[\|X\|^{\log n}]^{1/\log n}$ by a trivial lower bound for $\mathbf{E}\|X\|$.

Lemma 4.11 *In the setting of this section, we have*

$$\mathbf{E}\|X\| \gtrsim \max_{i\leq n} \sqrt{\sum_{j=1}^{n} b_{ij}^2} + \mathbf{E} \max_{i,j\leq n} |X_{ij}|.$$

Proof The first term is a lower bound by Lemma 3.7, while the second term is a lower bound by the trivial pointwise inequality $\|X\| \geq \max_{i,j} |X_{ij}|$. □

The simplest possible upper bound on the maximum of centered Gaussian random variables is $\mathbf{E}\max_{i,j} |X_{ij}| \lesssim \max_{i,j} b_{ij}\sqrt{\log n}$, which is sharp for i.i.d. Gaussian variables. Thus the lower bound of Lemma 4.11 matches the upper bound of Theorem 4.7 under a minimal homogeneity assumption: it suffices to assume that the number of entries whose standard deviation b_{kl} is of the same order as $\max_{i,j} b_{ij}$ grows polynomially with dimension (which still allows for a vanishing fraction of entries of the matrix to possess large variance). For example, in the sparse Wigner matrix model of Example 4.9, every row of the matrix that does not correspond to an isolated vertex in G contains at least one entry of variance one. Therefore, if G possesses no isolated vertices, there are at least n entries of X with variance one, and it follows immediately from Lemma 4.11 that the bound of Theorem 4.7 is sharp for sparse Wigner matrices. (There is no loss of generality in assuming that G has no isolated vertices: any isolated vertex yields a row that is identically zero, so we can simply remove such vertices from the graph without changing the norm.)

However, when the variances of the entries of X possess many different scales, the dimension-dependent upper bound $\mathbf{E}\max_{i,j} |X_{ij}| \lesssim \max_{i,j} b_{ij}\sqrt{\log n}$ can fail to be sharp. To obtain a sharp bound on the maximum of Gaussian random variables, we must proceed in a dimension-free fashion as in Lemma 4.5. In particular, combining Remark 4.6 and Lemma 4.11 yields the following explicit lower bound:

$$\mathbf{E}\|X\| \gtrsim \max_{i\le n} \sqrt{\sum_{j=1}^{n} b_{ij}^2} + \max_{i,j\le n} b_{ij}\sqrt{\log i},$$

provided that $\max_j b_{1j} \geq \max_j b_{2j} \geq \cdots \geq \max_j b_{nj} > 0$ (there is no loss of generality in assuming the latter, as we can always permute the rows and columns of X to achieve this ordering without changing the norm of X). It will not have escaped the attention of the reader that the latter lower bound is tantalizingly close both to the dimension-dependent upper bound of Theorem 4.7 and to the dimension-free upper bound of Theorem 4.2. This leads us to the following very natural conjecture [31].

Conjecture 1 Assume without loss of generality that the rows and columns of X have been permuted such that $\max_j b_{1j} \geq \max_j b_{2j} \geq \cdots \geq \max_j b_{nj} > 0$. Then

$$\begin{aligned}\mathbf{E}\|X\| &\asymp \|\mathbf{E}X^2\|^{1/2} + \mathbf{E}\max_{i,j\le n}|X_{ij}| \\ &\asymp \max_{i\le n}\sqrt{\sum_{j=1}^{n} b_{ij}^2} + \max_{i,j\le n} b_{ij}\sqrt{\log i}.\end{aligned}$$

Conjecture 1 appears completely naturally from our results, and has a surprising interpretation. There are two simple mechanisms that would certainly force the random matrix X to have large expected norm $\mathbf{E}\|X\|$: the matrix X is can be large "on average" in the sense that $\|\mathbf{E}X^2\|$ is large (note that the expectation here is *inside* the norm), or the matrix X can have an entry that exhibits a large fluctuation in the sense that $\max_{i,j} X_{ij}$ is large. Conjecture 1 suggests that these two mechanisms are, in a sense, the *only* reasons why $\mathbf{E}\|X\|$ can be large.

Given the remarkable similarity between Conjecture 1 and Theorem 4.7, one might hope that a slight sharpening of the proof of Theorem 4.7 would suffice to yield the conjecture. Unfortunately, it seems that the moment method is largely useless for the purpose of obtaining dimension-free bounds: indeed, the Corollary 4.8 shows that the moment method is already exploited optimally in the proof of Theorem 4.7. While it is sometimes possible to derive dimension-free results from dimension-dependent results by a stratification procedure, such methods either fail completely to capture the correct structure of the problem (cf. [19]) or retain a residual dimension-dependence (cf. [31]). It therefore seems likely that random process methods will prove to be essential for progress in this direction.

While Conjecture 1 appears completely natural in the present setting, we should also discuss a competing conjecture that was proposed much earlier by R. Latała. Inspired by certain results of Seginer [21] for matrices with i.i.d. entries, Latała conjectured the following sharp bound in the general setting of this section.

Conjecture 2 In the setting of this section, we have

$$\mathbf{E}\|X\| \asymp \mathbf{E} \max_{i \le n} \sqrt{\sum_{j=1}^{n} X_{ij}^2}.$$

As $\|X\|^2 \ge \max_i \sum_j X_{ij}^2$ holds deterministically, the lower bound in Conjecture 2 is trivial: it states that a matrix that possesses a large row must have large spectral norm. Conjecture 2 suggests that this is the *only* reason why the matrix norm can be large. This is certainly not the case for an arbitrary matrix X, and so it is not at all clear *a priori* why this should be true. Nonetheless, no counterexample is known in the setting of the Gaussian random matrices considered in this section.

While Conjectures 1 and 2 appear to arise from different mechanisms, it is observed in [31] that these conjectures are actually equivalent: it is not difficult to show that the right-hand side in both inequalities is equivalent, up to the universal constant, to the explicit expression recorded in Conjecture 1. In fact, let us note that both conjectured mechanisms are essentially already present in the proof of Theorem 4.2: in the comparison process Y_v that arises in the proof, the first term is strongly reminiscent of Conjecture 2, while the second term is reminiscent of the second term in Conjecture 1. In this sense, the mechanism that is developed in the proof of Theorem 4.2 provides even stronger evidence for the validity of these conjectures. The remaining inefficiency in the proof of Theorem 4.2 is discussed in detail in [31].

We conclude by discussing briefly a much more speculative question. The noncommutative Khintchine inequalities developed in the previous section hold in a very general setting, but are almost always suboptimal. In contrast, the bounds in this section yield nearly optimal results under the additional assumption that the matrix entries are independent. It would be very interesting to understand whether the bounds of the present section can be extended to the much more general setting captured by noncommutative Khintchine inequalities. Unfortunately, independence is used crucially in the proofs of the results in this section, and it is far from clear what mechanism might give rise to analogous results in the dependent setting.

One might nonetheless speculate what such a result might potentially look like. In particular, we note that both parameters that appear in the sharp bound Theorem 4.7 have natural analogues in the general setting: in the setting of this section

$$\|\mathbf{E}X^2\| = \sup_{v \in B} \mathbf{E}\langle v, X^2 v\rangle = \max_i \sum_j b_{ij}^2, \qquad \sup_{v \in B} \mathbf{E}\langle v, Xv\rangle^2 = \max_{i,j} b_{ij}^2.$$

We have already encountered both these quantities also in the previous section: $\sigma = \|\mathbf{E}X^2\|^{1/2}$ is the natural structural parameter that arises in noncommutative Khintchine inequalities, while $\sigma_* := \sup_v \mathbf{E}[\langle v, Xv\rangle^2]^{1/2}$ controls the fluctuations of the spectral norm by Gaussian concentration (see the proof of Lemma 3.7). By analogy with Theorem 4.7, we might therefore speculatively conjecture:

Conjecture 3 Let $X = \sum_{k=1}^{s} g_k A_k$ as in Theorem 3.2. Then

$$\mathbf{E}\|X\| \lesssim \|\mathbf{E}X^2\|^{1/2} + \sup_{v\in B} \mathbf{E}[\langle v, Xv\rangle^2]^{1/2}\sqrt{\log n}.$$

Such a generalization would constitute a far-reaching improvement of the noncommutative Khintchine theory. The problem with Conjecture 3 is that it is completely unclear how such a bound might arise: the only evidence to date for the potential validity of such a bound is the vague analogy with the independent case, and the fact that a counterexample has yet to be found.

4.4 Seginer's Inequality

Throughout this chapter, we have focused attention on Gaussian random matrices. We depart briefly from this setting in this section to discuss some aspects of structured random matrices that arise under other distributions of the entries.

The main reason that we restricted attention to Gaussian matrices is that most of the difficulty of capturing the structure of the matrix arises in this setting; at the same time, all upper bounds we develop extend without difficulty to more general distributions, so there is no significant loss of generality in focusing on the Gaussian case. For example, let us illustrate the latter statement using the moment method.

Lemma 4.12 *Let X and Y be symmetric random matrices with independent entries (modulo symmetry). Assume that X_{ij} are centered and subgaussian, that is, $\mathbf{E}X_{ij} = 0$ and $\mathbf{E}[X_{ij}^{2p}]^{1/2p} \lesssim b_{ij}\sqrt{p}$ for all $p \geq 1$, and let $Y_{ij} \sim N(0, b_{ij}^2)$. Then*

$$\mathbf{E}[\mathrm{Tr}[X^{2p}]]^{1/2p} \lesssim \mathbf{E}[\mathrm{Tr}[Y^{2p}]]^{1/2p} \quad \textit{for all } p \geq 1.$$

Proof Let X' be an independent copy of X. Then $\mathbf{E}[\mathrm{Tr}[X^{2p}]] = \mathbf{E}[\mathrm{Tr}[(X - \mathbf{E}X')^{2p}]] \leq \mathbf{E}[\mathrm{Tr}[(X - X')^{2p}]]$ by Jensen's inequality. Moreover, $Z = X - X'$ a symmetric random matrix satisfying the same properties as X, with the additional property that the entries Z_{ij} have symmetric distribution. Thus $\mathbf{E}[Z_{ij}^p]^{1/p} \lesssim \mathbf{E}[Y_{ij}^p]^{1/p}$ for all $p \geq 1$ (for odd p both sides are zero by symmetry, while for even p this follows from the subgaussian assumption using $\mathbf{E}[Y_{ij}^{2p}]^{1/2p} \asymp b_{ij}\sqrt{p}$). It remains to note that

$$\begin{aligned}
\mathbf{E}[\mathrm{Tr}[X^{2p}]] &= \sum_{\text{cycle } \gamma \text{ of length } 2p} \prod_{1\leq i\leq j\leq n} \mathbf{E}[X_{ij}^{\#_{ij}(\gamma)}] \\
&\leq C^{2p} \sum_{\text{cycle } \gamma \text{ of length } 2p} \prod_{1\leq i\leq j\leq n} \mathbf{E}[Y_{ij}^{\#_{ij}(\gamma)}] = C^{2p}\,\mathbf{E}[\mathrm{Tr}[Y^{2p}]]
\end{aligned}$$

for a universal constant C, where $\#_{ij}(\gamma)$ denotes the number of times the edge (i, j) appears in the cycle γ. The conclusion follows immediately. □

Lemma 4.12 shows that to upper bound the moments of a subgaussian random matrix with independent entries, it suffices to obtain a bound in the Gaussian case. The reader may readily verify that the completely analogous approach can be applied in the more general setting of the noncommutative Khintchine inequality. On the other hand, Gaussian bounds using the random process method extend to the subgaussian setting by virtue of a general subgaussian comparison principle [24, Theorem 2.4.12]. Beyond the subgaussian setting, similar methods can be used for entries with heavy-tailed distributions, see, for example, [4].

The above observations indicate that, in some sense, Gaussian random matrices are the "worst case" among subgaussian matrices. One can go one step further and ask whether there is some form of universality: do all subgaussian random matrices behave like their Gaussian counterparts? The universality phenomenon plays a major role in recent advances in random matrix theory: it turns out that many properties of Wigner matrices do not depend on the distribution of the entries. Unfortunately, we cannot expect universal behavior for structured random matrices: while Gaussian matrices are the "worst case" among subgaussian matrices, matrices with subgaussian entries can sometimes behave much better. The simplest example is the case of diagonal matrices (Example 3.9) with i.i.d. entries on the diagonal: in the Gaussian case $\mathbf{E}\|X\| \asymp \sqrt{\log n}$, but obviously $\mathbf{E}\|X\| \asymp 1$ if the entries are uniformly bounded (despite that uniformly bounded random variables are obviously subgaussian). In view of such examples, there is little hope to obtain a complete understanding of structured random matrices for arbitrary distributions of the entries. This justifies the approach we have taken: we seek sharp bounds for Gaussian matrices, which give rise to powerful upper bounds for general distributions of the entries.

Remark 4.13 We emphasize in this context that Conjectures 1 and 2 in the previous section are fundamentally Gaussian in nature, and *cannot* hold as stated for subgaussian matrices. For a counterexample along the lines of Example 4.3, see [21].

Despite these negative observations, it can be of significant interest to go beyond the Gaussian setting to understand whether the bounds we have obtained can be systematically improved under more favorable assumptions on the distributions of the entries. To illustrate how such improvements could arise, we discuss a result of Seginer [21] for random matrices with independent *uniformly bounded* entries.

Theorem 4.14 (Seginer) *Let X be an $n \times n$ symmetric random matrix with independent entries (modulo symmetry) and $\mathbf{E}X_{ij} = 0$, $\|X_{ij}\|_\infty \lesssim b_{ij}$ for all i, j. Then*

$$\mathbf{E}\|X\| \lesssim \max_{i \le n} \sqrt{\sum_{j=1}^{n} b_{ij}^2}\,(\log n)^{1/4}.$$

The uniform bound $\|X_{ij}\|_\infty \lesssim b_{ij}$ certainly implies the much weaker subgaussian property $\mathbf{E}[X_{ij}^{2p}]^{1/2p} \lesssim b_{ij}\sqrt{p}$, so that the conclusion of Theorem 4.7 extends immediately to the present setting by Lemma 4.12. In many cases, the latter bound is much sharper than the one provided by Theorem 4.14; indeed, Theorem 4.14 is suboptimal even for Wigner matrices (it could be viewed of a variant of the noncommutative Khintchine inequality in the present setting with a smaller power in the logarithmic factor). However, the interest of Theorem 4.14 is that it *cannot* hold for Gaussian entries: for example, in the diagonal case $b_{ij} = \mathbf{1}_{i=j}$, Theorem 4.14 gives $\mathbf{E}\|X\| \lesssim (\log n)^{1/4}$ while any Gaussian bound must give at least $\mathbf{E}\|X\| \gtrsim \sqrt{\log n}$. In this sense, Theorem 4.14 illustrates that it is possible in some cases to exploit the effect of stronger distributional assumptions in order to obtain improved bounds for non-Gaussian random matrices. The simple proof that we will give (taken from [4]) shows very clearly how this additional distributional information enters the picture.

Proof (Proof of Theorem 4.14) The proof works by combining two very different bounds on the matrix norm. On the one hand, due to Lemma 4.12, we can directly apply the Gaussian bound of Theorem 4.7 in the present setting. On the other hand, as the entries of X are uniformly bounded, we can do something that is impossible for Gaussian random variables: we can *uniformly* bound the norm $\|X\|$ as

$$\begin{aligned}\|X\| &= \sup_{v\in B}\left|\sum_{i,j=1}^{n} v_i X_{ij} v_j\right| \le \sup_{v\in B}\sum_{i,j=1}^{n}(|v_i|\,|X_{ij}|^{1/2})(|X_{ij}|^{1/2}|v_j|) \\ &\le \sup_{v\in B}\sum_{i,j=1}^{n} v_i^2|X_{ij}| = \max_{i\le n}\sum_{j=1}^{n}|X_{ij}| \le \max_{i\le n}\sum_{j=1}^{n} b_{ij},\end{aligned}$$

where we have used the Cauchy-Schwarz inequality in going from the first to the second line. The idea behind the proof of Theorem 4.14 is roughly as follows. Many small entries of X can add up to give rise to a large norm; we might expect the cumulative effect of many independent centered random variables to give rise to Gaussian behavior. On the other hand, if a few large entries of X dominate the norm, there is no Gaussian behavior and we expect that the uniform bound provides much better control. To capture this idea, we partition the matrix into two parts $X = X_1 + X_2$, where X_1 contains the "small" entries and X_2 contains the "large" entries:

$$(X_1)_{ij} = X_{ij}\mathbf{1}_{b_{ij}\le u}, \qquad (X_2)_{ij} = X_{ij}\mathbf{1}_{b_{ij}>u}.$$

Applying the Gaussian bound to X_1 and the uniform bound to X_2 yields

$$\begin{aligned}\mathbf{E}\|X\| &\leq \mathbf{E}\|X_1\| + \mathbf{E}\|X_2\| \\ &\lesssim \max_{i\leq n}\sqrt{\sum_{j=1}^{n} b_{ij}^2 \mathbf{1}_{b_{ij}\leq u}} + u\sqrt{\log n} + \max_{i\leq n}\sum_{j=1}^{n} b_{ij}\mathbf{1}_{b_{ij}>u} \\ &\leq \max_{i\leq n}\sqrt{\sum_{j=1}^{n} b_{ij}^2} + u\sqrt{\log n} + \frac{1}{u}\max_{i\leq n}\sum_{j=1}^{n} b_{ij}^2.\end{aligned}$$

The proof is completed by optimizing over u. □

The proof of Theorem 4.14 illustrates the improvement that can be achieved by trading off between Gaussian and uniform bounds on the norm of a random matrix. Such tradeoffs play a fundamental role in the general theory that governs the suprema of bounded random processes [24, Chapter 5]. Unfortunately, this tradeoff is captured only very crudely by the suboptimal Theorem 4.14.

Developing a sharp understanding of the behavior of bounded random matrices is a problem of significant interest: the bounded analogue of sparse Wigner matrices (Example 4.9) has interesting connections with graph theory and computer science, cf. [1] for a review of such applications. Unlike in the Gaussian case, however, it is clear that the degree of the graph that defines a sparse Wigner matrix cannot fully explain its spectral norm in the present setting: very different behavior is exhibited in dense vs. locally tree-like graphs of the same degree [4, section 4.2]. To date, a deeper understanding of such matrices beyond the Gaussian case remains limited.

5 Sample Covariance Matrices

We finally turn our attention to a random matrix model that is somewhat different than the matrices we considered so far. The following model will be considered throughout this section. Let Σ be a given $d \times d$ positive semidefinite matrix, and let $X_1, X_2, \ldots, X_n$ be i.i.d. centered Gaussian random vectors in $\mathbb{R}^d$ with covariance matrix Σ. We consider in the following the $d \times d$ symmetric random matrix

$$Z = \frac{1}{n}\sum_{k=1}^{n} X_k X_k^* = \frac{XX^*}{n},$$

where we defined the $d \times n$ matrix $X_{ik} = (X_k)_i$. In contrast to the models considered in the previous sections, the random matrix Z is not centered: we have in fact $\mathbf{E}Z = \Sigma$. This gives rise to the classical statistical interpretation of this matrix. We can think of $X_1, \ldots, X_n$ as being i.i.d. data drawn from a centered Gaussian distribution with unknown covariance matrix Σ. In this setting, the random matrix Z, which depends only on the observed data, provides an unbiased estimator of the covariance matrix of the underlying data. For this reason, Z is known as the *sample covariance matrix*. Of primary interest in this setting is not so much the matrix norm $\|Z\| = \|X\|^2/n$ itself, but rather the deviation $\|Z - \Sigma\|$ of Z from its mean.

The model of this section could be viewed as being "semi-structured." On the one hand, the covariance matrix Σ is completely arbitrary, and it therefore allows for an arbitrary variance and dependence pattern within each column of the matrix X (as in the most general setting of the noncommutative Khintchine inequality). On the other hand, the columns of X are assumed to be i.i.d., so that no nontrivial structure among the columns is captured by the present model. While the latter assumption is limiting, it allows us to obtain a complete understanding of the structural parameters that control the expected deviation $\mathbf{E}\|Z - \Sigma\|$ in this setting [10].

Theorem 5.1 (Koltchinskii-Lounici) *In the setting of this section*

$$\mathbf{E}\|Z - \Sigma\| \asymp \|\Sigma\| \left(\sqrt{\frac{r(\Sigma)}{n}} + \frac{r(\Sigma)}{n} \right),$$

where $r(\Sigma) := \mathrm{Tr}[\Sigma]/\|\Sigma\|$ *is the* effective rank *of* Σ.

The remainder of this section is devoted to the proof of Theorem 5.1.

5.1 Upper Bound

The proof of Theorem 5.1 will use the random process method using tools that were already developed in the previous sections. It would be clear how to proceed if we wanted to bound $\|Z\|$: as $\|Z\| = \|X\|^2/n$, it would suffice to bound $\|X\|$ which is the supremum of a Gaussian process. Unfortunately, this idea does not extend directly to the problem of bounding $\|Z - \Sigma\|$: the latter quantity is not the supremum of a centered Gaussian process, but rather of a *squared* Gaussian process

$$\|Z - \Sigma\| = \sup_{v \in B} \left| \frac{1}{n} \sum_{k=1}^{n} \{\langle v, X_k \rangle^2 - \mathbf{E}\langle v, X_k \rangle^2\} \right|.$$

We therefore cannot directly apply a Gaussian comparison method such as the Slepian-Fernique inequality to control the expected deviation $\mathbf{E}\|Z - \Sigma\|$.

To surmount this problem, we will use a simple device that is widely used in the study of squared Gaussian processes (or *Gaussian chaos*), cf. [12, section 3.2].

Lemma 5.2 (Decoupling) *Let* $\tilde{X}$ *be an independent copy of* X. *Then*

$$\mathbf{E}\|Z - \Sigma\| \le \frac{2}{n} \mathbf{E}\|X\tilde{X}^*\|.$$

Proof By Jensen's inequality

$$\mathbf{E}\|Z-\Sigma\| = \frac{1}{n}\mathbf{E}\|\mathbf{E}[(X+\tilde{X})(X-\tilde{X})^*|X]\| \le \frac{1}{n}\mathbf{E}\|(X+\tilde{X})(X-\tilde{X})^*\|.$$

It remains to note that $(X+\tilde{X}, X-\tilde{X})$ has the same distribution as $\sqrt{2}\,(X,\tilde{X})$. □

Roughly speaking, the decoupling device of Lemma 5.2 allows us to replace the square XX^* of a Gaussian matrix by a product of two independent copies $X\tilde{X}^*$. While the latter is still not Gaussian, it becomes Gaussian if we condition on one of the copies (say, $\tilde{X}$). This means that $\|X\tilde{X}^*\|$ is the supremum of a Gaussian process *conditionally* on $\tilde{X}$. This is precisely what we will exploit in the sequel: we use the Slepian-Fernique inequality conditionally on $\tilde{X}$ to obtain the following bound.

Lemma 5.3 *In the setting of this section*

$$\mathbf{E}\|Z-\Sigma\| \lesssim \mathbf{E}\|X\|\frac{\sqrt{\mathrm{Tr}[\Sigma]}}{n} + \|\Sigma\|\sqrt{\frac{r(\Sigma)}{n}}.$$

Proof By Lemma 5.2 we have

$$\mathbf{E}\|Z-\Sigma\| \le \frac{2}{n}\mathbf{E}\bigg[\sup_{v,w\in B} Z_{v,w}\bigg], \qquad Z_{v,w} := \sum_{k=1}^{n}\langle v, X_k\rangle\langle w, \tilde{X}_k\rangle.$$

Writing for simplicity $\mathbf{E}_{\tilde{X}}[\cdot] = \mathbf{E}[\cdot|\tilde{X}]$, we can estimate

$$\begin{aligned}\mathbf{E}_{\tilde{X}}(Z_{v,w}-Z_{v',w'})^2 &\le 2\langle v-v', \Sigma(v-v')\rangle\sum_{k=1}^{n}\langle w,\tilde{X}_k\rangle^2 + 2\langle v', \Sigma v'\rangle\sum_{k=1}^{n}\langle w-w',\tilde{X}_k\rangle^2\\ &\le 2\|\tilde{X}\|^2\|\Sigma^{1/2}(v-v')\|^2 + 2\|\Sigma\|\,\|\tilde{X}^*(w-w')\|^2\\ &= \mathbf{E}_{\tilde{X}}(Y_{v,w}-Y_{v',w'})^2,\end{aligned}$$

where we defined

$$Y_{v,w} = \sqrt{2}\,\|\tilde{X}\|\,\langle v, \Sigma^{1/2}g\rangle + (2\|\Sigma\|)^{1/2}\,\langle w, \tilde{X}g'\rangle$$

with g, g' independent standard Gaussian vectors in $\mathbb{R}^d$ and $\mathbb{R}^n$, respectively. Thus

$$\begin{aligned}\mathbf{E}_{\tilde{X}}\bigg[\sup_{v,w\in B} Z_{v,w}\bigg] \le \mathbf{E}_{\tilde{X}}\bigg[\sup_{v,w\in B} Y_{v,w}\bigg] &\lesssim \|\tilde{X}\|\,\mathbf{E}\|\Sigma^{1/2}g\| + \|\Sigma\|^{1/2}\,\mathbf{E}_{\tilde{X}}\|\tilde{X}g\|\\ &\le \|\tilde{X}\|\sqrt{\mathrm{Tr}[\Sigma]} + \|\Sigma\|^{1/2}\,\mathrm{Tr}[\tilde{X}\tilde{X}^*]^{1/2}\end{aligned}$$

by the Slepian-Fernique inequality. Taking the expectation with respect to $\tilde{X}$ and using that $\mathbf{E}\|\tilde{X}\| = \mathbf{E}\|X\|$ and $\mathbf{E}[\mathrm{Tr}[\tilde{X}\tilde{X}^*]^{1/2}] \le \sqrt{n}\,\mathrm{Tr}[\Sigma]$ yields the conclusion. □

Lemma 5.3 has reduced the problem of bounding $\mathbf{E}\|Z - \Sigma\|$ to the much more straightforward problem of bounding $\mathbf{E}\|X\|$: as $\|X\|$ is the supremum of a Gaussian process, the latter is amenable to a direct application of the Slepian-Fernique inequality precisely as was done in the proof of Lemma 3.11.

Lemma 5.4 *In the setting of this section*

$$\mathbf{E}\|X\| \lesssim \sqrt{\mathrm{Tr}[\Sigma]} + \sqrt{n\|\Sigma\|}.$$

Proof Note that

$$\begin{aligned}
\mathbf{E}(\langle v, Xw\rangle - \langle v', Xw'\rangle)^2 &\le 2\,\mathbf{E}(\langle v - v', Xw\rangle)^2 + 2\,\mathbf{E}(\langle v', X(w-w')\rangle)^2 \\
&= 2\|\Sigma^{1/2}(v-v')\|^2\|w\|^2 + 2\|\Sigma^{1/2}v'\|^2\|w-w'\|^2 \\
&\le \mathbf{E}(X'_{v,w} - X'_{v',w'})^2
\end{aligned}$$

when $\|v\|, \|w\| \le 1$, where we defined

$$X'_{v,w} = \sqrt{2}\,\langle v, \Sigma^{1/2}g\rangle + \sqrt{2}\,\|\Sigma\|^{1/2}\,\langle w, g'\rangle$$

with g, g' independent standard Gaussian vectors in $\mathbb{R}^d$ and $\mathbb{R}^n$, respectively. Thus

$$\mathbf{E}\|X\| = \mathbf{E}\left[\sup_{v,w\in B}\langle v, Xw\rangle\right] \le \mathbf{E}\left[\sup_{v,w\in B} X'_{v,w}\right] \lesssim \mathbf{E}\|\Sigma^{1/2}g\| + \|\Sigma\|^{1/2}\mathbf{E}\|g\|$$

by the Slepian-Fernique inequality. The proof is easily completed. □

The proof of the upper bound in Theorem 5.1 is now immediately completed by combining the results of Lemma 5.3 and Lemma 5.4.

Remark 5.5 The proof of the upper bound given here reduces the problem of controlling the supremum of a Gaussian chaos process by decoupling to that of controlling the supremum of a Gaussian process. The original proof in [10] uses a different method that exploits a much deeper general result on the suprema of empirical processes of squares, cf. [24, Theorem 9.3.7]. While the route we have taken is much more elementary, the original approach has the advantage that it applies directly to subgaussian matrices. The result of [10] is also stated for norms other than the spectral norm, but proof given here extends readily to this setting.

5.2 Lower Bound

It remains to prove the lower bound in Theorem 5.1. The main idea behind the proof is that the decoupling inequality of Lemma 5.2 can be partially reversed.

Lemma 5.6 *Let $\tilde{X}$ be an independent copy of X. Then for every $v \in \mathbb{R}^d$*

$$\mathbf{E}\|(Z-\Sigma)v\| \geq \frac{1}{n}\,\mathbf{E}\|X\tilde{X}^*v\| - \frac{\|\Sigma v\|}{\sqrt{n}}.$$

Proof The reader may readily verify that the random matrix

$$X' = \left(I - \frac{\Sigma v v^*}{\langle v, \Sigma v\rangle}\right)X$$

is independent of the random vector X^*v (and therefore of $\langle v, Zv\rangle$). Moreover

$$(Z-\Sigma)v = \frac{XX^*v}{n} - \Sigma v = \frac{X'X^*v}{n} + \left(\frac{\langle v, Zv\rangle}{\langle v, \Sigma v\rangle} - 1\right)\Sigma v.$$

As the columns of X' are i.i.d. and independent of X^*v, the pair $(X'X^*v, X^*v)$ has the same distribution as $(X_1'\|X^*v\|, X^*v)$ where X_1' denotes the first column of X'. Thus

$$\mathbf{E}\|(Z-\Sigma)v\| = \mathbf{E}\left\|\frac{X_1'\|X^*v\|}{n} + \left(\frac{\langle v, Zv\rangle}{\langle v, \Sigma v\rangle} - 1\right)\Sigma v\right\| \geq \frac{1}{n}\,\mathbf{E}\|X^*v\|\,\mathbf{E}\|X_1'\|,$$

where we used Jensen's inequality conditionally on X'. Now note that

$$\mathbf{E}\|X_1'\| \geq \mathbf{E}\|X_1\| - \|\Sigma v\|\frac{\mathbf{E}|\langle v, X_1\rangle|}{\langle v, \Sigma v\rangle} \geq \mathbf{E}\|X_1\| - \frac{\|\Sigma v\|}{\langle v, \Sigma v\rangle^{1/2}}.$$

We therefore have

$$\mathbf{E}\|(Z-\Sigma)v\| \geq \frac{1}{n}\,\mathbf{E}\|X_1\|\,\mathbf{E}\|\tilde{X}^*v\| - \frac{1}{n}\,\mathbf{E}\|X^*v\|\frac{\|\Sigma v\|}{\langle v, \Sigma v\rangle^{1/2}} \geq \frac{1}{n}\,\mathbf{E}\|X\tilde{X}^*v\| - \frac{\|\Sigma v\|}{\sqrt{n}},$$

as $\mathbf{E}\|X^*v\| \leq \sqrt{n}\,\langle v, \Sigma v\rangle^{1/2}$ and as $X_1\|\tilde{X}^*v\|$ has the same distribution as $X\tilde{X}^*v$. □

As a corollary, we can obtain the first term in the lower bound.

Corollary 5.7 *In the setting of this section, we have*

$$\mathbf{E}\|Z-\Sigma\| \gtrsim \|\Sigma\|\sqrt{\frac{r(\Sigma)}{n}}.$$

Proof Taking the supremum over $v \in B$ in Lemma 5.6 yields

$$\mathbf{E}\|Z - \Sigma\| + \frac{\|\Sigma\|}{\sqrt{n}} \geq \sup_{v \in B} \frac{1}{n} \mathbf{E}\|X\tilde{X}^* v\| = \frac{1}{n} \mathbf{E}\|X_1\| \sup_{v \in B} \mathbf{E}\|\tilde{X}^* v\|.$$

Using Gaussian concentration as in the proof of Lemma 3.7, we obtain

$$\mathbf{E}\|X_1\| \gtrsim \mathbf{E}[\|X_1\|^2]^{1/2} = \sqrt{\mathrm{Tr}[\Sigma]}, \qquad \mathbf{E}\|\tilde{X}^* v\| \gtrsim \mathbf{E}[\|\tilde{X}^* v\|^2]^{1/2} = \sqrt{n \langle v, \Sigma v \rangle}.$$

This yields

$$\mathbf{E}\|Z - \Sigma\| + \frac{\|\Sigma\|}{\sqrt{n}} \gtrsim \|\Sigma\| \sqrt{\frac{r(\Sigma)}{n}}.$$

On the other hand, we can estimate by the central limit theorem

$$\frac{\|\Sigma\|}{\sqrt{n}} \lesssim \sup_{v \in B} \mathbf{E}|\langle v, (Z - \Sigma)v \rangle| \leq \mathbf{E}\|Z - \Sigma\|,$$

as $\langle v, (Z - \Sigma)v \rangle = \langle v, \Sigma v \rangle \frac{1}{n} \sum_{k=1}^{n} \{Y_k^2 - 1\}$ with $Y_k = \langle v, X_k \rangle / \langle v, \Sigma v \rangle^{1/2} \sim N(0, 1)$. □

We can now easily complete the proof of Theorem 5.1.

Proof (Proof of Theorem 5.1) The upper bound follows immediately from Lemmas 5.3 and 5.4. For the lower bound, suppose first that $r(\Sigma) \leq 2n$. Then $\sqrt{r(\Sigma)/n} \gtrsim r(\Sigma)/n$, and the result follows from Corollary 5.7. On the other hand, if $r(\Sigma) > 2n$,

$$\mathbf{E}\|Z - \Sigma\| \geq \mathbf{E}\|Z\| - \|\Sigma\| \geq \frac{\mathbf{E}\|X_1\|^2}{n} - \|\Sigma\| \frac{r(\Sigma)}{2n} = \|\Sigma\| \frac{r(\Sigma)}{2n},$$

where we used that $Z = \frac{1}{n} \sum_{k=1}^{n} X_k X_k^* \succeq \frac{1}{n} X_1 X_1^*$. □

Acknowledgements The author warmly thanks IMA for its hospitality during the annual program "Discrete Structures: Analysis and Applications" in Spring 2015. The author also thanks Markus Reiß for the invitation to lecture on this material in the 2016 spring school in Lübeck, Germany, which further motivated the exposition in this chapter. This work was supported in part by NSF grant CAREER-DMS-1148711 and by ARO PECASE award W911NF-14-1-0094.

References

1. Agarwal, N., Kolla, A., Madan, V.: Small lifts of expander graphs are expanding (2013). Preprint arXiv:1311.3268
2. Anderson, G.W., Guionnet, A., Zeitouni, O.: An introduction to random matrices, *Cambridge Studies in Advanced Mathematics*, vol. 118. Cambridge University Press, Cambridge (2010)
3. Bai, Z.D., Silverstein, J.W.: No eigenvalues outside the support of the limiting spectral distribution of large-dimensional sample covariance matrices. Ann. Probab. **26**(1), 316–345 (1998)
4. Bandeira, A.S., Van Handel, R.: Sharp nonasymptotic bounds on the norm of random matrices with independent entries. Ann. Probab. (2016). To appear
5. Boucheron, S., Lugosi, G., Massart, P.: Concentration inequalities. Oxford University Press, Oxford (2013)
6. Chevet, S.: Séries de variables aléatoires gaussiennes à valeurs dans $E\hat{\otimes}_\varepsilon F$. Application aux produits d'espaces de Wiener abstraits. In: Séminaire sur la Géométrie des Espaces de Banach (1977–1978), pp. Exp. No. 19, 15. École Polytech., Palaiseau (1978)
7. Dirksen, S.: Tail bounds via generic chaining. Electron. J. Probab. **20**, no. 53, 29 (2015)
8. Erdős, L., Yau, H.T.: Universality of local spectral statistics of random matrices. Bull. Amer. Math. Soc. (N.S.) **49**(3), 377–414 (2012)
9. Gordon, Y.: Some inequalities for Gaussian processes and applications. Israel J. Math. **50**(4), 265–289 (1985)
10. Koltchinskii, V., Lounici, K.: Concentration inequalities and moment bounds for sample covariance operators. Bernoulli (2016). To appear
11. Latała, R.: Some estimates of norms of random matrices. Proc. Amer. Math. Soc. **133**(5), 1273–1282 (electronic) (2005)
12. Ledoux, M., Talagrand, M.: Probability in Banach spaces, *Ergebnisse der Mathematik und ihrer Grenzgebiete*, vol. 23. Springer-Verlag, Berlin (1991). Isoperimetry and processes
13. Lust-Piquard, F.: Inégalités de Khintchine dans C_p $(1 < p < \infty)$. C. R. Acad. Sci. Paris Sér. I Math. **303**(7), 289–292 (1986)
14. Mackey, L., Jordan, M.I., Chen, R.Y., Farrell, B., Tropp, J.A.: Matrix concentration inequalities via the method of exchangeable pairs. Ann. Probab. **42**(3), 906–945 (2014)
15. Marcus, A.W., Spielman, D.A., Srivastava, N.: Interlacing families II: Mixed characteristic polynomials and the Kadison-Singer problem. Ann. of Math. (2) **182**(1), 327–350 (2015)
16. Pisier, G.: Probabilistic methods in the geometry of Banach spaces. In: Probability and analysis (Varenna, 1985), *Lecture Notes in Math.*, vol. 1206, pp. 167–241. Springer, Berlin (1986)
17. Pisier, G.: Introduction to operator space theory, *London Mathematical Society Lecture Note Series*, vol. 294. Cambridge University Press, Cambridge (2003)
18. Reed, M., Simon, B.: Methods of modern mathematical physics. II. Fourier analysis, self-adjointness. Academic Press, New York-London (1975)
19. Riemer, S., Schütt, C.: On the expectation of the norm of random matrices with non-identically distributed entries. Electron. J. Probab. **18**, no. 29, 13 (2013)
20. Rudelson, M.: Almost orthogonal submatrices of an orthogonal matrix. Israel J. Math. **111**, 143–155 (1999)
21. Seginer, Y.: The expected norm of random matrices. Combin. Probab. Comput. **9**(2), 149–166 (2000)
22. Sodin, S.: The spectral edge of some random band matrices. Ann. of Math. (2) **172**(3), 2223–2251 (2010)
23. Srivastava, N., Vershynin, R.: Covariance estimation for distributions with $2 + \varepsilon$ moments. Ann. Probab. **41**(5), 3081–3111 (2013)
24. Talagrand, M.: Upper and lower bounds for stochastic processes, vol. 60. Springer, Heidelberg (2014)
25. Tao, T.: Topics in random matrix theory, *Graduate Studies in Mathematics*, vol. 132. American Mathematical Society, Providence, RI (2012)

26. Tao, T., Vu, V.: Random matrices: the universality phenomenon for Wigner ensembles. In: Modern aspects of random matrix theory, *Proc. Sympos. Appl. Math.*, vol. 72, pp. 121–172. Amer. Math. Soc., Providence, RI (2014)
27. Tomczak-Jaegermann, N.: The moduli of smoothness and convexity and the Rademacher averages of trace classes $S_p(1 \leq p < \infty)$. Studia Math. **50**, 163–182 (1974)
28. Tropp, J.: An introduction to matrix concentration inequalities. Foundations and Trends in Machine Learning (2015)
29. Tropp, J.: Second-order matrix concentration inequalities (2015). Preprint arXiv:1504.05919
30. Van Handel, R.: Chaining, interpolation, and convexity. J. Eur. Math. Soc. (2016). To appear
31. Van Handel, R.: On the spectral norm of Gaussian random matrices. Trans. Amer. Math. Soc. (2016). To appear
32. Vershynin, R.: Introduction to the non-asymptotic analysis of random matrices. In: Compressed sensing, pp. 210–268. Cambridge Univ. Press, Cambridge (2012)

Rates of Convergence for Empirical Spectral Measures: A Soft Approach

Elizabeth S. Meckes and Mark W. Meckes

Abstract Understanding the limiting behavior of eigenvalues of random matrices is the central problem of random matrix theory. Classical limit results are known for many models, and there has been significant recent progress in obtaining more quantitative, non-asymptotic results. In this paper, we describe a systematic approach to bounding rates of convergence and proving tail inequalities for the empirical spectral measures of a wide variety of random matrix ensembles. We illustrate the approach by proving asymptotically almost sure rates of convergence of the empirical spectral measure in the following ensembles: Wigner matrices, Wishart matrices, Haar-distributed matrices from the compact classical groups, powers of Haar matrices, randomized sums and random compressions of Hermitian matrices, a random matrix model for the Hamiltonians of quantum spin glasses, and finally the complex Ginibre ensemble. Many of the results appeared previously and are being collected and described here as illustrations of the general method; however, some details (particularly in the Wigner and Wishart cases) are new.

Our approach makes use of techniques from probability in Banach spaces, in particular concentration of measure and bounds for suprema of stochastic processes, in combination with more classical tools from matrix analysis, approximation theory, and Fourier analysis. It is highly flexible, as evidenced by the broad list of examples. It is moreover based largely on "soft" methods, and involves little hard analysis.

The most fundamental problem in random matrix theory is to understand the limiting behavior of the empirical spectral distribution of large random matrices, as the size tends to infinity. The first result on this topic is the famous Wigner semi-circle law, the first version of which was proved by Wigner in 1955 [52, 53]. A random matrix is called a *Wigner matrix* if it is Hermitian, with independent

E.S. Meckes • M.W. Meckes (✉)
Department of Mathematics, Applied Mathematics, and Statistics, Case Western Reserve University, 10900 Euclid Ave., Cleveland, OH 44106, USA
e-mail: elizabeth.meckes@case.edu; mark.meckes@case.edu

E. Carlen et al. (eds.), *Convexity and Concentration*, The IMA Volumes in Mathematics and its Applications 161, DOI 10.1007/978-1-4939-7005-6_5

entries on and above the diagonal. Wigner showed that, under some conditions on the distributions of the entries, the limiting empirical spectral measure of a (normalized) Wigner matrix is the semi-circular law ρ_{sc}.

Wigner's first version of the semi-circle law gave convergence in expectation only; i.e., he showed that the expected number of eigenvalues of a Wigner matrix in an interval converged to the value predicted by the semi-circle law, as the size of the matrix tended to infinity. His second paper improved this to convergence "weakly in probability." The analog for random unitary matrices, namely that their spectral measures converge to the uniform measure on the circle, seems intuitively obvious; surprisingly, convergence in mean and weak convergence in probability were not proved until nearly 40 years after Wigner's original work [9].

While these results are fundamental, the limitations of limit theorems such as these are well known. Just as the Berry–Esseen theorem and Hoeffding-type inequalities provide real tools for applications where the classical central limit theorem only justifies heuristics, it is essential to improve the classical limit results of random matrix theory to quantitative approximation results which have content for large but finite random matrices. See [8, 49] for extended discussions of this so-called "non-asymptotic" random matrix theory and its applications.

In this paper, we describe a systematic approach to bounding rates of convergence and proving tail inequalities for the empirical spectral measures of a wide variety of random matrix ensembles. This approach makes use of techniques from probability in Banach spaces, in particular concentration of measure and bounds for suprema of stochastic processes, in combination with more classical tools from matrix analysis, approximation theory, and Fourier analysis. Our approach is highly flexible, and can be used for a wide variety of types of matrix ensembles, as we will demonstrate in the following sections. Moreover, it is based largely on "soft" methods, and involves little hard analysis. Our approach is restricted to settings in which there is a concentration of measure phenomenon; in this sense, it has rather different strengths than the methods used in, for example, [13, 17, 47] and many other works referred to in those papers. Those approaches achieve sharper results without requiring a measure concentration hypothesis, but they require many delicate estimates and are mainly restricted to random matrices constructed from independent random variables, whereas our methods have no independence requirements.

The following key observation, a consequence of the classical Hoffman–Wielandt inequality (see [2, Theorem VI.4.1]), underlies the approach.

Lemma 1 (see [36, Lemma 2.3]) *For an $n \times n$ normal matrix M over $\mathbb{C}$, let $\lambda_1, \ldots, \lambda_n$ denote the eigenvalues, and let μ_M denote the spectral measure of M; i.e.,*

$$\mu_M := \frac{1}{n} \sum_{j=1}^{n} \delta_{\lambda_j}.$$

Then

(a) if $f : \mathbb{C} \to \mathbb{R}$ is 1-Lipschitz, then the map

$$M \longmapsto \int f \, d\mu_M$$

is $\frac{1}{\sqrt{n}}$-Lipschitz, with respect to the Hilbert–Schmidt distance on the set of normal matrices; and

(b) if ν is any probability measure on $\mathbb{C}$ and $p \in [1, 2]$, the map

$$M \longmapsto W_p(\mu_M, \nu)$$

is $\frac{1}{\sqrt{n}}$-Lipschitz.

Here W_p denotes the L_p-Kantorovich (or Wasserstein) distance on probability measures on $\mathbb{C}$, defined by

$$W_p(\mu, \nu) = \left(\inf_{\pi} \int |x - y|^p \, d\pi(x, y) \right)^{1/p},$$

where the infimum ranges over probability measures π on $\mathbb{C} \times \mathbb{C}$ with marginals μ and ν. The Kantorovich–Rubinstein theorem (see [50, Theorem 1.14]) gives that

$$W_1(\mu, \nu) = \sup_{|f|_L \leq 1} \left(\int f \, d\mu - \int f \, d\nu \right),$$

where $|f|_L$ denotes the Lipschitz constant of f; this connects part (a) of Lemma 1 with estimates on W_1.

In many random matrix ensembles of interest there is a concentration of measure phenomenon, meaning that well-behaved functions are "essentially constant," in the sense that they are close to their means with high probability. A prototype is the following Gaussian concentration phenomenon (see [28]).

Proposition 2 *If $F : \mathbb{R}^n \to \mathbb{R}$ is a 1-Lipschitz function and Z is a standard Gaussian random vector in $\mathbb{R}^n$, then*

$$\mathbb{P}\left[F(Z) - \mathbb{E}F(Z) \geq t\right] \leq e^{-t^2/2}$$

for all $t > 0$.

Suppose now that M is a random matrix satisfying such a concentration property. Lemma 1 means that one can obtain a bound on $W_p(\mu_M, \nu)$ which holds with high probability if one can bound $\mathbb{E}W_p(\mu_M, \nu)$. That is, a bound on the *expected* distance to the limiting measure immediately implies an asymptotically almost sure bound.

The tail estimates coming from measure concentration are typically exponential or better, and therefore imply almost sure convergence rates via the Borel–Cantelli lemma.

We are thus left with the problem of bounding the expected distance from the empirical spectral measure μ_M to some deterministic reference measure ν. There are two different methods used for this step, depending on the properties of the ensemble:

(1) **Eigenvalue rigidity.** In some ensembles, each of the (ordered) individual eigenvalues can be assigned a predicted location based on the limiting spectral measure for the ensemble, such that all (or at least many) eigenvalues concentrate strongly near these predicted locations. In this case ν is taken to be a discrete measure supported on those predicted locations, and the concentration allows one to easily estimate $\mathbb{E}W_p(\mu_M, \nu)$.

(2) **Entropy methods.** If instead we set $\nu = \mathbb{E}\mu_M$, then the Kantorovich–Rubinstein theorem implies that

$$W_1(\mu_M, \nu) = \sup_{|f|_L \leq 1} \left(\int f \, d\mu_M - \mathbb{E} \int f \, d\mu_M \right),$$

so that $W_1(\mu_M, \nu)$ is the supremum of a centered stochastic process indexed by the unit ball of the space of Lipschitz functions on $\mathbb{C}$. In ensembles with a concentration phenomenon for Lipschitz functions, part (a) of Lemma 1 translates to an increment condition on this stochastic process, which gives a route to bounding its expected supremum via classical entropy methods.

Finally, it may still be necessary to estimate the distance from the measure ν to the limiting spectral measure for the random matrix ensemble. The techniques used to do this vary by the ensemble, but this is a more classical problem of convergence of a sequence of deterministic measures to a limit, and any of the many techniques for obtaining rates of convergence may be useful.

Applications of concentration of measure to random matrices date from at least as long ago as the 1970s; a version of the argument for the concentration of $W_1(\mu_M, \nu)$ essentially appears in the 2000 paper [22] of Guionnet and Zeitouni. See [8, 29, 48] for surveys of concentration methods in random matrix theory.

The method of eigenvalue rigidity to bound Kantorovich distances is particularly suited to situations in which the empirical spectrum is a determinantal point process; this was first observed in the work of Dallaporta [6, 7]. The entropy approach to random Kantorovich distances was introduced in the context of random projections in [33, 34]; it was first applied for empirical spectral measures in [35, 36]. A further abstraction was given by Ledoux [30].

Organization The rest of this paper is a series of sections sketching some version of the program described above for a number of random matrix ensembles. Section 1 and section 2 discuss Wigner and Wishart matrices, combining eigenvalue rigidity arguments of Dallaporta [6, 7] with measure concentration. Section 3 discusses random matrices drawn uniformly from classical compact matrix groups, and Section 4 discusses powers of such matrices; both those sections follow [37] and also use the eigenvalue rigidity approach. The next three sections use the entropy method: Sections 5 and 6 discuss randomized sums and random compressions of Hermitian matrices, following [36], and Section 7 discusses Hamiltonians of quantum spin glasses, following [3]. Finally, Section 8, following [38], demonstrates in case of the complex Ginibre ensemble, how eigenvalue rigidity alone allows one to carry our much of our program even without the use of a general concentration phenomenon together with Lemma 1.

1 Wigner Matrices

In this section we outline how our approach can be applied to the most central model of random matrix theory, that of Wigner matrices. We begin with the most classical case: the Gaussian Unitary Ensemble (GUE). Let M_n be a random $n \times n$ Hermitian matrix, whose entries $\{[M_n]_{jk} \mid 1 \leq j \leq k \leq n\}$ are independent random variables, such that each $[M_n]_{jj}$ has a $N(0, n^{-1})$ distribution, and each $[M_n]_{jk}$ for $j < k$ has independent real and imaginary parts, each with an $N(0, (2n)^{-1})$ distribution. Since M_n is Hermitian, it has real eigenvalues $\lambda_1 \leq \cdots \leq \lambda_n$. Wigner's theorem implies that the empirical spectral measure

$$\mu_n = \frac{1}{n}\sum_{j=1}^{n} \delta_{\lambda_j}$$

converges to the semicircle law ρ_{sc}. The following result quantifies this convergence.

Theorem 3 *Let M_n be as above, and let μ_n denote its spectral measure. Then*

(a) $\mathbb{E}W_2(\mu_n, \rho_{sc}) \leq C\dfrac{\sqrt{\log(n)}}{n}$,

(b) $\mathbb{P}\left[W_2(\mu_n, \rho_{sc}) \geq C\dfrac{\sqrt{\log(n)}}{n} + t\right] \leq e^{-n^2t^2/2}$ *for all* $t \geq 0$*, and*

(c) with probability 1, for sufficiently large n, $W_2(\mu_n, \rho_{sc}) \leq C'\dfrac{\sqrt{\log(n)}}{n}$.

Here and in what follows, symbols such as c, C, C' denote constants which are independent of dimension.

Part (a) of Theorem 3 was proved by Dallaporta in [6] using the eigenvalue rigidity approach; the proof is outlined below.

Lemma 1 and the Gaussian concentration of measure property (Proposition 2), implies that if F is a 1-Lipschitz function (with respect to the Hilbert–Schmidt distance) on the space of Hermitian matrices, then

$$\mathbb{P}\left[F(M_n) \geq \mathbb{E}F(M_n) + t\right] \leq e^{-nt^2/2} \tag{1}$$

for all $t \geq 0$. This fact, together with part (b) of Lemma 1 and part (a) of Theorem 3 now imply part (b). Finally, part (c) follows from part (b) by the Borel–Cantelli lemma. So it remains only to prove part (a).

Define $\gamma_j \in \mathbb{R}$ such that $\rho_{sc}((-\infty, \gamma_j]) = \frac{j}{n}$; this is the predicted location of the jth eigenvalue λ_j of M_n. The discretization ν_n of the semi-circle law ρ_{sc} is given by

$$\nu_n := \frac{1}{n} \sum_{j=1}^{n} \delta_{\gamma_j}.$$

It can be shown that that $W_2(\rho_{sc}, \nu_n) \leq \frac{C}{n}$. Furthermore, by the definition of W_2,

$$\mathbb{E} W_2^2(\mu_n, \nu_n) \leq \frac{1}{n} \sum_{j=1}^{n} \mathbb{E} \left| \lambda_j - \gamma_j \right|^2.$$

This reduces the proof of part (a) to estimating the latter expectations.

It is a classical fact that the eigenvalues of the GUE form a determinantal point process with kernel

$$K_n(x, y) = \sum_{j=0}^{n} h_j(x) h_j(y) e^{-(x^2+y^2)/2},$$

where the h_j are the orthonormalized Hermite polynomials [39, Section 6.2]. (The reader is referred to [24] for the definition of a determinantal point process.) The following is then a special case of some important general properties of determinantal point processes [24, Theorem 7], [23].

Proposition 4 *For each $x \in \mathbb{R}$, let $\mathcal{N}_x$ denote the number of eigenvalues of M_n which are less than or equal to x. Then*

$$\mathcal{N}_x \overset{d}{=} \sum_{i=1}^{n} \xi_i,$$

where the ξ_i are independent $\{0, 1\}$-valued Bernoulli random variables.

Moreover,

$$\mathbb{E} \mathcal{N}_x = \int_{-\infty}^{x} K_n(u, u) \, du \qquad \textit{and} \qquad \operatorname{Var} \mathcal{N}_x = \int_{-\infty}^{x} \int_{x}^{\infty} K_n(u, v)^2 \, du \, dv.$$

The first part of this result can be combined with the classical Bernstein inequality to deduce that for each $t > 0$,

$$\mathbb{P}\left[|\mathcal{N}_x - \mathbb{E}\mathcal{N}_x| > t \right] \leq 2 \exp\left(-\frac{t^2}{2\sigma_x^2 + t} \right),$$

where $\sigma_x^2 = \operatorname{Var} \mathcal{N}_x$. Using estimates on $\mathbb{E}\mathcal{N}_x$ due to Götze and Tikhomirov [18] and on σ_x^2 due to Gustavsson [23] (both of which can be deduced from the second part of Proposition 4), this implies that for $x \in (-2+\delta, 2-\delta)$,

$$\mathbb{P}\left[|\mathcal{N}_x - n\rho_{sc}((-\infty, x])| > t + C\right] \leq 2\exp\left(-\frac{t^2}{2c_\delta \log(n) + t}\right)$$

for each $t \geq 0$. Combining this with the observation that

$$\mathbb{P}\left[\lambda_j > \gamma_j + t\right] = \mathbb{P}\left[\mathcal{N}_{\gamma_j + t} < j\right],$$

one can deduce, upon integrating by parts, that

$$\mathbb{E}\left|\lambda_j - \gamma_j\right|^2 \leq C_\varepsilon \frac{\log(n)}{n^2}$$

for $j \in [\varepsilon n, (1-\varepsilon)n]$. This provides the necessary estimates in the bulk of the spectrum. Dallaporta established similar but weaker bounds for the soft edge of the spectrum using essentially the last part of Proposition 4, and for the hard edge using tail estimates due to Ledoux and Rider [31]. This completes the proof of Theorem 3.

The real symmetric counterpart of the GUE is the Gaussian Orthogonal Ensemble (GOE), whose entries $\left\{[M_n]_{jk} \mid 1 \leq j \leq k \leq n\right\}$ are independent real random variables, such that each $[M_n]_{jj}$ has an $N(0, n^{-1})$ distribution, and each $[M_n]_{jk}$ for $j < k$ has a $N(0, (\sqrt{2}n)^{-1})$ distribution. The spectrum of the GOE does not form a determinantal point process, but a close distributional relationship between the eigenvalue counting functions of the GOE and GUE was found in [16, 41]. Using this, Dallaporta showed that part (a) of Theorem 3 also applies to the GOE. Part (b) then follows from the Gaussian concentration of measure property as before, and part (c) from the Borel–Cantelli lemma.

To move beyond the Gaussian setting, Dallaporta invokes the Tao–Vu four moment theorem [46, 45] and a localization theorem due to Erdős, Yau, and Yin [14] to extend Theorem 3(a) to random matrices with somewhat more general entries. The proofs of these results involve the kind of hard analysis which it is our purpose to avoid in this paper. However, it is straightforward, under appropriate hypotheses, to extend the measure concentration argument for part (b) of Theorem 3, and we indicate briefly how this is done.

A probability measure μ on $\mathbb{R}$ is said to satisfy a quadratic transportation cost inequality (QTCI) with constant $C > 0$ if

$$W_2(\mu, \nu) \leq \sqrt{CH(\nu|\mu)}$$

for any probability measure ν which is absolutely continuous with respect to μ, where $H(\nu|\mu)$ denotes relative entropy.

Proposition 5 (see [28, Chapter 6]) *Suppose that $X_1, \ldots, X_n$ are independent random variables whose distributions each satisfy a QTCI with constant C. If $F : \mathbb{R}^n \to \mathbb{R}$ is a 1-Lipschitz function, then*

$$\mathbb{P}\left[F(X) - \mathbb{E}F(X) \geq t\right] \leq e^{-t^2/C}$$

for all $t > 0$.

A QTCI is the most general possible hypothesis which implies subgaussian tail decay, independent of n, for Lipschitz functions of independent random variables; see [19]. It holds in particular for any distribution satisfying a logarithmic Sobolev inequality, including Gaussian distributions, or a distribution with a density on a finite interval bounded above and below by positive constants. Using Dallaporta's arguments for part (a) and substituting Proposition 5 in place of the Gaussian concentration phenomenon, we arrive at the following generalization of Theorem 3.

Theorem 6 *Let M_n be a random Hermitian matrix whose entries satisfy each of the following:*

- *The random variables $\left\{\operatorname{Re} M_{jk}\right\}_{1 \leq j \leq k \leq n}$ and $\left\{\operatorname{Im} M_{jk}\right\}_{1 \leq j < k \leq n}$ are all independent.*
- *The first four moments of each of these random variables is the same as for the GUE (respectively, GOE).*
- *Each of these random variables satisfies a QTCI with constant $cn^{-1/2}$.*

Let μ_n denote the spectral measure of M_n. Then

(a) $\mathbb{E}W_2(\mu_n, \rho_{sc}) \leq C\frac{\sqrt{\log(n)}}{n}$,

(b) $\mathbb{P}\left[W_2(\mu_n, \rho_{sc}) \geq C\frac{\sqrt{\log(n)}}{n} + t\right] \leq e^{-cn^2t^2}$ *for all $t \geq 0$, and*

(c) with probability 1, for sufficiently large n, $W_2(\mu_n, \rho_{sc}) \leq C'\frac{\sqrt{\log(n)}}{n}$.

As mentioned above, a QTCI is a minimal assumption to reach exactly this result by these methods. A weaker and more classical assumption would be a Poincaré inequality, which implies subexponential decay for Lipschitz functions, and is the most general hypothesis implying any decay independent of n; see [20] and the references therein. If the third condition in Theorem 6 is replaced by the assumption of a Poincaré inequality with constant $cn^{-1/2}$, then the same kind of argument leads to an almost sure convergence rate of order $\frac{\log(n)}{n}$; we omit the details.

2 Wishart Matrices

In this section we apply the strategy described in the introduction to Wishart matrices (i.e., random sample covariance matrices). Let $m \geq n$, and let X be an $m \times n$ random matrix with i.i.d. entries, and define the Hermitian positive-semidefinite random matrix

$$S_{m,n} := \frac{1}{m} X^* X.$$

We denote the eigenvalues of $S_{m,n}$ by $0 \le \lambda_1 \le \cdots \le \lambda_n$ and the empirical spectral measure by

$$\mu_{m,n} = \frac{1}{n} \sum_{j=1}^{n} \delta_{\lambda_j}.$$

It was first proved in [32] that, under some moment conditions, if $\frac{n}{m} \to \rho > 0$ as $n, m \to \infty$, then $\mu_{m,n}$ converges to the Marchenko–Pastur law μ_ρ with parameter ρ, with compactly supported density given by

$$f_\rho(x) = \frac{1}{2\pi x} \sqrt{(b_\rho - x)(x - a_\rho)},$$

on (a_ρ, b_ρ), with $a_\rho = (1 - \sqrt{\rho})^2$ and $b_\rho = (1 + \sqrt{\rho})^2$. The following result quantifies this convergence for many distributions.

Theorem 7 *Suppose that for each n,* $0 < c \le \frac{n}{m} \le 1$, *and that X is an* $m \times n$ *random matrix whose entries satisfy each of the following:*

- *The random variables* $\{\mathrm{Re}\, X_{jk}\}_{\substack{1 \le j \le m \\ 1 \le k \le n}}$ *and* $\{\mathrm{Im}\, X_{jk}\}_{\substack{1 \le j \le m \\ 1 \le k \le n}}$ *are all independent.*
- *The first four moments of each of these random variables are the same as for a standard complex (respectively, real) normal random variable.*
- *Each of these random variables satisfies a QTCI with constant C.*

Let $\rho = \frac{n}{m}$ *and let* $\mu_{m,n}$ *denote the spectral measure of* $S_{m,n} = \frac{1}{m} X^* X$. *Then*

(a) $\mathbb{E} W_2(\mu_{m,n}, \mu_\rho) \le C \frac{\sqrt{\log(n)}}{n}$,

(b) $\mathbb{P}\left[W_2(\mu_{m,n}, \mu_\rho) \ge C \frac{\sqrt{\log(n)}}{n} + t \right] \le e^{-cm \min\{nt^2, \sqrt{n}t\}}$ *for all* $t \ge c \frac{\sqrt{\log(n)}}{n}$, *and*

(c) with probability 1, for sufficiently large n, $W_2(\mu_{m,n}, \mu_\rho) \le C' \frac{\sqrt{\log(n)}}{n}$.

Strictly speaking, part (c) does not, as stated, imply almost sure convergence of $\mu_{m,n}$, since ρ and hence μ_ρ itself depends on n. However, if $\rho = \rho(n)$ has a limiting value ρ^* as $n \to \infty$ (as in the original Marchenko–Pastur result), then the measures μ_ρ converge to μ_{ρ^*}. This convergence can easily be quantified, but we will not pursue the details here.

Proof Part (a) was proved by Dallaporta in [7], by the same methods as in Theorem 6(a) discussed in the last section. First, when the entries of X are complex normal random variables (in which $S_{m,n}$ is the unitary Laguerre ensemble), the eigenvalues of $S_{m,n}$ form a determinantal point process. This implies an analogue of Proposition 4, from which eigenvalue rigidity results can be deduced, leading to the estimate in part (a) in this case. The result is extended to real Gaussian random matrices using interlacing results, and to more general distributions using versions of the four moment theorem for Wishart random matrices. The reader is referred to [7] for the details.

The proof of part (b) is more complicated than in the previous section, because the random matrix $S_{m,n}$ depends quadratically on the independent entries of X. However, we can still apply the machinery of measure concentration by using the fact that $S_{m,n}$ possesses local Lipschitz behavior, combined with a truncation argument. Indeed, if X, Y are $m \times n$ matrices over $\mathbb{C}$,

$$\begin{aligned} \left\| \frac{1}{m} X^* X - \frac{1}{m} Y^* Y \right\|_{HS} &\leq \frac{1}{m} \|X^*(X-Y)\|_{HS} + \frac{1}{m} \|(X^* - Y^*)Y)\|_{HS} \\ &\leq \frac{1}{m} \left(\|X\|_{op} + \|Y\|_{op} \right) \|X-Y\|_{HS}, \end{aligned} \tag{2}$$

where we have used the facts that both the Hilbert–Schmidt norm $\|\cdot\|_{HS}$ and the operator norm $\|\cdot\|_{op}$ are invariant under conjugation and transposition, and that $\|AB\|_{HS} \leq \|A\|_{op} \|B\|_{HS}$.

Thus, for a given $K > 0$, the function

$$X \mapsto \frac{1}{m} X^* X$$

is $\frac{2K}{\sqrt{m}}$-Lipschitz on $\left\{ X \in \mathrm{M}_{m,n}(\mathbb{C}) \mid \|X\|_{op} \leq K\sqrt{m} \right\}$, and so by Lemma 1(b), the function

$$F : X \mapsto W_2(\mu_{m,n}, \mu_\rho)$$

is $\frac{2K}{\sqrt{mn}}$-Lipschitz on this set. We can therefore extend F to a $\frac{2K}{\sqrt{mn}}$-Lipschitz function $\widetilde{F} : \mathrm{M}_{m,n}(\mathbb{C}) \to \mathbb{R}$ (cf. [15, Theorem 3.1.2]); we may moreover assume that $\widetilde{F}(X) \geq 0$ and

$$\sup_{X \in \mathrm{M}_{m,n}(\mathbb{C})} \widetilde{F}(X) = \sup_{\|X\|_{op} \leq K\sqrt{m}} W_2(\mu_{m,n}, \mu_\rho). \tag{3}$$

Proposition 5 now allows us to control $\widetilde{F}(X)$ and $\|X\|_{op}$, which are both Lipschitz functions of X.

First, an elementary discretization argument using Proposition 5 (cf. [49, Theorem 5.39], or alternatively Lemma 15 below) shows that

$$\mathbb{P}\left[\|X\|_{op} > K\sqrt{m} \right] \leq 2e^{-cm} \tag{4}$$

for some $K, c > 0$. We will use this K in the following.

Next, Proposition 5 implies that

$$\mathbb{P}\left[\widetilde{F}(X) > t\right] \leq Ce^{-cmnt^2} \tag{5}$$

as long as $t \geq 2\mathbb{E}\widetilde{F}(X)$. Now

$$\begin{aligned}\mathbb{E}\widetilde{F}(X) &= \mathbb{E}W_2(\mu_{m,n},\mu_\rho) + \mathbb{E}\left[\left(\widetilde{F}(X) - W_2(\mu_{m,n},\mu_\rho)\right)\mathbb{1}_{\|X\|_{op}>K\sqrt{m}}\right] \\ &\leq C\frac{\sqrt{\log(n)}}{n} + \left(\sup_{\|X\|_{op}\leq K\sqrt{m}} W_2(\mu_{m,n},\mu_\rho)\right)\mathbb{P}[\|X\|_{op} > K\sqrt{m}]\end{aligned} \tag{6}$$

by part (a) and (3). Since μ_ρ is supported on $[a_\rho, b_\rho]$, and $\mu_{m,n}$ is supported on $\left[0, \left\|\frac{1}{m}XX^*\right\|_{op}\right] = \left[0, \frac{1}{m}\|X\|_{op}^2\right]$,

$$\sup_{\|X\|_{op}\leq K\sqrt{m}} W_2(\mu_{m,n},\mu_\rho) \leq \max\{b_\rho, K^2\} \leq C,$$

and so by (4) and (6),

$$\mathbb{E}\widetilde{F}(X) \leq C\frac{\sqrt{\log(n)}}{n} + Ce^{-cm} \leq C'\frac{\sqrt{\log(n)}}{n}.$$

Finally, we have

$$\begin{aligned}\mathbb{P}\left[W_2(\mu_{m,n},\mu_\rho) > t\right] &\leq \mathbb{P}\left[W_2(\mu_{m,n},\mu_\rho) > t, \|X\|_{op} \leq K\sqrt{m}\right] + \mathbb{P}\left[\|X\|_{op} > K\sqrt{m}\right] \\ &\leq \mathbb{P}\left[\widetilde{F}(X) > t\right] + \mathbb{P}\left[\|X\|_{op} > K\sqrt{m}\right] \\ &\leq C'e^{-cmnt^2}\end{aligned} \tag{7}$$

for $c_1\frac{\sqrt{\log(n)}}{n} \leq t \leq \frac{c_2}{\sqrt{n}}$ by (4) and (5). We omit the details of the similar argument to obtain a subexponential bound for $t > \frac{c_2}{\sqrt{n}}$. This concludes the proof of part (b).

Part (c) follows as before using the Borel–Cantelli lemma. □

An alternative approach to quantifying the limiting behavior of the spectrum of Wishart matrices is to consider the singular values $0 \leq \sigma_1 \leq \cdots \leq \sigma_n$ of $\frac{1}{\sqrt{m}}X$; that is, $\sigma_j = \sqrt{\lambda_j}$. Lemma 1 can be applied directly in that context, by using the fact that the eigenvalues of the Hermitian matrix $\begin{bmatrix} 0 & X \\ X^* & 0 \end{bmatrix}$ are $\{\pm\sigma_j\}$. However, if one is ultimately interested in the eigenvalues $\{\lambda_j\}$, then translating the resulting concentration estimates to eigenvalues ends up requiring the same kind of analysis carried out above.

3 Uniform Random Matrices from the Compact Classical Groups

Each of the compact classical matrix groups $\mathbb{O}(n)$, $\mathbb{SO}(n)$, $\mathbb{U}(n)$, $\mathbb{SU}(n)$, $\mathbb{Sp}(2n)$ possesses a uniform (Haar) probability measure which is invariant under translation by a fixed group element. Each of these uniform measures possesses a concentration of measure property making it amenable to the program laid out in the introduction; moreover, the eigenvalues of a random matrix from any of these groups is a determinantal point process, meaning that the eigenvalue rigidity approach used in Section 1 applies here as well. The limiting empirical spectral measure for all of these groups is the uniform probability measure on the circle, as first shown in [9]. This convergence is quantified in the following result, proved in [37].

Theorem 8 *Let M_n be uniformly distributed in any of $\mathbb{O}(n)$, $\mathbb{SO}(n)$, $\mathbb{U}(n)$, $\mathbb{SU}(n)$, $\mathbb{Sp}(2n)$, and let μ_n denote its spectral measure. Let μ denote the uniform probability measure on the unit circle $\mathbb{S}^1 \subseteq \mathbb{C}$. Then*

(a) $\mathbb{E}W_2(\mu_n, \mu) \leq C\frac{\sqrt{\log(n)}}{n}$,

(b) $\mathbb{P}\left[W_2(\mu_n, \mu) \geq C\frac{\sqrt{\log(n)}}{n} + t\right] \leq e^{-cn^2t^2}$, *and*

(c) with probability 1, for sufficiently large n, $W_2(\mu_n, \mu) \leq C\frac{\sqrt{\log(n)}}{n}$.

We briefly sketch the proof below; for full details, see [37].

Part (a) is proved using the eigenvalue rigidity approach described in Section 1 for the GUE. We first order the eigenvalues of M_n as $\{e^{i\theta_j}\}_{1\leq j\leq n}$ with $0 \leq \theta_1 \leq \cdots \leq \theta_n < 2\pi$, and define the discretization ν_n of μ by

$$\nu_n := \frac{1}{n}\sum_{j=1}^{n} \delta_{e^{2\pi ij/n}}.$$

It is easy to show that $W_2(\mu, \nu_n) \leq \frac{C}{n}$, and by the definition of W_2,

$$\mathbb{E}W_2^2(\mu_n, \nu_n) \leq \frac{1}{n}\sum_{j=1}^{n} \mathbb{E}\left|e^{i\theta_j} - e^{2\pi ij/n}\right|^2 \leq \frac{1}{n}\sum_{j=1}^{n} \mathbb{E}\left|\theta_j - \frac{2\pi j}{n}\right|^2,$$

so that part (a) can be proved by estimating the latter expectations.

For these estimates, as for the GUE, one can make use of the determinantal structure of the eigenvalue processes of uniformly distributed random matrices. For the case of the unitary group $\mathbb{U}(n)$, the eigenvalue angles $\{\theta_j\}$ form a determinantal point process on $[0, 2\pi)$ with kernel

$$K_n := \frac{\sin\left(\frac{n(x-y)}{2}\right)}{\sin\left(\frac{(x-y)}{2}\right)};$$

this was first proved by Dyson [11]. The determinantal structure provides an analogue of Proposition 4:

Proposition 9 *For each $0 \leq x < 2\pi$, let $\mathcal{N}_x$ denote the number of eigenvalues $e^{i\theta_j}$ of $M_n \in \mathbb{U}(n)$ such that $\theta_j \leq x$. Then*

$$\mathcal{N}_x \stackrel{d}{=} \sum_{i=1}^{n} \xi_i, \tag{8}$$

where ξ_i are independent $\{0, 1\}$-valued Bernoulli random variables.

Moreover,

$$\mathbb{E}\mathcal{N}_x = \int_0^x K_n(u, u)\, du \qquad and \qquad \operatorname{Var}\mathcal{N}_x = \int_0^x \int_x^{2\pi} K_n(u, v)^2 \%\infty du\, dv. \tag{9}$$

Appropriately modified versions of Proposition 9 hold for the other groups as well, due to determinantal structures in those contexts identified by Katz and Sarnak [26].

Using (9), one can estimate $\mathbb{E}\mathcal{N}_x$ and $\operatorname{Var}\mathcal{N}_x$, and then use (8) and Bernstein's inequality to deduce that

$$\mathbb{P}\left[\left|\mathcal{N}_x - \frac{nx}{2\pi}\right| > t + C\right] \leq 2\exp\left(-\frac{t^2}{2c\log(n) + t}\right) \tag{10}$$

for all $t > 0$. Combining this with the observation that

$$\mathbb{P}\left[\theta_j > \frac{2\pi j}{n} + t\right] = \mathbb{P}\left[\mathcal{N}_{\frac{2\pi j}{n}+t} < j\right],$$

one can deduce, upon integrating by parts, that

$$\mathbb{E}\left|\theta_j - \frac{2\pi j}{n}\right|^2 \leq C\frac{\log(n)}{n^2}$$

for each j, which completes the proof of part (a). Observe that this is made slightly simpler than the proof of Theorem 3(a) for the GUE by the fact that all of the eigenvalues of a unitary matrix behave like "bulk" eigenvalues.

Part (b) of Theorem 8 follows from part (a) and the following concentration of measure property of the uniform measure on the compact classical groups. (There is an additional subtlety in dealing with the two components of $\mathbb{O}(n)$, which can be handled by conditioning on $\det M_n$.)

Proposition 10 *Let G_n be one of $\mathbb{SO}(n)$, $\mathbb{U}(n)$, $\mathbb{SU}(n)$, or $\mathbb{Sp}(2n)$, and let $F : G_n \to \mathbb{R}$ be 1-Lipschitz, with respect to either the Hilbert–Schmidt distance or the geodesic distance on G_n. Let M_n be a uniformly distributed random matrix in G_n. Then*

$$\mathbb{P}\left[|F(M_n) - \mathbb{E}F(M_n)| > t\right] \leq e^{-cnt^2}$$

for every $t > 0$.

For $\mathbb{SO}(n)$, $\mathbb{SU}(n)$, and $\mathbb{Sp}(2n)$, this property goes back to the work of Gromov and Milman [21]; for the precise version stated here see [1, Section 4.4]. For $\mathbb{U}(n)$ (which was not covered by the results of [21] because its Ricci tensor is degenerate), the concentration in Proposition 10 was proved in [37].

Finally, part (c) follows from part (b) via the Borel-Cantelli lemma, thus completing the proof of Theorem 8.

4 Powers of Uniform Random Matrices

The approach used with random matrices from the compact classical groups in the previous section can be readily generalized to powers of such matrices, as follows.

Theorem 11 *Let M_n be uniformly distributed in any of $\mathbb{O}(n)$, $\mathbb{SO}(n)$, $\mathbb{U}(n)$, $\mathbb{SU}(n)$, $\mathbb{Sp}(2n)$. Let $m \geq 1$, and let $\mu_{m,n}$ denote the spectral measure of M_n^m. Let μ denote the uniform probability measure on the unit circle $\mathbb{S}^1 \subseteq \mathbb{C}$. There are universal constants C, c such that*

(a) $\mathbb{E}W_2(\mu_{m,n}, \mu) \leq C\dfrac{\sqrt{m\left(\log\left(\frac{n}{m}\right)+1\right)}}{n}$,

(b) $\mathbb{P}\left[W_2(\mu_{m,n}, \mu) \geq C\dfrac{\sqrt{m\left(\log\left(\frac{n}{m}\right)+1\right)}}{n} + t\right] \leq e^{-cn^2t^2}$, *and*

(c) with probability 1, for sufficiently large n, $W_2(\mu_{m,n}, \mu) \leq C\dfrac{\sqrt{m\left(\log\left(\frac{n}{m}\right)+1\right)}}{n}$.

In fact, the same proof works for $m > 1$ as in the previous section, because of the following result of Rains [42]. The result is stated in the unitary case for simplicity, but analogous results hold in the other compact classical matrix groups.

Proposition 12 *Let $m \leq n$ be fixed. If M_n is uniformly distributed in $\mathbb{U}(n)$, the eigenvalues of M_n^m are distributed as those of m independent uniform unitary matrices of sizes $\left\lfloor \frac{n}{m} \right\rfloor := \max\left\{k \in \mathbb{N} \mid k \leq \frac{n}{m}\right\}$ and $\left\lceil \frac{n}{m} \right\rceil := \min\left\{k \in \mathbb{N} \mid k \geq \frac{n}{m}\right\}$, such that the sum of the sizes of the matrices is n.*

As a consequence, if $\mathcal{N}_x$ is the number of eigenvalues of M_n^m lying in the arc from 1 to e^{ix}, then

$$\mathcal{N}_x \stackrel{d}{=} \sum_{j=0}^{m-1} \mathcal{N}_x^j,$$

where the $\mathcal{N}_\theta^j$ are the counting functions of m independent random matrices, each uniformly distributed in $\mathbb{U}\left(\left\lfloor \frac{n}{m} \right\rfloor\right)$ or $\mathbb{U}\left(\left\lceil \frac{n}{m} \right\rceil\right)$. In particular, by Proposition 9 $\mathcal{N}_x$ is equal in distribution to a sum of independent Bernoulli random variables, and its mean and variance can be estimated using the available estimates for the individual summands established in the previous section. One can thus again apply Bernstein's inequality to obtain eigenvalue rigidity, leading to a bound on $\mathbb{E}W_2(\mu_{m,n}, \mu)$.

Crucially, the concentration phenomenon on the compact classical groups tensorizes in a dimension-free way: the product of uniform measure on the m smaller unitary groups above has the same concentration property as any one of those groups. This is a consequence of the fact that the uniform measures on the compact classical groups satisfy logarithmic Sobolev inequalities; see [1, Section 4.4] and the Appendix of [37]. This allows for the full program laid out in the introduction to be carried out in this case, yielding Theorem 11 above.

5 Randomized Sums

In this section we show how our approach can be applied to randomized sums of Hermitian matrices. In this and the following two sections, we no longer have a determinantal structure allowing us to use eigenvalue rigidity. Instead we will use entropy methods to bound the expected distance between the empirical spectral measure and its mean.

Let A_n and B_n be fixed $n \times n$ Hermitian matrices, and let $U_n \in \mathbb{U}(n)$ be uniformly distributed. Define

$$M_n := U_n A_n U_n^* + B_n;$$

the random matrix M_n is the so-called randomized sum of A_n and B_n. This random matrix model has been studied at some length in free probability theory; the limiting spectral measure was studied first by Voiculescu [51] and Speicher [43], who showed that if $\{A_n\}$ and $\{B_n\}$ have limiting spectral distributions μ_A and μ_B respectively, then the limiting spectral distribution of M_n is given by the free convolution $\mu_A \boxplus \mu_B$.

The following sharpening of this convergence is a special case of Theorem 3.8 and Corollary 3.9 of [36]; we present below a slightly simplified version of the argument from that paper.

Theorem 13 *In the setting above, let μ_n denote the empirical spectral measure of M_n, and let $\nu_n = \mathbb{E}\mu_n$. Then*

(a) $\mathbb{E}W_1(\mu_n, \nu_n) \leq \dfrac{C\,\|A_n\|_{op}^{2/3}\,(\|A_n\|_{op} + \|B_n\|_{op})^{1/3}}{n^{2/3}},$

(b) $\mathbb{P}\left[W_1(\mu_n, \nu_n) \geq \dfrac{C\,\|A_n\|_{op}^{2/3}\,(\|A_n\|_{op} + \|B_n\|_{op})^{1/3}}{n^{2/3}} + t\right] \leq e^{-cn^2t^2/\|A_n\|_{op}^2}$, *and*

(c) with probability 1*, for sufficiently large n,*

$$W_1(\mu_n, \nu_n) \leq C'\,\|A_n\|_{op}^{2/3}\,(\|A_n\|_{op} + \|B_n\|_{op})^{1/3} n^{-2/3}.$$

In the most typical situations of interest, $\|A_n\|_{op}$ and $\|B_n\|_{op}$ are bounded independently of n. If $\{A_n\}$ and $\{B_n\}$ have limiting spectral distributions μ_A and μ_B respectively, then the rate of convergence of the (deterministic) measures ν_n to $\mu_A \boxplus \mu_B$ will depend strongly on the sequences $\{A_n\}$ and $\{B_n\}$; we will not address that question here.

The Lipschitz property which is a crucial ingredient of our approach to prove Theorem 13 is provided by the following lemma.

Lemma 14 *For each* 1*-Lipschitz function $f : \mathbb{R} \to \mathbb{R}$, the maps*

$$U_n \mapsto \int f\, d\mu_n \qquad \text{and} \qquad U_n \mapsto W_1(\mu_n, \nu_n)$$

are $\frac{2\|A_n\|_{op}}{\sqrt{n}}$-Lipschitz on $\mathbb{U}(n)$.

Proof Let A and B be $n \times n$ Hermitian matrices, and let $U, V \in \mathbb{U}(n)$. Then it is straightforward to show that

$$\left\|\left(UAU^* + B\right) - \left(VAV^* + B\right)\right\|_{HS} \leq 2\,\|A\|_{op}\,\|U - V\|_{HS}$$

(see [36, Lemma 3.2]). The lemma now follows by Lemma 1. □

Part (b) of Theorem 13 now follows from part (a) using Lemma 14 and the concentration of measure phenomenon for $\mathbb{U}(n)$ (Proposition 10), and part (c) follows as usual by the Borel–Cantelli lemma. It remains to prove part (a); as mentioned above, this is done using entropy techniques for bounding the supremum of a stochastic process.

The following lemma summarizes what is needed here. This fact is well known to experts, but we were not able to find an explicit statement in the literature.

Lemma 15 *Suppose that $(V, \|\cdot\|)$ be a finite-dimensional normed space with unit ball $\mathcal{B}(V)$, and that $\{X_v \mid v \in V\}$ is a family of centered random variables such that*

$$\mathbb{P}[|X_u - X_v| \geq t] \leq 2e^{-t^2/K^2\|u-v\|^2}$$

for every $t \geq 0$. *Then*

$$\mathbb{E} \sup_{v \in \mathcal{B}(V)} X_v \leq CK\sqrt{\dim V}.$$

Proof This can be proved via an elementary ε-net argument, but a quicker proof can be given using Dudley's entropy bound (see [44, p. 22] for a statement, and [44, p. 70] and [10] for discussions of the history of this result and its name).

By rescaling it suffices to assume that $K = 1$. Let $N(\varepsilon)$ denote the number of ε-balls in V needed to cover the unit ball $\mathcal{B}(V)$. A standard volumetric argument (see, e.g., [49, Lemma 5.2]) shows that $N(\varepsilon) \leq (3/\varepsilon)^{\dim V}$ for each $0 < \varepsilon < 1$; of course $N(\varepsilon) = 1$ for $\varepsilon \geq 1$. Then Dudley's bound yields

$$\mathbb{E} \sup_{v \in \mathcal{B}(V)} X_v \leq C \int_0^\infty \sqrt{\log(N(\varepsilon))}\, d\varepsilon \leq C\sqrt{\dim V} \int_0^1 \sqrt{\log(3/\varepsilon)}\, d\varepsilon \leq C'\sqrt{\dim V}.$$

□

To apply this lemma in our setting, denote by

$$\mathrm{Lip}_0 := \{f : \mathbb{R} \to \mathbb{R} \mid |f|_L < \infty \text{ and } f(0) = 0\},$$

so that Lip_0 is a Banach space with norm $|\cdot|_L$. For each $f \in \mathrm{Lip}_0$, define the random variable

$$X_f := \int f\, d\mu_n - \mathbb{E} \int f\, d\mu_n. \tag{11}$$

By the Kantorovich–Rubinstein theorem,

$$W_1(\mu_n, \nu_n) = \sup\left\{X_f : f \in \mathcal{B}(\mathrm{Lip}_0)\right\}. \tag{12}$$

Lemma 14 and Proposition 10 imply that

$$\mathbb{P}\left[\left|X_f - X_g\right| \geq t\right] = \mathbb{P}\left[\left|X_{f-g}\right| \geq t\right] \leq 2\exp\left[-\frac{cn^2t^2}{\|A_n\|_{op}^2 |f-g|_L^2}\right]. \tag{13}$$

We would like to appeal to Lemma 15, but unfortunately, Lip_0 is infinite-dimensional. We can get around this problem with an additional approximation argument.

Observing that μ_n is supported on $[-\|M_n\|_{op}, \|M_n\|_{op}]$ and $\|M_n\|_{op} \leq \|A_n\|_{op} + \|B_n\|_{op}$, we begin by replacing Lip_0 with

$$\mathrm{Lip}_0([-R, R]) := \{f : [-R, R] \to \mathbb{R} \mid |f|_L < \infty \text{ and } f(0) = 0\},$$

with $R = \|A_n\|_{op} + \|B_n\|_{op}$, for (11), (12), and (13) above. Now for an integer $m \geq 1$, let $\mathrm{Lip}_0^m([-R,R])$ be the $2m$-dimensional space of piecewise affine functions $f \in \mathrm{Lip}_0([-R,R])$ such that f is affine on each interval $\left[-R + \frac{(k-1)R}{m}, -R + \frac{kR}{m}\right]$ for $k = 1, \ldots, 2m$. Given $f \in \mathrm{Lip}_0([-R,R])$, there is a unique function $g \in \mathrm{Lip}_0^m([-R,R])$ such that $g(\frac{jR}{m}) = f(\frac{jR}{m})$ for each integer $j \in [-m,m]$; and this g satisfies

$$|g|_L \leq |f|_L \qquad \text{and} \qquad \|f-g\|_\infty \leq \frac{|f|_L R}{2m}.$$

Thus by (12),

$$W_1(\mu_n, \nu_n) \leq \frac{R}{2m} + \sup\left\{X_g \mid g \in \mathcal{B}(\mathrm{Lip}_0^m([-R,R]))\right\}.$$

Now by (13) and Lemma 15,

$$\mathbb{E}W_1(\mu_n, \nu_n) \leq \frac{R}{2m} + C\frac{\|A_n\|_{op}\sqrt{m}}{n}.$$

Part (a) now follows by optimizing over m. This completes the proof of Theorem 13.

An additional conditioning argument allows one to consider the case that A_n and B_n are themselves random matrices in Theorem 13, assuming a concentration of measure property for their distributions. We refer to [36] for details.

It seems that the entropy method does not usually result in sharp rates; for example, in [36], we used the entropy approach for Wigner and Haar-distributed matrices, and the results were not as strong as those in Sections 1 and 3. On the other hand, the entropy method is more widely applicable than the determinantal point process methods which yielded the results of Sections 1 and 3. In addition to the randomized sums treated in this section, we show in Sections 6 and 7 how the entropy method can be used for random compressions and for the Hamiltonians of quantum spin glasses. The paper [36] also used the entropy approach to prove convergence rates for the empirical spectral measures of the circular orthogonal ensemble and the circular symplectic ensemble, which we have omitted from this paper.

6 Random Compressions

Let A_n be a fixed $n \times n$ Hermitian (respectively, real symmetric) matrix, and let U_n be uniformly distributed in $\mathbb{U}(n)$ (respectively, $\mathbb{O}(n)$). Let P_k denote the projection of $\mathbb{C}^n$ (respectively $\mathbb{R}^n$) onto the span of the first k standard basis vectors. Finally, define a random matrix M_n by

$$M := P_k U_n A_n U_n^* P_k^*. \tag{14}$$

Then M_n is a compression of A_n to a random k-dimensional subspace. In the case that $\{A_n\}_{n\in\mathbb{N}}$ has a limiting spectral distribution and $\frac{k}{n} \to \alpha$, the limiting spectral distribution of M_n can be determined using techniques of free probability (see [43]); the limit is given by a free-convolution power related to the limiting spectral distribution of A_n and the value α.

For this random matrix model, the program laid out in the introduction produces the following (cf. Theorem 3.5 and Corollary 3.6 in [36]).

Theorem 16 *In the setting above, let μ_n denote the empirical spectral distribution of M_n, and let $\nu_n = \mathbb{E}\mu_n$. Then*

(a) $\mathbb{E}W_1(\mu_n, \nu_n) \le \dfrac{C\,\|A_n\|_{op}}{(kn)^{1/3}}$,

(b) $\mathbb{P}\left[W_1(\mu_n, \nu_n) \ge \dfrac{C\,\|A_n\|_{op}}{(kn)^{1/3}} + t\right] \le e^{-cknt^2/\|A_n\|_{op}^2}$, *and*

(c) with probability 1, *for sufficiently large n,* $W_1(\mu_n, \nu_n) \le C'\,\|A_n\|_{op}\,(kn)^{-1/3}$.

The proof is essentially identical to the one in the previous section; the k-dependence in the bounds is a consequence of the fact that k, not n, is the size of the matrix when Lemma 1 is applied. As with Theorem 13, an additional conditioning argument allows one to consider the case that A_n is random, with distribution satisfying a concentration of measure property.

7 Hamiltonians of Quantum Spin Glasses

In this section we consider the following random matrix model for the Hamiltonian of a quantum spin glass: let $\{Z_{a,b,j}\}_{\substack{1\le a,b\le 3\\ 1\le j\le n}}$ be independent standard Gaussian random variables, and define the $2^n \times 2^n$ random Hermitian matrix H_n by

$$H_n := \frac{1}{\sqrt{9n}} \sum_{j=1}^{n} \sum_{a,b=1}^{3} Z_{a,b,j}\sigma_j^{(a)}\sigma_{j+1}^{(b)}, \tag{15}$$

where for $1 \le a \le 3$,

$$\sigma_j^{(a)} := I_n^{\otimes(j-1)} \otimes \sigma^{(a)} \otimes I_2^{\otimes(n-j)},$$

with I_2 denoting the 2×2 identity matrix, $\sigma^{(a)}$ denoting the 2×2 matrices

$$\sigma^{(1)} := \begin{bmatrix} 0 & 1 \\ 1 & 0 \end{bmatrix} \qquad \sigma^{(2)} := \begin{bmatrix} 0 & -i \\ i & 0 \end{bmatrix} \qquad \sigma^{(3)} := \begin{bmatrix} 1 & 0 \\ 0 & -1 \end{bmatrix},$$

and the labeling cyclic so that $\sigma_{n+1}^{(b)} := \sigma_1^{(b)}$. The random matrix H_n acts on the space $(\mathbb{C}^2)^{\otimes n}$ of n distinguishable qubits; the specific structure of H_n above corresponds to nearest neighbor interaction on a circle of qubits.

If μ_n denotes the empirical spectral measure of H_n, then the ensemble average $\nu_n = \mathbb{E}\mu_n$ is known in this context as the density of states measure μ_n^{DOS}. Recently, Keating, Linden, and Wells [27] showed that μ_n^{DOS} converges weakly to Gaussian, as $n \to \infty$; i.e., they showed that the empirical spectral measure of H_n converges to Gaussian in expectation. The paper [27] gives a similar treatment for more general collections of (still independent) coupling coefficients, and more general coupling geometries than that of nearest-neighbor interactions. In more recent work, Erdös and Schröder [12] have considered still more general coupling geometries, and found a sharp transition in the limiting behavior of the density of states measure depending on the size of the maximum degree of the underlying graph, relative to its number of edges.

The following result, essentially proved in [3], quantifies this convergence.

Theorem 17 *Let μ_n be the spectral measure of H_n and let γ denote the standard Gaussian distribution on $\mathbb{R}$. Then*

(a) $\mathbb{E}W_1(\mu_n, \gamma) \leq \dfrac{C}{n^{1/6}}$,

(b) $\mathbb{P}\left[W_1(\mu_n, \gamma) \geq \dfrac{C}{n^{1/6}} + t\right] \leq e^{-9nt^2/2}$, *and*

(c) with probability 1, for all sufficiently large n,

$$W_1(\mu_n, \gamma) \leq \frac{C'}{n^{1/6}}.$$

Because the coefficients $Z_{a,b,j}$ in (15) are taken to be i.i.d. Gaussian random variables, the Gaussian concentration of measure phenomenon (Proposition 2) can be combined with Lemma 1 to carry out a version of the approach used in the cases of random sums and random compressions (Sections 5 and 6). The following lemma provides the necessary link between Lemma 1 and Proposition 2 for this random matrix model.

Lemma 18 *Let $\mathbf{x} = \{x_{a,b,j}\} \in \mathbb{R}^{9n}$ (with, say, lexicographic ordering), and assume that $n \geq 3$. Define $H_n(\mathbf{x})$ by*

$$H_n(\mathbf{x}) := \frac{1}{3\sqrt{n}} \sum_{a,b=1}^{3} \sum_{j=1}^{n} x_{a,b,j} \sigma_j^{(a)} \sigma_{j+1}^{(b)}.$$

Then the map $\mathbf{x} \mapsto H_n$ is $\frac{2^{n/2}}{3\sqrt{n}}$-Lipschitz.

Lemma 18 and Lemma 1(b) together show that

$$\mathbf{x} \mapsto W_1(\mu_n, \gamma)$$

is a $\frac{1}{3\sqrt{n}}$-Lipschitz function of $\mathbf{x}$. Part (b) of Theorem 17 then follows from part (a) and Proposition 2, and part (c) follows by the Borel–Cantelli lemma.

The proof of part (a) has two main components. First, $W_1(\mu_n, \mathbb{E}\mu_n)$ is estimated via the approach used in Sections 5 and 6: Lemma 18, Lemma 1(a), and Proposition 2 show that the stochastic process

$$X_f := \int f \, d\mu_n - \mathbb{E} \int f \, d\mu_n$$

satisfies a subgaussian increment condition as in Lemma 15, which can then be used to show that

$$\mathbb{E} W_1(\mu_n, \mathbb{E}\mu_n) \leq \frac{C}{n^{1/6}}.$$

Second, the convergence in expectation proved in [27] was done via a pointwise estimate of the difference between the characteristic functions of $\mathbb{E}\mu_n$ and γ; this estimate can be parlayed into an estimate on $W_1(\mathbb{E}\mu_n, \gamma)$ via Fourier analysis. This is carried out in detail in [3] for the bounded-Lipschitz distance; a similar argument shows that

$$W_1(\mathbb{E}\mu_n, \gamma) \leq \frac{C}{n^{1/6}},$$

completing the proof of Theorem 17.

8 The Complex Ginibre Ensemble

Let G_n be an $n \times n$ random matrix with i.i.d. standard complex Gaussian entries; G_n is said to belong to the *complex Ginibre ensemble*. It was first established by Mehta that if μ_n is the empirical spectral measure of $\frac{1}{\sqrt{n}} G_n$, then as $n \to \infty$, μ_n converges to the circular law; i.e., to the uniform measure μ on the unit disc $D := \{z \in \mathbb{C} \mid |z| \leq 1\}$.

This is the one ensemble we treat in which the general concentration of measure approach does not apply. The issue is that while there is a concentration phenomenon for the i.i.d. Gaussian entries of G_n, the spectral measure of a nonnormal matrix (G_n is nonnormal with probability 1) is not a Lipschitz function of the matrix. Nevertheless, the eigenvalue process of G_n is a determinantal point process, and so some of the techniques used above are still available. We sketch the basic idea below; full details can be found in [38]

The eigenvalues of G_n form a determinantal point process on $\mathbb{C}$ with the kernel

$$K(z, w) = \frac{1}{\pi} e^{-(|z|^2+|w|^2)/2} \sum_{k=0}^{n-1} \frac{(z\overline{w})^k}{k!}. \tag{16}$$

This means that in principle, the determinantal approach to eigenvalue rigidity used in the case of the GUE (Section 1) and of the compact classical groups (Section 3) can be used for this model. A challenge, however, is the lack of an obvious order on the eigenvalues of an arbitrary matrix over $\mathbb{C}$; without one, there is no hope to assign predicted locations around which the individual eigenvalues concentrate. We therefore impose an order on $\mathbb{C}$ which is well adapted for our purposes; we refer to this as the *spiral order*. Specifically, the linear order $\prec$ on $\mathbb{C}$ is defined by making 0 initial, and for nonzero $w, z \in \mathbb{C}$, we declare $w \prec z$ if any of the following holds:

- $\lfloor \sqrt{n}\,|w| \rfloor < \lfloor \sqrt{n}\,|z| \rfloor$.
- $\lfloor \sqrt{n}\,|w| \rfloor = \lfloor \sqrt{n}\,|z| \rfloor$ and $\arg w < \arg z$.
- $\lfloor \sqrt{n}\,|w| \rfloor = \lfloor \sqrt{n}\,|z| \rfloor$, $\arg w = \arg z$, and $|w| \geq |z|$.

Here we are using the convention that $\arg z \in (0, 2\pi]$.

We order the eigenvalues according to $\prec$: first the eigenvalues in the disc of radius $\frac{1}{\sqrt{n}}$ are listed in order of increasing argument, then the ones in the annulus with inner radius $\frac{1}{\sqrt{n}}$ and outer radius $\frac{2}{\sqrt{n}}$ in order of increasing argument, and so on. We then define predicted locations $\tilde{\lambda}_j$ for (most of) the eigenvalues based on the spiral order: $\tilde{\lambda}_1 = 0$, $\{\tilde{\lambda}_2, \tilde{\lambda}_3, \tilde{\lambda}_4\}$ are $\frac{1}{\sqrt{n}}$ times the 3rd roots of unity (in increasing order with respect to $\prec$), the next five are $\frac{2}{\sqrt{n}}$ times the 5th roots of unity, and so on. Letting ν_n denote the normalized counting measure supported on the $\{\tilde{\lambda}_j\}$, it is easy to show that

$$W_2(\nu_n, \mu) \leq \frac{C}{\sqrt{n}}.$$

(In fact, there is a slight modification for about $\sqrt{n \log(n)}$ of the largest eigenvalues, the details of which we will not discuss here.)

The same type of argument as in the earlier determinantal cases gives a Bernstein-type inequality for the eigenvalue counting function on an initial segment with respect to the spiral order, which in turn leads to eigenvalue rigidity for most of the eigenvalues. The largest eigenvalues can be treated with a more elementary argument, leading via the usual coupling argument to the bound

$$\mathbb{E} W_2(\mu_n, \nu_n) \leq C \left(\frac{\log(n)}{n} \right)^{1/4}.$$

(One can deduce a slightly tighter bound for $\mathbb{E} W_p(\mu_n, \nu_n)$ for $1 \leq p < 2$, and a weaker one for $p > 2$.)

In this setting we cannot argue that the concentration of $W_1(\mu_n, \mu)$ is immediate from general concentration properties of the ensemble, but the eigenvalue rigidity itself can be used as a substitute. Indeed,

$$W_2(\mu_n, \nu_n)^2 \leq \frac{1}{n} \sum_{j=1}^{n} \left| \lambda_j - \tilde{\lambda}_j \right|^2,$$

and so

$$\mathbb{P}\left[W_2(\mu_n, \nu_n)^2 > t\right] \leq \mathbb{P}\left[\sum_{j=1}^{n}\left|\lambda_j - \tilde{\lambda}_j\right|^2 > nt\right] \leq \sum_{j=1}^{n}\mathbb{P}\left[\left|\lambda_j - \tilde{\lambda}_j\right|^2 > t\right].$$

For most of the eigenvalues the eigenvalue rigidity about $\tilde{\lambda}_j$ is strong enough to bound this quite sharply; as before, for about $\sqrt{n\log(n)}$ of the largest eigenvalues a more trivial bound is used. Since this approach does not produce a particularly clean tail inequality for $W_2(\mu_n, \nu_n)$, we will instead simply state the almost-sure convergence rate which follows by the Borel–Cantelli lemma.

Theorem 19 *Let μ_n denote the empirical spectral measure of $\frac{1}{\sqrt{n}}G_n$, and let μ denote the uniform measure on the unit disc in $\mathbb{C}$. Then with probability* 1*, for sufficiently large n,*

$$W_2(\mu_n, \mu) \leq C\frac{\sqrt{\log n}}{n^{1/4}}.$$

Acknowledgements This research was partially supported by grants from the U.S. National Science Foundation (DMS-1308725 to E.M.) and the Simons Foundation (#315593 to M.M.). This paper is an expansion of the first-named author's talk at the excellent workshop "Information Theory and Concentration Phenomena" at the Institute for Mathematics and its Applications, as part of the IMA Thematic Year on Discrete Structures: Analysis and Applications. The authors thank the IMA for its hospitality.

References

1. G. W. Anderson, A. Guionnet, and O. Zeitouni. *An Introduction to Random Matrices*, volume 118 of *Cambridge Studies in Advanced Mathematics*. Cambridge University Press, Cambridge, 2010.
2. R. Bhatia. *Matrix Analysis*, volume 169 of *Graduate Texts in Mathematics*. Springer-Verlag, New York, 1997.
3. D. Buzinski and E. S. Meckes. Almost sure convergence in quantum spin glasses. *J. Math. Phys.*, 56(12), 2015.
4. S. Chatterjee. Concentration of Haar measures, with an application to random matrices. *J. Funct. Anal.*, 245(2):379–389, 2007.
5. S. Chatterjee and M. Ledoux. An observation about submatrices. *Electron. Commun. Probab.*, 14:495–500, 2009.
6. S. Dallaporta. Eigenvalue variance bounds for Wigner and covariance random matrices. *Random Matrices Theory Appl.*, 1(3):1250007, 28, 2012.
7. S. Dallaporta. Eigenvalue variance bounds for covariance matrices. *Markov Process. Related Fields*, 21(1):145–175, 2015.
8. K. R. Davidson and S. J. Szarek. Local operator theory, random matrices and Banach spaces. In *Handbook of the Geometry of Banach Spaces, Vol. I*, pages 317–366. North-Holland, Amsterdam, 2001.
9. P. Diaconis and M. Shahshahani. On the eigenvalues of random matrices. *J. Appl. Probab.*, 31A:49–62, 1994. Studies in applied probability.

10. R. Dudley. V. N. Sudakov's work on expected suprema of Gaussian processes. In Proceedings of High Dimensional Probability VII: The Cargèse Volume, volume 71 of Progress in Probability, pages 37–43. Birkhäuser, Basel, 2016.
11. F. J. Dyson. Correlations between eigenvalues of a random matrix. *Comm. Math. Phys.*, 19:235–250, 1970.
12. L. Erdős and D. Schröder. Phase transition in the density of states of quantum spin glasses. *Math. Phys. Anal. Geom.*, 17(3–4):441–464, 2014.
13. L. Erdős and H.-T. Yau. Universality of local spectral statistics of random matrices. *Bull. Amer. Math. Soc. (N.S.)*, 49(3):377–414, 2012.
14. L. Erdős, H.-T. Yau, and J. Yin. Rigidity of eigenvalues of generalized Wigner matrices. *Adv. Math.*, 229(3):1435–1515, 2012.
15. L. C. Evans and R. F. Gariepy. *Measure theory and fine properties of functions*. Studies in Advanced Mathematics. CRC Press, Boca Raton, FL, 1992.
16. P. J. Forrester and E. M. Rains. Interrelationships between orthogonal, unitary and symplectic matrix ensembles. In *Random matrix models and their applications*, volume 40 of *Math. Sci. Res. Inst. Publ.*, pages 171–207. Cambridge Univ. Press, Cambridge, 2001.
17. F. Götze and A. Tikhomirov. Optimal bounds for convergence of expected spectral distributions to the semi-circular law. To appear in *Probab. Theory Related Fields*.
18. F. Götze and A. Tikhomirov. The rate of convergence for spectra of GUE and LUE matrix ensembles. *Cent. Eur. J. Math.*, 3(4):666–704 (electronic), 2005.
19. N. Gozlan. A characterization of dimension free concentration in terms of transportation inequalities. *Ann. Probab.*, 37(6):2480–2498, 2009.
20. N. Gozlan, C. Roberto, and P.-M. Samson. From dimension free concentration to the Poincaré inequality. *Calc. Var. Partial Differential Equations*, 52(3–4):899–925, 2015.
21. M. Gromov and V. D. Milman. A topological application of the isoperimetric inequality. *Amer. J. Math.*, 105(4):843–854, 1983.
22. A. Guionnet and O. Zeitouni. Concentration of the spectral measure for large matrices. *Electron. Comm. Probab.*, 5:119–136 (electronic), 2000.
23. J. Gustavsson. Gaussian fluctuations of eigenvalues in the GUE. *Ann. Inst. H. Poincaré Probab. Statist.*, 41(2):151–178, 2005.
24. J. B. Hough, M. Krishnapur, Y. Peres, and B. Virág. Determinantal processes and independence. *Probab. Surv.*, 3:206–229, 2006.
25. V. Kargin. A concentration inequality and a local law for the sum of two random matrices. *Probab. Theory Related Fields*, 154(3–4):677–702, 2012.
26. N. M. Katz and P. Sarnak. *Random Matrices, Frobenius Eigenvalues, and Monodromy*, volume 45 of *American Mathematical Society Colloquium Publications*. American Mathematical Society, Providence, RI, 1999.
27. J. P. Keating, N. Linden, and H. J. Wells. Spectra and eigenstates of spin chain Hamiltonians. *Comm. Math. Phys.*, 338(1):81–102, 2015.
28. M. Ledoux. *The Concentration of Measure Phenomenon*, volume 89 of *Mathematical Surveys and Monographs*. American Mathematical Society, Providence, RI, 2001.
29. M. Ledoux. Deviation inequalities on largest eigenvalues. In *Geometric Aspects of Functional Analysis*, volume 1910 of *Lecture Notes in Math.*, pages 167–219. Springer, Berlin, 2007.
30. M. Ledoux. γ_2 and Γ_2. Unpublished note, available at http://perso.math.univ-toulouse.fr/ledoux/files/2015/06/gGamma2.pdf, 2015.
31. M. Ledoux and B. Rider. Small deviations for beta ensembles. *Electron. J. Probab.*, 15:no. 41, 1319–1343, 2010.
32. V. A. Marčenko and L. A. Pastur. Distribution of eigenvalues in certain sets of random matrices. *Mat. Sb. (N.S.)*, 72 (114):507–536, 1967.
33. E. Meckes. Approximation of projections of random vectors. *J. Theoret. Probab.*, 25(2): 333–352, 2012.
34. E. Meckes. Projections of probability distributions: a measure-theoretic Dvoretzky theorem. In *Geometric Aspects of Functional Analysis*, volume 2050 of *Lecture Notes in Math.*, pages 317–326. Springer, Heidelberg, 2012.

35. E. S. Meckes and M. W. Meckes. Another observation about operator compressions. *Proc. Amer. Math. Soc.*, 139(4):1433–1439, 2011.
36. E. S. Meckes and M. W. Meckes. Concentration and convergence rates for spectral measures of random matrices. *Probab. Theory Related Fields*, 156(1–2):145–164, 2013.
37. E. S. Meckes and M. W. Meckes. Spectral measures of powers of random matrices. *Electron. Commun. Probab.*, 18:no. 78, 13, 2013.
38. E. S. Meckes and M. W. Meckes. A rate of convergence for the circular law for the complex Ginibre ensemble. *Ann. Fac. Sci. Toulouse Math. (6)*, 24(1):93–117, 2015.
39. M. L. Mehta. *Random Matrices*, volume 142 of *Pure and Applied Mathematics (Amsterdam)*. Elsevier/Academic Press, Amsterdam, third edition, 2004.
40. S. Ng and M. Walters. A method to derive concentration of measure bounds on Markov chains. *Electron. Comm. Probab.*, 20(95), 2015.
41. S. O'Rourke. Gaussian fluctuations of eigenvalues in Wigner random matrices. *J. Stat. Phys.*, 138(6):1045–1066, 2010.
42. E. M. Rains. Images of eigenvalue distributions under power maps. *Probab. Theory Related Fields*, 125(4):522–538, 2003.
43. R. Speicher. Free convolution and the random sum of matrices. *Publ. Res. Inst. Math. Sci.*, 29(5):731–744, 1993.
44. M. Talagrand. *Upper and Lower Bounds for Stochastic Processes: Modern Methods and Classical Problems*, volume 60 of *Ergebnisse der Mathematik und ihrer Grenzgebiete. 3. Folge.* Springer, Heidelberg, 2014.
45. T. Tao and V. Vu. Random matrices: universality of local eigenvalue statistics up to the edge. *Comm. Math. Phys.*, 298(2):549–572, 2010.
46. T. Tao and V. Vu. Random matrices: universality of local eigenvalue statistics. *Acta Math.*, 206(1):127–204, 2011.
47. T. Tao and V. Vu. Random matrices: sharp concentration of eigenvalues. *Random Matrices Theory Appl.*, 2(3):1350007, 31, 2013.
48. J. A. Tropp. An introduction to matrix concentration inequalities. *Foundations and Trends in Machine Learning*, 8(1–2), 2015.
49. R. Vershynin. Introduction to the non-asymptotic analysis of random matrices. In *Compressed Sensing*, pages 210–268. Cambridge Univ. Press, Cambridge, 2012.
50. C. Villani. *Topics in Optimal Transportation*, volume 58 of *Graduate Studies in Mathematics*. American Mathematical Society, Providence, RI, 2003.
51. D. Voiculescu. Limit laws for random matrices and free products. *Invent. Math.*, 104(1): 201–220, 1991.
52. E. Wigner. Characteristic vectors of bordered matrices with infinite dimensions. *Ann. of Math. (2)*, 62:548–564, 1955.
53. E. Wigner. On the distribution of the roots of certain symmetric matrices. *Ann. of Math. (2)*, 67:325–327, 1958.

Concentration of Measure Without Independence: A Unified Approach Via the Martingale Method

Aryeh Kontorovich and Maxim Raginsky

Abstract The concentration of measure phenomenon may be summarized as follows: a function of many weakly dependent random variables that is not too sensitive to any of its individual arguments will tend to take values very close to its expectation. This phenomenon is most completely understood when the arguments are mutually independent random variables, and there exist several powerful complementary methods for proving concentration inequalities, such as the martingale method, the entropy method, and the method of transportation inequalities. The setting of dependent arguments is much less well understood. This chapter focuses on the martingale method for deriving concentration inequalities without independence assumptions. In particular, we use the machinery of so-called Wasserstein matrices to show that the Azuma-Hoeffding concentration inequality for martingales with almost surely bounded differences, when applied in a sufficiently abstract setting, is powerful enough to recover and sharpen several known concentration results for nonproduct measures. Wasserstein matrices provide a natural formalism for capturing the interplay between the metric and the probabilistic structures, which is fundamental to the concentration phenomenon.

1 Introduction

At its most abstract, the concentration of measure phenomenon may be summarized as follows: a function of several weakly dependent random variables that is not too sensitive to any of the individual arguments will tend to take values very close

A. Kontorovich
Department of Computer Science, Ben-Gurion University, Beer-Sheva, Israel
e-mail: karyeh@cs.bgu.ac.il

M. Raginsky (✉)
Department of Electrical and Computer Engineering and the Coordinated Science Laboratory, University of Illinois, Urbana, IL, USA
e-mail: maxim@illinois.edu

E. Carlen et al. (eds.), *Convexity and Concentration*, The IMA Volumes in Mathematics and its Applications 161, DOI 10.1007/978-1-4939-7005-6_6

to its expectation. This phenomenon is most completely understood in the case of independent arguments, and the recent book [2] provides an excellent survey (see also [30] for an exposition from the viewpoint of, and with applications to, information theory).

The case of dependent arguments has yet to mature into such a unified, overarching theory. The earliest concentration results for nonproduct measures were established for Haar measures on various groups, and relied strongly on the highly symmetric nature of the Haar measure in question. These results include Lévy's classic isoperimetric inequality on the sphere [20] and Maurey's concentration inequality on the permutation group [28]. To the best of our knowledge, the first concentration result for a nonproduct, non-Haar measure is due to Marton [22], where she proved a McDiarmid-type bound for contracting Markov chains. A flurry of activity followed. Besides Marton's own follow-up work [23, 24, 25], the transportation method she pioneered was extended by Samson [33], and martingale techniques [32, 6, 17], as well as methods relying on the Dobrushin interdependence matrix [19, 4, 36], have been employed in obtaining concentration results for nonproduct measures. The underlying theme is that the independence assumption may be relaxed to one of *weak* dependence, the latter being quantified by various mixing coefficients.

This chapter is an attempt at providing an abstract unifying framework that generalizes and sharpens some of the above results. This framework combines classical martingale techniques with the method of *Wasserstein matrices* [10]. In particular, we rely on Wasserstein matrices to obtain general-purpose quantitative estimates of the local variability of a function of many dependent random variables after taking a conditional expectation with respect to a subset of the variables. A concentration inequality in a metric space must necessarily capture the interplay between the metric and the distribution, and, in our setting, Wasserstein matrices provide the ideal analytical tool for this task. As an illustration, we recover (and, in certain cases, sharpen) some results of [19, 6, 17] by demonstrating all of these to be special cases of the Wasserstein matrix method.

The remainder of the chapter is organized as follows. Section 2 is devoted to setting up the basic notation and preliminary definitions. A brief discussion of the concentration of measure phenomenon in high-dimensional spaces is presented in Section 3, together with a summary of key methods to establish concentration under the independence assumption. Next, in Section 4, we present our abstract martingale technique and then demonstrate its wide scope in Section 5 by deriving many of the previously published concentration inequalities as special cases. We conclude in Section 6 by listing some open questions.

2 Preliminaries and Notation

2.1 Metric Probability Spaces

A *metric probability space* is a triple (Ω, μ, d), where Ω is a Polish space equipped with its Borel σ-field, μ is a Borel probability measure on Ω, and d is metric on Ω, assumed to be a measurable function on the product space $\Omega \times \Omega$. We do not assume that d is the same metric that metrizes the Polish topology on Ω.

2.2 Product Spaces

Since concentration of measure is a high-dimensional phenomenon, a natural setting for studying it is that of a product space. Let T be a finite index set, which we identify with the set $[n] \triangleq \{1, \dots, n\}$, where $n = |T|$ (this amounts to fixing some linear ordering of the elements of T). We will use the following notation for subintervals of T: $[i] \triangleq \{1, \dots, i\}$; $[i,j] \triangleq \{i, i+1, \dots, j\}$ for $i \neq j$; $(i,j] \triangleq \{i+1, \dots, j\}$ for $i < j$; $(i,j) \triangleq \{i+1, \dots, j-1\}$ for $i < j-1$; etc.

With each $i \in T$, we associate a measurable space $(\mathsf{X}_i, \mathcal{B}_i)$, where X_i is a Polish space and $\mathcal{B}_i$ is its Borel σ-field. For each $I \subseteq T$, we will equip the product space $\mathsf{X}^I \triangleq \prod_{i \in I} \mathsf{X}_i$ with the product σ-field $\mathcal{B}^I \triangleq \bigotimes_{i \in I} \mathcal{B}_i$. When $I = T$, we will simply write X and $\mathcal{B}$. We will write x^I and x for a generic element of X^I and X, respectively. Given two sets $I, J \subset T$ with $I \cap J = \varnothing$, the *concatenation* of $x^I \in \mathsf{X}^I$ and $z^J \in \mathsf{X}^J$ is defined as $y = x^I z^J \in \mathsf{X}^{I \cup J}$ by setting

$$y_i = \begin{cases} x_i, & i \in I \\ z_i, & i \in J \end{cases}.$$

Given a random object $X = (X_i)_{i \in T}$ taking values in X according to a probability law μ, we will denote by $\mathbb{P}_\mu[\cdot]$ and $\mathbb{E}_\mu[\cdot]$ the probability and expectation with respect to μ, by $\mu^I(\mathrm{d}x^I | x^J)$ the regular conditional probability law of X^I given $X^J = x^J$, and by $\mu^I(\mathrm{d}x^I)$ the marginal probability law of X^I. When $I = \{i\}$, we will write $\mu_i(\cdot)$ and $\mu_i(\cdot | x^J)$.

For each $i \in T$, we fix a metric on X_i, which is assumed to be measurable with respect to the product σ-field $\mathcal{B}_i \otimes \mathcal{B}_i$. For each $I \subseteq T$, equip X^I with the product metric ρ^I, where

$$\rho^I(x^I, z^I) \triangleq \sum_{i \in I} \rho_i(x_i, z_i), \qquad \forall x^I, z^I \in \mathsf{X}^I.$$

When $I \equiv T$, we will simply write ρ instead of ρ^T. In this way, for any Borel probability measure μ on X, we can introduce a "global" metric probability space (X, μ, ρ), as well as "local" metric probability spaces $(\mathsf{X}^I, \mu^I, \rho^I)$ and $(\mathsf{X}^I, \mu^I(\cdot | x^J), \rho_I)$ for all $I, J \subset T$ and all $x^J \in \mathsf{X}^J$.

2.3 Couplings and Transportation Distances

Let Ω be a Polish space. A *coupling* of two Borel probability measures μ and ν on Ω is a Borel probability measure $\mathbf{P}$ on the product space $\Omega \times \Omega$, such that $\mathbf{P}(\cdot \times \Omega) = \mu$ and $\mathbf{P}(\Omega \times \cdot) = \nu$. We denote the set of all couplings of μ and ν by $C(\mu, \nu)$. Let d be a lower-semicontinuous metric on Ω. We denote by $\mathrm{Lip}(\Omega, d)$ the space of all functions $\Omega \to \mathbb{R}$ that are Lipschitz with respect to d, and by $\mathrm{Lip}_c(\Omega, d)$ the subset of $\mathrm{Lip}(\Omega, d)$ consisting of c-Lipschitz functions. The L^1 *Wasserstein* (or *transportation*) *distance* between μ and ν is defined as

$$W_d(\mu, \nu) \triangleq \inf_{\mathbf{P} \in C(\mu,\nu)} \mathbb{E}_{\mathbf{P}}[d(X, Y)], \tag{1}$$

where (X, Y) is a random element of $\Omega \times \Omega$ with law $\mathbf{P}$. The transportation distance admits a dual (Kantorovich–Rubinstein) representation

$$W_d(\mu, \nu) = \sup_{f \in \mathrm{Lip}_1(\Omega,d)} \left| \int_\Omega f \mathrm{d}\mu - \int_\Omega f \mathrm{d}\nu \right|. \tag{2}$$

For example, when we equip Ω with the trivial metric $d(\omega, \omega') = \mathbf{1}\{\omega \neq \omega'\}$, the corresponding Wasserstein distance coincides with the total variation distance:

$$W_d(\mu, \nu) = \|\mu - \nu\|_{\mathrm{TV}} = \sup_A |\mu(A) - \nu(A)|,$$

where the supremum is over all Borel subsets of Ω.

In the context of the product space (X, ρ) defined earlier, we will use the shorthand W_i for W_{ρ_i}, W^I for W_{ρ^I}, and W for W_ρ.

2.4 Markov Kernels and Wasserstein Matrices

A *Markov kernel* on X is a mapping $K : \mathsf{X} \times \mathcal{B} \to [0, 1]$, such that $x \mapsto K(x, A)$ is measurable for each $A \in \mathcal{B}$, and $K(x, \cdot)$ is a Borel probability measure on X for each $x \in \mathsf{X}$. Given a Markov kernel K and a bounded measurable function $f : \mathsf{X} \to \mathbb{R}$, we denote by Kf the bounded measurable function

$$Kf(x) \triangleq \int_{\mathsf{X}} f(y) K(x, \mathrm{d}y), \qquad x \in \mathsf{X}.$$

Likewise, given a Borel probability measure μ on X, we denote by μK the Borel probability measure

$$\mu K(A) \triangleq \int_{\mathsf{X}} K(x, A) \mu(\mathrm{d}x), \qquad A \in \mathcal{B}.$$

It is not hard to see that $\int_{\mathsf{X}} f \mathrm{d}(\mu K) = \int_{\mathsf{X}} (Kf) \mathrm{d}\mu$.

Given a measurable function $f : \mathsf{X} \to \mathbb{R}$, we define the *local oscillation of f at $i \in T$* as

$$\delta_i(f) \triangleq \sup_{\substack{x,z \in \mathsf{X} \\ x^{T\setminus\{i\}} = z^{T\setminus\{i\}}}} \frac{|f(x) - f(z)|}{\rho_i(x_i, z_i)},$$

where we follow the convention $0/0 = 0$. This quantity measures the variability of f in its ith argument when all other arguments are held fixed. As will become evident later on, our martingale technique for establishing concentration inequalities for a given function $f : \mathsf{X} \to \mathbb{R}$ requires controlling the local oscillations $\delta_i(Kf)$ in terms of the local oscillations $\delta_i(f)$ for appropriately chosen Markov kernels K.

To get an idea of what is involved, let us consider the simple case when each X_i is endowed with the scaled trivial metric $\rho_i(x_i, z_i) \triangleq \alpha_i \mathbf{1}\{x_i \neq z_i\}$, where $\alpha_i > 0$ is some fixed constant. Then

$$\delta_i(f) = \frac{1}{\alpha_i} \sup \left\{ |f(x) - f(z)| : x, z \in \mathsf{X},\ x^{T\setminus\{i\}} = z^{T\setminus\{i\}} \right\}.$$

The corresponding metric ρ on X is the weighted Hamming metric

$$\rho_{\alpha}(x, z) \triangleq \sum_{i \in T} \alpha_i \mathbf{1}\{x_i \neq z_i\}. \tag{3}$$

Fix a Markov kernel K on X. The *Dobrushin contraction coefficient* of K (also associated in the literature with Doeblin's name) is the smallest $\theta \geq 0$ for which $\|K(x, \cdot) - K(z, \cdot)\|_{\mathrm{TV}} \leq \theta$ holds for all $x, z \in \mathsf{X}$. The term *contraction* is justified by the well-known inequality (apparently going back to Markov himself [21, §5])

$$\|\mu K - \nu K\|_{\mathrm{TV}} \leq \theta \|\mu - \nu\|_{\mathrm{TV}}, \tag{4}$$

which holds for all probability measures μ, ν on X. Then we have the following estimate:

Proposition 2.1 *If K is a Markov kernel on X with Dobrushin coefficient θ, then for every $i \in T$ and for every $f \in \mathrm{Lip}(\mathsf{X}, \rho_{\alpha})$, we have*

$$\delta_i(Kf) \leq \frac{\theta}{\alpha_i} \sum_{j \in T} \alpha_j \delta_j(f). \tag{5}$$

Proof Fix an index $i \in T$ and any two $x, z \in \mathsf{X}$ that differ only in the ith coordinate: $x^{T\setminus\{i\}} = z^{T\setminus\{i\}}$ and $x_i \neq z_i$. Pick an arbitrary coupling $\mathbf{P}_{x,z} \in C(K(x, \cdot), K(z, \cdot))$. Then

$$|Kf(x) - Kf(z)| = \left| \int_{\mathsf{X}} K(x, \mathrm{d}u) f(u) - \int_{\mathsf{X}} K(z, \mathrm{d}y) f(y) \right|$$

$$
\begin{aligned}
&= \left| \int_{\mathsf{X}\times\mathsf{X}} \mathbf{P}_{x,z}(\mathrm{d}u, \mathrm{d}y)\big(f(u) - f(y)\big) \right| \\
&\le \sum_{j\in T} \delta_j(f) \int_{\mathsf{X}\times\mathsf{X}} \mathbf{P}_{x,z}(\mathrm{d}u, \mathrm{d}y)\rho_j(u_j, y_j) \\
&= \sum_{j\in T} \alpha_j\delta_j(f) \int_{\mathsf{X}\times\mathsf{X}} \mathbf{P}_{x,z}(\mathrm{d}u, \mathrm{d}y)\mathbf{1}\{u_j \neq y_j\} \\
&\le \sum_{j\in T} \alpha_j\delta_j(f) \cdot \int_{\mathsf{X}\times\mathsf{X}} \mathbf{P}_{x,z}(\mathrm{d}u, \mathrm{d}y)\mathbf{1}\{u \neq y\},
\end{aligned}
$$

where the first inequality is by the definition of $\delta_i(f)$, while the second one follows from the obvious implication $u_j \neq y_j \Rightarrow u \neq y$. Taking the infimum of both sides over all couplings $\mathbf{P}_{x,z} \in \mathcal{C}(K(x,\cdot), K(z,\cdot))$ yields

$$
\begin{aligned}
|Kf(x) - Kf(z)| &\le \sum_{j\in T} \alpha_j\delta_j(f) \cdot \|K(x,\cdot) - K(z,\cdot)\|_{\mathrm{TV}} \\
&\le \theta \sum_{j\in T} \alpha_j\delta_j(f).
\end{aligned}
$$

Finally, dividing both sides of the above inequality by α_i and taking the supremum over all choices of x, z that differ only in the ith coordinate, we obtain (5). □

One shortcoming of the above result (which is nontrivial only under the rather strong condition

$$
\theta < \frac{\alpha_i}{\alpha_j} < \theta^{-1} \tag{6}
$$

for all $i, j \in T$) is that it gives only a very rough idea of the influence of $\delta_j(f)$ for $j \in T$ on $\delta_i(Kf)$. For example, if $\alpha_1 = \ldots = \alpha_n = 1$, then the condition (6) reduces to the Dobrushin contraction condition $\theta < 1$, and the inequality (5) becomes

$$
\delta_i(Kf) \le \theta \sum_{j\in T} \delta_j(f),
$$

suggesting that all of the $\delta_j(f)$'s influence $\delta_i(Kf)$ equally. However, this picture can be refined. To that end, we introduce the notion of a *Wasserstein matrix* following Föllmer [10]. Let us denote by $\boldsymbol{\delta}(f)$ the vector $(\delta_i(f))_{i\in T}$. We say that a nonnegative matrix $V = (V_{ij})_{i,j\in T}$ is a Wasserstein matrix for K if, for every $f \in \mathrm{Lip}(\mathsf{X}, \rho)$ and for every $i \in T$,

$$
\delta_i(Kf) \le \sum_{j\in T} V_{ij}\delta_j(f), \tag{7}
$$

or, in vector form, if $\boldsymbol{\delta}(Kf) \preceq V\boldsymbol{\delta}(f)$.

One of our main objectives will be to show that concentration inequalities for functions f of $X \sim \mu$ can be obtained using Wasserstein matrices for certain Markov kernels K related to μ. In order to motivate the introduction of Wasserstein matrices, we record a couple of contraction estimates for Markov kernels that may be of independent interest. To that end, we introduce another coupling-based distance between probability measures [2, Chap. 8]: for two Borel probability measures on X, define

$$\bar{W}(\mu, \nu) \triangleq \inf_{\mathbf{P} \in C(\mu,\nu)} \sqrt{\sum_{i \in T} (\mathbb{E}_{\mathbf{P}}[\rho_i(X_i, Y_i)])^2}, \tag{8}$$

where (X, Y) is a random element of $\mathsf{X} \times \mathsf{X}$. Even though $\bar{W}$ is not a Wasserstein distance, we can use the inequality $\sqrt{a+b} \leq \sqrt{a} + \sqrt{b}$ for $a, b \geq 0$ to show that $\bar{W}(\mu, \nu) \leq W(\mu, \nu)$.

Proposition 2.2 *Let V be a Wasserstein matrix for a Markov kernel K on X. Then for any Lipschitz function $f : \mathsf{X} \to \mathbb{R}$,*

$$\left|\mathbb{E}_{\mu K}[f(X)] - \mathbb{E}_{\nu K}[f(X)]\right| \leq \|V\boldsymbol{\delta}(f)\|_{\ell^2(T)} \, \bar{W}(\mu, \nu). \tag{9}$$

Proof Fix an arbitrary coupling $\mathbf{P} \in C(\mu, \nu)$ and let (X, Y) be a random element of $\mathsf{X} \times \mathsf{X}$ with law $\mathbf{P}$. Then

$$\begin{aligned}
\left|\mathbb{E}_{\mu K}[f(X)] - \mathbb{E}_{\nu K}[f(X)]\right| &= \left|\mathbb{E}_{\mu}[Kf(X)] - \mathbb{E}_{\nu}[Kf(X)]\right| \\
&= |\mathbb{E}_{\mathbf{P}}[Kf(X) - Kf(Y)]| \\
&\leq \sum_{i \in T} \delta_i(Kf) \cdot \mathbb{E}_{\mathbf{P}}[\rho_i(X_i, Y_i)] \\
&\leq \sum_{i \in T} \sum_{j \in T} V_{ij}\delta_j(f) \cdot \mathbb{E}_{\mathbf{P}}[\rho_i(X_i, Y_i)].
\end{aligned}$$

where in the last step we have used the definition of the Wasserstein matrix. Using the Cauchy–Schwarz inequality, we obtain

$$\begin{aligned}
\left|\mathbb{E}_{\mu K}[f(X)] - \mathbb{E}_{\nu K}[f(X)]\right| &\leq \sqrt{\sum_{i \in T} \left|\sum_{j \in T} V_{ij}\delta_j(f)\right|^2 \cdot \sum_{i \in T} (\mathbb{E}_{\mathbf{P}}[\rho_i(X_i, Y_i)])^2} \\
&= \|V\boldsymbol{\delta}(f)\|_{\ell^2(T)} \cdot \sqrt{\sum_{i \in T} (\mathbb{E}_{\mathbf{P}}[\rho_i(X_i, Y_i)])^2}.
\end{aligned}$$

Taking the infimum of both sides over all $\mathbf{P} \in C(\mu, \nu)$, we obtain (9). $\square$

Corollary 2.3 *Let V be a Wasserstein matrix for a Markov kernel K on X. Then, for any two Borel probability measures μ and ν on X,*

$$W(\mu K, \nu K) \leq \|V\mathbf{1}\|_{\ell^2(T)} \bar{W}(\mu, \nu), \tag{10}$$

where $\mathbf{1} \in \mathbb{R}^T$ is the vector of all ones, and therefore

$$W(\mu K, \nu K) \leq \|V\mathbf{1}\|_{\ell^2(T)} W(\mu, \nu).$$

Proof A function $f : \mathsf{X} \to \mathbb{R}$ belongs to $\mathrm{Lip}_1(\mathsf{X}, \rho)$ if and only if $\boldsymbol{\delta}(f) \in [0, 1]^T$. Using the dual representation (2) of W and applying Proposition 2.2, we can write

$$\begin{aligned} W(\mu K, \nu K) &= \sup_{f \in \mathrm{Lip}_1(\mathsf{X}, \rho)} \left| \mathbb{E}_{\mu K}[f(X)] - \mathbb{E}_{\nu K}[f(X)] \right| \\ &\leq \sup_{\xi \in [0,1]^T} \|V\xi\|_{\ell^2(T)} \bar{W}(\mu, \nu). \end{aligned}$$

Since V is a nonnegative matrix, the supremum is achieved by $\xi = \mathbf{1}$. □

2.5 Relative Entropy

Finally, we will need some key notions from information theory. The *relative entropy* (or *information divergence*) between two probability measures μ, ν on a space Ω is defined as

$$D(\nu \| \mu) \triangleq \begin{cases} \int_\Omega \mathrm{d}\mu f \log f, & \text{if } \nu \ll \mu \text{ with } f = \mathrm{d}\nu/\mathrm{d}\mu \\ +\infty, & \text{otherwise} \end{cases}.$$

We use natural logarithms throughout the chapter. The relative entropy is related to the total variation distance via Pinsker's inequality[1]

$$\|\mu - \nu\|_{\mathrm{TV}} \leq \sqrt{\frac{1}{2} D(\mu \| \nu)}. \tag{11}$$

[1]Though commonly referred to as *Pinsker's inequality*, (11) as given here (with the optimal constant $\frac{1}{2}$) was proven by Csiszár [7] and Kullback [18] in 1967.

3 Concentration of Measure and Sufficient Conditions

In this section, we give a precise definition of the concentration of measure phenomenon, review several sufficient conditions for it to hold, and briefly discuss how it can be established under the independence assumption via tensorization. For more details and further references, the reader can consult [2] or [30].

We say that the metric probability space (X, μ, ρ) has the concentration of measure property if there exists a positive constant $c > 0$, such that, for every Lipschitz function $f : \mathsf{X} \to \mathbb{R}$,

$$\mathbb{P}_\mu \left\{ f(X) - \mathbb{E}_\mu[f(X)] \geq t \right\} \leq e^{-t^2/2c\|f\|^2_{\mathrm{Lip}}}, \qquad \forall t > 0 \tag{12}$$

where

$$\|f\|_{\mathrm{Lip}} \triangleq \sup_{\substack{x,y \in \mathsf{X} \\ x \neq y}} \frac{|f(x) - f(y)|}{\rho(x,y)}$$

is the Lipschitz constant of f. A sufficient (and, up to constants, necessary) condition for (12) is that, for every $f \in \mathrm{Lip}_1(\mathsf{X}, \rho)$, the random variable $f(X)$ with $X \sim \mu$ is *c-subgaussian*, i.e.,

$$\log \mathbb{E}_\mu \left[e^{\lambda(f(X) - \mathbb{E}_\mu[f(X)])} \right] \leq \frac{c\lambda^2}{2}, \qquad \forall \lambda \in \mathbb{R}. \tag{13}$$

A fundamental result of Bobkov and Götze [1] states that the subgaussian estimate (13) holds for all $f \in \mathrm{Lip}_1(\mathsf{X}, \rho)$ if and only if μ satisfies the so-called *transportation-information inequality*

$$W(\mu, \nu) \leq \sqrt{2c\, D(\nu \| \mu)}, \tag{14}$$

where ν ranges over all Borel probability measures on X. We will use the shorthand $\mu \in T_\rho(c)$ to denote the fact that the inequality (14) holds for all ν. The key role of transportation-information inequalities in characterizing the concentration of measure phenomenon was first recognized by Marton in a breakthrough paper [22], with further developments in [23, 24, 25].

The *entropy method* (see, e.g., [2, Chap. 6] and [30, Chap. 3]) provides another route to establishing (13). Its underlying idea can be briefly described as follows. Given a measurable function $f : \mathsf{X} \to \mathbb{R}$, consider the logarithmic moment-generating function

$$\psi_f(\lambda) \triangleq \log \mathbb{E}_\mu \left[e^{\lambda(f(X) - \mathbb{E}_\mu[f(X)])} \right]$$

of the centered random variable $f(X) - \mathbb{E}_\mu[f(X)]$. For any $\lambda \neq 0$, introduce the *tilted probability measure* $\mu^{(\lambda f)}$ with

$$\frac{d\mu^{(\lambda f)}}{d\mu} = \frac{e^{\lambda f}}{\mathbb{E}_\mu[e^{\lambda f}]} = \frac{e^{\lambda(f-\mathbb{E}_\mu f)}}{e^{\psi_f(\lambda)}}.$$

Then a simple calculation shows that the relative entropy $D(\mu^{(\lambda f)}\|\mu)$ can be expressed as

$$D(\mu^{(\lambda f)}\|\mu) = \lambda\psi_f'(\lambda) - \psi_f(\lambda) \equiv \lambda^2\left(\frac{\psi_f(\lambda)}{\lambda}\right)'$$

where the prime denotes differentiation with respect to λ. Using the fact that $\psi_f(0) = 0$ and integrating, we obtain the following formula for $\psi_f(\lambda)$:

$$\psi_f(\lambda) = \lambda \int_0^\lambda \frac{D(\mu^{(tf)}\|\mu)}{t^2} dt. \tag{15}$$

This representation is at the basis of the so-called *Herbst argument*, which for our purposes can be summarized as follows:

Lemma 3.1 (Herbst) *The metric probability space* (X, μ, ρ) *has the concentration property with constant c if, for any* $f \in \mathrm{Lip}_1(\mathsf{X}, \rho)$,

$$D(\mu^{(tf)}\|\mu) \le \frac{ct^2}{2}, \qquad \forall t > 0. \tag{16}$$

Remark 3.2 Up to a constant, the converse is also true [34, Prob. 3.12]: if the subgaussian estimate (13) holds for every $f \in \mathrm{Lip}_1(\mathsf{X}, \rho)$, then

$$D(\mu^{(tf)}\|\mu) \le 2ct^2, \qquad \forall t > 0$$

for every $f \in \mathrm{Lip}_1(\mathsf{X}, \rho)$.

In this way, the problem of establishing the concentration phenomenon reduces to showing that (16) holds for every $f \in \mathrm{Lip}_1(\mathsf{X}, \rho)$, typically via logarithmic Sobolev inequalities or other functional inequalities.

3.1 Concentration of Measure Under the Independence Assumption

To set the stage for the general treatment of the concentration phenomenon in high dimensions, we first consider the independent case, i.e., when coordinates X_i, $i \in T$, of the random object $X \sim \mu$ are mutually independent. In other words, the probability measure μ is equal to the product of its marginals: $\mu = \mu_1 \otimes \ldots \otimes \mu_n$. The key to establishing the concentration property in such a setting is *tensorization*, which is an umbrella term for any result that allows one to derive the "global" concentration property of the high-dimensional product space $(\mathsf{X}_1 \otimes \ldots \otimes \mathsf{X}_n, \rho, \mu_1 \otimes \ldots \otimes \mu_n)$ from "local" concentration properties of the coordinate spaces $(\mathsf{X}_i, \rho_i, \mu_i)$, $i \in T$.

Below, we list two such tensorization results, one for the transportation-information inequalities and one for the relative entropy. Both of these results are deep consequences of the interplay between the independence structure of μ and the metric structure of ρ. Indeed, a function $f : \mathsf{X} \to \mathbb{R}$ belongs to $\mathrm{Lip}_1(\mathsf{X}, \rho)$ if and only if $\delta_i(f) \leq 1$ for all $i \in T$, i.e., if and only if, for every $i \in T$ and every $x^{T\backslash\{i\}} \in \mathsf{X}^{T\backslash\{i\}}$, the function $f_i : \mathsf{X}_i \to R$ given by $f_i(y_i) \triangleq f(y_i x^{T\backslash\{i\}})$ is 1-Lipschitz with respect to the metric ρ_i on X_i. With this in mind, it is reasonable to expect that if one can establish a concentration property for all 1-Lipschitz functions on the coordinate spaces X_i, then one can deduce a concentration property for functions on the product space X that are 1-Lipschitz in each coordinate.

Lemma 3.3 (Tensorization of transportation-information inequalities) *Suppose that there exist constants $c_1, \ldots, c_n \geq 0$, such that*

$$\mu_i \in T_{\rho_i}(c_i), \qquad \forall i \in T.$$

Then $\mu = \mu_1 \otimes \ldots \otimes \mu_n \in T_\rho(c)$ with $c = \sum_{i=1}^n c_i$.

For example, by an appropriate rescaling of Pinsker's inequality (11), we see that, if each coordinate space X_i is endowed with the scaled trivial metric $\rho_i(x_i, z_i) = \alpha_i \mathbf{1}\{x_i \neq z_i\}$ for some $\alpha_i > 0$, then any Borel probability measure μ_i on X_i satisfies the transportation-information inequality with $c_i = \alpha_i^2/4$. By the above tensorization lemma, any product measure $\mu_1 \otimes \ldots \otimes \mu_n$ on the product space $\mathsf{X}_1 \otimes \ldots \otimes \mathsf{X}_n$ equipped with the weighted Hamming metric ρ_α defined in (3) satisfies $T_{\rho_\alpha}(c)$ with $c = \frac{1}{4}\sum_{i \in T} \alpha_i^2$. Consequently, by the Bobkov–Götze theorem, the subgaussian estimate (13) holds for any function $f \in \mathrm{Lip}_1(\mathsf{X}, \rho_\alpha)$, which in turn implies, via the Chernoff bound, that

$$\mathbb{P}_\mu\Big\{f - \mathbb{E}_\mu f \geq t\Big\} \leq \exp\left(-\frac{2t^2}{\sum_{i \in T} \alpha_i^2}\right), \qquad \forall t \geq 0.$$

This provides an alternative derivation of McDiarmid's inequality (with the sharp constant in the exponent), which was originally proved using the martingale method.

Lemma 3.4 (Tensorization of relative entropy) *Consider a product measure $\mu = \mu_1 \otimes \ldots \otimes \mu_n$. Then for any other probability measure ν on X we have*

$$D(\nu \| \mu) \leq \sum_{i \in T} \mathbb{E}_\nu D\Big(\nu_i(\cdot | X^{T\backslash\{i\}}) \| \mu_i\Big). \tag{17}$$

The idea is to apply this lemma to $\nu = \mu^{(tf)}$ for some $t \geq 0$ and an arbitrary $f \in \mathrm{Lip}_1(\mathsf{X}, \rho)$. In that case, a simple calculation shows that the conditional probability measure $\nu_i(\mathrm{d}x_i|x^{T\setminus\{i\}}) = \mu_i^{(tf)}(\mathrm{d}x_i|x^{T\setminus\{i\}})$ is equal to the tilted distribution $\mu_i^{(tf_i)}$ with $f_i(x_i) = f(x_i x^{T\setminus\{i\}})$, and therefore

$$D(\mu^{(tf)}\|\mu) \leq \sum_{i=1}^{n} \mathbb{E}_{\mu^{(tf)}} D\big(\mu_i^{(tf_i)}\big\|\mu_i\big).$$

If $f \in \mathrm{Lip}_1(\mathsf{X}, \rho)$, then $f_i \in \mathrm{Lip}_1(\mathsf{X}_i, \rho_i)$. Thus, if we can show that, for any $g \in \mathrm{Lip}_1(\mathsf{X}_i, \rho_i)$,

$$D(\mu_i^{(tg)}\|\mu_i) \leq \frac{c_i t^2}{2}, \qquad \forall t \geq 0,$$

then the estimate

$$D(\mu^{(tf)}\|\mu) \leq \frac{ct^2}{2}, \qquad \forall t \geq 0$$

holds with $c = \sum_{i=1}^{n} c_i$ for all $f \in \mathrm{Lip}(\mathsf{X}, \rho)$ by the tensorization lemma. Invoking the Herbst argument, we conclude that (X, μ, ρ) has the concentration property with the same c.

4 The Abstract Martingale Method

In this section, we present a general martingale-based scheme for deriving concentration inequalities for functions of many dependent random variables. Let $f : \mathsf{X} \to \mathbb{R}$ be the function of interest, and let $X = (X_i)_{i \in T}$ be a random element of the product space X with probability law μ. Let $\mathcal{F}_0 \subset \mathcal{F}_1 \subset \ldots \subset \mathcal{F}_m$ be a filtration (i.e., an increasing sequence of σ-fields) on X, such that $\mathcal{F}_0$ is trivial and $\mathcal{F}_m = \mathcal{B}$. The idea is to decompose the centered random variable $f(X) - \mathbb{E}_\mu[f(X)]$ as a sum of martingale differences

$$M^{(j)} \triangleq \mathbb{E}_\mu[f(X)|\mathcal{F}_j] - \mathbb{E}_\mu[f(X)|\mathcal{F}_{j-1}], \qquad j = 1, \ldots, m.$$

By construction, $\mathbb{E}_\mu[f(X)|\mathcal{F}_m] = f(X)$ and $\mathbb{E}_\mu[f(X)|\mathcal{F}_0] = \mathbb{E}_\mu[f(X)]$, so the problem of bounding the probability $\mathbb{P}_\mu\{|f - \mathbb{E}_\mu f| \geq t\}$ for a given $t \geq 0$ reduces to bounding the probability

$$\mathbb{P}_\mu\left\{\left|\sum_{j=1}^{m} M^{(j)}\right| \geq t\right\}.$$

The latter problem hinges on being able to control the martingale differences $M^{(j)}$. In particular, if each $M^{(j)}$ is a.s. bounded, we have the following:

Theorem 4.1 (Azuma–Hoeffding inequality) *Let $\{M^{(j)}\}_{j=1}^m$ be a martingale difference sequence with respect to a filtration $\{\mathcal{F}_j\}_{j=0}^m$. Suppose that, for each j, there exist $\mathcal{F}_{j-1}$-measurable random variables $A^{(j)}$ and $B^{(j)}$, such that $A^{(j)} \le M^{(j)} \le B^{(j)}$ a.s. Then*

$$\mathbb{E}\left[\exp\left(\lambda \sum_{j=1}^m M^{(j)}\right)\right] \le \exp\left(\frac{\lambda^2 \sum_{j=1}^m \|B^{(j)} - A^{(j)}\|_\infty^2}{8}\right), \quad \forall \lambda \in \mathbb{R}. \tag{18}$$

Consequently, for any $t \ge 0$,

$$\mathbb{P}\left\{\left|\sum_{j=1}^m M^{(j)}\right| \ge t\right\} \le 2\exp\left(-\frac{2t^2}{\sum_{j=1}^m \|B^{(j)} - A^{(j)}\|_\infty^2}\right). \tag{19}$$

The most straightforward choice of the filtration is also the most natural one: take $m = |T| = n$, and for each $i \in T$ take $\mathcal{F}_i = \sigma(X^{[i]})$. For $i \in T$, define a Markov kernel $K^{(i)}$ on X by

$$K^{(i)}(x, \mathrm{d}y) \triangleq \delta_{x^{[i-1]}}(\mathrm{d}y^{[i-1]}) \otimes \mu^{[i,n]}(\mathrm{d}y^{[i,n]}|x^{[i-1]}). \tag{20}$$

Then, for any $f \in L^1(\mu)$ we have

$$\begin{aligned} K^{(i)}f(x) &\triangleq \int_{\mathsf{X}} f(y) K^{(i)}(x, \mathrm{d}y) \\ &= \int_{\mathsf{X}^{[i,n]}} f(x^{[i-1]} y^{[i,n]}) \mu^{[i,n]}(\mathrm{d}y^{[i,n]}|x^{[i-1]}) \\ &= \mathbb{E}_\mu[f(X)|X^{[i-1]} = x^{[i-1]}]; \end{aligned}$$

in particular, $K^{(1)}f = \mathbb{E}_\mu f$. We extend this definition to $i = n+1$ in the obvious way:

$$K^{(n+1)}(x, \mathrm{d}y) = \delta_x(\mathrm{d}y),$$

so that $K^{(n+1)}f = f$. Then, for each $i \in T$, we can write $M^{(i)} = K^{(i+1)}f - K^{(i)}f$. With this construction, we can state the following theorem that applies to the case when each coordinate space X_i is endowed with a bounded measurable metric ρ_i:

Theorem 4.2 *Assume that, for all i,*

$$\|\rho_i\| \triangleq \sup_{x_i, z_i \in \mathsf{X}_i} \rho_i(x_i, z_i) < \infty.$$

For each $i \in \{1, \dots, n+1\}$, let $V^{(i)}$ be a Wasserstein matrix for the Markov kernel $K^{(i)}$ defined in (20), in the sense that $\delta(K^{(i)}f) \preceq V^{(i)}\delta(f)$ holds for each

$f \in \mathrm{Lip}(\mathsf{X}, \rho)$ as in (7). Define the matrix $\Gamma = (\Gamma_{ij})_{i,j \in T}$ with entries

$$\Gamma_{ij} \triangleq \|\rho_i\| V_{ij}^{(i+1)}.$$

Then, for any $f \in \mathrm{Lip}(\mathsf{X}, \rho)$ and for any $t \geq 0$, we have

$$\mathbb{P}_\mu\Big\{|f(X) - \mathbb{E}_\mu[f(X)]| \geq t\Big\} \leq 2\exp\left(-\frac{2t^2}{\|\Gamma\boldsymbol{\delta}(f)\|^2_{\ell^2(T)}}\right). \tag{21}$$

Proof For each $i \in T$, using the tower property of conditional expectations, we can write

$$\begin{aligned} M^{(i)} &= \mathbb{E}_\mu[f(X)|X^{[i]} = x^{[i]}] - \mathbb{E}_\mu[f(X)|X^{[i-1]} = x^{[i-1]}] \\ &= \mathbb{E}_\mu[f(X)|X^{[i]} = x^{[i]}] - \mathbb{E}_\mu\big[\mathbb{E}_\mu[f(X)|X^{[i-1]} = x^{[i-1]}, X_i]\big|X^{[i-1]} = x^{[i-1]}\big] \\ &= \int_{\mathsf{X}^{[i,n]}} \mu^{[i,n]}(\mathrm{d}y^{[i,n]}|x^{[i-1]})\Big(\int_{\mathsf{X}^{(i,n]}} \mu^{(i,n]}(\mathrm{d}z^{(i,n]}|x^{[i]})f(x^{[i-1]}x_i z^{(i,n]}) \\ &\quad - \int_{\mathsf{X}^{(i,n]}} \mu^{(i,n]}(\mathrm{d}z^{(i,n]}|x^{[i-1]}y_i)f(x^{[i-1]}y_i z^{(i,n]})\Big) \\ &= \int_{\mathsf{X}^{[i,n]}} \mu^{[i,n]}(\mathrm{d}y^{[i,n]}|x^{[i-1]})\left(K^{(i+1)}f(x^{[i-1]}x_i y^{(i,n]}) - K^{(i+1)}f(x^{[i-1]}y_i y^{(i,n]})\right). \end{aligned}$$

From this, it follows that $A^{(i)} \leq M^{(i)} \leq B^{(i)}$ a.s., where

$$\begin{aligned} A^{(i)} &\triangleq \int_{\mathsf{X}^{[i,n]}} \mu^{[i,n]}(\mathrm{d}y^{[i,n]}|x^{[i-1]}) \inf_{x_i \in \mathsf{X}_i} \left(K^{(i+1)}f(x^{[i-1]}x_i y^{(i,n]}) - K^{(i+1)}f(x^{[i-1]}y_i y^{(i,n]})\right) \\ B^{(i)} &\triangleq \int_{\mathsf{X}^{[i,n]}} \mu^{[i,n]}(\mathrm{d}y^{[i,n]}|x^{[i-1]}) \sup_{x_i \in \mathsf{X}_i} \left(K^{(i+1)}f(x^{[i-1]}x_i y^{(i,n]}) - K^{(i+1)}f(x^{[i-1]}y_i y^{(i,n]})\right), \end{aligned}$$

and

$$\|B^{(i)} - A^{(i)}\|_\infty \leq \|\rho_i\|\delta_i\big(K^{(i+1)}f\big). \tag{22}$$

By definition of the Wasserstein matrix, we have

$$\delta_i\left(K^{(i+1)}f\right) \leq \sum_{j \in T} V_{ij}^{(i+1)}\delta_j(f).$$

Substituting this estimate into (22), we get

$$\sum_{i=1}^n \|B^{(i)} - A^{(i)}\|^2_\infty \leq \sum_{i=1}^n |(\Gamma\boldsymbol{\delta}(f))_i|^2 \equiv \|\Gamma\boldsymbol{\delta}(f)\|^2_{\ell^2(T)}. \tag{23}$$

The probability estimate (21) then follows from the Azuma–Hoeffding inequality (19). □

We can also use the martingale method to obtain a tensorization result for transportation inequalities without independence assumptions. This result, which generalizes a theorem of Djellout, Guillin, and Wu [8, Thm. 2.11], can be used even when the metrics ρ_i are not necessarily bounded.

Theorem 4.3 *Suppose that there exist constants* $c_1, \ldots, c_n \geq 0$, *such that*

$$\mu_i(\cdot|x^{[i-1]}) \in T_{\rho_i}(c_i), \qquad \forall i \in T, \, x^{[i-1]} \in \mathsf{X}^{[i-1]}. \tag{24}$$

For each $i \in \{1, \ldots, n+1\}$, *let* $V^{(i)}$ *be a Wasserstein matrix for* $K^{(i)}$. *Then* $\mu \in T_\rho(c)$ *with*

$$c = \sum_{i \in T} c_i \Big(\sum_{j \in T} V_{ij}^{(i+1)} \Big)^2. \tag{25}$$

Proof By the Bobkov–Götze theorem [1], it suffices to show that, for every $f : \mathsf{X} \to \mathbb{R}$ with $\|f\|_{\mathrm{Lip}} \leq 1$, the random variable $f(X)$ with $X \sim \mu$ is c-subgaussian, with c given by (25). To that end, we again consider the martingale decomposition

$$f - \mathbb{E}_\mu[f] = \sum_{i \in T} M^{(i)}$$

with $M^{(i)} = K^{(i+1)}f - K^{(i)}f$. We will show that, for every i,

$$\log \mathbb{E}_\mu \Big[e^{\lambda M^{(i)}} \Big| X^{[i-1]} \Big] \leq \frac{c_i \Big(\sum_{j \in T} V_{ij}^{(i+1)} \Big)^2 \lambda^2}{2}, \qquad \forall \lambda \in \mathbb{R}. \tag{26}$$

This, in turn, will yield the desired subgaussian estimate

$$\begin{aligned} \mathbb{E}_\mu \left[e^{\lambda (f - \mathbb{E}_\mu[f])} \right] &= \mathbb{E}_\mu \left[\exp \left(\lambda \sum_{i \in T} M^{(i)} \right) \right] \\ &\leq \exp \left(\frac{c\lambda^2}{2} \right) \end{aligned}$$

for every $\lambda \in \mathbb{R}$.

To proceed, note that, for a fixed realization $x^{[i-1]}$ of $X^{[i-1]}$, $M^{(i)} = K^{(i+1)}f - K^{(i)}f$ is $\sigma(X_i)$-measurable, and

$$\|M^{(i)}\|_{\mathrm{Lip}} \leq \sup_{\substack{x,y \in \mathsf{X} \\ x^{T \setminus \{i\}} = y^{T \setminus \{i\}}}} \frac{|K^{(i+1)}f(x) - K^{(i+1)}f(y)|}{\rho_i(x_i, y_i)}$$

$$\begin{aligned}&\equiv \delta_i\big(K^{(i+1)}f\big)\\&\le \sum_{j\in T} V_{ij}^{(i+1)}\delta_j(f)\\&\le \sum_{j\in T} V_{ij}^{(i+1)},\end{aligned}$$

where we have used the definition of the Wasserstein matrix, as well as the fact that $\|f\|_{\rm Lip} \le 1$ is equivalent to $\delta_j(f) \le 1$ for all $j \in T$. Since $\mu_i(\cdot|x^{[i-1]}) \in T_{\rho_i}(c)$ by hypothesis, we obtain the estimate (26) by the Bobkov–Götze theorem. □

As a sanity check, let us confirm that, in the case when μ is a product measure and the product space X is endowed with the weighted Hamming metric ρ_α defined in (3), Theorems 4.2 and 4.3 both reduce to McDiarmid's inequality. To see this, we first note that, when the X_i's are independent, we can write

$$K^{(i)}f(x) = \int_{\mathsf{X}^{[i,n]}} f(x^{[i-1]}y^{[i,n]})\mu_i(\mathrm{d}y_i)\mu_{i+1}(\mathrm{d}y_{i+1})\dots\mu_n(\mathrm{d}y_n)$$

for each $i \in T, f \in L^1(\mu)$, and $x \in \mathsf{X}$. This, in turn, implies that

$$\begin{aligned}\delta_i(K^{(i+1)}f) &= \alpha_i^{-1} \sup_{\substack{x,z\in\mathsf{X}\\ x^{T\setminus\{i\}}=z^{T\setminus\{i\}}}} \left|K^{(i+1)}f(x) - K^{(i+1)}f(z)\right|\\ &= \alpha_i^{-1} \sup_{\substack{x,z\in\mathsf{X}\\ x^{T\setminus\{i\}}=z^{T\setminus\{i\}}}} \left|\int_{\mathsf{X}^{(i,n]}} f(x^{[i]}y^{(i,n]})\mu_{i+1}(\mathrm{d}y_{i+1})\dots\mu_n(\mathrm{d}y_n)\right.\\ &\qquad\qquad \left. - \int_{\mathsf{X}^{(i,n]}} f(z^{[i]}y^{(i,n]})\mu_{i+1}(\mathrm{d}y_{i+1})\dots\mu_n(\mathrm{d}y_n)\right|\\ &\le \alpha_i^{-1} \sup_{\substack{x,z\in\mathsf{X}\\ x^{T\setminus\{i\}}=z^{T\setminus\{i\}}}} |f(x)-f(z)|\\ &= \delta_i(f),\end{aligned}$$

where we have used the fact that, with $\rho_i(x_i, z_i) = \alpha_i \mathbf{1}\{x_i \neq z_i\}$, $\|\rho_i\| = \alpha_i$ for every $i \in T$. Therefore, for each $i \in T$, we can always choose a Wasserstein matrix $V^{(i+1)}$ for $K^{(i+1)}$ in such a way that its ith row has zeroes everywhere except for the ith column, where it has a 1. Now, for any function $f : \mathsf{X} \to \mathbb{R}$ which is 1-Lipschitz with respect to ρ_α, we can take $\boldsymbol{\delta}(f) = \mathbf{1}$. Therefore, for any such f Theorem 4.2 gives

$$\mathbb{P}_\mu\left\{|f(X) - \mathbb{E}_\mu[f(X)]| \ge t\right\} \le 2\exp\left(-\frac{2t^2}{\sum_{i=1}^n \alpha_i^2}\right), \qquad \forall t \ge 0$$

which is precisely McDiarmid's inequality. Since the constant 2 in McDiarmid's inequality is known to be sharp, this shows that the coefficient 2 in the exponent in (21) is likewise optimal. Moreover, with our choice of ρ_i, condition (24) of Theorem 4.3 holds with $c_i = \alpha_i^2/4$, and, in light of the discussion above, we can arrange $\sum_j V_{ij}^{(i+1)} = 1$. Therefore, by Theorem 4.3, any function $f : \mathsf{X} \to \mathbb{R}$ which is 1-Lipschitz with respect to ρ_α is c-subgaussian with constant

$$c = \sum_{i \in T} c_i \left(\sum_{j \in T} V_{ij}^{(i+1)} \right)^2 = \frac{1}{4} \sum_{i \in T} \alpha_i^2,$$

which is just another equivalent statement of McDiarmid's inequality.

It is also possible to consider alternative choices of the filtration $\{\mathcal{F}_j\}_{j=0}^m$. For example, if we partition the index set T into m disjoint subsets (blocks) $T_1, \ldots, T_m$, we can take

$$\mathcal{F}_j \triangleq \sigma\left(X_i : i \in \Lambda_j\right), \qquad \forall i \in T$$

where $\Lambda_j \triangleq T_1 \cup \ldots \cup T_j$. Defining for each $j \in [m]$ the Markov kernel

$$\tilde{K}^{(j)}(x, \mathrm{d}y) \triangleq \delta_{x^{\Lambda_{i-1}}}(\mathrm{d}y^{\Lambda_{i-1}}) \otimes \mu^{T \setminus \Lambda_{i-1}}(\mathrm{d}y^{T \setminus \Lambda_{i-1}} | x^{\Lambda_{i-1}}),$$

we can write

$$M^{(j)} = \mathbb{E}_\mu[f(X)|\mathcal{F}_j] - \mathbb{E}_\mu[f(X)|\mathcal{F}_{j-1}] = K^{(j+1)}f - K^{(j)}f$$

for every $j \in [m]$. As before, we take $K^{(1)}f = \mathbb{E}_\mu[f]$ and $K^{(m+1)}f = f$. Given a measurable function $f : \mathsf{X} \to \mathbb{R}$, we can define the oscillation of f in the jth block $T_j, j \in [m]$, by

$$\tilde{\delta}_j(f) \triangleq \sup_{\substack{x,z \in \mathsf{X} \\ x^{T \setminus T_j} = z^{T \setminus T_j}}} \frac{|f(x) - f(z)|}{\rho^{T_j}(x^{T_j}, z^{T_j})}.$$

The definition of a Wasserstein matrix is modified accordingly: we say that a nonnegative matrix $\tilde{V} = (\tilde{V}_{jk})_{j,k \in [m]}$ is a Wasserstein matrix for a Markov kernel K on X with respect to the partition $\{T_j\}_{j=1}^m$ if, for any Lipschitz function $f : \mathsf{X} \to \mathbb{R}$,

$$\tilde{\delta}_j(Kf) \le \sum_{k=1}^m \tilde{V}_{jk} \tilde{\delta}_k(f)$$

for all $j \in [m]$. With these definitions at hand, the following theorem, which generalizes a result of Paulin [29, Thm. 2.1], can be proved in the same way as Theorem 4.2:

Theorem 4.4 *For each $j \in [m+1]$, let $\tilde{V}^{(j)} = (\tilde{V}^{(j)}_{k\ell})_{k,\ell \in [m]}$ be a Wasserstein matrix for $\tilde{K}^{(j)}$ with respect to the partition $\{T_j\}$. Define the matrix $\tilde{\Gamma} = (\tilde{\Gamma}_{k\ell})_{k,\ell \in [m]}$ with entries*

$$\tilde{\Gamma}_{k\ell} \triangleq \|\rho^{T_k}\| \tilde{V}^{(k+1)}_{k\ell},$$

where

$$\|\rho^{T_k}\| \triangleq \sup_{x^{T_k}, z^{T_k}} \rho^{T_k}(x^{T_k}, z^{T_k})$$

is the diameter of the metric space $(\mathsf{X}^{T_k}, \rho^{T_k})$. Then, for any $f \in \mathrm{Lip}(\mathsf{X}, \rho)$ and for any $t \geq 0$,

$$\mathbb{P}_\mu \Big\{ |f(X) - \mathbb{E}_\mu[f(X)]| \geq t \Big\} \leq 2 \exp \left(- \frac{2t^2}{\| \tilde{\Gamma} \tilde{\delta}(f) \|^2_{\ell^2(m)}} \right).$$

5 The Martingale Method in Action

We now show that several previously published concentration inequalities for functions of dependent random variables arise as special cases of Theorem 4.2 by exploiting the freedom to choose the Wasserstein matrices $V^{(i)}$. In fact, careful examination of the statement of Theorem 4.2 shows that, for each $i \in T$, we only need to extract the ith row of $V^{(i+1)}$.

5.1 Concentration Inequalities Under the Dobrushin Uniqueness Condition

One particularly clean way of constructing the desired Wasserstein matrices is via the classical comparison theorem of Dobrushin for Gibbs measures [9]. For our purposes, we give its formulation due to Föllmer [11]:

Lemma 5.1 *Let ν and $\tilde{\nu}$ be two Borel probability measures on X. Define the matrix $C^\nu = (C^\nu_{ij})_{i,j \in T}$ and the vector $b^{\nu,\tilde{\nu}} = (b^{\nu,\tilde{\nu}}_i)_{i \in T}$ by*

$$C^\nu_{ij} = \sup_{\substack{x,z \in \mathsf{X} \\ x^{T \setminus \{j\}} = z^{T \setminus \{j\}}}} \frac{W_i\big(\nu_i(\cdot | x^{T \setminus \{i\}}), \nu_i(\cdot | z^{T \setminus \{i\}})\big)}{\rho_j(x_j, z_j)} \tag{27}$$

and

$$b_i^{\nu,\tilde{\nu}} = \int_{\mathsf{X}^{T\setminus\{i\}}} \tilde{\nu}^{T\setminus\{i\}}(\mathrm{d}x^{T\setminus\{i\}}) W_i\big(\nu_i(\cdot|x^{T\setminus\{i\}}), \tilde{\nu}_i(\cdot|x^{T\setminus\{i\}})\big). \tag{28}$$

Suppose that the spectral radius of C^ν is strictly smaller than unity. Then, for any $f \in L^1(\mu)$,

$$|\mathbb{E}_\nu f - \mathbb{E}_{\tilde{\nu}} f| \le \sum_{j,k \in T} \delta_j(f) D^\nu_{jk} b_k^{\nu,\tilde{\nu}}, \tag{29}$$

where $D^\nu \triangleq \sum_{m=0}^\infty (C^\nu)^m$.

Remark 5.2 The matrix C^ν is called the *Dobrushin interdependence matrix* of ν. When the spectral radius of C^ν is strictly smaller than unity, we say that ν satisfies the *Dobrushin uniqueness condition*. This condition is used in statistical physics to establish the absence of phase transitions, which is equivalent to uniqueness of a global Gibbs measure consistent with a given local specification (see the book of Georgii [12] for details).

Given an index $i \in T$, we will extract the ith row of a Wasserstein matrix for $K^{(i+1)}$ by applying the Dobrushin comparison theorem to a particular pair of probability measures on X. Let $x, z \in \mathsf{X}$ be two configurations that differ only in the ith coordinate: $x^{T\setminus\{i\}} = z^{T\setminus\{i\}}$ and $x_i \neq z_i$. Thus, we can write $z = x^{[i-1]} z_i x^{(i,n]}$, and

$$\begin{aligned}
&K^{(i+1)} f(x) - K^{(i+1)} f(z) \\
&= \mathbb{E}_\mu[f(X)|X^{[i]} = x^{[i-1]}x_i] - \mathbb{E}_\mu[f(X)|X^{[i]} = x^{[i-1]}z_i] \\
&= \int_{\mathsf{X}^{(i,n]}} f(x^{[i-1]}x_i y^{(i,n]}) \mu^{(i,n]}(\mathrm{d}y^{(i,n]}|x^{[i-1]}x_i) - \int_{\mathsf{X}^{(i,n]}} f(x^{[i-1]}z_i y^{(i,n]}) \mu^{(i,n]}(\mathrm{d}y^{(i,n]}|x^{[i-1]}z_i) \\
&= \int_{\mathsf{X}^{(i,n]}} \big(f(x^{[i-1]}x_i y^{(i,n]}) - f(x^{[i-1]}z_i y^{(i,n]})\big)\, \mu^{(i,n]}(\mathrm{d}y^{(i,n]}|x^{[i-1]}x_i) \\
&\qquad + \int_{\mathsf{X}^{(i,n]}} f(x^{[i-1]}z_i y^{(i,n]}) \mu^{(i,n]}(\mathrm{d}y^{(i,n]}|x^{[i-1]}x_i) \\
&\qquad - \int_{\mathsf{X}^{(i,n]}} f(x^{[i-1]}z_i y^{(i,n]}) \mu^{(i,n]}(\mathrm{d}y^{(i,n]}|x^{[i-1]}z_i).
\end{aligned} \tag{30}$$

By definition of the local oscillation, the first integral in (30) is bounded by $\delta_i(f)\rho_i(x_i, z_i)$. To handle the remaining terms, define two probability measures $\nu, \tilde{\nu}$ on X by

$$\nu(\mathrm{d}y) \triangleq \delta_{x^{[i-1]}z_i}(\mathrm{d}y^{[i]}) \otimes \mu^{(i,n]}(\mathrm{d}y^{(i,n]}|x^{[i-1]}x_i)$$

$$\tilde{\nu}(\mathrm{d}y) \triangleq \delta_{x^{[i-1]}z_i}(\mathrm{d}y^{[i]}) \otimes \mu^{(i,n]}(\mathrm{d}y^{(i,n]}|x^{[i-1]}z_i).$$

Using this definition and Lemma 5.1, we can write

$$\begin{aligned}
&\int_{\mathsf{X}^{(i,n]}} f(x^{[i-1]} z_i y^{(i,n]}) \mu^{(i,n]}(\mathrm{d}y^{(i,n]} | x^{[i-1]} x_i) - \int_{\mathsf{X}^{(i,n]}} f(x^{[i-1]} z_i y^{(i,n]}) \mu^{(i,n]}(\mathrm{d}y^{(i,n]} | x^{[i-1]} z_i) \\
&\qquad = \int f \mathrm{d}\nu - \int f \mathrm{d}\tilde{\nu} \\
&\qquad \le \sum_{j,k \in T} \delta_j(f) D^{\nu}_{jk} b^{\nu,\tilde{\nu}}_k. \qquad (31)
\end{aligned}$$

It remains to obtain explicit upper bounds on the entries of D^{ν} and $b^{\nu,\tilde{\nu}}$. To that end, we first note that, for a given $j \in T$ and for any $u, y \in \mathsf{X}$,

$$\begin{aligned}
&W_j\big(\nu_j(\cdot|u^{T\setminus\{j\}}), \nu_j(\cdot|y^{T\setminus\{j\}})\big) \\
&\qquad = \begin{cases} 0, & j \le i \\ W_j\big(\mu_j(\cdot|x^{[i-1]} x_i u^{(i,n]\setminus\{j\}}), \mu_j(\cdot|x^{[i-1]} z_i u^{(i,n]\setminus\{j\}})\big), & j > i \end{cases}.
\end{aligned}$$

Therefore, $C^{\nu}_{jk} \le C^{\mu}_{jk}$. Likewise, for a given $k \in T$ and for any $y \in \mathsf{X}$,

$$\begin{aligned}
&W_k\big(\nu_k(\cdot|y^{T\setminus\{k\}}), \tilde{\nu}_k(\cdot|y^{T\setminus\{k\}})\big) \\
&\qquad = \begin{cases} 0, & k \le i \\ W_k\big(\mu_k(\cdot|x^{[i-1]} x_i y^{(i,n]\setminus\{k\}}), \mu_k(\cdot|x^{[i-1]} z_i y^{(i,n]\setminus\{k\}})\big), & k > i \end{cases}
\end{aligned}$$

Therefore, $b^{\nu,\tilde{\nu}}_k \le C^{\mu}_{ki} \rho_i(x_i, z_i)$. Since the matrices C^{ν} and C^{μ} are nonnegative, $D^{\nu}_{jk} \le D^{\mu}_{jk}$. Consequently, we can write

$$\begin{aligned}
\int f \mathrm{d}\nu - \int f \mathrm{d}\tilde{\nu} &\le \sum_{j,k \in T} \delta_j(f) D^{\mu}_{jk} C^{\mu}_{ki} \rho_i(x_i, z_i) \\
&= \sum_{j \in T} \delta_j(f) (D^{\mu} C^{\mu})_{ji} \rho_i(x_i, z_i) \\
&= \sum_{j \in T} \delta_j(f) (D^{\mu} - \mathrm{id})_{ji} \rho_i(x_i, z_i). \qquad (32)
\end{aligned}$$

Therefore, from (31) and (32), we have

$$\begin{aligned}
\frac{K^{(i+1)} f(x) - K^{(i+1)} f(z)}{\rho_i(x_i, z_i)} &\le \delta_i(f) + \sum_{j \in T} (D^{\mu} - \mathbf{1})^{\mathsf{T}}_{ij} \delta_j(f) \\
&= \sum_{j \in T} (D^{\mu})^{\mathsf{T}}_{ij} \delta_j(f). \qquad (33)
\end{aligned}$$

We have thus proved the following:

Corollary 5.3 *Suppose that the probability measure μ satisfies the Dobrushin uniqueness condition, i.e., the spectral radius of its Dobrushin interdependence matrix C^{μ} is strictly smaller than unity. Then, for any $t \geq 0$, the concentration inequality* (21) *holds with*

$$\Gamma_{ij} = \|\rho_i\|(D^{\mu})^{T}_{ij}, \qquad i,j \in T. \tag{34}$$

For example, when each X_i is equipped with the trivial metric $\rho_i(x_i, z_i) = \mathbf{1}\{x_i \neq z_i\}$, we have $\|\rho_i\| = 1$ for all i, and consequently obtain the concentration inequality

$$\mathbb{P}_{\mu}\Big\{|f(X) - \mathbb{E}_{\mu}[f(X)]| \geq t\Big\} \leq 2\exp\left(-\frac{2t^2}{\|(D^{\mu})^{\mathsf{T}}\boldsymbol{\delta}(f)\|^2_{\ell^2(T)}}\right). \tag{35}$$

The same inequality, but with a worse constant in the exponent, was obtained by Külske [19, p. 45].

5.2 *Concentration Inequalities Via Couplings*

Another method for constructing Wasserstein matrices for the Markov kernels $K^{(i)}$ is via couplings. One notable advantage of this method is that it does not explicitly rely on the Dobrushin uniqueness condition; however, some such condition is typically necessary in order to obtain good bounds for the norm $\|\Gamma\boldsymbol{\delta}(f)\|_{\ell^2(T)}$.

Fix an index $i \in T$ and any two $x, z \in \mathsf{X}$ that differ only in the ith coordinate: $x^{T\setminus\{i\}} = z^{T\setminus\{i\}}$ and $x_i \neq z_i$. Let $\mathbf{P}^{[i]}_{x,z}$ be any coupling of the conditional laws $\mu^{(i,n]}(\cdot|x^{[i]})$ and $\mu^{(i,n]}(\cdot|z^{[i]})$. Then for any $f \in L^1(\mu)$ we can write

$$\begin{aligned}
&K^{(i+1)}f(x) - K^{(i+1)}f(z) \\
&\quad= \int_{\mathsf{X}^{(i,n]}\times\mathsf{X}^{(i,n]}} \mathbf{P}^{[i]}_{x,z}(\mathrm{d}u^{(i,n]}, \mathrm{d}y^{(i,n]})\left(f(x^{[i]}, u^{(i,n]}) - f(z^{[i]}, y^{(i,n]})\right) \\
&\quad\leq \delta_i(f)\rho_i(x_i, z_i) + \sum_{j\in T:j>i}\delta_j(f)\int_{\mathsf{X}^{(i,n]}\times\mathsf{X}^{(i,n]}} \mathbf{P}^{[i]}_{x,z}(\mathrm{d}u^{(i,n]}, \mathrm{d}y^{(i,n]})\rho_j(u_j, y_j).
\end{aligned}$$

Therefore,

$$\begin{aligned}
\frac{|K^{(i+1)}f(x) - K^{(i+1)}f(z)|}{\rho_i(x_i, z_i)} &\leq \delta_i(f) + \sum_{j\in T:j>i}\frac{\int\rho_j\mathrm{d}\mathbf{P}^{[i]}_{x,z}}{\rho_i(x_i, z_i)}\delta_j(f) \\
&\leq \delta_i(f) + \sum_{j\in T:j>i}\sup_{\substack{x,z\in\mathsf{X} \\ x^{T\setminus\{i\}} = z^{T\setminus\{i\}}}}\frac{\int\rho_j\mathrm{d}\mathbf{P}^{[i]}_{x,z}}{\rho_i(x_i, z_i)}\delta_j(f).
\end{aligned}$$

Remembering that we only need the ith row of a Wasserstein matrix for $K^{(i+1)}$, we may take

$$V_{ij}^{(i+1)} = \begin{cases} 0, & i > j \\ 1, & i = j \\ \sup\limits_{\substack{x,z \in \mathsf{X} \\ x^{T\setminus\{i\}} = z^{T\setminus\{i\}}}} \dfrac{\int \rho_j \mathrm{d}\mathbf{P}_{x,z}^{[i]}}{\rho_i(x_i, z_i)}, & i < j \end{cases}. \tag{36}$$

We have thus proved the following:

Corollary 5.4 *For each index $i \in T$ and for each pair $x, z \in \mathsf{X}$ of configurations with $x^{T\setminus\{i\}} = z^{T\setminus\{i\}}$, pick an arbitrary coupling $\mathbf{P}_{x,z}^{[i]}$ of the conditional laws $\mu^{\{i\}}(\cdot|x^{[i]})$ and $\mu^{\{i\}}(\cdot|z^{[i]})$. Then, for any $t \geq 0$, the concentration inequality* (21) *holds with*

$$\Gamma_{ij} = \|\rho_i\| V_{ij}^{(i+1)}, \qquad i, j \in T \tag{37}$$

where the entries $V_{ij}^{(i+1)}$ are given by (36).

In the case when each X_i is equipped with the trivial metric $\rho_i(x_i, z_i) = \mathbf{1}\{x_i \neq z_i\}$, the entries Γ_{ij} for $j > i$ take the form

$$\Gamma_{ij} = \sup_{\substack{x,z \in \mathsf{X} \\ x^{T\setminus\{i\}} = z^{T\setminus\{i\}}}} \mathbf{P}_{x,z}^{[i]} \left\{ Y_j^{(0)} \neq Y_j^{(1)} \right\}, \tag{38}$$

where $(Y^{(0)}, Y^{(1)}) = \left((Y_{i+1}^{(0)}, \ldots, Y_n^{(0)}), (Y_{i+1}^{(1)}, \ldots, Y_n^{(1)})\right)$ is a random object taking values in $\mathsf{X}^{(i,n]} \times \mathsf{X}^{(i,n]}$. A special case of this construction, under the name of *coupling matrix*, was used by Chazottes et al. [6]. In that work, each $\mathbf{P}_{x,z}^{[i]}$ was chosen to minimize

$$\mathbf{P}\{Y^{(0)} \neq Y^{(1)}\},$$

over all couplings $\mathbf{P}$ of $\mu^{(i,n]}(\cdot|x^{[i]})$ and $\mu^{(i,n]}(\cdot|z^{[i]})$, in which case we have

$$\begin{aligned} \mathbf{P}_{x,z}^{[i]}\{Y^{(0)} \neq Y^{(1)}\} &= \inf_{\mathbf{P} \in C(\mu^{(i,n]}(\cdot|x^{[i]}), \mu^{(i,n]}(\cdot|z^{[i]}))} \mathbf{P}\{Y^{(0)} \neq Y^{(1)}\} \\ &= \left\| \mu^{(i,n]}(\cdot|x^{[i]}) - \mu^{(i,n]}(\cdot|z^{[i]}) \right\|_{\mathrm{TV}}. \end{aligned}$$

However, it is not clear how to relate the quantities $\mathbf{P}_{x,z}^{[i]} \left\{ Y_j^{(0)} \neq Y_j^{(1)} \right\}$ and $\mathbf{P}_{x,z}^{[i]} \left\{ Y^{(0)} \neq Y^{(1)} \right\}$, apart from the obvious bound

$$\mathbf{P}_{x,z}^{[i]}\{Y_j^{(0)} \neq Y_j^{(1)}\} \leq \mathbf{P}_{x,z}^{[i]}\{Y^{(0)} \neq Y^{(1)}\} = \left\| \mu^{(i,n]}(\cdot|x^{[i]}) - \mu^{(i,n]}(\cdot|z^{[i]}) \right\|_{\mathrm{TV}},$$

which gives

$$\Gamma_{ij} \leq \sup_{\substack{x,z \in \mathsf{X} \\ x^{T\setminus\{i\}} = z^{T\setminus\{i\}}}} \left\| \mu^{(i,n]}(\cdot|x^{[i]}) - \mu^{(i,n]}(\cdot|z^{[i]}) \right\|_{\mathrm{TV}}.$$

An alternative choice of coupling is the so-called *maximal coupling* due to Goldstein [13], which for our purposes can be described as follows: let $U = (U_\ell)_{\ell=1}^m$ and $Y = (Y_\ell)_{\ell=1}^m$ be two random m-tuples taking values in a product space $\mathsf{E} = \mathsf{E}_1 \times \ldots \times \mathsf{E}_m$, where each E_ℓ is Polish. Then there exists a coupling $\mathbf{P}$ of the probability laws $\mathcal{L}(U)$ and $\mathcal{L}(Y)$, such that

$$\mathbf{P}\left\{U^{[\ell,m]} \neq Y^{[\ell,m]}\right\} = \left\| \mathcal{L}(U^{[\ell,m]}) - \mathcal{L}(Y^{[\ell,m]}) \right\|_{\mathrm{TV}}, \qquad \ell \in \{1,\ldots,m\}. \tag{39}$$

Thus, for each $i \in T$ and for every pair $x, z \in \mathsf{X}$ with $x^{T\setminus\{i\}} = z^{T\setminus\{i\}}$, let $\mathbf{P}_{x,z}^{[i]}$ be the Goldstein coupling of $\mu^{(i,n]}(\cdot|x^{[i]})$ and $\mu^{(i,n]}(\cdot|z^{[i]})$. Then for each $j \in \{i+1,\ldots,n\}$, using (39) we have

$$\begin{aligned}\mathbf{P}_{x,z}^{[i]}\{Y_j^{(0)} \neq Y_j^{(1)}\} &\leq \mathbf{P}_{x,z}^{[i]}\{(Y_j^{(0)},\ldots,Y_n^{(0)}) \neq (Y_j^{(1)},\ldots,Y_n^{(1)})\} \\ &= \left\| \mu^{[j,n]}(\cdot|x^{[i]}) - \mu^{[j,n]}(\cdot|z^{[i]}) \right\|_{\mathrm{TV}}.\end{aligned}$$

This choice of coupling gives rise to the upper-triangular matrix $\Gamma = (\Gamma_{ij})_{i,j \in T}$ with

$$\Gamma_{ij} = \begin{cases} 0, & i > j \\ 1, & i = j \\ \sup\limits_{\substack{x,z \in \mathsf{X} \\ x^{T\setminus\{i\}} = z^{T\setminus\{i\}}}} \left\| \mu^{[j,n]}(\cdot|x^{[i]}) - \mu^{[j,n]}(\cdot|z^{[i]}) \right\|_{\mathrm{TV}}, & i < j \end{cases}. \tag{40}$$

Substituting this matrix into (21), we recover the concentration inequality of Kontorovich and Ramanan [17], but with an improved constant in the exponent.

Remark 5.5 It was erroneously claimed in [16, 14, 15] that the basic concentration inequalities of Chazottes et al. [6] and Kontorovich and Ramanan [17] are essentially the same, only derived using different methods. As the discussion above elucidates, the two methods use different couplings (the former, explicitly, and the latter, implicitly) — which yield quantitatively different and, in general, incomparable mixing coefficients.

Remark 5.6 Kontorovich and Ramanan obtained the matrix (40) using analytic methods without constructing an explicit coupling. In 2012, S. Shlosman posed the following question: could this matrix have been derived using a suitable coupling? We can now answer his question in the affirmative: the coupling is precisely Goldstein's maximal coupling.

As an illustration, let us consider two specific types of the probability law μ: a directed Markov model (i.e., a Markov chain) and an undirected Markov model (i.e., a Gibbsian Markov random field). In the directed case, suppose that the elements of T are ordered in such a way that μ can be disintegrated in the form

$$\mu(\mathrm{d}x) = \mu_1(\mathrm{d}x_1) \otimes K_1(x_1, \mathrm{d}x_2) \otimes K_2(x_2, \mathrm{d}x_3) \otimes \ldots \otimes K_{n-1}(x_{n-1}, \mathrm{d}x_n), \tag{41}$$

where μ_0 is a Borel probability measure on $(\mathsf{X}_1, \mathscr{B}_1)$, and, for each $i \in [1, n-1]$, K_i is a Markov kernel from X_i to X_{i+1}. For each $i \in [1, n)$, let

$$\theta_i \triangleq \sup_{x_i, z_i \in \mathsf{X}_i} \|K_i(x_i, \cdot) - K_i(z_i, \cdot)\|_{\mathrm{TV}}$$

be the Dobrushin contraction coefficient of K_i. Fix $1 \le i < j \le n$ and $x^{[i-1]} \in \mathsf{X}^{[i-1]}$, $y_i, y_i' \in \mathsf{X}_i$. An easy calculation [14] shows that, defining the signed measures η_i on X_{i+1} by $\eta(\mathrm{d}x_{i+1}) = K_i(y_i, \mathrm{d}x_{i+1}) - K_i(y_i', \mathrm{d}x_{i+1})$ and ζ_j on X_j by

$$\zeta_j = \eta_i K_i K_{i+1} K_{i+2} \ldots K_{j-1},$$

we have

$$\left\| \mu^{[j,n]}(\cdot|x^{[i-1]}y_i) - \mu^{[j,n]}(\cdot|x^{[i-1]}y_i') \right\|_{\mathrm{TV}} = \left\| \zeta_j \right\|_{\mathrm{TV}} \le \theta_i \theta_{i+1} \ldots \theta_{j-1}, \tag{42}$$

where (4) was repeatedly invoked to obtain the last inequality. The above yields an upper bound on the Γ_{ij} in (40) and hence in the corresponding concentration inequality in (21). When more delicate (e.g., spectral [15, 29]) estimates on $\left\| \zeta_j \right\|_{\mathrm{TV}}$ are available, these translate directly into tighter concentration bounds.

In the undirected case, μ is a Gibbsian Markov random field induced by pair potentials [12]. To keep things simple, we assume that the local spaces X_i are all finite. Define an undirected graph with vertex set $T = [n]$ and edge set $E = \{[i, i+1] : 1 \le i < n\}$ (i.e., a chain graph with vertex set T). Associate with each edge $(i,j) \in E$ a potential function $\psi_{ij} : \mathsf{X}_i \times \mathsf{X}_j \to [0, \infty)$. Together, these define a probability measure μ on X via

$$\mu(x) = \frac{\prod_{(i,j)\in E} \psi_{ij}(x_i, x_j)}{\sum_{y \in \mathsf{X}} \prod_{(i,j)\in E} \psi_{ij}(y_i, y_j)}.$$

Since μ is a Markov measure on X, there is a sequence of Markov kernels $K_1, \ldots, K_{n-1}$ generating μ in the sense of (41). It is shown in [14] that the contraction coefficient θ_i of the kernel K_i is bounded by

$$\theta_i \le \frac{R_i - r_i}{R_i + r_i},$$

where

$$R_i = \sup_{(x_i,x_{i+1}) \in \mathsf{X}_i \times \mathsf{X}_{i+1}} \psi_{i,i+1}(x_i, x_{i+1}), \qquad r_i = \inf_{(x_i,x_{i+1}) \in \mathsf{X}_i \times \mathsf{X}_{i+1}} \psi_{i,i+1}(x_i, x_{i+1}).$$

The estimate above implies a concentration result, either via (42) or via (35). To apply the latter, recall that $D^\mu = \sum_{k=1}^{\infty}(C^\mu)^k$, where C^μ is the Dobrushin interdependence matrix defined in (27). Assuming that ρ is the unweighted Hamming metric (i.e., $\rho_i(x_i, z_i) = \mathbf{1}\{x_i \neq z_i\}$ for all i) and that the θ_i's are all majorized by some $\theta < 1$, it is easy to see that $(C^\mu)_{ij} \leq \theta^{|i-j|}$.

6 Open Questions

Our focus in this chapter has been on the martingale method for establishing concentration inequalities. In the case of product measures, other techniques, such as the entropy method or transportation-information inequalities, often lead to sharper bounds. However, these alternative techniques are less developed in the dependent setting, and there appears to be a gap between what is achievable using the martingale method and what is achievable using other means. We close the chapter by listing some open questions that are aimed at closing this gap:

- **(Approximate) tensorization of entropy.** In the independent case, it is possible to derive the same concentration inequality (e.g., McDiarmid's inequality) using either the martingale method or the entropy method, often with the same sharp constants. However, once the independence assumption is dropped, the situation is no longer so simple. Consider, for example, tensorization of entropy. Several authors (see, e.g., [26, 3, 27]) have obtained so-called *approximate* tensorization inequalities for the relative entropy in the case of weakly dependent random variables: under certain regularity conditions on μ, there exists a constant $A_\mu \geq 1$, such that, for any other probability measure ν,

$$D(\nu \| \mu) \leq A_\mu \cdot \sum_{i \in T} \mathbb{E}_\nu D\big(\nu_i(\cdot | X^{T \setminus \{i\}}) \big\| \mu_i(\cdot | X^{T \setminus \{i\}})\big). \tag{43}$$

  Having such an inequality in hand, one can proceed to prove concentration for Lipschitz functions in exactly the same way as in the independent case. However, it seems that the constants A_μ in (43) are not sharp in the sense that the resulting concentration inequalities are typically worse than what one can obtain using Theorems 4.2 or 4.3 under the same assumptions on μ and f. This motivates the following avenue for further investigation: Derive sharp inequalities of the form (43) by relating the constant A_μ to appropriately chosen Wasserstein matrices.
- **General Wasserstein-type matrices.** Using the techniques pioneered by Marton, Samson proved the following concentration of measure result: Consider a

function $f : \mathsf{X} \to \mathbb{R}$ satisfying an "asymmetric" Lipschitz condition of the form

$$f(x) - f(y) \le \sum_{i \in T} \alpha_i(x) \mathbf{1}\{x_i \neq y_i\}, \qquad \forall x, y \in \mathsf{X}$$

for some functions $\alpha_i : \mathsf{X} \to \mathbb{R}$, such that $\sum_{i \in T} \alpha_i^2(x) \le 1$ for all $x \in \mathsf{X}$. Then, for any Borel probability measure μ on X, we have

$$\mathbb{P}_\mu \Big\{ f(X) - \mathbb{E}_\mu[f(X)] \ge t \Big\} \le \exp\left(- \frac{t^2}{2\|\Delta\|_2^2} \right), \tag{44}$$

where the matrix Δ has entries $\Delta_{ij} = \sqrt{\Gamma_{ij}}$ with Γ_{ij} given by (40), and

$$\|\Delta\|_2 \triangleq \sup_{v \in \mathbb{R}^T \setminus \{0\}} \frac{\|\Delta v\|_{\ell^2(T)}}{\|v\|_{\ell^2(T)}}$$

is the operator norm of Δ. A more general result in this vein was derived by Marton [24], who showed that an inequality of the form (44) holds with Δ computed in terms of any matrix Γ of the form (36), where each ρ_i is the trivial metric. Samson's proof relies on a fairly intricate recursive coupling argument. It would be interesting to develop analogs of (44) for arbitrary choices of the metrics ρ_i and with full freedom to choose the Wasserstein matrices $V^{(i)}$ for each $i \in T$. A recent paper by Wintenberger [35] pursues this line of work.

- **The method of exchangeable pairs and Wasserstein matrices.** An alternative route towards concentration inequalities in the dependent setting is via Stein's method of exchangeable pairs [4, 5]. Using this method, Chatterjee obtained the following result [4, Chap. 4]: Let $f : \mathsf{X} \to \mathbb{R}$ be a function which is 1-Lipschitz with respect to the weighted Hamming metric ρ_α defined in (3). Let μ be a Borel probability measure on X, whose Dobrushin interdependence matrix C^μ satisfies the condition $\|C^\mu\|_2 < 1$. Then, for any $t \ge 0$,

$$\mathbb{P}_\mu \Big\{ |f(X) - \mathbb{E}_\mu[f(X)] \ge t| \Big\} \le 2 \exp\left(- \frac{(1 - \|C^\mu\|_2) t^2}{\sum_{i \in T} \alpha_i^2} \right). \tag{45}$$

The key ingredient in the proof of (45) is the so-called Gibbs sampler, i.e., the Markov kernel $\bar{K}$ on X given by

$$\bar{K}(x, \mathrm{d}y) \triangleq \frac{1}{|T|} \sum_{i \in T} \delta_{x^{T \setminus \{i\}}}(\mathrm{d}y^{T \setminus \{i\}}) \otimes \mu_i(\mathrm{d}y_i | x^{T \setminus \{i\}}).$$

This kernel leaves μ invariant, i.e., $\mu = \mu \bar{K}$, and it is easy to show (see, e.g., [27]) that it contracts the $\bar{W}$ distance: for any other probability measure ν on X,

$$\bar{W}(\mu\bar{K}, \nu\bar{K}) \le \left(1 - \frac{1 - \|C^{\mu}\|_2}{|T|}\right)\bar{W}(\mu, \nu).$$

Since one can obtain contraction estimates for Markov kernels using Wasserstein matrices, it is natural to ask whether Chatterjee's result can be derived as a special case of a more general method, which would let us freely choose an arbitrary Markov kernel K that leaves μ invariant and control the constants in the resulting concentration inequality by means of a judicious choice of a Wasserstein matrix for K. Such a method would most likely rely on general comparison theorems for Gibbs measures [31].

Acknowledgements The second author would like to thank IMA for an invitation to speak at the workshop on Information Theory and Concentration Phenomena in Spring 2015, which was part of the annual program "Discrete Structures: Analysis and Applications." The authors are grateful to the anonymous referee for several constructive suggestions, and to Dr. Naci Saldi for spotting an error in an earlier version of the manuscript. A. Kontorovich was partially supported by the Israel Science Foundation (grant No. 1141/12) and a Yahoo Faculty award. M. Raginsky would like to acknowledge the support of the U.S. National Science Foundation via CAREER award CCF–1254041.

References

1. Sergey G. Bobkov and Friedrich Götze. Exponential integrability and transportation cost related to logarithmic Sobolev inequalities. *J. Funct. Anal.*, 163:1–28, 1999.
2. Stéphane Boucheron, Gábor Lugosi, and Pascal Massart. *Concentration Inequalities: A Nonasymptotic Theory of Independence*. Oxford University Press, 2013.
3. Pietro Caputo, Georg Menz, and Prasad Tetali. Approximate tensorization of entropy at high temperature. *Annales de la Faculté des Sciences de Toulouse Sér. 6*, 24(4):691–716, 2015.
4. Sourav Chatterjee. *Concentration inequalities with exchangeable pairs*. PhD thesis, Stanford University, 2005.
5. Sourav Chatterjee. Stein's method for concentration inequalities. *Probability Theory and Related Fields*, 138:305–321, 2007.
6. Jean-René Chazottes, Pierre Collet, Christof Külske, and Frank Redig. Concentration inequalities for random fields via coupling. *Probability Theory and Related Fields*, 137 (1-2):201–225, 2007.
7. Imre Csiszár. Information-type measures of difference of probability distributions and indirect observations. *Studia Sci. Math. Hungar.*, 2:299–318, 1967.
8. H. Djellout, A. Guillin, and L. Wu. Transportation cost-information inequalities and applications to random dynamical systems and diffusions. *Ann. Probab.*, 32(3B):2702–2732, 2004.
9. Roland L. Dobrushin. Prescribing a system of random variables by conditional distributions. *Theory of Probability and Its Applications*, 15(3):458–486, 1970. Translated from Russian.
10. Hans Föllmer. Tail structure of Markov chains on infinite product spaces. *Z. Wahrscheinlichkeitstheorie und Verw. Gebiete*, 50:273–285, 1979.
11. Hans Föllmer. A covariance estimate for Gibbs measures. *Journal of Functional Analsysi*, 46:387–395, 1982.
12. Hans-Otto Georgii. *Gibbs Measures and Phase Transitions*, volume 9 of *de Gruyter Studies in Mathematics*. Walter de Gruyter & Co., 2nd edition, 2011.
13. Sheldon Goldstein. Maximal coupling. *Z. Wahrscheinlichkeitstheorie und Verw. Gebiete*, 46(193–204), 1979.

14. Aryeh Kontorovich. Obtaining measure concentration from Markov contraction. *Markov Processes and Related Fields*, 4:613–638, 2012.
15. Aryeh Kontorovich and Roi Weiss. Uniform Chernoff and Dvoretzky-Kiefer-Wolfowitz-type inequalities for Markov chains and related processes. *Journal of Applied Probability*, 51:1–14, 2014.
16. Aryeh (Leonid) Kontorovich. *Measure Concentration of Strongly Mixing Processes with Applications*. PhD thesis, Carnegie Mellon University, 2007.
17. Leonid (Aryeh) Kontorovich and Kavita Ramanan. Concentration Inequalities for Dependent Random Variables via the Martingale Method. *Ann. Probab.*, 36(6):2126–2158, 2008.
18. Solomon Kullback. A lower bound for discrimination information in terms of variation. *IEEE Trans. Inform. Theory*, 13:126–127, 1967. Correction, volume 16, p. 652, 1970.
19. Christof Külske. Concentration inequalities for functions of Gibbs fields with application to diffraction and random Gibbs measures. *Commun. Math. Phys.*, 239:29–51, 2003.
20. Paul Lévy. *Problèmes concrets d'analyse fonctionnelle*. Gauthier-Villars, Paris, 1951. 2d ed.
21. Andrei A. Markov. Extension of the law of large numbers to dependent quantities. *Izvestiia Fiz.-Matem. Obsch. Kazan Univ.*, 15:135–156, 1906.
22. Katalin Marton. Bounding $\bar{d}$-distance by informational divergence: a method to prove measure concentration. *Ann. Probab.*, 24(2):857–866, 1996.
23. Katalin Marton. Measure concentration for a class of random processes. *Probability Theory and Related Fields*, 110(3):427–439, 1998.
24. Katalin Marton. Measure concentration and strong mixing. *Studia Scientiarum Mathematicarum Hungarica*, 19(1-2):95–113, 2003.
25. Katalin Marton. Measure concentration for Euclidean distance in the case of dependent random variables. *Ann. Probab.*, 32(3):2526–2544, 2004.
26. Katalin Marton. An inequality for relative entropy and logarithmic Sobolev inequalities in Euclidean spaces. *Journal of Functional Analysis*, 264(34–61), 2013.
27. Katalin Marton. Logarithmic Sobolev inequalities in discrete product spaces: a proof by a transportation cost distance. arXiv.org preprint 1507.02803, July 2015.
28. Bernard Maurey. Construction de suites symétriques. *C. R. Acad. Sci. Paris Sér. A-B 288*, (14):A679–A681, 1979.
29. Daniel Paulin. Concentration inequalities for Markov chains by Marton couplings and spectral methods. *Electronic Journal of Probability*, 20:1–32, 2015.
30. Maxim Raginsky and Igal Sason. *Concentration of Measure Inequalities in Information Theory, Communications, and Coding*. Foundations and Trends in Communications and Information Theory. Now Publishers, 2nd edition, 2014.
31. Patrick Rebeschini and Ramon van Handel. Comparison theorems for Gibbs measures. *J. Stat. Phys.*, 157:234–281, 2014.
32. Emmanuel Rio. Inégalités de Hoeffding pour les fonctions lipschitziennes de suites dépendantes. *C. R. Acad. Sci. Paris Sér. I Math.*, 330(10):905–908, 2000.
33. Paul-Marie Samson. Concentration of measure inequalities for Markov chains and Φ-mixing processes. *Ann. Probab.*, 28(1):416–461, 2000.
34. Ramon van Handel. Probability in high dimension. ORF 570 Lecture Notes, Princeton University, June 2014.
35. Olivier Wintenberger. Weak transport inequalities and applications to exponential and oracle inequalities. *Electronic Journal of Probability*, 20(114):1–27, 2015.
36. Liming Wu. Poincaré and transportation inequalities for Gibbs measures under the Dobrushin uniqueness condition. *Ann. Probab.*, 34(5):1960–1989, 2006.

Strong Data-Processing Inequalities for Channels and Bayesian Networks

Yury Polyanskiy and Yihong Wu

Abstract The data-processing inequality, that is, $I(U;Y) \leq I(U;X)$ for a Markov chain $U \to X \to Y$, has been the method of choice for proving impossibility (converse) results in information theory and many other disciplines. Various channel-dependent improvements (called strong data-processing inequalities, or SDPIs) of this inequality have been proposed both classically and more recently. In this note we first survey known results relating various notions of contraction for a single channel. Then we consider the basic extension: given SDPI for each constituent channel in a Bayesian network, how to produce an end-to-end SDPI?

Our approach is based on the (extract of the) Evans-Schulman method, which is demonstrated for three different kinds of SDPIs, namely, the usual Ahlswede-Gács type contraction coefficients (mutual information), Dobrushin's contraction coefficients (total variation), and finally the F_I-curve (the best possible non-linear SDPI for a given channel). Resulting bounds on the contraction coefficients are interpreted as probability of site percolation. As an example, we demonstrate how to obtain SDPI for an n-letter memoryless channel with feedback given an SDPI for $n = 1$.

Finally, we discuss a simple observation on the equivalence of a linear SDPI and comparison to an erasure channel (in the sense of "less noisy" order). This leads to a simple proof of a curious inequality of Samorodnitsky (2015), and sheds light on how information spreads in the subsets of inputs of a memoryless channel.

1 Introduction

Multiplication of a componentwise non-negative vector by a stochastic matrix results in a vector that is "more uniform". This observation appears in several classical works [Mar06, Doe37, Bir57] differing in their particular way of making

Y. Polyanskiy (✉)
Department of EECS, MIT, Cambridge, MA, USA
e-mail: yp@mit.edu

Y. Wu
Department of Statistics, Yale University, New Haven, CT, USA
e-mail: yihong.wu@yale.edu

E. Carlen et al. (eds.), *Convexity and Concentration*, The IMA Volumes in Mathematics and its Applications 161, DOI 10.1007/978-1-4939-7005-6_7

quantitative estimates. For example, Birkhoff's work [Bir57] initiated a study (sometimes known as geometric ergodicity) of contraction of the projective distance $d_P(x, y) \triangleq \log \max_i \frac{x_i}{y_i} - \log \min_i \frac{x_i}{y_i}$ between vectors in $\mathbb{R}^n_+$. Here, instead, we will be interested in contraction of statistical distances and information measures involving probability distributions, which we define next.

Fix a transition probability kernel (channel) $P_{Y|X} : \mathcal{X} \to \mathcal{Y}$ acting between two measurable spaces. We denote by $P_{Y|X} \circ P$ the distribution on $\mathcal{Y}$ induced by the push-forward of the distribution P, which is the distribution of the output Y when the input X is distributed according to P, and by $P \times P_{Y|X}$ the joint distribution P_{XY} if $P_X = P$. We also denote by $P_{Z|Y} \circ P_{Y|X}$ the serial composition of channels.[1]

We define three quantities that will play key role in our discussion: the total variation, the Kullback-Leibler (KL) divergence and the mutual information

$$d_{\mathrm{TV}}(P, Q) \triangleq \sup_E |P[E] - Q[E]| = \frac{1}{2} \int |\mathrm{d}P - \mathrm{d}Q|, \tag{1}$$

$$D(P\|Q) \triangleq \int \log \frac{\mathrm{d}P}{\mathrm{d}Q} \, \mathrm{d}P, \tag{2}$$

$$I(A; B) \triangleq D(P_{AB}\|P_A P_B). \tag{3}$$

The purpose of this paper is to give exposition to the phenomenon that upon passing through a non-degenerate noisy channel distributions become strictly closer and this leads to a loss of information. Namely we have three effects:

1. Total-variation (or Dobrushin) contraction:

$$d_{\mathrm{TV}}(P_{Y|X} \circ P, P_{Y|X} \circ Q) < d_{\mathrm{TV}}(P, Q) .$$

2. Divergence contraction:

$$D(P_{Y|X} \circ P\|P_{Y|X} \circ Q) < D(P\|Q)$$

3. Information loss: For any Markov chain[2] $U \to X \to Y$ we

$$I(U; Y) < I(U; X) .$$

[1] More formally, we should have written $P_{Y|X} : \mathcal{P}(\mathcal{X}) \to \mathcal{P}(\mathcal{Y})$ as a map between spaces of probability measures $\mathcal{P}(\cdot)$ on respective bases. The rationale for our notation $P_{Y|X} : \mathcal{X} \to \mathcal{Y}$ is that we view Markov kernels as randomized functions. Then, a single distribution P on $\mathcal{X}$ is a randomized function acting from a space of a single point, i.e. $P : [1] \to \mathcal{X}$, and that in turn explains our notation $P_{Y|X} \circ P$ for denoting the induced marginal distribution.

[2] The notation $A \to B \to C$ simply means that $A \perp\!\!\!\perp C|B$.

These strict inequalities are collectively referred to as *strong data-processing inequalities* (SDPIs). The goal of this paper is to show intricate interdependencies between these effects, as well as introducing tools for quantifying how strict these SDPIs are.

Organization In Section 2 we overview the case of a single channel. Notably, most of the results in the literature are proved for finite alphabets, i.e., $|\mathcal{X}||\mathcal{Y}| < \infty$, with a few exceptions such as [CKZ98, PW16b]. We provide in Appendix A a self-contained proof of some of these results for general alphabets.

From then on we focus on the question: *Given a multi-terminal network with a single source and multiple sinks, and given SDPIs for each of the channels comprising the network, how do we obtain an SDPI for the composite channel from source to sinks?* It turns out that this question has been addressed implicitly in the work of Evans and Schulman [ES99] on redundancy required in circuits of noisy gates. Rudiments also appeared in Dawson [Daw75] as well as Boyen and Koller [BK98].

In Section 3 we present the essence of the Evans-Schulman method and derive upper bounds on the mutual information contraction coefficient η_{KL} for Bayesian networks (directed graphical models). We also interpret the resulting bounds as probabilities of disrupting end-to-end connectivity under independent removals of graph vertices (site percolation). Then in Section 4 we derive analogous estimates for Dobrushin's coefficient η_{TV} that governs the contraction of the total variation on networks. While the results exactly parallel those for mutual information, the proof relies on new arguments using coupling. Finally, Section 5 extends the technique to bounding the F_I-curves (the non-linear SDPIs). Section 6 concludes with an alternative point of view on mutual information contraction, namely that of comparison to an erasure channel. As an example we give a short proof of a result of Samorodnitsky [Sam15] about distribution of information in subsets of channel outputs.

Notation Elements of the Cartesian product $\mathcal{X}^n$ are denoted $x^n \triangleq (x_1, \ldots, x_n)$ to emphasize their dimension. Given a transition probability kernel from $P_{Y|X} : \mathcal{X} \to \mathcal{Y}$ we denote $P^n_{Y|X} = P_{Y^n|X^n}$ the kernel acting from $\mathcal{X}^n \to \mathcal{Y}^n$ componentwise independently:

$$P_{Y^n|X^n}(y^n|x^n) \triangleq \prod_{j=1}^{n} P_{Y|X}(y_j|x_j).$$

To demonstrate the general bounds we consider the running example of $P_{Y|X}$ being an n-letter binary symmetric channel (BSC), given by

$$Y = X + Z, \quad X, Y \in \mathbb{F}_2^n, \ Z \sim \mathrm{Bern}(\delta)^n \tag{4}$$

and denoted by $\mathsf{BSC}(\delta)^n$. Throughout this paper $\bar{\delta} \triangleq 1 - \delta$.

2 SDPI for a Single Channel

2.1 Contraction Coefficients for *f*-Divergence and Mutual Information

Let $f : (0, \infty) \to \mathbb{R}$ be a convex function that is strictly convex at 1 and $f(1) = 0$. Let $D_f(P\|Q) \triangleq \mathbb{E}_Q[f(\frac{dP}{dQ})]$ denote the f-divergence of P and Q with $P \ll Q$, cf. [Csi67].[3] For example, the total variation (1) and the KL divergence (2) correspond to $f(x) = \frac{1}{2}|x-1|$ and $f(x) = x \log x$, respectively; taking $f(x) = (x-1)^2$ we obtain the χ^2-divergence: $\chi^2(P\|Q) \triangleq \int (\frac{dP}{dQ})^2 dQ - 1$.

For any Q that is not a point mass, define:

$$\eta_f(P_{Y|X}, Q) \triangleq \sup_{P: 0 < D_f(P\|Q) < \infty} \frac{D_f(P_{Y|X} \circ P \| P_{Y|X} \circ Q)}{D_f(P\|Q)}, \tag{5}$$

$$\eta_f(P_{Y|X}) \triangleq \sup_Q \eta_f(Q). \tag{6}$$

It is easy to show that the supremum is over a non-empty set whenever Q is not a point mass (see Appendix A). For notational simplicity when the channel is clear from context we abbreviate $\eta_f(P_{Y|X})$ as η_f. For contraction coefficients of total variation, χ^2 and KL divergence, we write $\eta_{\mathrm{TV}}, \eta_{\chi^2}$ and η_{KL}, respectively, which play prominent roles in this exposition.

One of the main tools for studying ergodicity property of Markov chains as well as Gibbs measures, $\eta_{\mathrm{TV}}(P_{Y|X})$ is known as the *Dobrushin's coefficient* of the kernel $P_{Y|X}$. Dobrushin [Dob56] showed that the supremum in the definition of η_{TV} can be restricted to point masses, namely,

$$\eta_{\mathrm{TV}}(P_{Y|X}) = \sup_{x,x'} d_{\mathrm{TV}}(P_{Y|X=x}, P_{Y|X=x'}), \tag{7}$$

thus providing a simple criterion for strong ergodicity of Markov processes. Later [CKZ98, Proposition II.4.10(i)] (see also [CIR+93, Theorem 4.1] for finite alphabets) demonstrated that all other contraction coefficients are upper bounded by the Dobrushin's coefficient, with inequality being typically strict (cf. the BSC example below):

Theorem 1 ([CKZ98, Proposition II.4.10]) *For every f-divergence, we have*

$$\eta_f(P_{Y|X}) \le \eta_{\mathrm{TV}}(P_{Y|X}). \tag{8}$$

---

[3]More generally, $D_f(P\|Q) \triangleq \mathbb{E}_\mu \left[f\left(\frac{dP/d\mu}{dQ/d\mu} \right) \right]$, where μ is a dominating probability measure of P and Q, e.g., $\mu = (P+Q)/2$, with the understanding that $f(0) = f(0+)$, $0f(\frac{0}{0}) = 0$ and $0f(\frac{a}{0}) = \lim_{x \downarrow 0} xf(\frac{a}{x})$ for $a > 0$.

For the opposite direction, lower bounds on η_f typically involves η_{χ^2}, the contraction coefficient of the χ^2-divergence. It is well known, e.g. Sarmanov [Sar58], that $\eta_{\chi^2}(P_{Y|X}, P_X)$ is the squared second largest eigenvalue of the conditional expectation operator, which in turn equals the *maximal correlation* coefficient of the joint distribution P_{XY}:

$$S(X;Y) \triangleq \sup_{f,g} \rho(f(X), g(Y)) = \sqrt{\eta_{\chi^2}(P_{Y|X}, P_X)}\,, \tag{9}$$

where $\rho(\cdot,\cdot)$ denotes the correlation coefficient and the supremum is over real-valued functions f, g such that $f(X)$ and $g(Y)$ are square integrable.

The relationship between η_{KL} and η_{χ^2} on finite alphabets has been systematically studied by Ahlswede and Gács [AG76]. In particular, [AG76] proved

$$\eta_{\chi^2}(P_{Y|X}, P_X) \le \eta_{\mathrm{KL}}(P_{Y|X}, P_X), \tag{10}$$

and noticed that the inequality is frequently strict.[4] Furthermore, for finite alphabets, the following equivalence is demonstrated in [AG76]:

$$\eta_{\chi^2}(P_X, P_{Y|X}) < 1 \iff \eta_{\mathrm{KL}}(P_X, P_{Y|X}) < 1 \tag{11}$$

$$\iff \text{graph } \{(x,y) : P_X(x) > 0, P_{Y|X}(y|x) > 0\} \text{ is connected.} \tag{12}$$

As a criterion for $\eta_f(P_{Y|X}, P_X) < 1$, this is an improvement of (8) only for channels with $\eta_{\mathrm{TV}}(P_{Y|X}) = 1$. The lower bound (10) can in fact be considerably generalized:

Theorem 2 *Let f be twice continuously differentiable on $(0,\infty)$ with $f''(1) > 0$. Then for any P_X that is not a point mass,*

$$\eta_{\chi^2}(P_{Y|X}, P_X) \le \eta_f(P_{Y|X}, P_X)\,, \tag{13}$$

and

$$\eta_{\chi^2}(P_{Y|X}) \le \eta_f(P_{Y|X})\,. \tag{14}$$

See Appendix A.1 for a proof of (13) for the general case, which yields (14) by taking suprema over P_X on both sides. Note that (14) (resp. (13)) have been proved in [CKZ98, Proposition II.6.15] for the general alphabet (resp. in [Rag14, Theorem 3.3] for finite alphabets).

[4] See [AG76, Theorem 9] and [AGKN13] for examples.

Moreover, (14) in fact holds with equality for all nonlinear and operator convex f, e.g., for KL divergence and for squared Hellinger distance; see [CRS94, Theorem 1] and [CKZ98, Proposition II.6.13 and Corollary II.6.16]. Therefore, we have:

Theorem 3

$$\eta_{\chi^2}(P_{Y|X}) = \eta_{\mathrm{KL}}(P_{Y|X}) \,. \tag{15}$$

See Appendix A.1 for a self-contained proof. This result was first obtained in [AG76] using different methods for discrete space. Rather naturally, we also have [CKZ98, Proposition II.4.12]:

$$\eta_f(P_{Y|X}) = 1 \quad \iff \quad \eta_{\mathrm{TV}}(P_{Y|X}) = 1$$

for any non-linear f.

As an illustrating example, for $\mathsf{BSC}(\delta)$ defined in (4), we have cf. [AG76]

$$\eta_{\chi^2} = \eta_{\mathrm{KL}} = (1-2\delta)^2 < \eta_{\mathrm{TV}} = |1-2\delta|. \tag{16}$$

Appendix B presents general results on the contraction coefficients for binary-input arbitrary-output channels, which can be bounded using Hellinger distance within a factor of two.

We next discuss the fixed-input contraction coefficient $\eta_{\mathrm{KL}}(P_{Y|X}, Q)$. Unfortunately, there is no simple reduction to the χ^2-case as in (15). Besides the lower bound (10), there is a variety of upper bounds relating η_{KL} and η_{χ^2}. We quote [MZ15, Theorem 11], who show for finite input-alphabet case:

$$\eta_{\mathrm{KL}}(P_{Y|X}, Q) \le \frac{1}{\min_x Q(x)} \eta_{\chi^2}(P_{Y|X}, Q) \,.$$

Another bound (which also holds for all η_f with operator-convex f) is in [Rag14, Theorem 3.6]:

$$\eta_{\mathrm{KL}}(P_{Y|X}, Q) \le \max\left(\eta_{\chi^2}(P_{Y|X}, Q),\ \sup_{0<\beta<1} \eta_{\mathrm{LC}_\beta}(P_{Y|X}, Q)\right),$$

where η_{LC_β} denotes contraction coefficient of an f-divergence $\mathrm{LC}_\beta(P\|Q) = \beta\bar\beta \int \frac{(P-Q)^2}{\beta P + \bar\beta Q}$ with $\beta \in (0,1)$ and $\bar\beta = 1-\beta$ (see also Appendix B).

We also note in passing that SDPIs are intimately related to hypercontractivity and maximal correlation, as discovered by Ahlswede and Gács [AG76] and recently improved by Anantharam et al. [AGKN13] and Nair [Nai14]. Indeed, the main result of [AG76] characterizes $\eta_{\mathrm{KL}}(P_{Y|X}, P_X)$ as the maximal ratio of hyper-contractivity of the conditional expectation operator $\mathbb{E}[\cdot|X]$.

The fixed-input contraction coefficient $\eta_{\mathrm{KL}}(Q)$ is closely related to the (modified) log-Sobolev inequalities. Indeed, if $\eta_{\mathrm{KL}}(Q) < 1$ where Q is the invariant measure for the Markov kernel $P_{Y|X}$, i.e., $P_{Y|X} \circ Q = Q$, then any initial distribution P such that $D(P\|Q) < \infty$ converges to Q exponentially fast since

$$D(P^n_{Y|X} \circ P\|Q) \leq \eta^n_{\mathrm{KL}}(P_{Y|X}, Q)D(P\|Q),$$

where the exponent $\eta_{\mathrm{KL}}(P_{Y|X}, Q)$ can in turn be estimated from log-Sobolev inequalities, e.g. [Led99]. When Q is not invariant, it was shown [DMLM03] that

$$1 - \alpha(Q) \leq \eta_{\mathrm{KL}}(P_{Y|X}, Q) \leq 1 - C\alpha(Q)$$

holds for some universal constant C, where $\alpha(Q)$ is a modified log-Sobolev (also known as 1-log-Sobolev) constant:

$$\alpha(Q) = \inf_{f \neq 1, \|f\|_2 = 1} \frac{\mathbb{E}\left[f^2(X) \log \frac{f^2(X)}{f^2(X')}\right]}{\mathbb{E}[f^2(X) \log f^2(X)]}, \qquad P_{XX'} = Q \times (P_{X|Y} \circ P_{Y|X}).$$

For further connections between η_{KL} and log-Sobolev inequalities on finite alphabets, see [Rag13, Rag14].

There exist several other characterizations of η_{KL}, such as the following one in terms of the contraction of mutual information (cf. [CK81, Exercise III.3.12, p. 350] for finite alphabet):

$$\eta_{\mathrm{KL}}(P_{Y|X}) = \sup \frac{I(U;Y)}{I(U;X)}, \tag{17}$$

where the supremum is over all Markov chains $U \to X \to Y$ with fixed $P_{Y|X}$ (or equivalently, over all joint distributions P_{XU}) such that $I(U;X) < \infty$. This result is an immediate consequence of the following input-dependent version (see Appendix A.3 for a proof in the general case; the finite alphabet case has been shown in [AGKN13])

Theorem 4 *For any P_X that is not a point mass,*

$$\eta_{\mathrm{KL}}(P_{Y|X}, P_X) = \sup \frac{I(U;Y)}{I(U;X)}, \tag{18}$$

where the supremum is taken over all Markov chains $U \to X \to Y$ with fixed $P_{XY} = P_X \circ P_{Y|X}$ such that $0 < I(U;X) < \infty$.

Another characterization of η_{KL}, in view of (15) and (9), is

$$\eta_{\mathrm{KL}}(P_{Y|X}) = \sup \rho^2(f(X), g(Y)),$$

where the supremum is over all P_X and real-valued square-integrable $f(X)$ and $g(Y)$.

2.2 Non-linear SDPI

How to quantify the information loss if $\eta_{\mathrm{KL}} = 1$ for the channel of interest? In fact this situation can arise in very basic settings, such as the additive-noise Gaussian channel under the moment constraint on the input distributions (cf. [PW16b, Theorem 9, Section 4.5]), where the mutual information does not contract linearly as in (17), but can still contract *non-linearly*. In such cases, establishing a strong-data processing inequality can be done by following the joint-range idea of Harremoës and Vajda [HV11]. Namely, we aim to find (or bound) the *best possible data-processing function* F_I defined as follows.

Definition 1 (F_I-curve) Fix $P_{Y|X}$ and define

$$F_I(t, P_{Y|X}) \triangleq \sup_{P_{UX}} \left\{ I(U;Y) \colon I(U;X) \leq t, P_{UXY} = P_{UX} P_{Y|X} \right\}. \tag{19}$$

Equivalently, the supremum is taken over all joint distributions P_{UXY} with a given conditional $P_{Y|X}$ and satisfying $U \to X \to Y$. The upper concave envelope of F_I is denoted by F_I^c:

$$F_I^c(t, P_{Y|X}) \triangleq \inf\{f(t) : \forall t' \geq 0 \; F_I(t', P_{Y|X}) \leq f(t'), f\text{–concave}\}.$$

Equivalently, we have

$$F_I^c(t, P_{Y|X}) = \sup_{P_{VUX}} \left\{ I(U;Y|V) \colon I(U;X|V) \leq t, P_{VUXY} = P_{VUX} P_{Y|X} \right\}, \tag{20}$$

where $I(A;B|C) \triangleq I(A,C;B) - I(C;B)$ is the conditional mutual information, and averaging over V serves the role of concavification (so that V can be taken binary). Whenever it does not lead to confusion we will write $F_{Y|X}(t)$ instead of $F_I(t, P_{Y|X})$.

The operational significance of the F_I-curve is that it gives the optimal input-independent strong data processing inequality:

$$I(U;Y) \leq F_I(I(U;X)),$$

which generalizes (17) since $F_I'(0) = \eta_{\mathrm{KL}}(P_{Y|X})$ and $t \mapsto \frac{1}{t} F_I(t)$ is decreasing (see, e.g., [CPW15, Section I]). See [CPW15] for bounds and expressions for BSC and Gaussian channels.

Frequently it is more convenient to work with the concavified version F_I^c as it allows for some natural extension of the results about contraction coefficients. Proposition 18 shows that F_I may not be concave.

2.3 Some Applications: Classical and New

The main example of a strong data-processing inequality (SDPI) was discovered by Ahlswede and Gács [AG76]. They have shown, using the characterization (11), that whenever $P_{Y|X}$ is a discrete memoryless channel that does not admit zero-error communication, we have $\eta_{\mathrm{KL}}(P_{Y|X}) \leq \eta < 1$ and

$$I(W;Y) \leq \eta I(W;X) \tag{21}$$

for all Markov chains $W \to X \to Y$.

SDPIs have been popular for establishing lower (impossibility) bounds in various setups, in both classical and more recent works. We mention only a few of these applications:

- By Dobrushin for showing non-existence of multiple phases in Ising models at high temperatures [Dob70];
- By Erkip and Cover in portfolio theory [EC98];
- By Evans and Schulman in analysis of noise-resistant circuits [ES99];
- By Evans, Kenyon, Peres, and Schulman in the analysis of inference on trees and percolation [EKPS00];
- By Courtade in distributed data-compression [Cou12];
- By Duchi, Wainwright, and Jordan in statistical limitations of differential privacy [DJW13];
- By the authors to quantify optimal communication and optimal control in line networks [PW16b];
- By Liu, Cuff, and Verdú in key generation [LCV15];
- By Xu and Raginsky in distributed estimation [XR15].

All of the applications above use SDPI (21) to prove negative (impossibility) statements. A notable exception is the work of Boyen and Koller [BK98], who considered the basic problem of computing the posterior-belief vector of a hidden Markov model: that is, given a Markov chain $\{X_j\}$ observed over a memoryless channel $P_{Y|X}$, one aims to recompute $P_{X_j|Y_{-\infty}^j}$ as each new observation Y_j arrives. The problem arises when X is of large dimension and then for practicality one is constrained to approximate (quantize) the posterior. However, due to the recursive nature of belief computations, the cumulative effect of these approximations may become overwhelming. Boyen and Koller [BK98] proposed to use the SDPI similar to (21) with $\eta < 1$ for the Markov chain $\{X_j\}$ and show that this cumulative effect stays bounded since $\sum \eta^n < \infty$. Similar considerations also enable one to provide provable guarantees for simulation of inter-dependent stochastic processes.

3 Contraction of Mutual Information in Networks

We start by defining a *Bayesian network* (also known as a *directed graphical model*). Let G be a finite directed acyclic graph with set of vertices $\{Y_v : v \in \mathcal{V}\}$ denoting random variables taking values in a fixed finite alphabet.[5] We assume that each vertex Y_v is associated with a conditional distribution $P_{Y_v|Y_{\text{pa}(v)}}$ where $\text{pa}(v)$ denotes parents of v, with the exception of one special "source" node X that has no inbound edges (there may be other nodes without inbound edges, but those have to have their marginals specified). Notice that if $V \subset \mathcal{V}$ is an arbitrary set of nodes we can progressively chain together all the random transformations and unequivocally compute $P_{V|X}$ (here and below we use V and $Y_V = \{Y_v : v \in V\}$ interchangeably). We assume that vertices in $\mathcal{V}$ are topologically sorted so that $v_1 > v_2$ implies there is no path from v_1 to v_2. Associated to each node we also define

$$\eta_v \triangleq \eta_{\text{KL}}(P_{Y_v|Y_{\text{pa}(v)}}) .$$

See the excellent book of Lauritzen [Lau96] for a thorough introduction to a graphical model language of specifying conditional independencies.

The following result can be distilled from [ES99]:

Theorem 5 *Let $W \in \mathcal{V}$ and $V \subset \mathcal{V}$ such that $W > V$. Then*

$$\eta_{\text{KL}}(P_{V,W|X}) \le \eta_W \cdot \eta_{\text{KL}}(P_{V,\text{pa}(W)|X}) + (1 - \eta_W) \cdot \eta_{\text{KL}}(P_{V|X}) . \tag{22}$$

Furthermore, let $\text{perc}(V)$ *denote the probability that there is a path from X to V*[6] *in the graph if each node v is removed independently with probability $1 - \eta_v$ (site percolation). Then, we have for every $V \subset \mathcal{V}$*

$$\eta_{\text{KL}}(P_{V|X}) \le \text{perc}(V) . \tag{23}$$

In particular, if $\eta_v < 1$ for all $v \in \mathcal{V}$, then $\eta_{\text{KL}}(P_{V|X}) < 1$.

Proof Consider an arbitrary random variable U such that

$$U \to X \to (V, W) .$$

Let $A = \text{pa}(W) \setminus V$. Without loss of generality we may assume A does not contain X: indeed, if A includes X, then we can introduce an artificial node X' such that

---

[5]At the expense of technical details, the alphabet can be replaced with any countably generated (e.g. Polish) measurable space. For clarity of presentation we focus here on finite alphabets.

[6]More formally, $\text{perc}(V)$ equals probability that there exists a sequence of nodes $v_1, \ldots, v_n$ with $v_1 = X$, $v_n \in V$ satisfying two conditions: 1) for each $i \in [n-1]$ the pair (v_i, v_{i+1}) is a directed edge in G; and 2) each v_i is not removed.

$X' = X$ and include X' into A instead of X. Relevant conditional independencies are encoded in the following graph:

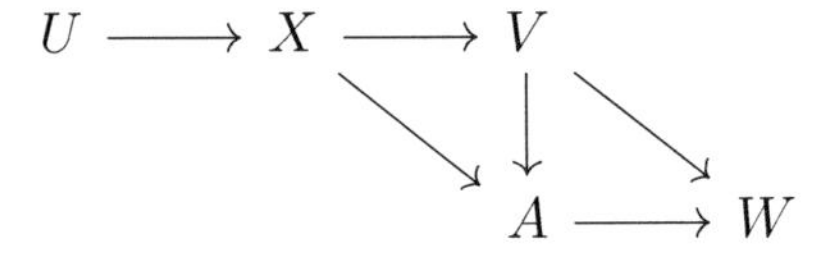

From the characterization (17) it is sufficient to show

$$I(U;V,W) \leq (1-\eta_W)I(U;V) + \eta_W I(U;V,A)\,. \tag{24}$$

Denote $B = V \backslash \mathrm{pa}(W)$ and $C = V \cap \mathrm{pa}(W)$. Then $\mathrm{pa}(W) = (A,C)$ and $V = (B,C)$. To verify (24) notice that by assumption we have

$$U \to X \to (V,A) \to W\,.$$

Therefore conditioned on V we have the Markov chain

$$U \to X \to A \to W \qquad |V$$

and the channel $A \to W$ is a restriction of the original $P_{W|\mathrm{pa}(W)}$ to a subset of the inputs. Indeed, $P_{W|A,V} = P_{W|\mathrm{pa}(W),B} = P_{W|\mathrm{pa}(W)}$ by the assumption of the graphical model. Thus, for every realization $v = (b,c)$ of V, we have $P_{W|A=a,V=v} = P_{W|A=a,C=c}$ and therefore

$$I(U;W|V=v) \leq \eta(P_{W|A,C=c})I(U;A|V=v) \leq \eta(P_{W|A,C})I(U;A|V=v), \tag{25}$$

where the last inequality uses the following property of the contraction coefficient which easily follows from either (6) or (17):

$$\sup_c \eta(P_{W|A,C=c}) \leq \eta(P_{W|A,C}). \tag{26}$$

Averaging both sides of (25) over $v \sim P_V$ and using the definition $\eta_W = \eta(P_{W|\mathrm{pa}(W)}) = \eta(P_{W|A,C})$, we have

$$I(U;W|V) \leq \eta_W I(U;A|V)\,. \tag{27}$$

Adding $I(U;V)$ to both sides yields (24).

We now move to proving the percolation bound (23). First, notice that if a vertex W satisfies $W > V$, then letting $\{\exists\,\pi : X \to V\}$ be the event that there exists a directed path from X to (any element of) the set V under the site percolation model, we notice that $\{W \text{ removed}\}$ is independent from $\{\exists\,\pi : X \to V\}$ and $\{\exists\,\pi : X \to V \cup \mathrm{pa}(W)\}$. Thus we have

$$\begin{aligned}
\mathrm{perc}(V \cup \{W\}) &\triangleq \mathbb{P}[\exists\, \pi : X \to V \cup \{W\}] \\
&= \mathbb{P}[\exists\, \pi : X \to V \cup \{W\}, W \text{ removed}] + \mathbb{P}[\exists\, \pi : X \to V \cup \{W\}, W \text{ kept}] \\
&= \mathbb{P}[\exists\, \pi : X \to V, W \text{ removed}] + \mathbb{P}[\exists\, \pi : X \to V \cup \mathrm{pa}(W), W \text{ kept}] \\
&= \mathbb{P}[\exists\, \pi : X \to V](1 - \eta_W) + \eta_W \mathbb{P}[\exists\, \pi : X \to V \cup \mathrm{pa}(W)] \\
&= (1 - \eta_W)\mathrm{perc}(V) + \eta_W \mathrm{perc}(V \cup \mathrm{pa}(W)) .
\end{aligned}$$

That is, the set-function perc$(\cdot)$ satisfies the recursion given by the right-hand side of (22). Now notice that (23) holds trivially for $V = \{X\}$, since both sides are equal to 1. Then, by induction on the maximal element of V and applying (22) we get that (23) holds for all V. □

Theorem 5 allows us to estimate contraction coefficients in arbitrary (finite) networks by peeling off last nodes one by one. Next we derive a few corollaries:

Corollary 6 *Consider a fixed (single-letter) channel $P_{Y|X}$ and assume that it is used repeatedly and with perfect feedback to send information from W to $(Y_1, \dots, Y_n)$. That is, we have for some encoder functions f_j*

$$P_{Y^n|W}(y^n|w) = \prod_{j=1}^{n} P_{Y|X}(y_j|f_j(w, y^{j-1})),$$

which corresponds to the graphical model:

$$W \longrightarrow Y_1 \longrightarrow Y_2 \longrightarrow Y_3 \cdots$$

Then

$$\eta_{\mathrm{KL}}(P_{Y^n|W}) \le 1 - (1 - \eta_{\mathrm{KL}}(P_{Y|X}))^n < n \cdot \eta_{\mathrm{KL}}(P_{Y|X})$$

Proof Apply Theorem 5 n times. □

Let us call a path $\pi = (X, \cdots, v)$ with $v \in V$ to be *shortcut-free from X to V*, denoted $X \overset{sf}{\to} V$, if there does not exist another path π' from X to any node in V such that π' is a subset of π. (In particular v necessarily is the first node in V that π visits.) Also for every path $\pi = (X, v_1, \dots, v_m)$ we define

$$\eta^{\pi} \triangleq \prod_{j=1}^{m} \eta_{v_j} .$$

Corollary 7 *For any subset V we have*

$$\eta_{\mathrm{KL}}(P_{V|X}) \le \sum_{\pi: X \overset{sf}{\to} V} \eta^{\pi} \,. \tag{28}$$

In particular, we have the estimate of Evans-Schulman [ES99]:

$$\eta_{\mathrm{KL}}(P_{V|X}) \le \sum_{\pi: X \to V} \eta^{\pi} \,. \tag{29}$$

Proof Both results are simple consequence of union-bounding the right-hand side of (23). But for completeness, we give an explicit proof. First, notice the following two self-evident observations:

1. If A and B are disjoint sets of nodes, then

$$\sum_{\pi: X \overset{sf}{\to} A \cup B} \eta^{\pi} = \sum_{\pi: X \overset{sf}{\to} A, \text{ avoid } B} \eta^{\pi} + \sum_{\pi: X \overset{sf}{\to} B, \text{ avoid } A} \eta^{\pi}. \tag{30}$$

2. Let $\pi : X \to V$ and π_1 be π without the last node, then

$$\pi : X \overset{sf}{\to} V \quad \iff \quad \pi_1 : X \overset{sf}{\to} \{\mathrm{pa}(V) \setminus V\}. \tag{31}$$

Now represent $V = (V', W)$ with $W > V'$, denote $P = \mathrm{pa}(W) \setminus V$ and assume (by induction) that

$$\eta_{\mathrm{KL}}(P_{V'|X}) \le \sum_{\pi: X \overset{sf}{\to} V} \eta^{\pi} \tag{32}$$

$$\eta_{\mathrm{KL}}(P_{V',P|X}) \le \sum_{\pi: X \overset{sf}{\to} \{V',P\}} \eta^{\pi} \,. \tag{33}$$

By (30) and (31) we have

$$\sum_{\pi: X \overset{sf}{\to} V} \eta^{\pi} = \sum_{\pi: X \overset{sf}{\to} V'} \eta^{\pi} + \sum_{\pi: X \overset{sf}{\to} W, \text{ avoid } V'} \eta^{\pi} \tag{34}$$

$$= \sum_{\pi: X \overset{sf}{\to} V'} \eta^{\pi} + \eta_W \sum_{\pi: X \overset{sf}{\to} P, \text{ avoid } V'} \eta^{\pi} \tag{35}$$

Then by Theorem 5 and induction hypotheses (32)–(33) we get

$$\eta_{\mathrm{KL}}(P_{V|X}) \le \eta_W \sum_{\pi : X \xrightarrow{sf} \{V', P\}} \eta^\pi + (1 - \eta_W) \sum_{\pi : X \xrightarrow{sf} V'} \eta^\pi \tag{36}$$

$$= \eta_W \left(\sum_{\pi : X \xrightarrow{sf} P,\ \text{avoid } V'} \eta^\pi - \sum_{\pi : X \xrightarrow{sf} V',\ \text{pass } P} \eta^\pi \right) + \sum_{\pi : X \xrightarrow{sf} V'} \eta^\pi \tag{37}$$

$$\le \eta_W \sum_{\pi : X \xrightarrow{sf} P,\ \text{avoid } V'} \eta^\pi + \sum_{\pi : X \xrightarrow{sf} V'} \eta^\pi \tag{38}$$

where in (37) we applied (30) and split the summation over $\pi : X \xrightarrow{sf} V'$ into paths that avoid and pass nodes in P. Comparing (35) and (38) the conclusion follows. □

Both estimates (28) and (29) are compared to that of Theorem 5 in Table 1 in various graphical models.

Evaluation for the BSC We consider the contraction coefficient for the n-letter binary symmetric channel $\mathsf{BSC}(\delta)^n$ defined in (4). By (16), for $n = 1$ we have $\eta_{\mathrm{KL}} = (1 - 2\delta)^2$. Then by Corollary 6 we have for arbitrary n:

Table 1 Comparing bounds on the contraction coefficient $\eta_{\mathrm{KL}}(P_{Y|X})$. For simplicity, we assume that the η_{KL} coefficients of all constituent kernels are bounded from above by η.

Name	Graph	Theorem 5	Estimate (28) via shortcut-free paths	Original Evans-Schulman estimate (29)
Markov chain 1	$X \to Y_1 \to B \to Y_2$	η	η	$\eta + \eta^3$
Markov chain 2	$X \to A$, $A \to B$, $X \to B \to Y$	η^2	η^2	$\eta^2 + \eta^3$
Parallel channels	$X \to Y_1$, $X \to Y_2$	$2\eta - \eta^2$	2η	2η
Parallel channels with feedback	$X \to Y_1$, $Y_1 \to Y_2$, $X \to Y_2$	$2\eta - \eta^2$	2η	3η

$$\eta_{\mathrm{KL}} \le 1 - (4\delta(1-\delta))^n . \tag{39}$$

A simple lower bound for η_{KL} can be obtained by considering (17) and taking $U \sim \mathrm{Bern}(1/2)$ and $U \to X$ being an n-letter repetition code, namely, $X = (U, \dots, U)$. Let[7] $\epsilon = \mathbb{P}[|Z| \ge n/2]$ be the probability of error for the maximal likelihood decoding of U based on Y, which satisfies the Chernoff bound $\epsilon \le (4\delta(1-\delta))^{n/2}$. We have from Jensen's inequality

$$I(U;Y) = H(U) - H(U|Y) \ge 1 - h(\epsilon) = 1 - (4\delta(1-\delta))^{\frac{n}{2}+O(\log n)} ,$$

where we used the fact that the binary entropy $h(x) = -x\log x - (1-x)\log(1-x) = -x\log x + O(x^2)$ as $x \to 0$. Consequently, we get

$$\eta_{\mathrm{KL}} \ge 1 - (4\delta(1-\delta))^{\frac{n}{2}+O(\log n)} . \tag{40}$$

Comparing (39) and (40) we see that $\eta_{\mathrm{KL}} \to 1$ exponentially fast. To get the exact exponent we need to replace (39) by the following improvement:

$$\eta_{\mathrm{KL}} \le \eta_{\mathrm{TV}} \le 1 - (4\delta(1-\delta))^{\frac{n}{2}+O(\log n)} ,$$

where the first inequality is from (8) and the second is from (48) below. Thus, all in all we have for $\mathsf{BSC}(\delta)^n$ as $n \to \infty$

$$\eta_{\mathrm{KL}}, \eta_{\mathrm{TV}} = 1 - (4\delta(1-\delta))^{\frac{n}{2}+O(\log n)} . \tag{41}$$

4 Dobrushin's Coefficients in Networks

The proof of Theorem 5 relies on the characterization (17) of η_{KL} via mutual information, which satisfies the chain rule. Neither of these two properties is enjoyed by the total variation. Nevertheless, the following is an exact counterpart of Theorem 5 for total variation.

Theorem 8 *Under the same assumption of Theorem 5,*

$$\eta_{\mathrm{TV}}(P_{V,W|X}) \le (1-\eta_W)\eta_{\mathrm{TV}}(P_{V|X}) + \eta_W \eta_{\mathrm{TV}}(P_{\mathrm{pa}(W),V|X}) , \tag{42}$$

where $\eta_W = \eta_{\mathrm{TV}}(P_{W|\mathrm{pa}(W)})$. Furthermore, let $\mathrm{perc}(V)$ denote the probability that there is a path from X to V in the graph if each node v is removed independently with probability $1-\eta_v$ (site percolation). Then, we have for every $V \subset \mathcal{V}$

$$\eta_{\mathrm{TV}}(P_{V|X}) \le \mathrm{perc}(V) . \tag{43}$$

In particular, if $\eta_v < 1$ for all $v \in V$, then $\eta_{\mathrm{TV}}(P_{V|X}) < 1$.

[7]For elements of $\mathbb{F}_2^n$, $|\cdot|$ is the Hamming weight.

Proof Fix $x, \tilde{x}$ and denote by P (resp. Q) the distribution conditioned on $X = x$ (resp. x'). Denote $Z = \mathrm{pa}(W)$. The goal is to show

$$d_{\mathrm{TV}}(P_{VW}, Q_{VW}) \le (1 - \eta_W) d_{\mathrm{TV}}(P_V, Q_V) + \eta_W d_{\mathrm{TV}}(P_{ZV}, Q_{ZV}) . \tag{44}$$

which, by the arbitrariness of x, x' and in view of the characterization of η in (7), yields the desired (42). By Lemma 22 in Appendix C, there exists a coupling of P_{ZV} and Q_{ZV}, denoted by $\pi_{ZVZ'V'}$, such that

$$\begin{aligned} \pi[(Z, V) \neq (Z', V')] &= d_{\mathrm{TV}}(P_{ZV}, Q_{ZV}) , \\ \pi[V \neq V'] &= d_{\mathrm{TV}}(P_V, Q_V) \end{aligned}$$

simultaneously (that is, this coupling is jointly optimal for the total variation of the joint distributions and one pair of marginals).

Conditioned on $Z = z$ and $Z' = z'$ and independently of VV', let WW' be distributed according to a maximal coupling of the conditional laws $P_{W|Z=z}$ and $P_{W|Z=z'}$ (recall that $Q_{W|Z} = P_{W|Z} = P_{W|\mathrm{pa}(W)}$ by definition). This defines a joint distribution $\pi_{ZVWZ'V'W'}$, under which we have the Markov chain $VV' \to ZZ' \to WW'$. Then

$$\pi[W \neq W' | ZVZ'V'] = \pi[W \neq W' | ZZ'] = d_{\mathrm{TV}}(P_{W|\mathrm{pa}(W)=Z}, P_{W|\mathrm{pa}(W)=Z'}) \le \eta_W \mathbf{1}_{\{Z \neq Z'\}}.$$

Therefore we have

$$\begin{aligned} \pi[W \neq W' | V = V'] &= \mathbb{E}[\pi[W \neq W' | ZZ'] | V = V'] \\ &\le \eta_W \pi[Z \neq Z' | V = V']. \end{aligned}$$

Multiplying both sides by $\pi[V = V']$ and then adding $\pi[V \neq V']$, we obtain

$$\begin{aligned} \pi[(W, V) \neq (W', V')] &\le (1 - \eta_W)\pi[V \neq V'] + \eta_W \pi[(Z, V) \neq (Z', V')] \\ &= (1 - \eta_W) d_{\mathrm{TV}}(P_V, Q_V) + \eta_W d_{\mathrm{TV}}(P_{ZV}, Q_{ZV}), \end{aligned}$$

where the LHS is lower bounded by $d_{\mathrm{TV}}(P_{WV}, Q_{WV})$ and the equality is due to the choice of π. This yields the desired (44), completing the proof of (42). The rest of the proof is done as in Theorem 5. ⊔

As a consequence of Theorem 8, both Corollary 6 and 7 extend to total variation verbatim with η_{KL} replaced by η_{TV}:

Corollary 9 *In the setting of Corollary 6 we have*

$$\eta_{\mathrm{TV}}(P_{Y^n|W}) \le 1 - (1 - \eta_{\mathrm{TV}}(P_{Y|X}))^n < n \cdot \eta_{\mathrm{KL}}(P_{Y|X}) . \tag{45}$$

Corollary 10 *In the setting of Corollary 7 we have*

$$\eta_{\mathrm{TV}}(P_{V|X}) \le \sum_{\pi : X \xrightarrow{sf} V} \eta_{\mathrm{TV}}^{\pi} \le \sum_{\pi : X \to V} \eta_{\mathrm{TV}}^{\pi} ,$$

where for any path $\pi = (X, v_1, \ldots, v_m)$ *we denoted* $\eta_{\mathrm{TV}}^{\pi} \triangleq \prod_{j=1}^{m} \eta_{\mathrm{TV}}(P_{v_j|\mathrm{pa}(v_j)})$.

Evaluation for the BSC Consider the n-letter BSC defined in (4), where $Y = X+Z$ with $Z \sim \mathrm{Bern}(\delta)^n$ and $|Z| \sim \mathrm{Binom}(n, \delta)$. By Dobrushin's characterization (7), we have

$$\begin{aligned}
\eta_{\mathrm{TV}} &= \max_{x,x' \in \mathbb{F}_2^n} d_{\mathrm{TV}}(P_{Y|X=x}, P_{Y|X=x'}) \\
&= d_{\mathrm{TV}}(\mathrm{Bern}(\delta)^n, \mathrm{Bern}(1-\delta)^n) \\
&= d_{\mathrm{TV}}(\mathrm{Binom}(n,\delta), \mathrm{Binom}(n, 1-\delta)) && (46) \\
&= 1 - 2\mathbb{P}[|Z| > n/2] - \mathbb{P}[|Z| = n/2] && (47) \\
&= 1 - (4\delta(1-\delta))^{\frac{n}{2} + O(\log n)} , && (48)
\end{aligned}$$

where (46) follows from the sufficiency of $|Z|$ for testing the two distributions, (47) follows from $d_{\mathrm{TV}}(P, Q) = 1 - \int P \wedge Q$ and (48) follows from standard binomial tail estimates (see, e.g., [Ash65, Lemma 4.7.2]). The above sharp estimate should be compared to the bound obtained by applying Corollary 9:

$$\eta_{\mathrm{TV}} \le 1 - (2\delta)^n . \tag{49}$$

Although (49) correctly predicts the exponential convergence of $\eta_{\mathrm{TV}} \to 1$ whenever $\delta < \frac{1}{2}$, the exponent estimated is not optimal.

5 Bounding F_I-Curves in Networks

In this section our goal is to produce upper bound bounds on the F_I-curve of a Bayesian network $F_{V|X}$ in terms of those of the constituent channels. For any vertex v of the network, denote the F_I-curve of the channel $P_{v|\mathrm{pa}(v)}$ by $F_{v|\mathrm{pa}(v)}$, abbreviated by F_v, and the concavified version by F_v^c.

Theorem 11 *In the setting of Theorem 5,*

$$F_{V,W|X} \le F_{V|X} + F_W^c \circ (F_{\mathrm{pa}(W),V|X} - F_{V|X}) , \tag{50}$$

$$F_{V,W|X}^c \le F_{V|X}^c + F_W^c \circ (F_{\mathrm{pa}(W),V|X}^c - F_{V|X}^c) . \tag{51}$$

Furthermore, the right-hand side of (51) is non-negative, concave, nondecreasing and upper bounded by the identity mapping id.

Remark 1 The F_I-curve estimate in Theorem 11 implies that of contraction coefficients of Theorem 5. To see this, note that since $F_{\mathrm{pa}(W),V|X} \le \mathrm{id}$, the following is a relaxation of (50):

$$\mathrm{id} - F_{V,W|X} \ge (\mathrm{id} - F_W) \circ (\mathrm{id} - F_{V|X}). \tag{52}$$

Consequently, if each channel in the network satisfies an SDPI, then the end-to-end SDPI is also satisfied. That is, if each vertex has a non-trivial F_I-curve, i.e., $F_v < \mathrm{id}$ for all $v \in \mathcal{V}$, then the channel $X \to V$ also has a strict contractive property, i.e., $F_{V|X} < \mathrm{id}$.

Furthermore, since $F_W^c(t) \le \eta_W t$, noting the fact that $F'_{V|X}(0) = \eta_{\mathrm{KL}}(P_{V|X})$ and taking the derivative on both sides of (50) we see that the latter implies (22).

Proof We first show that for any channel $P_{Y|X}$, its $F_{Y|X}$-curve satisfies that $t \mapsto t - F_{Y|X}(t)$ is nondecreasing. Indeed, it is known, cf. [CPW15, Section I], that $t \mapsto \frac{F_{Y|X}(t)}{t}$ is non-increasing. Thus, for $t_1 < t_2$ we have

$$\begin{aligned} t_2 - F_{Y|X}(t_2) &\ge t_2 - \frac{t_2}{t_1} F_{Y|X}(t_1) \\ &= \frac{t_2}{t_1}\left(t_1 - F_{Y|X}(t_1)\right) \\ &\ge t_1 - F_{Y|X}(t_1), \end{aligned}$$

where the last step follows from the fact that $F_{Y|X}(t) \le t$. Similarly, for any concave function $\Phi : \mathbb{R}_+ \to \mathbb{R}_+$ s.t. $\Phi(0) = 0$ we have $\frac{\Phi(t_2)}{t_2} \le \frac{\Phi(t_1)}{t_1}$. Therefore, the argument above implies $t \mapsto t - \Phi(t)$ is nondecreasing and, in particular, so is $t \mapsto t - F_W^c(t)$.

Let P_{UX} be such that $I(U;X) \le t$ and $I(U;W,V) = F_{V,W|X}(t)$. By the same argument that leads to (27) we obtain

$$\begin{aligned} I(U;W|V=v_0) &\le F_W(I(U;A|V=v_0)) \\ &\le F_W^c(I(U;A|V=v_0)). \end{aligned}$$

Averaging over $v_0 \sim P_V$ and applying Jensen's inequality we get

$$I(U;W,V) \le F_W^c(I(U;\mathrm{pa}(W),V) - I(U;V)) + I(U;V).$$

Therefore,

$$\begin{aligned} F_{V,W|X}(t) &\le F_W^c(I(U;\mathrm{pa}(W),V) - I(U;V)) + I(U;V) \\ &\le F_W^c(F_{\mathrm{pa}(W),V|X}(t) - I(U;V)) + I(U;V) \qquad (53) \\ &= F_{\mathrm{pa}(W),V|X}(t) - (\mathrm{id} - F_W^c)(F_{\mathrm{pa}(W),V|X}(t) - I(U;V)) \\ &\le F_{\mathrm{pa}(W),V|X}(t) - (\mathrm{id} - F_W^c)(F_{\mathrm{pa}(W),V|X}(t) - F_{V|X}(t)) \qquad (54) \end{aligned}$$

$$= F_{V|X}(t) + F^c_W(F_{\mathrm{pa}(W),V|X}(t) - F_{V|X}(t))$$

$$\leq F^c_{V|X}(t) + F^c_W(F^c_{\mathrm{pa}(W),V|X}(t) - F^c_{V|X}(t)) \tag{55}$$

where (53) and (54) follow from the facts that $t \mapsto F_W(t)$ and $t \mapsto t - F_W(t)$ are both nondecreasing, and (55) follows from that $a + F^c_W(b-a)$ is nondecreasing in both a and b.

Finally, we need to show that the right-hand side of (55) is nondecreasing and concave (this automatically implies that (55) is an upper-bound to the concavification $F^c_{V|X}$). To that end, denote $t_\lambda = \lambda t_1 + (1-\lambda)t_0$, $f_\lambda = F^c_{V|X}(t_\lambda)$, $g_\lambda = F^c_{\mathrm{pa}(W),V|X}(t_\lambda)$ and notice the chain

$$f_\lambda + F^c_W(g_\lambda - f_\lambda) \geq \lambda f_1 + (1-\lambda)f_0 + F^c_W(\lambda(g_1 - f_1) + (1-\lambda)(g_0 - f_0)) \tag{56}$$

$$\geq \lambda(f_1 + F^c_W(g_1 - f_1)) + (1-\lambda)(f_0 + F^c_W(g_0 - f_0)) \tag{57}$$

where (56) is from concavity of $F^c_{V|X}$, $F^c_{\mathrm{pa}(W),V|X}$ and monotonicity of $(a,b) \mapsto a + F^c_W(b-a)$, and (57) is from concavity of F^c_W. □

Corollary 12 *In the setting of Corollary 6 we have*

$$F_{Y^n|W}(t) \leq t - \psi^{(n)}(t)\,,$$

where $\psi^{(1)} = \psi$, $\psi^{(k+1)} = \psi^{(k)} \circ \psi$ and $\psi : \mathbb{R}_+ \to \mathbb{R}_+$ is a convex function such that

$$F_{Y|X}(t) \leq t - \psi(t)\,.$$

Proof The case of $n = 1$ follows from the assumption on ψ. The case of $n > 1$ is proved by induction, with the induction step being an application of Theorem 11 with $V = Y^{n-1}$ and $W = Y_n$. □

Generally, the bound of Corollary 12 cannot be improved in the vicinity of zero. As an example where this is tight, consider a parallel erasure channel, whose F_I-curve for $t \leq \log q$ is computed in Theorem 17 below.

Evaluation for the BSC To ease the notation, all logarithms are with respect to base two in this section. Let $h(y) = y \log \frac{1}{y} + (1-y)\log \frac{1}{1-y}$ denote the binary entropy function and $h^{-1} : [0,1] \to [0, \frac{1}{2}]$ its functional inverse. Let $p * q \triangleq p(1-q) + q(1-p)$ for $p, q \in [0,1]$ denote binary convolution and define

$$\psi(t) \triangleq t - 1 + h(\delta * h^{-1}(\max(1-t, 0))) \tag{58}$$

which is convex and increasing in t on $\mathbb{R}_+$. For $n = 1$ it was shown in [CPW15, Section 2] that the F_I-curve of $\mathsf{BSC}(\delta)$ is given by

$$F_I(t, \mathsf{BSC}(\delta)) = F_I^c(t, \mathsf{BSC}(\delta)) = t - \psi(t) .$$

Applying Corollary 12 we obtain the following bound on the F_I-curve of BSC of blocklength n (even with feedback):

Proposition 13 *Let $Z_1, \ldots, Z_n \overset{i.i.d.}{\sim} Bern(\delta)$ be independent of U. For any (encoder) functions $f_j, j = 1, \ldots, n$, define*

$$X_j = f_j(U, Y^{j-1}), \quad Y_j = X_j + Z_j .$$

Then

$$I(U; Y^n) \le I(U; X^n) - \psi^{(n)}(I(U; X^n)) , \tag{59}$$

where $\psi^{(1)} = \psi$, $\psi^{(k+1)} = \psi^{(k)} \circ \psi$ and ψ is defined in (58).

Remark 2 The estimate (59) was first shown by A. Samorodnitsky (private communication) under extra technical constraints on the joint distribution of (X^n, W) and in the absence of feedback. We have then observed that Evans-Schulman type of technique yields (59) generally.

Since $\psi(t) = 4\delta(1-\delta)t + o(t)$ as $t \to 0$ we get

$$F_I^c(t, \mathsf{BSC}(\delta)^n) \le t - t(4\delta(1-\delta))^{n+o(n)}$$

as $n \to \infty$ for any fixed t. A simple lower bound, for comparison purposes, can be inferred from (40) after noticing that there we have $I(U; X) = 1$, and so

$$F_I^c(1, \mathsf{BSC}(\delta)^n) \ge 1 - (4\delta(1-\delta))^{\frac{n}{2}+O(\log n)} ,$$

This shows that the bound of Proposition 13 is order-optimal: $F(t) \to t$ exponentially fast. Exact exponent is given by (41).

As another point of comparison, we note the following. Existence of capacity-achieving error-correcting codes then easily implies

$$\lim_{n\to\infty} \frac{1}{n} F_I^c(n\theta, \mathsf{BSC}(\delta)^n) = \min(\theta, C) ,$$

where $C = 1 - h(\delta)$ is the Shannon capacity of $\mathsf{BSC}(\delta)$. Since for $t > 1$ we have $\psi(t) = t - C$ one can show that

$$\lim_{n\to\infty} \frac{1}{n} \psi^{(n)}(n\theta) = |\theta - C|^+ ,$$

and therefore we conclude that in this sense the bound (59) is asymptotically tight.

6 SDPI via Comparison to Erasure Channels

So far our leading example has been the binary symmetric channel (4). We now consider another important example:

Example 1 For any set $\mathcal{X}$, the *erasure channel* on $\mathcal{X}$ with erasure probability δ is a random transformation from $\mathcal{X}$ to $\mathcal{X} \cup \{?\}$, where $? \notin \mathcal{X}$ defined as

$$P_{E|X}(e|x) = \begin{cases} \delta, & e = ? \\ 1-\delta, & e = x \end{cases}.$$

For $\mathcal{X} = [q]$, we call it the *q-ary erasure channel* denoted by $\mathsf{EC}_q(\delta)$. In the binary case, we denote the binary erasure channel by $\mathsf{BEC}(\delta) \triangleq \mathsf{EC}_2(\delta)$. A simple calculation shows that for every P_{UX} we have

$$I(U;E) = (1-\delta)I(U;X) \tag{60}$$

and therefore for $\mathsf{EC}_q(\delta)$ we have $\eta_{\mathrm{KL}}(P_{E|X}) = 1-\delta$ and $F_I(t) = \min((1-\delta)t, \log q)$.

Next we recall a standard information-theoretic ordering on channels, cf. [EGK11, Section 5.6]:

Definition 2 Given two channels with common input alphabet, $P_{Y|X}$ and $P_{Y'|X}$, we say that $P_{Y'|X}$ is less noisy than $P_{Y|X}$, denoted by $P_{Y|X} \le_{l.n.} P_{Y'|X}$ if for all joint distributions P_{UX} we have

$$I(U;Y) \le I(U;Y')\,. \tag{61}$$

We also have an equivalent formulation in terms of divergence:

Proposition 14 $P_{Y|X} \le_{l.n.} P_{Y'|X}$ *if and only if for all* P_X, Q_X *we have*

$$D(Q_Y\|P_Y) \le D(Q_{Y'}\|P_{Y'}) \tag{62}$$

where $P_Y, P_{Y'}, Q_Y, Q_{Y'}$ *are the output distributions induced by* P_X, Q_X *over* $P_{Y|X}$ *and* $P_{Y'|X}$*, respectively.*

See Appendix A.4 for the proof.[8]

---

[8]It is tempting to put forward a fixed-P_X version of the previous criterion (similar to Theorem 4). That would, however, require some extra assumptions on P_X. Indeed, knowing that $I(W;Y) \le I(W;Y')$ for all $P_{W,X}$ with a given fixed P_X tells us nothing about how distributions $P_{Y|X=x}$ and $P_{Y'|X=x}$ compare outside the support of P_X. (For discrete channels and strictly positive P_X, however, it is easy to argue that indeed (62) holds for all Q_X if and only if (61) holds for all $P_{U,X}$ with a given marginal P_X.)

The following result shows that the contraction coefficient of KL divergence can be equivalently formulated as being less noisy than the corresponding erasure channel:[9]

Proposition 15 *For an arbitrary channel $P_{Y|X}$ we have*

$$\eta_{\mathrm{KL}}(P_{Y|X}) \le \eta \quad \iff \quad P_{Y|X} \le_{l.n.} P_{E|X}\,, \tag{63}$$

where $P_{E|X}$ is the erasure channel on the same input alphabet and erasure probability $1-\eta$.

Proof The definition of $\eta_{\mathrm{KL}}(P_{Y|X})$ guarantees for every P_{UX}

$$I(U;Y) \le (1-\delta)I(U;X), \tag{64}$$

where the right-hand side is precisely $I(U;E)$ by (60). □

It turns out that the notion of less-noisiness tensorizes:

Proposition 16 *If $P_{Y_1|X_1} \le_{l.n.} P_{Y_1'|X_1}$ and $P_{Y_2|X_2} \le_{l.n.} P_{Y_2'|X_2}$ then*

$$P_{Y_1|X_1} \times P_{Y_2|X_2} \le_{l.n.} P_{Y_1'|X_1} \times P_{Y_2'|X_2}$$

In particular,

$$\eta_{\mathrm{KL}}(P_{Y|X}) \le \eta \quad \implies \quad P^n_{Y|X} \le_{l.n.} P^n_{E|X}\,. \tag{65}$$

where $P_{E|X}$ is the erasure channel on the same input alphabet and erasure probability $1-\eta$.

Proof Construct a relevant joint distribution $U \to X^2 \to (Y^2, Y'^2)$ and consider

$$I(U;Y_1,Y_2) = I(U;Y_1) + I(U;Y_2|Y_1)\,. \tag{66}$$

Now since $U \perp\!\!\!\perp Y_2|Y_1$ we have by $P_{Y_2|X_2} \le_{l.n.} P_{Y_2'|X_2}$

$$I(U;Y_2|Y_1) \le I(U;Y_2'|Y_1)$$

and putting this back into (66) we get

$$I(U;Y_1,Y_2) \le I(U;Y_1) + I(U;Y_2'|Y_1) = I(U;Y_1,Y_2')\,.$$

[9] Note that another popular partial order for random transformations – that of stochastic degradation – may also be related to contraction coefficients, see [Rag14, Remark 3.2].

Repeating the same argument, but conditioning on Y_2' we get

$$I(U; Y_1, Y_2) \leq I(U; Y_1', Y_2'),$$

as required. The last claim of the proposition follows from Proposition 15. □

Consequently, everything that has been said in this paper about $\eta_{\mathrm{KL}}(P_{Y|X})$ can be restated in terms of seeking to compare a given channel in the sense of the $\leq_{l.n.}$ order to an erasure channel. It seems natural, then, to consider erasure channel in somewhat greater details.

6.1 F_I-Curve of Erasure Channels

Theorem 17 *Consider the q-ary erasure channel of blocklength n and erasure probability δ. Its F_I-curve is bounded by*

$$F_I^c(t, \mathsf{EC}_q(\delta)^n) \leq \mathbb{E}[\min(B \log q, t)], \qquad B \sim \mathrm{Binom}(n, 1-\delta). \tag{67}$$

The bound is tight in the following cases:

1. *at $t = k \log q$ with integral $k \leq n$ if and only if an $(n, k, n-k+1)_q$ MDS code exists*[10]
2. *for $t \leq \log q$ and $t \geq (n-1)\log q$;*
3. *for all t when $n = 1, 2, 3$.*

Remark 3 Introducing $B' \sim \mathrm{Binom}(n-1, 1-\delta)$ and using the identity $\mathbb{E}[B\mathbf{1}_{\{B\leq a\}}] = n(1-\delta)\mathbb{P}[B' \leq a-1]$, we can express the right-hand side of (67) in terms of binomial CDFs:

$$\mathbb{E}[\min(B, x)] = x + \mathbb{P}[B' \leq \lfloor x \rfloor - 1](1-\delta)(n-x) - x\delta\mathbb{P}[B' \leq \lfloor x \rfloor]$$

This implies that the upper bound (67) is piecewise-linear, increasing and concave.

Proof Consider arbitrary $U \to X^n \to E^n$ with $P_{E^n|X^n} = \mathsf{EC}_q(\delta)^n$. Let S be random subset of $[n]$ which includes each $i \in [n]$ independently with probability $1-\delta$. A direct computation shows that

$$I(U; E^n) = I(U; X_S, S) = \sum_{\sigma \subset [n]} \mathbb{P}[S = \sigma] I(U; X_\sigma) \tag{68}$$

$$\leq \sum_{\sigma \subset [n]} \mathbb{P}[S = \sigma] \min(|\sigma| \log q, t) = \mathbb{E}[\min(B \log q, t)]. \tag{69}$$

From here (67) follows by taking supremum over P_{U,X^n}.

[10] We remind that a subset $\mathcal{C}$ of $[q]^n$ is called an $(n, k, d)_q$ code if $|\mathcal{C}| = q^k$ and Hamming distance between any two points from $\mathcal{C}$ is at least d. A code is called maximum-distance separable (MDS) if $d = n - k + 1$. This is equivalent to the property that projection of $\mathcal{C}$ onto any subset of k coordinates is bijective.

Claims about tightness follow by constructing $U = X^n$ and taking X^n to be the output of the MDS code (so that $H(X_\sigma) = \min(|\sigma| \log q, t))$ and invoking the concavity of $F_I(t)$. One also notes that $[n, 1, n]_q$ (repetition code) and $[n, n-1, 2]$ (single parity check code) show tightness at $t = \log q$ and $t = (n-1)\log q$.

Finally, we prove that when $t = k \log q$ and the bound (67) is tight then a (possibly non-linear) $(n, k, n-k+1)_q$ MDS code must exist. First, notice that the right-hand side of (67) is a piecewise-linear and concave function. Thus the bound being tight for $F_I(t)$ (that is a concave-envelope of $F_I(t)$) should also be tight as a bound for $F_I(t)$. Consequently, there must exist $U \to X^n \to E^n$ such that the bound (69) is tight with $t = I(U; X^n)$. This implies that we should have

$$I(U; X_\sigma) = \min(\sigma \log q, t) \tag{70}$$

for all $\sigma \subset [n]$. In particular, we have $I(U; X_i) = \log q$ and thus $H(X_i|U) = 0$ and without loss of generality we may assume that $U = X^n$. Again from (70) we have that $H(X^n) = H(X^k) = k \log q$. This implies that X^n is a uniform distribution on a set of size q^k and projection on any k coordinates is injective. This is exactly the characterization of an MDS code (possibly non-linear) with parameters $(n, k, n-k+1)_q$. □

We also formulate some interesting observations for binary erasure channels:

Proposition 18 *For* $\mathsf{BEC}(n, \delta)$ *we have:*

1. *For* $n \geq 3$ *we have that* $F_I(t)$ *is not concave. More exactly,* $F_I(t) < F_I^c(t)$ *for* $t \in (1, 2)$.
2. *For arbitrary* n *and* $t \leq \log 2$ *or* $t \geq (n-1)\log 2$ *we have* $F_I(t) = F_I^c(t) = \mathbb{E}[\min(B \log 2, t)]$ *with* B *defined in in (67).*
3. *For* $t = 2, n = 4$ *the bound (67) is not tight and* $F_I^c(t) < \mathbb{E}[\min(B \log 2, t)]$.

Proof First note that in Definition 1 of $F_I(t)$ the supremum is a maximum and and U can be restricted to alphabet of size $|\mathcal{X}| + 2$. So in particular, $F_I(t) = f$ if and only if there exists $I(U; Y^n) = f, I(U; X^n) \leq t$.

Now consider $t \in (1, 2)$ and $n = 3$ and suppose (U, X^n) achieves the bound. For the bound to be tight we must have $I(U; X^3) = t$. For the bound to be tight we must have $I(U; X_i) = 1$ for all i, that is $H(X_i) = 1$, $H(X_i|U) = 0$ and $H(X^n|U) = 0$. Consequently, without loss of generality we may take $U = X^n$. So for the bound to be tight we need to find a distribution s.t.

$$H(X^3) = H(X_1, X_2) = H(X_2, X_3) = H(X_1, X_3) = t, H(X_1) = H(X_2) = H(X_3) = 1. \tag{71}$$

It is straightforward to verify that this set of entropies satisfies Shannon inequalities (i.e. submodularity of entropy checks), so the main result of [ZY97] shows that there does exist a sequence of triples X^3 (over large alphabets) which attains this point. We will show, however, that this is impossible for binary-valued random variables. First, notice that the set of achievable entropy vectors by binary triplets is a closed subset of $\mathbb{R}_+^7$ (as a continuous image of a compact set). Thus, it is sufficient to show that (71) itself is not achievable.

Second, note that for any pair A, B of binary random variables with uniform marginals we must have

$$A = B + Z, \qquad B \perp\!\!\!\perp Z \sim \mathrm{Bern}(p)\,.$$

Without loss of generality, assume that $X_2 = X_1 + Z$ where $H(Z) = t - 1 > 0$. Moreover, $H(X_3|X_1, X_2) = 0$ implies that $X_3 = f(X_1, X_2)$ for some function f.

Given X_1 we have $H(X_3|X_1 = x) = H(X_3|X_2 = x) = t - 1 > 0$. So the function $X_1 \mapsto f(X_1, x)$ should not be constant for either choice of $x \in \{0, 1\}$ and the same holds for $X_2 \mapsto f(x, X_2)$. Eliminating cases leaves us with $f = X_1 + X_2$ or $f = X_1 + X_2 + 1$. But then $X_3 = X_1 + X_2 = Z$ and $H(X_3) < 1$, which is a contradiction.

Since by Theorem 17 we know that the bound (67) is tight for $F_I(t)$ we conclude that

$$F_I(t) < F_I^c(t), \quad \forall t \in (1, 2)\,.$$

To show the second claim consider $U = X^n$ and $X_1 = \cdots = X_n \sim \mathrm{Bern}(p)$ for $t \le \log 2$. For $t \ge (n-1)\log 2$ take X^{n-1} to be iid $\mathrm{Bern}(\frac{1}{2})$ and

$$X_n = X_1 + \cdots + X_{n-1} + Z\,,$$

where $Z \sim \mathrm{Bern}(p)$. This yields $I(U; X_\sigma) = H(X_\sigma) = |\sigma| \log 2$ for every subset $\sigma \subset [n]$ of size up to $n - 1$. Consequently, the bound (67) must be tight.

Finally, third claim follows from Theorem 17 and the fact that there is no [4, 2, 3] binary code, e.g. [MS77, Corollary 7, Chapter 11]. □

Putting together (65) and (67) we get the following upper bound on the concavified F_I-curve of n-letter product channels in terms of the contraction coefficient of the single-letter channel.

Corollary 19 *If* $\eta_{\mathrm{KL}}(P_{Y|X}) = \eta$, *then*

$$F_I^c(t, P_{Y|X}^n) \le \mathbb{E}[\min(B \log q, t)], \qquad B \sim \mathrm{Binom}(n, 1-\delta)\,.$$

This gives an alternative proof of Corollary (6) for the case of no feedback.

6.2 *Samorodnitsky's SDPI*

So far, we have been concerned with bounding the "output" mutual information in terms of a certain "input" one. However, frequently, one is interested in bounding some "output" information given knowledge of several input ones. For example, for the parallel channel we have shown that

$$I(W;Y^n) \le (1-(1-\eta_{\mathrm{KL}}(P_{Y|X}))^n) I(W;X^n) .$$

But it turns out that a stronger bound can be given if we have finer knowledge about the joint distribution of W and X^n.

The following bound can be distilled from [Sam15]:

Theorem 20 (Samorodnitsky) *Consider the Bayesian network*

$$U \to X^n \to Y^n ,$$

where $P_{Y^n|X^n} = \prod_{i=1}^n P_{Y_i|X_i}$ is a memoryless channel with $\eta_i \triangleq \eta_{KL}(P_{Y_i|X_i})$. Then we have

$$I(U;Y^n) \le I(U;X_S|S) = I(U;X_S,S) , \tag{72}$$

where $S \perp\!\!\!\perp (U,X^n,Y^n)$ is a random subset of $[n]$ generated by independently sampling each element i with probability η_i. In particular, if $\eta_i = \eta$ for all i, then

$$I(U;Y^n) \le \sum_{\sigma \subset [n]} \eta^{|\sigma|}(1-\eta)^{n-|\sigma|} I(U;X_\sigma) \tag{73}$$

Proof Just put together characterization (63), tensorization property Proposition 16 to get $I(U;Y^n) \le I(U;E^n)$, where E^n is the output of the product of erasure channels with erasure probabilities $1-\eta_i$. Then the calculation (68) completes the proof. □

Remark 4 Let us say that "total" information $I(U;X^n)$ is distributed among subsets of $[n]$ as given by the following numbers:

$$I_k \triangleq \binom{n}{k}^{-1} \sum_{T \in \binom{[n]}{k}} I(U;X_T) .$$

Then bound (73) says (replacing $\mathrm{Binom}(n,\eta)$ by its mean value ηn):

$$I(U;Y^n) \lesssim I_{\eta n} .$$

Informally: the only kind of information about U that has a chance to be inferred on the basis of Y^n is one that is contained in subsets of X of size at most ηn.

Remark 5 Another implication of the Theorem is a strengthening of the Mrs. Gerber's Lemma. Fix a single-letter channel $P_{Y|X}$ and suppose that for some increasing *convex* function $m(\cdot)$ and all random variables X we have

$$H(Y) \ge m(H(X)) .$$

Then, in the setting of the Theorem we have

$$H(Y^n) \geq m\left(\frac{1}{\eta n}H(X_S|S)\right). \tag{74}$$

Note that by Han's inequality (74) is strictly better than the simple consequence of the chain rule: $H(Y^n) \geq nm(H(X^n)/n)$. For the case of $P_{Y|X} = \mathsf{BSC}(\delta)$ the bound (74) is a sharpening of the Mrs. Gerber's Lemma, and has been the focus of [Sam15], see also [Ord16]. To prove (74) let $X^n \to E^n$ be $\mathsf{EC}(1-\eta)$. Then, by Theorem 20 applied to $U = X_i$, $n = i-1$ we have

$$H(X_i|Y^{i-1}) \geq H(X_i|E^{i-1}).$$

Thus, from the chain rule and convexity of $m(\cdot)$ we obtain

$$H(Y^n) = \sum_i H(Y_i|Y^{i-1}) \geq nm\left(\frac{1}{n}\sum_i H(X_i|E^{i-1})\right),$$

and the proof is completed by computing $H(E^n)$ in two ways:

$$\begin{aligned} nh(\eta) + H(X_S|S) &= H(E^n) \\ &= \sum_i H(E_i|E^{i-1}) = \sum_i h(\eta) + \eta H(X_i|E^{i-1}). \end{aligned}$$

Remark 6 Using Proposition 14 we may also state a divergence version of the Theorem: In the setting of Theorem 20 for any pair of distributions P_{X^n} and Q_{X^n} we have

$$D(P_{Y^n}\|Q_{Y^n}) \leq D(P_{X_S|S}\|Q_{X_S|S}|P_S).$$

Similarly, we may extend the argument in the previous remark: If for a fixed Q_X, Q_Y (not necessarily related by $P_{Y|X}$) there exists an increasing concave function f such that for all P_X and $P_Y = P_{Y|X} \circ P_X$ we have

$$D(P_X\|Q_X) \leq f(D(P_Y\|Q_Y)) \qquad \forall P_X$$

then

$$D(P_{Y^n}\|(Q_Y)^n) \leq nf\left(\frac{1}{\eta n}D(P_{X_S|S}\|\prod_{i\in S} Q_X|P_S)\right).$$

Acknowledgements We thank Prof. M. Raginsky for references [BK98, Daw75, Gol79] and Prof. A. Samorodnitsky for discussions on Proposition 13 with us. We also thank Aolin Xu for pointing out (41). We are grateful to an anonymous referee for helpful comments.

A Contraction Coefficients on General Spaces

A.1 *Proof of Theorem 2*

We show that

$$\eta_f(P_{Y|X}, P_X) = \sup_{Q_X} \frac{D_f(Q_Y\|P_Y)}{D_f(Q_X\|P_X)} \geq \eta_{\chi^2}(P_{Y|X}, P_X) = \sup_{Q_X} \frac{\chi^2(Q_Y\|P_Y)}{\chi^2(Q_X\|P_X)}, \tag{75}$$

where both suprema are over all Q_X such that the respective denominator is in $(0, \infty)$. With the assumption that P_X is not a point mass, namely, there exists a measurable set E such that $P_X(E) \in (0, 1)$, it is clear that such Q_X always exists. For example, let $Q_X = \frac{1}{2}(P_X + P_{X|X\in E})$, where $P_{X|X\in E}(\cdot) \triangleq \frac{P_X(\cdot\cap E)}{P_X(E)}$. Then $\frac{1}{2} \leq \frac{dQ_X}{dP_X} \leq \frac{1}{2}(1 + \frac{1}{P_X(E)})$ and hence $D_f(Q_X\|P_X) < \infty$ since f is continuous. Furthermore, $Q_X \neq P_X$ implies that $D_f(Q_X\|P_X) \neq 0$ [Csi67].

The proof follows that of [CIR+93, Theorem 5.4] using the local quadratic behavior of f-divergence; however, in order to deal with general alphabets, additional approximation steps are needed to ensure the likelihood ratio is bounded away from zero and infinity.

Fix Q_X such that $\chi^2(Q_X\|P_X) < \infty$. Let $A = \{x : \frac{dQ_X}{dP_X}(x) < a\}$ where $a > 0$ is sufficiently large such that $Q_X(A) \geq 1/2$. Let $Q'_X = Q_{X|X\in A}$ and $Q'_Y = P_{Y|X} \circ Q'_X$. Then $\frac{dQ'_Y}{dP_Y} \leq \frac{a}{Q_X(A)} \leq 2a$. Let $Q''_X = \frac{1}{a}P_X + (1 - \frac{1}{a})Q'_X$ and $Q''_Y = P_{Y|X} \circ Q'_X = \frac{1}{a}P_Y + (1 - \frac{1}{a})Q'_Y$. Then we have

$$\frac{1}{a} \leq \frac{dQ''_X}{dP_X} \leq 2a + \frac{1}{a}, \quad \frac{1}{a} \leq \frac{dQ''_Y}{dP_Y} \leq 2a + \frac{1}{a}. \tag{76}$$

Note that $\chi^2(Q'_X\|P_X) = \frac{1}{Q(X\in A)}\mathbb{E}_P[(\frac{dQ_X}{dP_X})^2 \mathbf{1}_{\{X\in A\}}] - 1$. By dominated convergence theorem, $\chi^2(Q'_X\|P_X) \to \chi^2(Q_X\|P_X)$ as $a \to \infty$. On the other hand, since $Q'_Y \to Q_Y$ pointwise, the weak lower-semicontinuity of χ^2-divergence yields $\liminf_{a\to\infty} \chi^2(Q'_Y\|P_Y) \geq \chi^2(Q_Y\|P_Y)$. Furthermore, using the simple fact that $\chi^2(\epsilon P + (1-\epsilon)Q\|P) = (1-\epsilon)^2\chi^2(Q\|P)$, we have $\frac{\chi^2(Q''_X\|P_X)}{\chi^2(Q''_Y\|P_Y)} = \frac{\chi^2(Q'_X\|P_X)}{\chi^2(Q'_Y\|P_Y)}$. Therefore, to prove (75), it suffices to show for each fixed a, for any $\delta > 0$, there exists $\tilde{P}_X$ such that $\frac{D_f(\tilde{P}_X\|\tilde{P}_Y)}{D_f(Q_X\|P_X)} \geq \frac{\chi^2(Q''_X\|P_X)}{\chi^2(Q''_Y\|P_Y)} - \delta$.

For $0 < \epsilon < 1$, let $\tilde{P}_X = \bar{\epsilon}P_X + \epsilon Q''_X$, which induces $\tilde{P}_Y = P_{Y|X} \circ \tilde{P}_X = \bar{\epsilon}P_Y + \epsilon Q''_Y$. Then $D_f(\tilde{P}_X\|P_X) = \mathbb{E}_{P_X}[f(1 + \epsilon(\frac{dQ''_X}{dP_X} - 1))]$. Recall from (76) that $\frac{dQ''_X}{dP_X} \in [\frac{1}{a}, \frac{1}{a} + 2a]$. Since f'' is continuous and $f''(1) = 1$, by Taylor's theorem and dominated convergence theorem, we have $D_f(\tilde{P}_X\|P_X) = \frac{\epsilon^2}{2}\chi^2(Q''_X\|P_X)(1 + o(1))$. Analogously, $D_f(\tilde{P}_Y\|P_Y) = \frac{\epsilon^2}{2}\chi^2(Q''_Y\|P_Y)(1 + o(1))$. This completes the proof of $\eta_f(P_X) \geq \eta_{\chi^2}(P_X)$.

Remark 7 In the special case of KL divergence, we can circumvent the step of approximating by bounded likelihood ratio: By [PW16a, Lemma 4.2], since $\chi^2(Q_Y\|P_Y) \le \chi^2(Q_X\|P_X) < \infty$, we have $D(\tilde{P}_X\|P_X) = \epsilon^2\chi^2(Q_X\|P_X)/2 + o(\epsilon^2)$ and $D(\tilde{P}_Y\|P_Y) = \epsilon^2\chi^2(Q_Y\|P_Y)/2 + o(\epsilon^2)$, as $\epsilon \to 0$. Therefore $\frac{\chi^2(Q_Y\|P_Y)}{\chi^2(Q_X\|P_X)} \le \lim_{\epsilon\to 0}\frac{D(\tilde{P}_Y\|P_Y)}{D(\tilde{P}_X\|P_X)} \le \eta_{\mathrm{KL}}(P_X)$. Therefore $\eta_{\mathrm{KL}}(P_X) \ge \eta_{\chi^2}(P_X)$

A.2 *Proof of Theorem 3*

We prove

$$\eta_{\mathrm{KL}} = \eta_{\chi^2}. \tag{77}$$

First of all, $\eta_{\mathrm{KL}} \ge \eta_{\chi^2}$ follows from Theorem 2. For the other direction we closely follow the argument of [CRS94, Theorem 1]. Below we prove the following integral representation:

$$D(Q\|P) = \int_0^\infty \chi^2(Q\|P^t)\mathrm{d}t, \tag{78}$$

where $P^t \triangleq \frac{tQ+P}{1+t}$. Then

$$\begin{aligned} D(Q_Y\|P_Y) &= \int_0^\infty \chi^2(Q_Y\|P_Y^t)\mathrm{d}t \\ &\le \int_0^\infty \eta_{\chi^2}\cdot\chi^2(Q_X\|P_X^t)\mathrm{d}t = \eta_{\chi^2}D(Q_X\|P_X). \end{aligned}$$

where we used $P_Y^t = P_{Y|X}\circ P_X^t$. It remains to check (78). Note that

$$-\log x = \int_0^\infty \frac{1-x}{(x+t)(1+t)}\mathrm{d}t$$

Therefore

$$D(Q\|P) = \int_0^\infty \frac{1}{1+t}\mathbb{E}_Q\left[\frac{\mathrm{d}Q-\mathrm{d}P}{\mathrm{d}P+t\mathrm{d}Q}\right]\mathrm{d}t$$

Note that $t\mathbb{E}_Q\left[\frac{\mathrm{d}Q-\mathrm{d}P}{\mathrm{d}P+t\mathrm{d}Q}\right] = -\mathbb{E}_P\left[\frac{\mathrm{d}Q-\mathrm{d}P}{\mathrm{d}P+t\mathrm{d}Q}\right]$. Therefore $\mathbb{E}_Q\left[\frac{\mathrm{d}Q-\mathrm{d}P}{\mathrm{d}P+t\mathrm{d}Q}\right] = \frac{1}{1+t}\int\frac{(\mathrm{d}Q-\mathrm{d}P)^2}{\mathrm{d}P+t\mathrm{d}Q}$ $= (1+t)\chi^2(Q\|P^t)$, completing the proof of (78).

It is instructive to remark how this result was established for finite alphabets originally in [AG76]. Consider the map

$$P_X \mapsto V_r(P_X, Q_X) \triangleq D(P_{Y|X} \circ P_X \| P_{Y|X} \circ Q_X) - rD(P_X \| Q_X).$$

A simple differentiation shows that Hessian of this map at P_X is negative-definite if and only if $r > \eta_{\chi^2}(P_{Y|X}, P_X)$ and negative semidefinite if and only if $r \geq \eta_{\chi^2}(P_{Y|X}, P_X)$ (note that this does not depend on Q_X). Thus, taking $r = \eta_{\chi^2}(P_{Y|X})$ the map $P_X \mapsto V_r(P_X, Q_X)$ is concave in P_X for all Q_X. Thus, its local extremum at $P_X = Q_X$ is a global maximum and hence $V_r(P_X, Q_X) \leq 0$.

A.3 Proof of Theorem 4

We shall assume that P_X is not a point mass, namely, there exists a measurable set E such that $P_X(E) \in (0, 1)$. Define

$$\eta_{\mathrm{KL}}(P_X) = \sup_{Q_X} \frac{D(Q_Y \| P_Y)}{D(Q_X \| P_X)}$$

where the supremum is over all Q_X such that $0 < D(Q_X \| P_X) < \infty$. It is clear that such Q_X always exists (e.g., $Q_X = P_{X|X \in E}$ and $D(Q_X \| P_X) = \log \frac{1}{P_X(E)} \in (0, \infty)$). Let

$$\eta_I(P_X) = \sup \frac{I(U;Y)}{I(U;X)}$$

where the supremum is over all Markov chains $U \to X \to Y$ with fixed P_{XY} such that $0 < I(U;X) < \infty$. Such Markov chains always exist, e.g., $U = \mathbf{1}_{\{X \in E\}}$ and then $I(U;X) = h(P_X(E)) \in (0, \log 2)$. The goal of this appendix is to prove (18), namely

$$\eta_{\mathrm{KL}}(P_X) = \eta_I(P_X).$$

The inequality $\eta_I(P_X) \leq \eta_{\mathrm{KL}}(P_X)$ follows trivially:

$$I(U;Y) = D(P_{Y|U} \| P_Y | P_U) \leq \eta_{\mathrm{KL}}(P_X) D(P_{X|U} \| P_X | P_U) = \eta_{\mathrm{KL}}(P_X) I(X;U).$$

For the other direction, fix Q_X such that $0 < D(Q_X \| P_X) < \infty$. First, consider the case where $\frac{\mathrm{d}Q_X}{\mathrm{d}P_X}$ is bounded, namely, $\frac{\mathrm{d}Q_X}{\mathrm{d}P_X} \leq a$ for some $a > 0$ Q_X-a.s. For any $\epsilon \leq \frac{1}{2a}$, let $U \sim \mathrm{Bern}(\epsilon)$ and define the probability measure $\tilde{P}_X = \frac{P_X - \epsilon Q_X}{1-\epsilon}$. Let $P_{X|U=0} = \tilde{P}_X$ and $P_{X|U=1} = Q_X$, which defines a Markov chain $U \to X \to Y$ such that X, Y is distributed as the desired P_{XY}. Note that

$$\frac{I(U;Y)}{I(U;X)} = \frac{\bar{\epsilon} D(\tilde{P}_Y \| P_Y) + \epsilon D(Q_Y \| P_Y)}{\bar{\epsilon} D(\tilde{P}_X \| P_X) + \epsilon D(Q_X \| P_X)}$$

where $\bar{\epsilon} = 1 - \epsilon$ and $\tilde{P}_Y = P_{Y|X} \circ \tilde{P}_X$. We claim that

$$D(\tilde{P}_X \| P_X) = o(\epsilon), \tag{79}$$

which, in view of the data processing inequality $D(\tilde{P}_X \| P_X) \leq D(\tilde{P}_Y \| P_Y)$, implies $\frac{I(U;Y)}{I(U;X)} \xrightarrow{\epsilon \downarrow 0} \frac{D(Q_Y \| P_Y)}{D(Q_X \| P_X)}$ as desired. To establish (79), define the function

$$f(x,\epsilon) \triangleq \begin{cases} \frac{1-\epsilon x}{\epsilon(1-\epsilon)} \log \frac{1-\epsilon x}{1-\epsilon}, & \epsilon > 0 \\ (x-1)\log e, & \epsilon = 0\,. \end{cases}$$

One easily notices that f is continuous on $[0,a] \times [0, \frac{1}{2a}]$ and thus bounded. So we get, by bounded convergence theorem,

$$\frac{1}{\epsilon} D(\tilde{P}_X \| P_X) = \mathbb{E}_{P_X}\left[f\left(\frac{\mathrm{d}Q_X}{\mathrm{d}P_X}, \epsilon\right)\right] \to \mathbb{E}_{P_X}\left[\frac{\mathrm{d}Q_X}{\mathrm{d}P_X} - 1\right] \log e = 0\,.$$

To drop the boundedness assumption on $\frac{\mathrm{d}Q_X}{\mathrm{d}P_X}$ we simply consider the conditional distribution $Q'_X \triangleq Q_{X|X \in A}$ where $A = \{x : \frac{\mathrm{d}Q_X}{\mathrm{d}P_X}(x) < a\}$ and $a > 0$ is sufficiently large so that $Q_X(A) > 0$. Clearly, as $a \to \infty$, we have $Q'_X \to Q_X$ and $Q'_Y \to Q_Y$ pointwise (i.e. $Q'_Y(E) \to Q_Y(E)$ for every measurable set E), where $Q'_Y \triangleq P_{Y|X} \circ Q'_X$. Hence the lower-semicontinuity of divergence yields

$$\liminf_{a \to \infty} D(Q'_Y \| P_Y) \geq D(Q_Y \| P_Y)\,.$$

Furthermore, since $\frac{\mathrm{d}Q'_X}{\mathrm{d}P_X} = \frac{1}{Q_X(A)} \frac{\mathrm{d}Q_X}{\mathrm{d}P_X} \mathbf{1}_A$, we have

$$D(Q'_X \| P_X) = \log \frac{1}{Q_X(A)} + \frac{1}{Q_X(A)} \mathbb{E}_Q\left[\log \frac{\mathrm{d}Q_X}{\mathrm{d}P_X} \mathbf{1}\left\{\frac{\mathrm{d}Q_X}{\mathrm{d}P_X} \leq a\right\}\right]. \tag{80}$$

Since $Q_X(A) \to 1$, by dominated convergence (note: $\mathbb{E}_Q[|\log \frac{\mathrm{d}Q_X}{\mathrm{d}P_X}|] < \infty$) we have $D(Q'_X \| P_X) \to D(Q_X \| P_X)$. Therefore,

$$\liminf_{a \to \infty} \frac{D(Q'_Y \| P_Y)}{D(Q'_X \| P_X)} \geq \frac{D(Q_Y \| P_Y)}{D(Q_X \| P_X)},$$

completing the proof.

A.4 Proof of Proposition 14

First, notice the following simple result:

$$D(Q\|\lambda P + \bar{\lambda} Q) = o(\lambda), \lambda \to 0 \quad \iff \quad P \ll Q \tag{81}$$

Indeed, if $P \not\ll Q$, then there is a set E with $p = P[E] > 0 = Q[E]$. Denote the binary divergence by $d(p\|q) \triangleq D(\text{Bern}(p)\|\text{Bern}(q))$. Applying data-processing for divergence to $X \mapsto 1_E(X)$, we get

$$D(Q\|\lambda P + \bar{\lambda} Q) \geq d(0\|\lambda p) = \log \frac{1}{1-\lambda p}$$

and the derivative at $\lambda \to 0$ is non-zero. If $P \ll Q$, then let $f = \frac{dP}{dQ}$ and notice

$$\log \bar{\lambda} \leq \log(\bar{\lambda} + \lambda f) \leq \lambda(f-1)\log e\,.$$

Dividing by λ and assuming $\lambda < \frac{1}{2}$ we get

$$\left|\frac{1}{\lambda}\log(\bar{\lambda} + \lambda f)\right| \leq C_1 f + C_2\,,$$

for some absolute constants C_1, C_2. Thus, by the dominated convergence theorem we get

$$\frac{1}{\lambda} D(Q\|\lambda P + \bar{\lambda} Q) = -\int dQ\left(\frac{1}{\lambda}\log(\bar{\lambda} + \lambda f)\right) \to \int dQ(1-f) = 0\,.$$

Another observation is that

$$\lim_{\lambda \to 0} D(P\|\lambda P + \bar{\lambda} Q) = D(P\|Q)\,, \tag{82}$$

regardless of the finiteness of the right-hand side (this is a property of all convex lower-semicontinuous functions).

Now, we prove Proposition 14. One direction is easy: if $D(Q_Y\|P_Y) \leq D(Q_{Y'}\|P_{Y'})$, then

$$I(W;Y) = D(P_{Y|W}\|P_Y|P_W) \leq D(P_{Y'|W}\|P_{Y'}|P_W) = I(W;Y')\,.$$

For the other direction, consider an arbitrary pair (P_X, Q_X). Let $W = \text{Bern}(\epsilon)$ and $P_{X|W=0} = P_X, P_{X|W=1} = Q_X$. Then, we get

$$I(W;Y) = \bar{\epsilon} D(P_Y\|\bar{\epsilon} P_Y + \epsilon Q_Y) + \epsilon D(Q_Y\|\bar{\epsilon} P_Y + \epsilon Q_Y)\,,$$

and similarly for $I(W; Y')$. Assume that $D(Q_{Y'}\|P_{Y'}) < \infty$, for otherwise (62) holds trivially. Then $Q_{Y'} \ll P_{Y'}$ and we get from (81) and (82) that

$$I(W; Y') = \epsilon D(Q_{Y'}\|P_{Y'}) + o(\epsilon)\,. \tag{83}$$

On the other hand, again from (82)

$$I(W; Y) \geq \epsilon D(Q_Y\|\bar{\epsilon}P_Y + \epsilon Q_Y) = \epsilon D(Q_Y\|P_Y) + o(\epsilon)\,. \tag{84}$$

Since by assumption $I(W; Y) \leq I(W; Y')$ we conclude from comparing (83) to (84) that $D(Q_Y\|P_Y) \leq D(Q_{Y'}\|P_{Y'}) < \infty$, completing the proof.

B Contraction Coefficients for Binary-Input Channels

In this appendix we provide a tight characterization of the KL contraction coefficient for binary-input channel $P_{Y|X}$, where $X \in \{0, 1\}$ and Y is arbitrary. Clearly, $\eta_{\mathrm{KL}}(P_{Y|X})$ is a function of $P \triangleq P_{Y|X=0}$ and $Q \triangleq P_{Y|X=1}$, which we abbreviate as $\eta(\{P, Q\})$. The behavior of this quantity closely resembles that of divergence between distributions. Indeed, we expect $\eta(\{P, Q\})$ to be bigger if P and Q are more dissimilar and, furthermore, $\eta(\{P, Q\}) = 0$ (resp. 1) if and only if $P = Q$ (resp. $P \perp Q$). Next we show that $\eta(\{P, Q\})$ is essentially equivalent to Hellinger distance:

Theorem 21 *Consider a binary input channel $P_{Y|X} : \{0, 1\} \to \mathcal{Y}$ with $P_{Y|X=0} = P$ and $P_{Y|X=1} = Q$. Then, its contraction coefficient $\eta_{\mathrm{KL}}(P_{Y|X}) = \eta_{\chi^2}(P_{Y|X}) \triangleq \eta(\{P, Q\})$ satisfies*

$$\frac{H^2(P, Q)}{2} \leq \eta(\{P, Q\}) \leq H^2(P, Q) - \frac{H^4(P, Q)}{2}\,, \tag{85}$$

where Hellinger distance is defined as $H^2(P, Q) \triangleq 2 - 2\int \sqrt{dPdQ}$.

Remark 8 An obvious upper bound is $\eta(\{P, Q\}) \leq d_{\mathrm{TV}}(P, Q)$ by Theorem 1, which is worse than Theorem 21 since d_{TV} is smaller than the square-root of the right-hand side of (85). In fact it is straightforward to verify that the upper bound holds with equality when the output Y is also binary-valued. In particular, Theorem 21 implies that $\eta(\{P, Q\})$ is always within *a factor of two* of $H^2(P, Q)$.

Proof First notice the identities:

$$\chi^2(\mathrm{Bern}(\alpha)\|\mathrm{Bern}(\beta)) = \frac{(\alpha - \beta)^2}{\beta\bar{\beta}},$$

$$\chi^2(\alpha P + \bar{\alpha} Q \| \beta P + \bar{\beta} Q) = (\alpha - \beta)^2 \int \frac{(P-Q)^2}{\beta P + \bar{\beta} Q},$$

where we denote $\bar{\alpha} = 1 - \alpha$. Therefore the (input-dependent) χ^2-contraction coefficient is given by

$$\eta_{\chi^2}(\mathrm{Bern}(\beta), P_{Y|X}) = \sup_{\alpha \neq \beta} \frac{\chi^2(\alpha P + \bar{\alpha} Q \| \beta P + \bar{\beta} Q)}{\chi^2(\mathrm{Bern}(\alpha) \| \mathrm{Bern}(\beta))} = \beta\bar{\beta} \int \frac{(P-Q)^2}{\beta P + \bar{\beta} Q} \triangleq \mathrm{LC}_\beta(P\|Q),$$

where $\mathrm{LC}_\beta(P\|Q)$, clearly an f-divergence, is known as the Le Cam divergence (see, e.g., [Vaj09, p. 889]). In view of Theorem 3, the input-independent KL-contraction coefficient coincides with that of χ^2 and hence

$$\eta(\{P, Q\}) = \sup_{\beta \in (0,1)} \mathrm{LC}_\beta(P\|Q).$$

Thus the desired bound (85) follows from the characterization of the joint range between pairs of f-divergence [HV11], namely, H^2 versus LC_β, by taking the convex hull of their joint range restricted to Bernoulli distributions. Instead of invoking this general result, next we prove (85) using elementary arguments. Since $\mathrm{LC}_{1/2}(P\|Q) = 1 - 2\int \frac{dPdQ}{dP+dQ} \geq 1 - \int \sqrt{dPdQ} = \frac{1}{2}H^2(P,Q)$, the left inequality of (85) follows immediately. To prove the right inequality, by Cauchy-Schwartz, note that we have $(1 - \frac{1}{2}H^2(P,Q))^2 = (\int \sqrt{dPdQ})^2 = (\int \sqrt{\beta dP + \bar{\beta} dQ} \sqrt{\frac{dPdQ}{\beta dP + \bar{\beta} dQ}})^2 \leq \int \frac{dPdQ}{\beta dP + \bar{\beta} dQ} = 1 - \mathrm{LC}_\beta(P\|Q)$, for any $\beta \in (0,1)$. □

C Simultaneously Maximal Couplings

Lemma 22 *Let $\mathcal{X}$ and $\mathcal{Y}$ be Polish spaces. Given any pair of Borel probability measures P_{XY}, Q_{XY} on $\mathcal{X} \times \mathcal{Y}$, there exists a coupling π of P_{XY} and Q_{XY}, namely, a joint distribution of (X, Y, X', Y') such that $\mathcal{L}(X,Y) = P_{XY}$ and $\mathcal{L}(X',Y') = Q_{XY}$ under π, such that*

$$\pi\{(X,Y) \neq (X',Y')\} = d_{\mathrm{TV}}(P_{XY}, Q_{XY}) \quad \text{and} \quad \pi\{X \neq X'\} = d_{\mathrm{TV}}(P_X, Q_X). \tag{86}$$

Remark 9 After submitting this manuscript, we were informed that this result is the main content of [Gol79]. For interested reader we keep our original proof which is different from [Gol79] by relying on Kantorovich's dual representation and, thus, is non-constructive.

Remark 10 A triply-optimal coupling achieving in addition to (86) also $\pi[Y \neq Y'] = d_{TV}(P_Y, Q_Y)$ need not exist. Indeed, consider the example where X, Y are $\{0, 1\}$-valued and

$$P_{XY} = \begin{pmatrix} \frac{1}{2} & 0 \\ 0 & \frac{1}{2} \end{pmatrix}, \quad Q_{XY} = \begin{pmatrix} 0 & \frac{1}{2} \\ \frac{1}{2} & 0 \end{pmatrix}.$$

In other words, $X, Y \sim \text{Bern}(1/2)$ under both P and Q; however, $X = Y$ under P and $X = 1 - Y$ under Q. Furthermore, since $d_{TV}(P_X, Q_X) = d_{TV}(P_Y, Q_Y) = 0$, under any coupling $\pi_{XYX'Y'}$ of P_{XY} and Q_{XY} that simultaneously couples P_X to Q_X and P_Y to Q_Y maximally, we have $X = X'$ and $Y = Y'$, which contradicts $X = Y$ and $X' = 1 - Y'$. On the other hand, it is clear that a doubly optimal coupling (as claimed by Lemma 22) exists: just take $X = X' = Y \sim \text{Bern}(1/2)$ and $Y' = 1 - X'$. It is not hard to show that such a coupling also attains the minimum

$$\min_{\pi} \pi[(X, Y) \neq (X', Y')] + \pi[X \neq X'] + \pi[Y \neq Y'] = 2.$$

Proof Define the cost function $c(x, y, x', y') \triangleq \mathbf{1}_{\{(x,y)\neq(x',y')\}} + \mathbf{1}_{\{x\neq x'\}} = 2\mathbf{1}_{\{x\neq x'\}} + \mathbf{1}_{\{x=x',y\neq y'\}}$. Since the indicator of any open set is lower semicontinuous, so is $(x, y, x', y') \mapsto c(x, y, x', y')$. Applying Kantorovich's duality theorem (see, e.g., [Vil03, Theorem 1.3]), we have

$$\min_{\pi\in\Pi(P_{XY}, Q_{XY})} \mathbb{E}_\pi c(X, Y, X', Y') = \max_{f,g} \mathbb{E}_P[f(X, Y)] - \mathbb{E}_Q[g(X, Y)]. \tag{87}$$

where $f \in L_1(P)$, $g \in L_1(Q)$ and

$$f(x, y) - g(x', y') \leq c(x, y, x', y'). \tag{88}$$

Since the cost function is bounded, namely, c takes values in $[0, 2]$, applying [Vil03, Remark 1.3], we conclude that it suffices to consider $0 \leq f, g \leq 2$. Note that constraint (88) is equivalent to

$$\begin{aligned} f(x, y) - g(x', y') &\leq 2, \forall x \neq x', \forall y \neq y' \\ f(x, y) - g(x, y') &\leq 1, \forall x, \forall y \neq y' \\ f(x, y) - g(x, y) &\leq 0, \forall x, \forall y \end{aligned}$$

where the first condition is redundant given the range of f, g. In summary, the maximum on the right-hand side of (87) can be taken over all f, g satisfying the following constraints:

$$\begin{aligned} 0 \leq f, g &\leq 2 \\ f(x, y) - g(x, y') &\leq 1, \forall x, y \neq y' \\ f(x, y) - g(x, y) &\leq 0, \forall x, y \end{aligned}$$

Then

$$\max_{f,g} \mathbb{E}_P[f(X,Y)] - \mathbb{E}_Q[g(X,Y)] = \int_{\mathcal{X}} \max_{\phi,\psi} \left\{ \int_{\mathcal{Y}} p(x,y)\phi(y) - q(x,y)\psi(y) \right\} \tag{89}$$

where the maximum on the right-hand side is over $\phi, \psi : \mathcal{Y} \to \mathbb{R}$ satisfying

$$\begin{aligned} 0 \le \phi, \psi \le 2 \\ \phi(y) - \psi(y') \le 1, \forall y \ne y' \\ \phi(y) - \psi(y) \le 0, \forall y \end{aligned} \tag{90}$$

The optimization problem in the bracket on the RHS of (89) can be solved using the following lemma:

Lemma 23 *Let* $p, q \ge 0$. *Let* $(x)_+ \triangleq \max\{x, 0\}$. *Then*

$$\max_{\phi,\psi} \left\{ \int_{\mathcal{Y}} p\phi - q\psi : 0 \le \phi \le \psi \le 2, \sup \phi \le 1 + \inf \psi \right\} = \int (p-q)_+ + \left(\int (p-q) \right)_+ . \tag{91}$$

Proof First we show that it suffices to consider $\phi = \psi$. Given any feasible pair (ϕ, ψ), set $\phi' = \max\{\phi, \inf \psi\}$. To check that (ϕ', ϕ') is a feasible pair, note that clearly ϕ' takes values in $[0, 2]$. Furthermore, $\sup \phi' \le \sup \phi \le 1 + \inf \psi \le 1 + \inf \phi'$. Therefore the maximum on the left-hand side of (91) is equal to

$$\max_{\phi} \left\{ \int_{\mathcal{Y}} (p-q)\phi : 0 \le \phi \le 2, \sup \phi \le 1 + \inf \phi \right\} .$$

Let $a = \inf \phi$. Then

$$\begin{aligned} &\max_{\phi} \left\{ \int (p-q)\phi : 0 \le \phi \le 2, \sup \phi \le 1 + \inf \phi \right\} \\ &= \sup_{0 \le a \le 2} \max_{\phi} \left\{ \int (p-q)\phi : a \le \phi \le 2 \wedge (1+a) \right\} \\ &= \sup_{0 \le a \le 1} \max_{\phi} \left\{ \int (p-q)\phi : a \le \phi \le 1 + a \right\} \\ &= \sup_{0 \le a \le 1} \left\{ (1+a) \int (p-q)_+ + a \int (p-q)_- \right\} \end{aligned}$$

$$= \sup_{0 \le a \le 1} \left\{ \int (p-q)_+ + a \int (p-q) \right\}$$

$$= \int (p-q)_+ + \left(\int (p-q) \right)_+ .$$

□

Applying Lemma 23 to (89) for fixed x, we have

$$\max_{f,g} \mathbb{E}_P[f(X,Y)] - \mathbb{E}_Q[g(X,Y)]$$

$$= \int_{\mathcal{X}} \left(\int_{\mathcal{Y}} (p(x,y) - q(x,y))_+ + (p(x) - q(x))_+ \right)$$

$$= \int_{\mathcal{X}} \int_{\mathcal{Y}} (p(x,y) - q(x,y))_+ + \int_{\mathcal{X}} (p(x) - q(x))_+ = d_{\mathrm{TV}}(P_{XY}, Q_{XY}) + d_{\mathrm{TV}}(P_X, Q_X)$$

Combining the above with (87), we have

$$\min_{\pi_{XYX'Y'}} \pi\{(X,Y) \neq (X',Y')\} + \pi\{X \neq X'\} = d_{\mathrm{TV}}(P_{XY}, Q_{XY}) + d_{\mathrm{TV}}(P_X, Q_X).$$

Since $\pi\{(X,Y) \neq (X',Y')\} \geq d_{\mathrm{TV}}(P_{XY}, Q_{XY})$ and $\pi\{X \neq X'\} \geq d_{\mathrm{TV}}(P_X, Q_X)$ for any π, the minimizer of the sum on the left-hand side achieves equality simultaneously for both terms, proving the theorem. □

Acknowledgements Yury Polyanskiy's research has been supported by the Center for Science of Information (CSoI), an NSF Science and Technology Center, under grant agreement CCF-09-39370 and by the NSF CAREER award under grant agreement CCF-12-53205.

Yihong Wu's research has been supported in part by NSF grants IIS-1447879, CCF-1423088 and the Strategic Research Initiative of the College of Engineering at the University of Illinois.

References

[AG76] R. Ahlswede and P. Gács. Spreading of sets in product spaces and hypercontraction of the Markov operator. *Ann. Probab.*, pages 925–939, 1976.

[AGKN13] Venkat Anantharam, Amin Gohari, Sudeep Kamath, and Chandra Nair. On maximal correlation, hypercontractivity, and the data processing inequality studied by Erkip and Cover. *arXiv preprint arXiv:1304.6133*, 2013.

[Ash65] Robert B. Ash. *Information Theory*. Dover Publications Inc., New York, NY, 1965.

[Bir57] G. Birkhoff. Extensions of Jentzsch's theorem. *Trans. of AMS*, 85:219–227, 1957.

[BK98] Xavier Boyen and Daphne Koller. Tractable inference for complex stochastic processes. In *Proceedings of the 14th Conference on Uncertainty in Artificial Intelligence—UAI 1998*, pages 33–42. San Francisco: Morgan Kaufmann, 1998. Available at http://www.cs.stanford.edu/~xb/uai98/.

[CIR+93] J.E. Cohen, Yoh Iwasa, Gh. Rautu, M.B. Ruskai, E. Seneta, and Gh. Zbaganu. Relative entropy under mappings by stochastic matrices. *Linear algebra and its applications*, 179:211–235, 1993.

[CK81] I. Csiszár and J. Körner. *Information Theory: Coding Theorems for Discrete Memoryless Systems*. Academic, New York, 1981.

[CKZ98] J. E. Cohen, J. H. B. Kempermann, and Gh. Zbăganu. *Comparisons of Stochastic Matrices with Applications in Information Theory, Statistics, Economics and Population*. Springer, 1998.

[Cou12] T. Courtade. *Two Problems in Multiterminal Information Theory*. PhD thesis, U. of California, Los Angeles, CA, 2012.

[CPW15] F. Calmon, Y. Polyanskiy, and Y. Wu. Strong data processing inequalities for input-constrained additive noise channels. *arXiv*, December 2015. arXiv:1512.06429.

[CRS94] M. Choi, M.B. Ruskai, and E. Seneta. Equivalence of certain entropy contraction coefficients. *Linear algebra and its applications*, 208:29–36, 1994.

[Csi67] I. Csiszár. Information-type measures of difference of probability distributions and indirect observation. *Studia Sci. Math. Hungar.*, 2:229–318, 1967.

[Daw75] DA Dawson. Information flow in graphs. *Stoch. Proc. Appl.*, 3(2):137–151, 1975.

[DJW13] John C Duchi, Michael Jordan, and Martin J Wainwright. Local privacy and statistical minimax rates. In *Foundations of Computer Science (FOCS), 2013 IEEE 54th Annual Symposium on*, pages 429–438. IEEE, 2013.

[DMLM03] P. Del Moral, M. Ledoux, and L. Miclo. On contraction properties of Markov kernels. *Probab. Theory Relat. Fields*, 126:395–420, 2003.

[Dob56] R. L. Dobrushin. Central limit theorem for nonstationary Markov chains. I. *Theory Probab. Appl.*, 1(1):65–80, 1956.

[Dob70] R. L. Dobrushin. Definition of random variables by conditional distributions. *Theor. Probability Appl.*, 15(3):469–497, 1970.

[Doe37] Wolfgang Doeblin. *Le cas discontinu des probabilités en chaîne*. na, 1937.

[EC98] Elza Erkip and Thomas M. Cover. The efficiency of investment information. *IEEE Trans. Inf. Theory*, 44(3):1026–1040, 1998.

[EGK11] Abbas El Gamal and Young-Han Kim. *Network information theory*. Cambridge university press, 2011.

[EKPS00] William Evans, Claire Kenyon, Yuval Peres, and Leonard J Schulman. Broadcasting on trees and the Ising model. *Ann. Appl. Probab.*, 10(2):410–433, 2000.

[ES99] William S Evans and Leonard J Schulman. Signal propagation and noisy circuits. *IEEE Trans. Inf. Theory*, 45(7):2367–2373, 1999.

[Gol79] Sheldon Goldstein. Maximal coupling. *Probability Theory and Related Fields*, 46(2):193–204, 1979.

[HV11] P. Harremoës and I. Vajda. On pairs of f-divergences and their joint range. *IEEE Trans. Inf. Theory*, 57(6):3230–3235, Jun. 2011.

[Lau96] Steffen L Lauritzen. *Graphical Models*. Oxford University Press, 1996.

[LCV15] Jingbo Liu, Paul Cuff, and Sergio Verdu. Secret key generation with one communicator and a zero-rate one-shot via hypercontractivity. *arXiv preprint arXiv:1504.05526*, 2015.

[Led99] M. Ledoux. Concentration of measure and logarithmic Sobolev inequalities. *Seminaire de probabilites XXXIII*, pages 120–216, 1999.

[Mar06] Andrey Andreyevich Markov. Extension of the law of large numbers to dependent quantities. *Izv. Fiz.-Matem. Obsch. Kazan Univ.(2nd Ser)*, 15:135–156, 1906.

[MS77] Florence Jessie MacWilliams and Neil James Alexander Sloane. *The theory of error correcting codes*. Elsevier, 1977.

[MZ15] Anuran Makur and Lizhong Zheng. Bounds between contraction coefficients. *arXiv preprint arXiv:1510.01844*, 2015.

[Nai14] C. Nair. Equivalent formulations of hypercontractivity using information measures. In *Proc. 2014 Zurich Seminar on Comm.*, 2014.

[Ord16] Or Ordentlich. Novel lower bounds on the entropy rate of binary hidden Markov processes. In *Proc. 2016 IEEE Int. Symp. Inf. Theory (ISIT)*, Barcelona, Spain, July 2016.

[PW16a] Y. Polyanskiy and Y. Wu. Lecture notes on information theory. 2016. http://people.lids.mit.edu/yp/homepage/data/itlectures_v4.pdf.

[PW16b] Yury Polyanskiy and Yihong Wu. Dissipation of information in channels with input constraints. *IEEE Trans. Inf. Theory*, 62(1):35–55, January 2016. also arXiv:1405.3629.

[Rag13] Maxim Raginsky. Logarithmic Sobolev inequalities and strong data processing theorems for discrete channels. In *2013 IEEE International Symposium on Information Theory Proceedings (ISIT)*, pages 419–423, 2013.

[Rag14] Maxim Raginsky. Strong data processing inequalities and ϕ-Sobolev inequalities for discrete channels. *arXiv preprint arXiv:1411.3575*, November 2014.

[Sam15] Alex Samorodnitsky. On the entropy of a noisy function. *arXiv preprint arXiv:1508.01464*, August 2015.

[Sar58] O. V. Sarmanov. Maximal correlation coefficient (non-symmetric case). *Dokl. Akad. Nauk SSSR*, 121(1):52–55, 1958.

[Vaj09] I. Vajda. On metric divergences of probability measures. *Kybernetika*, 45(6):885–900, 2009.

[Vil03] C. Villani. *Topics in optimal transportation.* American Mathematical Society, Providence, RI, 2003.

[XR15] Aolin Xu and Maxim Raginsky. Converses for distributed estimation via strong data processing inequalities. In *Proc. 2015 IEEE Int. Symp. Inf. Theory (ISIT)*, Hong Kong, CN, July 2015.

[ZY97] Zhen Zhang and Raymond W Yeung. A non-Shannon-type conditional inequality of information quantities. *IEEE Trans. Inf. Theory*, 43(6):1982–1986, 1997.

An Application of a Functional Inequality to Quasi-Invariance in Infinite Dimensions

Maria Gordina

Abstract One way to interpret smoothness of a measure in infinite dimensions is quasi-invariance of the measure under a class of transformations. Usually such settings lack a reference measure such as the Lebesgue or Haar measure, and therefore we cannot use smoothness of a density with respect to such a measure. We describe how a functional inequality can be used to prove quasi-invariance results in several settings. In particular, this gives a different proof of the classical Cameron-Martin (Girsanov) theorem for an abstract Wiener space. In addition, we revisit several more geometric examples, even though the main abstract result concerns quasi-invariance of a measure under a group action on a measure space.

Keywords and phrases Quasi-invariance • Group action • Functional inequalities

1991 *Mathematics Subject Classification*. Primary 58G32 58J35; Secondary 22E65 22E30 22E45 58J65 60B15 60H05.

1 Introduction

Our goal in this paper is to describe how a functional inequality can be used to prove quasi-invariance of certain measures in infinite dimensions. Even though the original argument was used in a geometric setting, we take a slightly different approach in this paper. Namely, we formulate a method that can be used to prove quasi-invariance of a measure under a group action.

Such methods are useful in infinite dimensions when usually there is no natural reference measure such as the Lebesgue measure. At the same time quasi-invariance of measures is a useful tool in proving regularity results when it is reformulated as an integration by parts formula. We do not discuss significance of such results, and

M. Gordina (✉)
Department of Mathematics, University of Connecticut, Storrs, CT 06269, USA
e-mail: maria.gordina@uconn.edu

E. Carlen et al. (eds.), *Convexity and Concentration*, The IMA Volumes in Mathematics and its Applications 161, DOI 10.1007/978-1-4939-7005-6_8

moreover we do not refer to the extensive literature on the subject, as it is beyond the scope of our paper.

We start by describing an abstract setting of how finite-dimensional approximations can be used to prove such a quasi-invariance. In [11] this method was applied to projective and inductive limits of finite-dimensional Lie groups acting on themselves by left or right multiplication. In that setting a functional inequality (integrated Harnack inequality) on the finite-dimensional approximations leads to a quasi-invariance theorem on the infinite-dimensional group space. As we pointed out in [11], this is an abstraction of results in [9] for loop groups, and while there were probably earlier results of similar flavor, the most relevant later publications include [1] and [13]. Similar methods were used in the elliptic setting on infinite-dimensional Heisenberg-like groups in [10], and on semi-infinite Lie groups in [17]. Note that the assumptions we make below in Section 3 have been verified in these settings, including the sub-elliptic case for infinite-dimensional Heisenberg group in [2]. Even though the integrated Harnack inequality we use in these situations has a distinctly geometric flavor, we show in this paper that it does not have to be.

The paper is organized as follows. The general setting is described in Sections 2 and 3, where Theorem 3.2 is the main result. One of the ingredients for this result is quasi-invariance for finite-dimensional approximations which is described in Section 3. We review the connection between an integrated Harnack inequality and Wang's Harnack inequality in Section 4. Finally, Section 5 gives several examples of how one can use Theorem 3.2. We describe in detail the case of an abstract Wiener space, where the group in question is identified with the Cameron-Martin subspace acting by translation on the Wiener space. In addition we discuss elliptic (Riemannian) and sub-elliptic (sub-Riemannian) infinite-dimensional groups which are examples of a subgroup acting on the group by multiplication.

2 Notation

Suppose G is a topological group with the identity e, X is a topological space, $(X, \mathcal{B}, \mu)$ is a measure space, where $\mathcal{B}$ is the Borel σ-algebra, and μ is a probability measure. We assume that G is endowed with the structure of a Hilbert Lie group (e.g., [8]), and further that its Lie algebra $\mathfrak{g} := Lie(G) = T_eG$ is equipped with a Hilbertian inner product, $\langle \cdot, \cdot \rangle$. The corresponding distance on G is denoted by $d(\cdot, \cdot)$. In addition, we assume that G is separable, and therefore we can use what is known about Borel actions of Polish groups [3, 4]. Once we have an inner product on the Lie algebra $\mathfrak{g}$, we can define the length of a path in G as follows. Suppose $k \in C^1([0,1], G)$, $k(0) = e$, then

$$l_G(k(\cdot)) := \int_0^1 |L_{k(t)^{-1}*}\dot{k}(t)|dt, \tag{2.1}$$

where L_g is the left translation by $g \in G$.

We assume that G acts measurably on X, that is, there is a (Borel) measurable map $\Phi : G \times X \longrightarrow X$ such that

$$\Phi(e, x) = x, \text{ for all } x \in X,$$
$$\Phi(g_1, \Phi(g_2, x)) = \Phi(g_1 g_2, x), \text{ for all } x \in X, g_1, g_2 \in G.$$

We often will use $\Phi_g := \Phi(g, \cdot)$ for $g \in G$.

Definition 2.1 Suppose Φ is a measurable group action of G on X.

(1) In this case we denote by $\left(\Phi_g\right)_* \mu$ the *pushforward measure* defined by

$$\left(\Phi_g\right)_* \mu(A) := \mu\left(\Phi\left(g^{-1}, A\right)\right), \text{ for all } A \in \mathcal{B}, g \in G;$$

(2) the measure μ is *invariant* under the action Φ if

$$\left(\Phi_g\right)_* \mu = \mu \text{ for all } g \in G;$$

(3) the measure μ is *quasi-invariant* with respect to the action Φ if $\left(\Phi_g\right)_* \mu$ and μ are mutually absolutely continuous for all $g \in G$.

Notation 2.2 *For a topological group G acting measurably on the measure space $(X, \mathcal{B}, \mu)$ in such a way that μ is quasi-invariant under the action by G, the Radon-Nikodym derivative of $\left(\Phi_g\right)_* \mu$ with respect to μ is denoted by*

$$J_g(x) := \frac{\left(\Phi_g\right)_* \mu(dx)}{\mu(dx)} \textit{ for all } g \in G, x \in X.$$

For a thorough discussion of the Radon-Nikodym derivative in this setting we refer to [6, Appendix D]

3 Finite-Dimensional Approximations and Quasi-Invariance

We start by describing approximations to both the group G and the measure space $(X, \mathcal{B}, \mu)$. At the end we also need to impose certain conditions to have consistency of the group action defined on these approximations. As X is a topological space, we denote by $C_b(X)$ the space of continuous bounded real-valued functions.

Assumption 1 (Lie group assumptions). *Suppose $\{G_n\}_{n \in \mathbb{N}}$ is a collection of finite-dimensional unimodular Lie subgroups of G such that $G_n \subset G_m$ for all $n < m$. We assume that there exists a smooth section $\{s_n : G \longrightarrow G_n\}_{n \in \mathbb{N}}$, that is, $s_n \circ i_n = id_{G_n}$, where $i_n : G_n \longrightarrow G$ is the smooth injection. We suppose that $\bigcup_{n \in \mathbb{N}} G_n$ is a dense subgroup of G. In addition, we assume that the length of a path in G can be approximated by the lengths in G_n, namely, if $k \in C^1([0, 1], G)$, $k(0) = e$, then*

$$l_G(k(\cdot)) = \lim_{n\to\infty} l_{G_n}(s_n(k(\cdot))). \tag{3.1}$$

Note that s_n does not have to be a group homomorphism.

Assumption 2 (Measure space assumptions). *We assume that X is a separable topological space with a sequence of topological spaces $X_n \subset X$ which come with corresponding continuous maps $\pi_n : X \longrightarrow X_n$ satisfying the following properties. For any $f \in C_b(X)$*

$$\int_X f d\mu = \lim_{n\to\infty} \int_{X_n} f \circ j_n d\mu_n, \tag{3.2}$$

where $j_n : X_n \longrightarrow X$ is the continuous injection map, and μ_n is the pushforward measure $(\pi_n)_* \mu$.

Our last assumption concerns the group action for these approximations.

Assumption 3 (Group action assumptions). *The approximations to group G and the measure space (X, μ) are consistent with the group action in the following way*

$$\Phi(G_n \times X_n) \subset X_n \text{ for each } n \in \mathbb{N},$$

$$\Phi_g : X \to X \text{ is a continuous map for each } g \in G.$$

We denote by Φ^n the restriction of Φ to $G_n \times X_n$. Observe that $\Phi^n = \Phi \circ (i_n, j_n)$ which together with Assumption 3, it is clear that Φ^n is a measurable group action of G_n on $(X_n, \mathcal{B}_n, \mu_n)$.

Suppose now that μ_n is quasi-invariant under the group action Φ^n, and let J_g^n be the Radon-Nikodym derivative $\left(\Phi_g^n\right)_* \mu_n$ with respect to μ_n. We assume that there is a positive constant $C = C(p)$ such that for any $p \in [1, \infty)$ and $g \in G_n$

$$\|J_g^n\|_{L^p(X_n, \mu_n)} \leqslant \exp\left(C(p)\, d_{G_n}^2(e, g)\right). \tag{3.3}$$

Note that the constant $C(p)$ does not depend on n.

Remark 3.1 The fact that this estimate is Gaussian (with the square of the distance) does not seem to be essential. But as we do not have examples with a different exponent, we leave (3.3) as is. Moreover, we could consider a more general function on the right-hand side than an exponential of the distance squared.

Theorem 3.2 (Quasi-invariance of μ) *Suppose we have a group G and a measure space $(X, \mathcal{B}, \mu)$ satisfying Assumptions 1, 2, and 3. In addition, we assume that the uniform estimate* (3.3) *on the Radon-Nikodym derivatives holds.*

Then for all $g \in G$ the measure μ is quasi-invariant under the action Φ_g. Moreover, for all $p \in (1, \infty)$,

$$\left\| \frac{d\left(\Phi_g\right)_* \mu}{d\mu} \right\|_{L^p(X,\mu)} \leqslant \exp\left(C(p)\, d_G^2(e,g)\right). \tag{3.4}$$

Proof Using (3.3) we see that for any bounded continuous function $f \in C_b(X)$, $n \in \mathbb{N}$, and $g \in G$

$$\int_{X_n} |(f \circ j_n)(\Phi^n_{s_n(g)}(x))| d\mu_n(x) = \int_{X_n} J^n_{s_n(g)}(x)|(f \circ j_n)(x)| d\mu_n(x)$$
$$\leqslant \|f \circ j_n\|_{L^{p'}(X_n,\mu_n)} \exp\left(C(p)\, d^2_{G_n}(e, s_n(g))\right),$$

where p' is the conjugate exponent to p. Note that by Assumption 3 and definitions of j_n and Φ^n for all $(g,x) \in G_n \times X_n$

$$j_n\left(\Phi^n_g(x)\right) = \Phi^n_g(x) = \Phi_g(x) = \Phi_g\left(j_n(x)\right)$$

and therefore

$$f \circ j_n(\Phi^n_g(x)) = f\left(\Phi^n_g(x)\right) = f \circ \Phi_g\left(j_n(x)\right), (g,x) \in G_n \times X_n.$$

Thus

$$\int_{X_n} |(f \circ j_n)\left(\Phi^n_g(x)\right)| d\mu_n(x) = \int_{X_n} |\left(f \circ \Phi_g\right)\left(j_n(x)\right)| d\mu_n(x).$$

Allowing $n \to \infty$ in the last identity and using (3.2) and the fact that $f \circ \Phi_g \in C_b(X)$ yields

$$\int_X |f\left(\Phi_g(x)\right)| d\mu(x) \leqslant \|f\|_{L^{p'}(X,\mu)} \exp\left(C(p)\, d^2_{G_n}(e,g)\right), \text{ for all } g \in G_n. \tag{3.5}$$

Thus, we have proved that (3.5) holds for $f \in C_b(X)$ and $g \in G_n$. Now we would like to prove (3.5) for the distance d_G instead of d_{G_n} with g still in G_n. Take any path $k \in C^1([0,1], G)$ such that $k(0) = e$ and $k(1) = g$, and observe that then $s_n \circ k \in C^1([0,1], G_n)$ and therefore (3.5) holds with $d^2_{G_n}(e,g)$ replaced by $l_{G_n}(s_n \circ k)(1)$. Now we can use (3.2) in Assumption 1 and optimizing over all such paths k to see that

$$\int_X |f\left(\Phi_g(x)\right)| d\mu(x) \leqslant \|f\|_{L^{p'}(X,\mu)} \exp\left(C(p)\, d^2_G(e,g)\right), \text{ for all } g \in \bigcup_{n \in \mathbb{N}} G_n. \tag{3.6}$$

By Assumption 1 this union is dense in G, therefore dominated convergence along with the continuity of $d_G(e,g)$ in g implies that (3.6) holds for all $g \in G$. Since the bounded continuous functions are dense in $L^{p'}(X,\mu)$ (see, for example, [14, Theorem A.1, p. 309]), the inequality in (3.6) implies that the linear functional $\varphi_g : C_b(X) \to \mathbb{R}$ defined by

$$\varphi_g(f) := \int_X f\left(\Phi_g(x)\right) d\mu(x)$$

has a unique extension to an element, still denoted by φ_g, of $L^{p'}(X,\mu)^*$ which satisfies the bound

$$|\varphi_g(f)| \leqslant \|f\|_{L^{p'}(X,\mu)} \exp\left(C(p)\, d_G^2(e,g)\right)$$

for all $f \in L^{p'}(X,\mu)$. Since $L^{p'}(X,\mu)^* \cong L^p(W,\mu)$, there exists a function $J_g \in L^p(X,\mu)$ such that

$$\varphi_g(f) = \int_X f(x) J_g(x) d\mu(x), \tag{3.7}$$

for all $f \in L^{p'}(X,\mu)$, and

$$\|J_g\|_{L^p(X,\mu)} \leqslant \exp\left(C(p)\, d_G^2(e,g)\right).$$

Now restricting (3.7) to $f \in C_b(X)$, we may rewrite this equation as

$$\int_X f\left(\Phi_g(x)\right) d\mu(x) = \int_W f(x) J_g(x) d\mu(x). \tag{3.8}$$

Then a monotone class argument (again use [14, Theorem A.1]) shows that (3.8) is valid for all bounded measurable functions f on W. Thus, $d\left(\Phi_g\right)_* \mu / d\mu$ exists and is given by J_g, which is in L^p for all $p \in (1,\infty)$ and satisfies the bound (3.4). □

4 A Functional Inequality

In this section we would like to revisit an observation made in [11]. Namely, [11, Lemma D.1] connects Wang's Harnack inequality with an estimate similar to (3.3). It is easy to transfer this argument from the setting of Riemannian manifolds to a more general situation.

We start with an integral operator on $L^2(X,\nu)$, where (X,ν) is a σ-finite measure space. Namely, let

$$Tf(x) := \int_X p(x,y) f(y)\, d\nu(y), f \in L^2(X,\nu),$$

where the integral kernel $p(x, y)$ is assumed to satisfy the following properties.

positive	$p(x, y) > 0$ for all $x, y \in X$,
conservative	$\int_X p(x, y)\, d\nu(y) = 1$ for all $x \in X$,
symmetric	$p(x, y) = p(y, x)$ for all $x, y \in X$,
continuous	$p(\cdot, \cdot) : X \times X \longrightarrow \mathbb{R}$ is continuous.

Some of these assumptions might not be needed for the proof of Proposition 4.1, but we make them to simplify the exposition. Note that in our applications this integral kernel is the heat kernel for a strongly continuous, symmetric, Markovian semigroup in $L^2(X, \nu)$, therefore the corresponding heat kernel is positive, symmetric with the total mass not exceeding 1, in addition to having the semigroup property or being the approximate identity in $L^2(X, \nu)$. In our examples this heat semigroup is also conservative, therefore the heat kernel is conservative (stochastically complete), and thus $p(x, y)\, d\nu(y)$ is a probability measure.

The following proposition is a generalization of [11, Lemma D.1], and it simply reflects the fact that $(L^p)^*$ and $L^{p'}$ are isometrically isomorphic Banach spaces for $1 < p < \infty$ and $p' = p/(p-1)$, the conjugate exponent to p.

Proposition 4.1 *Let $x, y \in X$, $p \in (1, \infty)$ and $C \in (0, \infty]$ which might depend on x and y. Then*

$$[(Tf)(x)]^p \leqslant C^p (Tf^p)(y) \text{ for all } f \geqslant 0 \tag{4.1}$$

if and only if

$$\left(\int_X \left[\frac{p(x, z)}{p(y, z)} \right]^{p'} p(y, z)\, d\nu(z) \right)^{1/p'} \leqslant C. \tag{4.2}$$

Proof Since $p(\cdot, \cdot)$ is positive, we can write

$$(Tf)(x) = \int_X \frac{p(x, z)}{p(y, z)} f(z)\, p(y, z)\, d\nu(z).$$

We denote $d\mu_y(\cdot) := p(y, \cdot)\, d\nu(\cdot)$ and $g_{x,y}(\cdot) := \frac{p(x,\cdot)}{p(y,\cdot)}$, then

$$(Tf)(x) = \int_X f(z)\, g_{x,y}(z)\, d\mu_y(z). \tag{4.3}$$

Since $g_{x,y} \geqslant 0$ and $L^p(\mu)^*$ is isomorphic to $L^{p'}(\mu)$, the pairing in (4.3) implies that

$$\left\| g_{x,y} \right\|_{L^{p'}(\mu_y)} = \sup_{f \geqslant 0} \frac{\int_X f(z)\, g_{x,y}(z)\, d\mu_y(z)}{\|f\|_{L^p(\mu_y)}} = \sup_{f \geqslant 0} \frac{(Tf)(x)}{\left[(Tf^p)(y)\right]^{1/p}}.$$

The last equation may be written more explicitly as

$$\left(\int_X \left[\frac{p(x,z)}{p(y,z)} \right]^{p'} p(y,z)\, d\nu(z) \right)^{1/p'} = \sup_{f \geqslant 0} \frac{(Tf)(x)}{\left[(Tf^p)(y)\right]^{1/p}},$$

and from this equation the result follows. □

Remark 4.2 In the case when T is a Markov semigroup P_t and $p(\cdot,\cdot) = p_t(\cdot,\cdot)$ is the corresponding integral kernel, Proposition 4.1 shows that a Wang's Harnack inequality is equivalent to an integrated Harnack inequality. Subsection 5.2 gives more details on this equivalence for Riemannian manifolds, see Corollary 5.10.

Remark 4.3 The connection between Proposition 4.1 and (3.3) can be seen if we choose x and y in (4.2) as the endpoints of the group action as follows. Let $x, y \in X$ and $g \in G$ be such that $\Phi_e(x) = x$ and $\Phi_g(x) = y$, then to apply Proposition 4.1 we can take the constant in (4.2) to be equal to

$$\exp\left(C(p-1)\, d_{G_n}^2(e,g) \right).$$

Here the measure on X is $d\mu_x(z) = p(x,z)\, d\nu(z)$.

5 Examples

5.1 Abstract Wiener Space

Standard references on basic facts on the Gaussian measures include [5, 15]. Let (H, W, μ) be an abstract Wiener space, that is, H is a real separable Hilbert space densely continuously embedded into a real separable Banach space W, and μ is the Gaussian measure defined by the characteristic functional

$$\int_W e^{i\varphi(x)} d\mu(x) = \exp\left(-\frac{|\varphi|_{H^*}^2}{2} \right)$$

for any $\varphi \in W^* \subset H^*$. We will identify W^* with a dense subspace of H such that for any $h \in W^*$ the linear functional $\langle \cdot, h \rangle$ extends continuously from H to W. We will usually write $\langle \varphi, w \rangle := \varphi(w)$ for $\varphi \in W^*$, $w \in W$. More details can be found in [5]. It is known that μ is a Borel measure, that is, it is defined on the Borel σ-algebra $\mathcal{B}(W)$ generated by the open subsets of W.

We would like to apply the material from Sections 4 3 with $(X,\mu)=(W,\mu)$ and the group $G=E_W$ being the group of (measurable) rotations and translations by the elements from the Cameron-Martin subspace H. We can view this group as an infinite-dimensional analogue of the Euclidean group.

Notation 5.1 *We call an orthogonal transformation of H which is a topological homeomorphism of W^* a* **rotation** *of W^*. The space of all such rotations is denoted by $O(W)$. For any $R\in O(W)$ its adjoint, R^*, is defined by*

$$\langle \varphi, R^* w\rangle := \langle R^{-1}\varphi, w\rangle,\ w\in W, \varphi\in W^*.$$

Proposition 5.2 *For any $R\in O(W)$ the map R^* is a $\mathcal{B}(W)$-measurable map from W to W and*

$$\mu\circ\left(R^*\right)^{-1}=\mu.$$

Proof The measurability of R^* follows from the fact that R is continuous on H. For any $\varphi\in W^*$

$$\int_W e^{i\varphi(x)}d\mu\left(\left(R^*\right)^{-1}x\right)=\int_W e^{i\langle\varphi,x\rangle}d\mu\left(\left(R^*\right)^{-1}x\right)=\int_W e^{i\langle\varphi,R^*x\rangle}d\mu(x)=$$
$$\exp\left(-\frac{|R^{-1}\varphi|^2_{H^*}}{2}\right)=\exp\left(-\frac{|\varphi|^2_{H^*}}{2}\right)=\int_W e^{i\varphi(x)}d\mu(x)$$

since R is an isometry. □

Corollary 5.3 *Any $R\in O(W)$ extends to a unitary map on $L^2(W,\mu)$.*

Definition 5.4 The *Euclidean group* E_W is a group generated by measurable rotations $R\in O(W)$ and translation $T_h: W\to W$, $T_h(w):=w+h$.

To describe finite-dimensional approximations as in Section 3 we need to give more details on the identification of W^* with a dense subspace of H. Let $i: H\to W$ be the inclusion map, and $i^*: W^*\to H^*$ be its transpose, i.e. $i^*\ell:=\ell\circ i$ for all $\ell\in W^*$. Also let

$$H_*:=\{h\in H:\langle\cdot,h\rangle_H\in \mathrm{Ran}(i^*)\subset H^*\}$$

or in other words, $h\in H$ is in H_* iff $\langle\cdot,h\rangle_H\in H^*$ extends to a continuous linear functional on W. We will continue to denote the continuous extension of $\langle\cdot,h\rangle_H$ to W by $\langle\cdot,h\rangle_H$. Because H is a dense subspace of W, i^* is injective and because i is injective, i^* has a dense range. Since $h\mapsto\langle\cdot,h\rangle_H$ as a map from H to H^* is a conjugate linear isometric isomorphism, it follows from the above comments that for any $h\in H$ we have $h\mapsto\langle\cdot,h\rangle_H\in W^*$ is a conjugate linear isomorphism too, and that H_* is a dense subspace of H.

Now suppose that $P : H \to H$ is a finite rank orthogonal projection such that $PH \subset H_*$. Let $\{e_j\}_{j=1}^n$ be an orthonormal basis for PH and $\ell_j = \langle \cdot, e_j \rangle_H \in W^*$. Then we may extend P to a (unique) continuous operator from $W \to H$ (still denoted by P) by letting

$$P_n w := \sum_{j=1}^n \langle w, e_j \rangle_H e_j = \sum_{j=1}^n \ell_j (w) e_j \text{ for all } w \in W. \tag{5.1}$$

As we pointed out in [10, Equation 3.43] there exists $C < \infty$ such that

$$\|Pw\|_H \leqslant C \|w\|_W \text{ for all } w \in W. \tag{5.2}$$

Notation 5.5 *Let* $\operatorname{Proj}(W)$ *denote the collection of finite rank projections on* W *such that* $PW \subset H_*$ *and* $P|_H : H \to H$ *is an orthogonal projection, i.e.* P *has the form given in Equation* (5.1).

Also let $\{e_j\}_{j=1}^\infty \subset H_*$ be an orthonormal basis for H. For $n \in \mathbb{N}$, define $P_n \in \operatorname{Proj}(W)$ as in Notation 5.5, i.e.

$$P_n (w) = \sum_{j=1}^n \langle w, e_j \rangle_H e_j = \sum_{j=1}^n \ell_j (w) e_j \text{ for all } w \in W. \tag{5.3}$$

Then we see that $P_n|_H \uparrow Id_H$.

Proposition 5.6 *The Gaussian measure* μ *is quasi-invariant under the translations from* H *and invariant under orthogonal transformations of* H.

Proof The second part of the statement is the content of Proposition 5.2. We now prove quasi-invariance of μ under translation by elements in H. Let $\{P_n\}_{n\in\mathbb{N}}$ be a collection of operators defined by (5.3)for an orthonormal basis $\{e_j\}_{j=1}^\infty$ of H such that $\{e_j\}_{j=1}^\infty \subseteq H_*$. Then $H_n := P_n (H) \cong \mathbb{R}^n$, and the pushforward measure $(P_n)_* \mu$ is simply the standard Gaussian measure $p_n (x)\, dx$ on H_n. So if we identify the group of translation G with H and $s_n := P_n|_H$, then the group action is given by $\Phi_h (w) := w + h, w \in W, h \in H$. Note that Assumptions 1, 2, and 3 are satisfied, where $j_n = P_n : W \longrightarrow H_n$, etc. In particular, if we denote $h_n := P_n (h) \in \mathbb{R}^n, h \in H$, then for any measurable function $f : W \longrightarrow \mathbb{R}$ we see that

$$f \circ P_n (w + h) = f \circ P_n \circ \Phi_{h_n} (w) = f \circ \Phi_{h_n} \circ P_n (w).$$

Therefore

$$\int_W f \circ P_n (w + h)\, d\mu (w) = \int_{H_n} f (x + P_n h)\, p_n (x)\, dx =$$

$$\int_{H_n} f(x) p_n(x-h_n)\,dx = \int_{H_n} f(x) \frac{p_n(x-h_n)}{p_n(x)} p_n(x)\,dx$$

$$= \int_{H_n} f(x) J_{h_n}(x) p_n(x)\,dx.$$

Using an explicit form of the Radon-Nikodym derivative $J_{h_n}(x)$, we see that for any $f \in L^{p'}(W, \mu)$

$$\int_W |f \circ P_n(w+h)|\,d\mu(w) \leqslant \|f\|_{L^{p'}(p_n(x)dx)} \left\| \frac{p_n(x-h_n)}{p_n(x)} \right\|_{L^p(p_n(x)dx)}$$

$$\leqslant \|f \circ P_n\|_{L^{p'}(p_n(x)dx)} \exp\left(\frac{(p-1)\|h_n\|_H^2}{2}\right).$$

Thus (3.3) is satisfied, and therefore Theorem 3.2 is applicable, which proves the quasi-invariance with the Radon-Nikodym derivative satisfying

$$\|J_h\|_{L^p(W,\mu)} \leqslant \exp\left(\frac{(p-1)\|h\|_H^2}{2}\right). \tag{5.4}$$

□

Remark 5.7 The statement of Proposition 5.6 of course follows from the Cameron-Martin theorem which states that μ is quasi-invariant under translations by elements in H with the Radon-Nikodym derivative given by

$$\frac{d(T_h)_*\mu}{d\mu}(w) = \frac{d(\mu \circ T_h^{-1})}{d\mu}(w) = \frac{d(\mu \circ T_{-h})}{d\mu}(w) = e^{-\langle h,w\rangle - \frac{|h|^2}{2}}, \quad w \in W, h \in H.$$

Thus (5.4) is sharp.

Remark 5.8 Following [12] we see that quasi-invariance of the Gaussian measure μ induces the Gaussian regular representation of the Euclidean group E_W on $L^2(W, \mu)$ by

$$(U_{R,h}f)(w) := \left(\frac{d(\mu \circ (T_h R^*))}{d\mu}(w)\right)^{1/2} f\left((T_h R^*)^{-1}(w)\right) =$$

$$\left(\frac{d(\mu \circ T_h)}{d\mu}(w)\right)^{1/2} f\left((R^*)^{-1}(w-h)\right) =$$

$$e^{\langle h,w\rangle - \frac{|h|^2}{2}} f\left((R^*)^{-1}(w-h)\right), \quad w \in W$$

which is well defined by Corollary 5.3. It is clear that this is a unitary representation.

5.2 Wang's Harnack Inequality

This follows [11, Appendix D]. The following theorem appears in [18, 19] with $k = -K$, $V \equiv 0$. We will use the following notation

$$c(t) := \begin{cases} \frac{t}{e^t - 1} & t \neq 0, \\ 1 & t = 0. \end{cases} \tag{5.5}$$

Theorem 5.9 (Wang's Harnack inequality) *Suppose that M is a complete connected Riemannian manifold such that* Ric $\geqslant kI$ *for some $k \in \mathbb{R}$. Then for all $p > 1, f \geqslant 0, t > 0$, and $x, y \in M$ we have*

$$(P_t f)^p(y) \leqslant (P_t f^p)(z) \exp\left(\frac{p'k}{e^{kt} - 1} d^2(y, z)\right). \tag{5.6}$$

Corollary 5.10 *Let (M, g) be a complete Riemannian manifold such that* Ric $\geqslant kI$ *for some $k \in \mathbb{R}$. Then for every $y, z \in M$, and $p \in [1, \infty)$*

$$\left(\int_M \left[\frac{p_t(y, x)}{p_t(z, x)}\right]^p p_t(z, x)\, dV(x)\right)^{1/p} \leqslant \exp\left(\frac{c(kt)(p-1)}{2t} d^2(y, z)\right) \tag{5.7}$$

where $c(\cdot)$ is defined by (5.5), $p_t(x, y)$ is the heat kernel on M and $d(y, z)$ is the Riemannian distance from x to y for $x, y \in M$.

Proof From Lemma 4.1 and Theorem 5.9 with

$$C = \exp\left(\frac{p'}{p} \frac{k}{e^{kt} - 1} d^2(y, z)\right) = \exp\left(\frac{1}{p-1} \frac{k}{e^{kt} - 1} d^2(y, z)\right),$$

it follows that

$$\left(\int_M \left[\frac{p_t(x, z)}{p_t(y, z)}\right]^{p'} p_t(y, z)\, dV(z)\right)^{1/p'} \leqslant \exp\left(\frac{1}{p-1} \frac{k}{e^{kt} - 1} d^2(y, z)\right).$$

Using $p - 1 = (p' - 1)^{-1}$ and then interchanging the roles of p and p' gives (5.7). □

The reason we call (4.2) an integrated Harnack inequality on a d-dimensional manifold M is as follows. Recall the classical Li–Yau Harnack inequality ([16] and [7, Theorem 5.3.5]) which states that if $\alpha > 1$, $s > 0$, and Ric $\geqslant -K$ for some $K \geqslant 0$, then

$$\frac{p_t(y, x)}{p_{t+s}(z, x)} \leqslant \left(\frac{t+s}{t}\right)^{d\alpha/2} \exp\left(\frac{\alpha d^2(y, z)}{2s} + \frac{d\alpha K s}{8(\alpha - 1)}\right), \tag{5.8}$$

for all $x, y, z \in M$ and $t > 0$. However, when $s = 0$, (5.8) gives no information on $p_t(y, x)/p_t(z, x)$ when $y \neq z$. This inequality is based on the Laplacian $\Delta/2$ rather than Δ, t and s should be replaced by $t/2$ and $s/2$ when applying the results in [16, 7].

5.3 *Infinite-Dimensional Heisenberg-Like Groups: Riemannian and Sub-Riemannian Cases*

These examples represent infinite-dimensional versions of the group action of a Lie group on itself by left or right multiplication. The difference is in geometry of the space on which the group acts on: Riemannian and sub-Riemannian. In both cases we proved (3.3), where the constant C depends on the geometry, and the distance used is Riemannian or Carnot-Carathéodory.

Let (W, H, μ) be an abstract Wiener space and let $\mathbf{C}$ be a finite-dimensional inner product space. Define $\mathfrak{g} := W \times \mathbf{C}$ to be an infinite-dimensional Heisenberg-like Lie algebra, which is constructed as an infinite-dimensional step 2 nilpotent Lie algebra with continuous Lie bracket. Namely, let $\omega : W \times W \to \mathbf{C}$ be a continuous skew-symmetric bilinear form on W. We will also assume that ω is surjective.

Let $\mathfrak{g}$ denote $W \times \mathbf{C}$ when thought of as a Lie algebra with the Lie bracket given by

$$[(X_1, V_1), (X_2, V_2)] := (0, \omega(X_1, X_2)). \tag{5.9}$$

Let G denote $W \times \mathbf{C}$ when thought of as a group with multiplication given by

$$g_1 g_2 := g_1 + g_2 + \frac{1}{2}[g_1, g_2],$$

where g_1 and g_2 are viewed as elements of $\mathfrak{g}$. For $g_i = (w_i, c_i)$, this may be written equivalently as

$$(w_1, c_1) \cdot (w_2, c_2) = \left(w_1 + w_2, c_1 + c_2 + \frac{1}{2}\omega(w_1, w_2) \right). \tag{5.10}$$

Then G is a Lie group with Lie algebra $\mathfrak{g}$, and G contains the subgroup $G_{CM} = H \times \mathbf{C}$ which has Lie algebra $\mathfrak{g}_{CM}$. In terms of Section 2 the Cameron-Martin (Hilbertian) subgroup G_{CM} is the group that is acting on the Heisenberg group G by left or right multiplication.

Using Notation 5.5 we can define finite-dimensional approximations to G by using $P \in \mathrm{Proj}(W)$. We assume in addition that PW is sufficiently large to satisfy Hörmander's condition (that is, $\{\omega(A, B) : A, B \in PW\} = \mathbf{C}$). For each $P \in \mathrm{Proj}(W)$, we define $G_P := PW \times \mathbf{C} \subset H_* \times \mathbf{C}$ and a corresponding projection $\pi_P : G \to G_P$

$$\pi_P(w, x) := (Pw, x).$$

We will also let $\mathfrak{g}_P = \mathrm{Lie}(G_P) = PW \times \mathbf{C}$. For each $P \in \mathrm{Proj}(W)$, G_P is a finite-dimensional connected unimodular Lie group.

Notation 5.11 (Riemannian and horizontal distances on G_{CM})

(1) For $x = (A, a) \in G_{CM}$, let

$$|x|^2_{\mathfrak{g}_{CM}} := \|A\|^2_H + \|a\|^2_{\mathbf{C}}.$$

The *length* of a C^1-path $\sigma : [0, 1] \to G_{CM}$ is defined as

$$\ell(\sigma) := \int_0^1 |L_{\sigma^{-1}(s)*}\dot{\sigma}(s)|_{\mathfrak{g}_{CM}}\, ds.$$

By C^1_{CM} we denote the set of paths $\sigma : [0, 1] \to G_{CM}$.

(2) A C^1-path $\sigma : [0, 1] \to G_{CM}$ is *horizontal* if $L_{\sigma(t)^{-1}*}\dot{\sigma}(t) \in H \times \{0\}$ for a.e. t. Let $C^{1,h}_{CM}$ denote the set of horizontal paths $\sigma : [0, 1] \to G_{CM}$.

(3) The *Riemannain distance* between $x, y \in G_{CM}$ is defined by

$$d(x, y) := \inf\{\ell(\sigma) : \sigma \in C^1_{CM} \text{ such that } \sigma(0) = x \text{ and } \sigma(1) = y\}.$$

(4) The *horizontal distance* between $x, y \in G_{CM}$ is defined by

$$d^h(x, y) := \inf\{\ell(\sigma) : \sigma \in C^{1,h}_{CM} \text{ such that } \sigma(0) = x \text{ and } \sigma(1) = y\}.$$

The Riemannian and horizontal distances are defined analogously on G_P and will be denoted by d_P and d^h_P correspondingly. In particular, for a sequence $\{P_n\}_{n=1}^{\infty} \subset \mathrm{Proj}(W)$, we will let $G_n := G_{P_n}$, $d_n := d_{P_n}$, and $d^h_n := d^h_{P_n}$.

Now we are ready to define the corresponding heat kernel measures on G. We start by considering two Brownian motions on $\mathfrak{g}$

$$b_t := (B(t), B_0(t)), t \geqslant 0,$$
$$b^h_t := (B(t), 0(t)), t \geqslant 0,$$

with variance determined by

$$\mathbb{E}\left[\langle (B(s), B_0(s)), (A, a)\rangle_{\mathfrak{g}_{CM}} \langle (B(t), B_0(t)), (C, c)\rangle_{\mathfrak{g}_{CM}}\right]$$
$$= \mathrm{Re}\,\langle (A, a), (C, c)\rangle_{\mathfrak{g}_{CM}} \min(s, t)$$

for all $s, t \in [0, \infty)$, $A, C \in H_*$ and $a, c \in \mathbf{C}$.

A (Riemannian) *Brownian motion* on G is the continuous G-valued process defined by

$$g(t) = \left(B(t), B_0(t) + \frac{1}{2} \int_0^t \omega(B(\tau), dB(\tau)) \right). \tag{5.11}$$

Further, for $t > 0$, let $\mu_t = \mathrm{Law}(g(t))$ be a probability measure on G. We refer to μ_t as the time t *heat kernel measure on* G.

Similarly a *horizontal Brownian motion* on G is the continuous G-valued process defined by

$$g^h(t) = \left(B(t), \frac{1}{2} \int_0^t \omega(B(\tau), dB(\tau)) \right). \tag{5.12}$$

Then for $t > 0$, let $\mu_t^h = \mathrm{Law}\left(g^h(t)\right)$ be a probability measure on G. We refer to μ_t as the time t *horizontal heat kernel measure on* G.

As the proof of [10, Theorem 8.1] explains, in this case Assumptions 1, 2, and 3 are satisfied, and moreover, (3.3) is satisfied as follows. Namely, [10, Corollary 7.3] says that the Ricci curvature is bounded from below by $k(\omega)$ uniformly for all G_n, so (3.3) holds as follows.

$$\left\| J_k^n \right\|_{L^p(\mu^n)} \leqslant \exp\left(\frac{c(k(\omega)t)(p-1)}{2t} d_n^2(\mathbf{e}, k) \right), k \in G_n, \tag{5.13}$$

where $c(\cdot)$ is defined by (5.5).

In the sub-Riemannian case we have

$$\left\| J_k^{h,n} \right\|_{L^p(\mu_h^n)} \leqslant \exp\left(\left(1 + \frac{8\|\omega\|_{2,n}^2}{\rho_{2,n}} \right) \frac{(1+p)\left(d_n^h(e,k)\right)^2}{4t} \right), \tag{5.14}$$

where the geometric constants are defined as in [2, p. 25].

Acknowledgements The author is grateful to Sasha Teplyaev and Tom Laetsch for useful discussions and helpful comments.

This research was supported in part by NSF Grant DMS-1007496.

References

[1] Hélène Airault and Paul Malliavin. Quasi-invariance of Brownian measures on the group of circle homeomorphisms and infinite-dimensional Riemannian geometry. *J. Funct. Anal.*, 241(1):99–142, 2006.

[2] Fabrice Baudoin, Maria Gordina, and Tai Melcher. Quasi-invariance for heat kernel measures on sub-Riemannian infinite-dimensional Heisenberg groups. *Trans. Amer. Math. Soc.*, 365(8):4313–4350, 2013.

[3] Howard Becker. Polish group actions: dichotomies and generalized elementary embeddings. *J. Amer. Math. Soc.*, 11(2):397–449, 1998.
[4] Howard Becker and Alexander S. Kechris. Borel actions of Polish groups. *Bull. Amer. Math. Soc. (N.S.)*, 28(2):334–341, 1993.
[5] Vladimir I. Bogachev. *Gaussian measures*, volume 62 of *Mathematical Surveys and Monographs*. American Mathematical Society, Providence, RI, 1998.
[6] Theo Bühler. *On the algebraic foundation of bounded cohomology*. PhD thesis, ETH, 2008.
[7] E. B. Davies. *Heat kernels and spectral theory*, volume 92 of *Cambridge Tracts in Mathematics*. Cambridge University Press, Cambridge, 1989.
[8] Pierre de la Harpe. *Classical Banach-Lie algebras and Banach-Lie groups of operators in Hilbert space*. Lecture Notes in Mathematics, Vol. 285. Springer-Verlag, Berlin-New York, 1972.
[9] Bruce K. Driver. Integration by parts and quasi-invariance for heat kernel measures on loop groups. *J. Funct. Anal.*, 149(2):470–547, 1997.
[10] Bruce K. Driver and Maria Gordina. Heat kernel analysis on infinite-dimensional Heisenberg groups. *J. Funct. Anal.*, 255(9):2395–2461, 2008.
[11] Bruce K. Driver and Maria Gordina. Integrated Harnack inequalities on Lie groups. *J. Differential Geom.*, 83(3):501–550, 2009.
[12] Bruce K. Driver and Brian C. Hall. The energy representation has no non-zero fixed vectors. In *Stochastic processes, physics and geometry: new interplays, II (Leipzig, 1999)*, volume 29 of *CMS Conf. Proc.*, pages 143–155. Amer. Math. Soc., Providence, RI, 2000.
[13] Shizan Fang. Integration by parts for heat measures over loop groups. *J. Math. Pures Appl. (9)*, 78(9):877–894, 1999.
[14] Svante Janson. *Gaussian Hilbert spaces*, volume 129 of *Cambridge Tracts in Mathematics*. Cambridge University Press, Cambridge, 1997.
[15] Hui Hsiung Kuo. *Gaussian measures in Banach spaces*. Springer-Verlag, Berlin, 1975. Lecture Notes in Mathematics, Vol. 463.
[16] Peter Li and Shing-Tung Yau. On the parabolic kernel of the Schrödinger operator. *Acta Math.*, 156(3–4):153–201, 1986.
[17] Tai Melcher. Heat kernel analysis on semi-infinite Lie groups. *J. Funct. Anal.*, 257(11): 3552–3592, 2009.
[18] Feng-Yu Wang. Logarithmic Sobolev inequalities on noncompact Riemannian manifolds. *Probab. Theory Related Fields*, 109(3):417–424, 1997.
[19] Feng-Yu Wang. Equivalence of dimension-free Harnack inequality and curvature condition. *Integral Equations Operator Theory*, 48(4):547–552, 2004.

Borell's Formula on a Riemannian Manifold and Applications

Joseph Lehec

Abstract Borell's formula is a stochastic variational formula for the log-Laplace transform of a function of a Gaussian vector. We establish an extension of this to the Riemannian setting and give a couple of applications, including a new proof of a convolution inequality on the sphere due to Carlen, Lieb and Loss.

1 Introduction

Throughout the article $(\Omega, \mathcal{A}, \mathsf{P})$ is a probability space, equipped with a filtration $(\mathcal{F}_t)_{t\leq T}$ and carrying a standard n–dimensional Brownian motion $(B_t)_{t\leq T}$. The time horizon T shall vary along the article, most of the time it will be finite. By *standard* we mean that (B_t) starts from 0 and has quadratic covariation given by $[B]_t = tI_n$ for all $t \leq T$. Let $\mathbb{H}$ be the Cameron-Martin space, namely the space of absolutely continuous paths $u\colon [0, T] \to \mathbb{R}^n$, starting from 0 and equipped with the norm

$$\|u\|_{\mathbb{H}} = \left(\int_0^T |\dot{u}_s|^2 \, ds \right)^{1/2},$$

where $|\dot{u}_s|$ denotes the Euclidean norm of the derivative of u at time s. In this context a *drift* is a process which is adapted to the filtration $(\mathcal{F}_t)$ and which belongs to $\mathbb{H}$ almost surely. Let γ_n be the standard Gaussian measure on $\mathbb{R}^n$. The starting point of the present article is the so-called Borell formula: If $f\colon \mathbb{R}^n \to \mathbb{R}$ is measurable and bounded from below, then

$$\log\left(\int_{\mathbb{R}^n} e^f \, d\gamma_n \right) = \sup_U \left\{ \mathsf{E}\left[f(B_1 + U_1) - \frac{1}{2}\|U\|_{\mathbb{H}}^2 \right] \right\} \tag{1}$$

where the supremum is taken over all drifts U (here the time horizon is $T = 1$). Actually a more general formula holds true, where the function f is allowed to depend on the whole trajectory of (B_t) rather than just B_1. More precisely, let

J. Lehec (✉)
Ceremade (UMR CNRS 7534), Université Paris-Dauphine, Place de Lattre de Tassigny, 75016 Paris, France
e-mail: lehec@ceremade.dauphine.fr

E. Carlen et al. (eds.), *Convexity and Concentration*, The IMA Volumes in Mathematics and its Applications 161, DOI 10.1007/978-1-4939-7005-6_9

$(\mathbb{W}, \mathcal{B}, \gamma)$ be the n–dimensional Wiener space: $\mathbb{W}$ is the space of continuous paths from $[0, T]$ to $\mathbb{R}^n$, $\mathcal{B}$ is the Borel σ–field associated to the topology given by the uniform convergence (uniform convergence on compact sets if $T = +\infty$) and γ is the law of the standard Brownian motion. If $F\colon \mathbb{W} \to \mathbb{R}$ is measurable and bounded from below, then

$$\log\left(\int_{\mathbb{W}} \mathrm{e}^F \, d\gamma\right) = \sup_U \left\{ \mathsf{E}\left[F(B+U) - \frac{1}{2}\|U\|_{\mathbb{H}}^2\right]\right\} \tag{2}$$

This is due to Boué and Dupuis [6]. Of course, applying this formula to a functional F of the form $F(w) = f(w_1)$ we recover Borell's formula (1).

Recall that if μ is a probability measure on $\mathbb{R}^n$ the relative entropy of μ with respect to the Gaussian measure γ_n is by definition

$$\mathrm{H}(\mu \mid \gamma_n) = \int_{\mathbb{R}^n} \log\left(\frac{d\mu}{d\gamma_n}\right) d\mu,$$

if μ is absolutely continuous with respect to γ_n and $\mathrm{H}(\mu \mid \gamma_n) = +\infty$ otherwise. It is well known that there is a convex duality between the Laplace transform and the relative entropy. Namely for every function f we have

$$\log\left(\int_{\mathbb{R}^n} \mathrm{e}^f \, d\gamma_n\right) = \sup_\mu \left\{ \int_{\mathbb{R}^n} f \, d\mu - \mathrm{H}(\mu \mid \gamma_n)\right\} \tag{3}$$

where the supremum is taken on every probability measure μ on $\mathbb{R}^n$. In our previous work [13], we established the following dual version of Borell's formula (1). If μ is a probability measure on $\mathbb{R}^n$ satisfying certain mild technical assumptions, then

$$\mathrm{H}(\mu \mid \gamma_n) = \inf \left\{ \frac{1}{2} \mathsf{E}\left[\|U\|_{\mathbb{H}}^2\right]\right\} \tag{4}$$

where the infimum is taken over all drifts U such that $B_1 + U_1$ has law μ. Informally this says that the minimal energy needed to constrain the Brownian motion to have a prescribed law at time 1 coincides with the relative entropy of this law with respect to γ_n. Note that combining this with the convex duality (3) we easily retrieve Borell's formula (1):

$$\begin{aligned} \log\left(\int_{\mathbb{R}^n} \mathrm{e}^f \, d\gamma_n\right) &= \sup_\mu \left\{ \int_{\mathbb{R}^n} f \, d\mu - \mathrm{H}(\mu \mid \gamma_n)\right\} \\ &= \sup_\mu \left\{ \int_{\mathbb{R}^n} f \, d\mu - \inf_{U : B_1 + U_1 \sim \mu} \left\{ \frac{1}{2} \mathsf{E}\left[\|U\|_{\mathbb{H}}^2\right]\right\}\right\} \\ &= \sup_U \left\{ \mathsf{E}\left[f(B_1 + U_1) - \frac{1}{2}\|U\|_{\mathbb{H}}^2\right]\right\}. \end{aligned}$$

The article [13] also contains a path space version of (4) allowing to recover the Boué–Dupuis formula (2), but we shall not spell this out here.

We are interested in the use of such formulas to prove functional inequalities. This was initiated by Borell in [5], in which he proved (1) and showed that it yields the Prékopa–Leindler inequality very easily. This was further developed in the author's works [13, 14] where many other functional inequalities were derived from (1), (2) and (4). The main purpose of the present article is to establish a version of Borell's formula (1) for the Brownian motion on a Riemannian manifold and to give a couple of applications, including a new proof of the Brascamp–Lieb inequality on the sphere of Carlen, Lieb and Loss. We also give a Riemannian version of the dual formula (4) and we apply it to recover the log–Sobolev inequality under a curvature condition.

2 Borell's Formula for a Diffusion

Let $\sigma\colon \mathbb{R}^n \to M_n(\mathbb{R})$, let $b\colon \mathbb{R}^n \to \mathbb{R}^n$ and assume that the stochastic differential equation

$$dX_t = \sigma(X_t)\, dB_t + b(X_t)\, dt \tag{5}$$

has a unique strong solution. Assume also for simplicity that the explosion time is $+\infty$. Then there exists a measurable functional

$$G\colon \mathbb{R}^n \times \mathbb{W} \to \mathbb{W}$$

(it is probably safer to complete the σ–field $\mathcal{B}$ at this stage) such that for every $x \in \mathbb{R}^n$ the process $X = G(x, B)$ is the unique solution of (5) starting from x. This hypothesis is satisfied in particular if σ and b are locally Lipschitz and grow at most linearly, see, for instance, [11, Chapter IV]. The process (X_t) is then a diffusion with generator L given by

$$Lf = \frac{1}{2}\langle \sigma\sigma^T, \nabla^2 f\rangle + \check{h}ab$$

for every $\mathcal{C}^2$–smooth function f. We denote the associated semigroup by (P_t): for any test function f

$$P_t f(x) = \mathsf{E}_x\left[f(X_t)\right],$$

where as usual the subscript x denotes the starting point of (X_t). Fix a finite time horizon T. Fix $x \in \mathbb{R}^n$, let $f\colon R^n \to \mathbb{R}$ and assume that f is bounded from below. Applying the representation formula (2) to the functional

$$F\colon w \in \mathbb{W} \mapsto f\left(G(x, w)_T\right)$$

we get

$$\log \int_{\mathbb{W}} e^{f(G(x,w)_T)}\, \gamma(dw) = \sup_U \left\{ \mathsf{E}\left[f\left(G(x, B+U)_T\right) - \frac{1}{2}\|U\|_{\mathbb{H}}^2 \right] \right\}$$

where the supremum is taken on all drifts U. By definition of the semigroup (P_t) we have

$$\log \int_{\mathbb{W}} e^{f(G(x,w)_T)}\, \gamma(dw) = \log P_T(e^f)(x).$$

Also, we have the following lemma.

Lemma 1 *Let $(U_t)_{t\leq T}$ be a drift. The process $X^U = G(x, B+U)$ is the unique process satisfying*

$$X_t^U = x + \int_0^t \sigma(X_s^U)\,(dB_s + dU_s) + \int_0^t b(X_s^U)\,ds, \quad t \leq T \tag{6}$$

almost surely.

Proof Assume first that $\|U\|_{\mathbb{H}}$ is bounded. Then by Novikov's criterion the process (D_t) given by

$$D_t = \exp\left(-\int_0^t \langle \dot{U}_s, dB_s \rangle - \frac{1}{2}\int_0^t |\dot{U}_s|^2\, ds \right)$$

is a uniformly integrable martingale. Moreover, according to Girsanov's formula, under the measure Q given by $d\mathsf{Q} = D_T\, d\mathsf{P}$, the process $B+U$ is a standard Brownian motion on $[0,T]$, see, for instance, [12, section 3.5] for more details. Now since the stochastic differential equation (5) is assumed to have a unique strong solution and by definition of G, almost surely for Q, the unique process satisfying (6) is $X^U = G(x, B+U)$. Since Q and P are equivalent this is the result. For general U, the result follows by applying the bounded case to $(U_t) = (U_{t\wedge T_n})$ where T_n is the stopping time

$$T_n = \inf\left\{ t \geq 0\colon \int_0^t |\dot{U}_s|^2\, ds \geq n \right\},$$

and letting n tend to $+\infty$. □

To sum up, we have established the following result.

Theorem 2 *For any function $f\colon \mathbb{R}^n \to \mathbb{R}$ bounded from below we have*

$$\log P_T(e^f)(x) = \sup_U \left\{ \mathsf{E}\left[f(X_T^U) - \frac{1}{2}\|U\|_{\mathbb{H}}^2 \right] \right\},$$

where the supremum is taken over all drifts U and the process X^U is the unique solution of (6).

Remarks This direct consequence of the representation formula (2) was already noted by Boué and Dupuis. They used it to recover Freidlin and Wentzell's large deviation principle as the diffusion coefficient tends to 0. Let us also note that the non-explosion hypothesis is not essential. One can consider $\mathbb{R}^n \cup \{\infty\}$, the one point compactification of $\mathbb{R}^n$, set $X_t = \infty$ after explosion time, and restrict to functions f that tend to 0 at infinity. In the same way we could also deal with a Dirichlet boundary condition.

3 Borell's Formula on a Riemannian Manifold

Let (M, g) be a complete Riemannian manifold of dimension n. In this section we wish to establish a Borell type formula for the Brownian motion on M. To do so we need to recall first the intrinsic construction of the Brownian motion on M.

Let us start with some definitions from Riemannian geometry. Recall that the orthonormal frame bundle $\mathcal{O}(M)$ is the set of $(n+1)$–tuples

$$\phi = (x, \phi^1, \dots, \phi^n)$$

where x is in M and $(\phi^1, \dots, \phi^n)$ is an orthonormal basis of $T_x(M)$. Given an element $\phi = (x, \phi^1, \dots, \phi^n)$ of $\mathcal{O}(M)$ and a vector $v \in T_x(M)$, the *horizontal lift* of v at ϕ, denoted $\mathcal{H}(v)$, is an element of $T_\phi(\mathcal{O}(M))$ defined as follows: Choose a curve (x_t) starting from x with speed v and for $i \le n$ let ϕ_t^i be the parallel translation of ϕ^i along (x_t). Since parallel translation preserves the inner product, we thus obtain a smooth curve (ϕ_t) on $\mathcal{O}(M)$ and we can set

$$\mathcal{H}(v) = \dot{\phi}_0.$$

This is a lift of v in the sense that for any smooth f on M

$$\mathcal{H}(v)(f \circ \pi) = v(f),$$

where $\pi : \mathcal{O}(M) \to M$ is the canonical projection. Now for $i \le n$ we define a vector field on $\mathcal{O}(M)$ by setting

$$\mathcal{H}^i(x, \phi^1, \dots, \phi^n) = \mathcal{H}(\phi^i).$$

The operator

$$\Delta_{\mathcal{H}} = \sum_{i=1}^n (\mathcal{H}^i)^2$$

is called the *horizontal Laplacian*. It is related to the Laplace–Beltrami operator on M, denoted Δ, through the following commutation property: for any smooth f on M we have

$$\Delta_{\mathcal{H}}(f \circ \pi) = \Delta(f) \circ \pi. \tag{7}$$

Note that the horizontal Laplacian is by definition a sum of squares of vector fields, and that this is typically not the case for the Laplace–Beltrami operator. We are now in a position to define the horizontal Brownian motion on $\mathcal{O}(M)$. Let

$$B_t = (B_t^1, \dots, B_t^n)$$

be a standard n–dimensional Brownian motion. We consider the following stochastic differential equation on $\mathcal{O}(M)$

$$d\Phi_t = \sum_{i=1}^n \mathcal{H}^i(\Phi_t) \circ dB_t^i. \tag{8}$$

Throughout the rest of the article, the notation $H \circ dM$ denotes the Stratonovitch integral. The equation (8) is a short way of saying that for any smooth function g on $\mathcal{O}(M)$ we have

$$g(\Phi_t) = g(\Phi_0) + \sum_{i=1}^n \int_0^t \mathcal{H}^i(g)(\Phi_t) \circ dB_t^i.$$

This always has a strong solution, see [11, Theorem V.1.1.]. Let us assume additionally that it does not explode in finite time. This is the case in particular if the Ricci curvature of M is bounded from below, see, for instance, [10, section 4.2], where a more precise criterion is given. Translating the equation above in terms of Itô increments we easily get

$$dg(\Phi_t) = \sum_{i=1}^n \mathcal{H}^i(g)(\Phi_t)\, dB_t^i + \frac{1}{2}\Delta_{\mathcal{H}} g(\Phi_t)\, dt.$$

Let (X_t) be the process given by $X_t = \pi(\Phi_t)$. Applying the previous formula and (7) we see that for any smooth f on M

$$df(X_t) = \sum_{i=1}^n \Phi_t^i(f)(X_t)\, dB_t^i + \frac{1}{2}\Delta f(X_t)\, dt. \tag{9}$$

In particular

$$f(X_t) - \frac{1}{2}\int_0^t \Delta f(X_s)\, ds, \quad t \geq 0$$

is a local martingale. This shows that (X_t) is a Brownian motion on M. The process (X_t) is called the *stochastic development* of (B_t). In the sequel, it will be convenient to identify the orthogonal basis $\Phi_t^1, \dots, \Phi_t^n$ with the orthogonal map

$$x \in \mathbb{R}^n \to \sum_{i=1}^{n} x_i \Phi_t^i \in T_{X_t}(M).$$

Then the equation (9) can be rewritten

$$df(X_t) = \langle \nabla f(X_t), \Phi_t \, dB_t \rangle + \frac{1}{2} \Delta f(X_t) \, dt.$$

Similarly the equation (8) can be rewritten in a more concise form

$$d\Phi_t = \mathcal{H}(\Phi_t) \circ dB_t. \tag{10}$$

To sum up, the process (Φ_t) is an orthonormal basis above (X_t) which is used to map the Brownian increment dB_t from $\mathbb{R}^n$ to the tangent space of M at X_t.
Now we establish a Borell type formula for the process (X_t). We know that there exists a measurable functional

$$G: \mathcal{O}(M) \times \mathbb{W} \to \mathcal{C}(\mathbb{R}, \mathcal{O}(M))$$

such that the process $\Phi = G(\phi, B)$ is the unique solution of (10) starting from ϕ.
Let $\phi \in \mathcal{O}(M)$, let $T > 0$, let $f: M \to \mathbb{R}$ and assume that f is bounded from below. Applying the representation formula (2) to the functional

$$F: w \in \mathbb{W} \mapsto f \circ \pi \, (G(\phi, w)_T)$$

we get

$$\log \left(\int_{\mathbb{W}} e^{f \circ \pi (G(\phi, B)_T)} \, d\gamma \right) = \sup_U \left\{ \mathsf{E} \left[f \circ \pi (G(\phi, B + U)_T) - \frac{1}{2} \|U\|_{\mathbb{H}}^2 \right] \right\}.$$

Let $x = \pi(\phi)$. Since $\pi(G(\phi, B))$ is a Brownian motion on M starting from x we have

$$\log \left(\int_{\mathbb{W}} e^{f \circ \pi (G(\phi, B)_T)} \, d\gamma \right) = \log P_T(e^f)(x),$$

where (P_t) is the heat semigroup on M. Also, letting $\Phi^U = G(\phi, B + U)$ and reasoning along the same lines as in the proof of Lemma 1, we obtain that Φ^U is the only solution to

$$d\Phi_t^U = \mathcal{H}(\Phi_t^U) \circ (dB_t + dU_t)$$

starting from ϕ. We also let $X^U = \pi(\Phi^U)$ and call this process the stochastic development of $B+U$ starting from ϕ. To sum up, we have established the following result.

Theorem 3 *Let $f: M \to \mathbb{R}$, let $\phi \in \mathcal{O}(M)$, let $x = \pi(\phi)$ and let $T > 0$. If f is bounded from below, then*

$$\log P_T\left(\mathrm{e}^f\right)(x) = \sup_U \left\{ \mathsf{E}\left[f(X_T^U) - \frac{1}{2}\|U\|_{\mathbb{H}}^2 \right] \right\},$$

where the supremum is taken on all drifts U and where given a drift U, the process X^U is the stochastic development of $B + U$ starting from ϕ.

4 Brascamp–Lieb Inequality on the Sphere

In the article [14], we explained how to derive the Brascamp–Lieb inequality and its reversed version from Borell's formula. In this section we extend this to the sphere, and give a proof based on Theorem 3 of the spherical version of the Brascamp–Lieb inequality, due to Carlen, Lieb and Loss in [7].

Theorem 4 *Let $g_1, \ldots, g_{n+1}$ be non–negative functions on the interval $[-1, 1]$. Let σ_n be the Haar measure on the sphere $\mathbb{S}^n$, normalized to be a probability measure. We have*

$$\int_{\mathbb{S}^n} \prod_{i=1}^{n+1} g_i(x_i)\, \sigma_n(dx) \leq \prod_{i=1}^{n+1} \left(\int_{\mathbb{S}^n} g_i(x_i)^2\, \sigma_n(dx) \right)^{1/2}$$

Remark The inequality does not hold if we replace the L^2 norm in the right–hand side by a smaller L^p norm, like the L^1 norm. Somehow this 2 accounts for the fact that the coordinates of a uniform random vector on $\mathbb{S}^n$ are not independent. We refer to the introduction of [7] for a deeper insight on this inequality.

In addition to the Borell type formulas established in the previous two sections, our proof relies on a sole inequality, spelled out in the lemma below. Let $P_i: \mathbb{S}^n \to [-1; 1]$ be the application that maps x to its i–th coordinate x_i. The spherical gradient of P_i at x is the projection of the coordinate vector e_i onto $x^\perp$:

$$\nabla P_i(x) = e_i - x_i x.$$

Lemma 5 *Let $x \in \mathbb{S}^n$ and let $y \in x^\perp$. For $i \leq n + 1$, if $\nabla P_i(x) \neq 0$ let*

$$\theta^i = \frac{\nabla P_i(x)}{|\nabla P_i(x)|}$$

and let θ^i be an arbitrary unit vector of $x^\perp$ otherwise. Then for any $y \in x^\perp$ we have

$$\sum_{i=1}^{n+1} \langle \theta^i, y \rangle^2 \leq 2|y|^2.$$

Proof Assume first that $\nabla P_i(x) \neq 0$ for every i. Since $\nabla P_i(x) = e_i - x_i x$ and y is orthogonal to x we then have

$$\langle \theta^i, y \rangle^2 = y_i^2 + x_i^2 \langle \theta^i, y \rangle^2 \leq y_i^2 + x_i^2 |y|^2.$$

Summing this over $i \leq n + 1$ yields the result. On the other hand, if there exists i such that $\nabla P_i(x) = 0$ then $x = \pm e_i$ and it is almost immediate to check that the desired inequality holds true. □

Proof of Theorem 4. Let us start by describing the behaviour of a given coordinate of a Brownian motion on $\mathbb{S}^n$. Let (B_t) be a standard Brownian motion on $\mathbb{R}^n$, let ϕ be a fixed element of $\mathcal{O}(\mathbb{S}^n)$ and let (Φ_t) be the horizontal Brownian motion given by

$$\Phi_0 = \phi \quad \text{and} \quad d\Phi_t = \mathcal{H}(\Phi_t) \circ dB_t.$$

We also let $X_t = \pi(\Phi_t)$ be the stochastic development of (B_t) and $X_t^i = P_i(X_t)$, for every $i \leq n + 1$. We have

$$dX_t^i = \langle \nabla P_i(X_t), \Phi_t dB_t \rangle + \frac{1}{2} \Delta P_i(X_t)\, dt. \tag{11}$$

Let θ be an arbitrary unit vector of $\mathbb{R}^n$ and define a process (θ_t^i) by

$$\theta_t^i = \begin{cases} \Phi_t^* \left(\frac{\nabla P_i(X_t)}{|\nabla P_i(X_t)|} \right), & \text{if } \nabla P_i(X_t) \neq 0, \\ \theta & \text{otherwise.} \end{cases}$$

Since Φ_t is an orthogonal map θ_t belongs to the unit sphere of $\mathbb{R}^n$. Consequently the process (W_t^i) defined by $dW_t^i = \langle \theta_t^i, dW_t \rangle$ is a one dimensional standard Brownian motion. Observe that $|\nabla P_i| = (1 - P_i^2)^{1/2}$ and recall that P_i is an eigenfunction for the spherical Laplacian: $\Delta P_i = -nP_i$. Equality (11) becomes

$$dX_t^i = \left(1 - (X_t^i)^2\right)^{1/2} dW_t^i - \frac{n}{2} X_t^i\, dt.$$

This stochastic differential equation is usually referred to as the Jacobi diffusion in the literature, see, for instance, [2, section 2.7.4]. What matters for us is that it does possess a unique strong solution. Indeed the drift term is linear and although the diffusion factor $(1 - x^2)^{1/2}$ is not locally Lipschitz, it is Hölder continuous with exponent $1/2$, which is sufficient to insure strong uniqueness in dimension 1, see [11, Chapter 4, Theorem 3.2]. Let (Q_t) be the semigroup associated to the process (X_t^i). The stationary measure ν_n is easily seen to be given by

$$\nu_n(dt) = c_n \mathbb{1}_{[-1,1]}(t)(1-t^2)^{\frac{n}{2}-1}\, dt,$$

where c_n is the normalization constant. Obviously ν_n coincides with the pushforward of σ_n by P_i.

We now turn to the actual proof of the theorem. Let $g_1, \dots, g_{n+1}$ be non-negative functions on $[-1,1]$ and assume (without loss of generality) that they are bounded away from 0. Let $f_i = \log(g_i)$ for all i and let

$$f \colon x \in \mathbb{S}^n \mapsto \sum_{i=1}^{n+1} f_i(x_i).$$

The functions f_i, f are bounded from below. Fix a time horizon T, let U be a drift and let (Φ_t^U) be the process given by

$$\begin{cases} \Phi_0^U = \phi \\ d\Phi_t^U = \mathcal{H}(\Phi_t^U) \circ (dB_t + dU_t), \quad t \leq T. \end{cases}$$

We also let $X_t^U = \pi(\Phi_t^U)$ be the stochastic development of $B + U$. These processes are well defined by the results of the previous section. We want to bound $f(X_T^U) - \frac{1}{2}\|U\|_{\mathbb{H}}^2$ from above. By definition

$$f(X_T^U) = \sum_{i=1}^{n+1} f_i(P_i X_T^U)).$$

Let (θ_t^i) be the process given by

$$\theta_t^i = \Phi_t^* \left(\frac{\nabla P_i(X_t^U)}{|\nabla P_i(X_t^U)|} \right)$$

(again replace this by an arbitrary fixed unit vector if $\nabla P_i(X_t^U) = 0$). Then let (W_t^i) be the one dimensional Brownian motion given by $dW_t^i = \langle \theta_t^i, dW_t \rangle$ and let (U_t^i) be the one dimensional drift given by $dU_t^i = \langle \theta_t^i, dU_t \rangle$. The process $(P_i(X_t^U))$ then satisfies

$$dP_i(X_t^U) = \left(1 - P_i(X_t^U)^2\right)^{1/2} \left(dW_t^i + dU_t^i\right) - \frac{n}{2} P_i(X_t^U)\, dt. \tag{12}$$

Applying Lemma 5, we easily get

$$\sum_{i=1}^{n+1} \|U^i\|_{\mathbb{H}}^2 \leq 2\|U\|_{\mathbb{H}}^2,$$

almost surely (note that in the left-hand side of the inequality $\mathbb{H}$ is the Cameron–Martin space of $\mathbb{R}$ rather than $\mathbb{R}^n$). Therefore

$$f(X_T^U) - \frac{1}{2}\|U\|_{\mathbb{H}}^2 \leq \sum_{i=1}^{n+1} \left(f_i(P_i(X_T^U)) - \frac{1}{4}\|U^i\|_{\mathbb{H}}^2 \right). \tag{13}$$

Recall (12) and apply Theorem 2 to the semigroup (Q_t) and to the function $2f_i$ rather than f_i. This gives

$$\mathsf{E}\left[f_i(P_i(X_t^U)) - \frac{1}{4}\|U^i\|_{\mathbb{H}}^2 \right] \leq \frac{1}{2} \log Q_T(\mathrm{e}^{2f_i})(x_i),$$

for every $i \leq n+1$. Taking expectation in (13) thus yields

$$\mathsf{E}\left[f(X_T^U) - \frac{1}{2}\|U\|_{\mathbb{H}}^2 \right] \leq \frac{1}{2} \sum_{i=1}^{n+1} \log Q_T\left(\mathrm{e}^{2f_i}\right)(x_i).$$

Taking the supremum over all drifts U and using Theorem 3 we finally obtain

$$P_T(\mathrm{e}^f)(x) \leq \prod_{i=1}^{n+1} \left(Q_T(\mathrm{e}^{2f_i})(x_i) \right)^{1/2}.$$

The semigroup (P_t) is ergodic and converges to σ_n as t tends to $+\infty$. Similarly (Q_t) converges to ν_n. So letting T tend to $+\infty$ in the previous inequality gives

$$\int_{\mathbb{S}^n} \mathrm{e}^f \, d\sigma_n \leq \prod_{i=1}^{n+1} \left(\int_{[-1,1]} \mathrm{e}^{2f_i} \, d\nu_n \right)^{1/2},$$

which is the result. □

Remark Barthe, Cordero–Erausquin and Maurey in [3] and together with Ledoux in [4] gave several extensions of Theorem 4. The method exposed here also allows to recover most of their results. We chose to stick to the original statement of Carlen, Lieb and Loss for simplicity.

5 The Dual Formula

Recall the dual version of Borell's formula: If μ is an absolutely continuous measure on $\mathbb{R}^n$ satisfying some (reasonable) technical assumptions, then

$$\mathrm{H}(\mu \mid \gamma_n) = \inf\left\{\frac{1}{2}\mathsf{E}\left[\|U\|_{\mathbb{H}}^2\right]\right\} \tag{14}$$

where the infimum is taken over all drifts U such that $B_1 + U_1$ has law μ. Moreover the infimum is attained and the optimal drift can be described as follows. Let f be the density of μ with respect to γ_n and (P_t) be the heat semigroup on $\mathbb{R}^n$. The following stochastic differential equation

$$\begin{cases} X_0 = 0 \\ dX_t = dB_t + \nabla \log P_{1-t}f(X_t)\, dt, \quad t \leq 1. \end{cases}$$

has a unique strong solution. The solution satisfies $X_1 = \mu$ in law and is optimal in the sense that there is equality in (14) for the drift U given by $\dot{U}_t = \nabla \log P_{1-t}f(X_t)$. We refer to [13] for a proof of these claims. The purpose of this section is to generalize this to the case of a Brownian motion on a Riemannian manifold. Diffusions such as the ones considered in section 2 could be treated in a very similar way but we shall omit this here.

The setting of this section is thus the same as that of section 3: (M, g) is a complete Riemannian manifold of dimension n whose Ricci curvature is bounded from below and (B_t) is a standard Brownian motion on $\mathbb{R}^n$. We denote the heat semigroup on M by (P_t).

Theorem 6 *Fix $x \in M$ and a time horizon T. Let μ be a probability measure on M, assume that μ is absolutely continuous with respect to $\delta_x P_T$ and let f be its density. If f is Lipschitz and bounded away from 0, then*

$$\mathrm{H}(\mu \mid \delta_x P_T) = \inf\left\{\frac{1}{2}\mathsf{E}\left[\|U\|_{\mathbb{H}}^2\right]\right\}$$

where the infimum is taken on all drifts U such that the stochastic development of $B + U$ starting from x has law μ at time T.

Proving that any drift satisfying the constraint has energy at least as large as the relative entropy of μ is a straightforward adaptation of Proposition 1 from [13], and we shall leave this to the reader. Alternatively, one can use Theorem 3 and combine it with the following variational formula for the entropy:

$$\mathrm{H}(\mu \mid \delta_x P_T) = \sup_f \left\{\int_M f\, d\mu - \log P_T(\mathrm{e}^f)(x)\right\}.$$

Moreover, as in the Euclidean case, there is actually an optimal drift, whose energy is exactly the relative entropy of μ. This is the purpose of the next result.

Theorem 7 *Let ϕ be a fixed element of $\mathcal{O}(M)$. Let $x = \pi(\phi)$ and let T be a time horizon. Let μ have density f with respect to $\delta_x P_T$, and assume that f is Lipschitz and bounded away from 0. The stochastic differential equation*

$$\begin{cases} \Phi_0 = \phi \\ d\Phi_t = \mathcal{H}(\Phi_t) \circ \big(dB_t + \Phi_t^* \nabla \log P_{T-t} f(Y_t)\big) \\ Y_t = \pi(\Phi_t) \end{cases} \tag{15}$$

has a unique strong solution on $[0, T]$. The law of the process (Y_t) is given by the following formula: For every functional $H : \mathcal{C}([0, T]; M) \to \mathbb{R}$ we have

$$\mathsf{E}\,[H(Y)] = \mathsf{E}\,[H(X) f(X_1)]\,, \tag{16}$$

where (X_t) is a Brownian motion on M starting from x. Moreover, letting U be the drift given by

$$U_t = \int_0^t \Phi_s^* \nabla \log P_{T-s} f(Y_s)\, ds, \quad t \le T, \tag{17}$$

we have

$$\mathrm{H}(\mu \mid \delta_x P_T) = \frac{1}{2} \mathsf{E}\left[\|U\|_{\mathbb{H}}^2\right].$$

Remark Equality (16) can be reformulated as: The vector Y_T has law μ and the conditional laws of X and Y given their endpoints coincide. So in some sense Y is as close to being a Brownian motion as possible while having law μ at time T.

Proof Since $\mathrm{Ric} \ge -\lambda\, g$, we have the following estimate for the Lipschitz norm of f:

$$\|P_t f\|_{\mathrm{Lip}} \le \mathrm{e}^{\lambda t/2} \|f\|_{\mathrm{Lip}}.$$

One way to see this is to use Kendall's coupling for Brownian motions on a manifold, see, for instance, [10, section 6.5]. Alternatively, it is easily derived from the commutation property $|\nabla P_t f|^2 \le \mathrm{e}^{\lambda t} P_t(|\nabla f|^2)$ which, in turn, follows from Bochner's formula, see [2, Theorem 3.2.3]. Recall that f is assumed to be bounded away from 0, and for every $t \le T$ let $F_t = \log P_{T-t} f$. Then $(t, x) \mapsto \nabla F_t(x)$ is smooth and bounded on $[0, T[\times M$, which is enough to insure the existence of a unique strong solution to (15). Besides an easy computation shows that

$$\partial_t F_t = -\frac{1}{2}(\Delta F_t + |\nabla F_t|^2). \tag{18}$$

Then using (15) and Itô's formula we get

$$\begin{aligned} dF_t(Y_t) &= \langle \nabla F_t(Y_t), \Phi_t dB_t \rangle + \frac{1}{2} |\nabla F_t(Y_t)|^2 \, dt. \\ &= \langle \dot{U}_t, dB_t \rangle + \frac{1}{2} |\dot{U}_t|^2 \, dt \end{aligned}$$

(recall the definition (17) of U). Therefore

$$\frac{1}{f(Y_T)} = \mathrm{e}^{-F_T(Y_T)} = \exp\left(- \int_0^T \langle \dot{U}_t, dB_t \rangle - \frac{1}{2} \|U\|_{\mathbb{H}}^2 \right). \tag{19}$$

Observe that the variable $\|U\|_{\mathbb{H}}$ is bounded (just because ∇F is bounded). So Girsanov's formula applies: $1/f(Y_T)$ has expectation 1 and under the measure Q given by $d\mathsf{Q} = (1/f(Y_T))\, d\mathsf{P}$ the process $B + U$ is a standard Brownian motion on $\mathbb{R}^n$. Since Y is the stochastic development of $B+U$ starting from x, this shows that under Q the process Y is a Brownian motion on M starting from x. This is a mere reformulation of (16). For the entropy equality observe that since Y_T has law μ, we have

$$\mathrm{H}(\mu \mid \delta_x P_T) = \mathsf{E}[\log f(Y_T)].$$

Using (19) again and the fact that $\int \langle \dot{U}_t, dB_t \rangle$ is a martingale we get the desired equality. □

To conclude this article, let us derive from this formula the log–Sobolev inequality for a manifold having a positive lower bound on its Ricci curvature. This is of course well known, but our point is only to illustrate how the previous theorem can be used to prove inequalities. Recall the definition of the Fisher information: if μ is a probability measure on M having Lipschitz and positive density f with respect to some reference measure m, the relative Fisher information of μ with respect to m is defined by

$$\mathrm{I}(\mu \mid m) = \int_M \frac{|\nabla f|^2}{f} \, dm = \int_M |\nabla \log f|^2 \, d\mu.$$

By Bishop's Theorem, if $\mathrm{Ric} \geq \kappa\, g$ pointwise for some positive κ, then the volume measure on M is finite. We let m be the volume measure normalized to be a probability measure.

Theorem 8 *If* $\mathrm{Ric} \geq \kappa\, g$ *pointwise for some* $\kappa > 0$, *then for any probability measure* μ *on* M *having a Lipschitz and positive density with respect to* m *we have*

$$\mathrm{H}(\mu \mid m) \leq \frac{n}{2} \log\left(1 + \frac{\mathrm{I}(\mu \mid m)}{n\kappa} \right). \tag{20}$$

Remarks Since $\log(1+x) \leq x$ this inequality is a dimensional improvement of the more familiar inequality

$$\mathrm{H}(\mu \mid m) \leq \frac{1}{2\kappa}\, \mathrm{I}(\mu \mid m). \tag{21}$$

Theorem 8 is not new, it is due to Bakry, see [1, Corollaire 6.8]. Note that in the same survey (Proposition 6.6) he obtains another sharp form of (21) that takes the dimension into account, namely

$$\mathrm{H}(\mu \mid m) \leq \frac{n-1}{2\kappa\, n}\, \mathrm{I}(\mu \mid m). \tag{22}$$

Unfortunately we were not able to recover this one with our method. Note also that depending on the measure μ, the right-hand side of (22) can be smaller than that of (20), or the other way around.

Proof By the Bonnet–Myers theorem M is compact. Fix $x \in M$ and a time horizon T. Let $p_T(x,\cdot)$ be the density of the measure $\delta_x P_T$ with respect to m (in other words let (p_t) be the heat kernel on M). If $d\mu = \rho\, dm$, then μ has density $f = \rho / p_T(x,\cdot)$ with respect to $\delta_x P_T$. Since $p_T(x,\cdot)$ is smooth and positive (see, for instance, [8, chapter 6]) f satisfies the technical assumptions of the previous theorem. Let $F_t = \log P_{T-t} f$ and let (Y_t) be the process given by (15). We know from the previous theorem that

$$\mathrm{H}(\mu \mid \delta_x P_T) = \frac{1}{2}\,\mathsf{E}\left[\int_0^T |\nabla F_t(Y_t)|^2\, dt\right]. \tag{23}$$

Using (18) we easily get

$$\partial_t(|\nabla F|^2) = -\langle \nabla \Delta F, \nabla F\rangle - \langle \nabla |\nabla F|^2, \nabla F\rangle.$$

Applying Itô's formula we obtain after some computations (omitting variables in the right-hand side)

$$d|\nabla F(t,Y_t)|^2 = \langle \nabla |\nabla F|^2, \Phi_t dB_t\rangle - \langle \nabla \Delta F, \nabla F\rangle\, dt + \frac{1}{2}\Delta |\nabla F|^2\, dt.$$

Now recall Bochner's formula

$$\frac{1}{2}\Delta |\nabla F|^2 = \langle \nabla \Delta F, \nabla F\rangle + \|\nabla^2 F\|_{HS}^2 + \mathrm{Ric}(\nabla F, \nabla F).$$

So that

$$d|\nabla F(t,Y_t)|^2 = \langle \nabla |\nabla F|^2, \Phi_t dB_t\rangle + \|\nabla^2 F\|_{HS}^2\, dt + \mathrm{Ric}(\nabla F, \nabla F)\, dt. \tag{24}$$

Since ∇F is bounded and the Ricci curvature non-negative, the local martingale part in the above equation is bounded from above. So by Fatou's lemma it is a sub–martingale, and its expectation is non-decreasing. So taking expectation in (24) and using the hypothesis $\mathrm{Ric} \geq \kappa\, g$ we get

$$\frac{d}{dt}\mathsf{E}\left[|\nabla F_t(Y_t)|^2\right] \geq \mathsf{E}\left[\|\nabla^2 F_t(Y_t)\|_{HS}^2\right] + \kappa\,\mathsf{E}\left[|\nabla F_t(Y_t)|^2\right]. \tag{25}$$

Throwing away the Hessian term would lead us to the inequality (21). Let us exploit this term instead. Using Cauchy–Schwarz and Jensen's inequalities we get

$$\mathsf{E}\left[\|\nabla^2 F_t(Y_t)\|_{HS}^2\right] \geq \frac{1}{n}\mathsf{E}\left[\Delta F_t(Y_t)^2\right] \geq \frac{1}{n}\,\mathsf{E}[\Delta F_t(Y_t)]^2.$$

Also, by (16) and recalling that (P_t) is the heat semigroup on M we obtain

$$\begin{aligned}\mathsf{E}\left[\Delta F_t(Y_t) + |\nabla F_t(Y_t)|^2\right] &= \mathsf{E}\left[\frac{\Delta P_{T-t}f(Y_t)}{P_{T-t}f(Y_t)}\right] = \mathsf{E}\left[\frac{\Delta P_{T-t}f(X_t)}{P_{T-t}f(X_t)}f(X_T)\right]\\ &= \mathsf{E}\left[\Delta P_{T-t}f(X_t)\right] = \Delta P_T(f)(x).\end{aligned}$$

Letting $\alpha(t) = \mathsf{E}\left[|\nabla F_t(Y_t)|^2\right]$ and $C_T = \Delta P_T f(x)$ we thus get from (25)

$$\begin{aligned}\alpha'(t) &\geq \kappa\alpha(t) + \frac{1}{n}(\alpha(t) - C_T)^2\\ &\geq \frac{1}{n}\,\alpha(t)\,(\alpha(t) + n\,\kappa - 2C_T)\end{aligned}$$

Since $P_T f$ tends to a constant function as T tends to $+\infty$, C_T tends to 0. So if T is large enough $n\kappa - 2C_T$ is positive and the differential inequality above yields

$$\alpha(t) \leq \frac{n\,\kappa(T)\,\alpha(T)}{\mathrm{e}^{\kappa(T)(T-t)}\,(n\,\kappa(T) + \alpha(T)) - \alpha(T)}, \quad t \leq T$$

where $\kappa(T) = \kappa - 2C_T/n$. Integrating this between 0 and T we get

$$\int_0^T \alpha(t)\,dt \leq n\,\log\left(1 + \frac{\alpha(T)(1 - \mathrm{e}^{-\kappa(T)T})}{n\,\kappa(T)}\right). \tag{26}$$

Observe that $\kappa(T) \to \kappa$ as T tends to $+\infty$. By (23) and since $(\delta_x P_t)$ converges to m measure on M we have

$$\int_0^T \alpha(t)\,dt \to 2\,\mathrm{H}(\mu \mid m),$$

as T tends to $+\infty$. Also, since Y_T has law μ

$$\alpha(T) = \mathsf{E}\left[|\nabla \log f(Y_T)|^2\right] = \mathrm{I}(\mu \mid \delta_x P_T) \to \mathrm{I}(\mu \mid m).$$

Therefore, letting T tend to $+\infty$ in (26) yields

$$\mathrm{H}(\mu \mid m) \leq \frac{n}{2} \log\left(1 + \frac{\mathrm{I}(\mu \mid m)}{n\kappa}\right),$$

which is the result. □

Let us give an open problem to finish this article. We already mentioned that Borell recovered the Prékopa–Leindler inequality from (1). It is natural to ask whether there is probabilistic proof of the Riemannian Prékopa–Leindler inequality of Cordero, McCann and Schmuckenschlger [9] based on Theorem 3. Copying naively Borell's argument, we soon face the following difficulty: If X and Y are two Brownian motions on a manifold coupled by parallel transport, then unless the manifold is flat, the midpoint of X and Y is not a Brownian motion. We believe that there is a way around this but we could not find it so far.

References

[1] D. Bakry. L'hypercontractivité et son utilisation en théorie des semigroupes. (French) [Hypercontractivity and its use in semigroup theory] Lectures on probability theory (Saint-Flour, 1992), 1–114, Lecture Notes in Math., 1581, Springer, Berlin, 1994.

[2] D. Bakry, I. Gentil and M. Ledoux. Analysis and geometry of Markov diffusion operators. Grundlehren der Mathematischen Wissenschaften [Fundamental Principles of Mathematical Sciences], 348. Springer, Cham, 2014.

[3] F. Barthe, D. Cordero–Erausquin and B. Maurey. *Entropy of spherical marginals and related inequalities.* J. Math. Pures Appl. (9) 86 (2006), no. 2, 89–99.

[4] F. Barthe, D. Cordero–Erausquin, M. Ledoux and B. Maurey. *Correlation and Brascamp–Lieb inequalities for Markov semigroups.* Int. Math. Res. Not. IMRN 2011, no. 10, 2177–2216.

[5] C. Borell. *Diffusion equations and geometric inequalities*, Potential Anal. 12 (2000), no 1, 49–71.

[6] M. Boué and P. Dupuis. *A variational representation for certain functionals of Brownian motion.* Ann. Probab. 26 (1998), no 4, 1641–1659.

[7] E.A. Carlen, E.H. Lieb and M. Loss. *A sharp analog of Young's inequality on* $\mathbb{S}^N$ *and related entropy inequalities.* J. Geom. Anal. 14 (2004), no. 3, 487–520.

[8] I. Chavel. Eigenvalues in Riemannian geometry. Including a chapter by Burton Randol. With an appendix by Jozef Dodziuk. Pure and Applied Mathematics, 115. Academic Press, Inc., Orlando, FL, 1984.

[9] D. Cordero–Erausquin, R.J McCann and M. Schmuckenschlger. *A Riemannian interpolation inequality à la Borell, Brascamp and Lieb.* Invent. Math. 146 (2001), no. 2, 219–257.

[10] E.P. Hsu. Stochastic analysis on manifolds. Graduate Studies in Mathematics, 38. American Mathematical Society, Providence, RI, 2002.

[11] N. Ikeda and S. Watanabe. Stochastic differential equations and diffusion processes. Second edition. North–Holland Mathematical Library, 24. North–Holland Publishing Co., Amsterdam; Kodansha, Ltd., Tokyo, 1989.

[12] I. Karatzas and S.E. Shreve. Brownian motion and stochastic calculus. Second edition. Graduate Texts in Mathematics, 113. Springer–Verlag, New York, 1991.

[13] J. Lehec, *Representation formula for the entropy and functional inequalities*. Ann. Inst. Henri Poincaré Probab. Stat. 49 (2013), no. 3, 885–899.

[14] J. Lehec, *Short probabilistic proof of the Brascamp–Lieb and Barthe theorems*. Canad. Math. Bull. 57 (2014), no. 3, 585–597.

Fourth Moments and Products: Unified Estimates

Ivan Nourdin and Giovanni Peccati

Abstract We provide a unified discussion, based on the properties of eigenfunctions of the generator of the Ornstein-Uhlenbeck semigroup, of *quantitative fourth moment theorems* and of the *weak Gaussian product conjecture*. In particular, our approach illustrates the connections between moment estimates for non-linear functionals of Gaussian fields, and the general semigroup approach towards fourth moment theorems, recently initiated by Ledoux and further investigated by Poly et al.

Keywords Fourth moment theorem • Gaussian fields • Gaussian vectors • Moment inequalities • Ornstein-Uhlenbeck semigroup • Polarization conjecture • Probabilistic approximations • Variance inequalities • Wiener chaos

1 Introduction

1.1 Overview

The concept of a *fourth moment theorem* for sequences of non-linear functionals of a Gaussian field has been first introduced in [14] (see also [13]), where it was proved that, if $\{F_n : n \geqslant 1\}$ is a sequence of normalized random variables living in a fixed Wiener chaos of a Gaussian field, then F_n converges in distribution towards a Gaussian random variable if and only if the fourth moment of F_n converges to 3 (note that the value 3 equals the fourth moment of a standard centered Gaussian random variable). A quantitative version of such a result was then obtained in [10], by combining the Malliavin calculus of variations with the so-called Stein's method for normal approximations (see Section 3 below, as well as [11, Chapter 5]). Since then, the results from [10, 14] have been the seeds of a large number of applications and generalizations to many fields, ranging from non-commutative probability, to

I. Nourdin (✉) • G. Peccati
Université du Luxembourg, Esch-sur-Alzette, Luxembourg
e-mail: ivan.nourdin@uni.lu

E. Carlen et al. (eds.), *Convexity and Concentration*, The IMA Volumes
in Mathematics and its Applications 161, DOI 10.1007/978-1-4939-7005-6_10

random fields on homogeneous spaces, stochastic geometry, and computer science. See the constantly updated webpage [17] for an overview of this area.

A particularly fruitful direction of research has been recently opened in the references [1, 2, 7], where these results have been generalized to the framework of sequences of random variables that are eigenfunctions of a diffusive Markov semi-group generator, thus paving the way for novel functional estimates connected to logarithmic Sobolev and transport inequalities—see [8, 12]. An in-depth discussion of some recent developments in this direction can be found, e.g., in [5].

The aim of this paper is to provide a unified discussion of the main inequalities leading to the results of [11, 14], and of one of the crucial estimates used in [9] in order to prove the following remarkable estimate: let γ_n denote the n-dimensional standard Gaussian measure, and let $F_1, \ldots, F_d \in L^2(\gamma_n)$ be such that each F_i is an eigenfunction of the associated generator $\mathcal{L}$ of the Ornstein-Uhlenbeck semigroup. One then has

$$\int_{\mathbb{R}^n} (F_1 \cdots F_d)^2 \, d\gamma_n \geqslant \prod_{i=1}^{d} \int_{\mathbb{R}^n} F_i^2 \, d\gamma_n. \tag{1.1}$$

As discussed in [9], relation (1.1) represents a weak form of the (still open) *Gaussian product conjecture* discussed in [6], that would in turn imply a complete solution of the (still unsolved) *real polarization problem* introduced in [3, 16]. We recall that the Gaussian product conjecture (yet unproved, we stress, even in the case $d = 3$) states that, for every collection of linear mappings $l_i : \mathbb{R}^n \to \mathbb{R}$, $i = 1, \ldots, d$, and for every positive integer $k \geqslant 1$,

$$\int_{\mathbb{R}^n} l_1^{2k} \cdots l_d^{2k} d\gamma_n \geqslant \int_{\mathbb{R}^n} l_1^{2k} d\gamma_n \cdots \int_{\mathbb{R}^n} l_d^{2k} d\gamma_n \tag{1.2}$$

As we will see below, the main estimates discussed in this paper rely on the fundamental property that, if the random variable F_i is an eigenfunction of $\mathcal{L}$ with eigenvalue $-k_i$, $i = 1, \ldots, d$, then the product $F_1 \cdots F_d$ is an element of the direct sum of the first $r := k_1 + \cdots + k_d$ eigenspaces of $\mathcal{L}$, that is:

$$F_1 \cdots F_d \in \bigoplus_{k=0}^{r} \mathrm{Ker}(\mathcal{L} + kI),$$

where I is the identity operator. This property of eigenfunctions (which can be encountered in a large number of probabilistic structures—much beyond the Gaussian framework) is central in the theory of probabilistic approximations developed in [1, 2, 7].

We shall now formally present our general framework, together with some well-known preliminary facts. For the rest of the paper, we focus for simplicity on the case of a finite dimensional underlying Gaussian space. It is easily shown that, at the cost of some slightly heavier notation, the arguments developed below carry on almost verbatim to an infinite-dimensional setting. See, e.g., [11] for proofs and further details.

1.2 Setting

For $n \geqslant 1$, let γ_n denote the standard Gaussian measure on $\mathbb{R}^n$, given by

$$d\gamma_n(x) = (2\pi)^{-n/2} \exp\{-\|x\|^2/2\}dx,$$

with $\|\cdot\|$ the Euclidean norm on $\mathbb{R}^n$. In what follows, we shall denote by $\{P_t : t \geqslant 0\}$ the *Ornstein-Uhlenbeck semigroup* on $\mathbb{R}^n$, with infinitesimal generator

$$\mathcal{L}f = \Delta f - \langle x, \nabla f\rangle = \sum_{i=1}^{n} \frac{\partial^2 f}{\partial x_i^2} - \sum_{i=1}^{n} x_i \frac{\partial f}{\partial x_i}. \tag{1.3}$$

(Note that $\mathcal{L}$ acts on smooth functions f as an invariant and symmetric operator with respect to γ_n.)

We denote by $\{H_k : k = 0, 1, \ldots\}$ the collection of Hermite polynomials on the real line, defined recursively as $H_0 \equiv 1$, and $H_{k+1} = \delta H_k$, where $\delta f(x) := xf(x) - f'(x)$. It is a well-known fact that the family $\{k!^{-1/2} H_k : k = 0, 1, ..\}$ constitutes an orthonormal basis of $L^2(\gamma_1) := L^2(\mathbb{R}, \mathcal{B}(\mathbb{R}), \gamma_1)$.

Another well-known fact is that the spectrum of $\mathcal{L}$ coincides with the set of negative integers, that is, $\mathrm{Sp}(\mathcal{L}) = \mathbb{Z}_- = \{0, -1, -2, \ldots\}$. Also, the kth eigenspace of $\mathcal{L}$, corresponding to the vector space $\mathrm{Ker}(\mathcal{L} + kI)$ (with I the identity operator) and known as the kth *Wiener chaos* associated with γ_n, coincides with the class of those polynomial functions $F(x_1, \cdots, x_n)$ having the form

$$F(x_1, \cdots, x_n) = \sum_{i_1+i_2+\cdots+i_n=k} \alpha(i_1, \cdots, i_n) \prod_{j=1}^{n} H_{i_j}(x_j), \tag{1.4}$$

for some collection of real weights $\{\alpha(i_1, \cdots, i_n)\}$. We will use many times the fact that the generator $\mathcal{L}$ satisfies the integration by parts formula

$$\int_{\mathbb{R}^n} f\, \mathcal{L}g\, d\gamma_n = -\int_{\mathbb{R}^n} \langle \nabla f, \nabla g\rangle d\gamma_n \tag{1.5}$$

for every pair of smooth functions $f, g : \mathbb{R}^n \to \mathbb{R}$.

1.3 Key Estimates

Let $F_1, \cdots, F_d$ be eigenfunctions of $\mathcal{L}$ corresponding to eigenvalues $-k_1, \cdots, -k_d$, respectively. Using (1.4) we note that each F_i is a multivariate polynomial of degree k_i. Hence, the product $F_1 \cdots F_d$ is a multivariate polynomial of degree $r = k_1 + \cdots + k_d$. As a result, and after expanding $F_1 \cdots F_d$ over the basis of multivariate Hermite polynomials (that is, the complete orthogonal system constituted by $x \mapsto \prod_{i=1}^n H_{i_j}(x_j)$ for any $n \in \mathbb{N}$ and $i_j \geqslant 1$), we obtain that $F_1 \cdots F_d$ has a finite expansion over the first eigenspaces of $\mathcal{L}$, that is,

$$F_1 \cdots F_d \in \bigoplus_{k=0}^{r} \mathrm{Ker}(\mathcal{L} + kI). \tag{1.6}$$

From this, we deduce in particular that

$$\int_{\mathbb{R}^n} F_1 \cdots F_d \ (\mathcal{L} + rI)(F_1 \cdots F_d)\, d\gamma_n \geqslant 0 \tag{1.7}$$

and

$$\int_{\mathbb{R}^n} \mathcal{L}(F_1 \cdots F_d)\ (\mathcal{L} + rI)(F_1 \cdots F_d)\, d\gamma_n \leqslant 0. \tag{1.8}$$

The two estimates (1.7)–(1.8) are the key of the entire paper. In Section 3, relation (1.8) will be fruitfully combined with the following *Malliavin-Stein inequality*, according to which, for every sufficiently smooth $F \in L^2(\gamma_n)$ (for instance, one can take F to be an element of the direct sum of a finite collection of Wiener chaoses), one has the following bound:

$$\sup_{A \in \mathcal{B}(\mathbb{R})} \left| \int_{\mathbb{R}^n} \mathbf{1}_{\{F \in A\}} d\gamma_n - \gamma_1(A) \right| \leqslant 2 \sqrt{\int_{\mathbb{R}^n} \left(1 - \langle \nabla F, -\nabla \mathcal{L}^{-1} F \rangle\right)^2 d\gamma_n}, \tag{1.9}$$

where $\mathcal{L}^{-1}$ is the pseudo-inverse of $\mathcal{L}$. See, e.g., [11, Theorem 5.1.3] for a proof.

1.4 Plan

The rest of the paper is organized as follows. In Section 2 we provide a self-contained and alternate proof on the main estimate from [9]. Section 3 contains a proof of the quantitative fourth moment theorem in dimension one, whereas a multidimensional version is discussed in Section 4.

2 A Weak (Wick) Form of the Gaussian Product Conjecture

We start by presenting an alternate proof of the main results from [9], that is crucially based on the inequality (1.7).

Theorem 2.1 *Let $F_1, \ldots, F_d$ be eigenfunctions of $\mathcal{L}$. Then*

$$\int_{\mathbb{R}^n} (F_1 \cdots F_d)^2 \, d\gamma_n \geqslant \int_{\mathbb{R}^n} {F_1}^2 d\gamma_n \cdots \int_{\mathbb{R}^n} {F_d}^2 \, d\gamma_n. \tag{2.10}$$

Proof For $i = 1, \ldots, d$, denote by $-k_i$ the eigenvalue associated with F_i. We proceed by induction on the integer $r = k_1 + \cdots + k_d$. Since inequality is clear when $r = 0$ and $r = 1$, we only have to explain how to pass from $r - 1$ to r. As anticipated, to do so, we will make an important use of (1.7). Set $G := F_1 \cdots F_d$ and let us compute $\mathcal{L}G$ using the very definition of $\mathcal{L}$. In what follows, a hat " $\widehat{\ }$ " over a symbol indicates that the corresponding term is omitted (that is, replaced by 1) in a product. We have

$$\begin{aligned} \mathcal{L}G &= \sum_{i=1}^{d} F_1 \cdots \widehat{F_i} \cdots F_d \, \mathcal{L}F_i + \sum_{1 \leqslant i \neq j \leqslant d} F_1 \cdots \widehat{F_i} \cdots \widehat{F_j} \cdots F_d \, \langle \nabla F_i, \nabla F_j \rangle \\ &= -rG + \sum_{1 \leqslant i \neq j \leqslant d} F_1 \cdots \widehat{F_i} \cdots \widehat{F_j} \cdots F_d \, \langle \nabla F_i, \nabla F_j \rangle, \end{aligned}$$

the last equality coming from $\mathcal{L}F_i = -k_i F_i$. It thus follows from (1.7) that

$$\sum_{1 \leqslant i \neq j \leqslant d} \int_{\mathbb{R}^n} {F_1}^2 \cdots \widehat{F_i}^2 \cdots \widehat{F_j}^2 \cdots {F_d}^2 \, F_i F_j \langle \nabla F_i, \nabla F_j \rangle \, d\gamma_n \geqslant 0. \tag{2.11}$$

On the other hand, one has also that

$$\mathcal{L}(G^2) = \sum_{i=1}^{d} {F_1}^2 \cdots \widehat{F_i}^2 \cdots {F_d}^2 \, \mathcal{L}({F_i}^2) + \sum_{1 \leqslant i \neq j \leqslant d} {F_1}^2 \cdots \widehat{F_i}^2 \cdots \widehat{F_j}^2 \cdots {F_d}^2 \, \langle \nabla(F_i^2), \nabla(F_j^2) \rangle.$$

Since $\int_{\mathbb{R}^n} \mathcal{L}(G^2) d\gamma_n = 0$ and $\langle \nabla(F_i^2), \nabla(F_j^2) \rangle = 4F_i F_j \langle \nabla F_i, \nabla F_j \rangle$, we deduce from (2.11) that

$$\sum_{i=1}^{d} \int_{\mathbb{R}^n} {F_1}^2 \cdots \widehat{F_i}^2 \cdots {F_d}^2 \, \mathcal{L}({F_i}^2) \, d\gamma_n \leqslant 0.$$

But $\mathcal{L}(F_i^2) = 2F_i\mathcal{L}F_i + 2\|\nabla F_i\|^2 = -2k_iF_i^2 + 2\|\nabla F_i\|^2$. As a result,

$$r\int_{\mathbb{R}^n} G^2\, d\gamma_n \geqslant \sum_{i=1}^{d}\int_{\mathbb{R}^n} {F_1}^2\cdots\widehat{F_i}^2\cdots{F_d}^2\ \|\nabla F_i\|^2\, d\gamma_n$$

$$= \sum_{i=1}^{d}\sum_{\ell=1}^{d}\int_{\mathbb{R}^n} {F_1}^2\cdots\widehat{F_i}^2\cdots{F_d}^2\ (\nabla_\ell F_i)^2\, d\gamma_n,$$

where we used the vector notation $\nabla = (\nabla_1,\ldots,\nabla_d)$. For any fixed ℓ, $\nabla_\ell F_i$ is an eigenfunction of $\mathcal{L}$ associated with eigenvalue $-(k_i-1)$. Thus, by the induction assumption applied to the family $F_1,\cdots,F_{i-1},\nabla_\ell F_i,F_{i+1},\cdots,F_d$ (which corresponds to integer $r-1$), we deduce that

$$r\int_{\mathbb{R}^n} G^2 d\gamma_n \geqslant \sum_{i=1}^{d}\int_{\mathbb{R}^n} {F_1}^2 d\gamma_n\cdots\int_{\mathbb{R}^n} {F_i}^2 d\gamma_n\cdots\int_{\mathbb{R}^n} F_d^2 d\gamma_n\ \int_{\mathbb{R}^n}\|\nabla F_i\|^2 d\gamma_n$$

$$= r\int_{\mathbb{R}^n} {F_1}^2 d\gamma_n\cdots\int_{\mathbb{R}^n} {F_d}^2 d\gamma_n,$$

the last equality being a consequence of $\int_{\mathbb{R}^n}\|\nabla F_i\|^2 d\gamma_n = k_i\int_{\mathbb{R}^n} {F_i}^2 d\gamma_n$. The proof of Theorem 2.1 is concluded. ∎

3 A Proof of the Fourth Moment Theorem Based on Integration by Parts

Following the approach developed in [1] (see also [2]), we now provide a self-contained proof, based on (1.7), of one of the main estimates from [10]. As anticipated, this is a quantitative version of the fourth moment theorem first proved in [14].

Theorem 3.1 *Let F be a non-constant eigenfunction of $\mathcal{L}$. Assume further that $\int_{\mathbb{R}^n} F^2 d\gamma_n = 1$. We have*

$$\sup_{A\in\mathcal{B}(\mathbb{R})}\left|\int_{\mathbb{R}^n} \mathbf{1}_{\{F\in A\}} d\gamma_n - \gamma_1(A)\right| \leqslant \frac{2}{\sqrt{3}}\sqrt{\int_{\mathbb{R}^n} F^4 d\gamma_n - 3}.$$

Proof Denote by $-k$ the eigenvalue associated with F, and note that, since F is non-constant, then necessarily $k \geqslant 1$. Using the definition of $\mathcal{L}$ we have that

$$\mathcal{L}(F^2) = -2kF^2 + 2\|\nabla F\|^2. \tag{3.12}$$

Relation (1.8) (in the case $d = 2$ and $F_1 = F_2 = F$) yields that

$$\int_{\mathbb{R}^n} \mathcal{L}(F^2)\ (\mathcal{L} + 2kI)(F^2)\, d\gamma_n \leqslant 0. \tag{3.13}$$

From this, we deduce that

$$\begin{aligned}\int_{\mathbb{R}^n} \|\nabla F\|^4 d\gamma_n &= \frac{1}{4}\int_{\mathbb{R}^n} (\mathcal{L}(F^2) + 2kF^2)^2 d\gamma_n \\ &= \frac{1}{4}\int_{\mathbb{R}^n} \mathcal{L}(F^2)\ (\mathcal{L} + 2kI)(F^2)\, d\gamma_n + \frac{k}{2}\int_{\mathbb{R}^n} F^2\ (\mathcal{L} + 2kI)(F^2)\, d\gamma_n \\ &\leqslant k\int_{\mathbb{R}^n} F^2\, \|\nabla F\|^2\, d\gamma_n.\end{aligned}$$

But $F = -\frac{1}{k}\mathcal{L}F$, so we have

$$1 = \int_{\mathbb{R}^n} F^2 d\gamma_n = -\frac{1}{k}\int_{\mathbb{R}^n} F\mathcal{L}F d\gamma_n = \frac{1}{k}\int_{\mathbb{R}^n} \|\nabla F\|^2 d\gamma_n,$$

as well as

$$\int_{\mathbb{R}^n} F^4 d\gamma_n = -\frac{1}{k}\int_{\mathbb{R}^n} F^3\mathcal{L}F d\gamma_n = \frac{1}{k}\int \langle\nabla(F^3), \nabla F\rangle d\gamma_n = \frac{3}{k}\int_{\mathbb{R}^n} F^2\|\nabla F\|^2 d\gamma_n.$$

Combining the previous estimates, one deduces that

$$3\int_{\mathbb{R}^n} \left(1 - \frac{1}{k}\|\nabla F\|^2\right)^2 d\gamma_n \leqslant \int_{\mathbb{R}^n} F^4 d\gamma_n - 3. \tag{3.14}$$

The conclusion now follows from (1.9).

■

4 Multivariate Fourth Moment Theorems

To conclude the paper, we now provide a proof of a multidimensional CLT taken from [15]—see also the discussion contained in [11, Chapter 6]. We follow the approach developed in [4]. To avoid the use of contractions, we will restrict ourselves only to the case of pairwise distinct eigenvalues. In what follows, we denote by $N(0, 1)$ and $N_d(0, I_d)$, respectively, a one dimensional standard Gaussian random variable, and a d dimensional centered Gaussian vector with identity covariance matrix. As usual, the symbol $\xrightarrow{d}$ stands for convergence in distribution of random variables with values in a Euclidean space, whereas Var indicates a variance.

Theorem 4.1 *Fix d integers $k_1, \cdots, k_d \geqslant 1$ and suppose they are pairwise distinct; consider also a sequence $\{n(m) : m \geqslant 1\}$ of positive integers such that $n(m) \to \infty$, as $m \to \infty$. For every $m \geqslant 1$ let $F_m = (F_{1,m}, \dots, F_{d,m})$ be a vector of random variables defined on the probability space $(\mathbb{R}^{n(m)}, \mathcal{B}(\mathbb{R}^{n(m)}), \gamma_{n(m)})$, such that $F_{i,m} \in \ker(\mathcal{L} + k_i I)$, for $1 \leqslant i \leqslant d$, where $\mathcal{L}$ is the generator of the Ornstein-Uhlenbeck semigroup associated with $\gamma_{n(m)}$. For simplicity, assume also that $\int_{\mathbb{R}^{n(m)}} F_{i,m}^2 \, d\gamma_{n(m)} = 1$ for any i and m. If $F_{i,m} \xrightarrow{d} N(0,1)$ for every $i = 1, \dots, d$, then $F_m \xrightarrow{d} N_d(0, I_d)$, as $m \to \infty$.*

Proof We write $n(m) = n$ in order to simplify the discussion. According to [11, Theorem 6.1.1], it suffices to show that, if $F_{i,m} \xrightarrow{d} N(0,1)$ as $m \to \infty$ for every $i = 1, \dots, d$ then, for $1 \leqslant i, j \leqslant d$ and as $m \to \infty$,

$$\int_{\mathbb{R}^n} \left(\langle \nabla F_{i,m}, -\nabla \mathcal{L}^{-1} F_{j,m} \rangle - \delta_{ij} \right)^2 d\gamma_n \to 0,$$

where δ_{ij} stands for the Kronecker symbol. In our situation, it is equivalent to show that

$$\int_{\mathbb{R}^n} \left(\langle \nabla F_{i,m}, \nabla F_{j,m} \rangle - k_j \delta_{ij} \right)^2 d\gamma_n \to 0. \tag{4.15}$$

For $i = j$, this follows from (3.14) since, by hypercontractivity, $F_{i,m} \xrightarrow{d} N(0,1)$ implies that $\int_{\mathbb{R}^n} {F_{i,m}}^4 d\gamma_n \to 3$. Let us thus assume that $i \neq j$. For the sake of readability, we temporarily suppress the index m from $F_{i,m}$ and $F_{j,m}$. We can write

$$\langle \nabla F_i, \nabla F_j \rangle = \frac{1}{2} \left(\mathcal{L} + (k_i + k_j) I \right) \left(F_i F_j \right).$$

Thus, using the symmetry of $\mathcal{L}$,

$$\begin{aligned} \int_{\mathbb{R}^n} \langle \nabla F_i, \nabla F_j \rangle^2 d\gamma_n &= \frac{1}{4} \int_{\mathbb{R}^n} \left(\left(\mathcal{L} + (k_i + k_j) I \right) \left(F_i F_j \right) \right)^2 d\gamma_n \\ &= \frac{1}{4} \int_{\mathbb{R}^n} F_i F_j \left(\mathcal{L} + (k_i + k_j) I \right)^2 \left(F_i F_j \right) d\gamma_n. \end{aligned}$$

Now, due to (1.6), observe that $F_i F_j \in \bigoplus_{r=1}^{k_i + k_j} \operatorname{Ker}(\mathcal{L} + rI)$. As a result, noting π_r the projection onto $\operatorname{Ker}(\mathcal{L} + rI)$ and with $s_{ij} = k_i + k_j$,

$$\begin{aligned} &\int_{\mathbb{R}^n} F_i F_j \left(\mathcal{L} + s_{ij} I \right)^2 \left(F_i F_j \right) d\gamma_n \\ &= \int_{\mathbb{R}^n} F_i F_j \left(\mathcal{L} \left(\mathcal{L} + s_{ij} I \right) \right) \left(F_i F_j \right) d\gamma_n + s_{ij} \int_{\mathbb{R}^n} F_i F_j \left(\mathcal{L} + s_{ij} I \right) \left(F_i F_j \right) d\gamma_n \end{aligned}$$

$$= -\sum_{r=1}^{s_{ij}} r(s_{ij}-r)\int_{\mathbb{R}^n}\|\nabla\pi_r(F_iF_j)\|^2 d\gamma_n + s_{ij}\int_{\mathbb{R}^n} F_iF_j\left(\mathcal{L}+s_{ij}I\right)\left(F_iF_j\right)d\gamma_n$$
$$\leqslant s_{ij}\int_{\mathbb{R}^n} F_iF_j\left(\mathcal{L}+s_{ij}I\right)\left(F_iF_j\right)d\gamma_n.$$

Therefore

$$\int_{\mathbb{R}^n}\langle\nabla F_i,\nabla F_j\rangle^2 d\gamma_n \leqslant \frac{k_i+k_j}{4}\int_{\mathbb{R}^n} F_iF_j\left(\mathcal{L}+(k_i+k_j)I\right)\left(F_iF_j\right)d\gamma_n$$
$$= \frac{k_i+k_j}{2}\int_{\mathbb{R}^n} F_iF_j\,\langle\nabla F_i,F_j\rangle d\gamma_n. \tag{4.16}$$

Now, using the definition of $\mathcal{L}$ and integrating by parts yields

$$\int_{\mathbb{R}^n} F_iF_j\langle\nabla F_i,\nabla F_j\rangle d\gamma_n = \frac{1}{4}\int_{\mathbb{R}^n}\langle\nabla(F_i^2),\nabla(F_j^2)\rangle d\gamma_n = -\frac{1}{4}\int_{\mathbb{R}^n} F_i^2\mathcal{L}(F_j^2)d\gamma_n$$
$$= -\frac{1}{2}\int_{\mathbb{R}^n} F_i^2\left(F_j\mathcal{L}F_j+\|\nabla F_j\|^2\right)d\gamma_n = \frac{k_j}{2}\int_{\mathbb{R}^n} F_i^2F_j^2 d\gamma_n - \frac{1}{2}\int_{\mathbb{R}^n} F_i^2\|\nabla F_j\|^2 d\gamma_n$$
$$= \frac{k_j}{2}\int_{\mathbb{R}^n} F_i^2\left(F_j^2-1\right)d\gamma_n - \frac{1}{2}\int_{\mathbb{R}^n} F_i^2\left(\|\nabla F_j\|^2-k_j\right)d\gamma_n$$
$$\leqslant \frac{k_j}{2}\int_{\mathbb{R}^n} F_i^2\left(F_j^2-1\right)d\gamma_n + \frac{1}{2}\sqrt{\int_{\mathbb{R}^n} F_i^4 d\gamma_n}\,\sqrt{\mathrm{Var}(\|\nabla F_j\|^2)}$$
$$= \frac{k_j}{2}\left(\int_{\mathbb{R}^n} F_i^2F_j^2 d\gamma_n - 1\right) + \frac{1}{2}\sqrt{\int_{\mathbb{R}^n} F_i^4 d\gamma_n}\,\sqrt{\mathrm{Var}(\|\nabla F_j\|^2)}.$$

Plugging such an estimate into (4.16) and reintroducing the index m, we can thus write

$$\int_{\mathbb{R}^n}\langle\nabla F_{i,m},\nabla F_{j,m}\rangle^2 d\gamma_n \leqslant \frac{k_i+k_j}{2}\left(\frac{1}{2}\sqrt{\int_{\mathbb{R}^n} F_{i,m}^4 d\gamma_n}\,\sqrt{\mathrm{Var}(\|\nabla F_{j,m}\|^2)} + R_{i,j}(m)\right),$$

where

$$R_{ij}(m) = \frac{k_j}{2}\left(\int_{\mathbb{R}^n} F_{i,m}^2F_{j,m}^2 d\gamma_n - 1\right).$$

By a classical hypercontractivity argument, the fact that $F_{i,m}\xrightarrow{d}N(0,1)$ for every i implies that $\int_{\mathbb{R}^n} {F_{i,m}}^4 d\gamma_n \to 3$ as $m\to\infty$ for every i. In particular, due to (3.14), we have that

$$\mathrm{Var}(\|\nabla F_{i,m}\|^2)\to 0 \quad \text{as } m\to\infty \text{ for every } i. \tag{4.17}$$

Thus, in order to show (4.15) we are left to check that $R_{ij}(m) \to 0$, as $m \to \infty$. By symmetry, we can assume without loss of generality that $k_j < k_i$. Now, we note that, by (3.12), it holds that

$$\begin{aligned}2\|\nabla F_{i,m}\|^2 - 2E\|\nabla F_{i,m}\|^2 &= (\mathcal{L} + 2\lambda_{k_i} I)(F_{i,m}^2) - E[(\mathcal{L} + 2\lambda_{k_i} I)(F_{i,m}^2)] \\ &= \sum_{r=1}^{2k_i} (2k_i - r)\pi_r(F_{i,m}^2),\end{aligned}$$

where π_r denotes the projection onto $\mathrm{Ker}(\mathcal{L} + rI)$. By orthogonality of the projections corresponding to different eigenvalues and by using (4.17), we deduce that

$$\int_{\mathbb{R}^n} \pi_r(F_{i,m}^2)^2 d\gamma_n \to 0 \qquad \text{for } r \in \{1, \cdots, 2k_i - 1\}. \tag{4.18}$$

We exploit this fact by writing

$$\int_{\mathbb{R}^n} F_{i,m}^2 F_{j,m}^2 d\gamma_n = 1 + \sum_{r=1}^{2k_i - 1} \int_{\mathbb{R}^n} \pi_r\left(F_{i,m}^2\right) F_{j,m}^2 d\gamma_n + \int_{\mathbb{R}^n} \pi_{2k_i}\left(F_{i,m}^2\right) F_{j,m}^2 d\gamma_n.$$

By Cauchy-Schwarz and (4.18), all integrals $\int_{\mathbb{R}^n} \pi_r(F_{i,m}^2) F_{j,m}^2 d\gamma_n$ inside the sum in the middle vanish in the limit. Finally, the fact that $k_i > k_j$ ensures that the third term is actually equal to zero. Thus, one has that $R_{ij}(m) \to 0$ as $m \to \infty$, and the proof of the multivariate fourth moment theorem for different orders is complete. ■

Acknowledgements We thank an anonymous referee for insightful comments.
IN is partially supported by the grant F1R-MTH-PUL-15CONF (CONFLUENT) at Luxembourg University.
GP is partially supported by the grant F1R-MTH-PUL-15STAR (STARS) at Luxembourg University.

References

[1] E. Azmoodeh, S. Campese and G. Poly (2014). Fourth Moment Theorems for Markov diffusion generators. *J. Funct. Anal.* **266**, no. 4, pp. 2341–2359.
[2] E. Azmoodeh, D. Malicet, G. Mijoule and G. Poly (2016): Generalization of the Nualart-Peccati criterion. *Ann. Probab.* **44**, no. 2, pp. 924–954.
[3] C. Benítem, Y. Sarantopolous and A.M. Tonge (1998): Lower bounds for norms of products of polynomials. *Math. Proc. Camb. Phil. Soc.* **124**, pp. 395–408.
[4] S. Campese, I. Nourdin, G. Peccati and G. Poly (2015): Multivariate Gaussian approximations on Markov chaoses. *Electron. Commun. Probab.* **21**, paper no. 48.
[5] L.H.Y. Chen and G. Poly (2015): Stein's method, Malliavin calculus, Dirichlet forms and the fourth moment Theorem. In: *Festschrift Masatoshi Fukushima* (Z-Q Chen, N. Jacob, M. Takeda and T. Uemura, eds.), Interdisciplinary Mathematical Sciences Vol. 17, World Scientific, pp. 107–130

[6] P.E. Frenkel (2007): Pfaffians, hafnians and products of real linear functionals. *Math. Res. Lett.* **15**, no. 2, pp. 351–358.
[7] M. Ledoux (2012): Chaos of a Markov operator and the fourth moment condition. *Ann. Probab.* **40**, no. 6, pp. 2439–2459.
[8] M. Ledoux, I. Nourdin and G. Peccati (2015): Stein's method, logarithmic Sobolev and transport inequalities. *Geom. Funct. Anal.* **25**, pp. 256–30
[9] D. Malicet, I. Nourdin, G. Peccati and G. Poly (2016): Squared chaotic random variables: new moment inequalities with applications. *J. Funct. Anal.* **270**, no. 2, pp. 649–670.
[10] I. Nourdin and G. Peccati (2009): Stein's method on Wiener chaos. *Probab. Theory Rel. Fields* **145**, no. 1, pp. 75–118.
[11] I. Nourdin and G. Peccati (2012): *Normal Approximations with Malliavin Calculus: From Stein's Method to Universality*. Cambridge University Press.
[12] I. Nourdin, G. Peccati and Y. Swan (2014): Entropy and the fourth moment phenomenon. *J. Funct. Anal.* **266**, pp. 3170–3207
[13] D. Nualart and S. Ortiz-Latorre (2008): Central limit theorems for multiple stochastic integrals and Malliavin calculus. *Stochastic Process. Appl.* **118**, no. 4, pp. 614–628.
[14] D. Nualart and G. Peccati (2005): Central limit theorems for sequences of multiple stochastic integrals. *Ann. Probab.* **33**, no. 1, pp. 177–193.
[15] G. Peccati and C.A. Tudor (2005): Gaussian limits for vector-valued multiple stochastic integrals. *Séminaire de Probabilités* **XXXVIII**, pp. 247–262.
[16] R. Ryan and B. Turett (1998): Geometry of spaces of polynomials. *J. Math. Anal. Appl.* **221**, pp. 698–711.
[17] A website about Stein's method and Malliavin calculus. Web address: https://sites.google.com/site/malliavinstein/home

Asymptotic Expansions for Products of Characteristic Functions Under Moment Assumptions of Non-integer Orders

Sergey G. Bobkov

Abstract This is mostly a review of results and proofs related to asymptotic expansions for characteristic functions of sums of independent random variables (known also as Edgeworth-type expansions). A number of known results is refined in terms of Lyapunov coefficients of non-integer orders.

Let $X_1, \dots, X_n$ be independent random variables with zero means, variances $\sigma_k^2 = \mathrm{Var}(X_k)$, such that $\sum_{k=1}^n \sigma_k^2 = 1$, and with finite absolute moments of some integer order $s \geq 2$. Introduce the Lyapunov coefficients

$$L_s = \sum_{k=1}^n \mathbb{E}\,|X_k|^s \qquad (s \geq 2).$$

If L_3 is small, the distribution F_n of the sum $S_n = X_1 + \cdots + X_n$ will be close in a weak sense to the standard normal law with density and distribution function

$$\varphi(x) = \frac{1}{\sqrt{2\pi}}\, e^{-x^2/2}, \quad \Phi(x) = \int_{-\infty}^x \varphi(y)\,dy \qquad (x \in \mathbb{R}).$$

This variant of the central limit theorem may be quantified by virtue of the classical Berry-Esseen bound

$$\sup_x |\,\mathbb{P}\{S_n \leq x\} - \Phi(x)| \leq cL_3$$

(where c is an absolute constant). Moreover, in case $s > 3$, in some sense the rate of approximation of F_n can be made much better – to be of order at most L_s, if we replace the normal law by a certain "corrected normal" signed measure μ_{s-1} on the real line. The density φ_{s-1} of this measure involves the cumulants γ_p of S_n of orders up to $s - 1$ (which are just the sums of the cumulants of X_k); for example,

---

S.G. Bobkov (✉)
University of Minnesota, Minneapolis, MN 55455, USA
e-mail: bobkov@math.umn.edu

E. Carlen et al. (eds.), *Convexity and Concentration*, The IMA Volumes in Mathematics and its Applications 161, DOI 10.1007/978-1-4939-7005-6_11

$$\varphi_3(x) = \varphi(x)\Big(1 + \frac{\gamma_3}{3!}\,H_3(x)\Big),$$

$$\varphi_4(x) = \varphi(x)\Big(1 + \frac{\gamma_3}{3!}\,H_3(x) + \frac{\gamma_4}{4!}\,H_4(x) + \frac{\gamma_3^2}{2!\,3!^2}\,H_6(x)\Big),$$

where H_k denotes the Chebyshev-Hermite polynomial of degree k. More generally,

$$\varphi_{s-1}(x) = \varphi(x)\sum \frac{1}{k_1!\dots k_{s-3}!}\,\Big(\frac{\gamma_3}{3!}\Big)^{k_1}\dots\Big(\frac{\gamma_{s-1}}{(s-1)!}\Big)^{k_{s-3}}\,H_k(x), \tag{0.1}$$

where $k = 3k_1 + \cdots + (s-1)k_{s-3}$ and where the summation is running over all collections of non-negative integers $k_1, \dots, k_{s-3}$ such that $k_1 + 2k_2 + \cdots + (s-3)k_{s-3} \leq s-3$.

When the random variables $X_k = \frac{1}{\sqrt{n}}\,\xi_k$ are identically distributed, the sum in (0.1) represents a polynomial in $\frac{1}{\sqrt{n}}$ of degree at most $s-3$ with free term 1. In that case, the Lyapunov coefficient

$$L_s = \mathbb{E}\,|\xi_1|^s\, n^{-\frac{s-2}{2}}$$

has a smaller order for growing n in comparison with all terms of the sum.

The closeness of the measures F_n and μ_{s-1} is usually studied with the help of Fourier methods. That is, as the first step, it is established that on a relatively long interval $|t| \leq T$ the characteristic function $f_n(t) = \mathbb{E}\,e^{itS_n}$ together with its first s derivatives are properly approximated by the Fourier-Stieltjes transform

$$g_{s-1}(t) = \int_{-\infty}^{\infty} e^{itx}\,d\mu_{s-1}(x)$$

and its derivatives. In particular, it is aimed to achieve relations such as

$$\big|f_n^{(p)}(t) - g_{s-1}^{(p)}(t)\big| \leq C_s L_s \min\{1, |t|^{s-p}\}\,e^{-ct^2}, \qquad p = 0, 1, \dots, s, \tag{0.2}$$

in which case one may speak about an asymptotic expansion for f_n by means of g_{s-1}. When it turns out possible to convert these relations to the statements about the closeness of the distribution function associated to F_n and μ_{s-1}, one obtains an Edgeworth expansion for F_n (or for density of F_n, when it exists). Basic results in this direction were developed by many researchers in the 1930–1970s, including Cramér, Esseen, Gnedenko, Petrov, Statulevičius, Bikjalis, Bhattacharya and Ranga Rao, Götze and Hipp among others (cf. [C, E, G1, G-K, P1, P2, P3, St1, St2, Bi1, Bi2, Bi3, B-C-G1, B-C-G2, B-C-G3, Pr1, Pr2, Bi1, Bi2, B-RR, G-H, Se, B1]).

In these notes, we focus on the questions that are only related to the first part of the problem, i.e., to the asymptotic expansions for f_n. We review several results, clarify basic technical ingredients of the proofs, and make some refinements where possible. In particular, the following questions are addressed: On which intervals do we have asymptotic expansions for the characteristic functions? How may the constants C_s depend on the growing parameter s? Another issue, which is well motivated, e.g., by limit problems about the normal approximation in terms of transport distances (cf. [B2]), is how to extend corresponding statements to the case of non-integer (or, fractional) values of s.

In a separate (first) part, we collect several results about the distributions of single random variables, including general inequalities on the moments, cumulants, and derivatives of characteristic functions, which lead to corresponding Taylor's expansions. In the second part, there have been collected some results on the behavior of Lyapunov's coefficients and moment inequalities for sums of independent random variables, with first applications to products of characteristic functions. Asymptotic expansions g_{s-1} for f_n are constructed and studied in the third part. In particular, in the interval $|t| \leq cL_s^{-1/3(s-2)}$ (in case L_s is small), we derive a sharper form of (0.2),

$$\left|f_n^{(p)}(t) - g_{s-1}^{(p)}(t)\right| \leq C^s L_s \max\{|t|^{s-p}, |t|^{3(s-2)+p}\} e^{-t^2/2}.$$

This interval of approximation, which we call moderate, appears in a natural way in many investigations, mostly focused on the case $p = 0$ and when X_k's are equidistributed. The fourth part is devoted to the extension of this interval to the size $|t| \leq 1/L_3$ which we call a long interval. This is possible at the expense of the constant in the exponent and with a different behavior of s-dependent factors, by showing that both $f_n^{(p)}(t)$ and $g_{s-1}^{(p)}(t)$ are small in absolute value outside the moderate interval. All results are developed for real values of the main parameter s. More precisely, we use the following plan.

PART I. Single random variables

1. Generalized chain rule formula.
2. Logarithm of the characteristic functions.
3. Moments and cumulants.
4. Bounds on the derivatives of the logarithm.
5. Taylor expansion for Fourier-Stieltjes transforms.
6. Taylor expansion for logarithm of characteristic functions.

PART II. Lyapunov coefficients and products of characteristic functions

7. Properties of Lyapunov coefficients.
8. Logarithm of the product of characteristic functions.
9. The case $2 < s \leq 3$.

PART III. "Corrected normal characteristic" functions

10. Polynomials P_m in the normal approximation.
11. Cumulant polynomials Q_m.
12. Relations between P_m and Q_m.
13. Corrected normal approximation on moderate intervals.
14. Signed measures μ_m associated with g_m.

PART I. Single random variables

1 Generalized Chain Rule Formula

The following calculus formula is frequently used in a multiple differentiation.

Proposition 1.1 *Suppose that a complex-valued function $y = y(t)$ is defined and has p derivatives in some open interval of the real line ($p \geq 1$). If $z = z(y)$ is analytic in the region containing all values of y, then*

$$\frac{d^p}{dt^p} z(y(t)) = p! \sum \frac{d^{s_p} z(y)}{dy^{s_p}} \Big|_{y=y(t)} \prod_{r=1}^{p} \frac{1}{k_r!} \left(\frac{1}{r!} \frac{d^r y(t)}{dt^r} \right)^{k_r}, \tag{1.1}$$

where $s_p = k_1 + \cdots + k_p$ and where the summation is performed over all non-negative integer solutions $(k_1, \ldots, k_p)$ to the equation $k_1 + 2k_2 + \cdots + pk_p = p$.

This formula can be used to develop a number of interesting identities and inequalities like the following ones given in the next lemma.

Lemma 1.2 *With the summation as before, for any $\lambda \in \mathbb{R}$ and any integer $p \geq 1$,*

$$\sum (s_p - 1)! \prod_{r=1}^{p} \frac{1}{k_r!} \lambda^{k_r} = \frac{(1+\lambda)^p - 1}{p} \tag{1.2}$$

$$\sum s_p! \prod_{r=1}^{p} \frac{1}{k_r!} \lambda^{k_r} = \lambda (1+\lambda)^{p-1}. \tag{1.3}$$

In particular, if $0 \leq \lambda \leq 2^{-p} \lambda_0$, then

$$\sum \prod_{r=1}^{p} \frac{1}{k_r!} \lambda^{k_r} \leq \lambda e^{\lambda_0/4}. \tag{1.4}$$

In addition,

$$\sum \prod_{r=1}^{p} \frac{1}{k_r!} \left(\frac{\lambda^r}{r} \right)^{k_r} = \lambda^p. \tag{1.5}$$

Proof First, apply Proposition 1.1 with $z(y) = -\log(1-y)$, in which case (1.1) becomes

$$-\frac{d^p}{dt^p}\log(1-y(t)) = p!\sum\frac{(s_p-1)!}{(1-y(t))^{s_p}}\prod_{r=1}^{p}\frac{1}{k_r!}\Big(\frac{1}{r!}\,\frac{d^r y(t)}{dt^r}\Big)^{k_r}. \tag{1.6}$$

Choosing $y(t) = \lambda\,\frac{t}{1-t} = -\lambda + \lambda(1-t)^{-1}$ so that $\frac{d^r y(t)}{dt^r} = r!\,\lambda(1-t)^{-(r+1)}$, the above sum on the right-hand side equals

$$\sum(s_p-1)!\,(1-y(t))^{-s_p}\,(1-t)^{-p-s_p}\,\lambda^{s_p}\prod_{r=1}^{p}\frac{1}{k_r!}.$$

On the other hand, writing $-\log(1-y(t)) = \log(1-t) - \log\big((1+\lambda)(1-t)-\lambda\big)$, we get

$$-\frac{d^p}{dt^p}\log(1-y(t)) = -\frac{(p-1)!}{(1-t)^p} + (1+\lambda)^p\,\frac{(p-1)!}{\big((1+\lambda)(1-t)-\lambda\big)^p}.$$

Therefore, (1.6) yields

$$(p-1)!\Big[\frac{(1+\lambda)^p}{\big((1+\lambda)(1-t)-\lambda\big)^p} - \frac{1}{(1-t)^p}\Big] = p!\sum\frac{(s_p-1)!\,\lambda^{s_p}}{(1-y(t))^{s_p}\,(1-t)^{p+s_p}}\prod_{r=1}^{p}\frac{1}{k_r!}.$$

Putting $t=0$, we obtain the identity (1.2). Differentiating it with respect to λ and multiplying by λ, we arrive at (1.3). In turn, using $s_p! \geq 1$ and the property that the function $p \to (p-1)2^{-p}$ is decreasing in $p \geq 2$, (1.3) implies that, for all $p \geq 2$,

$$(1+\lambda)^{p-1} \leq e^{(p-1)\lambda} \leq e^{\lambda_0 (p-1)2^{-p}} \leq e^{\lambda_0/4},$$

which obviously holds for $p = 1$ as well.

Finally, let us apply (1.1) with $z(y) = e^y$, when this identity becomes

$$\frac{d^p}{dt^p}\,e^{y(t)} = p!\,e^{y(t)}\sum\prod_{r=1}^{p}\frac{1}{k_r!}\Big(\frac{1}{r!}\,\frac{d^r y(t)}{dt^r}\Big)^{k_r}. \tag{1.7}$$

It remains to choose here $y(t) = -\log(1-\lambda t)$, so that $\frac{d^r y(t)}{dt^r} = \lambda^r\,(r-1)!\,(1-t)^{-r}$, and then this equality yields (1.5) at the point $t=0$. □

For an illustration, consider Gaussian functions $g(t) = e^{-t^2/2}$. By the definition,

$$g^{(p)}(t) = (-1)^{p-1} H_p(t) g(t),$$

where H_p denotes the Chebyshev–Hermite polynomial of degree p with leading term 1. From (1.7) with $y(t) = -t^2/2$, we have

$$g^{(p)}(t) = p!\, g(t) \sum_{k_1+2k_2=p} \frac{(-t)^{k_1}}{k_1!k_2!}\, 2^{-k_2}.$$

Using $|t|^{k_1} \leq \max\{1, |t|^p\}$ and applying the identity (1.5), we get a simple upper bound

$$|H_p(t)| \leq p! \max\{1, |t|^p\}. \tag{1.8}$$

2 Logarithm of the Characteristic Functions

If a random variable X has finite absolute moment $\beta_p = \mathbb{E}\,|X|^p$ for some integer $p \geq 1$, its characteristic function $f(t) = \mathbb{E}\, e^{itX}$ has continuous derivatives up to order p and is non-vanishing in some interval $|t| \leq t_0$. Hence, in this interval the principal value of the logarithm $\log f(t)$ is well defined and also has continuous derivatives up to order p, which actually can be expressed explicitly in terms of the first derivatives of f. More precisely, the chain rule formula of Proposition 1.1 with $z(y) = \log y$ immediately yields the following identity:

Proposition 2.1 *Let $\beta_p < \infty$ $(p \geq 1)$. In the interval $|t| \leq t_0$, where $f(t)$ is non-vanishing,*

$$\frac{d^p}{dt^p} \log f(t) = p! \sum \frac{(-1)^{s_p-1}\,(s_p-1)!}{f(t)^{s_p}} \prod_{r=1}^{p} \frac{1}{k_r!} \Big(\frac{1}{r!} f^{(r)}(t)\Big)^{k_r}, \tag{2.1}$$

where $s_p = k_1 + \cdots + k_p$ and the summation is running over all tuples $(k_1, \dots, k_p)$ of non-negative integers such that $k_1 + 2k_2 + \cdots + pk_p = p$.

As was shown by Sakovič [Sa], in the interval $\sqrt{\beta_2}\,|t| \leq \frac{\pi}{2}$ we necessarily have $\mathrm{Re}(f(t)) \geq 0$. This result was sharpened by Rossberg [G2] proving that

$$\mathrm{Re}(f(t)) \geq \cos(\sqrt{\beta_2}\,|t|) \quad \text{for} \quad \sqrt{\beta_2}\,|t| \leq \pi.$$

See also Shevtsova [Sh2] for a more detailed exposition of the question. Thus, the representation (2.1) holds true in the open interval $\sqrt{\beta_2}\,|t| < \frac{\pi}{2}$.

To quickly see that $f(t)$ is non-vanishing on a slightly smaller interval, one can just apply Taylor's formula. Indeed, if $\mathbb{E}X = 0$, $\mathbb{E}X^2 = \beta_2 = \sigma^2$ $(0 < \sigma < \infty)$, then $f(0) = 1, f'(0) = 0$, $|f''(t)| \leq \sigma^2$, and we get

$$|1 - f(t)| \leq \sup_{|z| \leq |t|} |f''(z)|\, \frac{t^2}{2} \leq \frac{\sigma^2 t^2}{2} < 1$$

for $\sigma|t| < \sqrt{2}$. In particular, $|f(t)| \geq \frac{1}{2}$ for $\sigma|t| \leq 1$, so that in this interval the principal value of the logarithm $\log f(t)$ is continuous and has continuous derivatives up to order p.

Let us mention several particular cases in (2.1). Clearly, $(\log f)' = f'f^{-1}$ and $(\log f)'' = f''f^{-1} - f'^2 f^{-2}$. The latter formula can be given in an equivalent form.

Proposition 2.2 *If the variance* $\sigma^2 = \mathrm{Var}(X)$ *is finite, then at any point* t *such that* $f(t) \neq 0$, *we have*

$$(\log f(t))'' = -\frac{1}{2f(t)^2}\,\mathbb{E}\,(X-Y)^2\,e^{it(X+Y)},$$

where Y *is an independent copy of* X. *In particular,*

$$|(\log f(t))''| \leq \frac{\sigma^2}{|f(t)|^2}. \tag{2.2}$$

Indeed, the right-hand side of the equality $f(t)^2\,(\log f(t))'' = f''(t)f(t) - f'(t)^2$ may be written as

$$\begin{aligned}
-\big(\mathbb{E}X^2 e^{it(X+Y)} - \mathbb{E}\,XY\,e^{it(X+Y)}\big) &= -\Big(\mathbb{E}\,\frac{X^2+Y^2}{2}\,e^{it(X+Y)} - \mathbb{E}\,XY\,e^{it(X+Y)}\Big) \\
&= -\frac{1}{2}\,\mathbb{E}\,(X-Y)^2\,e^{it(X+Y)}.
\end{aligned}$$

Therefore,

$$|f(t)|^2\,|(\log f(t))''| \leq \frac{1}{2}\,\mathbb{E}\,(X-Y)^2 = \mathrm{Var}(X).$$

For the next two derivatives, let us note that

$$f(t)^3\,(\log f(t))''' = f'''(t)f(t)^2 - 3f''(t)f'(t)f(t) + 2f'(t)^3, \tag{2.3}$$

$$\begin{aligned}
f(t)^4\,(\log f(t))'''' &= f''''(t)f(t)^3 - 4f'''(t)f'(t)f(t)^2 - 3f''(t)^2 f(t)^2 \\
&\quad + 12f''(t)f'(t)^2 f(t) - 6f'(t)^4.
\end{aligned} \tag{2.4}$$

3 Moments and Cumulants

Again, let a random variable X have a finite absolute moment $\beta_p = \mathbb{E}\,|X|^p$ for an integer $p \geq 1$. Since the characteristic function $f(t) = \mathbb{E}\,e^{itX}$ is non-vanishing in some interval $|t| \leq t_0$, and $\log f(t)$ has continuous derivatives up to order p, one may introduce the normalized derivatives at zero

$$\gamma_r = \gamma_r(X) = \frac{d^r}{i^p\, dt^r} \log f(t)\Big|_{t=0}, \qquad r = 0, 1, 2, \dots, p,$$

called the cumulants of X. Each γ_p is determined by the first moments $\alpha_r = \mathbb{E}X^r$, $r = 1, \dots, p$. Namely, at $t = 0$, the identity (2.1) of Proposition 1.1 gives:

Proposition 3.1 *Let $\beta_p < \infty$ $(p \geq 1)$. For $|t| \leq t_0$, we have*

$$\gamma_p = p! \sum (-1)^{s_p - 1} (s_p - 1)! \prod_{r=1}^{p} \frac{1}{k_r!} \Big(\frac{\alpha_r}{r!}\Big)^{k_r}, \tag{3.1}$$

where $s_p = k_1 + \dots + k_p$ and where the summation is running over all tuples $(k_1, \dots, k_p)$ of non-negative integers such that $k_1 + 2k_2 + \dots + pk_p = p$.

For example, $\gamma_1 = \alpha_1$, $\gamma_2 = \alpha_2 - \alpha_1^2$. Moreover, if $\alpha_1 = \mathbb{E}X = 0$, $\sigma^2 = \mathbb{E}X^2$, then

$$\gamma_1 = \alpha_1, \quad \gamma_2 = \alpha_2 = \sigma^2, \quad \gamma_3 = \alpha_3, \quad \gamma_4 = \alpha_4 - 3\alpha_2^2 = \beta_4 - 3\sigma^4.$$

One may reverse (3.1) by applying the generalized chain rule to the composition $f(t) = e^{\log f(t)}$, see (1.7). We then get a similar formula

$$\alpha_p = p! \sum \prod_{r=1}^{p} \frac{1}{k_r!} \Big(\frac{\gamma_r}{r!}\Big)^{k_r}. \tag{3.2}$$

Let us now turn to the question of bounding the cumulants in terms of the moments. By Markov's inequality, there are uniform bounds on the derivatives $|f^{(r)}(t)| \leq \beta_r \leq \beta_p^{r/p}$ for $r = 1, \dots, p$. Hence, the combination of identity (1.2) of Lemma 1.2 with $\lambda = \frac{1}{|f(t)|}$ and identity (2.1) of Proposition 2.1 leads to the bound

$$\Big|\frac{d^p}{dt^p} \log f(t)\Big| \leq \Big[\Big(1 + \frac{1}{|f(t)|}\Big)^p - 1\Big] (p-1)!\, \beta_p. \tag{3.3}$$

This inequality may be compared to the result of Bikjalis [Bi3], who showed that

$$\Big|\frac{d^p}{dt^p} \log f(t)\Big| \leq \frac{1}{|f(t)|^p}\, 2^{p-1} (p-1)!\, \beta_p. \tag{3.4}$$

In particular, when $|f(t)| \geq \frac{1}{2}$, it gives the relation $\big|\frac{d^p}{dt^p} \log f(t)\big| \leq 2^{2p-1} (p-1)!\, \beta_p$. However, in this case (3.3) yields a better bound

$$\Big|\frac{d^p}{dt^p} \log f(t)\Big| \leq (3^p - 1)\, (p-1)!\, \beta_p.$$

We will discuss further sharpenings in the next section, and now just note that at the point $t = 0$, (3.4) provides a bound on the cumulants, $|\gamma_p| \le (2^{p-1} - 1)\,(p-1)!\,\beta_p$. Another result of Bikjalis [Bi3] provides an improvement for mean zero random variables.

Proposition 3.2 *If $\beta_p = \mathbb{E}\,|X|^p < \infty$ for some integer $p \ge 1$, and $\mathbb{E}X = 0$, then*

$$|\gamma_p| \le (p-1)!\,\beta_p. \tag{3.5}$$

Proof The case $p = 1$ is obvious. Since $\gamma_2 = \alpha_2 = \beta_2$ in case $\mathbb{E}X = 0$, the desired bound also follows for $p = 2$. So, let $p \ge 3$. Differentiating the identity $f'(t) = f(t)\,(\log f(t))'$ near zero $p-1$ times in accordance with the binomial formula, one gets

$$\frac{d^p}{dt^p} f(t) = \sum_{r=0}^{p-1} C_{p-1}^r \, \frac{d^{p-1-r}}{dt^{p-1-r}} f(t)\, \frac{d^{r+1}}{dt^{r+1}} \log f(t),$$

where here and in the sequel we use the notation $C_n^k = \frac{n!}{k!(n-k)!}$ for the binomial coefficients. Equivalently,

$$\frac{d^p}{dt^p} \log f(t) = \frac{1}{f(t)} \frac{d^p}{dt^p} f(t) - \frac{1}{f(t)} \sum_{r=0}^{p-2} C_{p-1}^r \, \frac{d^{p-1-r}}{dt^{p-1-r}} f(t)\, \frac{d^{r+1}}{dt^{r+1}} \log f(t). \tag{3.6}$$

At $t = 0$, this identity becomes

$$\gamma_p = \alpha_p - \sum_{r=0}^{p-3} C_{p-1}^r \, \alpha_{p-1-r}\, \gamma_{r+1},$$

where we used the assumption $\alpha_1 = 0$. One can now proceed by induction on p. Since $|\alpha_{p-1-r}| \le \beta_p^{(p-1-r)/p}$ and $\gamma_{r+1} \le r!\beta_p^{(r+1)/p}$ (the induction hypothesis), we obtain that

$$|\gamma_p| \le \beta_p + \beta_p \sum_{r=0}^{p-3} C_{p-1}^r \, r! = (p-1)!\,\beta_p \left[\frac{1}{(p-1)!} + \sum_{r=0}^{p-3} \frac{1}{(p-1-r)!}\right].$$

The expression in the square brackets $\frac{1}{(p-1)!} + (\frac{1}{2!} + \cdots + \frac{1}{(p-1)!})$ is equal to 1 for $p = 2$ and is smaller than $\frac{1}{6} + (e-2) < 1$ for $p \ge 3$. □

The factorial growth of the constant in the inequality (3.5) is optimal, up to an exponentially growing factor, which was noticed by Bulinskii [Bu] in his study of upper bounds in a more general scheme of random vectors and associated mixed cumulants. To illustrate possible lower bounds, he considered the symmetric Bernoulli distribution assigning the mass $\frac{1}{2}$ to the points ± 1. In this case, the characteristic function is $f(t) = \cos t$, and one may use the Taylor expansion

$$\log f(t) = \log \cos t = -\sum_{p=1}^{\infty} \frac{2^{2p}\,(2^{2p}-1)}{(2p)!}\,B_p\,\frac{t^{2p}}{2p}, \qquad |t| < \frac{\pi}{4},$$

involving Bernoulli numbers $B_p = \frac{2\,(2p)!}{(2\pi)^{2p}}\,d_{2p}$, where $d_{2p} = \sum_{n=1}^{\infty} \frac{1}{n^{2p}}$. Thus, for even integer values of p,

$$|\gamma_p| = \frac{2^p\,(2^p-1)}{p}\,B_{p/2} = \frac{2\,(2^p-1)}{\pi^p}\,(p-1)!\,d_p.$$

From Stirling's formula, one gets $|\gamma_p| \geq (\frac{p}{\pi e})^p\,\sqrt{2\pi}$. To compare with the upper bound of Proposition 2.2, note that in this Bernoulli case, $\beta_p = 1$ for all p.

4 Bounds on the Derivatives of the Logarithm

We will now extend the Bikjalis argument, so as to obtain the following improvement of the bounds (3.3)–(3.4), assuming that X has mean zero and t is small enough. More precisely, we are going to derive the bound

$$\left|\frac{d^p}{dt^p}\,\log f(t)\right| \leq (p-1)!\,\beta_p \tag{4.1}$$

in the interval $\sigma|t| \leq \varepsilon = \frac{1}{5}$ (except for the value $p = 2$), where $\sigma^2 = \beta_2 = \mathbb{E}X^2$. This can be done with the help of the lower bound

$$|f(t)| \geq 1 - \frac{\sigma^2 t^2}{2} \geq 1 - \frac{\varepsilon^2}{2}, \qquad \sigma|t| \leq \varepsilon. \tag{4.2}$$

First let us check (4.1) for the first 4 values of p. Since $|f'(t)| \leq \sigma^2|t|$, we have

$$|(\log f(t))'| \leq \frac{\beta_2|t|}{|f(t)|} \leq \frac{0.2\,\beta_2^{1/2}}{1-\frac{\varepsilon^2}{2}} \leq 0.21\beta_2^{1/2}. \tag{4.3}$$

When $p = 2$, according to inequality (2.2) of Proposition 2.2,

$$|(\log f(t))''| \leq \frac{\beta_2}{|f(t)|^2} \leq \frac{\beta_2}{(1-\frac{\varepsilon^2}{2})^2} \leq 1.05\beta_2. \tag{4.4}$$

When $p = 3$, we use (2.3) giving

$$|(\log f(t))'''| \leq \frac{1+3\varepsilon+2\varepsilon^3}{|f(t)|^3}\,\beta_3 \leq \frac{1+3\varepsilon+2\varepsilon^3}{(1-\frac{\varepsilon^2}{2})^3}\,\beta_3 \leq 2\beta_3.$$

When $p = 4$, we use (2.4) giving similarly

$$|(\log f(t))''''| \le \frac{4 + 4\varepsilon + 12\varepsilon^2 + 6\varepsilon^4}{|f(t)|^4}\,\beta_4 \le \frac{4 + 4\varepsilon + 12\varepsilon^2 + 6\varepsilon^4}{(1 - \frac{\varepsilon^2}{2})^4}\,\beta_4 \le 6\beta_4.$$

In order to derive (4.1) for $p \ge 5$, we perform the induction step, applying (4.3)–(4.4) and assuming that, in the interval $\sigma|t| \le \varepsilon$,

$$|(\log f(t))^{(r)}| \le (r-1)!\,\beta_r \qquad \text{for } 3 \le r \le p-1. \tag{4.5}$$

By this hypothesis, using the recursive formula (3.6) and the bounds (4.3)–(4.4), we have

$$\begin{aligned}
|f(t)|\,|(\log f(t))^{(p)}| &\le |f^{(p)}(t)| + \sum_{r=0}^{p-2} C_{p-1}^r\,|f^{(p-1-r)}(t)|\,|(\log f(t))^{(r+1)}| \\
&= |f^{(p)}(t)| + (p-1)\,|f'(t)|\,|(\log f(t))^{(p-1)}| \\
&\quad + \sum_{r=2}^{p-3} C_{p-1}^r\,|f^{(p-1-r)}(t)|\,|(\log f(t))^{(r+1)}| \\
&\quad + |f^{(p-1)}(t)|\,|(\log f(t))'| + (p-1)\,|f^{(p-2)}(t)|\,|(\log f(t))''| \\
&\le \beta_p + (p-1)\,\beta_p^{1/p}\,\varepsilon \cdot \beta_{p-1}(p-2)! \\
&\quad + \sum_{r=2}^{p-3} C_{p-1}^r\,\beta_{p-1-r}\,\beta_{r+1}\,r! + \beta_{p-1} \cdot 0.21\beta_2^{1/2} \\
&\quad + (p-1)\beta_{p-2} \cdot 1.05\,\beta_2.
\end{aligned}$$

Here we apply again $\beta_r \le \beta_p^{r/p}$, giving

$$\begin{aligned}
\frac{|f(t)|}{\beta_p}\,|(\log f(t))^{(p)}| &\le 1 + (p-1)!\Big[\varepsilon + \sum_{r=2}^{p-3} \frac{1}{(p-1-r)!}\Big] + 0.21 + 1.05\,(p-1) \\
&\le 1 + (p-1)!\Big[\varepsilon + \sum_{k=2}^{p-1} \frac{1}{k!}\Big] + 0.05\,(p-1) \\
&\le 1 + 0.05\,(p-1) + (p-1)!\,(\varepsilon + e - 2).
\end{aligned}$$

Applying the lower bound (4.2), we obtain that

$$\frac{1}{\beta_p}\,|(\log f(t))^{(p)}| \le \frac{1}{1 - \frac{\varepsilon^2}{2}}\,\big(1 + 0.05\,(p-1) + (p-1)!\,(\varepsilon + e - 2)\big).$$

The latter expression does not exceed $(p-1)!$ (which is needed to make the induction step, i.e., to derive (4.5) for $r = p$), if and only if this is true for $p = 5$ (since after division by $(p-1)!$ the expression on the right will be decreasing in p). That is, we need to verify that $\frac{1}{1-\frac{\varepsilon^2}{2}}\,(1.2 + 24\,(\varepsilon + e - 2)) \leq 24$, which is indeed true for $\varepsilon = 0.2$. Hence, we have proved:

Proposition 4.1 *Let X be a random variable such that $\mathbb{E}X = 0$, $\mathbb{E}X^2 = \sigma^2$ $(\sigma > 0)$ and $\beta_p = \mathbb{E}\,|X|^p < \infty$ for some integer $p \geq 2$. Then, in the interval $\sigma|t| \leq \frac{1}{5}$, the characteristic function $f(t)$ of X is not vanishing and satisfies*

$$|(\log f(t))'| \leq 0.21\,\sigma, \qquad |(\log f(t))''| \leq 1.05\,\sigma.$$

Moreover, if $p \geq 3$, then

$$\left|\frac{d^p}{dt^p}\,\log f(t)\right| \leq (p-1)!\,\beta_p.$$

5 Taylor Expansion for Fourier-Stieltjes Transforms

Let X be a random variable with finite absolute moment $\beta_s = \mathbb{E}\,|X|^s$ of a real order $s > 0$, not necessarily integer. Put

$$\mathbb{E}X^k = \alpha_k, \quad \mathbb{E}\,|X|^k = \beta_k \quad (k = 0, 1, \dots, [s]).$$

In general, suitable expansions for the characteristic function $f(t) = \mathbb{E}\,e^{itX}$ can be developed according to the Taylor formula. Since f has $[s]$ continuous derivatives with $f^{(k)}(0) = i^k \alpha_k$, it admits the Taylor expansion

$$f(t) = \sum_{k=0}^{m} \alpha_k\,\frac{(it)^k}{k!} + \delta(t) \tag{5.1}$$

with $\delta(t) = o(t^m)$, where here and elsewhere we represent $s = m + \alpha$ with integer m and $0 < \alpha \leq 1$. The remainder term can be bounded in terms of β_s as follows:

Proposition 5.1 *For all t,*

$$\left|\frac{d^p}{dt^p}\,\delta(t)\right| \leq 2\beta_s\,\frac{|t|^{s-p}}{(m-p)!}, \qquad p = 0, 1, \dots, m. \tag{5.2}$$

Moreover, if $s = m + 1$ is integer, then

$$\left|\frac{d^p}{dt^p}\,\delta(t)\right| \leq \beta_s\,\frac{|t|^{s-p}}{(s-p)!}, \qquad p = 0, 1, \dots, s.$$

Proof By the very definition, $\delta(t) = \mathbb{E}\, R_m(tX)$, where $R_m(u) = e^{iu} - \sum_{l=0}^{m} \frac{(iu)^l}{l!}$, so that

$$\delta^{(p)}(t) = \mathbb{E}\,(iX)^p\, R_{m-p}(tX).$$

Given an integer number $k \geq 1$, note that $R_k^{(j)}(0) = 0$ for all $j = 0, \dots, k$ with $|R_k^{(k+1)}(u)| = 1$. In addition, $R_k^{(k)}(u) = i^k(e^{iu} - 1)$, so that $|R_k^{(k)}(u)| \leq 2$. Hence, by Taylor's formula,

$$|R_k(u)| \leq \frac{|u|^{k+1}}{(k+1)!} \qquad \text{and} \qquad |R_k(u)| \leq 2\,\frac{|u|^k}{k!}.$$

Although some other interesting bounds on the functions R_k are available (cf., e.g., [Sh1]), these two inequalities are sufficient to conclude that, for any $\alpha \in [0, 1]$,

$$\begin{aligned} |R_k(u)| &\leq \min\left\{2\,\frac{|u|^k}{k!},\, \frac{|u|^{k+1}}{(k+1)!}\right\} \\ &= \frac{|u|^k}{k!}\,\min\left\{2,\, \frac{|u|}{k+1}\right\} \;\leq\; \frac{|u|^k}{k!} \cdot \frac{2^{1-\alpha}}{(k+1)^\alpha}\,|u|^\alpha \;\leq\; \frac{2|u|^{k+\alpha}}{k!}. \end{aligned}$$

Therefore,

$$|\delta^{(p)}(t)| \leq \mathbb{E}\left[|X|^p\,\frac{2\,|tX|^{(m-p)+\alpha}}{(m-p)!}\right] = \frac{2\,|t|^{(s-p)}}{(m-p)!}\,\beta_s.$$

In case $s = m + 1$, the function $w(t) = \delta^{(p)}(t)$ has zero derivatives at $t = 0$ up to order $s - p - 1$, while $w^{(s-p)}(t) = \delta^{(s)}(t) = \mathbb{E}\,(iX)^s e^{itX}$ is bounded in absolute value by β_s. Hence, by Taylor's formula,

$$|w(t)| \;\leq\; \max_{|z| \leq |t|} |w^{(s-p)}(z)|\,\frac{|t|^{s-p}}{(s-p)!} \;\leq\; \beta_s\,\frac{|t|^{s-p}}{(s-p)!}.$$

□

More generally, consider the Fourier-Stieltjes transform $a(t) = \int_{-\infty}^{\infty} e^{itx}\,d\mu(x)$ of a Borel signed measure μ on the real line and introduce the corresponding absolute moment

$$\beta_s(\mu) = \int_{-\infty}^{\infty} |x|^s\,|\mu(dx)|,$$

where $|\mu|$ is the variation of μ treated as a positive measure on the line, and $s > 0$ is a real number. Clearly, a is $[s]$ times continuously differentiable on $\mathbb{R}$ with derivatives at the origin

$$a^{(p)}(0) = \int_{-\infty}^{\infty} (ix)^p \, d\mu(x), \qquad p = 0, 1, \dots, [s].$$

Here is a natural generalization of Proposition 5.1.

Proposition 5.2 *Let $s = m + \alpha$ with $m \geq 0$ integer and $0 < \alpha \leq 1$. If $a^{(p)}(0) = 0$ for all $p = 0, 1, \dots, m$, then for all $t \in \mathbb{R}$,*

$$|a^{(p)}(t)| \leq 2\beta_s(\mu) \, \frac{|t|^{s-p}}{(m-p)!}, \qquad p = 0, 1, \dots, m.$$

Moreover, if $s = m + 1$ is integer, then

$$|a^{(p)}(t)| \leq \beta_s(\mu) \, \frac{|t|^{s-p}}{(s-p)!}, \qquad p = 0, 1, \dots, [s].$$

Proof Note that $\mu(\mathbb{R}) = 0$ due to $a(0) = 0$. To prove the statement, one can repeat the arguments used in the proof of Proposition 5.1. By the moment assumption, $a(t) = \int_{-\infty}^{\infty} R_m(tx) \, d\mu(x)$, so

$$a^{(p)}(t) = \int_{-\infty}^{\infty} (ix)^p \, R_{m-p}(tx) \, d\mu(x).$$

Using the previous bound $|R_k(u)| \leq \frac{2|u|^{k+\alpha}}{k!}$ with $k = m - p$, we conclude that

$$|a^{(p)}(t)| \leq \int_{-\infty}^{\infty} \left[|x|^p \, \frac{2\,|tx|^{(m-p)+\alpha}}{(m-p)!} \right] |d\mu(x)| = \frac{2\,|t|^{(s-p)}}{(m-p)!} \, \beta_s(\mu).$$

The case $s = m + 1$ is similar. □

6 Taylor Expansion for Logarithm of Characteristic Functions

Our next task is to develop the Taylor expansion for $\log f(t)$ in analogy with the expansion (5.1) for the characteristic function $f(t)$ with a bound similar to (5.2), which would hold even if t is close to zero. Note that, in the most important case $p = m$, that bound yields

$$|f^{(m)}(t) - i^m \alpha_m| \;\leq\; 2\beta_s \, |t|^{\alpha}. \tag{6.1}$$

Hence, we need to derive a similar bound for $\log f(t)$, by replacing α_m with the cumulant γ_m.

We keep the same assumption as in the previous section: $\mathbb{E}X = 0$, $\beta_s = \mathbb{E}\,|X|^s < \infty$, $s = m + \alpha$ with $m \geq 2$ integer and $0 < \alpha \leq 1$. Let us return to the recursive formula

$$f(t)\,(\log f(t))^{(m)} = f^{(m)}(t) - \sum_{r=1}^{m-1} C_{m-1}^{r-1} f^{(m-r)}(t)\,(\log f(t))^{(r)}, \tag{6.2}$$

which at $t = 0$ becomes

$$i^m \gamma_m = i^m \alpha_m - \sum_{r=1}^{m-1} C_{m-1}^{r-1}\, i^{m-r} \alpha_{m-r}\, i^r \gamma_r. \tag{6.3}$$

Since $\alpha_1 = \gamma_1 = 0$, the last summation may be reduced to the values $2 \leq r \leq m-2$ for $m \geq 4$, while there is no sum for $m = 3$.

To argue by induction on m, our induction hypothesis will be

$$|(\log f(t))^{(r)} - i^r \gamma_r| \leq AB^r (r-1)!\, \beta_{r+\alpha} |t|^\alpha, \qquad r = 1, 2, \dots, m-1, \tag{6.4}$$

in the interval $\sigma\,|t| \leq \frac{1}{5}$, where the parameters $A, B \geq 1$ are to be chosen later on. Recall that Proposition 4.1 provides in this interval the bound

$$|(\log f(t))^{(r)}| \leq A_r (r-1)!\, \beta_r, \qquad r = 2, \dots, m, \tag{6.5}$$

with constants $A_2 = 1.05$ and $A_r = 1$ for $r \geq 3$. Now, let us apply (6.1) with $s = (m-r) + \alpha$. Then we have a similar relation

$$|f^{(m-r)}(t) - i^{m-r} \alpha_{m-r}| \leq 2\beta_{m-r+\alpha} |t|^\alpha, \qquad r = 0, 1, \dots, m-1, \tag{6.6}$$

which is valid for all t. Write

$$\begin{aligned} f^{(m-r)}(t)\,(\log f(t))^{(r)} &= (f^{(m-r)}(t) - i^{m-r} \alpha_{m-r})\,(\log f(t))^{(r)} \\ &\quad + i^{m-r} \alpha_{m-r} \left((\log f(t))^{(r)} - i^r \gamma_r \right) + i^{m-r} \alpha_{m-r}\, i^r \gamma_r. \end{aligned}$$

Applying the bounds (6.4)–(6.6) for $r = 2, \dots, m-1$, we get

$$\begin{aligned} |f^{(m-r)}(t)\,(\log f(t))^{(r)} - i^{m-r} \alpha_{m-r}\, i^r \gamma_r| &\leq 2\beta_{m-r+\alpha} |t|^\alpha \cdot A_r (r-1)!\, \beta_r \\ &\quad + \beta_{m-r} \cdot AB^r (r-1)!\, \beta_{r+\alpha} |t|^\alpha \\ &\leq (r-1)!\, \beta_s |t|^\alpha\, (2A_r + AB^r). \end{aligned}$$

When $r = 1$, we use a different bound based on the assumption that $\alpha_1 = \gamma_1 = 0$. Namely, by Proposition 4.1 in part concerning the first derivative, we have

$$|f^{(m-1)}(t)\,(\log f(t))'| \leq 2\beta_{m-1+\alpha} |t|^\alpha \cdot A_1 \beta_2^{1/2} \leq 2A_1\, \beta_s |t|^\alpha,$$

where $A_1 = 0.21$. Hence, subtracting the representation (6.3) from (6.2) and applying the bound (6.1), we get

$$|f(t)\,(\log f(t))^{(m)} - i^m\gamma_m| \leq 2\beta_s|t|^\alpha + (m-1)!\,\beta_s|t|^\alpha \sum_{r=1}^{m-1} \frac{1}{(m-r)!}\,(2A_r + AB^r)$$

$$= AB^m\,(m-1)!\,\beta_s|t|^\alpha \left[\frac{2}{AB^m} + \sum_{k=1}^{m-1} \frac{1}{k!}\left(\frac{2A_{m-k}}{AB^m} + B^{-k}\right)\right].$$

In addition, since $|f(t) - 1| \leq 2\beta_\alpha|t|^\alpha$, we have

$$|f(t)\,(\log f(t))^{(m)} - (\log f(t))^{(m)}| \leq A_m(m-1)!\,\beta_m \cdot 2\beta_\alpha|t|^\alpha \leq 2(m-1)!\,\beta_s|t|^\alpha.$$

Hence

$$|(\log f(t))^{(m)} - i^m\gamma_m| \leq AB^m\,(m-1)!\,\beta_s|t|^\alpha \left[\frac{4}{AB^m} + \sum_{k=1}^{m-1} \frac{1}{k!}\left(\frac{2A_{m-k}}{AB^m} + B^{-k}\right)\right],$$

and we can make an induction step by proving (6.4) for $r = m$, once the parameters satisfy

$$\frac{4}{AB^m} + \sum_{k=1}^{m-1} \frac{1}{k!}\left(\frac{2A_{m-k}}{AB^m} + B^{-k}\right) \leq 1.$$

To simplify, let us use a uniform bound $A_{m-k} \leq 1.05$, so that to estimate the above left-hand side from above by

$$\frac{4}{AB^m} + \sum_{k=1}^{\infty} \frac{1}{k!}\left(\frac{2.1}{AB^m} + B^{-k}\right) = \frac{4 + 2.1\,(e-1)}{AB^m} + (e^{1/B} - 1) < \frac{7.61}{AB^m} + (e^{1/B} - 1).$$

For example, for $B = 2$, the last term $\sqrt{e} - 1 < 0.65$. Hence, in case $m \geq 3$, we need $\frac{7.61}{8A} \leq 0.35$, where $A = 2.72$ fits well. Then we obtain (6.4) for $r = m$, i.e.,

$$|(\log f(t))^{(m)} - i^m\gamma_m| \leq A \cdot 2^m(m-1)!\,\beta_{m+\alpha}|t|^\alpha \tag{6.7}$$

for all $m \geq 1$ and with any $A \geq 2.72$, once we have this inequality for the first two values $m = 1$ and $m = 2$ (induction hypothesis).

When $m = 1$, according to (6.1) with $s = 1 + \alpha$, we have $|f'(t)| \leq 2\beta_{1+\alpha}|t|^\alpha$, so

$$|(\log f(t))'| = \frac{|f'(t)|}{|f(t)|} \leq \frac{2\beta_{1+\alpha}|t|^\alpha}{1 - \frac{\varepsilon^2}{2}} \leq 2.05\,\beta_{1+\alpha}|t|^\alpha,$$

so (6.7) is fulfilled. When $m = 2$,

$$(\log f(t))'' + \sigma^2 = \frac{f''(t)f(t) - f'(t)^2}{f(t)^2} + \sigma^2$$
$$= \frac{(f''(t) + \sigma^2)f(t) + \sigma^2 f(t)(f(t) - 1) - f'(t)^2}{f(t)^2}.$$

According to (6.1), $|f''(t) + \sigma^2| \le 2\beta_{2+\alpha}|t|^\alpha$ and $|f(t) - 1| \le 2\beta_\alpha|t|^\alpha$. Hence,

$$|(\log f(t))'' + \sigma^2| \le \frac{2\beta_{2+\alpha}|t|^\alpha + 2\sigma^2\beta_\alpha|t|^\alpha + 2\beta_1\beta_{1+\alpha}|t|^\alpha}{|f(t)|^2}$$
$$\le \frac{6\beta_{2+\alpha}|t|^\alpha}{(1 - \frac{\varepsilon^2}{2})^2} \le 6.25\,\beta_{2+\alpha}|t|^\alpha.$$

In both cases, (6.7) is fulfilled with $A = 2.72$. Thus, we have proved:

Lemma 6.1 *Let X be a random variable such that $\mathbb{E}X = 0$, $\mathbb{E}X^2 = \sigma^2$ ($\sigma > 0$), and $\beta_{m+\alpha} < \infty$ for some integer $m \ge 2$ and $0 < \alpha \le 1$. Then, in the interval $\sigma|t| \le \frac{1}{5}$, the characteristic function $f(t)$ of X is not vanishing and satisfies*

$$\left|\frac{d^m}{dt^m}\log f(t) - i^m\gamma_m\right| \le 2.72 \cdot 2^m\,(m-1)!\,\beta_{m+\alpha}|t|^\alpha.$$

This inequality remains to hold for $m = 1$ as well, if $\mathbb{E}X^2$ is finite.

Now, if s is integer, for any $p = 0, 1, \dots, s$, the function

$$w(t) = \frac{d^p}{dt^p}\log f(t) - \frac{d^p}{dt^p}\sum_{k=2}^{s-1}\gamma_k\,\frac{(it)^k}{k!}$$

has zero derivatives at $t = 0$ up to order $s - p - 1$, while $w^{(s-p)}(t) = \frac{d^s}{dt^s}\log f(t)$. Hence, by Proposition 4.1 and Taylor's formula,

$$|w(t)| \le \sup_{|z|\le|t|}|w^{(s-p)}(z)|\,\frac{|t|^{s-p}}{(s-p)!} \le (s-1)!\,\beta_s\,\frac{|t|^{s-p}}{(s-p)!}, \qquad \text{if } \sigma|t| \le \frac{1}{5}.$$

In the general case $s = m + \alpha$ with integer $m \ge 2$ and $0 < \alpha \le 1$, for any $p = 0, 1, \dots, m$, consider the function

$$w(t) = \frac{d^p}{dt^p}\log f(t) - \frac{d^p}{dt^p}\sum_{k=2}^{m}\gamma_k\,\frac{(it)^k}{k!}.$$

It has zero derivatives at $t = 0$ up to order $m-p-1$, while $w^{(m-p)}(t) = \frac{d^m}{dt^m} \log f(t) - \gamma_m i^m$. Hence, for $p \le m-1$, by Taylor's integral formula,

$$\begin{aligned} w(t) &= \frac{t^{m-p}}{(m-p-1)!} \int_0^1 (1-u)^{m-p-1} \, w^{(m-p)}(tu) \, du \\ &= \frac{t^{m-p}}{(m-p-1)!} \int_0^1 (1-u)^{m-p-1} \left((\log f)^{(m)}(tu) - \gamma_m i^m \right) du. \end{aligned}$$

Applying Lemma 6.1, we then get that

$$\begin{aligned} |w(t)| &\le \frac{|t|^{m-p}}{(m-p-1)!} \int_0^1 (1-u)^{m-p-1} \, 2.72 \cdot 2^m (m-1)! \, \beta_s \, |tu|^\alpha \, du \\ &= 2.72 \cdot 2^m \, (m-1)! \, \beta_s \, |t|^{s-p} \, \frac{\Gamma(\alpha+1)}{\Gamma(s-p+1)}. \end{aligned}$$

The obtained inequality is also true for $p = m$ (Lemma 6.1). Using $\Gamma(\alpha+1) \le 1$, we arrive at:

Proposition 6.2 *Let f be the characteristic function of a random variable X with $\mathbb{E}X = 0$ and $\beta_s = \mathbb{E}\,|X|^s < \infty$ for some $s > 2$. Put $s = m + \alpha$ with m integer and $0 < \alpha \le 1$. Then in the interval $\sigma|t| \le \frac{1}{5}$,*

$$\log f(t) = \sum_{k=2}^m \gamma_k \, \frac{(it)^k}{k!} + \varepsilon(t)$$

with

$$\left| \frac{d^p}{dt^p} \, \varepsilon(t) \right| \le 2.72 \cdot 2^m (m-1)! \, \beta_s \, \frac{|t|^{s-p}}{\Gamma(s-p+1)}$$

for all $p = 0, 1, \dots, m$. If $\alpha = 1$, in the same interval, for all $p = 0, 1, \dots, m+1$,

$$\left| \frac{d^p}{dt^p} \, \varepsilon(t) \right| \le m! \, \beta_s \, \frac{|t|^{s-p}}{\Gamma(s-p+1)}.$$

Let us state particular cases in this statement corresponding to the values $s = 3$ and $s = 4$.

Corollary 6.3 *Let $f(t)$ be the characteristic function of a random variable X with $\mathbb{E}X = 0$. If $\beta_3 = \mathbb{E}\,|X|^3 < \infty$, then in the interval $\sigma|t| \le \frac{1}{5}$,*

$$\log f(t) = -\frac{\sigma^2 t^2}{2} + \varepsilon(t) \quad \text{with} \quad \left| \frac{d^p}{dt^p} \, \varepsilon(t) \right| \le 6\beta_3 \, \frac{|t|^{3-p}}{(3-p)!}, \qquad p = 0, 1, 2, 3.$$

Moreover, if $\beta_4 = \mathbb{E}X^4 < \infty$, *then*

$$\log f(t) = -\frac{\sigma^2 t^2}{2} + \alpha_3 \frac{(it)^3}{6} + \varepsilon(t) \quad \textit{with} \quad \left|\frac{d^p}{dt^p}\,\varepsilon(t)\right| \leq 24\,\beta_4\,\frac{|t|^{4-p}}{(4-p)!},$$
$$p = 0, 1, 2, 3, 4.$$

PART II. Lyapunov coefficients and products of characteristic functions

7 Properties of Lyapunov Coefficients

From now on, we deal with a sequence $X_1, \dots, X_n$ of independent random variables such that $\mathbb{E}X_k = 0$, $\mathbb{E}X_k^2 = \sigma_k^2$ ($\sigma_k \geq 0$) and $\sum_{k=1}^n \sigma_k^2 = 1$. The latter insures that the sum

$$S_n = X_1 + \cdots + X_n$$

has the first two moments $\mathbb{E}S_n = 0$ and $\mathbb{E}S_n^2 = 1$. For $s \geq 2$, consider the absolute moments $\beta_{s,k} = \mathbb{E}\,|X_k|^s$ and the corresponding Lyapunov coefficients

$$L_s = \sum_{k=1}^n \mathbb{E}\,|X_k|^s.$$

First, below we state a few simple, but useful auxiliary results about these quantities.

Proposition 7.1 *The function* $L_s^{\frac{1}{s-2}}$ *is non-decreasing in* $s > 2$. *In particular,* $L_3 \leq L_s^{\frac{1}{s-2}}$ *for all* $s \geq 3$.

Proof Let F_k denote the distribution of X_k. By the basic assumption on the variances σ_k^2, the equality $d\mu(x) = \sum_{k=1}^n x^2\,dF_k(x)$ defines a probability measure on the real line. Moreover,

$$L_s = \sum_{k=1}^n \int_{-\infty}^{\infty} |x|^s\,dF_k(x) = \int_{-\infty}^{\infty} |x|^{s-2}\,d\mu(x) = \mathbb{E}\,|\xi|^{s-2},$$

where ξ is a random variable distributed according to μ. Hence, $L_s^{\frac{1}{s-2}} = (\mathbb{E}\,|\xi|^{s-2})^{\frac{1}{s-2}}$. Here the right-hand side represents a non-decreasing function in s. □

Proposition 7.2 *We have* $\max_k \sigma_k \leq L_s^{1/s}$ ($s \geq 2$). *In particular,* $L_3^{1/3} \geq \max_k \sigma_k$.

Proof Using $\sigma_k^s \leq \beta_{s,k}$, we have $\max_k \sigma_k \leq \big(\sum_{k=1}^n \sigma_k^s\big)^{1/s} \leq \big(\sum_{k=1} \beta_{s,k}\big)^{1/s} = L_s^{1/s}$. □

There is also a uniform lower bound on the Lyapunov coefficients depending upon n, only.

Proposition 7.3 *We have* $L_s \geq n^{-\frac{s-2}{2}}$ $(s \geq 2)$. *In particular,* $L_3 \geq \frac{1}{\sqrt{n}}$ *and* $L_4 \geq \frac{1}{n}$.

Proof Let $s > 2$. By Hölder's inequality with exponents $p = \frac{s}{s-2}$ and $q = \frac{s}{2}$,

$$1 = \sum_{k=1}^{n} \sigma_k^2 \leq n^{1/p} \Big(\sum_{k=1}^{n} \sigma_k^s\Big)^{1/q} \leq n^{1/p} \Big(\sum_{k=1}^{n} \beta_{s,k}\Big)^{1/q} = n^{1/p} L_s^{1/q}.$$

Hence, $L_s \geq n^{-q/p}$. □

Note that the finiteness of the moments $\beta_{s,k}$ for all $k \leq n$ is equivalent to the finiteness of the Lyapunov coefficient L_s. In this case, one may introduce the corresponding cumulants

$$\gamma_{p,k} = \gamma_p(X_k) = \frac{d^p}{i^p\, dt^p} \log v_k(t)\big|_{t=0}, \qquad p = 0, 1, 2, \dots, [s],$$

where $v_k = \mathbb{E}\, e^{itX_k}$ denote the characteristic functions of X_k. Since the characteristic function of S_n is given by the product

$$f_n(t) = \mathbb{E}\, e^{itS_n} = v_1(t) \dots v_n(t),$$

the cumulants of S_n exist for the same values of p and are given by

$$\gamma_p = \gamma_p(S_n) = \frac{d^p}{i^p\, dt^p} \log f_n(t)\big|_{t=0} = \sum_{k=1}^{n} \gamma_{p,k}.$$

The first values are $\gamma_0 = \gamma_1 = 0$, $\gamma_2 = 1$.

Applying Proposition 3.2 (Bikjalis inequality), we immediately obtain a similar relation between the Lyapunov coefficients and the cumulants of the sums.

Proposition 7.4 *For all* $p = 2, \dots, [s]$,

$$|\gamma_p| \leq (p-1)!\, L_p. \tag{7.1}$$

The Lyapunov coefficients may also be used to bound absolute moments of the sums S_n. In particular, there is the following observation due to Rosenthal [R].

Proposition 7.5 *With some constants* A_s *depending on* s, *only,*

$$\mathbb{E}\, |S_n|^s \leq A_s \max\{L_s, 1\}. \tag{7.2}$$

Moment inequalities of the form (7.2) are called Rosenthal's or Rosenthal-type inequalities. The study of the best value A_s has a long story, and here we only mention several results.

Define A_s^* to be an optimal constant in (7.2), when it is additionally assumed that the distributions of X_k are symmetric about the origin. By Jensen's inequality, for the optimal constant A_s there is a simple general relation

$$A_s^* \le A_s \le 2^{s-1} A_s^*,$$

which reduces in essence the study of Rosenthal-type inequalities to the symmetric case.

Johnson, Schechtman, and Zinn [J-S-Z] have derived the two-sided bounds

$$\frac{s}{\sqrt{2}\, e \, \log(\max(s,e))} \le (A_s^*)^{1/s} \le \frac{7.35\, s}{\log(\max(s,e))}.$$

Hence, asymptotically $A_s^{1/s}$ is of order $s/\log s$ for growing values of s. They have also obtained an upper bound with a better numerical factor, $(A_s^*)^{1/s} \le s/\sqrt{\log \max(s,e)}$, which implies a simple bound

$$A_s \le (2s)^s, \quad s > 2. \tag{7.3}$$

As for the best constant in the symmetric case, it was shown by Ibragimov and Sharakhmetov [I-S] that $A_s^* = \mathbb{E}\,|\xi - \eta|^s$ for $s > 4$, where ξ and η are independent Poisson random variables with parameter $\lambda = \frac{1}{2}$ (cf. also [Pi] for a similar description without the symmetry assumption). In particular, $(A_s^*)^{1/s} \sim \frac{s}{e \log s}$ as s tends to infinity. This result easily yields

$$A_s^* \le s! \qquad \text{for } s = 3, 4, 5, \dots,$$

and thus $A_s \le 2^{s-1} s!$ For even integers s, there is an alternative argument. Applying the expression (3.2) to S_n (for the cumulants in terms of the moments) and recalling (7.1), we get

$$\mathbb{E}\,|S_n|^s = \alpha_s(S_n) = s! \sum \prod_{r=1}^{s} \frac{1}{k_r!} \Big(\frac{\gamma_r(S_n)}{r!}\Big)^{k_r} \le s! \sum \prod_{r=1}^{s} \frac{1}{k_r!} \Big(\frac{L_{r^*}}{r}\Big)^{k_r}, \tag{7.4}$$

where $r^* = \max(r, 2)$, and where the summation is performed over all tuples $(k_1, \dots, k_s)$ of non-negative integers such that $k_1 + 2k_2 + \dots + sk_s = s$. (The left representation was emphasized in [P-U].) Now, by Proposition 7.1, $L_r \le L_s^{\frac{r-2}{s-2}} \le (\max(L_s, 1))^{r/s}$. Hence, by Lemma 1.2 (cf. (1.5)), the last sum in (7.4) does not exceed

$$\sum \prod_{r=2}^{s} \frac{1}{k_r!}\Big(\frac{(\max(L_s^{1/s},1))^r}{r}\Big)^{k_r} \leq \max(L_s,1).$$

Hence, $A_s \leq s!$ for $s = 4, 6, 8, \dots$

To involve real values of s, for our further purposes it will be sufficient to use the upper bound (7.3).

8 Logarithm of the Product of Characteristic Functions

We keep the same notations and assumptions as in the previous section. Let us return to the characteristic function

$$f_n(t) = \mathbb{E}\, e^{itS_n} = v_1(t)\dots v_n(t)$$

of the sum $S_n = X_1 + \dots + X_n$ in terms of the characteristic functions $v_k = \mathbb{E}\, e^{itX_k}$. To get the Taylor expansion for f_n, recall that, by Proposition 6.2, applied to each X_k, we have

$$v_k(t) = \exp\Big\{\sum_{l=2}^{m} \gamma_{l,k}\,\frac{(it)^l}{l!} + \varepsilon_k(t)\Big\}. \tag{8.1}$$

As we know, the function ε_k has $[s]$ continuous derivative, satisfying in the interval $\sigma_k|t| \leq \frac{1}{5}$

$$\Big|\frac{d^p}{dt^p}\,\varepsilon_k(t)\Big| \leq 2.72 \cdot 2^m (m-1)!\,\beta_{s,k}\,\frac{|t|^{s-p}}{\Gamma(s-p+1)}, \qquad p = 0, 1, \dots, m.$$

This assertion also extends to the case $p = m+1$, when $\alpha = 1$ (with better constants). Multiplying the expansions (8.1) and using $\gamma_2 = 1$, we arrive at a similar expansion for f.

Lemma 8.1 *Assume that $L_s < \infty$ for some $s = m + \alpha$ with $m \geq 2$ integer and $0 < \alpha \leq 1$. Then, in the interval $\max_k \sigma_k|t| \leq \frac{1}{5}$, we have*

$$e^{t^2/2} f_n(t) = \exp\big\{Q_m(it) + \varepsilon(t)\big\}, \qquad Q_m(it) = \sum_{l=3}^{m} \gamma_l\,\frac{(it)^l}{l!}, \tag{8.2}$$

where the function ε has $[s]$ continuous derivatives, satisfying for all $p = 0, 1, \dots, m$,

$$\Big|\frac{d^p}{dt^p}\,\varepsilon(t)\Big| \leq 2.72 \cdot 2^m (m-1)!\,L_s\,\frac{|t|^{s-p}}{\Gamma(s-p+1)}. \tag{8.3}$$

In addition, if $s = m + 1 \geq 3$, then in the same interval, for all $p = 0, 1, \dots, m + 1$,

$$\left| \frac{d^p}{dt^p} \varepsilon(t) \right| \leq m!\, L_s \frac{|t|^{s-p}}{\Gamma(s - p + 1)}. \tag{8.4}$$

Both bounds hold in the interval $L_s^{\frac{1}{s}}|t| \leq \frac{1}{5}$, since $L_s^{\frac{1}{s}} \geq \max_k \sigma_k$ (Proposition 8.2). In case $s \geq 3$, these bounds hold in the interval $L_3|t|^3 \leq 1$.

As a next natural step, we want to replace the term $e^{\varepsilon(t)}$ in (8.2) with a simpler one, $1 + \varepsilon(t)$, keeping similar bounds on the remainder term as in (8.3)–(8.4). To this aim, in the smaller interval $L_s^{1/s}|t| \leq \frac{1}{8}$, we consider the function

$$\delta(t) = e^{\varepsilon(t)} - 1.$$

By Proposition 1.1, for any $p = 1, \dots, m$,

$$\delta^{(p)}(t) = \frac{d^p}{dt^p} e^{\varepsilon(t)} = p!\, e^{\varepsilon(t)} \sum \prod_{r=1}^{p} \frac{1}{k_r!} \Big(\frac{1}{r!}\, \varepsilon^{(r)}(t) \Big)^{k_r}, \tag{8.5}$$

where the summation is performed over all non-negative integer solutions $k = (k_1, \dots, k_p)$ to $k_1 + 2k_2 + \dots + pk_p = p$. By (8.3) with $p = 0$,

$$|\varepsilon(t)| \leq 2.72 \cdot \frac{2^m}{m} L_s|t|^s \leq 2.72\, \frac{2^m}{m\, 8^s} \leq 1.36 \left(\frac{1}{4}\right)^s < 0.09,$$

since $s \geq m \geq 2$. Hence,

$$|\delta(t)| \leq e^{0.09}\, |\varepsilon(t)| \leq \frac{3 \cdot 2^m}{m} L_s|t|^s.$$

As for derivatives of order $1 \leq r \leq m$, applying (8.3) and the bound $C_m^r \leq 2^{m-1}$, we have

$$\begin{aligned} \frac{1}{r!}\, |\varepsilon^{(r)}(t)| &\leq 2.72 \cdot 2^m (m-1)!\, \frac{L_s|t|^{s-r}}{r!\, \Gamma(s - r + 1)} \\ &\leq 2.72 \cdot \frac{2^m}{m} \frac{m!}{r!\, (m-r)!} L_s|t|^{s-r} \leq 1.36 \cdot \frac{4^m}{m} L_s|t|^{s-r}. \end{aligned}$$

Here $\lambda \equiv 1.36 \cdot \frac{4^m}{m} L_s|t|^s \leq 0.68 \cdot 2^{-m} \leq 0.68 \cdot 2^{-p}$ whenever $1 \leq p \leq m$. Hence, by Lemma 1.2 with this value of λ and with $\lambda_0 = 0.68$ (cf. (1.4)), we have

$$\sum \prod_{r=1}^{p} \frac{1}{k_r!} \Big(1.36 \cdot \frac{4^m}{m} L_s|t|^{s-r} \Big)^{k_r} \leq e^{0.17}\, 1.36 \cdot \frac{4^m}{m} L_s|t|^{s-p}.$$

As a result, from (8.5) we get

$$\frac{1}{p!}\,|\delta^{(p)}(t)| \le e^{|\varepsilon(t)|} \sum \prod_{r=1}^{p} \frac{1}{k_r!}\,\Big|\frac{1}{r!}\,\varepsilon^{(r)}(t)\Big|^{k_r}$$

$$\le e^{0.09} \sum \prod_{r=1}^{p} \frac{1}{k_r!}\Big(1.36 \cdot \frac{4^m}{m}\,L_s|t|^{s-r}\Big)^{k_r}$$

$$= e^{0.09}\,e^{0.17}\,1.36 \cdot \frac{4^m}{m}\,L_s|t|^{s-p} \,\le\, 2 \cdot \frac{4^m}{m}\,L_s|t|^{s-p}.$$

As we have seen, the resulting bound also holds for $p = 0$ (with a better constant). More precisely, we thus get

$$\frac{1}{p!}\,|\delta^{(p)}(t)| \le 2 \cdot \frac{4^m}{m} L_s|t|^{s-p} \quad (1 \le p \le m), \qquad |\delta(t)| \le 3 \cdot \frac{2^m}{m} L_s|t|^s \quad (p = 0).$$

Scenario 2. In case $s = m + 1$ is integer, $m \ge 2$, one may involve an additional value $p = m + 1$. In case $p = 0$, (8.4) gives $|\varepsilon(t)| \le L_s|t|^s \le (\frac{1}{8})^3$, and then

$$|\delta(t)| \le e^{1/8^3}\,|\varepsilon(t)| \le 1.002\,L_s|t|^s.$$

For the derivatives of order $1 \le r \le m + 1$, we have

$$\frac{1}{r!}\,|\varepsilon^{(r)}(t)| \le m!\,\frac{L_s|t|^{s-r}}{r!\,\Gamma(s-r+1)}$$

$$= \frac{m!}{r!\,((m+1)-r)!}\,L_s|t|^{s-r} \,\le\, \frac{2^m}{m+1} L_s|t|^{s-r}.$$

Here $\frac{2^m}{m+1} L_s|t|^s \le \frac{1}{3}\,(\frac{2}{8})^m < \frac{1}{12}\,2^{-p}$, if $1 \le p \le m + 1$. Hence, by Lemma 1.2 with $\lambda_0 = \frac{1}{12}$,

$$\sum \prod_{r=1}^{p} \frac{1}{k_r!}\Big(\frac{2^m}{m+1} L_s|t|^{s-r}\Big)^{k_r} \le e^{1/48} \cdot \frac{2^m}{m+1} L_s|t|^{s-p}.$$

As a result, for any $p = 1, \dots, m + 1$,

$$\frac{1}{p!}\,|\delta^{(p)}(t)| \le e^{|\varepsilon(t)|} \sum \prod_{r=1}^{p} \frac{1}{k_r!}\Big|\frac{1}{r!}\,\varepsilon^{(r)}(t)\Big|^{k_r}$$

$$\le 1.002 \sum \prod_{r=1}^{p} \frac{1}{k_r!}\Big(\frac{2^m}{m+1} L_s|t|^{s-r}\Big)^{k_r}$$

$$= 1.002\, e^{1/48} \cdot \frac{2^m}{m+1} L_s |t|^{s-p} \le 1.1 \cdot \frac{2^m}{m+1} L_s |t|^{s-p}.$$

We thus get

$$\frac{1}{p!}\, |\delta^{(p)}(t)| \le 1.1 \cdot \frac{2^m}{m+1} L_s |t|^{s-p} \quad (1 \le p \le m+1), \qquad |\delta(t)| \le 1.1\, L_s |t|^s \quad (p = 0).$$

Let us summarize, replacing δ with ε (as the notation, only).

Proposition 8.2 *Assume that $L_s < \infty$ for $s = m + \alpha$ with $m \ge 2$ integer and $0 < \alpha \le 1$. Then in the interval $L_s^{\frac{1}{s}} |t| \le \frac{1}{8}$, we have*

$$e^{t^2/2} f_n(t) = e^{Q_m(it)}\, (1 + \varepsilon(t)), \qquad Q_m(it) = \sum_{l=3}^{m} \gamma_l \frac{(it)^l}{l!}, \tag{8.6}$$

where the function ε has $[s]$ continuous derivatives, satisfying

$$\frac{1}{p!} \left| \frac{d^p}{dt^p}\, \varepsilon(t) \right| \le C_m L_s |t|^{s-p}, \qquad p = 0, 1, \dots, m,$$

with $C_m = 2 \cdot \frac{4^m}{m}$. Moreover, if $s = m + 1$, one may take $C_m = 1.1 \cdot \frac{2^m}{m+1}$ for all $0 \le p \le m + 1$. If $p = 0$, this bound holds with $C_m = 3 \cdot \frac{2^m}{m}$. Moreover, one may take $C_m = 1.1$ when $s = m + 1$.

9 The Case $2 < s \le 3$

For the values $2 < s \le 3$, the cumulant sum in (8.2) and (8.6) does not contain any term, that is, $Q_m = 0$, so

$$f_n(t) = e^{-t^2/2}\, (1 + \varepsilon(t)).$$

Let us specify Proposition 8.2 in this case. If $L_s < \infty$ for $s = 2 + \alpha$, $0 < \alpha \le 1$, we obtain that in the interval $L_s^{1/s} |t| \le \frac{1}{8}$, the function $\varepsilon(t)$ has $[s]$ continuous derivatives satisfying

$$|\varepsilon(t)| \le 6 L_s |t|^s, \qquad \left| \frac{d^p}{dt^p}\, \varepsilon(t) \right| \le 16\, L_s |t|^{s-p} \quad (p = 1, 2).$$

Moreover, in case $s = 3$,

$$|\varepsilon(t)| \le 1.1\, L_s |t|^3, \qquad \left| \frac{d^p}{dt^p}\, \varepsilon(t) \right| \le 1.5\, L_3 |t|^{3-p} \quad (p = 1, 2, 3).$$

Using these representations, one may easily derive the following two propositions.

Proposition 9.1 *Let $L_s < \infty$ for $s = 2 + \alpha$ $(0 < \alpha < 1)$. Then in the interval $L_s^{\frac{1}{s}}|t| \leq \frac{1}{8}$,*

$$\left|f_n(t) - e^{-t^2/2}\right| \leq 6L_s|t|^s\, e^{-t^2/2},$$

$$\left|\frac{d}{dt}\left(f_n(t) - e^{-t^2/2}\right)\right| \leq 16\, L_s\left(|t|^{s-1} + |t|^{s+1}\right) e^{-t^2/2},$$

$$\left|\frac{d^2}{dt^2}\left(f_n(t) - e^{-t^2/2}\right)\right| \leq 32\, L_s\left(|t|^{s-2} + |t|^{s+2}\right) e^{-t^2/2}.$$

Proof Introduce the function $h(t) = f_n(t) - e^{-t^2/2} = e^{-t^2/2}\,\varepsilon(t)$. The first inequality is immediate. Next,

$$e^{t^2/2}\,|h'(t)| = |\varepsilon'(t) - t\varepsilon(t)| \leq 16L_s\,(|t|^{s-1} + |t|^{s+1}).$$

For the second derivative, we get

$$\begin{aligned} e^{t^2/2}\,|h''(t)| &\leq |\varepsilon''(t)| + 2|t|\,|\varepsilon'(t)| + |t^2 - 1|\,|\varepsilon(t)| \\ &\leq 16L_s\left(|t|^{s-2} + 2|t|\,|t|^{s-1} + |t^2 - 1|\,|t|^s\right) \\ &= 16L_s|t|^{s-2}\left(1 + 2t^2 + |t^2 - 1|\,t^2\right). \end{aligned}$$

If $|t| \leq 1$, then the expression in the last brackets is equal to $1 + 2t^2 - t^4 \leq 2\,(1 + t^4)$. If $|t| \geq 1$, it is equal to $1 + t^2 + t^4 \leq 2(1 + t^4)$. □

Proposition 9.2 *Let $L_3 < \infty$. Then in the interval $L_3^{1/3}|t| \leq \frac{1}{8}$,*

$$\left|f_n(t) - e^{-t^2/2}\right| \leq 1.1\, L_3|t|^3\, e^{-t^2/2},$$

$$\left|\frac{d}{dt}\left(f_n(t) - e^{-t^2/2}\right)\right| \leq 1.5\, L_3\,(t^2 + t^4)\, e^{-t^2/2},$$

$$\left|\frac{d^2}{dt^2}\left(f_n(t) - e^{-t^2/2}\right)\right| \leq 3L_3\,(|t| + |t|^5)\, e^{-t^2/2},$$

$$\left|\frac{d^3}{dt^3}\left(f_n(t) - e^{-t^2/2}\right)\right| \leq 12L_3\,(1 + t^6)\, e^{-t^2/2}.$$

Proof Again, consider the function $h(t) = f_n(t) - e^{-t^2/2} = e^{-t^2/2}\,\varepsilon(t)$. The case $p = 0$ is immediate. For $p = 1$,we have

$$e^{t^2/2}\,|h'(t)| = |\varepsilon'(t) - t\varepsilon(t)| \leq 1.5\, L_3\,(t^2 + t^4).$$

For $p = 2$, we get, using the previous arguments,

$$e^{t^2/2}\,|h''(t)| \le |\varepsilon''(t)| + 2|t|\,|\varepsilon'(t)| + |t^2-1|\,|\varepsilon(t)|$$
$$\le 1.5\,L_3\left(|t| + 2|t|\,t^2 + |t^2-1|\,|t|^3\right) \le 3L_3|t|\left(1+t^4\right).$$

Finally, for $p = 3$, using $|\varepsilon^{(p)}(t)| \le 2.2\,L_3|t|^{3-p}$ for $p = 0, 1, 2, 3$, we get

$$e^{t^2/2}\,|h'''(t)| \le |\varepsilon'''(t)| + 3|t|\,|\varepsilon''(t)| + 3\,|t^2-1|\,|\varepsilon'(t)| + |t^3-3t|\,|\varepsilon(t)|$$
$$\le 1.5\,L_3\left(1 + 3t^2 + 3\,|t^2-1|\,t^2 + |t^3-3t|\,|t|^3\right).$$

If $|t| \le 1$, the expression in the brackets equals and does not exceed $1 + 6t^2 - t^6 \le 1 + 4\sqrt{2} < 8$. If $|t| \ge 1$, it does not exceed $1 + 6t^4 + t^6 \le 8t^6$. □

PART III. "Corrected normal characteristic" functions

10 Polynomials P_m in the Normal Approximation

Let us return to the approximation given in Proposition 8.2, i.e.,

$$e^{t^2/2} f_n(t) = e^{Q_m(it)}\,(1+\varepsilon(t)), \quad \text{where} \quad Q_m(it) = \sum_{l=3}^{m} \gamma_l\,\frac{(it)^l}{l!} \quad (\gamma_l = \gamma_l(S_n)).$$

We are now going to simplify the expression $e^{Q_m(it)}\,(1+\varepsilon(t))$ to the form $1+P_m(it)+\varepsilon(t)$ with a certain polynomial P_m and with a new remainder term, which would be still as small as the Lyapunov coefficient L_s (including the case of derivatives). This may indeed be possible on a smaller interval in comparison with $L_s^{1/s}|t| \le 1$. In view of Propositions 9.1–9.2, one may naturally assume that $s > 3$, so that $s = m + \alpha$, $m \ge 3$ (integer), $0 < \alpha \le 1$.

Using Taylor's expansion for the exponential function, one can write

$$e^{Q_m(it)} = \sum_{k_1=0}^{\infty}\left(\frac{\gamma_3}{3!}\right)^{k_1}\frac{(it)^{3k_1}}{k_1!}\cdots\sum_{k_{s-3}=0}^{\infty}\left(\frac{\gamma_m}{m!}\right)^{k_{m-2}}\frac{(it)^{mk_{m-2}}}{k_{m-2}!}$$
$$= \sum_{k_1,\dots,k_{m-2}\ge 0}\frac{\gamma_3^{k_1}\cdots\gamma_m^{k_{m-2}}}{3!^{k_1}\cdots m!^{k_{m-2}}}\,\frac{(it)^{3k_1+\dots+mk_{m-2}}}{k_1!\dots k_{m-2}!} = \sum_{k=0}^{\infty} a_k\,(it)^k$$

with coefficients

$$a_k = \sum_{3k_1+\dots+mk_{m-2}=k}\frac{1}{k_1!\dots k_{m-2}!}\left(\frac{\gamma_3}{3!}\right)^{k_1}\cdots\left(\frac{\gamma_m}{m!}\right)^{k_{m-2}}.$$

Clearly, all these series are absolutely convergent for all t. A certain part of the last infinite series represents the desired polynomial P_m.

Definition 10.1 Put

$$P_m(it) = \sum \frac{1}{k_1! \dots k_{m-2}!} \left(\frac{\gamma_3}{3!}\right)^{k_1} \dots \left(\frac{\gamma_m}{m!}\right)^{k_{m-2}} (it)^{3k_1+\dots+mk_{m-2}},$$

where the summation runs over all collections of non-negative integers $(k_1, \dots, k_{m-2})$ that are not all zero and such that $d \equiv k_1 + 2k_2 + \dots + (m-2)k_{m-2} \leq m-2$.

Here the constraint $d \leq m-2$ has the aim to involve only those terms and coefficients in P_m that may not be small in comparison with L_s. Indeed, as we know from Proposition 7.4,

$$|\gamma_l| \leq (l-1)!\, L_l \leq (l-1)!\, L_s^{(l-2)/(s-2)}, \qquad 3 \leq l \leq [s],$$

which gives

$$\left| \left(\frac{\gamma_3}{3!}\right)^{k_1} \dots \left(\frac{\gamma_m}{m!}\right)^{k_{m-2}} \right| \leq \frac{L_s^{d/(s-2)}}{3^{k_1} \dots m^{k_{m-2}}}. \tag{10.1}$$

So, the left product is at least as small as L_s in case $d \geq m-1$, when L_s is small. Of course, this should be justified when comparing $e^{Q_m(it)}$ and $1 + P_m(it)$ on a proper interval of the t-axis. This will be done in the next two sections.

The index m for P indicates that all cumulants up to γ_m participate in the constructions of these polynomials. The power

$$k = 3k_1 + \dots + mk_{m-2} = d + 2(k_1 + k_2 + \dots + k_{m-2})$$

may vary from 3 to $3(m-2)$, with maximum $3(m-2)$ attainable when $k_1 = m-2$ and all other $k_r = 0$. Anyway, $\deg(P_m) \leq 3(m-2)$.

These observations imply a simple general bound on the growth of P_m, which will be needed in the sequel. First, $|t|^k \leq \max\left\{|t|^3, |t|^{3(m-2)}\right\}$. Hence, by Definition 10.1,

$$|P_m(it)| < \max\left\{|t|^3, |t|^{3(m-2)}\right\} \sum \frac{1}{k_1! \dots k_{m-2}!} \frac{L_s^{d/(s-2)}}{3^{k_1} \dots m^{k_{m-2}}},$$

Using the elementary bound

$$\sum \frac{1}{k_1! \dots k_{m-2}!} \frac{1}{3^{k_1} \dots m^{k_{m-2}}} < e^{1/3} \dots e^{1/m} < m, \tag{10.2}$$

we arrive at:

Proposition 10.2 *For all t real,*

$$|P_m(it)| \leq m \max\left\{|t|^3, |t|^{3(m-2)}\right\} \max\left\{L_s^{\frac{1}{s-2}}, L_s^{\frac{m-2}{s-2}}\right\}.$$

Let us describe the first three polynomials. Clearly, $P_3(it) = \gamma_3 \frac{(it)^3}{3!}$, while for $m = 4$,

$$P_4(it) = \sum_{0<k_1+2k_2\leq 2} \frac{1}{k_1!\,k_2!} \left(\frac{\gamma_3}{3!}\right)^{k_1} \left(\frac{\gamma_4}{4!}\right)^{k_2} (it)^{3k_1+4k_2} = \gamma_3 \frac{(it)^3}{3!} + \gamma_4 \frac{(it)^4}{4!} + \gamma_3^2 \frac{(it)^6}{2!\,3!^2}.$$

Correspondingly, for $m = 5$,

$$\begin{aligned} P_5(it) &= \sum_{0<k_1+2k_2+3k_3\leq 3} \frac{1}{k_1!\,k_2!\,k_3!} \left(\frac{\gamma_3}{3!}\right)^{k_1} \left(\frac{\gamma_4}{4!}\right)^{k_2} \left(\frac{\gamma_5}{5!}\right)^{k_3} (it)^{3k_1+4k_2+5k_3} \\ &= \gamma_3 \frac{(it)^3}{3!} + \gamma_4 \frac{(it)^4}{4!} + \gamma_5 \frac{(it)^5}{5!} + \gamma_3^2 \frac{(it)^6}{2!\,3!^2} + \gamma_3^3 \frac{(it)^9}{3!\,3!^3}. \end{aligned}$$

11 Cumulant Polynomials Q_m

Properties of the polynomials P_m will be explored via the study of the cumulant polynomials

$$Q_m(z) = \sum_{l=3}^{m} \frac{\gamma_l}{l!} z^l,$$

which will be treated as polynomials in the complex variable z. In this section we collect auxiliary facts, assuming that $L_s < \infty$ for some $s = m + \alpha$, $m \geq 3$, where m is integer and $0 < \alpha \leq 1$. In that case, the first term in Q_m is $\frac{\gamma_3}{3!} z^3$.

Lemma 11.1 *If* $|z| \max\left\{L_s^{\frac{1}{s-2}}, L_s^{\frac{1}{3(s-2)}}\right\} \leq \frac{1}{4}$, *then* $|Q_m(z)| < 0.007$. *Moreover,*

$$|Q_m(z)| \leq 0.42\, L_s^{\frac{1}{s-2}} |z|^3.$$

Proof Since $s \to L_s^{\frac{1}{s-2}}$ is non-decreasing, we have $|z| \max\left\{L_m^{\frac{1}{m-2}}, L_m^{\frac{1}{3(m-2)}}\right\} \leq \frac{1}{4}$. As we know, for any integer $3 \leq l \leq m$,

$$|\gamma_l| \leq (l-1)!\, L_l \leq (l-1)!\, L_m^{\frac{l-2}{m-2}}.$$

Hence,

$$|Q_m(z)| \le \sum_{l=3}^{m} \frac{1}{l} L_l |z|^l \le \sum_{l=3}^{m} \frac{1}{l} L_m^{\frac{l-2}{m-2}} |z|^l = L_m^{\frac{1}{m-2}} |z|^3 \sum_{l=3}^{m} \frac{1}{l} \left(L_m^{\frac{1}{m-2}} |z|\right)^{l-3}$$

$$\le 0.42\, L_m^{\frac{1}{m-2}} |z|^3 \le 0.42\, L_s^{\frac{1}{s-2}} |z|^3,$$

where we used $L_m^{\frac{1}{m-2}} |z| \le \frac{1}{4}$ together with $\sum_{l=3}^{\infty} \frac{4^{-(l-3)}}{l} = 64 \log \frac{4}{3} - 18 < 0.42$. This gives the second assertion. Finally, apply $L_s^{\frac{1}{s-2}} |z|^3 \le \frac{1}{64}$ to get the uniform bound on $|Q_m(z)|$. □

Lemma 11.2 *In the interval* $|t| \max\{L_s^{\frac{1}{s-2}}, L_s^{\frac{1}{3(s-2)}}\} \le \frac{1}{8}$, *we have*

$$e^{Q_m(it)} = \sum_{k=0}^{m-2} \frac{Q_m(it)^k}{k!} + \varepsilon(t)$$

with

$$\frac{1}{p!} \left| \frac{d^p}{dt^p} \varepsilon(t) \right| \le 4^{s-2} L_s |t|^{3(s-2)-p}, \qquad p = 0, 1, \dots$$

Proof Consider the function of the complex variable $\Psi(w) = e^w - \sum_{k=0}^{m-2} \frac{w^k}{k!} = \sum_{k=m-1}^{\infty} \frac{w^k}{k!}$. If $|w| \le 1$, then $|w|^k \le |w|^{m-1} \le |w|^{s-2}$ for all $k \ge m-1$, so,

$$|\Psi(w)| \le |w|^{s-2} \sum_{k=m-1}^{\infty} \frac{1}{k!} \le |w|^{s-2}.$$

This inequality will be used with $w = Q_m(z)$. The function $\Psi(Q_m(z))$ is analytic in the complex plane. So, we may apply Cauchy's contour integral formula

$$\frac{d^p}{dt^p} \Psi(Q_m(it)) = \frac{p!}{2\pi} \int_{|z-it|=\rho} \frac{\Psi(Q_m(z))}{(z-it)^{p+1}}\, dz$$

with an arbitrary $\rho > 0$, which gives

$$\left| \frac{d^p}{dt^p} \Psi(Q_m(it)) \right| \le \frac{p!}{\rho^p} \max_{|z-it|=\rho} |\Psi(Q_m(z))|.$$

Assume that $|t| > 0$ and choose $\rho = |t|$. Then on the circle $|z - it| = \rho$, necessarily $|z| \le 2|t|$ and, by the assumption on t,

$$|z| \max\{L_m^{\frac{1}{m-2}}, L_m^{\frac{1}{3(m-2)}}\} \le 2|t| \max\{L_m^{\frac{1}{m-2}}, L_m^{\frac{1}{3(m-2)}}\} \le \frac{1}{4}.$$

Hence, we may apply the uniform estimate of Lemma 11.1, $|Q_m(z)| \le 0.007 < 1$, so that, involving also the non-uniform estimate of the same lemma, we get

$$|\Psi(Q_m(z))| \le |Q_m(z)|^{s-2} \le \left(0.42\, L_s^{\frac{1}{s-2}}\, |z|^3\right)^{s-2}$$
$$\le (0.42)^{s-2} \cdot L_s \cdot (2|t|)^{3(s-2)} = 3.36^{s-2}\, L_s \cdot |t|^{3(s-2)}.$$

As a result,

$$\left|\frac{d^p}{dt^p}\Psi(Q_m(it))\right| \le \frac{p!}{|t|^p}\, 3.36^{s-2} L_s |t|^{3(s-2)}.$$

□

Note that, using $\sum_{k=m-1}^{\infty}\frac{1}{k!} \le \frac{1.5}{(m-1)!}$, the assertion of Lemma 11.2 could be sharpened to

$$\frac{1}{p!}\left|\frac{d^p}{dt^p}\,\varepsilon(t)\right| \le 3.2 \cdot \frac{4^s}{(m+1)!}\, L_s |t|^{3(s-2)-p}, \qquad p = 0, 1, \ldots$$

Lemma 11.3 *In the interval* $|t| \max\left\{L_s^{\frac{1}{s-2}}, L_s^{\frac{1}{3(s-2)}}\right\} \le \frac{1}{8}$, *we have*

$$\left|\frac{d^p}{dt^p}\, e^{Q_m(it)}\right| \le 1.01\, p!\, |t|^{-p}, \qquad p = 1, 2, \ldots$$

Proof By Cauchy's contour integral formula, for any $\rho > 0$,

$$\left|\frac{d^p}{dt^p}\, e^{Q_m(it)}\right| \le \frac{p!}{\rho^r}\, \exp\left\{\max_{|z-it|=\rho} |Q_m(z)|\right\}.$$

Assume $|t| > 0$ and choose again $\rho = |t|$. Then on the cicrle $|z - it| = \rho$ we have

$$|z| \max\left\{L_s^{\frac{1}{s-2}}, L_s^{\frac{1}{3(s-2)}}\right\} \le 2|t| \max\left\{L_s^{\frac{1}{s-2}}, L_s^{\frac{1}{3(s-2)}}\right\} \le \frac{1}{4}.$$

Hence, we may apply the uniform estimate of Lemma 11.1 and notice that $e^{0.007} < 1.01$. □

12 Relations Between P_m and Q_m

The basic relation between polynomials P_m and Q_m is described in the following statement.

Proposition 12.1 *If* $L_s < \infty$ $(s > 3)$, *then for* $|t| \max\left\{L_s^{\frac{1}{s-2}}, L_s^{\frac{1}{3(s-2)}}\right\} \le \frac{1}{8}$, *we have*

$$e^{Q_m(it)} = 1 + P_m(it) + \delta(t)$$

with

$$|\delta(t)| \le 0.2 \cdot 4^s L_s \max\left\{|t|^s, |t|^{3(s-2)}\right\}.$$

Moreover, for all $p = 1, \dots, [s]$,

$$\frac{1}{p!}\left|\frac{d^p}{dt^p}\,\delta(t)\right| \le 0.5 \cdot 7^s L_s \max\left\{|t|^{s-p}, |t|^{3(s-2)-p}\right\}.$$

Proof In view of Lemma 11.2, we may only be concerned with the remainder term

$$r(t) = \sum_{k=1}^{m-2} \frac{Q_m(it)^k}{k!} - P_m(it),$$

which we consider in the complex plane (by replacing it with $z \in \mathbb{C}$). Using the polynomial formula, let us represent the above sum as

$$\sum_{k=1}^{m-2} \frac{1}{k!}\left(\sum_{l=3}^{m} \gamma_l \frac{z^l}{l!}\right)^k = \sum_{k=1}^{m-2} \sum_{k_1+\dots+k_{m-2}=k} \frac{1}{k_1! \dots k_{m-2}!}\left(\frac{\gamma_3}{3!}\right)^{k_1} \dots \left(\frac{\gamma_m}{m!}\right)^{k_{m-2}} z^{3k_1+\dots+mk_{m-2}}.$$

Here the double sum almost defines $P_m(it)$ with the difference that Definition 10.1 contains the constraint $k_1 + 2k_2 + \dots + (m-2)k_{m-2} \le m-2$, while now we have a weaker constraint $k_1 + k_2 + \dots + k_{m-2} \le m-2$. Hence, all terms appearing in $P_m(it)$ are present in the above double sum, so

$$r(t) = \sum \frac{1}{k_1! \dots k_{m-2}!}\left(\frac{\gamma_3}{3!}\right)^{k_1} \dots \left(\frac{\gamma_m}{m!}\right)^{k_{m-2}} (it)^{3k_1+\dots+mk_{m-2}}$$

with summation subject to

$$k_1 + k_2 + \dots + k_{m-2} \le m-2, \qquad k_1 + 2k_2 + \dots + (m-2)k_{m-2} \ge m-1.$$

Necessarily, all $k_j \le m-2$ and at least one $k_j \ge 1$. Using $|\gamma_l| \le (l-1)!\, L_s^{\frac{l-2}{s-2}}$, we get

$$|r(z)| \le \sum \frac{1}{k_1! \dots k_{m-2}!} \prod_{l=3}^{m} L_s^{k_{l-2}\frac{l-2}{s-2}} |z|^N = \sum \frac{1}{k_1! \dots k_{m-2}!} L_s^M |z|^N,$$

where

$$M = M(k_1, \dots, k_{m-2}) = \frac{1}{s-2}\,(k_1 + 2k_2 + \dots + (m-2)k_{m-2}),$$

$$\begin{aligned} N = N(k_1,\dots,k_{m-2}) &= 3k_1 + \cdots + mk_{m-2} \\ &= (k_1 + 2k_2 + \cdots + (m-2)k_{m-2}) \\ &\quad +2\,(k_1 + k_2 + \cdots + k_{m-2}). \end{aligned}$$

Note that $m+1 \le N \le m(m-2)$, which actually will not be used, and $(s-2)M = N - 2k$. If $L_s^{\frac{1}{s-2}}|z| \le 1$, using the property $1 \le k \le s-2$, we have

$$L_s^{M-1}|z|^N \le |z|^{N-(s-2)(M-1)} = |z|^{(s-2)+2k} \le \max\left\{|z|^s, |z|^{3(s-2)}\right\}.$$

Hence

$$|r(z)| \le L_s \max\left\{|z|^s, |z|^{3(s-2)}\right\} \sum \prod_{l=3}^{m} \frac{1}{k_{l-2}!}\left(\frac{1}{l}\right)^{k_{l-2}}.$$

The latter sum is dominated by $e^{m-2} \le e^{s-2}$, so

$$|r(z)| \le e^{s-2} L_s \max\left\{|z|^s, |z|^{3(s-2)}\right\},$$

which can be used to prove Proposition 12.1 in case $p = 0$. Indeed, by Lemma 11.2 with its function $\varepsilon(t)$ for the interval $|t| \max\left\{L_s^{\frac{1}{s-2}}, L_s^{\frac{1}{3(s-2)}}\right\} \le \frac{1}{8}$, we have

$$|\delta(t)| \le |\varepsilon(t)| + |r(t)| \le 4^{s-2} L_s |t|^{3(s-2)} + e^{s-2} L_s \max\left\{|t|^s, |t|^{3(s-2)}\right\}.$$

Here $4^{-2} + e^{-2} < 0.2$, and we arrive at the first conclusion for $p = 0$.

In fact, one can a little sharpen the bound on $|r(z)|$, by noting that

$$\sum \prod_{l=3}^{m} \frac{1}{k_{l-2}!}\left(\frac{1}{l}\right)^{k_{l-2}} \le \exp\left\{\sum_{l=3}^{m} \frac{1}{l}\right\} - 1 \le e^{\log m - \log 2} - 1 = \frac{m-2}{2}.$$

Hence

$$|r(z)| \le \frac{s-2}{2} L_s \max\left\{|z|^s, |z|^{3(s-2)}\right\}.$$

This bound can be used for the remaining cases $1 \le p \le [s]$. One may apply the Cauchy contour integral formula to get that

$$|r^{(p)}(t)| \le \frac{p!}{\rho^p} \max_{|z-it|=\rho} |r(z)|.$$

Let us choose $\rho = \frac{1}{2}|t|$ and use the assumption $|t| \max\big\{L_s^{\frac{1}{s-2}}, L_s^{\frac{1}{3(s-2)}}\big\} \leq \frac{1}{8}$. On the circle $|z-it| = \rho$ it is necessary that $|z| \leq \frac{3}{2}|t|$ and thus $|z| \max\big\{L_s^{\frac{1}{s-2}}, L_s^{\frac{1}{3(s-2)}}\big\} \leq \frac{1}{4}$. Hence, we may apply the previous step with bounding $r(z)$ which was made under the weaker assumption $L_s^{\frac{1}{s-2}}|z| \leq 1$. This gives

$$\begin{aligned} |r^{(p)}(z)| &\leq \frac{p!}{|0.5\,t|^p}\,\frac{s-2}{2}\,L_s \max\big\{|z|^s, |z|^{3(s-2)}\big\} \\ &\leq \frac{p!}{|t|^p}\,2^s\,\frac{s-2}{2}\,L_s \max\big\{|1.5\,t|^s, |1.5\,t|^{3(s-2)}\big\}. \end{aligned}$$

This yields

$$|r^{(p)}(t)| \,\leq\, 2p!\,6.75^{s-2}(s-2)\,L_s \max\big\{|t|^{s-p}, |t|^{3(s-2)-p}\big\}.$$

Again, by Lemma 11.2 with its function $\varepsilon(t)$,

$$\begin{aligned} |\delta(t)| &\leq |\varepsilon(t)| + |r(t)| \\ &\leq 4^{-2}p!\,4^s L_s |t|^{3(s-2)} + 2p!\,6.75^{s-2}\,(s-2)\,L_s \max\big\{|t|^{s-p}, |t|^{3(s-2)-p}\big\}. \end{aligned}$$

Here $2 \cdot 6.75^{s-2}(s-2) \leq \frac{2}{e \log \frac{7}{6.75}}\,7^{s-2} < 0.413 \cdot 7^s$, and then we arrive at the desired conclusion. □

Corollary 12.2 *Let $L_s < \infty$ $(s \geq 3)$. In the interval $|t| \max\big\{L_s^{\frac{1}{s-2}}, L_s^{\frac{1}{3(s-2)}}\big\} \leq \frac{1}{8}$, we have $|P_m(it)| \leq 0.1$. Moreover, for all $p = 1, \dots, [s]$,*

$$\frac{1}{p!}\left|\frac{d^p}{dt^p}\,P_m(it)\right| \,\leq\, 1.4\,|t|^{-p}.$$

Proof First consider the case $p = 0$. By Lemma 11.1, $|Q_m(it)| \leq 0.007$, which implies, using the second estimate of Lemma 11.1 and our assumption,

$$\big|e^{Q_m(it)} - 1\big| \leq \frac{e^{0.007}-1}{0.007}\,|Q_m(it)| \leq 1.004{\cdot}0.42\,L_s^{\frac{1}{s-2}}|t|^3 \leq 1.004{\cdot}0.42{\cdot}\frac{1}{8^3} < 0.001.$$

In addition, by Proposition 12.1 (the obtained bound in case $p = 0$),

$$|\delta(t)| \leq 0.2 \cdot 4^s L_s \max\big\{|t|^s, |t|^{3(s-2)}\big\}.$$

By the assumption, $L_s|t|^s \,\leq\, \frac{1}{|8t|^{s-2}}\,|t|^s \,=\, \frac{1}{8^{s-2}}\,t^2$ and $L_s|t|^s \,\leq\, \frac{1}{|8t|^{3(s-2)}}\,|t|^s \,=\, \frac{1}{8^{3(s-2)}}\,t^{6-2s}$. Both estimates yield $L_s|t|^s \,\leq\, 8^{-s}$. Since also $L_s|t|^{3(s-2)} \,\leq\, 8^{-s}$, we have

$$L_s \max\big\{|t|^s, |t|^{3(s-2)}\big\} \leq 8^{-s},$$

so

$$|P_m(it)| \le \left|e^{Q_m(it)} - 1\right| + |\delta(t)| \le 0.001 + 0.2 \cdot \left(\frac{4}{8}\right)^s < 0.1,$$

which proves the corollary in this particular case.

Now, let $1 \le p \le [s]$. Combining Lemma 11.3 and Proposition 12.1, we have, using the previous step and the assumption $s \ge 3$:

$$\begin{aligned}\left|\frac{d^p}{dt^p} P_m(it)\right| &\le \left|\frac{d^p}{dt^p} e^{Q_m(it)}\right| + \left|\frac{d^p}{dt^p} \delta(t)\right| \\ &\le 1.01\, p!\, |t|^{-p} + 0.5\, p!\, 7^s L_s \max\left\{|t|^{s-p}, |t|^{3(s-2)-p}\right\} \\ &\le p!\, |t|^{-p}\left[1.01 + 0.5\left(\frac{7}{8}\right)^s\right] \le p!\, |t|^{-p}\left[1.01 + 0.5\left(\frac{7}{8}\right)^3\right].\end{aligned}$$

□

13 Corrected Normal Approximation on Moderate Intervals

We are now prepared to prove several assertions about the corrected normal approximation for the characteristic function $f_n(t)$ of the sum $S_n = X_1 + \cdots + X_n$ of independent random variables X_k. As usual, we assume that $\mathbb{E}X_k = 0$, $\mathbb{E}X_k^2 = \sigma_k^2$ ($\sigma_k \ge 0$) with $\sum_{k=1}^n \sigma_k^2 = 1$. Recall that Lyapunov's coefficients are defined by

$$L_s = \sum_{k=1}^n \mathbb{E}\,|X_k|^s, \qquad s \ge 2.$$

As before, we write $s = m + \alpha$, where m is integer and $0 < \alpha \le 1$. The range $2 < s \le 3$ was considered in Propositions 9.1–9.2, so our main concern will be the case $s > 3$. As a preliminary step, let us prove the following statement, including the value $s = 3$ (a limit case).

Lemma 13.1 *Let $L_s < \infty$. In the interval $|t| \max\left\{L_s^{\frac{1}{s-2}}, L_s^{\frac{1}{3(s-2)}}\right\} \le \frac{1}{8}$, we have*

$$f_n(t) = e^{-t^2/2}\left(1 + P_m(it) + r(t)\right) \tag{13.1}$$

with

$$\left|\frac{d^p}{dt^p} r(t)\right| \le C_s L_s \max\left\{|t|^{s-p}, |t|^{3(s-2)-p}\right\}, \qquad p = 0, 1, \dots, [s], \tag{13.2}$$

where one may take $C_s = 0.4 \cdot 4^s$ in case $p = 0$ and $C_s = 1.8 \cdot 7^s$ for $1 \le p \le [s]$.

Proof Combining Proposition 12.1 and Corollary 12.2 with Proposition 8.2, we may write

$$f_n(t) = e^{-t^2/2} e^{Q_m(it)} (1 + \varepsilon(t)) = e^{-t^2/2} (1 + P_m(it) + \delta(t)) (1 + \varepsilon(t))$$

with

$$|\delta(t)| \leq 0.2 \cdot 4^s L_s \max \left\{|t|^s, |t|^{3(s-2)}\right\}, \qquad |\varepsilon(t)| \leq 2^s L_s |t|^s, \qquad |P_m(it)| \leq 0.1,$$

where the second inequality was derived under the assumption that $L_s^{\frac{1}{s}} |t| \leq \frac{1}{8}$. It is fulfilled, since in general $L_s^{\frac{1}{s}} \leq \max \left\{L_s^{\frac{1}{s-2}}, L_s^{\frac{1}{3(s-2)}}\right\}$. In particular, we get $L_s|t|^s \leq 8^{-s}$, so $|\varepsilon(t)| \leq 4^{-s}$.

Since

$$r(t) = (1 + P_m(it))\varepsilon(t) + \delta(t)(1 + \varepsilon(t)),$$

we obtain that

$$\begin{aligned} |r(t)| &\leq 1.1 \cdot 2^s L_s |t|^s + 0.2 \cdot 4^s L_s \max \left\{|t|^s, |t|^{3(s-2)}\right\} \cdot (1 + 4^{-s}) \\ &\leq 4^s L_s \max \left\{|t|^s, |t|^{3(s-2)}\right\} \left[1.1 \cdot \left(\frac{2}{4}\right)^s + 0.2 + 0.2 \cdot 4^{-s} \right]. \end{aligned}$$

The expression in square brackets does not exceed $1.1 \cdot (\frac{2}{4})^3 + 0.2 + 0.2 \cdot 4^{-3} < 0.4$, which proves the assertion in case $p = 0$.

Now, let us turn to the derivatives of order $p = 1, \dots, [s]$ and apply other bounds given in Proposition 12.1, Corollary 12.2, and Proposition 8.2,

$$\begin{aligned} \frac{1}{p!} |\delta^{(p)}(t)| &\leq 0.5 \cdot 7^s L_s \max \left\{|t|^{s-p}, |t|^{3(s-2)-p}\right\}, \\ \frac{1}{p!} |\varepsilon^{(p)}(t)| &\leq 2 \cdot \frac{4^s}{s} L_s |t|^{s-p}, \\ \frac{1}{p!} |P_m^{(p)}(it)| &\leq 1.4 |t|^{-p} \end{aligned}$$

(which remain to hold in case $p = 0$ as well). Differentiating the product $P_m(it)\varepsilon(t)$ according to the Newton binomial formula, let us write

$$\big((1 + P_m(it)) \cdot \varepsilon(t)\big)^{(p)} = \sum_{k=0}^{p} \frac{p!}{k! \,(p-k)!} \big(1 + P_m(it)\big)^{(k)} \varepsilon(t)^{(p-k)}.$$

Applying the above estimates, we then get

$$\begin{aligned}\left|\left((1+P_m(it))\cdot\varepsilon(t)\right)^{(p)}\right| &\leq \sum_{k=0}^{p} \frac{p!}{k!\,(p-k)!}\,k!\,1.4\cdot|t|^{-k}\cdot(p-k)!\,\frac{2\cdot 4^s}{s}\,L_s|t|^{s-(p-k)}\\ &= 2.8\,\frac{(p+1)!}{s}\,4^s\,L_s|t|^{s-p}\\ &\leq 2.8\cdot p!\,4^s\,L_s|t|^{s-p}.\end{aligned}$$

To derive a similar bound for the product $\delta(t)\varepsilon(t)$, we use $L_s|t|^{3(s-2)} \leq 8^{-3(s-2)}$ together with $L_s|t|^s \leq 8^{-s}$. Then, the estimate on the p-th derivative of δ implies

$$|\delta^{(p)}(t)| \leq p!\,0.5\cdot 7^s\,8^{-s}\,|t|^{-p} \leq 0.4\,p!\,|t|^{-p}.$$

Hence, again according to the binomial formula,

$$\begin{aligned}\left|\left(\delta(t)\varepsilon(t)\right)^{(p)}\right| &\leq \sum_{k=0}^{p} \frac{p!}{k!\,(p-k)!}\,0.4\,k!\,|t|^{-k}\cdot(p-k)!\,4^s L_s\,|t|^{s-(p-k)}\\ &= 0.4\,(p+1)!\,4^s\,L_s|t|^{s-p}.\end{aligned}$$

Collecting these estimates, we obtain that

$$\begin{aligned}|r^{(p)}(t)| &\leq \left|(P_m(it)\varepsilon(t))^{(p)}\right| + \left|(\delta(t)\varepsilon(t))^{(p)}\right| + \left|\varepsilon^{(p)}(t)\right| + \left|\delta^{(p)}(t)\right|\\ &\leq p!\,\left(2.8\cdot 4^s + 0.4\,(p+1)\,4^s + 4^s + 0.5\cdot 7^s\right)L_s\max\left\{|t|^{s-p},\,|t|^{3(s-2)-p}\right\}.\end{aligned}$$

Here, since the function $t \to te^{-\beta t}$ is decreasing for $t > 1/\beta$ ($\beta > 0$), we have

$$(p+1)\,4^s \leq \frac{7}{4}\,(s+1)\left(\frac{4}{7}\right)^{s+1} 7^s \leq 4\left(\frac{4}{7}\right)^3 7^s < 0.75\cdot 7^s.$$

In addition, $4^s = (\frac{4}{7})^s\,7^s < 0.2\cdot 7^s$. So the expression in the brackets in front of L_s is smaller than $(2.8\cdot 0.2 + 0.4\cdot 0.75 + 0.2 + 0.7)\,7^s < 1.8\cdot 7^s$. □

In the representation for $f_n(t)$ in (13.1), one can take the term $r(t)$ out of the brackets, and then we get a more convenient form (at the expense of a larger power of t). Thus, put

$$g_m(t) = e^{-t^2/2}\,(1 + P_m(it)),$$

which serves as the corrected normal "characteristic" function. For the first values of m, one may recall the formulas for P_m at the end of Section 10, which give $g_2(t) = e^{-t^2/2}$,

$$g_3(t) = e^{-t^2/2}\left(1 + \gamma_3 \frac{(it)^3}{3!}\right),$$

$$g_4(t) = e^{-t^2/2}\left(1 + \gamma_3 \frac{(it)^3}{3!} + \gamma_4 \frac{(it)^4}{4!} + \gamma_3^2 \frac{(it)^6}{2!\,3!^2}\right),$$

$$g_5(t) = e^{-t^2/2}\left(\gamma_3 \frac{(it)^3}{3!} + \gamma_4 \frac{(it)^4}{4!} + \gamma_5 \frac{(it)^5}{5!} + \gamma_3^2 \frac{(it)^6}{2!\,3!^2} + \gamma_3^3 \frac{(it)^9}{3!\,3!^3}\right).$$

Proposition 13.2 *Let $L_s < \infty$ $(s \geq 3)$. In the interval $|t| \max\{L_s^{\frac{1}{s-2}}, L_s^{\frac{1}{3(s-2)}}\} \leq \frac{1}{8}$, for every $p = 0, 1, \dots, [s]$,*

$$\left|\frac{d^p}{dt^p}\left(f_n(t) - g_m(t)\right)\right| \leq C_s L_s \max\left\{|t|^{s-p}, |t|^{3(s-2)+p}\right\} e^{-t^2/2}, \tag{13.3}$$

where one may take $C_s = 0.5 \cdot 4^s$ in case $p = 0$ and $C_s = 6 \cdot 8^s$ for $1 \leq p \leq [s]$.

Proof Using the remainder term in (13.1), consider the function

$$R(t) \equiv f_n(t) - e^{-t^2/2}\,(1 + P_m(it)) = e^{-t^2/2}\, r(t).$$

In case $p = 0$, (13.2) gives the bound

$$|r(t)| \leq 0.5 \cdot 4^s L_s \max\left\{|t|^s, |t|^{3(s-2)}\right\}.$$

Hence, the same uniform bound holds for $R(t)$ as well.

Turning to the derivatives, we use the bounds

$$\frac{1}{p!}\,|r^{(p)}(t)| \leq 1.8 \cdot 7^s L_s \max\left\{|t|^{s-p}, |t|^{3(s-2)-p}\right\}$$

together with $|g^{(p)}(t)| \leq p!\,\max\{1, |t|^p\}\, g(t)$ for the Gaussian function $g(t) = e^{-t^2/2}$ (cf. (1.8)). Differentiating the product according to the binomial formula,

$$R^{(p)}(t) = \sum_{k=0}^{p} \frac{p!}{k!\,(p-k)!}\, g^{(p-k)}(t) r^{(k)}(t),$$

we therefore obtain that the absolute value of the above sum is bounded by

$$g(t) \sum_{k=0}^{p} \frac{p!}{k!\,(p-k)!}\,(p-k)!\,\max\left\{1, |t|^{p-k}\right\} \cdot k!\; 1.8 \cdot 7^s L_s \max\left\{|t|^{s-k}, |t|^{3(s-2)-k}\right\}$$

$$\leq 1.8 \cdot 7^s p!\, L_s g(t) \sum_{k=0}^{p} \max\left\{1, |t|^{p-k}\right\} \max\left\{|t|^{s-k}, |t|^{3(s-2)-k}\right\}$$

$$\leq 1.8 \cdot 7^s p!\, L_s g(t)\, (p+1)\, \max\left\{|t|^{s-p}, |t|^{3(s-2)+p}\right\}.$$

Here

$$(p+1)\, 7^s \leq \frac{8}{7}\,(s+1)\left(\frac{7}{8}\right)^{s+1} 8^s \leq \frac{8}{7}\,\frac{1}{e\log\frac{8}{7}}\, 8^s \leq 3.15 \cdot 8^s,$$

while $3.15 \cdot 1.8 < 5.7$. □

Remarks In the literature one can find different variations of the inequality (13.3). For integer values $s = m + 1$ and for $p = 0$, it was proved by Statulevičius, cf. [St1, St2] (with a similar behavior of the constants). A somewhat more complicated formulation describing the multidimensional expansion was given by Bikjalis [Bi2] (in the same situation).

14 Signed Measures μ_m Associated with g_m

Once it is observed that the characteristic function $f_n(t)$ of S_n is close on a relatively long interval to the corrected normal "characteristic function" $g_m(t) = e^{-t^2/2}\,(1 + P_m(it))$, it is reasonable to believe that in some sense the distribution of S_n is close to the signed measure μ_m, whose Fourier-Stieltjes transform is exactly $g_m(t)$, that is, with

$$\int_{-\infty}^{\infty} e^{itx}\, d\mu_m(x) = g_m(t), \qquad t \in \mathbb{R}.$$

In order to describe μ_m, let us recall the Chebyshev-Hermite polynomials

$$H_k(x) = (-1)^k\, (e^{-x^2/2})^{(k)}\, e^{x^2/2}, \qquad k = 0, 1, 2, \dots \ (x \in \mathbb{R}),$$

or equivalently, $\varphi^{(k)}(x) = (-1)^k H_k(x)\varphi(x)$ in terms of the normal density $\varphi(x) = \frac{1}{\sqrt{2\pi}}\, e^{-x^2/2}$. Each H_k is a polynomial of degree k with leading coefficient 1. For example,

$$H_0(x) = 1, \qquad H_2(x) = x^2 - 1, \qquad H_4(x) = x^4 - 6x^2 + 3,$$

$$H_1(x) = x, \qquad H_3(x) = x^3 - 3x, \qquad H_5(x) = x^5 - 10x^3 + 15x,$$

$$H_6(x) = x^6 - 15x^4 + 45x^2 - 15,$$

and so on. These polynomials are orthogonal on the real line with weight $\varphi(x)$ and form a complete orthogonal system in the Hilbert space $L^2(\mathbb{R}, \varphi(x)\, dx)$. By the repeated integration by parts (with $t \neq 0$),

$$e^{-t^2/2} = \int_{-\infty}^{\infty} e^{itx}\varphi(x)\, dx = \frac{1}{-it}\int_{-\infty}^{\infty} e^{itx}\varphi'(x)\, dx = \frac{1}{(-it)^k}\int_{-\infty}^{\infty} e^{itx}\varphi^{(k)}(x)\, dx.$$

In other words, we have the identity $\int_{-\infty}^{\infty} e^{itx}\, H_k(x)\varphi(x)\,dx = (it)^k\, e^{-t^2/2}$. Equivalently, using the inverse Fourier transform, one may write

$$H_k(x)\,\varphi(x) = \frac{1}{2\pi}\int_{-\infty}^{\infty} e^{-itx}\,(it)^p\, e^{-t^2/2}\,dt,$$

which may be taken as another definition of H_k.

Returning to Definition 10.1, we therefore obtain:

Proposition 14.1 *Let $L_s < \infty$ for $s = m+\alpha$ with an integer $m \geq 2$ and $0 < \alpha \leq 1$. The measure μ_m with Fourier-Stieltjes transform $g_m(t) = e^{-t^2/2}\,(1 + P_m(it))$ has density*

$$\varphi_m(x) = \varphi(x) + \varphi(x)\sum \frac{1}{k_1!\dots k_{m-2}!}\left(\frac{\gamma_3}{3!}\right)^{k_1}\dots\left(\frac{\gamma_m}{m!}\right)^{k_{m-2}} H_k(x),$$

where $k = 3k_1 + \dots + mk_{m-2}$ and where the summation runs over all collections of non-negative integers $(k_1,\dots,k_{m-2})$ that are not all zero and such that $k_1 + 2k_2 + \dots + (m-2)k_{m-2} \leq m-2$.

Recall that the cumulants γ_p of S_n are well defined for $p = 1,\dots,m$ and also for $p = m+1$ when s is integer. However, in this case γ_{m+1} is not present in the construction of φ_m. By the definition, if $2 < s \leq 3$, the above sum is empty, that is, $\varphi_2 = \varphi$.

In a more compact form, one may write $\varphi_m(x) = \varphi(x)(1 + R_m(x))$, where R_m is a certain polynomial of degree at most $3(m-2)$, defined by

$$R_m(x) = \sum \frac{1}{k_1!\dots k_{m-2}!}\left(\frac{\gamma_3}{3!}\right)^{k_1}\dots\left(\frac{\gamma_m}{m!}\right)^{k_{m-2}} H_k(x),$$

where $k = 3k_1 + \dots + mk_{m-2}$ and the summation is as before. For $m = 3$, we have $R_3(x) = \frac{\gamma_3}{3!}\,H_3(x) = \frac{\gamma_3}{3!}\,(x^3 - 3x)$, while for $m = 4$,

$$\begin{aligned}
R_4(x) &= \frac{\gamma_3}{3!}\,H_3(x) + \frac{\gamma_4}{4!}\,H_4(x) + \frac{\gamma_3^2}{2!\,3!^2}\,H_6(x) \\
&= \frac{\gamma_3}{3!}\,(x^3 - 3x) + \frac{\gamma_4}{4!}\,(x^4 - 6x^2 + 3) + \frac{\gamma_3^2}{2!\,3!^2}\,(x^6 - 15x^4 + 45x^2 - 15).
\end{aligned}$$

Correspondingly, for $m = 5$,

$$R_5(x) = \frac{\gamma_3}{3!}\,H_3(x) + \frac{\gamma_4}{4!}\,H_4(x) + \frac{\gamma_5}{5!}\,H_5(x) + \frac{\gamma_3^2}{2!\,3!^2}\,H_6(x) + \frac{\gamma_3^3}{3!\,3!^3}\,H_9(x).$$

Let us briefly describe a few basic properties of the measures μ_m.

Proposition 14.2 *The moments of S_n and μ_m coincide up to order m, that is,*

$$f_n^{(p)}(0) = g_m^{(p)}(0), \qquad p = 0, 1, \dots, m.$$

In particular, $\mu_m(\mathbb{R}) = \int_{-\infty}^{\infty} \varphi_m(x)\,dx = 1$.

The latter immediately follows from the Fourier transform formula

$$\int_{-\infty}^{\infty} e^{itx}\varphi_m(x)\,dx = g_m(t) = e^{-t^2/2}\,(1 + P_m(it)),$$

applied at $t = 0$. The more general assertion immediately follows from Proposition 13.2, which gives $|f_n^{(p)}(t) - g_m^{(p)}(t)| = O(|t|^{s-p})$ as $t \to 0$.

Proposition 14.3 *If $L_s < \infty$ for $s = m + \alpha$ with $m \geq 2$ integer and $0 < \alpha \leq 1$, then the measure μ_m has a total variation norm satisfying*

$$\big|\|\mu_m\|_{\mathrm{TV}} - 1\big| \leq m\sqrt{(3(m-2))!}\,\max\{L_s^{\frac{1}{s-2}}, L_s^{\frac{m-2}{s-2}}\}. \tag{14.1}$$

In addition,

$$\int_{-\infty}^{\infty} |x|^s\,|\mu_m(dx)| \leq s^{2s}\max\{L_s, 1\}. \tag{14.2}$$

Proof In the definition of P_m, the tuples $(k_1, \dots, k_{m-2})$ participating in the sum satisfy $1 \leq d \leq m-2$, where $d = k_1 + 2k_2 + \cdots + (m-2)k_{m-2}$. Thus (cf. (10.1)),

$$\begin{aligned}
\Big|\Big(\frac{\gamma_3}{3!}\Big)^{k_1}\cdots\Big(\frac{\gamma_m}{m!}\Big)^{k_{m-2}}\Big| &\leq \frac{1}{3^{k_1}\dots m^{k_{m-2}}}\,L_s^{\frac{d}{s-2}} \\
&\leq \frac{1}{3^{k_1}\dots m^{k_{m-2}}}\,\max\{L_s^{\frac{1}{s-2}}, L_s^{\frac{m-2}{s-2}}\} \leq \frac{1}{3^{k_1}\dots m^{k_{m-2}}}\,\max\{L_s, 1\}.
\end{aligned}$$

Hence

$$\begin{aligned}
\big|\|\mu_m\|_{\mathrm{TV}} - 1\big| &= \int_{-\infty}^{\infty} |R_m(x)|\,\varphi(x)\,dx \\
&\leq \max\{L_s^{\frac{1}{s-2}}, L_s^{\frac{m-2}{s-2}}\}\sum \frac{1}{k_1!\dots k_{m-2}!}\,\frac{1}{3^{k_1}\dots m^{k_{m-2}}}\int_{-\infty}^{\infty} |H_k(x)|\,\varphi(x)\,dx,
\end{aligned}$$

where $k = 3k_1 + \cdots + mk_{m-2}$ (which may vary from 3 to $3(m-2)$). Let Z be a random variable with the standard normal distribution. As is well known,

$$\int_{-\infty}^{\infty} H_k(x)^2\,\varphi(x)\,dx = \mathbb{E}\,H_k(Z)^2 = k!$$

Hence, by the Cauchy inequality,

$$\int_{-\infty}^{\infty} |H_k(x)|\,\varphi(x)\,dx = \mathbb{E}\,|H_k(Z)| \le \sqrt{k!} \le \sqrt{(3(m-2))!}$$

implying that

$$\big|\|\mu_m\|_{\rm TV} - 1\big| \le \sqrt{(3(m-2))!}\,\max\{L_s, L_s^{(m-2)/(s-2)}\} \sum \frac{1}{k_1!\dots k_{m-2}!}\,\frac{1}{3^{k_1}\dots m^{k_{m-2}}}.$$

The latter sum does not exceed $e^{1/3}\dots e^{1/m} < m$, cf. (10.2), and we obtain (14.1).

Let us now turn to the second assertion. If $m = 2$, then $\varphi_m = \varphi$ and μ_m is the standard Gaussian measure on the real line. In this case,

$$\int_{-\infty}^{\infty} |x|^s\,|\mu_m(dx)| = \mathbb{E}\,|Z|^s = \frac{2^{s/2}}{\sqrt{\pi}}\,\Gamma\Big(\frac{s+1}{2}\Big) \le \frac{2^{3/2}}{\sqrt{\pi}}\,\Gamma(2) < 1.6 < s^{2s}.$$

In case $m \ge 3$, again by the Cauchy inequality, for the same value of k as before, we have

$$\begin{aligned}\int_{-\infty}^{\infty} |x|^s\,|H_k(x)|\,\varphi(x)\,dx = \mathbb{E}\,|Z|^s\,|H_k(Z)| &\le \sqrt{\mathbb{E}\,|Z|^{2s}}\sqrt{k!}\\ &\le \sqrt{\mathbb{E}\,|Z|^{2s}}\sqrt{(3(m-2))!}\end{aligned}$$

Hence, applying once more the inequality (10.2) together with the last bound on the product of the cumulants, we obtain that

$$\begin{aligned}\int_{-\infty}^{\infty} |x|^s\,|\mu_m(dx)| &\le \int_{-\infty}^{\infty} |x|^s\,\varphi(x)\,dx + \int_{-\infty}^{\infty} |x|^s\,|R_m(x)|\,\varphi(x)\,dx\\ &\le \mathbb{E}\,|Z|^s + \sqrt{\mathbb{E}\,|Z|^{2s}}\,m\sqrt{(3(m-2))!}\,\max\{L_s, 1\}\\ &\le 2\sqrt{\mathbb{E}\,|Z|^{2s}}\,m\sqrt{(3(m-2))!}\,\max\{L_s, 1\}.\end{aligned}$$

To simplify the right-hand side, one may use

$$(3(m-2))! \le \Gamma(3s-5) = \frac{\Gamma(3s+1)}{3s\,(3s-1)(3s-2)(3s-3)(3s-4)(3s-5)}.$$

Since $3s-1 \ge \frac{8}{3}\,s$, $3s-2 \ge \frac{7}{3}\,s$, $3s-3 \ge 6$, $3s-4 \ge 5$, $3s-5 \ge 4$, we have $\Gamma(3s-5) \le \frac{1}{280\,s^3}\,\Gamma(3s+1)$ and thus

$$\bigg(\frac{1}{\max(L_s, 1)}\int_{-\infty}^{\infty} |x|^s\,|\mu(dx)|\bigg)^2$$

$$\leq 4\,\mathbb{E}\,|Z|^{2s}\,\frac{m^2}{280\,s^3}\,\Gamma(3s+1) \;\leq\; \frac{1}{70\,s}\,\mathbb{E}\,|Z|^{2s}\,\Gamma(3s+1)$$

$$= \frac{1}{70\,s}\,\frac{2^s}{\sqrt{\pi}}\,\Gamma\Big(s+\frac{1}{2}\Big)\,\Gamma(3s+1) \;<\; \frac{1}{70\,s}\,\frac{2^s}{\sqrt{\pi}}\,\Gamma(s+1)\,\Gamma(3s+1).$$

By Stirling's formula, $\Gamma(x+1) \leq 2\,(\frac{x}{e})^x\sqrt{2\pi x}$ $(x \geq 3)$, which allows us to bound the above right-hand side by

$$\frac{1}{70\,s}\,\frac{2^s}{\sqrt{\pi}}\cdot 2\Big(\frac{s}{e}\Big)^s\sqrt{2\pi s}\cdot 2\Big(\frac{3s}{e}\Big)^{3s}\sqrt{6\pi s} \;=\; \frac{2\sqrt{12\,\pi}}{35}\,\Big(\frac{54}{e^4}\Big)^s s^{4s} \;<\; s^{4s}.$$

□

As a consequence, one can complement Proposition 14.3 with the following statement which is of a special interest when L_s is large (since in that case the interval of approximation in this proposition is getting small).

Corollary 14.4 *Let $L_s < \infty$ for $s = m + \alpha$ with $m \geq 2$ integer and $0 < \alpha \leq 1$. Then, for all $t \in \mathbb{R}$ and $p = 0, 1, \dots, [s]$,*

$$\Big|\frac{d^p}{dt^p}\,\big(f_n(t) - g_m(t)\big)\Big| \;\leq\; 4s^{2s}\,\max\{L_s, 1\}\frac{|t|^{s-p}}{([s]-p)!}.$$

Proof Let P_n denote the distribution of S_n. By Proposition 5.2 applied to the signed measure $\mu = P_n - \mu_m$, we have

$$\Big|\frac{d^p}{dt^p}\,\big(f_n(t) - g_m(t)\big)\Big| \;\leq\; 2C_s\,\frac{|t|^{s-p}}{(m-p)!}, \quad \text{where } C_s = \int_{-\infty}^{\infty} |x|^s\,|P_n(dx) - \mu_m(dx)|.$$

Here

$$C_s \leq \int_{-\infty}^{\infty} |x|^s\,P_n(dx) + \int_{-\infty}^{\infty} |x|^s|\mu_m(dx)| = \mathbb{E}\,|S_n|^s + \int_{-\infty}^{\infty} |x|^s|\mu_m(dx)|.$$

The last integral may be estimated with the help of the bound (14.2), while the s-th absolute moment of S_n is estimated with the help of Rosenthal's inequality $\mathbb{E}\,|S_n|^s \leq (2s)^s\,\max\{L_s, 1\}$, cf. (7.3). Since $(2s)^s \leq s^{2s}$, we get $C_s \leq 2s^{2s}$. □

PART IV. Corrected normal approximation on long intervals

15 Upper Bounds for Characteristic Functions f_n

Let $X_1, \dots, X_n$ be independent random variables with $\mathbb{E}X_k = 0$, $\mathbb{E}X_k^2 = \sigma_k^2$ ($\sigma_k \geq 0$), assuming that $\sum_{k=1}^n \sigma_k^2 = 1$. Recall that

$$L_3 = \sum_{k=1}^n \mathbb{E}\,|X_k|^3.$$

On long intervals of the t-axis, we are aimed to derive upper bounds on the absolute value of the characteristic function $f_n(t) = \mathbb{E}\,e^{itS_n}$ of the sum $S_n = X_1 + \cdots + X_n$. Assume that X_k have finite 3-rd moments, and put $\beta_{3,k} = \mathbb{E}|X_k|^3$. We will need:

Lemma 15.1 *Let X be a random variable with characteristic function $v(t)$. If $\mathbb{E}X = 0$, $\mathbb{E}X^2 = \sigma^2$, $\mathbb{E}\,|X|^3 = \beta_3 < \infty$, then for all $t \in \mathbb{R}$,*

$$|v(t)| \leq e^{-\frac{1}{2}\sigma^2 t^2 + \frac{1}{3}\beta_3 |t|^3}.$$

In addition, if $\beta_s = \mathbb{E}\,|X|^s$ is finite for $s \geq 3$, then for all $p = 1, \dots, [s]$,

$$|v^{(p)}(t)| \leq e^{1/6}\,\beta_{p^*} \max\{1, |t|\}\, e^{-\frac{1}{2}\sigma^2 t^2 + \frac{1}{3}\beta_3 |t|^3}, \qquad p^* = \max\{p, 2\}.$$

Proof Let X' be an independent copy of X. Since X has mean zero, $\mathbb{E}\,|X - X'|^3 \leq 4\beta_3$, cf. [B-RR], Lemma 8.8. Hence, by Taylor's expansion, for any t real,

$$|v(t)|^2 = \mathbb{E}\,e^{it(X-X')} = 1 - \sigma^2 t^2 + \frac{4\theta}{3!}\,\beta_3 |t|^3 \leq \exp\Big\{-\sigma^2 t^2 + \frac{4\theta}{3!}\,\beta_3 |t|^3\Big\}$$

with some $\theta = \theta(t)$ such that $|\theta| \leq 1$. The first inequality now easily follows.

Since $|v''(t)| \leq \sigma^2$ and $v'(0) = 0$, we also have $|v'(t)| \leq \sigma^2 |t|$. On the other hand, putting $x = \sigma|t|$ and using $\beta \geq \sigma^3$, we have

$$-\frac{1}{2}\sigma^2 t^2 + \frac{1}{3}\beta_3 |t|^3 \geq -\frac{1}{2}x^2 + \frac{1}{3}x^3 \geq -\frac{1}{6} \qquad (x \geq 0).$$

This proves the second inequality of the lemma in case $p = 1$. If $p \geq 2$, then we only need to apply $|v^{(p)}(t)| \leq \beta_p$. □

Denoting by v_k the characteristic function of X_k, by the first inequality of Lemma 15.1, $|v_k(t)| \leq \exp\{-\frac{1}{2}\sigma_k^2 t^2 + \frac{1}{3}\beta_{3,k}|t|^3\}$. Multiplying these inequalities, we get

$$|f_n(t)| \leq \exp\Big\{-\frac{1}{2}t^2 + \frac{1}{3}L_3 |t|^3\Big\}.$$

If $|t| \leq \frac{1}{L_3}$, then $L_3 |t|^3 \leq t^2$ for $|t| \leq \frac{1}{L_3}$. Hence, the above bound yields:

Proposition 15.2 *We have* $|f_n(t)| \leq e^{-t^2/6}$ *whenever* $|t| \leq \frac{1}{L_3}$.

One can sharpen the statement of Proposition 15.2 by developing Taylor's expansion for $v_k(t)$, rather than for $|v_k(t)|^2$. By Taylor's integral formula,

$$v_k(t) = 1 - \frac{\sigma_k^2 t^2}{2} + \frac{1}{2} \int_0^t v_k'''(\tau)(t-\tau)^2 \, d\tau,$$

so $\left|v_k(t) - (1 - \frac{\sigma_k^2 t^2}{2})\right| \leq \frac{\beta_{3,k}}{6} |t|^3$. Here the left-hand side dominates $|v_k(t)| - (1 - \frac{\sigma_k^2 t^2}{2})$ in case $\sigma_k |t| \leq \sqrt{2}$, and then we obtain that

$$|v_k(t)| \leq 1 - \frac{\sigma_k^2 t^2}{2} + \frac{\beta_{3,k} |t|^3}{6} \leq \exp\left\{ - \frac{\sigma_k^2 t^2}{2} + \frac{\beta_{3,k} |t|^3}{6} \right\}.$$

Multiplying these inequalities, we get:

Proposition 15.3 *If* $\max_k \sigma_k |t| \leq \sqrt{2}$, *we have*

$$|f_n(t)| \leq \exp\left\{ - \frac{t^2}{2} + \frac{L_3 |t|^3}{6} \right\}.$$

Hence, if additionally $|t| \leq \frac{1}{L_3}$, *then* $|f_n(t)| \leq e^{-t^2/3}$.

This statement has an advantage over Proposition 15.2 in case of i.i.d. summands.

Now let us consider the case of the finite L_s with $2 < s \leq 3$ and define $\beta_{s,k} = \mathbb{E}\,|X_k|^s$. Here is an adaptation of Lemma 15.1.

Lemma 15.4 *Let X be a random variable with characteristic function* $v(t)$. *If* $\mathbb{E}X = 0$, $\mathbb{E}X^2 = \sigma^2$, $\mathbb{E}\,|X|^s = \beta_s < \infty$ *for* $2 < s \leq 3$, *then, for all* $t \in \mathbb{R}$,

$$|v(t)| \leq e^{-\frac{1}{2}\sigma^2 t^2 + 2\beta_s |t|^s}.$$

In addition,

$$|v'(t)| \leq e^{1/24}\, \sigma^2 |t|\, e^{-\frac{1}{2}\sigma^2 t^2 + 2\beta_s |t|^s}, \qquad |v''(t)| \leq e^{1/24}\, \sigma^2 e^{-\frac{1}{2}\sigma^2 t^2 + 2\beta_s |t|^s}.$$

Proof Let X' be an independent copy of X. Then $\mathrm{Var}(X - X') = 2\sigma^2$. Write

$$\begin{aligned} |X - X'|^s &= (X - X')^2\, |X - X'|^{s-2} \\ &\leq (X - X')^2 \left(|X|^{s-2} + |X'|^{s-2}\right) = (X^2 - 2XX' + X')\left(|X|^{s-2} + |X'|^{s-2}\right), \end{aligned}$$

implying that

$$\begin{aligned} \mathbb{E}\,|X - X'|^s &\leq \mathbb{E}\,|X|^s + \mathbb{E}\,|X'|^s + \mathbb{E}X^2\, \mathbb{E}\,|X'|^{s-2} + \mathbb{E}X'^2\, \mathbb{E}\,|X|^{s-2} \\ &= 2\,\mathbb{E}\,|X|^s + 2\,\mathbb{E}X^2\, \mathbb{E}\,|X|^{s-2}. \end{aligned}$$

Here $\mathbb{E}X^2 \leq \beta_s^{2/s}$ and $\mathbb{E}\,|X|^{s-2} \leq \beta_s^{(s-2)/s}$, so that we obtain $\mathbb{E}\,|X - X'|^s \leq 4\beta_s$.

Now, by Proposition 5.1 with $p = 0$, $m = 2$, applied to $X - X'$,

$$|v(t)|^2 = \mathbb{E}\,e^{it(X-X')} = 1 - \sigma^2 t^2 + \delta(t), \qquad |\delta(t)| \leq 4\beta_s |t|^s.$$

Hence, for any t real,

$$|v(t)|^2 \leq 1 - \sigma^2 t^2 + 4\beta_s |t|^s \leq \exp\big\{-\sigma^2 t^2 + 4\beta_s |t|^s\big\},$$

proving the first inequality. Since $|v''(t)| \leq \sigma^2$ and $v'(0) = 0$, we also have $|v'(t)| \leq \sigma^2|t|$, $|v''(t)| \leq \sigma^2$. On the other hand, putting $x = \sigma\,|t|$ and using $\beta_s \geq \sigma^s$, we have

$$-\frac{1}{2}\,\sigma^2 t^2 + 2\beta_s |t|^s \geq -\frac{1}{2}\,x^2 + 2x^s = \psi(x).$$

On the positive half-axis the function ψ attains minimum at the point $x_s = (2s)^{-\frac{1}{s-2}}$, at which

$$\psi(x_s) \;=\; -\frac{1}{2}\,(2s)^{-\frac{2}{s-2}} + 2\,(2s)^{-\frac{s}{s-2}} \;=\; -\frac{s-2}{2s}\left(\frac{1}{2s}\right)^{\frac{2}{s-2}} \;\geq\; -\frac{1}{24}.$$

□

Now, returning to the random variables X_k, by the first inequality of this lemma, we have $|v_k(t)| \leq \exp\big\{-\frac{1}{2}\,\sigma_k^2 t^2 + 2\beta_{s,k}|t|^s\big\}$. Multiplying them, we get $|f_n(t)| \leq \exp\{-\frac{t^2}{2}\,(1 - 4L_s|t|^{s-2})\}$, which yields:

Proposition 15.4 *If $2 < s \leq 3$, then $|f_n(t)| \leq e^{-t^2/6}$ in the interval $|t| \leq (6L_s)^{-\frac{1}{s-2}}$.*

Remarks The first inequality in Lemma 15.1 first appeared apparently in the work by Zolotarev [Z1]. Later in [Z2] he sharpened this bound to

$$\log|v(t)| \leq -\frac{1}{2}\,\sigma^2 t^2 + 2\kappa_3\,\beta_3 |t|^3, \qquad \kappa_3 = \sup_{x>0}\,\frac{\cos x - 1 + \frac{x^2}{2}}{x^3} = 0.099\ldots$$

Further refinements are due to Prawitz [Pr1, Pr2]. Sharper forms of Lemma 15.4, including s-dependent constants in front of $|t|^s$ for the values $2 < s \leq 3$, were studied by Ushakov and Shevtsova, cf. [U], [Sh2].

16 Bounds on the Derivatives of Characteristic Functions

Keeping notations of the previous section together with basic assumptions on the random variables X_k's, here we extend upper bounds on the characteristic function $f_n(t) = \mathbb{E}\,e^{itS_n}$ to its derivatives up to order $[s]$. Put $p^* = \max(p, 2)$.

Proposition 16.1 *Let $L_s < \infty$, for some $s \ge 3$. Then, for all $p = 0, \dots, [s]$,*

$$\left|\frac{d^p}{dt^p} f_n(t)\right| \le 2.03^p\, p!\, \max\{L_{p^*}, 1\}\, \max\{1, |t|^p\}\, e^{-t^2/6}, \qquad \text{if } |t| \le \frac{1}{L_3}. \tag{16.1}$$

Proof The case $p = 0$ follows from Proposition 15.2. For $p \ge 1$, denote by $v_k(t)$ the characteristic functions of X_k. We use the polynomial formula

$$f_n^{(p)}(t) = \sum \binom{p}{q_1 \ \dots \ q_n} v_1^{(q_1)}(t) \dots v_n^{(q_n)}(t)$$

with summation running over all integers $q_k \ge 0$ such that $q_1 + \dots + q_n = p$. By Lemma 15.1,

$$|v_k(t)| \le e^{-\frac{1}{2}\sigma_k^2 t^2 + \frac{1}{3}\beta_{3,k}|t|^3},$$

$$|v_k^{(q_k)}(t)| \le e^{1/6}\, \beta_{q_k^*,k}\, \max\{1, |t|\}\, e^{-\frac{1}{2}\sigma_k^2 t^2 + \frac{1}{3}\beta_{3,k}|t|^3}, \qquad q_k \ge 1,$$

where $\beta_{q,k} = \mathbb{E}\,|X_k|^q$ and $q_k^* = \max\{q_k, 2\}$. Applying these inequalities and noting that the number

$$l = \mathrm{card}\{k \le n : q_k \ge 1\}$$

is smaller than or equal to p, we get

$$\prod_{k=0}^n |v_k^{(q_k)}(t)| \le e^{p/6} \max\{1, |t|^p\}\, e^{-\frac{1}{2}t^2 + \frac{1}{3}L_3|t|^3}\, \beta_{q_1^*,1} \dots \beta_{q_n^*,n}.$$

Write $(q_1, \dots, q_n) = (0, \dots, q_{k_1}, \dots, q_{k_l}, \dots, 0)$, specifying all indexes k for which $q_k \ge 1$. Put $p_1 = q_{k_1}, \dots, p_l = q_{k_l}$. Thus, $p_j \ge 1$, $p_1 + \dots + p_l = p$, so $1 \le l \le p$, and the above bound takes the form

$$\prod_{k=0}^n |v_k^{(q_k)}(t)| \le e^{p/6} \max\{1, |t|^p\}\, e^{-\frac{1}{2}t^2 + \frac{1}{3}L_3|t|^3}\, \beta_{p_1^*,k_1} \dots \beta_{p_l^*,k_l}.$$

Using it in the polynomial formula and performing summation over all k_j's, we arrive at

$$|f_n^{(p)}(t)| \le e^{p/6} \max\{1, |t|^p\}\, e^{-\frac{1}{2}t^2 + \frac{1}{3}L_3|t|^3}\, \widetilde{L}_p$$

with

$$\widetilde{L}_p = \sum \binom{p}{p_1 \ \dots \ p_l} L_{p_1^*} \dots L_{p_l^*},$$

where the sum runs over all integers $l = 1, \dots, p$ and $p_1, \dots, p_l \geq 1$ such that $p_1 + \dots + p_l = p$.

Clearly, $\widetilde{L}_1 = 1$ and $\widetilde{L}_2 = 2$. If $p \geq 3$, using the property that the function $q \to L_q^{1/(q-2)}$ is not decreasing in $q > 2$ (Proposition 7.1), we get

$$L_{p_1^*} \dots L_{p_l^*} = \prod_{j:\, p_j \geq 2} L_{p_j} \leq \prod_{j:\, p_j \geq 2} L_p^{\frac{p_j-2}{p-2}} = L_p^{\nu}.$$

Here

$$(p-2)\nu = \sum_{j=1}^{l} (p_j - 2)\, 1_{\{p_j \geq 2\}} = p - 2l + \sum_{j:\, p_j = 1} 1 \leq p - 2$$

with the last inequality holding for $l \geq 2$. Also, when $l = 1$, necessarily $\sum_{j:p_j=1} 1 = 0$, so $\nu \leq 1$ in all cases. But then $L_p^{\nu} \leq \max\{L_p, 1\}$, which implies

$$\widetilde{L}_p \leq \max\{L_p, 1\} \sum_{l=1}^{p} \sum_{p_1 + \dots + p_l = p} \binom{p}{p_1 \ \dots \ p_l}$$

$$\leq \max\{L_p, 1\}\, p! \prod_{l=1}^{p} \sum_{p_l=1}^{\infty} \frac{1}{p_l!} \leq \max\{L_p, 1\}\, (e-1)^p\, p!$$

This inequality remains to hold for $p = 1$ and $p = 2$. Thus, for all $p \geq 1$,

$$|f_n^{(p)}(t)| \leq \big((e-1)\, e^{1/6}\big)^p p! \, \max\{L_{p^*}, 1\} \, \max\{1, |t|^p\}\, e^{-\frac{1}{2}t^2 + \frac{1}{3}L_3|t|^3}.$$

Here $(e-1)e^{1/6} < 2.03$. Also, if $|t| \leq \frac{1}{L_3}$, then $L_3|t|^3 \leq t^2$. □

Let us now turn to the case $2 < s < 3$ with finite Lyapunov coefficient L_s rather than L_3. In terms of the characteristic functions $v_k(t)$, the first derivative of $f_n(t)$ is just the sum

$$f_n'(t) = \sum_{k=1}^{n} v_1(t) \dots v_{k-1}(t)\, v_k'(t)\, v_{k+1}(t) \dots v_n(t).$$

Here, by Lemma 15.4, the k-th term is dominated by $e^{1/24}\,\sigma_k^2 |t|\, e^{-\frac{1}{2}t^2 + 2L_s|t|^s}$. Performing summation over all $k \leq n$, we then arrive at

$$|f_n'(t)| \leq e^{1/24}\, |t|\, e^{-\frac{1}{2}t^2 + 2L_s|t|^s}.$$

Now, let us turn to the second derivative. Assuming that $n \geq 2$, first write

$$f_n''(t) = \sum_{k=1}^{n} v_1(t) \dots v_{k-1}(t)\, v_k''(t)\, v_{k+1}(t) \dots v_n(t)$$
$$+ 2 \sum_{1 \leq k < l \leq n} v_1(t) \dots v_{k-1}(t)\, v_k'(t)\, v_{k+1}(t) \dots v_{l-1}(t)\, v_l'(t)\, v_{l+1}(t) \dots v_n(t).$$

Again by Lemma 15.4, we get

$$|v_1(t) \dots v_{k-1}(t)\, v_k''(t)\, v_{k+1}(t) \dots v_n(t)| \leq e^{1/24}\, \sigma_k^2\, e^{-\frac{1}{2}t^2 + 2L_s|t|^s}$$

and

$$|v_1(t) \dots v_{k-1}(t)\, v_k'(t)\, v_{k+1}(t) \dots v_{l-1}(t)\, v_l'(t)\, v_{l+1}(t) \dots v_n(t)|$$
$$\leq e^{1/12}\, \sigma_k^2 \sigma_l^2\, t^2\, e^{-\frac{1}{2}t^2 + 2L_s|t|^s}.$$

Performing summation in the representation for $f_n''(t)$ we arrive at

$$|f_n''(t)| \leq \big(e^{1/24} + e^{1/12}\, t^2\big)\, e^{-\frac{1}{2}t^2 + 2L_s|t|^s}.$$

If $n = 1$, the estimate is simplified to $|f_1''(t)| \leq e^{1/24}\, e^{-\frac{1}{2}t^2 + 2L_s|t|^s}$. One can summarize.

Proposition 16.2 *If* $2 < s < 3$, *then in the interval* $|t| \leq (6L_s)^{-\frac{1}{s-2}}$,

$$|f_n(t)| \leq e^{-t^2/6}, \qquad |f_n'(t)| \leq e^{1/24}\, |t|\, e^{-t^2/6}, \qquad |f_n''(t)| \leq e^{1/12}(1 + t^2)\, e^{-t^2/6}.$$

17 Upper Bounds for Approximating Functions $g_m(t)$

Our next step is to get bounds, similar to the ones in Sections 15–16, for the corrected normal "characteristic function"

$$g_m(t) = e^{-t^2/2}\,(1 + P_m(it))$$

with large values of $|t|$, more precisely – outside the interval of Proposition 13.2.

Proposition 17.1 *Let* $s \geq 3$. *In the region* $|t|\, \max\{L_s^{\frac{1}{s-2}}, L_s^{\frac{1}{3(s-2)}}\} \geq \frac{1}{8}$, *we have*

$$|g_m(t)| \leq (142\, s)^{3s/2}\, L_s\, e^{-t^2/8}. \tag{17.1}$$

Moreover, for every $p = 1, 2, \dots, [s]$,

$$|g_m^{(p)}(t)| \le (573\, s)^{2s}\, L_s\, e^{-t^2/8}. \tag{17.2}$$

Recall that, for real values $s = m + \alpha$, where $m \ge 2$ is integer and $0 < \alpha \le 1$,

$$P_m(it) = \sum \frac{1}{k_1! \dots k_{m-2}!} \left(\frac{\gamma_3}{3!}\right)^{k_1} \dots \left(\frac{\gamma_m}{m!}\right)^{k_{m-2}} (it)^k,$$

where the summation runs over all collections of non-negative integers $(k_1, \dots, k_{m-2})$ that are not all zero and such that

$$k \equiv 3k_1 + \dots + mk_{m-2}, \qquad d \equiv k_1 + 2k_2 + \dots + (m-2)k_{m-2} \le m - 2.$$

Note that all tuples that are involved satisfy $1 \le d \le s-2$ and $1 \le k \le 3d \le 3(s-2)$.

Proof of Proposition 17.1 We use the bound (10.1), implying that, for all complex t,

$$|P_m(it)| \le \sum \frac{1}{k_1! \dots k_{m-2}!} \frac{1}{3^{k_1} \dots m^{k_{m-2}}} L_s^{\frac{d}{s-2}} |t|^k.$$

If $L_s \ge 1$, then $L_s^{\frac{d}{s-2}} \le L_s$. In this case, using a simple inequality

$$x^\beta e^{-x} \le (\beta e^{-1})^\beta \qquad (x, \beta \ge 0) \tag{17.3}$$

together with the property $k \le 3(s-2)$, we have

$$|t|^k\, e^{-3t^2/8} \le \left(\frac{8k}{3e}\right)^{k/2} \le \left(\frac{8s}{e}\right)^{\frac{3}{2}(s-2)} < (3s)^{\frac{3}{2}(s-2)}.$$

Hence $L_s^{\frac{d}{s-2}} |t|^k\, e^{-t^2/2} \le (3s)^{\frac{3}{2}(s-2)}\, L_s\, e^{-t^2/8}$. Using the inequality (10.2), we then get

$$|g_m(t)| \le (1 + |P_m(it)|)\, e^{-t^2/2} \le m(3s)^{\frac{3(s-2)}{2}}\, L_s\, e^{-t^2/8} \le (3s)^{3s/2}\, L_s\, e^{-t^2/8},$$

which provides the desired estimate (17.1).

In the (main) case $L_s \le 1$, it will be sufficient to bound the products $L_s^{\frac{d}{s-2}-1} |t|^k\, e^{-3t^2/8}$ by the s-dependent constants uniformly for all admissible tuples. Put $x = L_s^{-\frac{1}{s-2}}$. Using the hypothesis $|t| \ge \frac{1}{8}\, x^{1/3}$, let us rewrite every such product and then estimate it as follows:

$$L_s^{\frac{d}{s-2}-1} |t|^k\, e^{-3t^2/8} = x^{(s-2)-d}\, e^{-t^2/4} \cdot |t|^k\, e^{-t^2/8}$$

$$\leq x^{(s-2)-d}\, e^{-\frac{1}{256}x^{2/3}} \cdot |t|^k\, e^{-t^2/8}$$
$$= (256\, y)^{\frac{3}{2}((s-2)-d)}\, e^{-y} \cdot (8u)^{k/2}\, e^{-u},$$

where we changed the variables $x = (256\, y)^{3/2}$, $t = (8u)^{1/2}$. Next, again we apply inequality (17.3), which allows us to bound the last expression by

$$\Big(256 \cdot \frac{3}{2e}\,(s-2-d)\Big)^{\frac{3}{2}(s-2-d)} \cdot \Big(\frac{8k}{2e}\Big)^{k/2} \leq \Big(\frac{384\, s}{e}\Big)^{\frac{3}{2}(s-2-d)} \cdot \Big(\frac{12\, s}{e}\Big)^{k/2}$$
$$\leq \Big(\frac{384\, s}{e}\Big)^{\frac{1}{2}(3(s-2-d)+k)}.$$

Here $3(s-2-d)+k = 3(s-2)-(3d-k) \leq 3(s-2)$. Hence, the last quantity may further be estimated by $\big(\frac{384\, s}{e}\big)^{\frac{3(s-2)}{2}} < (142\, s)^{\frac{3(s-2)}{2}}$, so

$$L_s^{\frac{d}{s-2}}\, |t|^k\, e^{-t^2/2} \leq (142\, s)^{\frac{3(s-2)}{2}}\, L_s\, e^{-t^2/8}.$$

This inequality remains to hold, when all $k_j = 0$ as well. Thus, similarly to the previous case,

$$(1 + |P_m(it)|)\, e^{-t^2/2} \leq m(142\, s)^{\frac{3(s-2)}{2}}\, L_s\, e^{-t^2/8} \leq (142\, s)^{3s/2}\, L_s\, e^{-t^2/8},$$

proving the first part of the proposition, i.e. for $p = 0$.

To treat the case of derivatives of an arbitrary order $p \geq 1$, one may use the property that g_m is an entire function and apply Cauchy's contour integral formula. This would reduce our task to bounding $|g_m|$ in a strip of the complex plane. Indeed, first consider the functions of the complex variable

$$R_k(z) = z^k\, e^{-z^2/2}, \qquad z = t + u, \;\; (t \neq 0 \text{ real}), \;\; |u| \leq \frac{|t|}{4} \;\; (u \text{ complex}).$$

We have $|z| \leq \frac{5}{4}\,|t|$ and $\mathrm{Re}(z^2) \geq t^2 - 2\,|t|\,|u| - |u|^2 \geq \frac{7}{16}\, t^2$, implying that

$$|R_k(z)| = |z|^k\, e^{-\mathrm{Re}(z^2)/2} \leq \Big(\frac{5}{4}\,|t|\Big)^k\, e^{-7t^2/32}.$$

For any $\rho > 0$, by Cauchy's integral formula, $|R_k^{(p)}(t)| \leq p!\, \rho^{-p} \max_{|z-t|=\rho} |R_k(z)|$. Choosing $\rho = \frac{|t|}{4}$ and applying the constraints $p \leq s+1$, $k \leq 3(s-2)$, we get

$$|R_k^{(p)}(t)| \leq p!\Big(\frac{4}{|t|}\Big)^p \cdot \Big(\frac{5}{4}\,|t|\Big)^k\, e^{-7t^2/32} \leq p!\, 4^{s+1}\Big(\frac{5}{4}\Big)^{3(s-2)} \cdot |t|^{k-p}\, e^{-7t^2/32}. \tag{17.4}$$

Case 1. First assume that $k \geq p$.

If $L_s \leq 1$, putting $x = L_s^{-\frac{1}{s-2}}$ as before and using the hypothesis $|t| \geq \frac{1}{8}\,x^{1/3}$, we have:

$$\begin{aligned} p!\, L_s^{\frac{d}{s-2}-1}\, |t|^{k-p}\, e^{-3t^2/32} &= p!\, x^{(s-2)-d}\, e^{-t^2/16} \cdot |t|^{k-p}\, e^{-t^2/32} \\ &\leq p!\, x^{(s-2)-d}\, e^{-\frac{1}{8^2 \cdot 16}\, x^{2/3}} \cdot |t|^{k-p}\, e^{-t^2/32} \\ &= p!\, (8^2 \cdot 16\, y)^{\frac{3}{2}\,((s-2)-d)}\, e^{-y} \cdot (32\, u)^{\frac{1}{2}\,(k-p)}\, e^{-u}. \end{aligned}$$

Again using the general inequality (17.3), one can bound the last expression by

$$\begin{aligned} & p!\, \Big(8^2 \cdot 16 \cdot \frac{3}{2e}\,(s-2-d)\Big)^{\frac{3}{2}\,(s-2-d)} \cdot \Big(32\, \frac{k-p}{2e}\Big)^{\frac{1}{2}\,(k-p)} \\ & \leq p!\, (566\, s)^{\frac{3}{2}\,((s-2)-d)} \cdot \Big(\frac{48s}{e}\Big)^{\frac{1}{2}\,(k-p)} \\ & \leq s^p\, (566\, s)^{\frac{1}{2}\,(3(s-2-d)+(k-p))}, \end{aligned}$$

where we applied elementary relations $p! \leq m^p \leq s^p$ for the values $p \leq m+1$ on the last step. Also note that

$$3(s-2-d) + (k-p) = 3(s-2) - (3d-k) - p \leq 3(s-2) - p.$$

Hence,

$$\begin{aligned} s^p\, (566\, s)^{\frac{1}{2}\,(3(s-2-d)+(k-p))} &\leq 566^{\frac{3}{2}\,(s-2)}\, s^{\frac{1}{2}\,(3(s-2)+p)} \\ &\leq 116^{2(s-2)}\, s^{2s-2} < (116\, s)^{2s-2}, \end{aligned}$$

and thus

$$p!\, L_s^{\frac{d}{s-2}-1}\, |t|^{k-p}\, e^{-3t^2/32} \leq (116\, s)^{2s-2}. \tag{17.5}$$

If $L_s \geq 1$, the argument is similar and leads to a better constant. Since now $|t| \geq \frac{1}{8}\,x$,

$$\begin{aligned} p!\, L_s^{\frac{d}{s-2}-1}\, |t|^{k-p}\, e^{-3t^2/32} &= p!\, x^{(s-2)-d}\, e^{-t^2/16} \cdot |t|^{k-p}\, e^{-t^2/32} \\ &\leq p!\, x^{(s-2)-d}\, e^{-\frac{1}{8^2 \cdot 16}\, x^2} \cdot |t|^{k-p}\, e^{-t^2/32} \\ &= p!\, (8^2 \cdot 16\, y)^{\frac{1}{2}\,((s-2)-d)}\, e^{-y} \cdot (32\, u)^{\frac{1}{2}\,(k-p)}\, e^{-u}. \end{aligned}$$

The last expression is bounded by

$$
\begin{aligned}
& p! \left(8^2 \cdot 16 \cdot \frac{1}{2e}\,(s-2-d)\right)^{\frac{1}{2}\,(s-2-d)} \cdot \left(32\,\frac{k-p}{2e}\right)^{\frac{1}{2}\,(k-p)} \\
& \leq p!\,(189\,s)^{\frac{1}{2}\,((s-2)-d)} \cdot \left(\frac{48s}{e}\right)^{\frac{1}{2}\,(k-p)} \\
& \leq s^p\,(189\,s)^{\frac{1}{2}\,((s-2-d)+(k-p))}.
\end{aligned}
$$

Replacing here $s-2-d$ with the larger value $3(s-2-d)$, we return to the previous step with constant 189 in place of 566. So, the bound (17.5) remains to hold in this case as well.

Case 2. Assume that $k < p$ and $L_s \leq 1$. In this case, the function $|t|^{k-p}\,e^{-3t^2/32}$ is decreasing in $|t|$. Using again $|t| \geq \frac{1}{8}\,x^{1/3}$ with $x = L_s^{-\frac{1}{s-2}}$, we have, by (17.3), for any $\beta \geq 0$,

$$
\begin{aligned}
p!\,L_s^{\frac{d}{s-2}-1}\,|t|^{k-p}\,e^{-3t^2/32} & \leq p!\,x^{(s-2)-d}\left(\frac{1}{8}\,x^{1/3}\right)^{k-p} e^{-\frac{3}{8^2\cdot 32}\,x^{2/3}} \\
& \leq p!\,8^s\,x^{(s-2)-d+\frac{1}{3}\,(k-p)}\,e^{-\frac{3}{8^2\cdot 32}\,x^{2/3}} \\
& = p!\,8^s\left(\frac{8^2\cdot 32\,\beta}{3e}\right)^{\beta} x^{(s-2)-d+\frac{1}{3}\,(k-p)-\frac{2}{3}\,\beta}.
\end{aligned}
$$

Here we choose β such that the power of x would be zero, that is, $\beta = \frac{3}{2}\,(s-2-d) + \frac{1}{2}\,(k-p)$. Let us verify that this number is indeed non-negative, that is, $(3d-k)+p \leq 3(s-2)$. This is obvious, when all $k_j = 0$. From the definition, it also follows that, when at least one $k_j > 0$,

$$
3d-k = 2\sum_{j=1}^{m-2}(j-1)k_j = 2d - 2\sum_{j=1}^{m-2}k_j \leq 2(m-2)-2.
$$

If $p \leq m$, we conclude that $(3d-k)+p \leq 2(m-2)-2+m = 3(m-2) < 3(s-2)$, which was required. If $s = m+1$ is integer, and $p = m+1$, we also have

$$
(3d-k)+p \leq 2(m-2)-2+(m+1) = 3m-5 < 3(s-2).
$$

Thus, one may use the chosen value of β. Since $\beta = \frac{1}{2}\,(3(s-2)-(3d-k)-p) \leq \frac{3(s-2)-p}{2}$, we then get that

$$
\begin{aligned}
p!\,L_s^{\frac{d}{s-2}-1}\,|t|^{k-p}\,e^{-3t^2/32} & \leq p!\,8^s\left(\frac{8^2\cdot 32\,\beta}{3e}\right)^{\beta} \\
& \leq s^p\,8^s\left(\frac{8^2\cdot 32\,s}{2e}\right)^{\frac{3(s-2)-p}{2}}
\end{aligned}
$$

$$\leq 8^s \left(\frac{8^2 \cdot 32}{2e}\right)^{\frac{3(s-2)}{2}} \cdot s^{\frac{3(s-2)+p}{2}} < (242\, s)^{2s-2}.$$

Case 3. Assume that $k < p$ and $L_s \geq 1$, $|t| \geq 1$. In this case one may just write

$$p!\, L_s^{\frac{d}{s-2}-1}\, |t|^{k-p}\, e^{-3t^2/32} \leq p! \leq s^p \leq s^{2s-2}.$$

Thus, in all these three cases,

$$p!\, L_s^{\frac{d}{s-2}-1}\, |t|^{k-p}\, e^{-3t^2/32} \leq (242\, s)^{2s-2},$$

and therefore, according to (17.4),

$$\begin{aligned} |R_k^{(p)}(t)| &\leq 4^{s+1} \left(\frac{5}{4}\right)^{3(s-2)} e^{-t^2/8} \cdot p!\, |t|^{k-p}\, e^{-3t^2/32} \\ &\leq 4^{s+1} \left(\frac{5}{4}\right)^{3(s-2)} (242\, s)^{2s-2}\, L_s^{1-\frac{d}{s-2}}\, e^{-t^2/8} \\ &< 573^{2s}\, s^{2s-2}\, L_s^{1-\frac{d}{s-2}}\, e^{-t^2/8}. \end{aligned} \tag{17.6}$$

Case 4. Assume that $k < p$, $L_s \geq 1$ and $|t| \leq 1$.

Returning to the Cauchy integral formula, let us now choose $\rho = 1$. For $z = t+u$, $|u| \leq 1$ (u complex), we have $|z| \leq 2$ and $\mathrm{Re}(z^2) \geq t^2 - 2\,|t|\,|u| - |u|^2 \geq \frac{1}{2}\, t^2 - 3$. Hence

$$|R_k(z)| = |z|^k\, e^{-\mathrm{Re}(z^2)/2} \leq e^{-3}\, 2^k\, e^{-t^2/2} \leq e^{-3}\, 2^{s+1}\, e^{-t^2/2}$$

and

$$|R_k^{(p)}(t)| \leq p! \max_{|z-t|=1} |R_k(z)| \leq p!\, e^{-3}\, 2^{s+1}\, e^{-t^2/2} \leq e^{-3}\, (2s)^{2s-2}\, e^{-t^2/2}.$$

This is better than the bound (17.6) obtained for the previous cases (note that the above right-hand side may be multiplied by the factor $L_s^{1-\frac{d}{s-2}}$ which is larger than 1).

As result, in all cases,

$$|R_k^{(p)}(t)| \leq 573^{2s}\, s^{2s-2}\, L_s^{1-\frac{d}{s-2}}\, e^{-t^2/8},$$

so

$$|g_m^{(p)}(t)| \leq \sum \frac{1}{k_1! \dots k_{m-2}!} \left| \left(\frac{\gamma_3}{3!}\right)^{k_1} \cdots \left(\frac{\gamma_m}{m!}\right)^{k_{m-2}} \right| |R_k^{(p)}(t)|$$

$$\leq \sum \frac{1}{k_1! \dots k_{m-2}!} \frac{1}{3^{k_1} \dots m^{k_{m-2}}} L_s^{\frac{d}{s-2}} \cdot 573^{2s} s^{2s-2} L_s^{1-\frac{d}{s-2}} e^{-t^2/8}$$

$$\leq m \cdot 573^{2s} s^{2s-2} L_s e^{-t^2/8}.$$

□

18 Approximation of f_n and Its Derivatives on Long Intervals

Again, let $X_1, \dots, X_n$ be independent random variables with $\mathbb{E}X_k = 0$, $\sigma_k^2 = \mathbb{E}X_n^2$ ($\sigma_k \geq 0$) such that $\sum_{k=1}^n \sigma_k^2 = 1$, and finite Lyapunov coefficient L_s. On a relatively long (moderate) interval I_s, Proposition 13.2 (for $s \geq 3$) and Propositions 9.1–9.2 (for $2 < s \leq 3$) provide an approximation for the characteristic function $f_n(t)$ of the sum $S_n = X_1 + \dots + X_n$ by the corrected normal "characteristic function"

$$g_m(t) = e^{-t^2/2} (1 + P_m(it)).$$

This approximation also includes closeness of the derivatives of f_n and g_m up to order $[s]$. On the other hand, according to Propositions 15.2 and 16.1–16.2, $f_n(t)$ and their derivatives are very small in absolute value outside the interval I_s, although still inside $|t| \leq \frac{1}{L_3}$ when $s \geq 3$. Since $g_m(t)$ is also small (section 17), one can enlarge the interval I_s and thus simplify these approximations at the expense of a constant in the exponent appearing in the bounds.

As before, let $s = m + \alpha$, where $m \geq 2$ is integer and $0 < \alpha \leq 1$.

Theorem 18.1 *Let $L_s < \infty$ for $s \geq 3$. In the interval $|t| \leq \frac{1}{L_3}$,*

$$\left|f_n(t) - g_m(t)\right| \leq (Cs)^{3s/2} L_s \min\left\{1, |t|^s\right\} e^{-t^2/8}. \tag{18.1}$$

Moreover, for all $p = 0, 1, \dots, [s]$,

$$\left|\frac{d^p}{dt^p} (f_n(t) - g_m(t))\right| \leq (Cs)^{3s} L_s \min\left\{1, |t|^{s-p}\right\} e^{-t^2/8}, \tag{18.2}$$

where C is an absolute constant. One may take $C = 990$ in (18.1) *and $C = 70$ in* (18.2).

Proof We distinguish between several cases.

Case 1a. Moderate interval $I_s : |t| \max\{L_s^{\frac{1}{s-2}}, L_s^{\frac{1}{3(s-2)}}\} \leq \frac{1}{8}$. By Proposition 13.2, in this interval

$$|f_n(t) - g_m(t)| \leq 4^s L_s \max\left\{|t|^s, |t|^{3(s-2)}\right\} e^{-t^2/2}.$$

If $|t| \leq 1$, the above maximum is equal to $|t|^s$, and we are done with $C = 4$. If $|t| \geq 1$, the above maximum is equal to $|t|^{3(s-2)}$, and then one may use a general inequality $x^\beta e^{-x} \leq (\frac{\beta}{e})^\beta$ $(x, \beta > 0)$. For $x = 3t^2/8$ it gives

$$|t|^{3(s-2)}\, e^{-3t^2/8} = \Big(\frac{8x}{3}\Big)^{\frac{3(s-2)}{2}} e^{-x} \leq \Big(\frac{4\,(s-2)}{e}\Big)^{\frac{3(s-2)}{2}} < (1.48\,s)^{3s/2},$$

so

$$|t|^{3(s-2)}\, e^{-t^2/8} = |t|^{3(s-2)}\, e^{-3t^2/8}\, e^{-t^2/8} < (1.48\,s)^{3s/2}\, e^{-t^2/8}.$$

Since also $4^s \leq 2.52^{3s/2}$ and $2.52 \cdot 1.48 < 4$, we conclude that

$$|f_n(t) - g_m(t)| \leq (4s)^{3s/2}\, L_s \min\{1, |t|^s\}\, e^{-t^2/8}, \qquad t \in I_s,$$

which is the required inequality (18.1) with $C = 4$.

This bound may serve as a simplified version of Proposition 13.2 in the case $p = 0$. This is achieved at the expense of a worse constant in the exponent, although it contains a much larger s-dependent factor in front of L_s.

Case 2a. Large region I_s' : $|t| \max\{L_s^{\frac{1}{s-2}}, L_s^{\frac{1}{3(s-2)}}\} \geq \frac{1}{8}$ with $1 \leq |t| \leq \frac{1}{L_3}$. In this case, we bound both $f_n(t)$ and $g_m(t)$ in absolute value by appropriate quantities.

First, we involve the bound of Proposition 15.2, $|f_n(t)| \leq e^{-t^2/6}$, which is valid for $|t| \leq \frac{1}{L_3}$, and derive an estimate of the form

$$e^{-t^2/24} \leq C_s L_s.$$

If $L_s \geq 8^{-3(s-2)}$, it holds with $C_s = 8^{3(s-2)}$. If $L_s \leq 8^{-3(s-2)}$, then necessarily $|t| \geq \frac{1}{8}\, L_s^{-\frac{1}{3(s-2)}}$, and therefore one may take

$$C_s = \frac{1}{L_s}\, \exp\Big\{-\frac{1}{24 \cdot 8^2}\, L_s^{-\frac{2}{3(s-2)}}\Big\}.$$

Putting $L_s^{-\frac{2}{3(s-2)}} = 1536\,x$, the right-hand side equals and may be bounded with the help of (17.3) by

$$(1536\,x)^{\frac{3(s-2)}{2}}\, e^{-x} \leq \Big(\frac{1536 \cdot 3(s-2)}{2e}\Big)^{\frac{3(s-2)}{2}} < (848\,s)^{3s/2}.$$

As a result, we arrive at the upper bound

$$|f_n(t)| \leq e^{-t^2/24} \cdot e^{-t^2/8} \leq (848\,s)^{3s/2}\, L_s \min\{1, |t|^s\}\, e^{-t^2/8}.$$

A similar bound also holds for the approximating function $g_m(it) = e^{-t^2/2} + P_m(it)e^{-t^2/2}$. Recall that, by Proposition 17.1, whenever $|t| \geq 1$,

$$|g_m(t)| \leq (142\,s)^{3s/2}\,L_s\,e^{-t^2/8} \leq (142\,s)^{3s/2}\,L_s\,\min\{1, |t|^s\}\,e^{-t^2/8},$$

implying

$$|f_n(t) - g_m(t)| \leq \big((848\,s)^{3s/2} + (142\,s)^{3s/2}\big)\,L_s\,\min\{1, |t|^s\}\,e^{-t^2/8}.$$

Since $s > 3$, the constant in front of L_s is smaller than $990^{3s/2}$.

Case 3a. Consider the region I_s' : $|t|\,\max\{L_s^{\frac{1}{s-2}}, L_s^{\frac{1}{3(s-2)}}\} \geq \frac{1}{8}$ with $|t| \leq \min\{1, \frac{1}{L_3}\}$. Necessarily $L_s \geq 8^{-3(s-2)}$, so $\max\{L_s, 1\} \leq 8^{3(s-2)}L_s$. Hence, by Corollary 14.4 with $p = 0$,

$$|f_n(t) - g_m(t)| \leq 4s^{2s}\,8^{3(s-2)}\,L_s\,\frac{|t|^s}{m!}$$

$$\leq \frac{4}{8^6}\,\frac{s^{2s}}{(m/e)^m}\,8^{3s}\,L_s|t|^s \leq \frac{4}{8^6}\,\frac{s^{2s}}{(s/2)^{s/2}}\,e^s\,8^{3s}\,L_s|t|^s < \frac{1}{2}\,(158\,s)^{3s/2}\,L_s|t|^s.$$

This implies (18.1), since $e^{-t^2/8} \geq e^{-1/8}$.

The first assertion (18.1) is thus proved, and we now extend this inequality to the case of derivatives, although with a different dependence of the constants in s indicated in (18.2). We distinguish between several cases in analogy with the previous steps.

Case 1b. By Proposition 13.2, in the interval I_s,

$$|f_n^{(p)}(t) - g_m^{(p)}(t)| \leq 6 \cdot 8^s L_s \max\left\{|t|^{s-p}, |t|^{3(s-2)+p}\right\} e^{-t^2/2}.$$

If $|t| \leq 1$, the above maximum is equal to $|t|^{s-p}$, and we are done.
If $|t| \geq 1$, the above maximum is equal to $|t|^{3(s-2)+p}$. Using once more (17.3) with $x = 3t^2/8$, we have

$$|t|^{3(s-2)+p}\,e^{-3t^2/8} = \Big(\frac{8x}{3}\Big)^{\frac{3(s-2)+p}{2}} e^{-x} \leq \Big(\frac{4\,(3(s-2)+p)}{3e}\Big)^{\frac{3(s-2)+p}{2}} < (2s)^{2s},$$

so

$$|t|^{3(s-2)}\,e^{-t^2/8} = |t|^{3(s-2)}\,e^{-3t^2/8}\,e^{-t^2/8} < (2s)^{2s}\,e^{-t^2/8}.$$

Since also $6 \cdot 8^s \leq 4^{2s}$, we conclude that

$$|f_n^{(p)}(t) - g_m^{(p)}(t)| \leq (8s)^{2s} L_s \min\{1, |t|^{s-p}\}\,e^{-t^2/8}, \qquad t \in I_s,$$

which implies the required inequality (18.2) with $C = 8$.

Case 2b. Large region I_s' with $1 \le |t| \le \frac{1}{L_3}$. Let us involve Proposition 16.1. Using $p! \le s^s$ and $\max\{L_{p^*}, 1\} \le \max\{L_s, 1\}$, the bound (16.1) of this proposition readily implies

$$|f_n^{(p)}(t)| \le (2.03\, s)^s \max\{L_s, 1\}\, |t|^s\, e^{-t^2/6}, \qquad 1 \le |t| \le \frac{1}{L_3}.$$

Thus, we need to derive an estimate of the form

$$(2.03\, s)^s \max\{L_s, 1\}\, |t|^s\, e^{-t^2/24} \le C_s L_s.$$

If $L_s \ge 8^{-3(s-2)}$, the latter inequality holds with

$$C_s = (2.03\, s)^s\, 8^{3(s-2)} \max_t |t|^s\, e^{-t^2/24} = (2.03\, s)^s\, 8^{3(s-2)} \left(\frac{12s}{e}\right)^{s/2} < (13s)^{3s}.$$

If $L_s \le 8^{-3(s-2)}$, then necessarily $|t| \ge \frac{1}{8}\, L_s^{-\frac{1}{3(s-2)}}$, i.e., $\frac{1}{L_s} \le (8t)^{3(s-2)}$. Hence,

$$\begin{aligned}
\frac{1}{L_s}\, (2.03\, s)^s\, |t|^s\, e^{-t^2/24} &\le (8t)^{3(s-2)}\, (2.03\, s)^s\, |t|^s\, e^{-t^2/24} \\
&= 8^{-6}\, (8^3 \cdot 2.03\, s)^s\, (24y)^{2s}\, e^{-y} \\
&\le 8^{-6}\, (8^3 \cdot 2.03\, s)^s \left(\frac{48\, s}{e}\right)^{2s} < 8^{-6}\, 69^{3s}.
\end{aligned}$$

As a result, we arrive at the upper bound

$$|f_n^{(p)}(t)| \le (69\, s)^{3s}\, L_s \min\{1, |t|^s\}\, e^{-t^2/8}.$$

As we know, a better bound holds for the function $g_m(it) = e^{-t^2/2} + P_m(it)e^{-t^2/2}$. By Proposition 17.1, whenever $|t| \ge 1$,

$$|g_m^{(p)}(t)| \le (573\, s)^{2s}\, L_s\, e^{-t^2/8} \le (69\, s)^{3s}\, L_s \min\{1, |t|^s\}\, e^{-t^2/8},$$

implying

$$|f_n^{(p)}(t) - g_m^{(p)}(t)| \le (1 + 8^{-6})\, (69\, s)^{3s}\, L_s \min\{1, |t|^s\}\, e^{-t^2/8}.$$

Since $s > 3$, the constant in front of L_s is smaller than 70^{3s}.

Case 3b. The region I_s' with $|t| \le \min\{1, \frac{1}{L_3}\}$. Necessarily $L_s \ge 8^{-3(s-2)}$, so $\max\{L_s, 1\} \le 8^{3(s-2)} L_s$. Hence, by Corollary 14.4, for all $p \le [s]$,

$$|f_n^{(p)}(t) - g_m^{(p)}(t)| \le 4s^{2s}\, 8^{3(s-2)} L_s\, |t|^{s-p}$$

$$\leq \frac{4}{8^6}\, s^{2s}\, 8^{3s}\, L_s |t|^{s-p} \leq (8s)^{3s}\, L_s |t|^{s-p}.$$

Clearly, this bound is better than what was obtained on the previous step. □

Finally, let us include an analog of Theorem 18.1 for the case $2 < s < 3$. The following statement can be proved with similar arguments on the basis of Propositions 9.1 and 16.2.

Theorem 18.2 *Let $L_s < \infty$ for $2 < s < 3$. In the interval $|t| \leq (6L_s)^{-\frac{1}{s-2}}$, we have*

$$\left|\frac{d^p}{dt^p}\left(f_n(t) - e^{-t^2/2}\right)\right| \leq CL_s \min\left\{1, |t|^{s-p}\right\} e^{-t^2/8}, \qquad p = 0, 1, 2,$$

where C is an absolute constant.

Remarks In the literature, inequalities similar to (18.1)–(18.2) can be found for integer values $s = m + 1 \geq 3$, often for identically distributed summands $X_k = \xi_k/\sqrt{n}$, only, when $L_s = \beta_s\, n^{-(n-2)/2}$, $\beta_s = \mathbb{E}\,|\xi_1|^s$. In the book by Petrov [P2], (18.2) is proved without the derivative of the maximal order $p = m + 1$ and with an indefinite constant C_s (cf. Lemma 4, p. 140, which is attributed to Osipov [O]). Bikjalis derived a more precise statement (cf. [Bi3]). In case $p = 0$, he proved that, in the interval $|t| \leq \frac{1}{10}\, \beta_s^{-\frac{1}{s-2}} \sqrt{n}$,

$$\left|f_n(t) - g_m(t)\right| \leq \frac{2^{s-1}}{0.99^s}\, \beta_s\, n^{-\frac{s-2}{2}}\, |t|^s\, e^{-t^2/4}, \tag{18.3}$$

while for $p = 1, \dots, s$, $|t| \leq \frac{1}{16e}\, \beta_s^{-\frac{1}{s-2}} \sqrt{n}$, we have

$$\left|\frac{d^p}{dt^p}\left(f_n(t) - g_m(t)\right)\right| \leq \frac{p!^2\, 64^{s+p-2}}{s-2}\, \beta_s\, n^{-\frac{s-2}{2}}\, |t|^{s-p}\, e^{-t^2/6}. \tag{18.4}$$

It is interesting that the right-hand side in (18.3) provides a sharper growth of the constant in s in comparison with (18.1). Similarly, for the critical value $p = s$, the right-hand side in (18.4) may be replaced with $(Cs)^{2s}\, L_s\, \min\{1, |t|^s\}\, e^{-t^2/8}$ which also gives some improvement over (18.2). On the other hand, inequalities (18.1)–(18.2) are applicable in the non-i.i.d. situation and for real values of s.

In the general non-i.i.d. case, some similar versions of (18.1) were studied in [Bi1, Bi2]. A variant of (18.2) can be found in the book by Bhattacharya and Ranga Rao [B-RR], who considered multidimensional summands. Their Theorem 9.9 covers the interval $|t| \leq c\, L_s^{-\frac{1}{s-2}}$, although it does not specify constants as functions of s. Note that the interval $|t| \leq 1/L_3$ as in Theorem 18.1 is longest possible (up to a universal factor), but we leave open the question on the worst growth rates of the s-dependent constants in such inequalities.

Acknowledgements Sergey G. Bobkov would like to thank Irina Shevtsova for reading of the manuscript, valuable comments, and additional references.

Partially supported by the Alexander von Humboldt Foundation and NSF grant DMS-1612961.

References

[B-RR] Bhattacharya, R. N.; Ranga Rao, R. Normal approximation and asymptotic expansions. John Wiley & Sons, Inc. 1976. Also: Soc. for Industrial and Appl. Math., Philadelphia, 2010.

[Bi1] Bikjalis, A. On multivariate characteristic functions. (Russian) Litovsk. Mat. Sbornik 8 (1968), 21–39.

[Bi2] Bikjalis, A. On asymptotic expansion for products of multidimensional characteristic functions. (Russian) Teor. Verojatnost. i Primenen. 14 (1969), 508–511.

[Bi3] Bikjalis, A. Remainder terms in asymptotic expansions for characteristic functions and their derivatives. (Russian) Litovsk. Mat. Sb. 7 (1967) 571–582 (1968). Selected Transl. in Math. Statistics and Probability, 11 (1973), 149–162.

[B1] Bobkov, S. G. Closeness of probability distributions in terms of Fourier-Stieltjes transforms. (Russian) Uspekhi Matem. Nauk, vol. 71, issue 6 (432), 2016, 37–98. English translation in: Russian Math. Surveys.

[B2] Bobkov, S. G. Berry-Esseen bounds and Edgeworth expansions in the central limit theorem for transport distances. Probab. Theory Related Fields. Published online 19 January 2017.

[B-C-G1] Bobkov, S. G.; Chistyakov, G. P.; Götze, F. Non-uniform bounds in local limit theorem in case of fractional moments. I–II. Math. Methods of Statistics, 20 (2011), no. 3, 171–191; no. 4, 269–287.

[B-C-G2] Bobkov, S. G.; Chistyakov, G. P.; Götze, F. Rate of convergence and Edgeworth-type expansion in the entropic central limit theorem. Ann. Probab. 41 (2013), no. 4, 2479–2512.

[B-C-G3] Bobkov, S. G.; Chistyakov, G. P.; Götze, F. Fisher information and the central limit theorem. Probab. Theory Related Fields 159 (2014), no. 1–2, 1–59.

[Bu] Bulinskii, A. V. Estimates for mixed semi-invariants and higher covariances of bounded random variables. (Russian) Teor. Verojatnost. i Premenen. 19 (1974), 869–873.

[C] Cramér, H. Random variables and probability distributions. Cambridge Tracts in Math. and Math. Phys., no. 36, Cambridge Univ. Press, New York, 1937.

[E] Esseen, C.-G. Fourier analysis of distribution functions. A mathematical study of the Laplace-Gaussian law. Acta Math. 77 (1945), 1–125.

[G1] Gnedenko, B. V. A local limit theorem for densities. (Russian) Doklady Akad. Nauk SSSR (N.S.) 95, (1954), 5–7.

[G2] Gnedenko, B. W. Positiv definite Verteilungsdichten. Einführung in die Wahrscheinlichkeitstheorie [Introduction to Probability Theory]. Translated from the sixth Russian edition by Hans-Joachim Rossberg and Gabriele Laue. Edited and with a foreword and an appendix by H.-J. Rossberg. AkademieVerlag, Berlin, 1991.

[G-K] Gnedenko, B. V.; Kolmogorov, A. N. Limit distributions for sums of independent random variables. Translated and annotated by K. L. Chung. With an Appendix by J. L. Doob. Addison-Wesley Publishing Company, Inc., Cambridge, Mass., 1954. ix+264 pp.

[G-H] Götze, F.; Hipp, C. Asymptotic expansions in the central limit theorem under moment conditions. Z. Wahrscheinlichkeitstheorie und Verw. Gebiete 42 (1978), no. 1, 67–87.

[I-S] Ibragimov, R.; Sharakhmetov, Sh. On an exact constant for the Rosenthal inequality. (Russian) Teor. Veroyatnost. i Primenen. 42 (1997), no. 2, 341–350; translation in Theory Probab. Appl. 42 (1997), no. 2, 294–302 (1998).

[J-S-Z] Johnson, W. B.; Schechtman, G.; Zinn, J. Best constants in moment inequalities for linear combinations of independent and exchangeable random variables. Ann. Probab. 13 (1985), no. 1, 234–253.

[O] Osipov, L. V. Asymptotic expansions in the central limit theorem. (Russian) Vestnik Leningrad. Univ. 22 (1967), no. 19, 45–62.

[P1] Petrov, V. V. Local limit theorems for sums of independent random variables. (Russian) Teor. Verojatnost. i Primenen. 9 (1964), 343–352.

[P2] Petrov, V. V. Sums of independent random variables. Translated from the Russian by A. A. Brown. Ergebnisse der Mathematik und ihrer Grenzgebiete, Band 82. Springer-Verlag, New York-Heidelberg, 1975. x+346 pp. Russian ed.: Moscow, Nauka, 1972, 414 pp.

[P3] Petrov, V. V. Limit theorems for sums of independent random variables. Moscow, Nauka, 1987 (in Russian), 318 pp.

[Pi] Pinelis, I. Exact Rosenthal-type bounds. Ann. Probab. 43 (2015), no. 5, 2511–2544.

[P-U] Pinelis, I. F.; Utev, S. A. Estimates of moments of sums of independent random variables. (Russian) Teor. Veroyatnost. i Primenen. 29 (1984), no. 3, 554–557.

[Pr1] Prawitz, H. Ungleichungen für den absoluten Betrag einer charackteristischen Funktion. (German) Skand. Aktuarietidskr. 1973, 11–16.

[Pr2] Prawitz, H. Weitere Ungleichungen für den absoluten Betrag einer charakteristischen Funktion. (German) Scand. Actuar. J. 1975, 21–28.

[R] Rosenthal, H. P. On the subspaces of L_p $(p > 2)$ spanned by sequences of independent random variables. Israel J. Math. 8 (1970), 273–303.

[Sa] Sakovič, G. N. On the width of a spectrum. I. (Ukrainian) Dopovidi Akad. Nauk Ukrain. RSR, 11 (1965), 1427–1430.

[Se] Senatov, V. V. Central limit theorem. Exactness of approximation and asymptotic expansions. (Russian) TVP Science Publishers, Moscow, 2009.

[Sh1] Shevtsova, I. G. On the accuracy of the approximation of the complex exponent by the first terms of its Taylor expansion with applications. J. Math. Anal. Appl. 418 (2014), no. 1, 185–210.

[Sh2] Shevtsova, I. G. Moment-type estimates for characteristic functions with application to von Mises inequality. J. Math. Sci. (N.Y.) 214 (2016), no. 1, 119–131.

[St1] Statulevičius, V. A. On the asymptotic expansion of the characteristic function of a sum of independent random variables. (Russian) Litovsk. Mat. Sb. 2 (1962), no. 2, 227–232.

[St2] Statulevičius, V. A. Limit theorems for densities and the asymptotic expansions for distributions of sums of independent random variables. (Russian) Teor. Verojatnost. i Primenen. 10 (1965), 645–659.

[U] Ushakov N. G. Selected topics in characteristic functions. Utrecht: VSP, 1999.

[Z1] Zolotarev, V. M. On the closeness of the distributions of two sums of independent random variables. (Russian) Teor. Verojatnost. i Primenen. 10 (1965), 519–526.

[Z2] Zolotarev, V. M. Some inequalities from probability theory and their application to a refinement of A. M. Ljapunov's theorem. (Russian) Dokl. Akad. Nauk SSSR 177 (1967), 501–504.

Part II
Convexity and Concentration for Sets and Functions

Non-standard Constructions in Convex Geometry: Geometric Means of Convex Bodies

Vitali Milman and Liran Rotem

Abstract In this note we discuss new constructions of convex bodies. By thinking of the polarity map $K \mapsto K^{\circ}$ as the inversion $x \mapsto x^{-1}$ one may construct new bodies which were not previously considered in convex geometry. We illustrate this philosophy by describing a recent result of Molchanov, who constructed continued fractions of convex bodies.

Our main construction is the geometric mean of two convex bodies. We define it using the above ideology, and discuss its properties and its structure. We also compare our new definition with the "logarithmic mean" of Böröczky, Lutwak, Yang and Zhang, and discuss volume inequalities. Finally, we discuss possible extensions of the theory to p-additions and to the functional case, and present a list of open problems.

An appendix to this paper, written by Alexander Magazinov, presents a 2-dimensional counterexample to a natural conjecture involving the geometric mean.

1 Introduction

A *convex body* in $\mathbb{R}^n$ is a compact, convex set $K \subseteq \mathbb{R}^n$. We will always make the additional assumption that 0 is in the interior of K, and denote the class of such convex bodies in $\mathbb{R}^n$ by $\mathcal{K}^n_{(0)}$. We also denote the (Lebesgue) volume of K by $|K|$, and the unit ball of ℓ^n_p by B^n_p.

V. Milman
School of Mathematical Sciences, Tel-Aviv University, Ramat Aviv, Tel Aviv 69978, Israel
e-mail: milman@post.tau.ac.il

L. Rotem (✉)
School of Mathematics, University of Minnesota, Minneapolis, MN, USA
e-mail: lrotem@umn.edu

The research was conducted while the author worked in Tel-Aviv University.

E. Carlen et al. (eds.), *Convexity and Concentration*, The IMA Volumes in Mathematics and its Applications 161, DOI 10.1007/978-1-4939-7005-6_12

The goal of this paper is to discuss constructions of new convex bodies out of old ones. The most well-known such construction is the *Minkowski addition*. For $K, T \in \mathcal{K}^n_{(0)}$ we define

$$K + T = \{x + y : x \in K, \ y \in T\}.$$

If $\lambda > 0$ and $K \in \mathcal{K}^n_{(0)}$, then the *homothety* λK is defined in the obvious way as $\lambda K = \{\lambda x : x \in K\}$. Once the addition and the homothety are defined, we may of course define the arithmetic mean of K and T as $A(K, T) = \frac{1}{2}(K + T)$.

Another standard construction in convex geometry is the polarity transform. The *polar* (or dual) of a body $K \in \mathcal{K}^n_{(0)}$ is defined by

$$K^\circ = \{y \in \mathbb{R}^n : \langle x, y \rangle \leq 1 \text{ for all } x \in K\},$$

where $\langle \cdot, \cdot \rangle$ denotes the standard Euclidean inner product on $\mathbb{R}^n$.

For an equivalent description of the polar body, remember that to every $K \in \mathcal{K}^n_{(0)}$ one may associate two standard functions $h_K, r_K : \mathbb{R}^n \to (0, \infty)$. The *support function* h_K is defined by

$$h_K(\theta) = \max_{x \in K} \langle x, \theta \rangle,$$

and the *radial function* r_K is defined by

$$r_K(\theta) = \max \{\lambda > 0 : \lambda\theta \in K\}.$$

As h_K is 1-homogeneous and r_K is (-1)-homogeneous, it is usually enough to think of them as functions on S^{n-1}, the unit Euclidean sphere in $\mathbb{R}^n$. We will often write $h_\theta(K)$ and $r_\theta(K)$ instead of $h_K(\theta)$ and $r_K(\theta)$, especially in situations where $\theta \in S^{n-1}$ is fixed and K changes. Each of the functions h_K and r_K determines the body K uniquely, and the polar body K° can be defined by the relation $r_\theta(K^\circ) = h_\theta(K)^{-1}$. Remember also that for every $K, T \in \mathcal{K}^n_{(0)}$ and every $\lambda > 0$ we have $h_\theta(\lambda K + T) = \lambda h_\theta(K) + h_\theta(T)$.

The polarity map $\circ : \mathcal{K}^n_{(0)} \to \mathcal{K}^n_{(0)}$ is an *abstract duality* in the sense of [3] (see also [22]). This means that it satisfies the following two properties:

- It is an *involution*: $(K^\circ)^\circ = K$ for all $K \in \mathcal{K}^n_{(0)}$.
- It is *order reversing*: If $K \supseteq T$, then $K^\circ \subseteq T^\circ$.

In fact, the polarity map is essentially the only duality on $\mathcal{K}^n_{(0)}$:

Theorem 1 *Let $\mathcal{T} : \mathcal{K}^n_{(0)} \to \mathcal{K}^n_{(0)}$ be an order reversing involution. Then there exists a symmetric and invertible linear map $u : \mathbb{R}^n \to \mathbb{R}^n$ such that $\mathcal{T}K = u(K^\circ)$.*

This theorem essentially appears in the work of Böröczky and Schneider [6]. A similar theorem on a different class of convex sets was proved by Artstein-Avidan and Milman [2]. On yet another class of convex sets, the theorem can also be deduced from the work of Gruber [13].

There is another famous duality in mathematics: the inverse map. The map $x \mapsto x^{-1}$ defined on $\mathbb{R}_+$ is a duality in the above sense. The same is true if one replaces $\mathbb{R}_+$ with the the class $\mathcal{M}_+^n$ of $n \times n$ positive-definite matrices. For the constructions described in this paper, it will be useful to think of K° as the inverse "K^{-1}".

Let us give one example of this point of view. Once we have an inverse map and an addition operation, we can easily construct the harmonic mean: The harmonic mean of K and T is simply

$$H(K,T) = \left(\frac{K^\circ + T^\circ}{2}\right)^\circ.$$

Naturally, we expect the harmonic mean to be smaller than the arithmetic mean. This is true, and was proved by Firey in [11]:

Theorem 2 (Firey) *For every $K, T \in \mathcal{K}_{(0)}^n$ one has*

$$\frac{K+T}{2} \supseteq \left(\frac{K^\circ + T^\circ}{2}\right)^\circ.$$

Since we will rely heavily on this result, we reproduce its short proof in a more modern notation:

Proof Fix $\theta \in S^{n-1}$. Since $r_\theta(K) \cdot \theta \in K$ and $r_\theta(T) \cdot \theta \in T$ we have

$$\frac{r_\theta(K) + r_\theta(T)}{2} \cdot \theta \in \frac{K+T}{2},$$

and hence by definition

$$r_\theta\left(\frac{K+T}{2}\right) \geq \frac{r_\theta(K) + r_\theta(T)}{2}.$$

On the other hand, we have

$$r_\theta\left(\left(\frac{K^\circ + T^\circ}{2}\right)^\circ\right) = \left(h_\theta\left(\frac{K^\circ + T^\circ}{2}\right)\right)^{-1}$$
$$= \left(\frac{h_\theta(K^\circ) + h_\theta(T^\circ)}{2}\right)^{-1} = \frac{2}{\frac{1}{r_\theta(K)} + \frac{1}{r_\theta(T)}}.$$

The result now follows from the arithmetic mean-harmonic mean inequality for real numbers. □

The construction of the harmonic mean is not terribly exciting, but it emerged naturally from the same philosophy as the rest of this paper. In the next section we will follow the work of Molchanov, and describe a more interesting construction – continued fractions of convex bodies. The next several sections are devoted to the geometric mean of convex bodies, the main construction of this paper. Section 8 is devoted to a possible extension of the theory to the functional case. Finally, in Section 9 we list several open problems.

2 Continued Fractions of Convex Bodies

For a sequence of positive real numbers $\{x_m\}_{m=1}^{\infty}$, the continued fraction $[x_1, x_2, x_3, \dots]$ is simply

$$\cfrac{1}{x_1 + \cfrac{1}{x_2 + \frac{1}{x_3 + \frac{1}{\dots}}}}.$$

More formally, we define $[x_1] = \frac{1}{x_1}$ and

$$[x_1, x_2, \dots, x_m] = (x_1 + [x_2, x_3, \dots, x_m])^{-1},$$

and we set

$$[x_1, x_2, x_3, \dots] = \lim_{m\to\infty} [x_1, x_2, \dots, x_m].$$

It is not hard to see that this sequence indeed converges if $x_m > \epsilon$ for all m and some fixed $\epsilon > 0$. In particular, the continued fraction converges whenever the x_i's are all integers.

In [18], Molchanov generalizes the construction of continued fractions to the general setting of a partially ordered abelian semigroup equipped with an abstract duality (i.e. an order reversing involution). We will only state his results for the class $\mathcal{K}_{(0)}^n$, where the duality is of course the polarity transform.

For a sequence of convex bodies $\{K_m\}_{m=1}^{\infty} \subseteq \mathcal{K}_{(0)}^n$ we set $[K_1] = K_1^\circ$ and

$$[K_1, K_2, \dots, K_m] = (K_1 + [K_2, K_3, \dots, K_m])^\circ.$$

In order to discuss the convergence of $\lim_{m\to\infty} [K_1, K_2, \dots, K_m]$ we need a suitable metric on $\mathcal{K}_{(0)}^n$. The obvious choice, and the one used by Molchanov, is the Hausdorff distance:

$$d(K, T) = \min\{r > 0 : K \subseteq T + rB_2^n \text{ and } T \subseteq K + rB_2^n\}.$$

We can now state Molchanov's theorem:

Theorem 3 (Molchanov) *Let $\{K_m\}_{m=1}^{\infty} \subseteq \mathcal{K}_{(0)}^n$ be a family of convex bodies. Assume that one of the following three conditions hold:*

1. $K_m \supseteq rB_2^n$ *for all m and for some* $r > 1$.
2. $B_2^n \subseteq K_m \subseteq R \cdot B_2^n$ *for all m and for some* $R < \infty$.
3. $rB_2^n \subseteq K_m \subseteq R \cdot B_2^n$ *for all m for some such* $r < 1$ *and* $R \le r/(1-r)$.

Then

$$[K_1, K_2, \ldots] = \lim_{m\to\infty} [K_1, K_2, \ldots, K_m]$$

exists in the Hausdorff sense.

As a corollary of the above theorem, one can deduce the following result:

Proposition 4 (Molchanov) *For every convex body* $K \in \mathcal{K}_{(0)}^n$ *such that* $K \supseteq B_2^n$ *there exists a unique body* $Z \in \mathcal{K}_{(0)}^n$ *such that* $Z^\circ = Z + K$.

Notice that if we think of Z° as the inverse "Z^{-1}", the equation $Z^\circ = Z + K$ is a "quadratic equation" of convex bodies. Its solution can be written in a continued fraction form, $Z = [K, K, K, \ldots]$, and the convergence of this fraction follows from Theorem 3. The uniqueness part of Proposition 4 does not appear in Molchanov's paper, but follows easily from his techniques.

We will now give a self-contained proof of Proposition 4. The proof is essentially Molchanov's, but since we do not strive for generality we can present the proof in a more transparent form. We begin with a lemma, also taken from Molchanov's paper:

Lemma 5 (Molchanov) *If* $K, T \supseteq rB_2^n$, *then* $d(K^\circ, T^\circ) \le r^{-2} \cdot d(K, T)$.

Proof Write $d = d(K, T)$. By definition of the Hausdorff distance we have

$$K \subseteq T + d \cdot B_2^n \subseteq T + \frac{d}{r}T = \frac{r+d}{r}T,$$

and since polarity is order reversing it follows that

$$K^\circ \supseteq \left(\frac{r+d}{r}T\right)^\circ = \frac{r}{r+d}T^\circ.$$

Since $K \supseteq rB_2^n$ we also have $B_2^n \supseteq rK^\circ$, so

$$K^\circ + \frac{d}{r^2}B_2^n \supseteq K^\circ + \frac{d}{r^2}rK^\circ = \left(\frac{r+d}{r}\right)K^\circ \supseteq \frac{r+d}{r} \cdot \frac{r}{r+d} \cdot T^\circ = T^\circ.$$

By exchanging the roles of K and T we also have $T^\circ + \frac{d}{r^2}B_2^n \supseteq T^\circ$, so $d(K^\circ, L^\circ) \le \frac{d}{r^2} = r^{-2}d(K, T)$. □

We may now proof Proposition 4:

Proof Define a sequence of convex bodies by

$$Z_1 = K^\circ$$
$$Z_{m+1} = (K + Z_m)^\circ .$$

Our first goal is to prove that $Z_m \supseteq \epsilon B_2^n$ for all m and some fixed $\epsilon > 0$. Indeed, K is assumed to be compact so $B_2^n \subseteq K \subseteq R \cdot B_2^n$ for some $R > 0$. If we now define two sequences of real numbers $\{a_m\}_{m=1}^\infty$, $\{b_m\}_{m=1}^\infty$ by

$$a_1 = \frac{1}{R} \qquad b_1 = 1$$
$$a_{m+1} = \frac{1}{R + b_m} \quad b_{m+1} = \frac{1}{1 + a_m},$$

it is trivial to prove by induction that $a_m B_2^n \subseteq Z_m \subseteq b_m B_2^n$ for all m. Since

$$\lim_{m\to\infty} a_m = [R, 1, R, 1, R, 1, \ldots] > 0,$$

it follows that $a_m > \epsilon$ for all m and some fixed $\epsilon > 0$, which proves our claim.

Using the above fact and the lemma, we deduce that for every $m > 1$ we have

$$d(Z_{m+1}, Z_m) = d\left((K + Z_m)^\circ, (K + Z_{m-1})^\circ\right) \le \left(\frac{1}{1+\epsilon}\right)^2 d\left(K + Z_m, K + Z_{m-1}\right)$$
$$= \left(\frac{1}{1+\epsilon}\right)^2 d(Z_m, Z_{m-1}).$$

Hence the sequence $\{Z_m\}$ is a Cauchy sequence, so the limit $Z = \lim_{m\to\infty} Z_m$ exists. In general the limit of bodies in $\mathcal{K}_{(0)}^n$ does not have to be in $\mathcal{K}_{(0)}^n$, as it may have an empty interior. In our case, however, we have $Z_m \supseteq \epsilon B_2^n$ for all m, so $Z \supseteq \epsilon B_2^n$ and $Z \in \mathcal{K}_{(0)}^n$. Sending $m \to \infty$ in the relation $Z_{m+1} = (K + Z_m)^\circ$ and using the continuity of the polarity transform we obtain $Z = (K + Z)^\circ$, so the existence part of the proposition is proved.

For the uniqueness, assume $Z, W \in \mathcal{K}_{(0)}^n$ satisfy both $Z^\circ = Z + K$ and $W^\circ = W + K$. Fix some $\epsilon > 0$ such that $Z, W \supseteq \epsilon B_2^n$. Then

$$d(Z, W) = d\left((Z + K)^\circ, (W + K)^\circ\right) \le \left(\frac{1}{1+\epsilon}\right)^2 d\left(Z + K, W + K\right)$$
$$= \left(\frac{1}{1+\epsilon}\right)^2 d(Z, W),$$

so $d(Z, W) = 0$ and $Z = W$. □

Denote the unique solution of $Z^\circ = Z + K$ by $Z(K)$. Notice that $Z(rB_2^n) = [r, r, r, \ldots] \cdot B_2^n$. In particular, for $r = 1$ we have $Z(B_2^n) = \frac{1}{\varphi} B_2^n$, where $\varphi = \frac{1+\sqrt{5}}{2}$ is the golden ratio. However, for other choices of K (say the unit cube), the body $Z(K)$ is completely mysterious, and we know very little about its properties. It appears to be a genuinely new construction in convexity.

3 The Geometric Mean of Convex Bodies

Over the recent years, there were several attempts to define the geometric mean of two convex bodies K and T. Let us recall some of these ideas, not in chronological order:

In [5], Böröczky, Lutwak, Yang and Zhang construct the following "0-mean", or "logarithmic mean", of convex bodies:

$$L_\lambda(K, T) = \left\{x \in \mathbb{R}^n : \langle x, \theta \rangle \le h_K(\theta)^{1-\lambda} h_T(\theta)^\lambda \text{ for all } \theta \in S^{n-1}\right\}.$$

In other words, the support function h_L of $L = L_\lambda(K, T)$ is the largest convex function such that $h_L(\theta) \le h_K(\theta)^{1-\lambda} h_T(\theta)^\lambda$ for all $\theta \in S^{n-1}$.

The authors of [5] conjecture that $L_\lambda(K, T)$ satisfy a Brunn-Minkowski type inequality. To describe their conjecture, let us denote by $\mathcal{K}_s^n$ the class of origin-symmetric convex bodies, i.e. the sets $K \in \mathcal{K}_{(0)}^n$ such that $K = -K$. The *log Brunn Minkowski conjecture* then states that for every $K, T \in \mathcal{K}_s^n$ and every $\lambda \in [0, 1]$ we have $|L_\lambda(K, T)| \ge |K|^{1-\lambda} |T|^\lambda$. It is still unknown whether this conjecture is true – it was proven in [5] in dimension $n = 2$, and in [23] by Saroglou for unconditional convex bodies in $\mathbb{R}^n$.

To explain its name, notice that the log-Brunn-Minkowski conjecture is a strengthening of the classic Brunn-Minkowski inequality. Indeed, by the arithmetic mean-geometric mean inequality we have

$$h_K(\theta)^{1-\lambda} h_T(\theta)^\lambda \le (1 - \lambda) h_K(\theta) + \lambda h_T(\theta) = h_{(1-\lambda)K+\lambda T}(\theta),$$

so the log-Brunn-Minkowski inequality implies that

$$|(1 - \lambda)K + \lambda T| \ge |L_\lambda(K, T)| \ge |K|^{1-\lambda} |T|^\lambda,$$

which is exactly the Brunn-Minkowski inequality in its dimension free form.

Let us mention that one can also consider the "dual" construction to L_λ, where instead of the support functions one take the geometric average of the radial functions. The body obtained is simply $L_\lambda (K^\circ, T^\circ)^\circ$, and Saroglou proved in [24] that $|L_\lambda(K^\circ, T^\circ)^\circ| \le |K|^{1-\lambda} |T|^\lambda$ for every $K, T \in \mathcal{K}_s^n$. By the Blaschke-Santaló inequality and Bourgain-Milman theorem [7], it follows that

$$|L_\lambda(K,T)| \geq c^n |K|^{1-\lambda} |T|^\lambda \tag{3.1}$$

for some universal constant $c > 0$.

For complex bodies, the situation is much clearer. Notice that, by our definition, a convex body $K \in \mathcal{K}_s^n$ is simply the unit ball of a norm on $\mathbb{R}^n$. Similarly, a complex convex body $K \subseteq \mathbb{C}^n$ is the unit ball of a norm on $\mathbb{C}^n$. By identifying $\mathbb{C}^n \simeq \mathbb{R}^{2n}$ we see that every complex body is also a real body, but not vice versa. In fact, a complex body $K \subseteq \mathbb{C}^n$ is a real body which is also symmetric with respect to complex rotations, i.e. $z \in K$ implies that $e^{i\theta}z \in K$ for all $\theta \in \mathbb{R}$.

There is a standard method in the literature to interpolate between complex norms, or equivalently, between complex bodies. This method is known simply as "complex interpolation" and is described, for example, in chapter 7 of [19]. In [8], Cordero-Erausquin proves that for every complex bodies K and T and every $\lambda \in [0,1]$, we have the relation $|C_\lambda(K,T)| \geq |K|^{1-\lambda}|T|^\lambda$ where C_λ denotes the complex interpolation. From here he deduces an extension of the Blaschke-Santaló inequality: Since

$$C_{1/2}(K \cap T, K^\circ \cap T) \subseteq B_2^{2n} \cap T$$

we must have $|K \cap T|\,|K^\circ \cap T| \leq \left|B_2^{2n} \cap T\right|^2$. Cordero-Erausquin asks whether this inequality also holds for real convex bodies, and this question is still open. A partial answer was given in [14] by Klartag, who proved in the real case a functional version of the inequality. As a corollary he proved that for every $K, T \in \mathcal{K}_s^n$ we have

$$|K \cap T|\,|K^\circ \cap T| \leq 2^n |B_2^n \cap T|^2 \tag{3.2}$$

It is also true that for complex bodies $C_\lambda(K,T) \subseteq L_\lambda(K,T)$, so the log-Brunn-Minkowski conjecture is true for complex bodies (see [20]).

Finally, let us briefly mention a third possible "geometric mean". The construction was studied by Cordero-Erausquin and Klartag in [9], following a previous work of Semmes [25]. Let $u_0, u_1 : \mathbb{R}^n \to \mathbb{R}$ be (sufficiently smooth) convex functions. A p-interpolation between u_0 and u_1 is a function $u : [0,1] \times \mathbb{R}^n \to \mathbb{R}$ such that $u(0,x) = u_0(x)$, $u(1,x) = u_1(x)$, and $u(t,x)$ satisfies the PDE

$$\partial_{tt}^2 u = \frac{1}{p}\left\langle (\mathrm{Hess}_x u)^{-1}\, \nabla\partial_t u, \nabla\partial_t u\right\rangle.$$

Here we will care about the case $p = 2$. Given u_0 and u_1 it is not clear that this PDE has a solution, let alone a unique solution. However, it is not hard to check that if $u_0 = \frac{1}{2}h_K^2$ and $u_1 = \frac{1}{2}h_L^2$ for some bodies K and L, then $u_t = \frac{1}{2}h_{R_t}^2$ (assuming it exists) for some family of convex bodies $R_t = R_t(K,L)$. Similarly to the previous two constructions, the authors conjectured that $|R_\lambda(K,L)| \geq |K|^{1-\lambda}|L|^\lambda$ for $K, T \in \mathcal{K}_s^n$. However, after the publication of [9], Cordero-Erausquin and Klartag found a counterexample to this inequality.

We will now present a new definition for the geometric mean of two convex bodies in $\mathbb{R}^n$, which seems to satisfy some natural properties. As a first step, let us consider the geometric mean of positive numbers. Given two numbers $x, y > 0$, we build two sequences by the recurrence relations

$$\begin{aligned} a_0 &= x & h_0 &= y \\ a_{n+1} &= \frac{a_n + h_n}{2} & h_{n+1} &= \left(\frac{a_n^{-1} + h_n^{-1}}{2}\right)^{-1}. \end{aligned}$$

It is an easy exercise to see that $\{a_n\}$ is decreasing, $\{h_n\}$ is increasing, and $\lim_{n\to\infty} a_n = \lim_{n\to\infty} h_n = \sqrt{xy}$.

A similar result is known to hold for positive definite matrices. Given two such matrices u and v, we define

$$\begin{aligned} A_0 &= u & H_0 &= v \\ A_{n+1} &= \frac{A_n + H_n}{2} & H_{n+1} &= \left(\frac{A_n^{-1} + H_n^{-1}}{2}\right)^{-1}. \end{aligned}$$

It is known that $\{A_n\}$ is decreasing (in the sense of matrices), $\{H_n\}$ is increasing, and the limits $\lim_{n\to\infty} A_n$ and $\lim_{n\to\infty} H_n$ exist and are equal. This joint limit is known as the geometric mean of u and v, and is often written as $u\#v$. It shares many of the properties of the geometric mean of numbers - see, e.g., [15] for a survey of such properties. An explicit formula for $u\#v$ is

$$u\#v = u^{1/2}\left(u^{-1/2}vu^{-1/2}\right)^{1/2}u^{1/2},$$

but it may be better to think of $u\#v$ as the unique solution of the matrix equation $xu^{-1}x = v$

Since we already understand the arithmetic mean and harmonic mean of convex bodies, we may simply repeat the same process. For $K, T \in \mathcal{K}^n_{(0)}$ we define

$$\begin{aligned} A_0 &= K & H_0 &= T \\ A_{n+1} &= \frac{A_n + H_n}{2} & H_{n+1} &= \left(\frac{A_n^\circ + H_n^\circ}{2}\right)^\circ. \end{aligned} \tag{3.3}$$

Theorem 6 *Fix $K, T \in \mathcal{K}^n_{(0)}$ and define sequences $\{A_n\}$ and $\{H_n\}$ according to (3.3). Then $\{A_n\}$ is decreasing and $\{H_n\}$ is increasing with respect to set inclusion, and the limits $\lim_{n\to\infty} A_n$ and $\lim_{n\to\infty} H_n$ exist (in the Hausdorff sense) and are equal.*

Proof By Theorem 2 we see that $A_n \supseteq H_n$ for every $n \geq 1$. If follows that

$$A_{n+1} = \frac{A_n + H_n}{2} \subseteq \frac{A_n + A_n}{2} = A_n,$$

and

$$H_{n+1} = \left(\frac{A_n^\circ + H_n^\circ}{2}\right)^\circ \supseteq \left(\frac{H_n^\circ + H_n^\circ}{2}\right)^\circ = H_n.$$

Hence $\{A_n\}$ is a decreasing sequence of convex bodies. It is also bounded from below by a "proper" convex body (with non-empty interior), since

$$A_n \supseteq H_n \supseteq H_1$$

for all $n \geq 1$. It follows that there exists a body $G_1 \in \mathcal{K}^n_{(0)}$ such that $A_n \to G_1$ in the Hausdorff sense. Similarly, $\{H_n\}$ is increasing and bounded from above by A_1, so it converges to some G_2.

Finally, taking the equation

$$A_{n+1} = \frac{A_n + H_n}{2}$$

and sending $n \to \infty$, we see that $G_1 = \frac{1}{2}(G_1 + G_2)$. Hence $G_1 = G_2$ and the proof is complete. □

Definition 7 The joint limit from the previous theorem is called the geometric mean of K and T:

$$G(K,T) = \lim_{n\to\infty} A_n = \lim_{n\to\infty} H_n.$$

If we need to refer to the bodies A_n and H_n from the process defining $G(K,T)$, we will write $A_n(K,T)$ and $H_n(K,T)$. It is immediate that $H_n(K,T) \subseteq G(K,T) \subseteq A_n(K,T)$ for all n. In particular, we have the arithmetic mean - geometric mean - harmonic mean inequality $H_1(K,T) \subseteq G(K,T) \subseteq A_1(K,T)$.

Figure 1 depicts one planar example of two convex polygons K and T and their geometric mean.

Even though our motivation is very different, the above definition was also inspired by a similar construction for 2-homogeneous functions of Asplund [4]. We have also recently discovered a paper of Fedotov [10] with a similar construction.

4 Properties of the Geometric Mean

The following proposition summarizes some of the basic properties of the geometric mean:

Proposition 8 *1.* $G(K,K) = K$.
2. *$G(K,T)$ is monotone in its arguments: If $K_1 \subseteq K_2$ and $T_1 \subseteq T_2$, then $G(K_1,T_1) \subseteq G(K_2,T_2)$.*

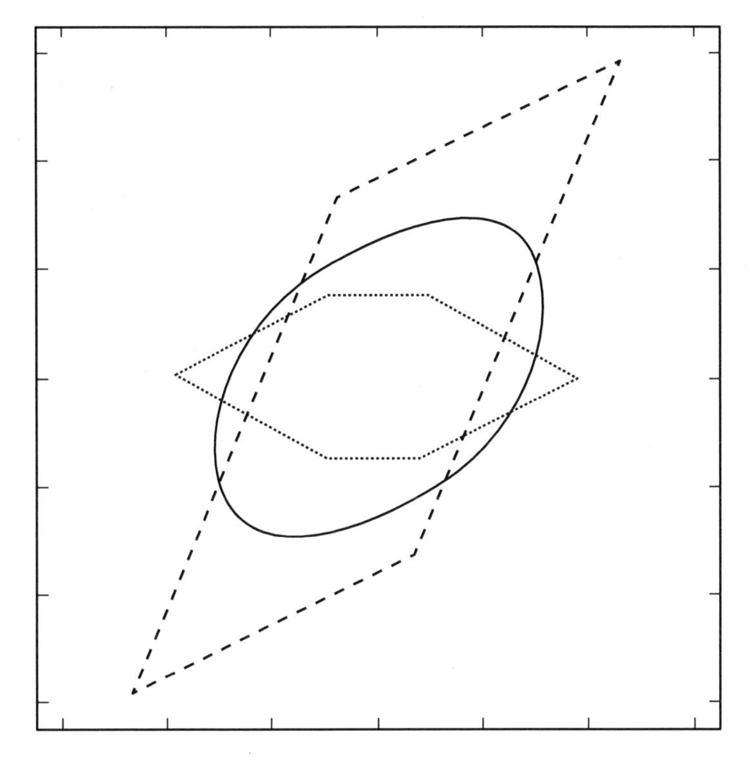

Fig. 1 K, T (dashed and dotted lines) and $G(K, T)$ (solid line)

3. $[G(K,T)]^\circ = G(K^\circ, T^\circ)$.
4. For any linear map u we have $G(uK, uT) = u(G(K,T))$.

Proof 1. This is obvious, as $A_n(K,K) = H_n(K,K) = K$ for all $n \geq 0$.
2. If $K_1 \subseteq K_2$ and $T_1 \subseteq T_2$, then easy induction of n shows that $A_n(K_1, T_1) \subseteq A_n(K_2, T_2)$ and $H_n(K_1, T_1) \subseteq H_n(K_2, T_2)$ for all $n \geq 0$. Sending $n \to \infty$ gives the result.
3. Again, use induction to show that $[A_n(K,T)]^\circ = H_n(K^\circ, T^\circ)$ and $[H_n(K,T)]^\circ = A_n(K^\circ, T^\circ)$ for all $n \geq 0$. Send $n \to \infty$ for the result.
4. Using yet another induction, $A_n(uK, uT) = u(A_n(K,T))$ and $H_n(uK, uT) = u(H_n(K,T))$ for all $n \geq 0$. Again, we obtain the required result in the limit.

□

Let us mention another easy but important property of the geometric mean: All of our means (the arithmetic mean, the harmonic mean, and the geometric mean) do not depend on the choice of a scalar product on $\mathbb{R}^n$. This is obvious for the arithmetic mean, but less so for the harmonic and the geometric mean, since the polarity map $K \mapsto K^\circ$ which appears in the definition *does* depend on this choice. However, remember from the proof of Theorem 2 that

$$r_\theta(H(K,T)) = \frac{2}{\frac{1}{r_\theta(K)} + \frac{1}{r_\theta(T)}}$$

for all $\theta \in \mathbb{R}^n$, so $H(K,T)$ may be constructed from K and T without mentioning polarity or any scalar product. It follows that the harmonic mean, and hence also the geometric mean, may be defined without fixing a scalar product on our space.

Our next goal is to compute the geometric mean in several important cases, which will give us a better intuition for it.

Proposition 9 *Let K be any convex body. Then:*

1. $G(K,K^\circ) = B_2^n$.
2. *For any positive definite linear map u we have* $G(K,uK^\circ) = u^{1/2}B_2^n$.
3. *For any* $\alpha,\beta > 0$ *we have* $G(\alpha K,\beta K^\circ) = \sqrt{\alpha\beta}B_2^n$.

Proof For (1), notice that

$$G(K,K^\circ)^\circ = G\left(K^\circ,\left(K^\circ\right)^\circ\right) = G(K^\circ,K) = G(K,K^\circ).$$

as the only body $\mathcal{K}_{(0)}^n$ to satisfy $X^\circ = X$ is $X = B_2^n$, the claim follows.

Part (2) follows from (1), as

$$G(K,uK^\circ) = u^{1/2}G\left(u^{-1/2}K,u^{1/2}K^\circ\right) = u^{1/2}G\left(u^{-1/2}K,\left(u^{-1/2}K\right)^\circ\right) = u^{1/2}B_2^n.$$

Finally, for (3) we take $u(x) = \frac{\beta}{\alpha}x$ in (2) and obtain

$$G(\alpha K,\beta K^\circ) = \alpha\cdot G\left(K,\frac{\beta}{\alpha}K^\circ\right) = \alpha\cdot\sqrt{\frac{\beta}{\alpha}}B_2^n = \sqrt{\alpha\beta}B_2^n.$$

□

This proposition gives us another way to think about the geometric mean, as an extension of the notion of polarity. We would like to say that T is polar to K with respect to Z if $g(K,T) = Z$. The above proposition says that K° is indeed polar to K with respect to the Euclidean ball. Several natural problems regarding the theory of "polarity with respect to a convex body" will appear in the final section of this paper.

One may also compare this proposition with its obvious counterparts for numbers and matrices: $G(x,x^{-1}) = 1$ for every $x > 0$ and $G(u,u^{-1}) = Id$ for every positive definite matrix u. We see that the ball B_2^n plays the same role as the number 1 for positive numbers or the identity matrix for positive definite matrices. Hence it makes sense to define $\sqrt{K} = G(K,B_2^n)$. For many of the open problems discussed in this paper one may first concentrate on this special case.

Proposition 10 *Let u, v be positive-definite matrices, and let*

$$\mathcal{E}_1 = \{x:\ \langle ux,x\rangle \le 1\}$$
$$\mathcal{E}_2 = \{x:\ \langle vx,x\rangle \le 1\}$$

be the corresponding ellipsoids. Then $G(\mathcal{E}_1, \mathcal{E}_2) = \{x : \langle wx, x\rangle \le 1\}$, where $w = u\#v$ is the matrix geometric mean of u and v.

Proof We have $\mathcal{E}_1 = u^{-1/2}B_2^n$, so $\mathcal{E}_1^\circ = u^{1/2}B_2^n = \{x : \langle u^{-1}x, x\rangle \le 1\}$.

Since $v = wu^{-1}w$, we see that

$$\mathcal{E}_2 = \{x : \langle wu^{-1}wx, x\rangle \le 1\} = \{x : \langle u^{-1}wx, wx\rangle \le 1\} = \{x : wx \in \mathcal{E}_1^\circ\} = w^{-1}\mathcal{E}_1^\circ.$$

Hence by Proposition 9 we have

$$G(\mathcal{E}_1, \mathcal{E}_2) = G(\mathcal{E}_1, w^{-1}\mathcal{E}_1^\circ) = w^{-1/2}B_2^n = \{x : \langle wx, x\rangle \le 1\}$$

like we wanted. □

The above result is somewhat surprising – For ellipsoids $\mathcal{E}_1, \mathcal{E}_2$ the intermediate sets $A_n(\mathcal{E}_1, \mathcal{E}_2)$ and $H_n(\mathcal{E}_1, \mathcal{E}_2)$ are not ellipsoids. Still, in the limit we obtain that $G(\mathcal{E}_1, \mathcal{E}_2)$ is an ellipsoid. Actually $\mathcal{E}_1$ and $\mathcal{E}_2$ are dual to each other with respect to the ellipsoid $G(\mathcal{E}_1, \mathcal{E}_2)$ (i.e., if the scalar product on $\mathbb{R}^n$ is chosen in such a way that $G(\mathcal{E}_1, \mathcal{E}_2)$ is the unit ball, then $\mathcal{E}_2 = \mathcal{E}_1^\circ$).

Proposition 10 has a nice corollary regarding the Banach-Mazur distance. For symmetric convex bodies $K, T \in \mathcal{K}_s^n$ the Banach-Mazur distance $d_{BM}(K, T)$ is defined by

$$d_{BM}(K, T) = \min\{\lambda > 0 : \text{There exists a linear map } u \text{ such that } uT \subseteq K \subseteq \lambda \cdot uT\}.$$

We have the following result:

Proposition 11 *For every* $K \in \mathcal{K}_s^n$ *one has* $d_{BM}(\sqrt{K}, B_2^n) \le \sqrt{d_{BM}(K, B_2^n)}$.

Proof Write $d = d_{BM}(K, B_2^n)$. By definition, there exists an ellipsoid $\mathcal{E}$ such that $\mathcal{E} \subseteq K \subseteq d \cdot \mathcal{E}$. By the monotonicity of the geometric mean it follows that $\sqrt{\mathcal{E}} \subseteq \sqrt{K} \subseteq \sqrt{d \cdot \mathcal{E}}$.

From Proposition 10 it follows that $\sqrt{\mathcal{E}} = G(\mathcal{E}, B_2^n)$ is an ellipsoid. Furthermore, the explicit formula given there immediately implies that $\sqrt{d \cdot \mathcal{E}} = \sqrt{d} \cdot \sqrt{\mathcal{E}}$. Hence we have $\sqrt{\mathcal{E}} \subseteq \sqrt{K} \subseteq \sqrt{d} \cdot \sqrt{\mathcal{E}}$ so

$$d_{BM}(\sqrt{K}, B_2^n) \le \sqrt{d} = \sqrt{d_{BM}(K, B_2^n)}.$$

□

We know from John's theorem that $d_{BM}(K, B_2^n) \le \sqrt{n}$ for all $K \in \mathcal{K}_s^n$, so we always have $d_{BM}(\sqrt{K}, B_2^n) \le n^{1/4}$. In particular, since it is known that $d_{BM}(B_p^n, B_2^n) = n^{|1/2 - 1/p|}$, it follows that there is no $K \in \mathcal{K}_s^n$ such that $\sqrt{K} = B_p^n$ if $p > 4$ or $p < 4/3$.

Finally, we conclude this section by computing an example in the plane that will be useful later:

Example 12 Fix $R > 1$ (that we will later send to ∞) and define

$$K = [-R, R] \times \left[-\frac{1}{R}, \frac{1}{R}\right] \subseteq \mathbb{R}^2,$$

$$T = \left[-\frac{1}{R}, \frac{1}{R}\right] \times [-R, R] \subseteq \mathbb{R}^2.$$

Notice that $K \supseteq T^\circ$, so $G(K,T) \supseteq G(T^\circ, T) = B_2^n$. For the opposite inclusion, let us follow one iteration. Define

$$A = \frac{K+T}{2}, \qquad B = \left(\frac{K^\circ + T^\circ}{2}\right)^\circ.$$

Obviously, $A = \frac{1}{2}\left(R + \frac{1}{R}\right) B_\infty^2$. For B we use the following estimate:

$$\begin{aligned} h_{B^\circ}(x,y) &= \frac{h_{K^\circ}(x,y) + h_{T^\circ}(x,y)}{2} = \frac{1}{2}\left(\max\left(\frac{|x|}{R}, R\,|y|\right) + \max\left(R\,|x|, \frac{|y|}{R}\right)\right) \\ &\geq \frac{1}{2}\left((R\,|y|) + (R\,|x|)\right) = \frac{R}{2}\left(|x| + |y|\right). \end{aligned}$$

Since $|x| + |y| = h_{B_\infty^2}(x,y)$, we have $B^\circ \supseteq \frac{R}{2} B_\infty^2$, so $B \subseteq \frac{2}{R} B_1^2$.

Hence

$$G(K,T) = G(A,B) \subseteq G\left(\frac{1}{2}\left(R + \frac{1}{R}\right) B_\infty^2, \frac{2}{R} B_1^2\right) = \sqrt{1 + \frac{1}{R^2}} B_2^2,$$

where the last step follows from Proposition 9. It follows that $\lim_{R\to\infty} G(K,T) = B_2^2$.

5 Structure of the Geometric Mean

We now turn our attention to finer questions regarding the geometric mean. First we prove a relation between $G(A,B)$ and the logarithmic mean $L_{1/2}(A,B)$ described in Section 3:

Proposition 13 *For $K, T \in \mathcal{K}_{(0)}^n$ we have $G(K,T) \subseteq L_{1/2}(K,T)$.*

Proof Define $\varphi_n = h\left(A_n(K,T)\right)$, and $\psi_n = h\left(H_n(K,T)\right)$, where h denotes the support function. We will also define another process by

$$\begin{aligned} \widetilde{\varphi}_0 &= h(K) & \widetilde{\psi}_0 &= h(T) \\ \widetilde{\varphi}_{n+1} &= \frac{1}{2}\left(\widetilde{\varphi}_n + \widetilde{\psi}_n\right) & \widetilde{\psi}_{n+1} &= \left[\frac{1}{2}\left(\widetilde{\varphi}_n^{-1} + \widetilde{\psi}_n^{-1}\right)\right]^{-1}. \end{aligned}$$

Notice that the functions $\widetilde{\varphi}_n, \widetilde{\psi}_n$ are not necessarily convex, unlike φ_n and ψ_n. Still, we claim that $\widetilde{\varphi}_n \geq \varphi_n$ and $\widetilde{\psi}_n \geq \psi_n$ for all n (here and everywhere else in the proof, inequalities between functions are meant in the pointwise sense). For $n = 0$ there is nothing to prove. If we assume the inequalities to be true for n, then for $n + 1$ we have

$$\varphi_{n+1} = h(A_{n+1}) = h\left(\frac{A_n + H_n}{2}\right) = \frac{h(A_n) + h(H_n)}{2} = \frac{\varphi_n + \psi_n}{2}$$
$$\leq \frac{\widetilde{\varphi}_n + \widetilde{\psi}_n}{2} = \widetilde{\varphi}_{n+1}.$$

We also have

$$\psi_{n+1} = h(H_{n+1}) = r\left(\frac{A_n^\circ + H_n^\circ}{2}\right)^{-1} \stackrel{(*)}{\leq} \left(\frac{r(A_n^\circ) + r(H_n^\circ)}{2}\right)^{-1} =$$
$$= \frac{2}{h(A_n)^{-1} + h(B_n)^{-1}} = \frac{2}{\varphi_n^{-1} + \psi_n^{-1}} \leq \frac{2}{\widetilde{\varphi}_n^{-1} + \widetilde{\psi}_n^{-1}} = \widetilde{\psi}_{n+1},$$

where $(*)$ was explained in the proof of Theorem 2. This completes the inductive proof.

It is a simple exercise in calculus that

$$\lim_{n\to\infty} \widetilde{\varphi}_n = \lim_{n\to\infty} \widetilde{\psi}_n = \sqrt{h(K)h(T)}.$$

Therefore, by taking the limit $n \to \infty$ in the inequality $\widetilde{\varphi}_n \geq \varphi_n$ we see that

$$h(G(K,T)) = \lim_{n\to\infty} \varphi_n \leq \lim_{n\to\infty} \widetilde{\varphi}_n = \sqrt{h(K)h(T)}.$$

This proves the result. □

For the "dual" logarithmic sum, we obtain an inclusion in the opposite direction:

Corollary 14 *For $K, T \subseteq \mathcal{K}^n_{(0)}$ and we have $G(K,T) \supseteq L_{1/2}(K^\circ, T^\circ)^\circ$.*

Proof Applying Proposition 13 to K° and T° we see that

$$G(K,T)^\circ = G\left(K^\circ, T^\circ\right) \subseteq L_{1/2}\left(K^\circ, T^\circ\right).$$

Taking polarity, we obtain the result. □

The inclusions in the last two results may be strict. For example, take $K = B^2_\infty \subseteq \mathbb{R}^2$ and $T = B^2_1 \subseteq \mathbb{R}^2$. Then $G(K,T) = B^2_2$ since $T = K^\circ$. However, a direct (yet tedious) computation shows that $L_{1/2}(K,T)$ and $L_{1/2}(K^\circ, T^\circ)^\circ$ are octagons. Figure 2 depicts the three bodies $G(K,T)$, $L_{1/2}(K,T)$ and $L_{1/2}(K^\circ, T^\circ)^\circ$.

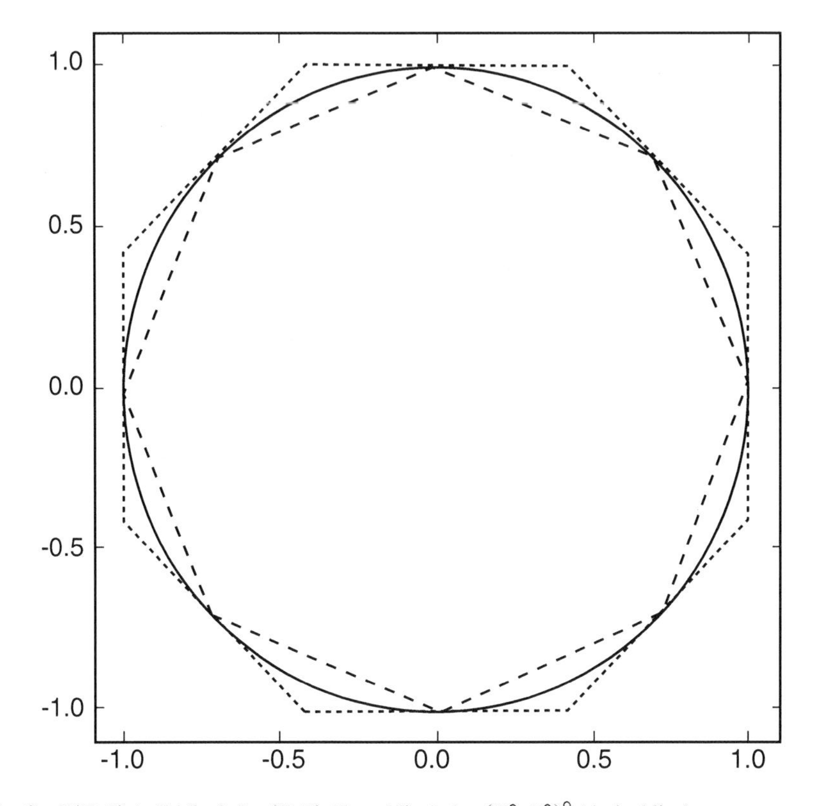

Fig. 2 $G(K, T)$ (solid line), $L_{1/2}(K, T)$ (dotted line), $L_{1/2}(K^{\circ}, T^{\circ})^{\circ}$ (dashed line)

In the figure we see that even though those three bodies are distinct, there are directions in which their radial functions coincide. This is not a coincidence, as the next proposition shows:

Proposition 15 *Let K and T be convex bodies. Assume that in direction η the bodies K and T have parallel supporting hyperplanes, with normal vector θ. Write $G = G(K, T)$. Then $h_G(\theta) = \sqrt{h_K(\theta)h_T(\theta)}$ and $r_G(\eta) = \sqrt{r_K(\eta)r_T(\eta)}$.*

Proof Write $a = r_K(\eta)\eta \in \partial K$ and $b = r_T(\eta)\eta \in \partial T$. Since the hyperplane $\{x : \langle x, \theta\rangle = \langle a, \theta\rangle\}$ is a supporting hyperplane for K we know that

$$h_K(\theta) = \langle a, \theta\rangle = r_K(\eta) \cdot \langle \eta, \theta\rangle ,$$

and similarly $h_T(\theta) = r_T(\eta) \cdot \langle \eta, \theta\rangle$.

On the one hand, by Proposition 13 we know that $h_G(\theta) \leq \sqrt{h_K(\theta)h_T(\theta)}$. On the other hand,

$$h_G(\theta) = \sup_{\alpha \in S^{n-1}} \left(\langle \alpha, \theta\rangle \cdot r_G(\alpha)\right) \geq \langle \eta, \theta\rangle \cdot r_G(\eta)$$

$$\geq \langle \eta, \theta \rangle \cdot \sqrt{r_K(\eta) r_T(\eta)} = \sqrt{h_K(\theta) h_T(\theta)},$$

where we used Corollary 14. Together we see that indeed $h_G(\theta) = \sqrt{h_K(\theta) h_T(\theta)}$. Hence we must also have $h_G(\theta) = \langle \eta, \theta \rangle \cdot r_G(\eta)$, so

$$r_G(\eta) = \frac{h_G(\theta)}{\langle \eta, \theta \rangle} = \sqrt{\frac{h_K(\theta)}{\langle \eta, \theta \rangle} \cdot \frac{h_T(\theta)}{\langle \eta, \theta \rangle}} = \sqrt{r_K(\eta) \cdot r_T(\eta)}.$$

□

Remember that if K is not smooth at a point, it may have many supporting hyperplanes at this point. The above proposition only requires that *one* supporting hyperplane of K in direction η is parallel to *one* supporting hyperplane of T in the same direction. Such directions η always exist, for any pair of convex bodies K and T. For example, if one defines a functional $\Phi : S^{n-1} \to \mathbb{R}$ by $\Phi(\eta) = \frac{r_K(\eta)}{r_T(\eta)}$, then it is enough to take the points where Φ attains its extrema.

Let us understand the relation between $G(K, T)$ and $L_{1/2}(K, T)$ in different terms. Formally we have only defined the mean $G(K, T)$ for compact sets. However, there is a natural extension of this notion to slabs: we will write $S_{\theta,c} = \{x : |\langle x, \theta \rangle| \leq c\}$, and set

$$G\left(S_{\theta,c}, S_{\eta,d}\right) = \begin{cases} S_{\theta,\sqrt{cd}} & \text{if } \theta = \eta \\ \mathbb{R}^n & \text{otherwise.} \end{cases}$$

One way to justify this formula is to think about a slab as a degenerated ellipsoid and take a suitable limit in Proposition 10.

From this definition it is immediate that

$$L_{1/2}(K, T) = \bigcap \left\{ G(S_{\theta,c}, S_{\eta,d}) : K \subseteq S_{\theta,c} \text{ and } T \subseteq S_{\eta,d} \right\}.$$

However, what happens if we allow arbitrary ellipsoids, and not only slabs?

Definition 16 Given convex bodies $K, T \subseteq \mathbb{R}^n$, we define the upper ellipsoidal envelope of K and T to be

$$\overline{G}(K, T) = \bigcap \{G(\mathcal{E}_1, \mathcal{E}_2) : K \subseteq \mathcal{E}_1 \text{ and } T \subseteq \mathcal{E}_2\},$$

Similarly we define the lower ellipsoidal envelope of K and T as

$$\underline{G}(K, T) = \overline{G}(K^\circ, T^\circ)^\circ = \mathrm{conv} \bigcup \{G(\mathcal{E}_1, \mathcal{E}_2) : K \supseteq \mathcal{E}_1 \text{ and } T \supseteq \mathcal{E}_2\}.$$

We obviously have $\underline{G}(K, T) \subseteq G(K, T) \subseteq \overline{G}(K, T)$. It will be interesting to know when is it true that $\overline{G}(K, T) = \underline{G}(K, T) = G(K, T)$.

To see why this may be interesting, remember that in Example 10 we proved an explicit formula for the geometric mean of ellipsoids. From this formula it is clear that

$$G(\alpha\mathcal{E}_1, \beta\mathcal{E}_2) = \sqrt{\alpha\beta}G(\mathcal{E}_1, \mathcal{E}_2)$$

for all ellipsoids $\mathcal{E}_1, \mathcal{E}_2$ and all $\alpha, \beta > 0$. Hence we also have

$$\overline{G}(\alpha K, \beta T) = \sqrt{\alpha\beta} \cdot \overline{G}(K, T),$$

and similarly for the lower envelope $\underline{G}$. It seems intuitive that this scaling property should hold for G as well, and it is obviously true whenever G coincides with one of its envelopes. However, recently Alexander Magazinov found a counterexample to this "scaling conjecture". In particular, his example shows that in general G does not have to coincide with its ellipsoidal envelopes. Magazinov's example appears as an appendix to this paper.

Another possible application of the equality $\overline{G}(K, T) = \underline{G}(K, T)$ (whenever it is true) will be given in Section (7), where we discuss a possible extension of the iteration process 3.3 to p-sums.

6 Volume Inequalities

Like in the case of the logarithmic mean or the complex interpolation, it is natural to ask whether we have an inequality of the form

$$|G(K, T)| \geq \sqrt{|K|\,|T|}$$

for $K, T \in \mathcal{K}_s^n$. Such an inequality will be intimately related to the Brunn-Minkowski and log-Brunn-Minkowski inequalities (remember the discussion in Section 3 and Proposition 13), as well as the Blaschke-Santaló inequality (take $T = K^{\circ}$ and remember Proposition 9).

Unfortunately, we have already seen in Example 12 that this inequality is false in general. In that example we had $|K| = |T| = 4$ for all values of R, but

$$\lim_{R\to\infty} |G(K, T)| = \left|B_2^2\right| = \pi < 4.$$

Still, it seems worthwhile to understand for what classes of bodies this inequality is true, and how far it is from being true in general. For example, the following is an immediate computation:

Proposition 17 *If $\mathcal{E}_1$ and $\mathcal{E}_2$ are ellipsoids, then $|G(\mathcal{E}_1, \mathcal{E}_2)| = \sqrt{|\mathcal{E}_1|\,|\mathcal{E}_2|}$.*

Proof Using the computations and notation of Example 10 we have

$$\begin{aligned}|G(\mathcal{E}_1, \mathcal{E}_2)| &= \left|w^{-1/2}B_2^n\right| = (\det w)^{-1/2}\,|B_2^n| = \det(u\#v)^{-1/2}\,|B_2^n| \\ &= (\det u \cdot \det v)^{-1/4}\,|B_2^n| = \sqrt{\left[(\det u)^{-1/2}\left|B_2^n\right|\right]\left[(\det v)^{-1/2}\left|B_2^n\right|\right]} \\ &= \sqrt{\left|u^{-1/2}B_2^n\right| \cdot \left|v^{-1/2}B_2^n\right|} = \sqrt{|\mathcal{E}_1|\,|\mathcal{E}_2|}.\end{aligned}$$

□

Corollary 18 *The log-Brunn-Minkowski conjecture is true for ellipsoids.*

Proof Combine the above Proposition with Proposition 13. □

In general, let us define the constant g_n to be the biggest constant such that

$$|G(K,T)| \geq g_n \cdot \sqrt{|K|\,|T|}.$$

for all $K, T \in \mathcal{K}_s^n$. Determining the asymptotics of g_n as $n \to \infty$ may have important consequences. For example, by Proposition 13 we see that

$$\left|L_{1/2}(K,T)\right| \geq g_n\sqrt{|K|\,|T|},$$

so this question is directly related to the log-Brunn-Minkowski conjecture. As another example, by monotonicity and Proposition 9 we have $G(K \cap T, K^\circ \cap T) \subseteq B_2^n \cap T$, so $|K \cap T| \cdot |K^\circ \cap T| \leq g_n^{-2} \cdot \left|B_2^n \cap T\right|^2$.

From proposition 17 and John's theorem one obtains the trivial bound $g_n \geq n^{-n/2}$. It seems possible that $g_n \geq c^n$ for some constant c. Such a result will essentially recover Klartag's result (3.2), perhaps with a different constant. It will also give a new proof of the inequality (3.1) which follows from the work of Saroglou.

7 *p*-Additions and *p*-Geometric Means

In the introduction to this paper we took time to explain the role of the polarity map, but we took for granted the fact that the "addition" of convex bodies is indeed the Minkowski sum. However, there are other interesting additions on convex sets, such as the p-additions. This notion was introduced by Firey [12] and studied extensively by Lutwak [16, 17]. For $K, T \in \mathcal{K}_{(0)}^n$ and $1 \leq p < \infty$, the p-sum $K +_p T$ is defined implicitly by the relation

$$h_\theta(K +_p T) = \left(h_\theta(K)^p + h_\theta(T)^p\right)^{1/p}.$$

The case $p = 1$ is of course the standard Minkowski addition.

Instead of the process (3.3), one may fix $1 \le p < \infty$ and look at the following process:

$$\begin{aligned} A_0 &= K & H_0 &= T \\ A_{n+1} &= \frac{A_n +_p H_n}{2^{1/p}} & H_{n+1} &= \left(\frac{A_n^\circ +_p H_n^\circ}{2^{1/p}}\right)^\circ. \end{aligned} \tag{7.1}$$

(the factor $2^{1/p}$ is the correct one, since $K +_p K = 2^{1/p} K$).

Theorem 19 *Fix $K, T \in \mathcal{K}^n_{(0)}$ and $1 \le p < \infty$, and define processes $\{A_n\}$, $\{H_n\}$ by (7.1). Then $\{A_n\}$ is decreasing, $\{H_n\}$ is increasing, and the limits $\lim_{n\to\infty} A_n$ and $\lim_{n\to\infty} H_n$ exist (in the Hausdorff sense) and are equal.*

The proof is almost identical to the proof of Theorem 6, so we omit the details. We call this joint limit the p-geometric mean of K and T and denote it by $G_p(K, T)$.

As a side note, one may also discuss the ∞-sum of convex bodies which is the limit of p-sums as $p \to \infty$. Explicitly

$$h_\theta(K +_\infty T) = \lim_{p\to\infty} \left(h_\theta(K)^p + h_\theta(T)^p\right)^{1/p} = \max\left\{h_\theta(K), h_\theta(T)\right\},$$

so $K +_\infty T = K \vee T$, the convex hull of the union $K \cup T$. For $p = \infty$ the process (7.1) becomes

$$\begin{aligned} A_0 &= K & H_0 &= T \\ A_{n+1} &= A_n \vee H_n & H_{n+1} &= A_n \cap H_n, \end{aligned} \tag{7.2}$$

but this process does not converge unless $K = T$. Indeed, for every $n \ge 1$ we have $A_n = K \vee T$ and $H_n = K \cap T$. Hence we will only discuss $1 \le p < \infty$.

All the results of this paper remain true when $G(K, T)$ is replaced by $G_p(K, T)$, with almost identical proofs. In particular $G_p(K, K^\circ) = B_2^n = G(K, K^\circ)$ for all $K \in \mathcal{K}_s^n$, and $G_p(\mathcal{E}_1, \mathcal{E}_2) = G(\mathcal{E}_1, \mathcal{E}_2)$. In fact, we have computed a few examples using a computer and did not find an example where $G_p(K, T) \ne G_q(K, T)$. Is it possible that $G_p(K, T)$ does not depend on p, at least on some non-trivial cases?

In this direction it is worth mentioning that since for ellipsoids $G_p(\mathcal{E}_1, \mathcal{E}_2)$ does not depend on p, the ellipsoidal envelopes $\overline{G}(K, T)$ and $\underline{G}(K, T)$ from Definition 16 also do not depend on p. In particular, we have $\underline{G}(K, T) \subseteq G_p(K, T) \subseteq \overline{G}(K, T)$ for all $1 \le p < \infty$. From here we see that if for some bodies $K, T \subset \mathcal{K}_s^n$ we have $\underline{G}(K, T) = \overline{G}(K, T)$, then $G_p(K, T)$ is indeed independent of p. As discussed in Section 5, we do not know when this is the case.

8 Functional Constructions

So far we only discussed constructions of new convex bodies out of old ones. However, similar constructions can be used for convex functions as well. We denote by $\mathrm{Cvx}(\mathbb{R}^n)$ the class of all convex and lower semi-continuous functions

$\varphi : \mathbb{R}^n \to (-\infty, \infty]$. The addition on $\mathrm{Cvx}(\mathbb{R}^n)$ is the regular pointwise addition, and the order is the pointwise order ($\varphi \le \psi$ if $\varphi(x) \le \psi(x)$ for all x). Like $\mathcal{K}^n_{(0)}$, the class $\mathrm{Cvx}(\mathbb{R}^n)$ also has an essentially unique duality, the Legendre transform

$$\varphi^*(y) = \sup_{x\in\mathbb{R}^n} [\langle x, y\rangle - \varphi(x)] .$$

More formally, we have the following theorem of Artstein-Avidan and Milman [1, 3]:

Theorem 20 *Every order reversing involution $\mathcal{T} : \mathrm{Cvx}(\mathbb{R}^n) \to \mathrm{Cvx}(\mathbb{R}^n)$ is the Legendre transform up to linear terms.*

Explicitly, there exist a constant $C \in \mathbb{R}$, a vector $v \in \mathbb{R}^n$, and an invertible symmetric linear transformation $B \in GL(n)$ such that

$$(\mathcal{T}\varphi)(x) = \varphi^*(Bx + v) + \langle x, v\rangle + C.$$

Hence we may think of φ^* as the inverse "φ^{-1}" and attempt to repeat the constructions of the previous sections. Notice that for functions the harmonic mean $H(\varphi, \psi) = \left[\frac{1}{2}(\varphi^* + \psi^*)\right]^*$ is exactly the inf-convolution, used by Asplund in [4]:

$$(H(\varphi, \psi))(x) = \frac{1}{2} \inf_{y\in\mathbb{R}^n} (\varphi(x+y) + \psi(x-y)) .$$

As a recent example of the benefits of thinking of φ^* as φ^{-1}, Rotem recently proved the following result [21]:

Theorem 21 *For every $\varphi \in \mathrm{Cvx}(\mathbb{R}^n)$ one has*

$$(\varphi + \delta)^* + (\varphi^* + \delta)^* = \delta,$$

where $\delta(x) = \frac{1}{2}|x|^2$ and $|\cdot|$ is the Euclidean norm on $\mathbb{R}^n$.

Notice that this theorem is the analogue for convex functions of the trivial identity $\frac{1}{x+1} + \frac{1}{1/x+1} = 1$ for positive real numbers. The function δ plays the role of the number 1 as δ is the unique function satisfying $\delta^* = \delta$. This theorem has applications for functional Blaschke-Santaló type inequalities and for the theory of summands.

By fixing a convex body $K \in \mathcal{K}^n_{(0)}$ and choosing $\varphi = \frac{1}{2}h_K^2$ in Theorem 21, it was shown in [21] that

$$\left(K +_2 B_2^n\right)^\circ +_2 \left(K^\circ +_2 B_2^n\right)^\circ = B_2^n$$

where the 2-sum $+_2$ was defined in the previous section. However, it turns out that the 2-sum cannot be replaced by the Minkowski sum, as the identity

$$\left(K + B_2^n\right)^\circ + \left(K^\circ + B_2^n\right)^\circ = B_2^n$$

is simply false. So in this example not only the theory can be extended to the functional case, but the functional case is better behaved than the classical case of convex bodies.

The theory of continued fractions can also be extended to this functional case. Specifically, Molchanov proves the following theorem in [18]:

Theorem 22 *Let* $\varphi \in \mathrm{Cvx}(\mathbb{R}^n)$ *be a non-negative function with* $\varphi(0) = 0$. *Assume that* $\frac{r}{2}|x|^2 \le \varphi(x) \le \frac{R}{2}|x|^2$ *for all* $x \in \mathbb{R}^n$, *for some constants* r, R *that satisfy* $r^2 + 4\frac{r}{R} > 4$. *Then the continued fraction* $[\varphi, \varphi, \varphi, \ldots]$ *converges to a function* $\zeta \in \mathrm{Cvx}(\mathbb{R}^n)$, *and* ζ *solves the functional equation* $\zeta^* = \zeta + \varphi$.

The convergence of $[\varphi, \varphi, \varphi, \ldots]$ is proved with respect to the metric

$$d(\varphi, \psi) = \min\{r > 0 : f \le g + r\delta \text{ and } g \le f + r\delta\}.$$

We will not give the details of the proof.

Finally, the construction of the geometric mean may also be carried out for convex functions. Given $\varphi, \psi \in \mathrm{Cvx}(\mathbb{R}^n)$ we define

$$\begin{aligned} \alpha_0 &= \varphi & \eta_0 &= \psi \\ \alpha_{n+1} &= \frac{\alpha_n + \eta_n}{2} & \eta_{n+1} &= \left(\frac{\alpha_n^* + \eta_n^*}{2}\right)^*. \end{aligned}$$

It is then possible to prove the following result:

Theorem 23 *Assume* $\varphi, \psi \in \mathrm{Cvx}(\mathbb{R}^n)$ *are everywhere finite. Then the pointwise limit*

$$\rho = \lim_{n\to\infty} \alpha_n = \lim_{n\to\infty} \beta_n$$

exists. We call ρ *the geometric mean of* φ *and* ψ *and write* $\rho = G(\varphi, \psi)$. *Furthermore, the functional geometric mean has the following properties:*

1. $G(\varphi, \varphi) = \varphi$.
2. $G(\varphi, \psi)$ *is monotone in its arguments: If* $\varphi_1 \subseteq \varphi_2$ *and* $\psi_1 \subseteq \psi_2$, *then* $G(\varphi_1, \psi_1) \subseteq G(\varphi_2, \psi_2)$.
3. $[G(\varphi, \psi)]^* = G(\varphi^*, \psi^*)$.
4. $G(\varphi, \varphi^*) = \delta$.
5. *For any linear map* u *we have* $G(\varphi \circ u, \psi \circ u) = G(\varphi, \psi) \circ u$.

We omit the details of the proof, as it is very similar to Theorem 6, Proposition 8 and Proposition 9.

Finally, let us note that in some ways the geometric mean of convex functions is even more well behaved than the geometric mean of convex bodies. For example, the following theorem is proved in [21]:

Theorem 24 *The geometric mean of convex functions is concave in its arguments. More explicitly, fix* $\varphi_0, \varphi_1, \psi_0, \psi_1 \in \mathrm{Cvx}(\mathbb{R}^n)$ *and* $0 < \lambda < 1$. *Define* $\varphi_\lambda = (1-\lambda)\varphi_0 + \lambda\varphi_1$ *and* $\psi_\lambda = (1-\lambda)\psi_0 + \lambda\psi_1$. *Then*

$$G(\varphi_\lambda, \psi_\lambda) \supseteq (1-\lambda) \cdot G(\varphi_0, \psi_0) + \lambda G(\varphi_1, \psi_1)$$

whenever all the geometric means in this expression are well defined.

This theorem is the natural extension of the fact that $f(x, y) = \sqrt{xy}$ is a concave function on $(\mathbb{R}_+)^2$. Like in Theorem 21, the functional version immediately implies a theorem for convex bodies *with the* 2*-sum*: For every convex bodies $K_0, K_1, T_0, T_1 \in \mathcal{K}^n_{(0)}$ one has

$$G_2(K_\lambda, T_\lambda) \supseteq \sqrt{1-\lambda}G_2(K_0, T_0) +_2 \sqrt{\lambda}G_2(K_1, T_1),$$

where $K_\lambda = \sqrt{1-\lambda}K_0 +_2 \sqrt{\lambda}K_1$ and $T_\lambda = \sqrt{1-\lambda}T_0 +_2 \sqrt{\lambda}T_1$.

However, it is also proved in [21] that the concavity property does not hold for the geometric mean of convex bodies with the regular Minkowski addition. So, like in Theorem 21, the functional theory is better behaved than the classical theory.

9 Open Problems

We conclude this paper by clearly listing the open problems that were mentioned in the previous sections, together with a few other.

1. As explained in Section 4, we would like to think of the relation $G(K, T) = Z$ as "T is polar to K with respect to Z". This ideology raises the following questions:
   (a) **Domain of polarity**: Fix $Z \in \mathcal{K}^n_{(0)}$. For which convex bodies K there exists a T such that $G(K, T) = Z$? In other words, what is the natural domain of this "extended polarity"?
   As a particular sub-problem, assume that for every $K \in \mathcal{K}^n_{(0)}$ there exists a $T \in \mathcal{K}^n_{(0)}$ such that $G(K, T) = Z$. Does it follow that Z is an ellipsoid?
   (b) **Uniqueness**: Is the polar body to K with respect to Z always unique? More explicitly, if $K, T_1, T_2 \in \mathcal{K}^n_{(0)}$ satisfy $G(K, T_1) = G(K, T_2)$, does it follow that $T_1 = T_2$?
   (c) **Order reversing**: Assume that $G(K_1, T_1) = G(K_2, T_2)$ and $K_1 \supseteq K_2$. Does it follow that $T_1 \subseteq T_2$? In other words, is the "extended polarity" transform order reversing? Notice that an affirmative answer to this question implies an affirmative answer to the previous question. It is also worth mentioning that $G(K, T_1) \supseteq G(K, T_2)$ does *not* imply that $T_1 \supseteq T_2$, and one can even construct a counterexample where K, T_1, T_2 are ellipsoids.

2. For which bodies $K, T \in \mathcal{K}_s^n$ we have $\overline{G}(K,T) = \underline{G}(K,T)$?
   Remember from the discussion in Sections 5 and 7 that this question is related to the next two. Also remember that by Magazinov's example in the appendix, the answer to this question is not "always".
3. Fix $K, T \in \mathcal{K}_{(0)}^n$ and $\alpha, \beta > 0$. When is it true that $G(\alpha K, \beta T) = \sqrt{\alpha\beta}G(K,T)$? Since this equality holds when $\alpha = \beta$, it is enough to assume that $\beta = 1$ or that $\alpha = \frac{1}{\beta}$. Again, the answer to this question is not "always".
4. Fix $K, T \in \mathcal{K}_{(0)}^n$. When is $G_p(K,T)$ as defined in Section 7 independent on $p \in [1, \infty)$?
5. As discussed in Section 6, what is the asymptotic behavior of

$$g_n = \inf_{K,T \in \mathcal{K}_s^n} \frac{|G(K,T)|}{\sqrt{|K|\,|T|}}$$

   as $n \to \infty$?
6. Does there exists an "exponential map" $E : \mathcal{K}_{(0)}^n \to \mathcal{K}_{(0)}^n$ such that

$$E\left(\frac{K+T}{2}\right) = G\left(E(K), E(T)\right)?$$

   What should be the image $\mathcal{I} \subseteq \mathcal{K}_{(0)}^n$ of E? Since $\exp([0,\infty)) = [1,\infty)$, it may be possible to take $\mathcal{I} = \left\{K \in \mathcal{K}_{(0)}^n : K \supseteq B_2^n\right\}$.
   It may be easier to construct the "logarithm" $L : \mathcal{I} \to \mathcal{K}_{(0)}^n$ with the property

$$G(L(K), L(T)) = \frac{L(K) + L(T)}{2}.$$

7. How to properly define the weighted geometric mean of two convex bodies? For numbers $x, y > 0$ and $\lambda \in [0,1]$, the λ-geometric mean of x and y is simply $x^{1-\lambda}y^{\lambda}$. For positive-definite matrices, the λ-geometric mean of u and v is usually defined as

$$u\#_\lambda v = u^{1/2}\left(u^{-1/2}vu^{-1/2}\right)^{\lambda} u^{1/2}.$$

   What should the definition be for convex bodies? It is possible to define, for example, $G_{1/4}(K,T) = G(G(K,T),T)$ and so on, but is there a more direct approach?

Acknowledgements The research was supported by ISF grant 826/13 and BSF grant 2012111. The second named author was also supported by the Adams Fellowship Program of the Israel Academy of Sciences and Humanities. We would like to thank the referee for his/her useful suggestions.

References

[1] Shiri Artstein-Avidan and Vitali Milman. A characterization of the concept of duality. *Electronic Research Announcements in Mathematical Sciences*, 14:42–59, 2007.

[2] Shiri Artstein-Avidan and Vitali Milman. The concept of duality for measure projections of convex bodies. *Journal of Functional Analysis*, 254(10):2648–2666, may 2008.

[3] Shiri Artstein-Avidan and Vitali Milman. The concept of duality in convex analysis, and the characterization of the Legendre transform. *Annals of Mathematics*, 169(2):661–674, mar 2009.

[4] Edgar Asplund. Averaged norms. *Israel Journal of Mathematics*, 5(4):227–233, oct 1967.

[5] Károly J. Böröczky, Erwin Lutwak, Deane Yang, and Gaoyong Zhang. The log-Brunn-Minkowski inequality. *Advances in Mathematics*, 231(3–4):1974–1997, oct 2012.

[6] Károly J. Böröczky and Rolf Schneider. A characterization of the duality mapping for convex bodies. *Geometric and Functional Analysis*, 18(3):657–667, aug 2008.

[7] Jean Bourgain and Vitali Milman. New volume ratio properties for convex symmetric bodies in $\mathbb{R}^n$. *Inventiones Mathematicae*, 88(2):319–340, jun 1987.

[8] Dario Cordero-Erausquin. Santaló's inequality on $\mathbb{C}^n$ by complex interpolation. *Comptes Rendus Mathematique*, 334(9):767–772, jan 2002.

[9] Dario Cordero-Erausquin and Bo'az Klartag. Interpolations , Convexity and Geometric Inequalities. In Bo'az Klartag, Shahar Mendelson, and Vitali Milman, editors, *Geometric Aspects of Functional Analysis, Israel Seminar 2006–2010*, volume 2050 of *Lecture Notes in Mathematics*, pages 151–168. Springer, Berlin, Heidelberg, 2012.

[10] V. P. Fedotov. Geometric mean of convex sets. *Journal of Soviet Mathematics*, 10(3):488–491, sep 1978.

[11] William J. Firey. Polar means of convex bodies and a dual to the Brunn-Minkowski theorem. *Canadian Journal of Mathematics*, 13:444–453, 1961.

[12] William J. Firey. p-means of convex bodies. *Mathematica Scandinavica*, 10:17–24, 1962.

[13] Peter M. Gruber. The endomorphisms of the lattice of norms in finite dimensions. *Abhandlungen aus dem Mathematischen Seminar der Universität Hamburg*, 62(1):179–189, 1992.

[14] Bo'az Klartag. Marginals of Geometric Inequalities. In Vitali Milman and Gideon Schechtman, editors, *Geometric Aspects of Functional Analysis, Israel Seminar 2004–2005*, volume 1910 of *Lecture Notes in Mathematics*, pages 133–166. Springer, Berlin, Heidelberg, 2007.

[15] Jimmie Lawson and Yongdo Lim. The Geometric Mean, Matrices, Metrics, and More. *The American Mathematical Monthly*, 108(9):797–812, 2001.

[16] Erwin Lutwak. The Brunn-Minkowski-Firey theory I: Mixed volumes and the Minkowski problem. *Journal of Differential Geometry*, 38(1):131–150, 1993.

[17] Erwin Lutwak. The Brunn-Minkowski-Firey theory II: Affine and geominimal surface areas. *Advances in Mathematics*, 118(2):244–294, 1996.

[18] Ilya Molchanov. Continued fractions built from convex sets and convex functions. *Communications in Contemporary Mathematics*, 17(05):1550003, oct 2015.

[19] Gilles Pisier. *The volume of convex bodies and Banach space geometry*, volume 94 of *Cambridge Tracts in Mathematics*. Cambridge University Press, Cambridge, 1989.

[20] Liran Rotem. A letter: The log-Brunn-Minkowski inequality for complex bodies. *arXiv:1412.5321*, dec 2014.

[21] Liran Rotem. Algebraically inspired results on convex functions and bodies. *Communications in Contemporary Mathematics*, 18(6):1650027, jun 2016.

[22] Alexander Rubinov. *Abstract Convexity and Global Optimization*, volume 44 of *Nonconvex Optimization and Its Applications*. Springer US, Boston, MA, 2000.

[23] Christos Saroglou. Remarks on the conjectured log-Brunn-Minkowski inequality. *Geometriae Dedicata*, jul 2014.

[24] Christos Saroglou. More on logarithmic sums of convex bodies. *Mathematika*, 62(03):818–841, may 2016.
[25] Stephen Semmes. Interpolation of Banach Spaces, Differential Geometry and Differential Equations. *Revista Matemática Iberoamericana*, 4(1):155–176, 1988.

Appendix: A Counterexample to the Scaling Property of the Geometric Mean

Alexander Magazinov*

Theorem A *Let K be a square with vertices $(\pm 4, 0)$ and $(0, \pm 4)$. Let T be a hexagon with vertices $(0, \pm 4)$, $(\pm 2, \pm 1)$ (the signs in the last expression are taken independently). Then*

$$G(K, T) \neq G\left((1+\varepsilon)K, \frac{1}{1+\varepsilon}T\right),$$

if $\varepsilon \neq 0$ and $|\varepsilon|$ is small enough.

The following lemma is the key to Theorem A. By a non-reflex angle in $\mathbb{R}^2$ we mean a closed convex cone $C \subseteq \mathbb{R}^2$ with vertex at $(0, 0)$ which has a non-empty interior, but that does not contain a full line. The notation $[u_1, u_2]$ for $u_1, u_2 \in \mathbb{R}^2$ denotes the closed interval with endpoints u_1 and u_2:

Lemma B *Let $C \subseteq \mathbb{R}^2$ be a non-relfex angle. Fix $K, T \in \mathcal{K}^2_{(0)}$ such that*

$$C \cap \partial K = [u_1, u_2], \quad C \cap \partial T = [\alpha u_1, \beta u_2]$$

for some $u_1, u_2 \in \mathbb{R}^2$, where $0 < \alpha < \beta$. Assume that the lines

$$\ell = \{u_1 + t(\alpha u_1 - \beta u_2) : t \in \mathbb{R}\}, \qquad \ell' = \{\beta u_2 + t(u_1 - u_2) : t \in \mathbb{R}\}$$

are support lines to K and T respectively. Then

1. There exists a unique Euclidean scalar product $Q(\cdot, \cdot)$ in $\mathbb{R}^2$ such that

$$Q(\alpha u_1 - \beta u_2, u_1) = Q(u_1 - u_2, u_2) = 0, \quad Q(u_1, \alpha u_1) = 1. \tag{A.1}$$

*School of Mathematical Sciences, Tel-Aviv University, magazinov-al@yandex.ru

2. *If Q is as above, then the curvilinear segment $\partial G(K,T)\cap C$ is an arc of the ellipse*

$$\{v \in \mathbb{R}^2 : Q(v,v) = 1\}.$$

Outline of the proof. Fix a linear map $f : \mathbb{R}^2 \to \mathbb{R}^2$ with the following properties:

$$|f(u_1)| = 1/\sqrt{\alpha}, \quad |f(u_2)| = 1/\sqrt{\beta}, \quad \angle(f(u_1), f(u_2)) = \arccos\sqrt{\alpha/\beta},$$

where $\angle(v,w)$ denotes the angle between the vectors v and w. For the existence in assertion 1 it is enough to set

$$Q(v,v') = \langle f(v), f(v')\rangle. \tag{A.2}$$

Uniqueness is immediate since (A.1) gives three linearly independent linear equations in the three variables $Q(u_1,u_1)$, $Q(u_1,u_2)$, $Q(u_2,u_2)$.

Now we prove assertion 2, keeping f from above. Note that the segment $[f(u_1), f(u_2)]$ is orthogonal to the vector $f(u_2)$, and the segment $[f(\alpha u_1), f(\beta u_2)]$ is orthogonal to the vector $f(u_1)$.

By construction $\alpha|f(u_1)|^2 = 1$. Then, since the triangles $(\mathbf{0}, f(u_1), f(u_2))$ and $(\mathbf{0}, \alpha f(u_1), \beta f(u_2))$ are both right-angled and have the same angle at $(0,0)$ they are similar, so we have $\beta|f(u_2)|^2 = 1$.

We say that a ray R from the origin is an *orthogonality ray* for a convex body $X \in \mathcal{K}^2_{(0)}$ if a line through the point $R \cap \partial X$ in the direction orthogonal to R is a support line to X. The condition that ℓ and ℓ' are support lines to K and T implies that the two rays

$$R_i = \{tf(u_i) : t > 0\} \quad (i = 1,2)$$

are orthogonality rays for both $f(K)$ and $f(T)$.

Note that

$$f(K)^\circ \cap f(C) = f(T) \cap f(C),$$

and the rays R_1 and R_2 are orthogonality rays for the body $f(K)^\circ$. Now we claim that

$$A_i(f(K), f(T)) \cap f(C) = A_i(f(K), f(K)^\circ) \cap f(C),$$

$$H_i(f(K), f(T)) \cap f(C) = H_i(f(K), f(K)^\circ) \cap f(C),$$

and R_1 and R_2 are orthogonality rays for each of the bodies $A_i(f(K), f(T))$, $H_i(f(K), f(T))$, $A_i(f(K), f(K)^\circ)$, $H_i(f(K), f(K)^\circ)$. Indeed, this can be checked straightforwardly by induction over i.

Passing to the limit, we have $G(f(K), f(T)) \cap f(C) = B_2^2 \cap f(C)$, so by Proposition 8 we have

$$G(K, T) \cap C = f^{-1}(B_2^2) \cap C.$$

Hence $\partial G(K, T) \cap C$ is indeed an arc of an ellipse $Q(v, v) = 1$, and assertion 2 is proved. □

Proof of Theorem A. We will prove the following claim. Let ε_1 and ε_2 be arbitrary real numbers such that $\max(|\varepsilon_1|, |\varepsilon_2|)$ is small enough. Let

$$C = \{(x, y) : x < 12y < 4x\}$$

be an open angle. We prove that the curve $\partial G\left((1 + \varepsilon_1)K, (1 + \varepsilon_2)T\right) \cap C$ contains exactly one non-smooth point, which lies on the line

$$\frac{x}{6 + 4\varepsilon_1 + 2\varepsilon_2} = \frac{y}{1 + \varepsilon_2},$$

pointing to the vertex $(3+2\varepsilon_1+\varepsilon_2, (1+\varepsilon_2)/2)$ of the body $\frac{1}{2}((1+\varepsilon_1)K+(1+\varepsilon_2)T)$.

We will give the proof for $\varepsilon_1 = \varepsilon_2 = 0$, as one can check that the argument is applicable in the general case.

Consider the angles

$$C_1 = \{(x, y) : 0 \leq 6y \leq x\},$$

$$C_2 = \{(x, y) : x \leq 6y \leq 3x\}.$$

We claim that for $i = 1, 2$ the curvilinear segments $\partial G(K, T) \cap C_i$ are elliptic arcs, and these arcs arise from distinct ellipses.

We have $G(K, T) = G(K_1, T_1)$, where

$$K_1 = \frac{K + T}{2}, \qquad T_1 = \left(\frac{K^\circ + T^\circ}{2}\right)^\circ.$$

In the positive quadrant the vertices of K_1 are

$$u_1 = (3, 0), \quad u_2 = (3, 1/2), \quad u_3 = (2, 2), \quad u_4 = (0, 4),$$

and the vertices of T_1 in the positive quadrant are

$$v_1 = (8/3, 0), \quad v_2 = (16/7, 8/7), \quad v_3 = (0, 4).$$

Clearly, Lemma B applies to K_1 , T_1 and each C_i. Hence $\partial G(K, T) \cap C_i$ are indeed elliptic arcs.

Assume these arcs belong to the same ellipse $Q(v, v) = 1$. Then

$$Q(v_1, u_1 - u_2) = Q(v_2, u_2 - u_3) = Q(u_2, v_1 - v_2) = 0.$$

But if

$$Q((x_1, y_1), (x_2, y_2)) = ax_1x_2 + b(x_1y_2 + x_2y_1) + cy_1y_2,$$

this would yield $a = b = c = 0$, a contradiction. Therefore the common point of the arcs $\partial G(K, T) \cap C_i$ is a non-smooth point of $\partial G(K, T)$.

Consequently, the only non-smooth point of the curve $\partial G((1 + \varepsilon_1)K, (1 + \varepsilon_2)T) \cap C$ changes its angular direction from the origin, even under the additional assumption $(1 + \varepsilon_1)(1 + \varepsilon_2) = 1$. This immediately implies

$$G(K, T) \neq G\left((1 + \varepsilon_1)K, \frac{1}{1 + \varepsilon_1}T\right).$$

□

Remark Nevertheless, the identity

$$G(K, T) = G(aK, a^{-1}T) \tag{A.3}$$

holds for some wide class of two-dimensional bodies. For instance, let K_0 be a regular n-gon and $T_0 = K_0^\circ$. Consider convex n-gons K and T that are obtained by arbitrary small enough perturbations of the vertices of K_0 and T_0, respectively. Then Lemma B allows one to reconstruct $\partial G(aK, a^{-1}T)$ completely and thus verify that (A.3) holds true for such K and T.

Randomized Isoperimetric Inequalities

Grigoris Paouris and Peter Pivovarov

Abstract We discuss isoperimetric inequalities for convex sets. These include the classical isoperimetric inequality and that of Brunn-Minkowski, Blaschke-Santaló, Busemann-Petty and their various extensions. We show that many such inequalities admit stronger randomized forms in the following sense: for natural families of associated random convex sets one has stochastic dominance for various functionals such as volume, surface area, mean width and others. By laws of large numbers, these randomized versions recover the classical inequalities. We give an overview of when such stochastic dominance arises and its applications in convex geometry and probability.

1 Introduction

The focus of this paper is on stochastic forms of isoperimetric inequalities for convex sets. To set the stage, we begin with two examples. Among the most fundamental isoperimetric inequalities is the Brunn-Minkowski inequality for the volume V_n of convex bodies $K, L \subseteq \mathbb{R}^n$,

$$V_n(K + L)^{1/n} \geqslant V_n(K)^{1/n} + V_n(L)^{1/n}, \tag{1.1}$$

where $K + L$ is the Minkowski sum $\{x + y : x \in K, y \in L\}$. The Brunn-Minkowski inequality is the cornerstone of the Brunn-Minkowski theory and its reach extends well beyond convex geometry; see Schneider's monograph [71] and Gardner's survey [27]. It is well known that (1.1) provides a direct route to the classical isoperimetric inequality relating surface area S and volume,

G. Paouris
Texas A&M University, Mailstop 3368, College Station, TX 77843-3368, USA
e-mail: grigoris@math.tamu.edu

P. Pivovarov (✉)
University of Missouri, 202 Math Sciences Building, 810 E. Rollins Street, Columbia, MO 65211, USA
e-mail: pivovarovp@missouri.edu

E. Carlen et al. (eds.), *Convexity and Concentration*, The IMA Volumes in Mathematics and its Applications 161, DOI 10.1007/978-1-4939-7005-6_13

$$\left(\frac{S(K)}{S(B)}\right)^{1/(n-1)} \geqslant \left(\frac{V_n(K)}{V_n(B)}\right)^{1/n}, \tag{1.2}$$

where B is the Euclidean unit ball. As equality holds in (1.1) if K and L are homothetic, it can be equivalently stated in isoperimetric form as follows:

$$V_n(K + L) \geqslant V_n(r_K B + r_L B), \tag{1.3}$$

where r_K, r_L denote the radii of Euclidean balls with the same volume as K, L, respectively, i.e., $r_K = (V_n(K)/V_n(B))^{1/n}$; for subsequent reference, with this notation, (1.2) reads

$$S(K) \geqslant S(r_K B). \tag{1.4}$$

Both (1.1) and (1.2) admit stronger empirical versions associated with random convex sets. Specifically, let $x_1, \ldots, x_N$ be independent random vectors (on some probability space $(\Omega, \mathcal{F}, \mathbb{P})$) distributed according to the uniform density on a convex body $K \subseteq \mathbb{R}^n$, say, $f_K = \frac{1}{V_n(K)} \mathbb{1}_K$, i.e., $\mathbb{P}(x_i \in A) = \int_A f_K(x)dx$ for Borel sets $A \subseteq \mathbb{R}^n$. For each such K and $N > n$, we associate a random polytope

$$K_N = \text{conv}\{x_1, \ldots, x_N\},$$

where conv denotes convex hull. Then the following stochastic dominance holds for the independent random polytopes K_{N_1}, L_{N_2} and $(r_K B)_{N_1}$, $(r_L B)_{N_2}$ associated with the bodies in (1.3): for all $\alpha \geqslant 0$,

$$\mathbb{P}\left(V_n(K_{N_1} + L_{N_2}) > \alpha\right) \geqslant \mathbb{P}\left(V_n((r_K B)_{N_1} + (r_L B)_{N_2}) > \alpha\right). \tag{1.5}$$

Integrating in α gives

$$\mathbb{E} V_n(K_{N_1} + L_{N_2}) \geqslant \mathbb{E} V_n((r_K B)_{N_1} + (r_L B)_{N_2}),$$

where $\mathbb{E}$ denotes expectation. By the law of large numbers, when $N_1, N_2 \to \infty$, the latter convex hulls converge to their ambient bodies and this leads to (1.3). Thus (1.1) is a global inequality which can be proved by a random approximation procedure in which stochastic dominance holds at each stage; for a different stochastic form of (1.1), see Vitale's work [76]. For the classical isoperimetric inequality, one has the following distributional inequality, for $\alpha \geqslant 0$,

$$\mathbb{P}\left(S(K_{N_1}) > \alpha\right) \geqslant \mathbb{P}\left(S((r_K B)_{N_1}) > \alpha\right). \tag{1.6}$$

The same integration and limiting procedure lead to (1.4). For fixed N_1 and N_2, the sets in the extremizing probabilities on the right-hand sides of (1.5) and (1.6) are not Euclidean balls, but rather sets that one generates using Euclidean balls. In particular, the stochastic forms are strictly stronger than the global inequalities (1.1) and (1.2).

The goal of this paper is to give an overview of related stochastic forms of isoperimetric inequalities. Both (1.1) and (1.2) hold for non-convex sets but we focus on stochastic dominance associated with convex sets. The underlying randomness, however, will not be limited to uniform distributions on convex bodies but will involve continuous distributions on $\mathbb{R}^n$. We will discuss a streamlined approach that yields stochastic dominance in a variety of inequalities in convex geometry and their applications. We pay particular attention to high-dimensional probability distributions and associated structures, e.g., random convex sets and matrices. Many of the results we discuss are from a series of papers [58, 59], along with D. Cordero-Erausquin, M. Fradelizi [24], S. Dann [25] and G. Livshyts [46]. We also present a few new results that fit in this framework and have not appeared previously.

Inequalities for the volume of random convex hulls in stochastic geometry have a rich history starting with Blaschke's resolution of Sylvester's famous four-point problem in the plane (see, e.g., [63, 18, 20, 28] for background and history). In particular, for planar convex bodies Blaschke proved that the random triangle K_3 (notation as above) satisfies

$$\mathbb{E}V_2(\Delta_3) \geqslant \mathbb{E}V_2(K_3) \geqslant \mathbb{E}V_2((r_K B_2)_3), \tag{1.7}$$

where Δ is a triangle in $\mathbb{R}^2$ with the same area as K and B_2 is the unit disk. Blaschke's proof of the lower bound draws on Steiner symmetrization, which is the basis for many related extremal inequalities, see, e.g,. [71, 28, 35]. More generally, shadow systems as put forth by Rogers and Shephard [72, 65] and developed by Campi and Gronchi, among others, play a fundamental role, e.g., [18, 21, 22], and will be defined and discussed further below. Finding maximizers in (1.7) for $n \geqslant 3$ has proved more difficult and is connected to the famous slicing problem, which we will not discuss here (see [13] for background).

A seminal result building on the lower bound in (1.7) is Busemann's random simplex inequality [16, 17]: for a convex body $K \subseteq \mathbb{R}^n$ and $p \geqslant 1$, the set $K_{o,n} = \mathrm{conv}\{o, x_1, \ldots, x_n\}$ (x_i's as above) satisfies

$$\mathbb{E}V_n(K_{o,n})^p \geqslant \mathbb{E}V_n((r_K B)_{o,n})^p. \tag{1.8}$$

This is a key ingredient in Busemann's intersection inequality,

$$\int_{S^{n-1}} V_{n-1}(K \cap \theta^{\perp})^n d\sigma(\theta) \leq \int_{S^{n-1}} V_{n-1}((r_K B) \cap \theta^{\perp})^n d\sigma(\theta), \tag{1.9}$$

where S^{n-1} is the unit sphere equipped with the Haar probability measure σ; (1.8) is also the basis for extending (1.9) to lower dimensional secitons as proved by Busemann and Straus [17] and Grinberg [33]; see also Gardner [29] for further extensions.

Inextricably linked to Busemann's random simplex inequality is the Busemann-Petty centroid inequality, proved by Petty [61]. The centroid body of a star body $K \subseteq \mathbb{R}^n$ is the convex body $Z(K)$ with support function given by

$$h(Z(K), y) = \frac{1}{V_n(K)} \int_K |\langle x, y\rangle|\, dx;$$

(star bodies and support functions are defined in §2) and it satisfies

$$V_n(Z(K)) \geqslant V_n(Z(r_K B)).$$

The latter occupies a special role in the theory of affine isoperimetric inequalities; see Lutwak's survey [47].

One can view (1.8) as a result about convex hulls or about the random parallelotope $\sum_{i=1}^n [-x_i, x_i]$ (since $n! V_n(K_{o,n}) = |\det[x_1, \ldots, x_n]|$). Both viewpoints generalize: for convex hulls K_N with $N > n$, this was done by Groemer [34] and for Minkowski sums of $N \geqslant n$ random line segments by Bourgain, Meyer, Milman and Pajor [11]; these are stated in §5, where we discuss various extensions for different functionals and underlying randomness. These are the starting point for a systematic study of many related quantities.

In particular, convex hulls and zonotopes are natural endpoint families of sets in L_p-Brunn-Minkowski theory and its recent extensions. In the last twenty years, this area has seen significant developments. L_p analogues of centroid bodies are important for affine isoperimetric inequalities, e.g., [48, 49, 36] and are fundamental in concentration of volume in convex bodies, e.g, [56, 42, 43]. The L_p-version of the Busemann-Petty centroid inequality, due to Lutwak, Yang and Zhang [48], concerns the convex body $Z_p(K)$ defined by its support function

$$h^p(Z_p(K), y) = \frac{1}{V_n(K)} \int_K |\langle x, y\rangle|^p\, dx \tag{1.10}$$

and states that

$$V_n(Z_p(K)) \geqslant V_n(Z_p(r_K B)). \tag{1.11}$$

A precursor to (1.11) is due to Lutwak and Zhang [53] who proved that when K is origin-symmetric,

$$V_n(Z_p(K)^\circ) \leqslant V_n(Z_p(r_K B)^\circ). \tag{1.12}$$

When $p \to \infty$, $Z_p(K)$ converges to $Z_\infty(K) = K$ and (1.12) recovers the classical Blaschke-Santaló inequality [68],

$$V_n(K^\circ) \leqslant V_n((r_K B)^\circ). \tag{1.13}$$

The latter holds more generally for non-symmetric bodies with an appropriate choice of center. The analogue of (1.12) in the non-symmetric case was proved by Haberl and Schuster [36], to which we refer for further references and background on L_p-Brunn-Minkowski theory.

Inequalities (1.11) and (1.12) are fundamental inequalities in the L_p Brunn-Minkowski theory. Recently, such inequalities have been placed in a general framework involving Orlicz functions by Lutwak, Yang, and Zhang, e.g., [50, 51] and a closely related concept, due to Gardner, Hug and Weil [30, 31], termed M-addition, which we discuss in §5; for further extensions and background, see [10]. We treat stochastic forms of fundamental related inequalities. For example, we show that in (1.5) one can replace Minkowski addition by M-addition. With the help of laws of large numbers, this leads to a streamlined approach to all of the above inequalities and others.

The notion of M-addition fits perfectly with the random linear operator point of view which we have used in our work on this topic [58, 59, 24]. For random vectors $x_1, \ldots, x_N$, we form the $n \times N$ random matrix $[x_1, \ldots, x_N]$ and view it as a linear operator from $\mathbb{R}^N$ to $\mathbb{R}^n$. If $C \subseteq \mathbb{R}^N$, then

$$[x_1, \ldots, x_N]C = \left\{ \sum_{i=1}^{N} c_i x_i : c = (c_i) \in C \right\}$$

is a random set in $\mathbb{R}^n$. In particular, if $C = \text{conv}\{e_1, \ldots, e_N\}$, where $e_1, \ldots, e_N$ is the standard unit vector basis for $\mathbb{R}^N$, then

$$[x_1, \ldots, x_N]\text{conv}\{e_1, \ldots, e_N\} = \text{conv}\{x_1, \ldots, x_N\}.$$

Let B_p^N denote the closed unit ball in ℓ_p^N. If $C = B_1^N$, then

$$[x_1, \ldots, x_N]B_1^N = \text{conv}\{\pm x_1, \ldots, \pm x_N\}.$$

If $C = B_\infty^N$, then one obtains Minkowski sums,

$$[x_1, \ldots, x_N]B_\infty^N = \sum_{i=1}^{N} [-x_i, x_i].$$

We define the empirical analogue $Z_{p,N}(K)$ of the L_p-centroid body $Z_p(K)$ by its (random) support function

$$h^p(Z_{p,N}(K), y) = \frac{1}{N} \sum_{i=1}^{N} |\langle x_i, y \rangle|^p, \tag{1.14}$$

where $x_1, \ldots, x_N$ are independent random vectors with density $\frac{1}{V_n(K)} \mathbb{1}_K$; this can be compared with (1.10); in matrix form, $Z_{p,N}(K) = N^{-1/p}[x_1, \ldots, x_N]B_q^N$, where $1/p + 1/q = 1$. In this framework, we will explain how uniform measures on Cartesian products of Euclidean balls arise as extremizers for

$$\mathbb{P}(\phi([X_1, \ldots, X_N]C) > \alpha) \tag{1.15}$$

and

$$\mathbb{P}(\phi(([X_1,\ldots,X_N]C)^\circ) > \alpha); \tag{1.16}$$

over the class of independent random vectors X_i with continuous distributions on $\mathbb{R}^n$ having bounded densities; here $C \subseteq \mathbb{R}^N$ is a compact convex set (sometimes with some additional symmetry assumptions) and ϕ an appropriate functional, e.g., volume, surface area, mean width, diameter, among others. Since the random sets in the extremizing probabilities are not typically balls but sets one generates using balls, there is no clear-cut path to reduce distributional inequalities for (1.15) and (1.16) from one another via duality; for comparison, note that the Lutwak-Yang-Zhang inequality for L_p centroid bodies (1.11) implies the Lutwak-Zhang result for their polars (1.12) by the Blaschke-Santaló inequality since the extremizers in each case are balls (or ellipsoids).

The random operator approach allows one to interpolate between inequalities for families of convex sets, but such inequalities in turn yield information about random operators. For example, recall the classical Bieberbach inequality on the diameter of a convex body $K \subseteq \mathbb{R}^n$,

$$\mathrm{diam}(K) \geqslant \mathrm{diam}(r_K B). \tag{1.17}$$

A corresponding empirical form is given by

$$\mathbb{P}(\mathrm{diam}(K_N) > \alpha) \geqslant \mathbb{P}(\mathrm{diam}((r_K B)_N) > \alpha). \tag{1.18}$$

The latter identifies the extremizers of the distribution of certain operator norms. Indeed, if K is an origin-symmetric convex body and we set $K_{N,s} = \mathrm{conv}\{\pm x_1,\ldots,\pm x_N\}$ ($x_i \in \mathbb{R}^n$), then (1.18) still holds and we have the following for the $\ell_1^N \to \ell_2^n$ operator norm,

$$\mathrm{diam}(K_{N,s}) = 2\left\|[x_1,\ldots,x_N] : \ell_1^N \to \ell_2^n\right\|.$$

We show in §6 that if $\mathbf{X} = [X_1,\ldots,X_N]$, where the X_i's are independent random vectors in $\mathbb{R}^n$ and have densities bounded by one, say, then for any N-dimensional normed space E, the quantity

$$\mathbb{P}\left(\|[X_1,\ldots,X_N] : E \to \ell_2^n\| > \alpha\right)$$

is minimized when the columns X_i are distributed uniformly in the Euclidean ball $\widetilde{B}$ of volume one, centered at the origin. This can be viewed as an operator analogue of the Bieberbach inequality (1.17). When $n = 1$, $\mathbf{X}$ is simply a $1 \times N$ row vector and the latter extends to semi-norms. Thus if F is a subspace of $\mathbb{R}^n$, we get the following for random vectors $x \in \mathbb{R}^N$ with independent coordinates with densities bounded by one: the probability

$$\mathbb{P}(\|P_F x\|_2 > \alpha) \tag{1.19}$$

is minimized when x is sampled in the unit cube $[-1/2, 1/2]^N$ - products of "balls" in one dimension (here $\|\cdot\|_2$ is the Euclidean norm and P_F is the orthogonal projection onto F). Combining (1.19) with a seminal result by Ball [4] on maximal volume sections of the cube, we obtain a new proof of a result of Rudelson and Vershynin [67] on small ball probabilities of marginal densities of product measures (which differs also from the proof in [46], our joint work G. Livshyts); this is explained in §6.

As mentioned above, Busemann's original motivation for proving the random simplex inequality (1.8) was to bound suitable averages of volumes of central hyperplane sections of convex bodies (1.9). If $V_n(K) = 1$ and $\theta \in S^{n-1}$, then $V_{n-1}(K \cap \theta^{\perp})$ is the value of the marginal density of $\mathbb{1}_K$ on $[\theta] = \text{span}\{\theta\}$ evaluated at 0, i.e. $\pi_{[\theta]}(\mathbb{1}_K)(0) = \int_{\theta^{\perp}} \mathbb{1}_K(x)dx$. Thus it is natural that marginal distributions of probability measures arise in this setting. One reason for placing Busemann-type inequalities in a probabilistic framework is that they lead to bounds for marginal distributions of random vectors not necessarily having independent coordinates, as in our joint work with S. Dann [25], which we discuss further in §5.

Lastly, we comment on some of the tools used to prove such inequalities. We make essential use of rearrangement inequalities such as that of Rogers [64], Brascamp, Lieb and Luttinger [12] and Christ [23]. These interface particularly well with Steiner symmetrization, shadow systems and other machinery from convex geometry. Another key ingredient is an inequality of Kanter [39] on stochastic dominance. In fact, we formulate the Rogers/Brascamp-Lieb-Luttinger inequality in terms of stochastic dominance using the notion of peaked measures as studied by Kanter [39] and Barthe [5, 6], among others. One can actually prove the minimization result for (1.19) directly using the Rogers/Brascamp-Lieb-Luttinger inequality and Kanter's theorem but we will show how these ingredients apply in a general framework for a variety of functionals. Similar techniques are used in proving analytic inequalities, e.g., for k-plane transform by Christ [23] and Baernstein and Loss [2]. Our focus is on phenomena in convex geometry and probability.

The paper is organized as follows. We start with definitions and background in §2. In §3, we discuss the rearrangement inequality of Rogers/Brascamp-Lieb-Luttinger and interpret it as a result about stochastic dominance for certain types of functions with a concavity property, called Steiner concavity, following Christ. In §4, we present examples of Steiner concave functions. In §5, we present general randomized inequalities. We conclude with applications to operator norms of random matrices and small deviations in §6.

2 Preliminaries

We work in Euclidean space $\mathbb{R}^n$ with the canonical inner-product $\langle \cdot, \cdot \rangle$ and Euclidean norm $\|\cdot\|_2$. As above, the unit Euclidean ball in $\mathbb{R}^n$ is $B = B_2^n$ and its volume is $\omega_n := V_n(B_2^n)$; S^{n-1} is the unit sphere, equipped with the Haar probability measure

σ. Let $G_{n,k}$ be the Grassmannian manifold of k-dimensional linear subspaces of $\mathbb{R}^n$ equipped with the Haar probability measure $\nu_{n,k}$.

A convex body $K \subseteq \mathbb{R}^n$ is a compact, convex set with non-empty interior. The set of all compact convex sets in $\mathbb{R}^n$ is denoted by $\mathcal{K}^n$. For a convex body K we write $\widetilde{K}$ for the homothet of K of volume one; in particular, $\widetilde{B} = \omega_n^{-1/n} B$. Let $\mathcal{K}_\circ^n$ denote the class of all convex bodies that contain the origin in their interior. For $K, L \in \mathcal{K}^n$, the Minkowski sum $K + L$ is the set $\{x + y : x \in K, y \in L\}$; for $\alpha > 0$, $\alpha K = \{\alpha x : x \in K\}$. We say that K is origin-symmetric (or simply 'symmetric'), if $-x \in K$ whenever $x \in K$. For $K \in \mathcal{K}^n$, the support function of K is given by

$$h_K(x) = \sup\{\langle y, x\rangle \ : \ y \in K\} \quad (x \in \mathbb{R}^n).$$

The mean width of K is

$$w(K) = \int_{S^{n-1}} h_K(\theta) + h_K(-\theta) d\sigma(\theta) = 2 \int_{S^{n-1}} h_K(\theta) d\sigma(\theta).$$

Recall that the intrinsic volumes $V_1, \ldots, V_n$ are functionals on convex bodies which can be defined via the Steiner formula: for any convex body $K \subseteq \mathbb{R}^n$ and $\varepsilon > 0$,

$$V_n(K + \varepsilon B) = \sum_{j=0}^{n} \omega_{n-j} V_j(K) \varepsilon^{n-j};$$

here $V_0 \equiv 1$, V_1 is a multiple of the mean width, $2V_{n-1}$ is the surface area and V_n is the volume; see [71].

For compact sets C_1, C_2 in $\mathbb{R}^n$, we let $\delta^H(C_1, C_2)$ denote the Hausdorff distance:

$$\delta^H(C_1, C_2) = \inf\{\varepsilon > 0 \ : \ C_1 \subseteq C_2 + \varepsilon B, C_2 \subseteq C_1 + \varepsilon B\}$$

A set $K \subseteq \mathbb{R}^n$ is star-shaped if it is compact, contains the origin in its interior and for every $x \in K$ and $\lambda \in [0, 1]$ we have $\lambda x \in K$. We call K a star-body if its radial function

$$\rho_K(\theta) = \sup\{t > 0 : t\theta \in K\} \quad (\theta \in S^{n-1})$$

is positive and continuous. Any positive continuous function $f : S^{n-1} \to \mathbb{R}$ determines a star body with radial function f.

Following Borell [8, 9], we say that a non-negative, non-identically zero, function ψ is γ-concave if: (i) for $\gamma > 0$, ϕ^γ is concave on $\{\psi > 0\}$, (ii) for $\gamma = 0$, $\log \psi$ is concave on $\{\psi > 0\}$; (iii) for $\gamma < 0$, ψ^γ is convex on $\{\psi > 0\}$. Let $s \in [-\infty, 1]$. A Borel measure μ on $\mathbb{R}^n$ is called s-concave if

$$\mu\left((1-\lambda)A + \lambda B\right) \geqslant \left((1-\lambda)\mu(A)^s + \lambda \mu(B)^s\right)^{\frac{1}{s}}$$

for all compact sets $A, B \subseteq \mathbb{R}^n$ such that $\mu(A)\mu(B) > 0$. For $s = 0$, one says that μ is log-concave and the inequality reads as

$$\mu((1-\lambda)A + \lambda B) \geqslant \mu(A)^{1-\lambda}\mu(B)^{\lambda}.$$

Also, for $s = -\infty$, the measure is called convex and the inequality is replaced by

$$\mu((1-\lambda)A + \lambda B) \geqslant \min\{\mu(A), \mu(B)\}.$$

An s-concave measure μ is always supported on some convex subset of an affine subspace E where it has a density. If μ is a measure on $\mathbb{R}^n$ absolutely continuous with respect to Lebesgue measure with density ψ, then it is s-concave if and only if its density ψ is γ-concave with $\gamma = \frac{s}{1-sn}$ (see [8, 9]).

Let A be a Borel subset of $\mathbb{R}^n$ with finite Lebesgue measure. The symmetric rearrangement A^* of A is the open ball with center at the origin, whose volume is equal to the measure of A. Since we choose A^* to be open, $\mathbf{1}_A^*$ is lower semicontinuous. The symmetric decreasing rearrangement of $\mathbf{1}_A$ is defined by $\mathbf{1}_A^* = \mathbf{1}_{A^*}$. We consider Borel measurable functions $f : \mathbb{R}^n \to \mathbb{R}_+$ which satisfy the following condition: for every $t > 0$, the set $\{x \in \mathbb{R}^n : f(x) > t\}$ has finite Lebesgue measure. In this case, we say that f vanishes at infinity. For such f, the symmetric decreasing rearrangement f^* is defined by

$$f^*(x) = \int_0^\infty \mathbb{1}_{\{f>t\}^*}(x)dt = \int_0^\infty \mathbb{1}_{\{f>t\}^*}(x)dt.$$

The latter should be compared with the "layer-cake representation" of f:

$$f(x) = \int_0^\infty 1_{\{f>t\}}(x)dt. \tag{2.1}$$

see [44, Theorem 1.13]. Note that the function f^* is radially symmetric, radially decreasing and equimeasurable with f, i.e., $\{f > a\}$ and $\{f^* > a\}$ have the same volume for each $a > 0$. By equimeasurability one has that $\|f\|_p = \|f^*\|_p$ for each $1 \leqslant p \leqslant \infty$, where $\|\cdot\|_p$ denote the $L_p(\mathbb{R}^n)$-norm.

Let $f : \mathbb{R}^n \to \mathbb{R}_+$ be a measurable function vanishing at infinity. For $\theta \in S^{n-1}$, we fix a coordinate system that $e_1 := \theta$. The Steiner symmetral $f(\cdot|\theta)$ of f with respect to $\theta^\perp := \{y \in \mathbb{R}^n : \langle y, \theta\rangle = 0\}$ is defined as follows: for $z := (x_2, \dots, x_n) \in \theta^\perp$, we set $f_{z,\theta}(t) = f(t, x_2, \dots, x_n)$ and define $f^*(t, x_2, \dots, x_n|\theta) := (f_{z,\theta})^*(t)$. In other words, we obtain $f^*(\cdot|\theta)$ by rearranging f along every line parallel to θ. We will use the following fact, proved in [4]: if $g : \mathbb{R}^n \to \mathbb{R}_+$ is an integrable function with compact support, there exists a sequence of functions g_k, where $g_0 = g$ and $g_{k+1} = g_k^*(\cdot|\theta_k)$, for some $\theta_k \in S^{n-1}$, such that $\lim_{k\to\infty} \|g_k - g^*\|_1 = 0$. We refer the reader to the books [44, 74] or the introductory notes [14] for further background material on rearrangement of functions.

3 Inequalities for Stochastic Dominance

We start with a seminal inequality now known as the Rogers/Brascamp-Lieb-Luttinger inequality. It was observed by Madiman and Wang in [77] that Rogers proved the inequality in [64] but it is widely known as the Brascamp-Lieb-Luttinger inequality [12]. We will state it only for integrable functions since this is the focus of our paper.

Theorem 3.1 *Let $f_1, \ldots, f_M$ be non-negative integrable functions on $\mathbb{R}$ and $u_1, \ldots, u_M \in \mathbb{R}^N$. Then*

$$\int_{\mathbb{R}^N} \prod_{i=1}^{M} f_i(\langle x, u_i \rangle) dx \leqslant \int_{\mathbb{R}^N} \prod_{i=1}^{M} f_i^*(\langle x, u_i \rangle) dx. \tag{3.1}$$

We will write the above inequality in an equivalent form using the notion of peaked measures. The ideas behind this definition can be tracked back to Anderson [1] and Kanter [39], among others, but here we follow the terminology and notation of Barthe in [5, 6]. Let μ_1, μ_2 be finite Radon measures on $\mathbb{R}^n$ with $\mu_1(\mathbb{R}^n) = \mu_2(\mathbb{R}^n)$. We say that μ_1 is more peaked than μ_2 (and we write $\mu_1 \succ \mu_2$ or $\mu_2 \prec \mu_1$) if

$$\mu_1(K) \geqslant \mu_2(K) \tag{3.2}$$

for all symmetric convex bodies K in $\mathbb{R}^n$. If X_1, X_2 are random vectors in $\mathbb{R}^n$ with distributions μ_1 and μ_2, respectively, we write $X_1 \succ X_2$ if $\mu_1 \succ \mu_2$. Let f_1, f_2 two non-negative integrable functions on $\mathbb{R}^n$ with $\int f_1 = \int f_2$. We write $f_1 \succ f_2$ if the measures μ_i with densities f_i satisfy $\mu_1 \succ \mu_2$. It follows immediately from the definition that the relation $\succ$ is transitive. Moreover if $\mu_i \succ \nu_i$ and $t_i > 0$, $1 \leqslant i \leqslant N$ then $\sum_i t_i \mu_i \succ \sum_i t_i \nu_i$. Another consequence of the definition is that if $\mu \succ \nu$ and E is a k-dimensional subspace then the marginal of μ on E, i.e. $\mu \circ P_E^{-1}$, is more peaked than the marginal of ν on E. To see this, take any symmetric convex body K in E and consider the infinite cylinder $C := K \times E^{\perp} \subseteq \mathbb{R}^n$. It is enough to check that $\mu(C) \geqslant \nu(C)$, and this is satisfied since C can be approximated from inside by symmetric convex bodies in $\mathbb{R}^n$. More generally, if $\mu \succ \nu$ then for every linear map T, we have

$$\mu \circ T \succ \nu \circ T, \tag{3.3}$$

where $\mu \circ T$ is the pushforward measure of μ through the map T.

Recall that $F : \mathbb{R}^n \to \mathbb{R}$ is quasi-concave (quasi-convex) if for all s the set $\{x : F(x) > s\}$ ($\{x : F(x) \leqslant s\}$) is convex.

Lemma 3.2 *Let* μ_1, μ_2 *be Radon measures on* $\mathbb{R}^n$ *with* $\mu_1(\mathbb{R}^n) = \mu_2(\mathbb{R}^n)$. *Then* $\mu_1 \succ \mu_2$ *if and only if*

$$\int_{\mathbb{R}^n} F(x)d\mu_1(x) \geqslant \int_{\mathbb{R}^n} F(x)d\mu_2(x) \tag{3.4}$$

for all even non-negative quasi-concave functions F.

Proof Assume first that $\mu_1 \succ \mu_2$ and that F is even and quasi-concave. Then by the layer-cake representation and Fubini's theorem,

$$\int_{\mathbb{R}^n} F(x)d\mu_1(x) = \int_0^\infty \int_{\{x:F(x)>s\}} d\mu_1(x)ds \geqslant$$

$$\int_0^\infty \int_{\{x:F(x)>s\}} d\mu_2(x)ds = \int_{\mathbb{R}^n} F(x)d\mu_2(x).$$

Conversely, if K is a symmetric convex body then $F := \mathbf{1}_K$ is even and quasi-concave and (3.4) becomes $\mu_1(K) \geqslant \mu_2(K)$ so (3.4) implies that $\mu_1 \succ \mu_2$. $\square$

We are now able to state the following equivalent form of the Rogers/Brascamp-Lieb-Luttinger inequality:

Proposition 3.3 *Let* $f_1, \ldots, f_N$ *be non-negative integrable functions on* $\mathbb{R}$. *Then as products on* $\mathbb{R}^N$,

$$\prod_{i=1}^N f_i \prec \prod_{i=1}^N f_i^*. \tag{3.5}$$

Let us explain why Theorem 3.1 implies Proposition 3.3. Note first that without loss of generality we can replace the assumption "integrable" with "having integral 1". Let K be a symmetric convex body in $\mathbb{R}^N$. Then it can be approximated by intersections of symmetric slabs of the form

$$K_m := \bigcap_{i=1}^m \{x \in \mathbb{R}^N : |\langle x, u_i \rangle| \leqslant 1\}$$

for suitable $u_1, \ldots, u_m \in \mathbb{R}^N$. Note that $\mathbf{1}_{K_m} = \prod_{i=1}^m \mathbf{1}_{[-1,1]}(\langle \cdot, u_i \rangle)$. Apply (3.1) with $M = m + N$ and $u_{m+i} := e_i$, $i = 1, \ldots, N$. Then (since $\mathbf{1}_{K_m} \to \mathbf{1}_K$ in L_1), we get that

$$\int_K \prod_{i=1}^N f_i(x_i)dx \leqslant \int_K \prod_{i=1}^N f_i^*(x_i)dx. \tag{3.6}$$

Since K is an arbitrary symmetric convex body in $\mathbb{R}^N$, we get (3.5). The latter is an extension of a theorem of Anderson [1] and it is the basis of Christ's extension of the Rogers/Brascamp-Lieb-Luttinger inequality [23]; see also the thesis of Pfiefer [62] and work of Baernstein and Loss [2].

In the other direction, consider non-negative integrable functions $f_1, \ldots, f_m$ and let $u_1, \ldots, u_m$ be vectors in $\mathbb{R}^N$. Write $F(x) := \prod_{i=1}^m f_i(x_i)$ and $F_*(x) := \prod_{i=1}^m f_i^*(x_i)$. Let T be the $m \times N$ matrix with rows $u_1, \ldots, u_m$. Note that (3.5) implies that $F \prec F_*$. By (3.3) we also have that $F \circ T \prec F_* \circ T$ so that for any symmetric convex body $K \subseteq \mathbb{R}^N$, $\int_K F \circ T(x)dx \leq \int_K F_* \circ T(x)dx$, hence

$$\int_{\mathbb{R}^N} \prod_{i=1}^m f_i(\langle x, u_i \rangle)dx \leqslant \int_{\mathbb{R}^N} \prod_{i=1}^m f_i^*(\langle x, u_i \rangle)dx$$

which is (3.1).

Actually we will use the Rogers/Brascamp-Lieb-Luttinger inequality in the following form [23].

Corollary 3.4 *Let $f_1, \ldots, f_m$ be non-negative integrable functions on $\mathbb{R}$. Let $u_1, \ldots, u_m$ be non-zero vectors in $\mathbb{R}^N$ and let $F_1, \ldots, F_M$ be non-negative, even, quasi-concave functions on $\mathbb{R}^N$. Then*

$$\int_{\mathbb{R}^N} \prod_{j=1}^M F_j(x) \prod_{i=1}^m f_i(\langle x, u_i \rangle)dx \leqslant \int_{\mathbb{R}^N} \prod_{j=1}^M F_j(x) \prod_{i=1}^m f_i^*(\langle x, u_i \rangle)dx. \tag{3.7}$$

Also, if F is a non-negative, even, quasi-convex function on $\mathbb{R}^N$, we have

$$\int_{\mathbb{R}^N} F(x) \prod_{i=1}^N f_i(x_i)dx \geqslant \int_{\mathbb{R}^N} F(x) \prod_{i=1}^m f_i^*(x_i)dx. \tag{3.8}$$

Proof (Sketch) Note that $\prod_{j=1}^M F_j(x)$ is again quasi-concave and even. So (3.7) follows from Proposition 3.3 and Lemma 3.2.

For the proof of (3.8) first notice that it is enough to prove in the case that $\int_{\mathbb{R}} f_i(t)dt = 1$, $1 \leqslant i \leqslant N$. Recall that for every $t > 0$, $\{F \leqslant t\}$ is convex and symmetric. Thus using Proposition 3.3 and Lemma 3.2, we get

$$\begin{aligned} &\int_{\mathbb{R}^N} F(x) \prod_{i=1}^N f_i(x_i)dx \\ &= \int_{\mathbb{R}^N} \left(\int_0^\infty \mathbf{1}_{\{F>t\}}(x)dt \right) \prod_{i=1}^N f_i(x_i)dx \end{aligned}$$

$$= \int_0^\infty \int_{\mathbb{R}^N} (1 - \mathbf{1}_{\{F \leqslant t\}}) \prod_{i=1}^N f_i(x_i) dx dt$$

$$= \int_0^\infty \left(\int_{\mathbb{R}^N} \prod_{i=1}^N f_i^*(x_i) dx - \int_{\mathbb{R}^N} \mathbf{1}_{\{F \leqslant t\}}) \prod_{i=1}^N f_i(x_i) dx \right) dt$$

$$\geqslant \int_0^\infty \left(\int_{\mathbb{R}^N} \prod_{i=1}^N f_i^*(x_i) dx - \int_{\mathbb{R}^N} \mathbf{1}_{\{F \leqslant t\}}) \prod_{i=1}^N f_i^*(x_i) dx \right) dt$$

$$= \int_{\mathbb{R}^N} F(x) \prod_{i=1}^N f_i^*(x_i) dx.$$

□

We say that a function f on $\mathbb{R}^n$ is unimodal if it is the increasing limit of a sequence of functions of the form,

$$\sum_{i=1}^m t_i \mathbf{1}_{K_i},$$

where $t_i \geqslant 0$ and K_i are symmetric convex bodies in $\mathbb{R}^n$. Even quasi-concave functions are unimodal and every even and non-increasing function on $\mathbb{R}_+$ is unimodal. In particular, for every integrable $f : \mathbb{R}^n \to \mathbb{R}_+$, f^* is unimodal. We will use the following lemma, which is essentially the bathtub principle (e.g., [44]).

Lemma 3.5 *Let $f : \mathbb{R}^n \to \mathbb{R}_+$ be an integrable function.*

1. *If $\beta := \int_0^\infty f(t) t^{n-1} dt < \infty$ and $\phi : \mathbb{R}_+ \to \mathbb{R}_+$ is a non-decreasing function, then*

$$\int_0^\infty \phi(t) f(t) t^{n-1} dt \geqslant \int_0^\infty \phi(t) h(t) t^{n-1} dt, \tag{3.9}$$

where $h := \mathbf{1}_{[0,(n\beta)^{\frac{1}{n}}]}$. If ϕ is non-increasing, then the inequality in (3.9) is reversed.

2. *If $n = 1$, $\|f\|_1 = 1$, $\|f\|_\infty \leqslant 1$ and f is even, then $f^* \prec \mathbf{1}_{[-\frac{1}{2},\frac{1}{2}]}$.*
3. *If f is rotationally invariant, $\|f\|_1 = 1$, and $\|f\|_\infty \leqslant 1$, then for every star-shaped set $K \subseteq \mathbb{R}^n$, $\int_K f(x) dx \leqslant \int_K \mathbf{1}_{\widetilde{B}}(x) dx$.*
4. *If $\|f\|_1 = 1$, $\|f\|_\infty \leqslant 1$, then $f^* \prec \mathbf{1}_{\widetilde{B}}$.*

Proof The proof of the first claim is standard, see, e.g., [58, Lemma 3.5]. The second claim follows from the first, by choosing $n = 1$, $\beta = \frac{1}{2}$ and $\phi := \mathbf{1}_{[0,a]}$, $a > 0$. The third claim follows by applying (3.9) after writing the desired inequality in polar coordinates. The last claim follows immediately from the third. □

A fundamental property of peaked measures is the following result of Kanter [39].

Theorem 3.6 *Let $f_1, f_2 : \mathbb{R}^{n_1} \to \mathbb{R}_+$ with $f_1 \succ f_2$ and f a unimodal function on $\mathbb{R}^{n_2}$. Then as products on $\mathbb{R}^{n_1} \times \mathbb{R}^{n_2}$,*

$$ff_1 \succ ff_2. \tag{3.10}$$

In particular, if f_i, g_i are unimodal functions on $\mathbb{R}^{n_i}$, $1 \leqslant i \leqslant M$ and $f_i \succ g_i$ for all i, then

$$\prod_{i=1}^{M} f_i \succ \prod_{i}^{M} g_i. \tag{3.11}$$

Proof (Sketch) Without loss of generality, assume $\int f_1 = \int f_2 = \int f = 1$. Consider first the case where $f := \mathbf{1}_L$ for some symmetric convex body L in $\mathbb{R}^{n_2}$. Let K be a symmetric convex body in $\mathbb{R}^{n_1} \times \mathbb{R}^{n_2}$. The Prékopa-Leindler inequality implies that the even function

$$F(x) := \int_{\mathbb{R}^{n_2}} \mathbf{1}_K(x,y)\mathbf{1}_L(y)dy$$

is log-concave. So, using Lemma 3.2,

$$\int_{\mathbb{R}^{n_1}} \int_{\mathbb{R}^{n_2}} \mathbf{1}_K(x,y)f_1(x)f(y)dxdy = \int_{\mathbb{R}^{n_1}} F(x)f_1(x)dx \geqslant$$

$$\int_{\mathbb{R}^{n_1}} F(x)f_2(x)dx = \int_{\mathbb{R}^{n_1}} \int_{\mathbb{R}^{n_2}} \mathbf{1}_K(x,y)f_2(x)f(y)dxdy,$$

hence $ff_1 \succ ff_2$. The general case follows easily. □

Theorem 3.6 and Lemma 3.5 immediately imply the following corollary.

Corollary 3.7 *Let $f_1, \ldots, f_m : \mathbb{R}^n \to \mathbb{R}_+$ be probability densities of continuous distributions such that $\max_{i \leqslant M} \|f_i\|_\infty \leqslant 1$. If $n = 1$, then*

$$\prod_{i=1}^{m} f_i^* \prec \mathbf{1}_{Q_m} \tag{3.12}$$

where Q_m is the m-dimensional cube of volume 1 centered at 0. In the general case we have that

$$\prod_{i=1}^{m} f_i^* \prec \prod_{i=1}^{m} \mathbf{1}_{\widetilde{B}}. \tag{3.13}$$

3.1 Multidimensional Case

Let f be a non-negative function on $\mathbb{R}^n$, $\theta \in S^{n-1}$ and $z \in \theta^\perp$. We write $f_{z,\theta}(t) := f_z(\theta) := f(z + t\theta)$. Let G be a non-negative function on the N-fold product $\mathbb{R}^n \times \ldots \times \mathbb{R}^n$. Let $\theta \in S^{n-1}$ and let $Y := \{y_1, \ldots, y_N\} \subseteq \theta^\perp := \{y \in \mathbb{R}^n : \langle y, \theta \rangle = 0\}$. We define a function $G_{Y,\theta} : \mathbb{R}^N \to \mathbb{R}_+$ as

$$G_{Y,\theta}(t_1, \ldots, t_N) := G(y_1 + t_1\theta, \ldots, y_N + t_N\theta).$$

We say that $G : \mathbb{R}^n \times \ldots \times \mathbb{R}^n \to \mathbb{R}_+$ is Steiner concave if for every θ and $Y \subseteq \theta^\perp$ we have that $G_{Y,\theta}$ is even and quasi-concave; similarly, we say G is Steiner convex if $G_{Y,\theta}$ is even and quasi-convex. For example, if $N = n$, then negative powers of the absolute value of the determinant of an $n \times n$ matrix are Steiner concave since the determinant is a multi-linear function of its columns (or rows). Our results depend on the following generalization of the Rogers and Brascamp-Lieb-Luttinger inequality due to Christ [23] (our terminology and presentation is suited for our needs and differs slightly from [23]).

Theorem 3.8 *Let $f_1, \ldots, f_N$ be non-negative integrable functions on $\mathbb{R}^n$, A an $N \times \ell$ matrix. Let $F^{(k)} : (\mathbb{R}^n)^\ell \to \mathbb{R}_+$ be Steiner concave functions $1 \leqslant k \leqslant M$ and let μ be a measure with a rotationally invariant quasi-concave density on $\mathbb{R}^n$. Then*

$$\int_{\mathbb{R}^n} \cdots \int_{\mathbb{R}^n} \prod_{k=1}^{M} F^{(k)}(x_1, \ldots, x_\ell) \prod_{i=1}^{N} f_i \left(\sum_{j=1}^{\ell} a_{ij} x_j \right) d\mu(x_\ell) \ldots d\mu(x_1) \leqslant$$

$$\int_{\mathbb{R}^n} \cdots \int_{\mathbb{R}^n} \prod_{k=1}^{M} F^{(k)}(x_1, \ldots, x_\ell) \prod_{i=1}^{N} f_i^* \left(\sum_{j=1}^{\ell} a_{ij} x_j \right) d\mu(x_\ell) \ldots d\mu(x_1). \tag{3.14}$$

Proof (Sketch) Note that in the case $n = 1$, (3.14) is just (3.7). We consider the case $n > 1$. Let $u_i \in \mathbb{R}^\ell$ be the rows of the matrix A. Fix a direction $\theta \in S^{n-1}$ and let $y_1, \ldots, y_\ell \in \theta^\perp$ the (unique) vectors such that $x_j = y_j + t_j\theta$. Consider the function

$$h_i(\langle u_i, t \rangle) := f_i \left(\sum_{j=1}^{\ell} a_{ij}(y_j + t_j\theta) \right), \quad 1 \leqslant i \leqslant N.$$

We defined the Steiner symmetral $f_i^*(\cdot|\theta) = h_i^*$ in the direction θ in §2. Then by Fubini's theorem we write each integral as an integral on $\theta^\perp$ and $[\theta] = \mathrm{span}\{\theta\}$, for each fixed $y_1, \ldots, y_\ell$ we apply (3.7) for the functions h_i and the quasi-concave functions $F^{(k)}_{Y,\theta}$. (Recall the definition of Steiner concavity). Using Fubini's theorem again, we have proved that

$$\int_{\mathbb{R}^n} \cdots \int_{\mathbb{R}^n} \prod_{k=1}^{M} F^{(k)}(x_1, \ldots, x_\ell) \prod_{i=1}^{N} f_i\left(\sum_{j=1}^{\ell} a_{ij}x_j\right) d\mu(x_\ell) \ldots d\mu(x_1) \leqslant$$

$$\int_{\mathbb{R}^n} \cdots \int_{\mathbb{R}^n} \prod_{k=1}^{M} F^{(k)}(x_1, \ldots, x_\ell) \prod_{i=1}^{N} f_i^*\left(\sum_{j=1}^{\ell} a_{ij}x_j|\theta\right) d\mu(x_\ell) \ldots d\mu(x_1). \quad (3.15)$$

In [12] it has been proved that the function f^* can be approximated (in the L_1 metric) by a suitable sequence of Steiner symmetrizations. This leads to (3.14). □

Let F be a Steiner concave function. Notice that the function $\tilde{F} := \mathbf{1}_{\{F>\alpha\}}$ is also Steiner concave. Indeed, if $\theta \in S^{n-1}$ and $Y \subseteq \theta^\perp$, notice that $\tilde{F}_{Y,\theta}(t) = 1$ if and only if $F_{Y,\theta}(t) > \alpha$. Since F is Steiner concave, $\tilde{F}_{Y,\theta}$ is the indicator function of a symmetric convex set. So $\tilde{F}$ is also Steiner concave. Thus we have the following corollary.

Corollary 3.9 *Let $F : \mathbb{R}^n \times \ldots \times \mathbb{R}^n \to \mathbb{R}_+$ be a Steiner concave function and let $f_i : \mathbb{R}^n \to \mathbb{R}_+$ be non-negative functions with $\|f_i\|_1 = 1$ for $1 \leqslant i \leqslant N$. Let ν be the (product) probability measure defined on $\mathbb{R}^n \times \ldots \times \mathbb{R}^n$ with density $\prod_i f_i$ and let ν^* have density $\prod_i f_i^*$. Then for each $\alpha > 0$,*

$$\nu\left(\{F(x_1, \ldots, x_N) > \alpha\}\right) \leqslant \nu^*\left(\{F(x_1, \ldots, x_N) > \alpha\}\right). \quad (3.16)$$

Moreover, if $G : \mathbb{R}^n \times \ldots \times \mathbb{R}^n \to \mathbb{R}_+$ is a Steiner convex function, then

$$\nu\left(\{G(x_1, \ldots, x_N) > \alpha\}\right) \geqslant \nu^*\left(\{G(x_1, \ldots, x_N) > \alpha\}\right). \quad (3.17)$$

Proof We apply (3.14) for μ the Lebesgue measure, $\ell = N$, A the identity matrix, $M = 1$ and for the function $\tilde{F}$ (as defined above). This proves (3.16). Working with the function $1 - \tilde{F}$ as in the proof of (3.8) we get (3.17). □

3.2 *Cartesian Products of Balls as Extremizers*

In the last section, we discussed how in the presence of Steiner concavity, one can replace densities by their symmetric decreasing rearrangements. Among products of bounded, radial, decreasing densities, the uniform measure on Cartesian products of balls arises in extremal inequalities under several conditions and we discuss two of them in this section.

We will say that a function $F : \mathbb{R}^n \times \ldots \times \mathbb{R}^n \to \mathbb{R}_+$ is coordinate-wise decreasing if for any $x_1, \ldots, x_N \in \mathbb{R}^n$, and $0 \leqslant s_i \leqslant t_i$, $1 \leqslant i \leqslant N$,

$$F(s_1x_1, \ldots, s_Nx_N) \geqslant F(t_1x_1, \ldots, t_Nx_N). \quad (3.18)$$

The next proposition can be proved by using Fubini's theorem iteratively and Lemma 3.5 (as in [24]).

Proposition 3.10 *Let $F : (\mathbb{R}^n)^N \to \mathbb{R}_+$ be a function that is coordinate-wise decreasing. If $g_1, \ldots, g_N : \mathbb{R}^n \to \mathbb{R}_+$ are rotationally invariant densities with* $\max_{i\leqslant N} \|g_i\|_\infty \leqslant 1$, *then*

$$\int_{\mathbb{R}^n} \cdots \int_{\mathbb{R}^n} F(x_1, \ldots, x_N) \prod_{i=1}^{N} g_i(x_i) dx_N \ldots dx_1 \tag{3.19}$$

$$\leqslant \int_{\mathbb{R}^n} \cdots \int_{\mathbb{R}^n} F(x_1, \ldots, x_N) \prod_{i=1}^{N} \mathbf{1}_{\widetilde{B}}(x_i) dx_N \ldots dx_1. \tag{3.20}$$

Using Corollary 3.7, we get the following.

Proposition 3.11 *Let $F : (\mathbb{R}^n)^N \to \mathbb{R}_+$ be quasi-concave and even. If $g_1, \ldots, g_N : \mathbb{R}^n \to \mathbb{R}_+$ are rotationally invariant densities with* $\max_{i\leqslant N} \|g_i\|_\infty \leqslant 1$, *then*

$$\int_{\mathbb{R}^n} \cdots \int_{\mathbb{R}^n} F(x_1, \ldots, x_N) \prod_{i=1}^{N} g_i(x_i) dx_N \ldots dx_1 \tag{3.21}$$

$$\leqslant \int_{\mathbb{R}^n} \cdots \int_{\mathbb{R}^n} F(x_1, \ldots, x_N) \prod_{i=1}^{N} \mathbf{1}_{\widetilde{B}}(x_i) dx_N \ldots dx_1. \tag{3.22}$$

4 Examples of Steiner Concave and Convex Functions

As discussed in the previous section, the presence of Steiner concavity (or convexity) allows one to prove extremal inequalities when the extremizers are rotationally invariant. The requisite Steiner concavity is present for many functionals associated with random structures. As we will see, in many important cases, verifying the Steiner concavity condition is not a routine matter but rather depends on fundamental inequalities in convex geometry. In this section we give several non-trivial examples of Steiner concave (or Steiner convex) functions and we describe the variety of tools that are involved.

4.1 Shadow Systems and Mixed Volumes

Shadow systems were defined by Shephard [73] and developed by Rogers and Shephard [65], and Campi and Gronchi, among others; see, e.g., [18, 20, 19, 21, 69] and the references therein. Let C be a closed convex set in $\mathbb{R}^{n+1}$. Let $(e_1, \ldots, e_{n+1})$

be an orthonormal basis of $\mathbb{R}^{n+1}$ and write $\mathbb{R}^{n+1} = \mathbb{R}^n \oplus \mathbb{R}e_{n+1}$ so that $\mathbb{R}^n = e_{n+1}^{\perp}$. Let $\theta \in S^{n-1}$. For $t \in \mathbb{R}$ let P_t be the projection onto $\mathbb{R}^n$ parallel to $e_{n+1} - t\theta$: for $x \in \mathbb{R}^n$ and $s \in \mathbb{R}$,

$$P_t(x + se_{n+1}) = x + ts\theta.$$

Set $K_t = P_t C \subseteq \mathbb{R}^n$. Then the family (K_t) is a shadow system of convex sets, where t varies in an interval on the real line. Shephard [72] proved that for each $1 \leqslant j \leqslant n$,

$$[0, 1] \ni t \mapsto V_j(P_t C)$$

is a convex function; see work of Campi and Gronchi, e.g., [22, 19] for further background and references. Here we consider the following N-parameter variation, which can be reduced to the one-parameter case.

Proposition 4.1 *Let n, N be postive integers and C be a compact convex set in $\mathbb{R}^n \times \mathbb{R}^N$. Let $\theta \in S^{n-1} \subseteq \mathbb{R}^n$. For $t \in \mathbb{R}^N$ and $(x, y) \in \mathbb{R}^n \times \mathbb{R}^N$, we define $P_t(x, y) = x + \langle y, t\rangle\theta$. Then for all $1 \leqslant j \leqslant n$,*

$$\mathbb{R}^N \ni t \mapsto V_j(P_t C)$$

is a convex function.

Proof (Sketch) Fix s and t in $\mathbb{R}^N$. It is sufficient to show that

$$[0, 1] \ni \lambda \mapsto V_j(P_{s+\lambda(t-s)} C)$$

is convex. Note that $\lambda \mapsto P_{s+\lambda(s-t)}C$ is a one-parameter shadow system and we can apply Shephard's result above; for an alternate argument, following Groemer [34], see [58]. □

Corollary 4.2 *Let C be a compact convex set in $\mathbb{R}^N$. Then for all $1 \leqslant j \leqslant n$,*

$$(\mathbb{R}^n)^N \ni (x_1, \ldots, x_N) \mapsto V_j([x_1, \ldots, x_N]C)$$

is Steiner convex on $\mathbb{R}^N$. Moreover, if C is 1-unconditional then the latter function is coordinate-wise increasing analogous to definition (3.18).

Proof Let $\theta \in S^{n-1}$ and $y_i \in \theta^{\perp}$ for $i = 1, \ldots, N$. Write $x_i = y_i + t_i\theta$. Let $\mathcal{C} = [y_1 + e_{n+1}, \ldots, y_N + e_{n+N}]C$. Then $\mathcal{C}$ is a compact convex set in $\mathbb{R}^n \times \mathbb{R}^N$ which is symmetric with respect to $\theta^{\perp}$ in $\mathbb{R}^{n+N}$ since $[y_1 + e_{n+1}, \ldots, y_N + e_{n+N}]C \subseteq \theta^{\perp}$. Let $P_t : \mathbb{R}^n \times \mathbb{R}^N \to \mathbb{R}^n$ be defined as in Proposition 4.1. Then

$$P_t([y_1 + e_{n+1}, \ldots, y_N + e_{n+N}]C = [y_1 + t_1\theta, \ldots, y_N + t_N\theta]C.$$

We apply the previous proposition to obtain the convexity claim. Now for each $\theta \in S^{n-1}$ and $y_1, \ldots, y_N \in \theta^{\perp}$, the sets $[y_1 + t_1\theta, \ldots, y_N + t_N\theta]C$ and $[y_1 - t_1\theta, \ldots, y_N - t_N\theta]C$ are reflections of one another and so the evenness condition (for Steiner convexity) holds as well. The coordinate-wise monotonicity holds since one has the following inclusion when C is 1-unconditional: for $0 \leqslant s_i \leqslant t_i$,

$$[s_1x_1, \ldots, s_Nx_N]C \subseteq [t_1x_1, \ldots, t_Nx_N]C.$$

□

4.2 Dual Setting

Here we discuss the following dual setting involving the polar dual of a shadow system. Rather than looking at projections of a fixed higher-dimensional convex set as in the previous section, this involves intersections with subspaces. We will invoke a fundamental inequality concerning sections of symmetric convex sets, known as Busemann's inequality [15]. This leads to a randomized version of an extension of the Blaschke-Santaló inequality to the class of convex measures (defined in §2). For this reason we will need the following extension of Busemann's inequality to convex measures from our joint work with D. Cordero-Erausquin and M. Fradelizi [24]; this builds on work by Ball [3], Bobkov [7], Kim, Yaskin and Zvavitch [40].

Theorem 4.3 (Busemann Theorem for convex measures) *Let ν be a convex measure with even density ψ on $\mathbb{R}^n$. Then the function Φ defined on $\mathbb{R}^n$ by $\Phi(0) = 0$ and for $z \neq 0$,*

$$\Phi(z) = \frac{\|z\|_2}{\int_{z^\perp} \psi(x)dx}$$

is a norm.

The latter result is key to the following theorem from [24] which extends work of Campi-Gronchi [21] to the setting of convex measures; the approach taken in [21] was the starting point for our work in this direction.

Proposition 4.4 *Let ν be a measure on $\mathbb{R}^n$ with a density ψ which is even and γ-concave on $\mathbb{R}^n$ for some $\gamma \geqslant -\frac{1}{n+1}$. Let $(K_t) := P_tC$ be an N-parameter shadow system of origin symmetric convex sets with respect to an origin symmetric body $C \subseteq \mathbb{R}^n \times \mathbb{R}^N$. Then the function $\mathbb{R}^N \ni t \mapsto \nu(K_t^\circ)^{-1}$ is convex.*

This result and the assumption on the symmetries of C and ν leads to the following corollary. The proof is similar to that given in [24].

Corollary 4.5 *Let $r \geqslant 0$, C be an origin-symmetric compact convex set in $\mathbb{R}^N$. Let ν be a radial measure on $\mathbb{R}^n$ with a density ψ which is $-1/(n+1)$-concave on $\mathbb{R}^n$.*

Then the function

$$G(x_1, \ldots, x_N) = \nu(([x_1 \ldots x_N]C + rB_2^N)^\circ)$$

is Steiner concave. Moreover if C is 1-unconditional then the function G is coordinate-wise decreasing.

Remark 1 The present setting is limited to origin-symmetric convex bodies. The argument of Campi and Gronchi [21] leading to the Blaschke-Santaló inequality has been extended to the non-symmetric case by Meyer and Reisner in [55]. It would be interesting to see an asymmetric version for random sets as it would give an empirical form of the Blaschke-Santaló inequality and related inequalities, e.g., [36] in the asymmetric case.

4.3 Minkowski Addition and Extensions

In this section, we recall several variations of Minkowski addition that are the basis of L_p-Brunn-Minkowski theory, $p \geqslant 1$, and its extensions. L_p-addition as originally defined by Firey [26] of convex sets K and L with the origin in their interior is given by

$$h^p_{K+_pL}(x) = h^p_K(x) + h^p_L(x).$$

The L_p-Brunn-Minkowski inequality of Firey states that

$$V_n(K +_p L)^{p/n} \geqslant V_n(K)^{p/n} + V_n(L)^{p/n}. \tag{4.1}$$

A more recent pointwise definition that applies to compact sets K and L is due to Lutwak, Yang and Zhang [52]

$$K +_p L = \{(1-t)^{1/q} + t^{1/q}y : x \in K, y \in L, 0 \leqslant t \leqslant 1\}, \tag{4.2}$$

where $1/p + 1/q = 1$; they proved that with the latter definition (4.1) extends to compact sets.

A general framework incorporating the latter as well as more general notions in the Orlicz setting initiated by Lutwak, Yang and Zhang [50, 51] was studied by Gardner, Hug and Weil [30, 31]. Let M be an arbitrary subset of $\mathbb{R}^m$ and define the *M-combination* $\oplus_M(K^1, \ldots, K^m)$ of arbitrary sets $K^1, \ldots, K^m$ in $\mathbb{R}^n$ by

$$\oplus_M(K^1, \ldots, K^m) = \left\{ \sum_{i=1}^{m} a_i x^{(i)} : x^{(i)} \in K^i, (a_1, \ldots, a_m) \in M \right\}$$

$$= \bigcup_{(a_i)\in M} (a_1 K^1 + \ldots + a_m K^m).$$

Gardner, Hug, and Weil [30] develop a general framework for addition operations on convex sets which model important features of the Orlicz-Brunn-Minkowski theory. The notion of M-addition is closely related to linear images of convex sets in this paper. In particular, if $C = M$ and $K^1 = \{x_1\}, \ldots, K^m = \{x_m\}$, where $x_1, \ldots, x_m \in \mathbb{R}^n$, then $[x_1, \ldots, x_m]C = \oplus_M(\{x_1\}, \ldots, \{x_m\})$.

As a sample result we mention just the following from [30] (see Theorem 6.1 and Corollary 6.4).

Theorem 4.6 *Let M be a convex set in $\mathbb{R}^m$, $m \geqslant 2$.*

i. If M is contained in the positive orthant and $K^1, \ldots, K^m$ are convex sets in $\mathbb{R}^n$, then $\oplus_M(K^1, \ldots, K^m)$ is a convex set.
ii. If M is 1-unconditional and $K^1, \ldots, K^m$ are origin-symmetric convex sets, then $\oplus_M(K^1, \ldots, K^m)$ is an origin symmetric convex set.

For several examples we mention the following:

(i) If $M = \{(1, 1)\}$ and K^1 and K^2 are convex sets, then $K^1 \oplus_M K^2 = K^1 + K^2$, i.e., $\oplus_M$ is the usual Minkowski addition.
(ii) If $M = B_q^N$ with $1/p + 1/q = 1$, and K^1 and K^2 are origin symmetric convex bodies, then $K^1 \oplus_M K^2 = K^1 +_p K^2$, i.e., $\oplus_M$ corresponds to L_p-addition as in (4.2).
(iii) There is a close connection between Orlicz addition as defined in [50, 51] and M-addition, as shown in [31]. In fact, we define Orlicz addition in terms of the latter as it interfaces well with our operator approach. As an example, let $\psi : [0, \infty)^2 \to [0, \infty)$ be convex, increasing in each argument, and $\psi(0, 0) = 0$, $\psi(1, 0) = \psi(0, 1) = 1$. Let K and L be origin-symmetric convex bodies and let $M = B_\psi^\circ$, where $B_\psi = \{(t_1, t_2) \in [-1, 1]^2 : \psi(|t_1|, |t_2|) \leqslant 1\}$. Then we define $K +_\psi L$ to be $K \oplus_M L$.

Let $N_1, \ldots, N_m$ be positive integers. For each $i = 1, \ldots, m$, consider collections of vectors $\{x_{i1}, \ldots, x_{iN_i}\} \subseteq \mathbb{R}^n$ and let $C_1, \ldots, C_m$ be compact, convex sets with $C_i \subseteq \mathbb{R}^{N_i}$. Then for any $M \subseteq \mathbb{R}^{N_1+\ldots+N_m}$,

$$\begin{aligned}
&\oplus_M([x_{11}, \ldots, x_{1N_1}]C_1, \ldots, [x_{m1}, \ldots, x_{mN_m}]C_m) \\
&= \left\{ \sum_{i=1}^m a_i \left(\sum_{j=1}^{N_i} c_{ij} x_{ij} \right) : (a_i)_i \in M, (c_{ij})_j \in C_i \right\} \\
&= \left\{ \sum_{i=1}^m \sum_{j=1}^{N_i} a_i c_{ij} x_{ij} : (a_i)_i \in M, (c_{ij})_j \in C_i \right\} \\
&= [x_{11}, \ldots, x_{1N_1}, \ldots, x_{m1}, \ldots, x_{mN_m}](\oplus_M(C_1', \ldots, C_m')),
\end{aligned}$$

where C_i' is the natural embedding of C_i into $\mathbb{R}^{N_1+\ldots+N_m}$. Thus the M-combination of families of sets of the form $[x_{i1},\ldots,x_{iN_i}]C_i$ fits exactly in the framework considered in this paper. In particular, if M is compact, convex and satisfies either of the assumptions of Theorem 4.6, then the j-th intrinsic volume of the latter set is a Steiner convex function by Corollary 4.2.

For subsequent reference we note one special case of the preceding identities. Let $C_1 = \operatorname{conv}\{e_1,\ldots,e_{N_1}\}$ and $C_2 = \operatorname{conv}\{e_1,\ldots,e_{N_2}\}$. Then we identify C_1 with $C_1' = \operatorname{conv}\{e_1,\ldots,e_{N_1}\}$ in $\mathbb{R}^{N_1+N_2}$, C_2 with $C_2' = \operatorname{conv}\{e_{N_1+1},\ldots,e_{N_1+N_2}\}$ in $\mathbb{R}^{N_1+N_2}$. If $x_1,\ldots,x_{N_1},x_{N_1+1},\ldots,x_{N_1+N_2}\in\mathbb{R}^n$, then

$$\begin{aligned}
&\operatorname{conv}\{x_1,\ldots,x_{N_1}\}\oplus_M \operatorname{conv}\{x_{N_1+1},\ldots,x_{N_1+N_2}\}\\
&= [x_1,\ldots,x_{N_1}]C_1\oplus_M [x_{N_1+1},\ldots,x_{N_1+N_2}]C_2\\
&= [x_1,\ldots,x_{N_1},x_{N_1+1},\ldots,x_{N_1+N_2}](C_1'\oplus_M C_2').
\end{aligned}$$

This will be used in §5.

4.4 Unions and Intersections of Euclidean Balls

Here we consider Euclidean balls $B(x_i,R) = \{x\in\mathbb{R}^n : |x-x_i|\leqslant r\}$ of a given radius $r>0$ with centers $x_1,\ldots,x_N\in\mathbb{R}^n$.

Theorem 4.7 *For each* $1\leqslant j\leqslant n$, *the function*

$$(\mathbb{R}^n)^N \ni (x_1,\ldots,x_N)\mapsto V_j\left(\bigcap_{i=1}^N B(x_i,r)\right) \tag{4.3}$$

is Steiner concave. Moreover, it is quasi-concave and even on $(\mathbb{R}^n)^N$.

Proof Let F be the function in (4.3). Let $\mathbf{u} = (u_1,\ldots,u_N)\in(\mathbb{R}^n)^N$ and $\mathbf{v} = (v_1,\ldots,v_N)\in(\mathbb{R}^n)^N$ belong to the support of F. One checks the following inclusion,

$$\bigcap_{i=1}^N B\left(\frac{u_i+v_i}{2},r\right)\supseteq \frac{1}{2}\bigcap_{i=1}^N B(u_i,r)+\frac{1}{2}\bigcap_{i=1}^N B(v_i,r),$$

and then applies the concavity of $K\mapsto V_j(K)^{1/j}$, which is a consequence of the Alexandrov-Fenchel inequalities. The evenness condition in the definition of Steiner concavity follows by using reflections, analogous to the proof of Corollary 4.2. □

Remark The latter theorem is also true when V_j is replaced by a function which is monotone with respect to inclusion, rotation-invariant and quasi-concave with respect to Minkowski addition; see [60].

The latter can be compared with the following result for the convex hull of unions of Euclidean balls.

Theorem 4.8 *The function*

$$(\mathbb{R}^n)^N \ni (x_1, \ldots, x_N) \mapsto V_j\left(\operatorname{conv}\left(\bigcup_{i=1}^N B(x_i, r)\right)\right)$$

is Steiner convex.

Proof Since

$$\operatorname{conv}\left(\bigcup_{i=1}^N B(x_i, r)\right) = \operatorname{conv}\{x_1, \ldots, x_N\} + B(0, r),$$

we can apply the same projection argument as in the proof of Corollary 4.2; see also work of Pfiefer [62] for a direct argument, extending Groemer's approach [34]. □

4.5 Operator Norms

Steiner convexity is also present for operator norms from an arbitrary normed space into ℓ_2^n.

Proposition 4.9 *Let E be an N-dimensional normed space. For $x_1, \ldots, x_N \in \mathbb{R}^n$, let $\mathbf{X} = [x_1, \ldots, x_N]$. Then the operator norm*

$$(\mathbb{R}^n)^N \ni \mathbf{X} \mapsto \|\mathbf{X} : E \to \ell_2^n\| \tag{4.4}$$

is Steiner convex.

Proof Denote the map in (4.4) by G. Then G is convex and hence the restriction to any line is convex. In particular, if $z \in S^{n-1}$ and $y_1, \ldots, y_N \in z^\perp$, then the function $G_Y : \mathbb{R}^N \to \mathbb{R}^+$ defined by

$$G_Y(t_1, \ldots, t_N) = G(y_1 + t_1 z_1, \ldots, y_N + t_N z_N)$$

is convex. To show that G_Y is even, we use the fact that $y_1, \ldots, y_N \in z^\perp$ to get for any $\lambda \in \mathbb{R}^N$,

$$\left\|\sum \lambda_i (y_i + t_i z)\right\|_2^2 = \left\|\sum \lambda_i (y_i - t_i z)\right\|_2^2,$$

hence $G_Y(t) = G_Y(-t)$. □

5 Stochastic Forms of Isoperimetric Inequalities

We now have all the tools to prove the randomized inequalities mentioned in the introduction and others. We will first prove two general theorems on stochastic dominance and then show how these imply a variety of randomized inequalities. At the end of the section, we discuss some examples of a different flavor.

For the next two theorems, we assume we have the following sequences of *independent* random vectors defined on a common probability space $(\Omega, \mathcal{F}, \mathbb{P})$; recall that $\widetilde{B} = \omega_n^{-1/n} B$.

1. $X_1, X_2, \ldots$, sampled according to densities $f_1, f_2, \ldots$ on $\mathbb{R}^n$, respectively (which will be chosen according to the functional under consideration).
2. $X_1^*, X_2^*, \ldots$, sampled according to $f_1^*, f_2^*, \ldots$, respectively.
3. $Z_1, Z_2 \ldots$ sampled uniformly in $\widetilde{B}$.

We use $\mathbf{X}$ to denote the $n \times N$ random matrix $\mathbf{X} = [X_1 \ldots X_N]$. Similarly, $\mathbf{X}^* = [X_1^* \ldots X_N^*]$ and $\mathbf{Z} = [Z_1 \ldots Z_N]$.

Theorem 5.1 *Let C be a compact convex set in $\mathbb{R}^N$ and $1 \leqslant j \leqslant n$. Then for each $\alpha \geqslant 0$,*

$$\mathbb{P}(V_j(\mathbf{X}C) > \alpha) \geqslant \mathbb{P}(V_j(\mathbf{X}^*C) > \alpha). \tag{5.1}$$

Moreover, if C is 1-unconditional and $\|f_i\|_\infty \leqslant 1$ for $i = 1, \ldots, N$, then for each $\alpha \geqslant 0$,

$$\mathbb{P}(V_j(\mathbf{X}C) > \alpha) \geqslant \mathbb{P}(V_j(\mathbf{Z}C) > \alpha). \tag{5.2}$$

Proof By Corollary 4.2, we have Steiner convexity. Thus we may apply Corollary 3.9 to obtain (5.1). If C is unconditional, then Proposition 3.10 applies so we can conclude (5.2). □

Theorem 5.2 *Let C be an origin symmetric convex body in $\mathbb{R}^N$. Let ν be a radial measure on $\mathbb{R}^n$ with a density ψ which is $-1/(n+1)$-concave on $\mathbb{R}^n$. Then for each $\alpha \geqslant 0$,*

$$\mathbb{P}(\nu((\mathbf{X}C)^\circ) > \alpha) \leqslant \mathbb{P}(\nu((\mathbf{X}^*C)^\circ) > \alpha). \tag{5.3}$$

Moreover, if C is 1-unconditional and $\|f_i\|_\infty \leqslant 1$ for $i = 1, \ldots, N$, then for each $\alpha \geqslant 0$,

$$\mathbb{P}(\nu((\mathbf{X}C)^\circ) > \alpha) \leqslant \mathbb{P}(\nu((\mathbf{Z}C)^\circ) > \alpha). \tag{5.4}$$

Proof By Corollary 4.5, the function is Steiner concave. Thus we may apply Corollary 3.9 to obtain (5.3). If C is unconditional, then Proposition 3.10 applies so we can conclude (5.4). □

We start by explicitly stating some of the results mentioned in the introduction. We will first derive consequences for points sampled in convex bodies or compact sets $K \subseteq \mathbb{R}^n$. In this case, we have immediate distributional inequalities as $(\frac{1}{V_n(K)}\mathbb{1}_K)^* = \frac{1}{V_n(r_K B)}\mathbb{1}_{r_K B}$, even without the unconditionality assumption on C. The case of compact sets deserves special mention for comparison to classical inequalities.

1. Busemann Random Simplex Inequality. As mentioned the Busemann random simplex inequality says that if $K \subseteq \mathbb{R}^n$ is a compact set with $V_n(K) > 0$ and $K_{o,n} = \mathrm{conv}\{o, X_1, \ldots, X_n\}$, where $X_1, \ldots, X_n$ are i.i.d. random vectors with density $f_i = \frac{1}{V_n(K)}\mathbb{1}_K$, then for $p \geqslant 1$,

$$\mathbb{E}V_n(K_{o,n})^p \geqslant \mathbb{E}V_n((r_K B)_{o,n})^p. \tag{5.5}$$

In our notation, $X_1^*, \ldots, X_n^*$ have density $\frac{1}{V_n(r_K B)}\mathbb{1}_{r_K B}$. For the set $C = \mathrm{conv}\{o, e_1, \ldots, e_n\}$, we have $K_{n,o} = \mathrm{conv}\{o, X_1, \ldots, X_n\}$. Thus the stochastic dominance of Theorem 5.1 implies (5.5) for all $p > 0$.

2. Groemer's Inequality for Random Polytopes. With the X_i's as in the previous example, set $K_N = \mathrm{conv}\{X_1, \ldots, X_N\}$. An inequality of Groemer [34] states that for $p \geqslant 1$,

$$\mathbb{E}V_n(K_N)^p \geqslant \mathbb{E}V_n((r_K B)_N)^p; \tag{5.6}$$

this was extended by Giannopoulos and Tsolomitis for $p \in (0, 1)$ in [32]. Let $C = \mathrm{conv}\{e_1, \ldots, e_N\}$ so that $K_N = [X_1, \ldots, X_N]C$ and $(r_K B)_N = [X_1^*, \ldots, X_N^*]C$. Then (5.6) follows from Theorem 5.1.

3. Bourgain-Meyer-Milman-Pajor Inequality for Random Zonotopes. Let $Z_{1,N}(K) = \sum_{i=1}^N [-X_i, X_i]$, with X_i as above. Bourgain, Meyer, Milman and Pajor [11] proved that for $p > 0$,

$$\mathbb{E}V_n(Z_{1,N}(K))^p \geqslant \mathbb{E}V_n(Z_{1,N}(r_K B))^p. \tag{5.7}$$

With the notation of the previous examples, $Z_{1,N}(K) = [X_1, \ldots, X_N]B_\infty^N$. Thus Theorem 5.1 implies (5.7).

4. Inequalities for Intrinsic Volumes. For completeness, we record here how one obtains the stochastic form of the isoperimetric inequality (1.6). In fact, we state a stochastic form of the following extended isoperimetric inequality for convex bodies $K \subseteq \mathbb{R}^n$: for $1 \leqslant j \leqslant n$,

$$V_j(K) \geqslant V_j(r_K B). \tag{5.8}$$

The latter is a particular case of the Alexandrov-Fenchel inequalities, e.g., [71]. With K_N as above, a stochastic form (5.8) is the following: for $\alpha \geqslant 0$,

$$\mathbb{P}(V_j(K_N) > \alpha) \geqslant \mathbb{P}(V_j((r_K B)_N) > \alpha), \tag{5.9}$$

which is immediate from Theorem 5.1. For expectations, results of this type for intrinsic volumes were proved by Pfiefer [62] and Hartzoulaki and the first named author [37].

For further information on the previous inequalities and others we refer the reader to the paper of Campi and Gronchi [20] and the references therein. We have singled out these four as particular examples of M-additions (defined in the previous section). For example, if $C = \mathrm{conv}\{e_1, \ldots, e_N\}$, we have

$$K_N = \oplus_C(\{X_1\}, \ldots, \{X_N\}).$$

Similarly, for $C = B_\infty^N$,

$$\sum_{i=1}^N [-X_i, X_i] = \oplus_C([-X_1, X_1], \ldots, [-X_N, X_N]).$$

One can also intertwine the above operations and others. For example, if $C = \mathrm{conv}\{e_1, e_1 + e_2, e_1 + e_2 - e_3\}$, then

$$[X_1, X_2, X_3]C = \mathrm{conv}\{X_1, X_1 + X_2, X_1 + X_2 - X_3\}$$

and Theorem 5.1 applies to such sets as well. The randomized Brunn-Minkowski inequality (1.5) is just one example of mixing two operations - convex hull and Minkowski summation. In the next example, we state a sample stochastic form of the Brunn-Minkowski inequality for M-addition in which (1.5) is just a special case; all of the previous examples also fit in this framework for additional summands. For other Brunn-Minkowski type inequalities for M-addition, see [30, 31].

5. Brunn-Minkowski Type Inequalities. Let K and L be convex bodies in $\mathbb{R}^n$ and let $M \subseteq \mathbb{R}^2$ be compact, convex and contained in the positive orthant. Then the following Brunn-Minkowski type inequality holds for each $1 \leqslant j \leqslant n$,

$$V_j(K \oplus_M L) \geqslant V_j(r_K B \oplus_M r_L B). \tag{5.10}$$

We first state a stochastic form of the latter. Let $K_{N_1} = \mathrm{conv}\{X_1, \ldots, X_{N_1}\}$, where $X_1, \ldots, X_{N_1}$ have density $f_i = \frac{1}{V_n(K)} \mathbb{1}_K$; similarly, we define $L_{N_2} = \mathrm{conv}\{X_{N_1+1}, \ldots, X_{N_1+N_2}\}$, where $X_{N_1+1}, \ldots, X_{N_1+N_2}$ have density $f_i = \frac{1}{V_n(L)} \mathbb{1}_L$. Then for $\alpha > 0$,

$$\mathbb{P}\left(V_j(K_{N_1} \oplus_M L_{N_2}) > \alpha\right) \geqslant \mathbb{P}\left(V_j((r_K B)_{N_1} \oplus_M (r_L B)_{N_2}) > \alpha\right). \tag{5.11}$$

To see that (5.11) holds, set

$$C_1 = \mathrm{conv}\{e_1, \ldots, e_{N_1}\}, \quad C_2 = \mathrm{conv}\{e_1, \ldots, e_{N_2}\}.$$

Identifying C_1 with $C_1' = \text{conv}\{e_1, \ldots, e_{N_1}\}$ in $\mathbb{R}^{N_1+N_2}$ and similarly C_2 with $C_2' = \text{conv}\{e_{N_1+1}, \ldots, e_{N_1+N_2}\}$ in $\mathbb{R}^{N_1+N_2}$ as in §4.3, we have

$$\begin{aligned} K_{N_1} \oplus_M L_{N_2} &= [X_1, \ldots, X_{N_1}]C_1 \oplus_M [X_{N_1+1}, \ldots, X_{N_1+N_2}]C_2 \\ &= [X_1, \ldots, X_{N_1}, X_{N_1+1}, \ldots, X_{N_1+N_2}](C_1' \oplus_M C_2'). \end{aligned}$$

Write $\mathbf{X_1} = [X_1, \ldots, X_{N_1}]$ and $\mathbf{X_2} = [X_{N_1+1}, \ldots, X_{N_1+N_2}]$, and $\mathbf{X_1^*} = [X_1^*, \ldots, X_{N_1}^*]$ and $\mathbf{X_2^*} = [X_{N_1+1}^*, \ldots, X_{N_1+N_2}^*]$. In block matrix form, we have

$$K_{N_1} \oplus_M L_{N_2} = [\mathbf{X_1},\ \mathbf{X_2}](C_1' \oplus_M C_2').$$

Similarly,

$$(r_K B)_{N_1} \oplus_M (r_L B)_{N_2} = [\mathbf{X_1^*},\ \mathbf{X_2^*}](C_1' \oplus_M C_2'),$$

and so Theorem 5.1 implies (5.11). To prove (1.5), we take $M = \{(1,1)\}$ and $j = n$ in (5.11). Inequality (5.10) follows from (5.11) when $N_1, N_2 \to \infty$. For simplicity of notation, we have stated this for only two sets and C_1, C_2 as above.

For another example involving a law of large numbers, we turn to the following, stated in the symmetric case for simplicity.

6. Orlicz-Busemann-Petty Centroid Inequality. Let $\psi : [0,\infty) \to [0,\infty)$ be a Young function, i.e., convex, strictly increasing with $\psi(0) = 0$. Let f be a bounded probability density of a continuous distribution on $\mathbb{R}^n$. Define the Orlicz centroid body $Z_\psi(f)$ associated to ψ by its support function

$$h(Z_\psi(f), y) = \inf\left\{\lambda > 0 : \int_{\mathbb{R}^n} \psi\left(\frac{|\langle x, y\rangle|}{\lambda}\right) f(x)dx \leqslant 1\right\}.$$

Let $r_f > 0$ be such that $\|f\|_\infty \mathbb{1}_{r_f B}$ is a probability density. Then

$$V_n(Z_\psi(f)) \geqslant V_n(Z_\psi(\|f\|_\infty \mathbb{1}_{r_f B})). \tag{5.12}$$

Here we assume that $h(Z_\psi(f), y)$ is finite for each $y \in S^{n-1}$ and so $h(Z_\psi(f), \cdot)$ defines a norm and hence is the support function of the symmetric convex body $Z_\psi(f)$. When f is the indicator of a convex body, (5.12) was proved by Lutwak, Yang and Zhang [50] (where it was also studied for more general functions ψ); it was extended to star bodies by Zhu [78]; the version for probability densities and the randomized version below is from [58]; an extension of (5.12) to the asymmetric case was carried out by Huang and He [38].

The empirical analogue of (5.12) arises by considering the following finite-dimensional origin-symmetric Orlicz balls

$$B_{\psi,N} := \left\{t = (t_1, \ldots, t_N) \in \mathbb{R}^N : \frac{1}{N}\sum_{i=1}^{N} \psi(|t_i|) \leqslant 1\right\}$$

with associated Orlicz norm $\|t\|_{B_{\psi/N}} := \inf\{\lambda > 0 : t \in \lambda B_{\psi,N}\}$, which is the support function for $B^{\circ}_{\psi,N}$. For independent random vectors $X_1, \ldots, X_N$ distributed according to f, we let

$$Z_{\psi,N}(f) = [X_1, \ldots, X_N]B^{\circ}_{\psi,N}.$$

Then for $y \in S^{n-1}$,

$$h(Z_{\psi,N}(f), y) = \|(\langle X_1, y\rangle, \ldots, \langle X_N, y\rangle)\|_{B_{\psi/N}}.$$

Applying Theorem 5.1 for $C = B^{\circ}_{\psi,N}$, we get that for $1 \leqslant j \leqslant n$ and $\alpha \geqslant 0$,

$$\mathbb{P}(V_j(Z_{\psi,N}(f)) > \alpha) \geqslant \mathbb{P}(V_j(Z_{\psi,N}(\|f\|_{\infty} \mathbb{1}_{r_f B})) > \alpha). \tag{5.13}$$

Using the law of large numbers, one may check that

$$Z_{\psi,N}(f) \to Z_{\psi}(f) \tag{5.14}$$

almost surely in the Hausdorff metric (see [58]); when $\psi(x) = x^p$ and $f = \frac{1}{V_n(K)}\mathbb{1}_K$, $Z_{\psi,N}(f) = Z_{p,N}(K)$ as defined in the introduction; in this case, the convergence in (5.14) is immediate by the classical strong law of large numbers (compare (1.10) and (1.14)). By integrating (5.13) and sending $N \to \infty$, we thus obtain (5.12).

We now turn to the dual setting.

7. Blaschke-Santaló Type Inequalities. The Blaschke-Santaló inequality states that if K is a symmetric convex body in $\mathbb{R}^n$, then

$$V_n(K^{\circ}) \leqslant V_n((r_K B)^{\circ}). \tag{5.15}$$

This was proved by Blaschke for $n = 2, 3$ and in general by Santaló [68]; see also Meyer and Pajor's proof by Steiner symmetrization [54] and [71, 28] for further background; origin symmetry in (5.15) is not needed but we discuss the randomized version only in the symmetric case. One can obtain companion results for all of the inequalities mentioned so far with suitable choices of symmetric convex bodies C. Let ν be a radially decreasing measure as in Theorem 5.2. Let $C = B_1^N$ and set $K_{N,s} = [X_1, \ldots, X_N]B_1^N$, where X_i has density $f_i = \frac{1}{V_n(K)}\mathbb{1}_K$. Then for $\alpha > 0$,

$$\mathbb{P}(\nu((K_{N,s})^{\circ}) > \alpha) \leqslant \mathbb{P}(\nu(((r_K B)_{N,s})^{\circ}) > \alpha).$$

Similarly, if K and L are origin-symmetric convex bodies and $M \subseteq \mathbb{R}^2$ is unconditional, then for $\alpha > 0$,

$$\mathbb{P}(\nu((K_{N_1,s} \oplus_M L_{N_1,s})^{\circ}) > \alpha) \leqslant \mathbb{P}(\nu(((r_K B)_{N_1,s} \oplus_M (r_L B)_{N_1,s})^{\circ}) > \alpha). \tag{5.16}$$

We also single out the polar dual of the last example on Orlicz-Busemann-Petty centroid bodies. Let ψ and $B_{\psi,N}$ be as above. Then

$$\mathbb{P}(\nu(Z^{\circ}_{\psi,N}(f)) > \alpha) \leqslant \mathbb{P}(\nu(Z^{\circ}_{\psi,N}(\|f\|_{\infty}\mathbb{1}_{r_f B})) > \alpha).$$

For a particular choice of ψ we arrive at the following example, which has not appeared in the literature before and deserves an explicit mention.

8. Level Sets of the Logarithmic Laplace Transform. For a continuous probability distribution with an even bounded density f, recall that the logarithmic Laplace transform is defined by

$$\Lambda(f, y) = \log \int_{\mathbb{R}^n} \exp(\langle x, y\rangle) f(x)dx.$$

For such f and $p > 0$, we define an origin-symmetric convex body $\Lambda_p(f)$ by

$$\Lambda_p(f) = \{y \in \mathbb{R}^n : \Lambda_f(y) \leqslant p\}.$$

The empirical analogue is defined as follows: for independent random vectors $X_1, \ldots, X_N$ with density f, set

$$\Lambda_{p,N}(f) = \left\{ y \in \mathbb{R}^n : \frac{1}{N}\sum_{i=1}^{N} \psi(|\langle X_i, y\rangle|) \leqslant e^p \right\}.$$

If we set $\psi_p(x) = e^{-p}(e^x - 1)$ then $([X_1, \ldots, X_N]B^{\circ}_{\psi_p,N})^{\circ} = \Lambda_{p,N}(f)$. Then we have the following stochastic dominance

$$\mathbb{P}(\nu(\Lambda_{p,N}(f)) > \alpha) \leqslant \mathbb{P}(\nu(\Lambda_{p,N}(\|f\|_{\infty}\mathbb{1}_{r_f B})) > \alpha),$$

where r_f satisfies $\|f\|_{\infty}\mathbb{1}_{r_f B} = 1$. When $N \to \infty$, we get

$$\nu(\Lambda_p(f)) \leqslant \nu(\Lambda_p(\|f\|_{\infty}\mathbb{1}_{r_f B}).$$

The latter follows from the law of large numbers as in [58, Lemma 5.4] and the argument given in [24, §5].

For log-concave densities, the level sets of the logarithmic Laplace transform are known to be isomorphic to the duals to the L_p-centroid bodies; see work of Latała and Wojtaszczyk [43], or Klartag and E. Milman [42]; these bodies are essential in establishing concentration properties of log-concave measures, e.g., [56, 41, 13].

9. Ball-Polyhedra. All of the above inequalities are volumetric in nature. For convex bodies, they all reduce to comparisons of bodies of a given volume. For an example of a different flavor, we have the following inequality involving random ball polyhedra: for $R > 0$,

$$\mathbb{P}\left(V_j\left(\bigcap_{i=1}^{N} B(X_i,R)\right) \geqslant \alpha\right) \leqslant \mathbb{P}\left(V_j\left(\bigcap_{i=1}^{N} B(Z_i,R)\right) > \alpha\right).$$

When the X_i's are sampled according to a particular density f associated with a convex body K, the latter leads to the following generalized Urysohn inequality,

$$V_j(K) \leqslant V_j((w(K))/2)B),$$

where $w(K)$ is the mean width of K, see [60]; the latter is not a volumetric inequality when $j < n$. The particular density f is the uniform measure on a star-shaped set $A(K,R)$ defined by specifying its radial function $\rho_{A(K,R)}(\theta) = R - h_K(-\theta)$; Steiner symmetrization of $A(K,R)$ preserves the mean-width of K (for large R) so the volumetric techniques here lead to a stochastic dominance inequality for mean width.

We have focused this discussion on stochastic dominance. It is sometimes useful to relax the probabilistic formulation and instead consider the quantities above in terms of bounded integrable functions. We give one such example.

10. Functional Forms. The following functional version of Busemann's random simplex inequality (1.8) is useful for marginal distributions of high-dimensional probability distributions; this is from joint work with S. Dann [25]. Let $f_1,\ldots,f_k$ be non-negative, bounded, integrable functions such that $\|f_i\|_1 > 0$ for each $i = 1,\ldots,k$. For $p \in \mathbb{R}$, set

$$g_p(f_1,\ldots,f_k) = \int_{\mathbb{R}^n}\cdots\int_{\mathbb{R}^n} V_k(\text{conv}\{0,x_1,\ldots,x_k\})^p \prod_{i=1}^{k} f_i(x_i)dx_1\ldots dx_k.$$

Then for $p > 0$,

$$g_p(f_1,\ldots,f_k) \geqslant \left(\prod_{i=1}^{k} \frac{\|f_i\|_1^{1+p/n}}{\omega_n^{1+p/n}\,\|f_i\|_\infty^{p/n}}\right) g_p(\mathbb{1}_{B_2^n},\ldots,\mathbb{1}_{B_2^n}).$$

The latter is just a special case of a general functional inequality [25]. Following Busemann's argument, we obtain the following. Let $1 \leqslant k \leqslant n-1$ and let f be a non-negative, bounded integrable function on $\mathbb{R}^n$. Then

$$\int_{G_{n,k}} \frac{\left(\int_E f(x)dx\right)^n}{\|f|_E\|_\infty^{n-k}} d\nu_{n,k}(E) \leqslant \frac{\omega_k^n}{\omega_n^k}\left(\int_{\mathbb{R}^n} f(x)dx\right)^k;$$

when $f = \mathbb{1}_K$ this recovers the inequality of Busemann and Straus [17] and Grinberg [33] extending (1.9). Schneider proved an analogue of the latter on the affine Grassmannian [70], which can also be extended to a sharp isoperimetric inequality for integrable functions [25]. The functional versions lead to small ball probabilities for projections of random vectors that need not have independent coordinates.

6 An Application to Operator Norms of Random Matrices

In the previous section we gave examples of functionals on random convex sets which are minorized or majorized for the uniform measure on the Cartesian product of Euclidean balls. In some cases the associated distribution function can be accurately estimated. For example, passing to complements in (5.2), we get for $\alpha \geqslant 0$,

$$\mathbb{P}(V_n(\mathbf{X}C) \leqslant \alpha) \leqslant \mathbb{P}(V_n(\mathbf{Z}C) \leqslant \alpha), \tag{6.1}$$

where $\mathbf{X}$ and $\mathbf{Z}$ are as in Theorem 5.1. When $C = B_1^N$, i.e., for random symmetric convex hulls, we have estimated the quantity on the right-hand side of (6.1) in [59] for all α less than an absolute contant (sufficiently small), at least when $N \leqslant e^n$. (The reason for the restriction is that we compute this for Gaussian matrices and the comparison to the uniform measure on the Cartesian products of balls is only valid in this range). This leads to sharp bounds for small deviation probabilities for the volume of random polytopes that were known before only for certain sub-gaussian distributions. The method of [59] applies more broadly. In this section we will focus on the case of the operator norm of a random matrix with independent columns. We refer readers interested in background on non-asymptotic random matrix theory to the article of Rudelson and Vershynin [66] and the references therein.

By combining Corollary 3.9, and Propositions 3.11 and 4.9, we get the following result, which is joint work with G. Livshyts [45].

Theorem 6.1 *Let $N, n \in \mathbb{N}$. Let E be an N-dimensional normed space. Then the random matrices $\mathbf{X}, \mathbf{X}^*$ and $\mathbf{Z}$ (as in §5) satisfy the following for each $\alpha \geqslant 0$,*

$$\mathbb{P}\left(\|\mathbf{X} : E \to \ell_2^n\| \leqslant \alpha\right) \leqslant \mathbb{P}\left(\|\mathbf{X}^* : E \to \ell_2^n\| \leqslant \alpha\right). \tag{6.2}$$

Moreover, if $\|f_i\|_\infty \leqslant 1$ for each $i = 1, \ldots, N$, then

$$\mathbb{P}\left(\|\mathbf{X} : E \to \ell_2^n\| \leqslant \alpha\right) \leqslant \mathbb{P}\left(\|\mathbf{Z} : E \to \ell_2^n\| \leqslant \alpha\right).$$

As before, the latter result reduces the small deviation problem to computations for matrices $\mathbf{Z}$ with independent columns sampled in the Euclidean ball of volume one. For the important case of the operator norm $\|\cdot\|_{2\to 2}$, i.e., $E := \ell_2^N$, we get the following bound.

Lemma 6.2 *For $\varepsilon > 0$,*

$$\mathbb{P}\left(\|\mathbf{Z}\|_{2\to 2} \leqslant \varepsilon\sqrt{N}\right) \leqslant (c\varepsilon)^{nN-1}, \tag{6.3}$$

where c is an absolute constant.

Proof Let C and K be symmetric convex bodies in $\mathbb{R}^d$, $V_d(K) = 1$ and $p < d$. By [57, Proposition 4.7]),

$$\left(\int_K \|x\|_C^{-p} dx\right)^{\frac{1}{p}} \leq \left(\frac{d}{d-p}\right)^{\frac{1}{p}} V_d(C)^{\frac{1}{d}}. \tag{6.4}$$

Let $d := nN$, $K := \widetilde{B}\times\cdots\times\widetilde{B} \subseteq \mathbb{R}^d$ and C be the unit ball in $\mathbb{R}^d$ for the operator norm $\left\|\cdot : \ell_2^N \to \ell_2^n\right\|$. Then the Hilbert-Schmidt norm $\|\cdot\|_{HS}$ satisfies $\|A\|_{HS} \leq \sqrt{n}\,\|A\|_{2\to 2}$ or $C \subseteq \sqrt{n}B_2^d$, which implies that $V_d(C)^{\frac{1}{d}} \leq \frac{c_1}{\sqrt{N}}$; in fact, arguing as in [75, Lemma 38.5] one can show that $V_d(C)^{\frac{1}{d}} \simeq \frac{1}{\sqrt{N}}$. Thus for $p = nN-1$, we get

$$\left(\mathbb{E}\,\|\mathbf{Z}\|_{2\to 2}^{-(nN-1)}\right)^{\frac{1}{nN-1}} \leqslant c_1 (nN)^{\frac{1}{nN-1}} N^{-1/2} \leqslant ec_1 N^{-1/2},$$

from which the lemma follows by an application of Markov's inequality. □

For $1 \times N$ matrices, Theorem 6.1 reduces to small-ball probabilities for norms of a random vector x in $\mathbb{R}^N$ distributed according to a density of the form $\prod_{i=1}^N f_i$ where each f_i is a density on the real line. In particular, if $\|f_i\|_\infty \leq 1$ for each $i = 1,\ldots,N$, then for any norm $\|\cdot\|$ on $\mathbb{R}^N$ (dual to that of E), we have for $\varepsilon > 0$,

$$\mathbb{P}\left(\|x\| \leqslant \varepsilon\right) \leqslant \mathbb{P}\left(\|z\| \leqslant \varepsilon\right), \tag{6.5}$$

where z is a random vector in the cube $[-1/2, 1/2]^N$ - the uniform measure on Cartesian products of "balls" in 1-dimension. In fact, by approximation from within, the same result holds if $\|\cdot\|$ is a semi-norm. Thus if x and z are as above, for each $\varepsilon > 0$ we have

$$\mathbb{P}(\|P_E x\|_2 \leqslant \varepsilon\sqrt{k}) \leqslant \mathbb{P}(\|P_E z\|_2 \leqslant \varepsilon\sqrt{k}) \leqslant (2\sqrt{\pi e}\varepsilon)^k, \tag{6.6}$$

where the last inequality uses a result of Ball [4]. In this way we recover the result of Rudelson and Vershynin from [67], who proved (6.6) with a bound of the form $(c\varepsilon)^k$ for some absolute constant c. Using the Rogers/Brascamp-Lieb-Luttinger inequality and Kanter's theorem, one can also obtain the sharp constant of $\sqrt{2}$ for the ℓ_∞-norm of marginal densities, which was first computed in [46] by adapting Ball's arguments from [4].

Acknowledgements We would like to thank Petros Valettas for useful discussions. We also thank Erwin Lutwak and Beatrice-Helen Vritsiou for helpful comments on earlier drafts of this paper. Part of this work was carried out during the Oberwolfach workshops *Convex Geometry and its Applications*, held December 6–12, 2015, and *Asymptotic Geometric Analysis*, held February 21–27, 2016. We thank the organizers of these meetings and the institute staff for their warm hospitality.

Grigoris Paouris is supported by US NSF grant CAREER-1151711 and BSF grant 2010288.

Peter Pivovarov is supported by US NSF grant DMS-1612936. This work was also partially supported by a grant from the Simons Foundation (#317733 to Peter Pivovarov).

References

[1] T. W. Anderson, *The integral of a symmetric unimodal function over a symmetric convex set and some probability inequalities*, Proc. Amer. Math. Soc. **6** (1955), 170–176.
[2] A. Baernstein and M. Loss, *Some conjectures about L^p norms of k-plane transforms*, Rend. Sem. Mat. Fis. Milano **67** (1997), 9–26 (2000).
[3] K. Ball, *Logarithmically concave functions and sections of convex sets in $\mathbf{R}^n$*, Studia Math. **88** (1988), no. 1, 69–84.
[4] ———, *Volumes of sections of cubes and related problems*, Geometric aspects of functional analysis (1987–88), Lecture Notes in Math., vol. 1376, Springer, Berlin, 1989, pp. 251–260.
[5] F. Barthe, *Mesures unimodales et sections des boules B_p^n*, C. R. Acad. Sci. Paris Sér. I Math. **321** (1995), no. 7, 865–868.
[6] ———, *Extremal properties of central half-spaces for product measures*, J. Funct. Anal. **182** (2001), no. 1, 81–107.
[7] S. G. Bobkov, *Convex bodies and norms associated to convex measures*, Probab. Theory Related Fields **147** (2010), no. 1-2, 303–332.
[8] C. Borell, *Convex measures on locally convex spaces*, Ark. Mat. **12** (1974), 239–252.
[9] ———, *Convex set functions in d-space*, Period. Math. Hungar. **6** (1975), no. 2, 111–136.
[10] K. J. Böröczky, E. Lutwak, D. Yang, and G. Zhang, *The log-Brunn-Minkowski inequality*, Adv. Math. **231** (2012), no. 3-4, 1974–1997.
[11] J. Bourgain, M. Meyer, V. Milman, and A. Pajor, *On a geometric inequality*, Geometric aspects of functional analysis (1986/87), Lecture Notes in Math., vol. 1317, Springer, Berlin, 1988, pp. 271–282.
[12] H. J. Brascamp, E. H. Lieb, and J. M. Luttinger, *A general rearrangement inequality for multiple integrals*, J. Functional Analysis **17** (1974), 227–237.
[13] S. Brazitikos, A. Giannopoulos, P. Valettas, and B. H. Vritsiou, *Geometry of isotropic convex bodies*, Mathematical Surveys and Monographs, vol. 196, American Mathematical Society, Providence, RI, 2014.
[14] A. Burchard, *A short course on rearrangement inequalities*, available at http://www.math.utoronto.ca/almut/rearrange.pdf, 2009.
[15] H. Busemann, *A theorem on convex bodies of the Brunn-Minkowski type*, Proc. Nat. Acad. Sci. U. S. A. **35** (1949), 27–31.
[16] ———, *Volume in terms of concurrent cross-sections*, Pacific J. Math. **3** (1953), 1–12.
[17] H. Busemann and E. G. Straus, *Area and normality*, Pacific J. Math. **10** (1960), 35–72.
[18] S. Campi, A. Colesanti, and P. Gronchi, *A note on Sylvester's problem for random polytopes in a convex body*, Rend. Istit. Mat. Univ. Trieste **31** (1999), no. 1-2, 79–94.
[19] S. Campi and P. Gronchi, *The L^p-Busemann-Petty centroid inequality*, Adv. Math. **167** (2002), no. 1, 128–141.
[20] ———, *Extremal convex sets for Sylvester-Busemann type functionals*, Appl. Anal. **85** (2006), no. 1-3, 129–141.
[21] ———, *On volume product inequalities for convex sets*, Proc. Amer. Math. Soc. **134** (2006), no. 8, 2393–2402 (electronic).
[22] ———, *Volume inequalities for sets associated with convex bodies*, Integral geometry and convexity, World Sci. Publ., Hackensack, NJ, 2006, pp. 1–15.
[23] M. Christ, *Estimates for the k-plane transform*, Indiana Univ. Math. J. **33** (1984), no. 6, 891–910.
[24] D. Cordero-Erausquin, M. Fradelizi, G. Paouris, and P. Pivovarov, *Volume of the polar of random sets and shadow systems*, Math. Ann. **362** (2015), no. 3-4, 1305–1325.
[25] S. Dann, G. Paouris, and P. Pivovarov, *Bounding marginal densities via affine isoperimetry*, Proc. Lond. Math. Soc. (3) **113** (2016), no. 2, 140–162.
[26] W. J. Firey, *p-means of convex bodies*, Math. Scand. **10** (1962), 17–24.
[27] R. J. Gardner, *The Brunn-Minkowski inequality*, Bull. Amer. Math. Soc. (N.S.) **39** (2002), no. 3, 355–405.

[28] ———, *Geometric Tomography*, second ed., Encyclopedia of Mathematics and its Applications, vol. 58, Cambridge University Press, New York, 2006.
[29] ———, *The dual Brunn-Minkowski theory for bounded Borel sets: dual affine quermassintegrals and inequalities*, Adv. Math. **216** (2007), no. 1, 358–386.
[30] R. J. Gardner, D. Hug, and W. Weil, *Operations between sets in geometry*, J. Eur. Math. Soc. (JEMS) **15** (2013), no. 6, 2297–2352.
[31] ———, *The Orlicz-Brunn-Minkowski theory: a general framework, additions, and inequalities*, J. Differential Geom. **97** (2014), no. 3, 427–476.
[32] A. Giannopoulos and A. Tsolomitis, *Volume radius of a random polytope in a convex body*, Math. Proc. Cambridge Philos. Soc. **134** (2003), no. 1, 13–21.
[33] E. L. Grinberg, *Isoperimetric inequalities and identities for k-dimensional cross-sections of convex bodies*, Math. Ann. **291** (1991), no. 1, 75–86.
[34] H. Groemer, *On the mean value of the volume of a random polytope in a convex set*, Arch. Math. (Basel) **25** (1974), 86–90.
[35] P. M. Gruber, *Convex and discrete geometry*, Grundlehren der Mathematischen Wissenschaften [Fundamental Principles of Mathematical Sciences], vol. 336, Springer, Berlin, 2007.
[36] C. Haberl and F. E. Schuster., *General L_p affine isoperimetric inequalities*, J. Differential Geom. **83** (2009), no. 1, 1–26.
[37] M. Hartzoulaki and G. Paouris, *Quermassintegrals of a random polytope in a convex body*, Arch. Math. (Basel) **80** (2003), no. 4, 430–438.
[38] Q. Huang and B. He, *An asymmetric Orlicz centroid inequality for probability measures*, Sci. China Math. **57** (2014), no. 6, 1193–1202.
[39] M. Kanter, *Unimodality and dominance for symmetric random vectors*, Trans. Amer. Math. Soc. **229** (1977), 65–85.
[40] J. Kim, V. Yaskin, and A. Zvavitch, *The geometry of p-convex intersection bodies*, Adv. Math. **226** (2011), no. 6, 5320–5337.
[41] B. Klartag, *A central limit theorem for convex sets*, Invent. Math. **168** (2007), no. 1, 91–131.
[42] B. Klartag and E. Milman, *Centroid bodies and the logarithmic Laplace transform—a unified approach*, J. Funct. Anal. **262** (2012), no. 1, 10–34.
[43] R. Latała and J. O. Wojtaszczyk, *On the infimum convolution inequality*, Studia Math. **189** (2008), no. 2, 147–187.
[44] E. H. Lieb and M. Loss, *Analysis*, second ed., Graduate Studies in Mathematics, vol. 14, American Mathematical Society, Providence, RI, 2001.
[45] G. Livshyts, G. Paouris, and P. Pivovarov, *Small deviations for operator norms*, work in progress.
[46] ———, *On sharp bounds for marginal densities of product measures*, Israel J. Math. **216** (2016), no. 2, 877–889.
[47] E. Lutwak, *Selected affine isoperimetric inequalities*, Handbook of convex geometry, Vol. A, B, North-Holland, Amsterdam, 1993, pp. 151–176.
[48] E. Lutwak, D. Yang, and G. Zhang, *L_p affine isoperimetric inequalities*, J. Differential Geom. **56** (2000), no. 1, 111–132.
[49] ———, *Sharp affine L_p Sobolev inequalities*, J. Differential Geom. **62** (2002), no. 1, 17–38.
[50] ———, *Orlicz centroid bodies*, J. Differential Geom. **84** (2010), no. 2, 365–387.
[51] ———, *Orlicz projection bodies*, Adv. Math. **223** (2010), no. 1, 220–242.
[52] ———, *The Brunn-Minkowski-Firey inequality for nonconvex sets*, Adv. in Appl. Math. **48** (2012), no. 2, 407–413.
[53] E. Lutwak and G. Zhang, *Blaschke-Santaló inequalities*, J. Differential Geom. **47** (1997), no. 1, 1–16.
[54] M. Meyer and A. Pajor, *On Santaló's inequality*, Geometric aspects of functional analysis (1987–88), Lecture Notes in Math., vol. 1376, Springer, Berlin, 1989, pp. 261–263.
[55] M. Meyer and S. Reisner, *Shadow systems and volumes of polar convex bodies*, Mathematika **53** (2006), no. 1, 129–148 (2007).

[56] G. Paouris, *Concentration of mass on convex bodies*, Geom. Funct. Anal. **16** (2006), no. 5, 1021–1049.
[57] ———, *Small ball probability estimates for log-concave measures*, Trans. Amer. Math. Soc. **364** (2012), no. 1, 287–308.
[58] G. Paouris and P. Pivovarov, *A probabilistic take on isoperimetric-type inequalities*, Adv. Math. **230** (2012), no. 3, 1402–1422.
[59] ———, *Small-ball probabilities for the volume of random convex sets*, Discrete Comput. Geom. **49** (2013), no. 3, 601–646.
[60] G. Paouris and P. Pivovarov, *Random ball-polyhedra and inequalities for intrinsic volumes*, Monatsh. Math. **182** (2017), no. 3, 709–729.
[61] C. M. Petty, *Isoperimetric problems*, Proceedings of the Conference on Convexity and Combinatorial Geometry (Univ. Oklahoma, Norman, Okla., 1971), Dept. Math., Univ. Oklahoma, Norman, Okla., 1971, pp. 26–41.
[62] R. E. Pfiefer, *The extrema of geometric mean values*, ProQuest LLC, Ann Arbor, MI, 1982, Thesis (Ph.D.)–University of California, Davis.
[63] ———, *The historical development of J. J. Sylvester's four point problem*, Math. Mag. **62** (1989), no. 5, 309–317.
[64] C. A. Rogers, *A single integral inequality*, J. London Math. Soc. **32** (1957), 102–108.
[65] C. A. Rogers and G. C. Shephard, *Some extremal problems for convex bodies*, Mathematika **5** (1958), 93–102.
[66] M. Rudelson and R. Vershynin, *Non-asymptotic theory of random matrices: extreme singular values*, Proceedings of the International Congress of Mathematicians. Volume III, Hindustan Book Agency, New Delhi, 2010, pp. 1576–1602.
[67] ———, *Small ball probabilities for linear images of high-dimensional distributions*, Int. Math. Res. Not. IMRN (2015), no. 19, 9594–9617.
[68] L. A. Santaló, *An affine invariant for convex bodies of n-dimensional space*, Portugaliae Math. **8** (1949), 155–161.
[69] C. Saroglou, *Shadow systems: remarks and extensions*, Arch. Math. (Basel) **100** (2013), no. 4, 389–399.
[70] R. Schneider, *Inequalities for random flats meeting a convex body*, J. Appl. Probab. **22** (1985), no. 3, 710–716.
[71] ———, *Convex bodies: the Brunn-Minkowski theory*, expanded ed., Encyclopedia of Mathematics and its Applications, vol. 151, Cambridge University Press, Cambridge, 2014.
[72] G. C. Shephard, *Shadow systems of convex sets*, Israel J. Math. **2** (1964), 229–236.
[73] ———, *Shadow systems of convex sets*, Israel J. Math. **2** (1964), 229–236.
[74] B. Simon, *Convexity*, Cambridge Tracts in Mathematics, vol. 187, Cambridge University Press, Cambridge, 2011, An analytic viewpoint.
[75] N. Tomczak-Jaegermann, *Banach-Mazur distances and finite-dimensional operator ideals*, Pitman Monographs and Surveys in Pure and Applied Mathematics, vol. 38, Longman Scientific & Technical, Harlow; copublished in the United States with John Wiley & Sons, Inc., New York, 1989.
[76] R. A. Vitale, *The Brunn-Minkowski inequality for random sets*, J. Multivariate Anal. **33** (1990), no. 2, 286–293.
[77] L. Wang and M. Madiman, *Beyond the entropy power inequality, via rearrangements*, IEEE Trans. Inform. Theory **60** (2014), no. 9, 5116–5137.
[78] G. Zhu, *The Orlicz centroid inequality for star bodies*, Adv. in Appl. Math. **48** (2012), no. 2, 432–445.

Forward and Reverse Entropy Power Inequalities in Convex Geometry

Mokshay Madiman, James Melbourne, and Peng Xu

Abstract The entropy power inequality, which plays a fundamental role in information theory and probability, may be seen as an analogue of the Brunn-Minkowski inequality. Motivated by this connection to Convex Geometry, we survey various recent developments on forward and reverse entropy power inequalities not just for the Shannon-Boltzmann entropy but also more generally for Rényi entropy. In the process, we discuss connections between the so-called functional (or integral) and probabilistic (or entropic) analogues of some classical inequalities in geometric functional analysis.

1 Introduction

The Brunn-Minkowski inequality plays a fundamental role not just in Convex Geometry, where it originated over 125 years ago, but also as an indispensable tool in Functional Analysis, and– via its connections to the concentration of measure phenomenon– in Probability. The importance of this inequality, and the web of its tangled relationships with many other interesting and important inequalities, is beautifully elucidated in the landmark 2002 survey of Gardner [70]. Two of the parallels that Gardner discusses in his survey are the Prékopa-Leindler inequality and the Entropy Power Inequality; since the time that the survey was written, these two inequalities have become the foundation and prototypes for two different but related analytic "liftings" of Convex Geometry. While the resulting literature is too vast for us to attempt doing full justice to in this survey, we focus on one

---

This work was supported in part by the U.S. National Science Foundation through grants DMS-1409504 (CAREER) and CCF-1346564. Some of the new results described in Section 3 were announced at the 2016 IEEE International Symposium on Information Theory [166] in Barcelona.

M. Madiman (✉) • J. Melbourne • P. Xu
Department of Mathematical Sciences, University of Delaware, Newark, DE, USA
e-mail: madiman@udel.edu; jamesm@udel.edu; xpeng@udel.edu

E. Carlen et al. (eds.), *Convexity and Concentration*, The IMA Volumes in Mathematics and its Applications 161, DOI 10.1007/978-1-4939-7005-6_14

particular strain of research– namely, the development of reverse entropy power inequalities– and using that as a narrative thread, chart some of the work that has been done towards these "liftings".

Let A, B be any nonempty Borel sets in $\mathbb{R}^d$. Write $A + B = \{x + y : x \in A,\ y \in B\}$ for the Minkowski sum, and $|A|$ for the d-dimensional volume (or Lebesgue measure) of A. The Brunn-Minkowski inequality (BMI) says that

$$\left|A + B\right|^{1/d} \geq |A|^{1/d} + |B|^{1/d}. \tag{1}$$

The BMI was proved in the late 19th century by Brunn for convex sets in low dimension ($d \leq 3$), and Minkowski for convex sets in $\mathbb{R}^d$; the reader may consult Kjeldsen [86, 87] for an interesting historical analysis of how the notion of convex sets in linear spaces emerged from these efforts (Minkowski's in particular). The extension of the BMI to compact– and thence Borel-measurable– subsets of $\mathbb{R}^d$ was done by Lusternik [99]. Equality holds in the inequality (1) for sets A and B with positive volumes if and only if they are convex and homothetic (i.e., one is a scalar multiple of the other, up to translation), possibly with sets of measure zero removed from each one. As of today, there are a number of simple and elegant proofs known for the BMI.

In the last few decades, the BMI became the starting point of what is sometimes called the Brunn-Minkowski theory, which encompasses a large and growing range of geometric inequalities including the Alexandrov-Fenchel inequalities for mixed volumes, and which has even developed important offshoots such as the L^p-Brunn-Minkowski theory [100]. Already in the study of the geometry of convex bodies (i.e., convex compact sets with nonempty interior), the study of log-concave functions turns out to be fundamental. One way to see this is to observe that uniform measures on convex bodies are not closed under taking lower-dimensional marginals, but yield log-concave densities, which do have such a closure property– while the closure property of log-concave functions under marginalization goes back to Prékopa [131, 132] and Brascamp-Lieb [38], their consequent fundamental role in the geometry of convex bodies was first clearly recognized in the doctoral work of K. Ball [10] (see also [11, 126]). Since then, the realization has grown that it is both possible and natural to state many questions and theorems in Convex Geometry directly for the category of log-concave functions or measures rather than for the category of convex bodies– V. Milman calls this the "Geometrization of Probability" program [125], although one might equally well call it the "Probabilitization of Convex Geometry" program. The present survey squarely falls within this program.

For the goal of embedding the geometry of convex sets in a more analytic setting, two approaches are possible:

1. *Functional (integral) lifting*: Replace sets by functions, and convex sets by log-concave or s-concave functions, and the volume functional by the integral. This is a natural extension because if we identify a convex body K with its indicator function 1_K (defined as being 1 on the set and 0 on its complement), then the integral of 1_K is just the volume of K. The earlier survey of V. Milman [125] is

entirely focused on this lifting of Convex Geometry; recent developments since then include the introduction and study of mixed integrals (analogous to mixed volumes) independently by Milman-Rotem [121, 120] and Bobkov-Colesanti-Fragala [31] (see also [17]). Colesanti [47] has an up-to-date survey of these developments in another chapter of this volume.

2. *Probabilistic (entropic) lifting*: Replace sets by random variables (or strictly speaking their distributions), and convex sets by random variables with log-concave or s-concave distributions, and the volume functional by the entropy functional (actually "entropy power", which we will discuss shortly). This is a natural analogue because if we identify a convex body K with the random variable U_K whose distribution is uniform measure on K, then the entropy of U_K is the logarithm of $|K|$. The parallels were observed early by Costa and Cover [51] (and perhaps also implicitly by Lieb [96]); subsequently, this analogy has been studied by many other authors, including by Dembo-Cover-Thomas [56] and in two series of papers by Lutwak-Yang-Zhang (see, e.g., [102, 101]) and Bobkov-Madiman (see, e.g., [24, 26]).

While this paper is largely focused on the probabilistic (entropic) lifting, we will also discuss how it is related to the functional (integral) lifting.

It is instructive at this point to state the integral and entropic liftings of the Brunn-Minkowski inequality itself, which are known as the Prékopa-Leindler inequality and the Entropy Power Inequality, respectively.

Prékopa-Leindler inequality (PLI): The Prékopa-Leindler inequality (PLI) [131, 92, 132] states that if $f, g, h : \mathbb{R}^d \to [0, \infty)$ are integrable functions satisfying, for a given $\lambda \in (0, 1)$,

$$h(\lambda x + (1 - \lambda)y) \geq f^{\lambda}(x)g^{1-\lambda}(y)$$

for every $x, y \in \mathbb{R}^d$, then

$$\int h \geq \left(\int f\right)^{\lambda} \left(\int g\right)^{1-\lambda}. \tag{2}$$

If one prefers, the PLI can also be written more explicitly as a kind of convolution inequality, as implictly observed in [38] and explicitly in [88]. Indeed, if one defines the Asplund product of two nonnegative functions by

$$(f \star g)(x) = \sup_{x_1+x_2=x} f(x_1)g(x_2),$$

and the scaling $(\lambda \cdot f)(x) = f^{\lambda}(x/\lambda)$, then the left side of (2) can be replaced by the integral of $[\lambda \cdot f] \star [(1 - \lambda) \cdot g]$.

To see the connection with the BMI, one simply has to observe that $f = 1_A, g = 1_B$ and $h = 1_{\lambda A+(1-\lambda)B}$ satisfy the hypothesis, and in this case, the conclusion is precisely the BMI in its "geometric mean" form $|\lambda A + (1 - \lambda)B| \geq |A|^{\lambda}|B|^{1-\lambda}$.

The equivalence of this inequality to the BMI in the form (1) is just one aspect of a broader set of equivalences involving the BMI. To be precise, for the class of Borel-measurable subsets of $\mathbb{R}^d$, the following are equivalent:

$$|A+B|^{\frac{1}{d}} \geq |A|^{\frac{1}{d}} + |B|^{\frac{1}{d}} \tag{3}$$

$$|\lambda A + (1-\lambda)B| \geq \left(\lambda |A|^{\frac{1}{d}} + (1-\lambda)|B|^{\frac{1}{d}}\right)^d \tag{4}$$

$$|\lambda A + (1-\lambda)B| \geq |A|^{\lambda} |B|^{1-\lambda} \tag{5}$$

$$|\lambda A + (1-\lambda)B| \geq \min\{|A|, |B|\}. \tag{6}$$

Let us indicate why the inequalities (3)–(6) are equivalent. Making use of the arithmetic mean-geometric mean inequality, we immediately have (4) $\Rightarrow$ (5) $\Rightarrow$ (6). Applying (3) to $\tilde{A} = \lambda A$, $\tilde{B} = (1-\lambda)B$ we have

$$\begin{aligned}
|\lambda A + (1-\lambda)B| &= |\tilde{A} + \tilde{B}| \\
&\geq (|\tilde{A}|^{\frac{1}{d}} + |\tilde{B}|^{\frac{1}{d}})^d \\
&= \left(|\lambda A|^{\frac{1}{d}} + |(1-\lambda)B|^{\frac{1}{d}}\right)^d \\
&= \left(\lambda |A|^{\frac{1}{d}} + (1-\lambda)|B|^{\frac{1}{d}}\right)^d,
\end{aligned}$$

where the last equality is by homogeneity of the Lebesgue measure. Thus (3) $\Rightarrow$ (4). It remains to prove that (6) $\Rightarrow$ (3). First notice that (6) is equivalent to

$$\begin{aligned}
|A+B| &\geq \min\{|A/\lambda|, |B/(1-\lambda)|\} \\
&= \min\{|A|/\lambda^d, |B|/(1-\lambda)^d\}.
\end{aligned}$$

It is easy to see that the right-hand side is maximized when $|A|/\lambda^d = |B|/(1-\lambda)^d$, or

$$\lambda = \frac{|A|^{\frac{1}{d}}}{|A|^{\frac{1}{d}} + |B|^{\frac{1}{d}}}.$$

Inserting λ into the above yields (3).

Entropy Power Inequality (EPI): In order to state the Entropy Power Inequality (EPI), let us first explain what is meant by entropy power. When random variable $X = (X_1, \ldots, X_d)$ has density $f(x)$ on $\mathbb{R}^d$, the *entropy* of X is

$$h(X) = h(f) := -\int_{\mathbb{R}^d} f(x) \log f(x) dx = \mathbf{E}[-\log f(X)]. \tag{7}$$

This quantity is sometimes called the Shannon-Boltzmann entropy or the differential entropy (to distinguish it from the discrete entropy functional that applies to probability distributions on a countable set). The *entropy power* of X is $N(X) = e^{\frac{2h(X)}{d}}$. As is usual, we abuse notation and write $h(X)$ and $N(X)$, even though these are functionals depending only on the density of X and not on its random realization. The entropy power $N(X) \in [0, \infty]$ can be thought of as a "measure of randomness". It is an (inexact) analogue of volume: if U_A is uniformly distributed on a bounded Borel set A, then it is easily checked that $h(U_A) = \log |A|$ and hence $N(U_A) = |A|^{2/d}$. The reason we don't define entropy power by $e^{h(X)}$ (which would yield a value of $|A|$ for the entropy power of U_A) is that the "correct" comparison is not to uniforms but to Gaussians. This is because just as Euclidean balls are special among subsets of $\mathbb{R}^d$, Gaussians are special among distributions on $\mathbb{R}^d$. Indeed, the reason for the appearance of the functional $|A|^{\frac{1}{d}}$ in the BMI is because this functional is (up to a universal constant) the radius of the ball that has the same volume as A, i.e., $|A|^{\frac{1}{d}}$ may be thought of as (up to a universal constant) the "effective radius" of A. To develop the analogy for random variables, observe that when $Z \sim N(0, \sigma^2 I)$ (i.e., Z has the Gaussian distribution with mean 0 and covariance matrix that is a multiple of the identity), the entropy power of Z is $N(Z) = (2\pi e)\sigma^2$. Thus the entropy power of X is (up to a universal constant) the variance of the isotropic normal that has the same entropy as X, i.e., if $Z \sim N(0, \sigma_Z^2 I)$ and $h(Z) = h(X)$, then

$$N(X) = N(Z) = (2\pi e)\sigma_Z^2.$$

Looked at this way, entropy power is the "effective variance" of a random variable, exactly as volume raised to $1/d$ is the effective radius of a set.

The EPI states that for any two independent random vectors X and Y in $\mathbb{R}^d$ such that the entropies of X, Y and $X + Y$ exist,

$$N(X + Y) \geq N(X) + N(Y).$$

The EPI was stated by Shannon [145] with an incomplete proof; the first complete proof was provided by Stam [148]. The EPI plays an extremely important role in the field of Information Theory, where it first arose and was used (first by Shannon, and later by many others) to prove statements about the fundamental limits of communication over various models of communication channels. Subsequently it has also been recognized as an extremely useful inequality in Probability Theory, with close connections to the logarithmic Sobolev inequality for the Gaussian distribution as well as to the Central Limit Theorem. We will not further discuss these other motivations for the study of the EPI in this paper, although we refer the interested reader to [85, 105] for more on the connections to central limit theorems.

It should be noted that one insightful way to compare the BMI and EPI is to think of the latter as a "99% analogue in high dimensions" of the former, in the sense that looking at most of the Minkowski sum of the supports of a large number of independent copies of the two random vectors effectively yields the EPI via a simple instance of the asymptotic equipartition property or Shannon-McMillan-Breiman theorem. A rigorous argument is given by Szarek and Voiculescu [153] (building on [152]), a short intuitive explanation of which can be found in an answer of Tao to a MathOverflow question[1]. The key idea of [153] is to use not the usual BMI but a "restricted" version of it where it is the exponent $2/d$ rather than $1/d$ that shows up[2].

Rényi entropies. Unified proofs can be given of the EPI and the BMI in two different ways, both of which may be thought of as providing extensions of the EPI to Rényi entropy. We will discuss both of these later; for now, we only introduce the notion of Rényi entropy. For a $\mathbb{R}^d$-valued random variable X with probability density function f, define its Rényi entropy of order p (or simply p-Rényi entropy) by

$$h_p(X) = h_p(f) := \frac{1}{1-p} \log \left(\int_{\mathbb{R}^d} f^p(x) dx \right), \tag{8}$$

if $p \in (0, 1) \cup (1, \infty)$. Observe that, defining h_1 "by continuity" and using l'Hospital's rule, $h_1(X) = h(X)$ is the (Shannon-Boltzmann) entropy. Moreover, by taking limits,

$$h_0(X) = \log |\mathrm{Supp}(f)|,$$
$$h_\infty(X) = -\log \|f\|_\infty,$$

where Supp(f) is the support of f (i.e., the smallest closed set such that f is zero outside it), and $\|f\|_\infty$ is the usual L^∞-norm of f (i.e., the essential supremum with respect to Lebesgue measure). We also define the p-Rényi entropy power by $N_p(X) = e^{\frac{2h_p(X)}{d}}$, so that the usual entropy power $N(X) = N_1(X)$ and for a random variable X whose support is A, $N_0(X) = |A|^{2/d}$.

Conventions. Throughout this paper, *we assume that all random variables considered have densities with respect to Lebesgue measure.* While the entropy of X can be meaningfully set to $-\infty$ when the distribution of X does not possess a density, for the most part we avoid discussing this case. Also, when X has probability density function f, we write $X \sim f$.

For real-valued functions A, B we will use the notation $A \lesssim B$ when $A(z) \leq CB(z)$ for some positive constant C independent of z. For our purposes this will be most interesting when A and B are in some way determined by dimension.

Organization. This survey is organized as follows. In Section 2, we review various statements and variants of the EPI, first for the usual Shannon-Boltzmann entropy in Section 2.2 and then for p-Rényi entropy in Section 2.3, focusing on the ∞-Rényi

---

[1] See http://mathoverflow.net/questions/167951/entropy-proof-of-brunn-minkowski-inequality.

[2] We mention in passing that Barthe [16] also proved a restricted version of the PLI. An analogue of "restriction" for the EPI would involve some kind of weak dependence between summands; some references to the literature on this topic are given later.

case in Section 2.4. In Section 3, we explore what can be said about inequalities that go the other way, under convexity constraints on the probability measures involved. We start by recalling the notions of κ-concave measures and functions in Section 3.1. In Section 3.2, we discuss reverse EPIs that require invoking a linear transformation (analogous to the reverse Brunn-Minkowski inequality of V. Milman), and explicit choices of linear transformations that can be used are discussed in Section 3.6. The three intermediate subsections focus on three different approaches to reverse Rényi EPIs that do not require invoking a linear transformation. Finally we discuss the relationship between integral and entropic liftings, in the context of the Blashke-Santaló inequality in Section 4, and end with some concluding remarks on nonlinear and discrete analogs in Section 5.

2 Entropy Power Inequalities

2.1 Some Basic Observations

Before we discuss more sophisticated results, let us recall some basic properties of Rényi entropy.

Theorem 2.1 *For independent $\mathbb{R}^d$-valued random variables X and Y, and any $p \in [0, \infty]$,*

$$N_p(X + Y) \geq \max\{N_p(X), N_p(Y)\}.$$

Proof Let $X \sim f$ and $Y \sim g$. For $p \in (1, \infty)$, we have the following with the inequality delivered by Jensen's inequality:

$$\begin{aligned}\int (f * g)^p(x)dx &= \int \left(\int f(x-y)g(y)dy\right)^p dx \\ &\leq \int\int f^p(x-y)g(y)dydx \\ &= \int \left(\int f^p(x-y)dx\right) g(y)dy \\ &= \int f^p(x)dx.\end{aligned}$$

Inserting the inequality into the order reversing function $\varphi(z) = z^{\frac{2}{d(1-p)}}$ we have our result.

The case that $p \in (0, 1)$ is similar, making note that now z^p is concave while $z^{\frac{2}{d(1-p)}}$ is order preserving. For $p = 1$, we can give a probabilistic proof: applying the nonnegativity of mutual information, which in particular implies that conditioning reduces entropy (see, e.g., [54]),

$$h(X+Y) \geq h(X+Y|Y) = h(X|Y) = h(X),$$

where we used translation-invariance of entropy for the first equality and independence of X and Y for the second. For $p = 0$, the conclusion simply follows by the fact that $|A + B| \geq \max\{|A|, |B|\}$ for any nonempty Borel sets A and B; this may be seen by translating B so that it contains 0, which does not affect any of the volumes and in which case $A + B \supset A$. For $p = \infty$, the conclusion follows from Hölder's inequality:

$$\int f(x-y)g(y)dy \leq \|g\|_\infty \|f\|_1 = \|g\|_\infty.$$

Thus we have the theorem for all values of $p \in [0, \infty]$. □

We now observe that for any fixed random vector, the Rényi entropy of order p is non-increasing in p.

Lemma 2.2 *For a $\mathbb{R}^d$-valued random variable X, and $0 \leq q < p \leq \infty$, we have*

$$N_q(X) \geq N_p(X).$$

Proof The result follows by expressing, for $X \sim f$,

$$h_p(X) = \frac{\log(\int f^p)}{1-p} = -\log \mathbb{E}\|f(X)\|_{p-1}$$

and using the "increasingness" of p-norms on probability spaces, which is nothing but an instance of Hölder's inequality. □

Definition 2.3 *A function $f : \mathbb{R}^d \to [0, \infty)$ is said to be log-concave if*

$$f(\alpha x + (1-\alpha)y) \geq f(x)^\alpha f(y)^{1-\alpha}, \tag{9}$$

for each $x, y \in \mathbb{R}^d$ and each $0 \leq \alpha \leq 1$.

If a probability density function f is log-concave, we will also use the adjective "log-concave" for a random variable X distributed according to f, and for the probability measure induced by it. Log-concavity has been deeply studied in probability, statistics, optimization and geometry, and is perhaps the most natural notion of convexity for probability density functions.

In general, the monotonicity of Lemma 2.2 relates two different Rényi entropies of the same distribution in one direction, but there is no reason for a bound to exist in the other direction. Remarkably, for log-concave random vectors, all Rényi entropies are comparable in both directions.

Lemma 2.4 ([115]) *If a random variable X in $\mathbb{R}^d$ has log-concave density f, then for $p \geq q > 0$,*

$$h_q(f) - h_p(f) \leq d\frac{\log q}{q-1} - d\frac{\log p}{p-1},$$

with equality iff $f(x) = e^{-\sum_{i=1}^{d} x_i}$ *on the positive orthant and 0 elsewhere.*

This lemma generalizes the following sharp inequality for log-concave distributions obtained in [25]:

$$h(X) \leq d + h_\infty(X). \tag{10}$$

In fact, Lemma 2.4 has an extension to the larger class (discussed later) of s-concave measures with $s < 0$; preliminary results in this direction are available in [25] and sharp results obtained in [22].

2.2 The Shannon-Stam EPI and Its Variants

2.2.1 The Basic EPI

The EPI has several equivalent formulations; we collect these together with minimal conditions below.

Theorem 2.5 *Suppose X and Y are independent* $\mathbb{R}^d$*-valued random variables such that* $h(X)$, $h(Y)$ *and* $h(X+Y)$ *exist. Then the following statements, which are equivalent to each other, are true:*

1. *We have*

$$N(X+Y) \geq N(X) + N(Y). \tag{11}$$

2. *For any* $\lambda \in [0, 1]$,

$$h(\sqrt{\lambda}X + \sqrt{1-\lambda}Y) \geq \lambda h(X) + (1-\lambda)h(Y). \tag{12}$$

3. *Denoting by* X^G *and* Y^G *independent, isotropic*[3], *Gaussian random variables with* $h(X^G) = h(X)$ *and* $h(Y^G) = h(Y)$, *one has*

$$h(X+Y) \geq h(X^G + Y^G). \tag{13}$$

In (11) *and* (13), *equality holds if and only if X and Y are Gaussian random variables with proportional covariance matrices; in* (12), *equality holds if and only if X and Y are Gaussian random variables with the same covariance matrix.*

[3] By isotropic here, we mean spherical symmetry, or equivalently, that the covariance matrix is taken to be a scalar multiple of the identity matrix.

Proof First let us show that we can assume $h(X), h(Y) \in (-\infty, \infty)$. By Theorem 2.1 we can immediately obtain $h(X+Y) \geq \max\{h(X), h(Y)\}$. It follows that all three inequalities hold immediately in the case that $\max\{h(X), h(Y)\} = \infty$. Now assume that neither $h(X)$ nor $h(Y)$ take the value $+\infty$ and consider $\min\{h(X), h(Y)\} = -\infty$. In this situation, the inequalities (11) and (12) are immediate. For (13), in the case that $h(X) = -\infty$ we interpret X^G as a Dirac point mass, and hence $h(X^G + Y^G) = h(Y^G) = h(Y) \leq h(X+Y)$.

We now proceed to prove the equivalences.

(11) ⇒ (12): Apply (11), substituting X by $\sqrt{\lambda}X$ and Y by $\sqrt{1-\lambda}Y$ and use the homogeneity of entropy power to obtain

$$N(\sqrt{\lambda}X + \sqrt{1-\lambda}Y) \geq \lambda N(X) + (1-\lambda)N(Y).$$

Apply the AM-GM inequality to the right-hand side and conclude by taking logarithms.

(12) ⇒ (13): Applying (12) in its exponentiated form $N(\sqrt{\lambda}X + \sqrt{1-\lambda}Y) \geq N^{\lambda}(X)N^{1-\lambda}(Y)$ after writing $X+Y = \sqrt{\lambda}(X/\sqrt{\lambda}) + \sqrt{1-\lambda}(Y/\sqrt{1-\lambda})$ we obtain

$$N(X+Y) \geq \left(N\left(\frac{X}{\sqrt{\lambda}}\right)\right)^{\lambda} \left(N\left(\frac{Y}{\sqrt{1-\lambda}}\right)\right)^{1-\lambda}.$$

Making use of the identity $N(X^G + Y^G) = N(X^G) + N(Y^G)$ and homogeneity again, we can evaluate the right-hand side at $\lambda = N(X^G)/N(X^G + Y^G)$ to obtain exactly $N(X^G + Y^G)$, recovering the exponentiated version of (13).

(13) ⇒ (11): Using the exponentiated version of (13),

$$N(X+Y) \geq N(X^G + Y^G) = N(X^G) + N(Y^G) = N(X) + N(Y).$$

Observe from the proof that a strict inequality in one statement implies a strict inequality in the rest.

What is left is to prove any of the 3 statements of the EPI when the entropies involved are finite. There are many proofs of this available in the literature (see, e.g., [148, 20, 96, 56, 153, 134]), and we will not detail any here, although we later sketch a proof via the sharp form of Young's convolution inequality. □

The conditions stated above cannot be relaxed, as observed by Bobkov and Chistyakov [30], who construct a distribution whose entropy exists but such that the entropy of the self-convolution does not exist. This, in particular, shows that the assumption for validity of the EPI stated, for example, in [56] is incomplete–existence of just $h(X)$ and $h(Y)$ is not sufficient. It is also shown in [30], however, that for any example where $h(X)$ exists but $h(X+X')$ does not (with X' an i.i.d. copy of X), necessarily $h(X) = -\infty$, so that it remains true that if the entropy exists and is a real number, then the entropy of the self-convolution also exists. They also have other interesting examples of the behavior of entropy on convolution: [30, Example 1] constructs a distribution with entropy $-\infty$ such that the entropy of the self-convolution is a real number, and [30, Proposition 5] constructs a distribution with finite entropy such that its convolution with any distribution of finite entropy has infinite entropy.

2.2.2 Fancier Versions of the EPI

Many generalizations and improvements of the EPI exist. For three or more independent random vectors X_i, the EPI trivially implies that

$$N(X_1 + \cdots + X_n) \geq \sum_{i=1}^{n} N(X_i), \tag{14}$$

with equality if and only if the random vectors are Gaussian and their covariance matrices are proportional to each other. In fact, it turns out that this can be refined, as shown by S. Artstein, K. Ball, Barthe and Naor [4]:

$$N\left(\sum_{i=1}^{n} X_i\right) \geq \frac{1}{n-1} \sum_{j=1}^{n} N\left(\sum_{i \neq j} X_i\right). \tag{15}$$

This implies the monotonicity of entropy in the Central Limit Theorem, which suggests that quantifying the Central Limit Theorem using entropy or relative entropy is a particularly natural approach. More precisely, if $X_1, \ldots, X_n$ are independent and identically distributed (i.i.d.) square-integrable random vectors, then

$$h\left(\frac{X_1 + \cdots + X_n}{\sqrt{n}}\right) \leq h\left(\frac{X_1 + \cdots + X_{n-1}}{\sqrt{n-1}}\right). \tag{16}$$

Simpler proofs of (15) were given independently by [106, 146, 160]. Generalizations of (15) to arbitrary collections of subsets on the right side were given by [107, 108], and some further fine properties of the kinds of inequalities that hold for the entropy power of a sum of independent random variables were revealed in [109]. Let us mention a key result of this type due to [108]. For a collcction $\mathcal{C}$ of nonempty subsets of $[n] := \{1, 2, \cdots, n\}$, a function $\beta : \mathcal{C} \to \mathbb{R}_+$ is called a *fractional partition*[4] if for each $i \in [n]$, we have $\sum_{s \in \mathcal{C}: i \in s} \beta_s = 1$. Then the entropy power of convolutions is fractionally superadditive, i.e., if $X_1, \ldots, X_n$ are independent $\mathbb{R}^d$-valued random variables, one has

$$N\left(\sum_{i=1}^{n} X_i\right) \geq \sum_{s \in \mathcal{C}} \beta_s N\left(\sum_{i \in s} X_i\right).$$

[4] If there exists a fractional partition β for $\mathcal{C}$ that is $\{0, 1\}$-valued, then β is the indicator function for a partition of the set $[n]$ using a subset of $\mathcal{C}$; hence the terminology.

This yields the usual EPI by taking $\mathcal{C}$ to be the collection of all singletons and $\beta_s \equiv 1$, and the inequality (15) by taking $\mathcal{C}$ to be the collection of all sets of size $n-1$ and $\beta_s = \frac{1}{n-1}$.

For i.i.d. summands in dimension 1, [3] and [82] prove an upper bound of the relative entropy between the distribution of the normalized sum and that of a standard Gaussian random variable. To be precise, suppose $X_1, \ldots, X_n$ are independent copies of a random variable X with $\mathrm{Var}(X) = 1$, and the density of X satisfies a Poincaré inequality with constant c, i.e., for every smooth function s,

$$c\mathrm{Var}(s(X)) \leq \mathbf{E}[\{s'(X)\}^2].$$

Then, for every $a \in \mathbb{R}^n$ with $\sum_{i=1}^n a_i^2 = 1$ and $\alpha(a) := \sum_{i=1}^n a_i^4$,

$$h(G) - h\left(\sum_{i=1}^n a_i X_i\right) \leq \frac{\alpha(a)}{\frac{c}{2} + (1 - \frac{c}{2})\alpha(a)} \left(h(G) - h(X)\right), \tag{17}$$

where G is a standard Gaussian random variable. Observe that this refines the EPI since taking $c = 0$ in the inequality (17) gives the EPI in the second form of Theorem 2.5. On the other hand, specializing (17) to $n = 2$ with $a_1 = a_2 = \frac{1}{\sqrt{2}}$, one obtains a lower bound for $h\big(\frac{X_1+X_2}{\sqrt{2}}\big) - h(X)$ in terms of the relative entropy $h(G) - h(X)$ of X from Gaussianity. Ball and Nguyen [13] develop an extension of this latter inequality to general dimension under the additional assumption of log-concavity.

It is natural to ask if the EPI can be refined by introducing an error term that quantifies the gap between the two sides in terms of how non-Gaussian the summands are. Such estimates are referred to as "stability estimates" since they capture how stable the equality condition for the inequality is, i.e., whether closeness to Gaussianity is guaranteed for the summands if the two sides in the inequality are not exactly equal but close to each other. For the EPI, the first stability estimates were given by Carlen and Soffer [42], but these are qualitative and not quantitative (i.e., they do not give numerical bounds on distance from Gaussianity of the summands when there is near-equality in the EPI, but they do assert that this distance must go to zero as the deficit in the inequality goes to zero). Recently Toscani [159] gave a quantitative stability estimate when the summands are restricted to have log-concave densities: For independent random vectors X and Y with log-concave densities,

$$N(X+Y) \geq (N(X) + N(Y))\, R(X, Y), \tag{18}$$

where the quantity $R(X, Y) \geq 1$ is a somewhat complicated quantity that we do not define here and can be interpreted as a measure of non-Gaussianity of X and Y. Indeed, [159] shows that $R(X, Y) = 1$ if and only if X and Y are Gaussian random vectors, but leaves open the question of whether $R(X, Y)$ can be related to some more familiar distance from Gaussianity. Even more recently, Courtade, Fathi and

Pananjady [53] showed that if X and Y are uniformly log-concave (in the sense that the densities of both are of the form e^{-V} with the Hessian of V bounded from below by a positive multiple of the identity matrix), then the deficit in the EPI is controlled in terms of the quadratic Wasserstein distances between the distributions of X and Y and Gaussianity.

There are also strengthenings of the EPI when one of the summands is Gaussian. Set $X^{(t)} = X + \sqrt{t}Z$, with Z a standard Gaussian random variable independent of X. Costa [50] showed that for any $t \in [0, 1]$,

$$N(X^{(t)}) \geq (1-t)N(X) + tN(X+Z). \tag{19}$$

This may be rewritten as $N(X^{(t)}) - N(X) \geq t[N(X+Z) - N(X)] = N(\sqrt{t}X + \sqrt{t}Z) - N(\sqrt{t}X)$. Setting $\beta = \sqrt{t}$, we have for any $\beta \in [0, 1]$ that $N(X + \beta Z) - N(X) \geq N(\beta X + \beta Z) - N(\beta X)$, substituting X by βX, we get

$$N(X+Z) - N(X) \geq N(\beta X + Z) - N(\beta X). \tag{20}$$

for any $\beta \in [0, 1]$. Therefore, for any $\beta, \beta' \in [0, 1]$ with $\beta > \beta'$, substitute X by βX and β by β'/β in (20), we have

$$N(\beta X + Z) - N(\beta X) \geq N(\beta' X + Z) - N(\beta' X).$$

In other words, Costa's result states that if $A(\beta) = N(\beta X + Z) - N(\beta X)$, then $A(\beta)$ is a monotonically increasing function for $\beta \in [0, 1]$. To see that this is a refinement of the EPI in the special case when one summand is Gaussian, note that the EPI in this case is the statement that $A(1) \geq A(0)$. An alternative proof of Costa's inequality was given by Villani [161]; for a generalization, see [128].

Very recently, a powerful extension of Costa's inequality was developed by Courtade [52], applying to a system in which $X, X+Z, V$ form a Markov chain (i.e., X and V are conditionally independent given $X+Z$) and Z is a Gaussian random vector independent of X. Courtade's result specializes in the case where $V = X + Z + Y$ to the following: If X, Y, Z be independent random vectors in $\mathbb{R}^d$ with Z being Gaussian, then

$$N(X+Z)N(Y+Z) \geq N(X)N(Y) + N(X+Y+Z)N(Z). \tag{21}$$

Applying the inequality (21) to X, $\sqrt{1-t}Z'$ and $\sqrt{t}Z$ where Z' is the independent copy of the standard normal distribution Z, we have

$$N(X^{(t)})N(\sqrt{1-t}Z' + \sqrt{t}Z) \geq N(X)N(\sqrt{1-t}Z') + N(X + \sqrt{1-t}Z' + \sqrt{t}Z)N(\sqrt{t}Z).$$

By the fact that $\sqrt{1-t}Z' + \sqrt{t}Z$ has the same distribution as Z, and by the fact that $N(Z) = 1$, we have $N(X^{(t)}) \geq (1-t)N(X) + tN(X+Z)$, which is Costa's inequality (19).

Motivated by the desire to prove entropic central limit theorems for statistical physics models, some extensions of the EPI to dependent summands have also been considered (see, e.g., [42, 154, 155, 79, 80]), although the assumptions tend to be quite restrictive for such results.

Finally there is an extension of the EPI that applies not just to sums but also to more general linear transformations applied to independent random variables. The main result of Zamir and Feder [171] asserts that if $X_1, \ldots, X_n$ are independent real-valued random variables, $Z_1, \ldots, Z_n$ are independent Gaussian random variables satisfying $h(Z_i) = h(X_i)$, and A is any matrix, then $h(AX) \geq h(AZ)$ where AX represents the left-multiplication of the vector X by the matrix A. As explained in [171], for this result to be nontrivial, the $m \times n$ matrix A must have $m < n$ and be of full rank. To see this, notice that if $m > n$ or if A is not of full rank, the vector AX does not have full support on $\mathbb{R}^m$ and $h(AX) = h(AZ) = -\infty$, while if $m = n$ and A is invertible, $h(AX) = h(AZ)$ holds with equality because of the conditions determining Z and the way entropy behaves under linear transformations.

2.3 Rényi Entropy Power Inequalities

2.3.1 First Rényi Interpolation of the EPI and BMI

Unified proofs can be given of the EPI and the BMI in different ways, each of which may be thought of as providing extensions of the EPI to Rényi entropy.

The first unified approach is via Young's inequality. Denote by L^p the Banach space $L^p(\mathbb{R}^d, dx)$ of measurable functions defined on $\mathbb{R}^d$ whose p-th power is integrable with respect to Lebesgue measure dx. In 1912, Young [167] introduced the fundamental inequality

$$\|f \star g\|_r \leq \|f\|_p \|g\|_q, \quad \frac{1}{p} + \frac{1}{q} = \frac{1}{r} + 1, \quad 1 < p, q, r < +\infty, \tag{22}$$

for functions $f \in L^p$ and $g \in L^q$, which implies that if two functions are in (possibly different) L^p-spaces, then their convolution is contained in a third L^p-space. In 1972, Leindler [91] showed the so-called reverse Young inequality, referring to the fact that the inequality (22) is reversed when $0 < p, q, r < 1$. The best constant that can be put on the right side of (22) or its reverse was found by Beckner [18]: the best constant is $(C_p C_q / C_r)^d$, where

$$C_p^2 = \frac{p^{\frac{1}{p}}}{|p'|^{\frac{1}{p'}}}, \tag{23}$$

and for any $p \in (0, \infty]$, p' is defined by

$$\frac{1}{p} + \frac{1}{p'} = 1. \tag{24}$$

Note that p' is positive for $p \in (1, \infty)$, and negative for $p \in (0, 1)$. Alternative proofs of both Young's inequality and the reverse Young inequality with this sharp constant were given by Brascamp and Lieb [37], Barthe [15], and Cordero-Erausquin and Ledoux [48].

We state the sharp Young and reverse Young inequalities now for later reference.

Theorem 2.6 ([18]) *Suppose $r \in (0, 1)$ and $p_i \in (0, 1)$ satisfy*

$$\sum_{i=1}^{n} \frac{1}{p_i} = n - \frac{1}{r'}. \tag{25}$$

Then, for any functions $f_j \in L^{p_j}$ ($j = 1, \ldots, n$),

$$\left\| \star_{j\in[n]} f_j \right\|_r \geq \frac{1}{C_r^d} \prod_{j\in[n]} [C_{p_j}^d \|f_j\|_{p_j}]. \tag{26}$$

The inequality is reversed if $r \in (1, \infty)$ and $p_i \in (1, \infty)$.

Dembo, Cover and Thomas [56] interpret the Young and reverse Young inequalities with sharp constant as EPIs for the Rényi entropy. If X_i are random vectors in $\mathbb{R}^d$ with densities f_i respectively, taking the logarithm of (26) and rewriting the definition of the Rényi entropy power as $N_p(X) = \|f\|_p^{-2p'/d}$, we have

$$\frac{d}{2r'} \log N_r\left(\sum_{i\in[n]} X_i \right) \leq d \log C_r - d \sum_{i\in[n]} \log C_{p_i} + \sum_{i\in[n]} \frac{d}{2p_i'} \log N_{p_i}(X_i). \tag{27}$$

Introduce two discrete probability measures λ and κ on $[n]$, with probabilities proportional to $1/p_i'$ and $1/p_i$, respectively. Setting $L_r = rn - r + 1 = r(n - 1/r')$, the condition (25), allows us to write explicitly

$$\kappa_i = \left(\frac{r}{L_r}\right)\frac{1}{p_i},$$

$$\lambda_i = \frac{r'}{p_i'},$$

for each $i \in [n]$, also using $1/p_i + 1/p_i' = 1$ for the latter. Then (27) reduces to

$$h_r(Y_{[n]}) \geq \frac{dr'}{2} \log C_r^2 - \frac{dr'}{2} \sum_{i\in[n]} \log C_{p_i}^2 + \sum_{i\in[n]} \lambda_i h_{p_i}(X_i).$$

Now, some straightforward calculations show that if we take the limit as $p_i, r \to 0$ from above, we get the BMI, while if we take the limit as $p_i, r \to 1$, we get the EPI (this was originally observed by Lieb [96]).

2.3.2 Second Rényi Interpolation of the EPI and BMI

Wang and Madiman [162] found a rearrangement-based refinement of the EPI that also applies to Rényi entropies. For a Borel set A, define its spherically decreasing symmetric rearrangement A^* by

$$A^* := B(0, r),$$

where $B(0, r)$ stands for the open ball with radius r centered at the origin and r is determined by the condition that $B(0, r)$ has volume $|A|$. Here we use the convention that if $|A| = 0$ then $A^* = \emptyset$ and that if $|A| = \infty$ then $A^* = \mathbb{R}^d$. Now for a measurable non-negative function f, define its spherically decreasing symmetric rearrangement f^* by

$$f^*(y) := \int_0^\infty \mathbb{1}_{\{y \in B_t^*\}} dt,$$

where $B_t := \{x : f(x) > t\}$. It is a classical fact (see, e.g., [39]) that rearrangement preserves L^p-norms, i.e., $\|f^*\|_p = \|f\|_p$. In particular, if f is a probability density function, so is f^*. If $X \sim f$, denote by X^* a random variable with density f^*, then the rearrangement-invariance of L^p-norms immediately implies that $h_p(X^*) = h_p(X)$ for each $p \in [0, \infty]$ (for $p = 1$, this is not done directly but via a limiting argument).

Theorem 2.7 ([162]) *Let $X_1, \ldots, X_n$ be independent $\mathbb{R}^d$-valued random vectors. Then*

$$h_p(X_1 + \ldots + X_n) \geq h_p(X_1^* + \ldots + X_n^*) \tag{28}$$

for any $p \in [0, \infty]$, provided the entropies exist.

In particular,

$$N(X + Y) \geq N(X^* + Y^*), \tag{29}$$

where X and Y are independent random vectors with density functions f and g, respectively, and X^* and Y^* are independent random vectors with density function f^* and g^*, respectively. Thanks to (29), we have effectively inserted an intermediate term in between the two sides of the formulation (13) of the EPI.

$$N(X + Y) \geq N(X^* + Y^*) \geq N(X^G + Y^G),$$

where the second inequality is by the fact that $h(X^G) = h(X^*) = h(X)$, combined with the third equivalent form of the EPI in Theorem 2.5. In fact, it is also shown in [162] that the EPI itself can be deduced from (29).

2.3.3 A Conjectured Rényi EPI

Let us note that neither of the above unifications of BMI and EPI via Rényi entropy directly gives a sharp bound on $N_p(X+Y)$ in terms of $N_p(X)$ and $N_p(Y)$. The former approach relates Rényi entropy powers of different indices, while the latter refines the third formulation in Theorem 2.1 (but not the first, because the equivalence that held for Shannon-Boltzmann entropy does not work in the Rényi case). The question of finding a sharp direct relationship between $N_p(X+Y)$ with $N_p(X)$ and $N_p(Y)$ remains open, with some non-sharp results for the $p > 1$ case obtained by Bobkov and Chistyakov [30], whose argument and results were recently tightened by Ram and Sason [133].

Theorem 2.8 ([133]) *For $p \in (1, \infty)$ and independent random vectors X_i with densities in $\mathbb{R}^d$,*

$$N_p(X_1 + \cdots + X_n) \geq c_p^{(n)} \sum_{i=1}^{n} N_p(X_i),$$

where $p' = p/(p-1)$ and

$$c_p^{(n)} = p^{\frac{1}{p-1}} \left(1 - \frac{1}{np'}\right)^{np'-1} \geq \frac{1}{e}.$$

We now discuss a conjecture of Wang and Madiman [162] about extremal distributions for Rényi EPIs of this sort. Consider the one-parameter family of distributions, indexed by a parameter $-\infty < \beta \leq \frac{2}{d+2}$, of the following form: g_0 is the standard Gaussian density in $\mathbb{R}^d$, and for $\beta \neq 0$,

$$g_\beta(x) = A_\beta \left(1 - \frac{\beta}{2}\|x\|^2\right)_+^{\frac{1}{\beta} - \frac{d}{2} - 1},$$

where A_β is a normalizing constant (which can be written explicitly in terms of gamma functions). We call g_β the *standard generalized Gaussian* of order β; any affine function of a standard generalized Gaussian yields a "generalized Gaussian". The densities g_β (apart from the obviously special value $\beta = 0$) are easily classified into two distinct ranges where they behave differently. *First*, for $\beta < 0$, the density is proportional to a negative power of $(1 + b\|x\|^2)$ for a positive constant b, and therefore corresponds to measures with full support on $\mathbb{R}^d$ that are heavy-tailed. For $\beta > 0$, note that $(1 - b\|x\|^2)_+$ with positive b is non-zero only for $\|x\| < b^{-\frac{1}{2}}$, and is concave in this region. Thus any density in the *second* class, corresponding to $0 < \beta \leq \frac{2}{d+2}$, is a positive power of $(1 - b\|x\|^2)_+$, and is thus a concave function supported on a centered Euclidean ball of finite radius. It is pertinent to note that although the first class includes many distributions from what one might call the "Cauchy family", it excludes the standard Cauchy distribution; indeed, not only do all the generalized Gaussians defined above have finite variance, but in fact the form has been chosen so that, for $Z \sim g_\beta$,

$$\mathbf{E}[\|Z\|^2] = d$$

for any β. The generalized Gaussians have been called by different names in the literature, including Barenblatt profiles, or the Student-r distributions ($\beta < 0$) and Student-t distributions ($0 < \beta \le \frac{2}{d+2}$).

For $p > \frac{d}{d+2}$, define β_p by

$$\frac{1}{\beta_p} = \frac{1}{p-1} + \frac{d+2}{2},$$

and write $Z^{(p)}$ for a random vector drawn from g_{β_p}. Note that β_p ranges from $-\infty$ to $\frac{2}{d+2}$ as p ranges from $\frac{d}{d+2}$ to ∞. The generalized Gaussians $Z^{(p)}$ arise naturally as the maximizers of the Rényi entropy power of order p under a variance constraint, as independently observed by Costa, Hero and Vignat [49] and Lutwak, Yang and Zhang [103]. They play the starring role in the conjecture of Wang and Madiman [162].

Conjecture 2.9 ([162]) *Let $X_1, \ldots, X_n$ be independent random vectors taking values in $\mathbb{R}^d$, and $p > \frac{d}{d+2}$. Suppose Z_i are independent random vectors, each a scaled version of $Z^{(p)}$. such that $h_p(X_i) = h_p(Z_i)$. Then*

$$N_p(X_1 + \ldots + X_n) \ge N_p(Z_1 + \ldots + Z_n).$$

Until very recently, this conjecture was only known to be true in the case where $p = 1$ (when it is the classical EPI) and the case where $p = \infty$ and $d = 1$ (which is due to Rogozin [137] and discussed in Section 2.4). In [114], we have very recently been able to prove Conjecture 2.9 for $p = \infty$ and any finite dimension d, generalizing Rogozin's inequality. All other cases remain open.

2.3.4 Other Work on Rényi Entropy Power Inequalities

Johnson and Vignat [84] also demonstrated what they call an "entropy power inequality for Rényi entropy", for any order $p \ge 1$. However, their inequality does not pertain to the usual convolution, but a new and somewhat complicated convolution operation (depending on p). This new operation reduces to the usual convolution for $p = 1$, and has the nice property that the convolution of affine transforms of independent copies of $Z^{(p)}$ is an affine transform of $Z^{(p)}$ (which fails for the usual convolution when $p > 1$).

As discussed earlier, Costa [50] proved a strengthening of the classical EPI when one of the summands is Gaussian. Savaré and Toscani [143] recently proposed a generalization of Costa's result to Rényi entropy power, but the notion of concavity they use based on solutions of a nonlinear heat equation does not have obvious probabilistic meaning. Curiously, it turns out that the definition of Rényi entropy power appropriate for the framework of [143] has a different constant in the exponent ($\frac{2}{d}+p-1$ as opposed to $\frac{2}{d}$). Motivated by [143], Bobkov and Marsiglietti [27] very recently proved Rényi entropy power inequalities with non-standard exponents. Their main result may be stated as follows.

Theorem 2.10 ([27]) *For $p \in (1,\infty)$ and independent random vectors X_i with densities in $\mathbb{R}^d$,*

$$\tilde{N}_p(X_1+\cdots+X_n) \geq \sum_{i=1}^{n} \tilde{N}_p(X_i),$$

where

$$\tilde{N}_p(X) = e^{\frac{p+1}{d}h_p(X)}.$$

It would be interesting to know if Theorem 2.10 is true for $p \in [0,1)$ (and hence all $p \geq 0$), since this would be a particularly nice interpolation between the BMI and EPI.

It is natural to look for Rényi entropy analogues of the refinements and generalizations of the EPI discussed in Section 2.2.2. While little has been done in this direction for general Rényi entropies (apart from the afore-mentioned work of [143]), the case of the Rényi entropy of order 0 (i.e., inequalities for volumes of sets)– which is, of course, of special interest– has attracted some study. For example, Zamir and Feder [172] demonstrated a nontrivial version of the BMI for sums of the form $v_1A_1 + \ldots v_kA_k$, where A_i are unit length subsets of $\mathbb{R}$ and v_i are vectors in $\mathbb{R}^d$, showing that the volume of the Minkowski sum is minimized when each A_i is an interval (i.e., the sum is a zonotope). This result was motivated by analogy with the "matrix version" of the EPI discussed earlier.

Indeed, the strong parallels between the BMI and the EPI might lead to the belief that every volume inequality for Minkowski sums has an analogue for entropy of convolutions, and vice versa. However, this turns out not to be the case. It was shown by Fradelizi and Marsiglietti [68] that the analogue of Costa's result (19) on concavity of entropy power, namely the assertion that $t \mapsto |A + tB_2^d|^{\frac{1}{d}}$ is concave for positive t and any given Borel set A, fails to hold[5] even in dimension 2. Another conjecture in this spirit that was made independently by V. Milman (as a generalization of Bergstrom's determinant inequality) and by Dembo, Cover and Thomas [56] (as an analogue of Stam's Fisher information inequality, which is closely related to the EPI) was disproved by Fradelizi, Giannopoulos and Meyer [63]. In [32], it was conjectured that analogues of fractional EPIs such as (15) hold

[5]They also showed some partial positive results– concavity holds in dimension 2 for connected sets, and in general dimension on a subinterval $[t_0,\infty)$ under some regularity conditions.

for volumes, and it was observed that this is indeed the case for convex sets. If this conjecture were true for general compact sets, it would imply that for any compact set, the volumes of the Minkowski self-averages (obtained by taking the Minkowski sum of k copies of the set, and scaling by $1/k$) are monotonically increasing[6] in k. However, [65] showed that this conjecture does not hold[7] in general– in fact, they showed that there exist many compact sets A in $\mathbb{R}^d$ for any $d \geq 12$ such that $|A + A + A| < (\frac{3}{2})^d |A + A|$. Finally while volumes of Minkowski sums of convex sets in $\mathbb{R}^d$ are supermodular (as shown in [66]), entropy powers of convolutions of log-concave densities fail to be supermodular even in dimension 1 (as shown in [109]). Thus the parallels between volume inequalities and entropy inequalities are not exact.

Another direction that has seen considerable exploration in recent years is stability of the BMI. This direction began with stability estimates for the BMI in the case where the two summands are convex sets [57, 73, 61, 62, 144][8], asserting that near-equality in the BMI implies that the summands are nearly homothetic. For general Borel sets, qualitative stability (i.e., that closeness to equality entails closeness to extremizers) was shown by Christ [45, 44], with the first quantitative estimates recently developed by Figalli and Jerison [60]. Qualitative stability for the more general Young's inequality has also been recently considered [43], but quantitative estimates are unknown to the extent of our knowledge.

2.4 An EPI for Rényi Entropy of Order ∞

In discussing Rényi entropy power inequalities, it is of particular interest to consider the case of $p = \infty$, because of close connections with the literature in probability theory on small ball estimates and the so-called Lévy concentration functions [127, 58], which in turn have applications to a number of areas including stochastic process theory [95] and random matrix theory [138, 158, 139].

Observe that by Theorem 2.1 we trivially have

$$N_\infty(X + Y) \geq \max\{N_\infty(X), N_\infty(Y)\} \geq \frac{1}{2}(N_\infty(X) + N_\infty(Y)). \tag{30}$$

[6]The significance of this arises from the fact that the Minkowski self-averages of any compact set converge in Hausdorff distance to the convex hull of the set, and furthermore, one also has convergence of the volumes if the original compact set had nonempty interior. Various versions of this fact were proved independently by Emerson and Greenleaf [59], and by Shapley, Folkmann and Starr [150]; a survey of such results including detailed historical remarks can be found in [66].

[7]On the other hand, partial positive results quantifying the convexifying effect of Minkowski summation were obtained in [65, 66].

[8]There is also a stream of work on stability estimates for other geometric inequalities related to the BMI, such as the isoperimetric inequality, but this would take us far afield.

In fact, the constant $\frac{1}{2}$ here is sharp, as uniform distributions on any symmetric convex set K (i.e., K is convex, and $x \in K$ if and only if $-x \in K$) of volume 1 are extremal: if X and X' are independently distributed according to $f = \mathbb{1}_K$, then denoting the density of $X - X'$ by u, we have

$$\|u\|_\infty = u(0) = \int f^2(x)dx = 1 = \|f\|_\infty,$$

so that $N_\infty(X + X') = N_\infty(X - X') = N_\infty(X) = \frac{1}{2}[N_\infty(X) + N_\infty(X')]$.

What is more, it is observed in [30] that when each X_i is real-valued, $1/2$ is the optimal constant for any number of summands.

Theorem 2.11 ([30]) *For independent, real-valued random variables $X_1, \ldots, X_n$,*

$$N_\infty\left(\sum_{i=1}^n X_i\right) \geq \frac{1}{2}\sum_{i=1}^n N_\infty(X_i).$$

The constant $1/2$ clearly cannot be improved upon (one can take $X_3, \ldots, X_n$ to be deterministic and the result follows from the $n = 2$ case). That one should have this sort of scaling in n for the lower bound (namely, linear in n when the summands are identically distributed with bounded densities) is not so obvious from the trivial maximum bound above. The proof of Theorem 2.11 draws on two theorems, the first due to Rogozin [137], which reduces the general case to the cube, and the second a geometric result on cube slicing due to K. Ball [9].

Theorem 2.12 ([137]) *Let $X_1, \ldots, X_n$ be independent $\mathbb{R}$-valued random variables with bounded densities. Then*

$$N_\infty(X_1 + \cdots + X_n) \geq N_\infty(Y_1 + \cdots + Y_n), \tag{31}$$

where $Y_1, \ldots, Y_n$ are a collection of independent random variables, with Y_i chosen to be uniformly distributed on a symmetric interval such that $N_\infty(Y_i) = N_\infty(X_i)$.

Theorem 2.13 ([9]) *Every section of the unit cube $[-\frac{1}{2}, \frac{1}{2}]^d$, denoted Q_d, by a $(d-1)$-dimensional subspace has volume bounded above by $\sqrt{2}$. This upper bound is attained iff the subspace contains a $(d-2)$-dimensional face of Q_d.*

Proof of Theorem 2.11. For X_i independent and $\mathbb{R}$-valued, with Y_i chosen as in Theorem 2.12,

$$N_\infty(X_1 + \cdots + X_n) \geq N_\infty(Y_1 + \cdots + Y_n).$$

Applying a sort of change of variables, and utilizing the degree 2 homogeneity of entropy powers, one can write

$$N_\infty(Y_1 + \cdots + Y_n) = \left(\sum_{i=1}^n N_\infty(Y_i)\right) N_\infty(\theta_1 U_1 + \cdots + \theta_n U_n),$$

where the U_i are independent uniform on $[-\frac{1}{2}, \frac{1}{2}]$ and θ is a unit vector (to be explicit, take $\theta_i = \sqrt{N_\infty(Y_i)/\sum_j N_\infty(Y_j)}$ and the above can be verified). Then utilizing the symmetry of $\theta_1 U_1 + \cdots + \theta_n U_n$ and the BMI, we see that the maximum of its density must occur at 0, yielding

$$N_\infty(\theta_1 U_1 + \cdots + \theta_n U_n) = \left|Q_d \cap \theta^\perp\right|_{d-1}^{-2} \geq \frac{1}{2}.$$

The result follows. □

Theorem 2.11 admits two natural generalizations. The first, also handled in [30] (and later recovered in [133] by taking the limit as $p \to \infty$ in Theorem 2.8), is the following.

Theorem 2.14 ([30]) *For independent random vectors $X_1, \ldots, X_n$ in $\mathbb{R}^d$.*

$$N_\infty(X_1 + \cdots + X_n) \geq \left(1 - \frac{1}{n}\right)^{n-1} [N_\infty(X_1) + \cdots + N_\infty(X_n)] \tag{32}$$

$$\geq \frac{1}{e}[N_\infty(X_1) + \cdots + N_\infty(X_n)]. \tag{33}$$

A second direction was pursued by Livshyts, Paouris and Pivovarov [97] in which the authors derive sharp bounds for the maxima of densities obtained as the projections of product measures. Specifically, [97, Theorem 1.1] shows that given probability density functions f_i on $\mathbb{R}$ with $\|f_i\|_\infty \leq 1$, with joint product density f defined by $f(x_1, \ldots, x_n) = \prod_{i=1}^n f_i(x_i)$, then

$$\|\pi_E(f)\|_\infty \leq \min\left(\left(\frac{n}{n-k}\right)^{(n-k)/2}, 2^{k/2}\right), \tag{34}$$

where $\pi_E(f)$ denotes the pushforward of the probability measure induced by f under orthogonal projection to a k-dimensional subspace E, i.e., $\pi_E(f)(x) = \int_{x+E^\perp} f(y)dy$. In addition, cubes are shown to be extremizers of the above inequality. In the language of information theory, this can be rewritten as follows.

Theorem 2.15 ([97]) *Let $X = (X_1, \ldots, X_n)$ where X_i are independent $\mathbb{R}$-valued random variables, and $N_\infty(X_i) \geq 1$. Then*

$$N_\infty(P_E X) \geq \max\left\{\frac{1}{2}, \left(1 - \frac{k}{n}\right)^{\frac{n}{k}-1}\right\}, \tag{35}$$

where P_E denotes the orthogonal projection to a k-dimensional subspace E, and equality can be achieved for X_i uniform on intervals.

In the $k = 1$ case, this implies Theorem 2.11 by applying the inequality (35) to

$$Y_i = X_i / \sqrt{N_\infty(X_i)},$$

and taking E to be the space spanned by the unit vector $\theta_i = \sqrt{N_\infty(X_i)/\sum_j N_\infty(X_j)}$. The Y_i defined satisfy the hypothesis so we have $N_\infty(P_E Y) \geq 1/2$, but

$$\begin{aligned} N_\infty(P_E Y) &= N_\infty(\langle \theta, Y \rangle) \\ &= N_\infty\left(\frac{X_1 + \cdots + X_n}{\sqrt{\sum_{j=1}^n N_\infty(X_j)}}\right) \\ &= \frac{N_\infty(X_1 + \cdots + X_n)}{\sum_{j=1}^n N_\infty(X_j)}, \end{aligned}$$

and the implication follows.

Conversely, for the one-dimensional subspace E spanned by the unit vector θ, and X_i satisfying $N_\infty(X_i) \geq 1$, if one applies Theorem 2.11 to $Y_i = \theta_i X_i$, we recover the one-dimensional case of the projection theorem as

$$\begin{aligned} N_\infty(P_E X) &= N_\infty(Y_1 + \cdots + Y_n) \\ &\geq \frac{1}{2}(N_\infty(Y_1) + \cdots + N_\infty(Y_n)) \\ &= \frac{1}{2}(\theta_1^2 N_\infty(X_1) + \cdots + \theta_n^2 N_\infty(X_n)) \\ &\geq \frac{1}{2}. \end{aligned}$$

Thus Theorem 2.15 can be seen as a k-dimensional generalization of the ∞-EPI for real random variables.

In recent work [114], we have obtained a generalization of Rogozin's inequality that allows us to prove multidimensional versions of both Theorems 2.14 and 2.15. Indeed, our extension of Rogozin's inequality reduces both the latter theorems to geometric inequalities about Cartesian products of Euclidean balls, allowing us to obtain sharp constants in Theorem 2.11 for any fixed dimension as well as to generalize Theorem 2.15 to the case where each X_i is a random vector.

3 Reverse Entropy Power Inequalities

3.1 κ-Concave Measures and Functions

κ-concave measures are measures that satisfy a generalized Brunn-Minkowski inequality, and were studied systematically by Borell [34, 35].

As a prerequisite, we define the t-weighted κ-mean of two numbers. For $a, b \in (0, \infty)$, $t \in (0, 1)$ and $\kappa \in (-\infty, 0) \cup (0, \infty)$, define

$$M_\kappa^t(a, b) = ((1-t)a^\kappa + tb^\kappa)^{\frac{1}{\kappa}}. \tag{36}$$

For $\kappa \in \{-\infty, 0, \infty\}$ define $M_\kappa^t(a, b) = \lim_{\kappa' \to \kappa} M_{\kappa'}^t(a, b)$ corresponding to

$$\{\min(a, b), a^{1-t}b^t, \max(a, b)\}$$

respectively. M_κ can be extended to $a, b \in [0, \infty)$ via direct evaluation when $\kappa \geq 0$ and again by limits when $\kappa < 0$ so that $M_\kappa(a, b) = 0$ whenever $ab = 0$.

Definition 3.1 *Fix $\kappa \in [-\infty, \frac{1}{d}]$. We say that a probability measure μ on $\mathbb{R}^d$ is κ-concave if the support of μ has non-empty interior*[9], *and*

$$\mu((1-t)A + tB) \geq M_\kappa^t(\mu(A), \mu(B))$$

for any Borel sets A, B, and any $t \in (0, 1)$.

We say that μ is a convex measure if it is κ-concave for some $\kappa \in [-\infty, \frac{1}{d}]$.

When the law of a random vector X is a κ-concave measure, we will refer to X as a κ-concave random vector.

Thus, the κ-concave measures are those that distribute volume in such a way that the vector space average of two sets is larger than the κ-mean of their respective volumes. Let us state some preliminaries. First notice that by Jensen's inequality μ being κ-concave implies μ is κ'-concave for $\kappa' \leq \kappa$. The support of a κ-concave measure is necessarily convex, and since we assumed that the support has nonempty interior, the dimension of the smallest affine subspace of $\mathbb{R}^d$ containing the support of μ is automatically d.

It is a nontrivial fact that concavity properties of a measure can equivalently be described pointwise in terms of its density.

Theorem 3.2 ([34]) *A measure μ on $\mathbb{R}^d$ is κ-concave if and only if it has a density (with respect to the Lebesgue measure on its support) that is a $s_{\kappa,d}$-concave function, in the sense that*

$$f((1-t)x + ty) \geq M_{s_{\kappa,d}}^t(f(x), f(y))$$

[9] We only assume this for simplicity of exposition– a more general theory not requiring absolute continuity of the measure μ with respect to Lebesgue measure on $\mathbb{R}^d$ is available in Borell's papers. Note that while the support of μ having nonempty interior in general is a weaker condition than absolute continuity, the two conditions turn out to coincide in the presence of a κ-concavity assumption.

whenever $f(x)f(y) > 0$ *and* $t \in (0, 1)$, *and where*

$$s_{\kappa,d} := \frac{\kappa}{1 - \kappa d}.$$

Examples

1. If X is the uniform distribution on a convex body K, it has an ∞-concave density function $f = |K|^{-1}\mathbb{1}_K$ and thus the probability measure is $1/d$-concave. Let us note that by our requirement that μ is "full-dimensional" (i.e., has support with nonempty interior), the only $1/d$-concave probability measures on $\mathbb{R}^d$ are of this type.
2. A measure that is 0-concave is also called a *log-concave measure*. Since $s_{0,d} = 0$ for any positive integer d, Theorem 3.2 implies that an absolutely continuous measure μ is log-concave if and only if its density is a log-concave function (as defined in Definition 2.3). In other words, X has a log-concave distribution if and only if its density function can be expressed on its support as $e^{-V(x)}$ for V convex. When $V(x) = \frac{1}{2}|x|^2 - \frac{d}{2}\log(2\pi)$, one has the standard Gaussian distribution; when $V(x) = x$ for $x \geq 0$ and $V(x) = \infty$ for $x < 0$, one has the standard exponential distribution; and so on.
3. If X is log-normal distribution with density function

$$f(x) := \frac{1}{x\sigma\sqrt{2\pi}} e^{-\frac{(\ln x - \mu)^2}{2\sigma^2}}$$

then the density function of X is $-\frac{\sigma}{4}$-concave, and for $\sigma < 4$, the probability measure is $\frac{-\sigma}{4-\sigma}$-concave.
4. If X is a Beta distribution with density function

$$\frac{x^{\alpha}(1-x)^{\beta}}{B(\alpha, \beta)}$$

with shape parameters $\alpha \geq 1$ and $\beta \geq 1$, then the density function of X is $\min(\frac{1}{\alpha-1}, \frac{1}{\beta-1})$-concave, and the probability measure is $\frac{1}{\max(\alpha,\beta)}$-concave.
5. If X is a d-dimensional Student's t-distribution with density function

$$f(x) := \frac{\Gamma(\frac{\nu+d}{2})}{\nu^{\frac{d}{2}}\pi^{\frac{d}{2}}\Gamma(\frac{\nu}{2})}\left(1 + \frac{|x|^2}{\nu}\right)^{-\frac{\nu+d}{2}}$$

with $\nu > 0$, then the density function of X is $-\frac{1}{\nu+d}$-concave, and the probability measure is $-\frac{1}{\nu}$-concave.
6. If X is a d-dimensional Pareto distribution of the first kind with density function

$$f(x) := a(a+1)\cdots(a+d-1)\left(\prod_{i=1}^{d}\theta_i\right)^{-1}\left(\sum_{i=1}^{d}\frac{x_i}{\theta_i} - d + 1\right)^{-(a+d)}$$

for $x_i > \theta_i > 0$ with $a > 0$, then the density function of X is $-\frac{1}{a+d}$-concave, and the probability measure is $-\frac{1}{a}$-concave.

The optimal κ for the distributions above can be found through direct computation on densities, let us also remind the reader that κ-concavity is an affine invariant. In other words, if X is κ-concave and T is affine, then TX is κ-concave as well, which supplies further examples through modification of the examples above.

We will also find useful an extension of Lemma 2.4 to convex measures (this was obtained in [25] under an additional condition, which was removed in [22]).

Lemma 3.3 *Let $\kappa \in (-\infty, 0]$. If X is a κ-concave random vector in $\mathbb{R}^d$, then*

$$h(X) - h_\infty(X) \leq \sum_{i=0}^{d-1} \frac{1-\kappa d}{1-\kappa i}, \tag{37}$$

with equality for the n-dimensional Pareto distribution.

To match notation with [25] notice that X being κ-concave is equivalent to X having a density function that can be expressed as $\varphi^{-\beta}$, for $\beta = d - \frac{1}{\kappa}$ and φ convex.

We now develop reverse Rényi entropy power inequalities for κ-concave measures, inspired by work on special cases (such as the log-concave case corresponding to $\kappa = 0$ in the terminology above, or the case of Shannon-Boltzmann entropy) in [26, 168, 33, 12].

3.2 Positional Reverse EPI's for Rényi Entropies

The reverse Brunn-Minkowski inequality (Reverse BMI) is a celebrated result in convex geometry discovered by V. Milman [122] (see also [123, 124, 130]). It states that given two convex bodies A and B in $\mathbb{R}^d$, one can find a linear volume-preserving map $u : \mathbb{R}^d \to \mathbb{R}^d$ such that with some absolute constant C,

$$|u(A) + B|^{1/d} \leq C(|A|^{1/d} + |B|^{1/d}). \tag{38}$$

The EPI may be formally strengthened by using the invariance of entropy under affine transformations of determinant ± 1, i.e., $N(u(X)) = N(X)$ whenever $|\det(u)| = 1$. Specifically,

$$\inf_{u_1,u_2} N(u_1(X) + u_2(Y)) \geq N(X) + N(Y), \tag{39}$$

where the maps $u_i : \mathbb{R}^d \to \mathbb{R}^d$ range over all affine entropy-preserving transformations. It was shown in [24] that in exact analogy to the Reverse BMI, the inequality (39) can be reversed with a constant not depending on dimension if we restrict to log-concave distributions. To state such results compactly, we adopt the following terminology.

Definition 3.4 *For each $d \in \mathbb{N}$, let $\mathcal{M}_d$ be a class of probability measures on $\mathbb{R}^d$, and write $\mathcal{M} = (\mathcal{M}_d : d \in \mathbb{N})$. Suppose that for every pair of independent random variables X and Y whose distributions lie in $\mathcal{M}_d$, there exist linear maps $u_1, u_2 : \mathbb{R}^d \to \mathbb{R}^d$ of determinant 1 such that*

$$N_p\big(u_1(X) + u_2(Y)\big) \le C_p\,(N_p(X) + N_p(Y)), \tag{40}$$

where C_p is a constant that depends only on p (and not on d or the distributions of X and Y). Then we say that a Positional Reverse p-EPI holds for $\mathcal{M}$.

Theorem 3.5 ([24]) *Let $\mathcal{M}_d^{LC}$ be the class of log-concave probability measures on $\mathbb{R}^d$, and $\mathcal{M}^{LC} = (\mathcal{M}_d^{LC} : d \in \mathbb{N})$. A Positional Reverse* 1*-EPI holds for $\mathcal{M}^{LC}$.*

Specializing to uniform distributions on convex bodies, it is shown in [26] that Theorem 3.5 recovers the Reverse BMI. Thus one may think of Theorem 3.5 as completing in a reverse direction the already extensively discussed analogy between the BMI and EPI.

Furthermore, [26] found[10] that Theorem 3.5 can be extended to larger subclasses of the class of convex measures.

Theorem 3.6 ([26]) *For $\beta_0 > 2$, let $\mathcal{M}_{d,\beta_0}$ be the class of probability measures whose densities of the form $f(x) = V(x)^{-\beta}$ for $x \in \mathbb{R}^d$, where $V : \mathbb{R}^d \to (0, \infty]$ is a positive convex function and $\beta \ge \beta_0 d$. Then a Positional Reverse* 1*-EPI holds for $\mathcal{M}_{\beta_0} = (\mathcal{M}_{d,\beta_0} : d \in \mathbb{N})$.*

In [33], it is shown that a Reverse EPI is not possible over all convex measures.

Theorem 3.7 ([33]) *For any constant C, there is a convex probability distribution μ on the real line with a finite entropy, such that*

$$\min\{N(X+Y), N(X-Y)\} \ge C N(X),$$

where X and Y are independent random variables distributed according to μ.

We have the following positional reverse p-Rényi EPI for log-concave random vectors; this does not seem to have explicitly observed before.

Theorem 3.8 *For any $p \in (0, \infty]$, a Positional Reverse p-Rényi EPI holds for $\mathcal{M}^{LC}$. Moreover, for $p \ge 1$, the constant $C_{\mathcal{M},p}$ in the corresponding inequality does not depend on p.*

Proof For any pair of independent log-concave random vectors X and Y, there exist linear maps u_1, u_2: $\mathbb{R}^d \to \mathbb{R}^d$ of determinant 1, such that for all $p > 1$, by Lemma 2.2, Theorem 3.5 and Lemma 2.4, one has

[10] Actually [26] only proved this under the additional condition that $\beta \ge 2d + 1$, but it turns out that this condition can be dispensed with, as explained in [115].

$$N_p(u_1(X)+u_2(Y)) \le N(u_1(X)+u_2(Y)) \lesssim N(X)+N(Y)$$
$$\lesssim N_\infty(X)+N_\infty(Y) \le N_p(X)+N_p(Y).$$

For $p < 1$, by Lemma 2.4 and Lemma 2.2, there exists a constant $C(p)$ depending solely on p such that

$$N_p(u_1(X)+u_2(Y)) \le C(p)N(u_1(X)+u_2(Y)) \le C(p)\left(N(X)+N(Y)\right)$$
$$\le C(p)\left(N_p(X)+N_p(Y)\right),$$

which provides the theorem. □

Later we will show that Theorem 3.8 can be used to recover the functional version of the reverse Brunn-Minkowski inequality proposed by Klartag and V. Milman [88].

3.3 Reverse ∞-EPI via a Generalization of K. Ball's Bodies

3.3.1 Busemann's Theorem for Convex Bodies

We first consider Bobkov's extension of K. Ball's convex bodies associated to log-concave measures. In this direction we associate a star shaped body to a density function via a generalization of the Minkowski functional of a convex body.

Definition 3.9 *For a probability density function f on $\mathbb{R}^d$ with the origin in the interior of the support of f, and $p \in (0,\infty)$, define $\Lambda_f^p : \mathbb{R}^d \to [0,\infty]$ by*

$$\Lambda_f^p(v) = \left(\int_0^\infty f(rv)dr^p\right)^{-1/p}$$

We will consider the class of densities $\mathcal{F}_p$ where $\Lambda_f^p(v) \in [0,\infty)$ for all $v \in \mathbb{R}^d$. For such densities, we can associate a body defined by

$$K_f^p = \{v \in \mathbb{R}^d : \Lambda_f^p(v) \le 1\}.$$

We can now state Bobkov's generalization [28] of the Ball-Busemann theorem.

Theorem 3.10 *If f is a s-concave density on $\mathbb{R}^d$, with $-\frac{1}{d} \le s \le 0$, then*

$$\Lambda_f^p((1-t)x+ty) \le (1-t)\Lambda_f^p(x) + t\Lambda_f^p(y), \tag{41}$$

for every $x, y \in \mathbb{R}^d$ and $t \in (0,1)$, provided $0 < p \le -1-1/s$.

Remark 3.11 Notice that, since Λ_f^p is positive homogeneous and (by Theorem 3.10) convex, it necessarily satisfies the triangle inequality. If we add the assumption that f is even, then Λ_f^p defines a norm.

There is remarkable utility in this type of association. In [11], Ball used the fact that one can directly pass from log-concave probability measures to convex bodies using this method to derive an analog of Hensley's theorem [76] for certain log-concave measures, demonstrating comparability of their slices by different hyperplanes. By generalizing this association to convex measures in [28], Bobkov derived analogs of Blaschke-Santalo inequalities, the Meyer-Reisner theorem [117] (this was proved independently in unpublished work, by Keith Ball, as discussed in [116]) for floating surfaces, and Hensley's theorem for convex measures. Thus this association of convex bodies with convex measures may be seen as a way to "geometrize" said measures.

Another application of this association of bodies to measures is to the study of the so-called intersection bodies.

Definition 3.12 *For any compact set K in $\mathbb{R}^d$ whose interior contains the origin, define $r : \mathbb{S}^{d-1} \to (0, \infty)$ by $r(\theta) = |K \cap \theta^{\perp}|_{d-1}$ (i.e., the volume of the $(d-1)$-dimensional slice of K by the subspace orthogonal to θ). The star-shaped body whose boundary is defined by the points $\theta r(\theta)$ is called the intersection body of K, and denoted $I(K)$.*

The most important fact about intersection bodies is the classical theorem of Busemann [40].

Theorem 3.13 ([40]) *If K be a symmetric convex body in $\mathbb{R}^d$, then $I(K)$ is a symmetric convex body as well.*

The symmetry is essential here; the intersection body of a non-symmetric convex body need not be convex[11]. Busemann's theorem is a fundamental result in convex geometry since it expresses a convexity property of volumes of central slices of a symmetric convex body, whereas Brunn's theorem (an easy implication of the BMI) asserts a concavity property of volumes of slices that are perpendicular to a given direction.

Busemann's theorem may be recast in terms of Rényi entropy, as implicitly recognized by K. Ball and explicitly described below.

Theorem 3.14 *If X is uniformly distributed on a symmetric convex body $K \subset \mathbb{R}^d$, then the mapping $M_\infty^X : \mathbb{R}^d \to \mathbb{R}$ defined by*

$$M_\infty^X(v) = \begin{cases} N_\infty^{1/2}(\langle v, X\rangle) & v \neq 0 \\ 0 & v = 0 \end{cases}$$

defines a norm on $\mathbb{R}^d$.

[11] There is a nontrivial way to extend the definition of intersection body to non-symmetric convex bodies so that the new definition results in a convex body; see [118] for details.

Before showing that Theorems 3.13 and 3.14 are equivalent, we need to recall the definition of the Minkowski functional.

Definition 3.15 *For a convex body L in $\mathbb{R}^d$ containing the origin, define $\rho_L : \mathbb{R}^d \to [0, \infty)$ by*

$$\rho_L(x) = \inf\{t \in (0, \infty) : x \in tL\}.$$

It is straightforward that ρ_L is positively homogeneous (i.e., $\rho_L(ax) = a\rho_L(x)$ for $a > 0$) and convex. When L is assumed to be symmetric, ρ_L defines a norm.

Proof of Theorem 3.13 ⇔ Theorem 3.14. Let K be a symmetric convex body and without loss of generality take $|K| = 1$. Let $X = X_K$ denote a random variable distributed uniformly on K.

For a unit vector $\theta \in \mathbb{S}^{d-1}$, as the pushforward of a symmetric log-concave measure under the linear map $x \mapsto \langle \theta, x \rangle$, the distribution of the real-valued random variable $\langle \theta, X \rangle$ is symmetric and log-concave. Denoting the symmetric, log-concave density of $\langle \theta, X \rangle$ by f_θ, we see that the mode of f_θ is 0, and consequently,

$$N_\infty^{1/2}(\langle \theta, X \rangle) = \frac{1}{f_\theta(0)} = \frac{1}{|K \cap \theta^\perp|_{d-1}} = \frac{1}{r(\theta)}.$$

By the definition of $I(K)$, we have $\rho_{I(K)}(r(\theta)\theta) = 1$. Thus, for any $\theta \in \mathbb{S}^{d-1}$,

$$\rho_{I(K)}(\theta) = \rho_{I(K)}\left(\frac{r(\theta)\theta}{r(\theta)}\right) = \frac{1}{r(\theta)} = M_\infty^X(\theta).$$

By homogeneity, this immediately extends to $\mathbb{R}^d$, establishing our result and also a pleasant duality; up to a constant factor, the Minkowski functional associated to $I(K)$ is a Rényi entropy power of the projections of X_K. □

3.3.2 A Busemann-Type Theorem for Measures

Theorem 3.14 is a statement about ∞-Rényi entropies associated to a $1/d$-concave random vector X (see Example 1 after Theorem 3.2). It is natural to wonder if Busemann's theorem can be extended to other p-Rényi entropies and more general classes of measures.

In [12], Ball-Nayar-Tkocz also give a simple argument, essentially going back to [11], that the information-theoretic statement of Busemann's theorem (namely Theorem 3.14) extends to log-concave measures. Interpreting in the language of Borell's κ-concave measures, [12] extends Theorem 3.14 to measures that are κ-concave with $\kappa \geq 0$. In what follows, we use the same argument as [12] to prove that Busemann's theorem can in fact be extended to all convex measures by invoking Theorem 3.10.

Theorem 3.16 *Let $\kappa \in [-\infty, 1/2]$. If (U, V) is a symmetric κ-concave random vector in $\mathbb{R}^2$, then*

$$e^{h_\infty(U+V)} \leq e^{h_\infty(U)} + e^{h_\infty(V)}.$$

Proof It is enough to prove the result for the weakest hypothesis $\kappa = -\infty$. We let φ denote the density function of (U, V) so that

$$\begin{aligned} U + V \sim w(x) &= \int_{\mathbb{R}} \varphi(x - t, t)dt \\ U \sim u(x) &= \int_{\mathbb{R}} \varphi(x, t)dt \\ V \sim v(x) &= \int_{\mathbb{R}} \varphi(t, x)dt. \end{aligned}$$

Since symmetry and the appropriate concavity properties of the densities force the maxima of u, v, w to occur at 0,

$$\begin{aligned} \frac{1}{\|w\|_\infty} &= \frac{1}{w(0)} \\ &= \left(\int_{\mathbb{R}} \varphi(-t, t)dt\right)^{-1} \\ &= \left(2\int_0^\infty \varphi(t(e_2 - e_1))dt\right)^{-1} \\ &= \frac{1}{2}\Lambda_\varphi^1(e_2 - e_1) \\ &\leq \frac{1}{2}\left(\Lambda_\varphi^1(e_2) + \Lambda_\varphi^1(e_1)\right) \\ &= \left(2\int_0^\infty \varphi(0, t)dt\right)^{-1} + \left(2\int_0^\infty \varphi(t, 0)dt\right)^{-1} \\ &= \frac{1}{u(0)} + \frac{1}{v(0)} \\ &= \frac{1}{\|u\|_\infty} + \frac{1}{\|v\|_\infty}, \end{aligned}$$

where the only inequality follows from Theorem 3.10 with $a = 1$ and $p = 1 = n - 1 - 1/\kappa$. By definition of h_∞, we have proved the desired inequality. □

As a nearly immediate consequence we have Busemann's theorem for convex measures.

Corollary 3.17 *For $\kappa \in [-\infty, \frac{1}{d}]$, if X is symmetric and κ-concave the function*

$$M_\infty^X(v) = \begin{cases} N_\infty^{1/2}(\langle v, X\rangle) & v \neq 0 \\ 0 & v = 0 \end{cases}$$

defines a norm.

Proof As we have observed $M = M_\infty^X$ is homogeneous. To prove the triangle inequality take vectors $u, v \in \mathbb{R}^d$ and define $(U, V) = (\langle X, u\rangle, \langle X, v\rangle)$, so that $U + V = \langle X, u + v\rangle$. Notice that (U, V) is clearly symmetric and as the affine pushforward of a κ-concave measure, is thus κ-concave as well. Thus by Theorem 3.16 we have

$$e^{h_\infty(U+V)} \leq e^{h_\infty(U)} + e^{h_\infty(V)}.$$

But this is exactly

$$N_\infty^{1/2}(\langle X, u + v\rangle) \leq N_\infty^{1/2}(\langle X, u\rangle) + N_\infty^{1/2}(\langle X, v\rangle),$$

which is what we sought to prove. □

3.3.3 Busemann-Type Theorems for Other Rényi Entropies

While the above extension deals with general measures, a further natural question relates to more general entropies. Ball-Nayar-Tkocz [12] conjecture that the Shannon entropy version holds for log-concave measures.

Conjecture 3.18 ([12]) *When X is a symmetric log-concave vector in $\mathbb{R}^d$ then the function*

$$M_1^X(v) = \begin{cases} N_1^{1/2}(\langle v, X\rangle) & v \neq 0 \\ 0 & v = 0 \end{cases}$$

defines a norm on $\mathbb{R}^d$.

As the homogeneity of M is immediate, the veracity of the conjecture depends on proving the triangle inequality

$$e^{h_1(\langle u+v, X\rangle)} \leq e^{h_1(\langle v, X\rangle)} + e^{h_1(\langle u, X\rangle)},$$

which is easily seen to be equivalent to the following modified Reverse EPI for symmetric log-concave measures on $\mathbb{R}^2$.

Conjecture 3.19 ([12]) *For a symmetric log-concave random vector in $\mathbb{R}^2$, with coordinates (U, V),*

$$N_1^{1/2}(U+V) \leq N_1^{1/2}(U) + N_1^{1/2}(V).$$

Towards this conjecture, it is proved in [12] that $e^{\alpha h_1(U+V)} \leq e^{\alpha h_1(U)} + e^{\alpha h_1(V)}$ when $\alpha = 1/5$. By extending the approach used by [12], we can obtain a family of Busemann-type results for p-Rényi entropies.

Theorem 3.20 *Fix $p \in [1, \infty]$. There exists a constant $\alpha_p > 0$ which depends only on the parameter p, such that for a symmetric log-concave random vector X in $\mathbb{R}^d$ and two vectors $u, v \in \mathbb{R}^d$, we have*

$$e^{\alpha_p h_p(\langle u+v, X\rangle)} \leq e^{\alpha_p h_p(\langle u, X\rangle)} + e^{\alpha_p h_p(\langle v, X\rangle)}.$$

Equivalently, for a symmetric log-concave random vector (X, Y) in $\mathbb{R}^2$ we have

$$e^{\alpha_p h_p(X+Y)} \leq e^{\alpha_p h_p(X)} + e^{\alpha_p h_p(Y)}.$$

In fact, if $p \in [1, \infty)$, one can take α_p above to be the unique positive solution α of

$$p^{\frac{\alpha}{p-1}} = \theta_p^\alpha + (1-\theta_p)^\alpha, \tag{42}$$

where

$$\theta_p := \left(\frac{\log p}{p-1}\right) \cdot \frac{1}{2(e+1)[2pe^2 + (4p+1)e + 1]},$$

with the understanding that the $p = 1$ case is understood by continuity (i.e., the left side of equation (42) is e^α in this case, and the pre-factor $\frac{\log p}{p-1}$ in θ_p is replaced by 1).

Remark 3.21 *If $p < \infty$, then $\theta_p > 0$, and on the other hand, trivially $\theta_p < \frac{1}{2(1+e)} < 1$. Denote the left and right sides of the equation (42) by $L_p(\alpha)$ and $R_p(\alpha)$ respectively. Then $1 = L_p(0) < R_p(0) = 2$, and since $p^{1/(p-1)} > 1$ for $p \in [1, \infty)$, we also have $\infty = \lim_{\alpha\to\infty} L_p(\alpha) > \lim_{\alpha\to\infty} R_p(\alpha) = 0$. Since L_p and R_p are continuous functions of α, equation (42) must have a positive solution α_p. Moreover, since L_p is an increasing function and R_p is a decreasing function, there must be a unique positive solution α_p. In particular, easy simulation gives $\alpha_1 \approx 0.240789 > 1/5$, and simulation also shows that the unique solution α_p is non-decreasing in p. Consequently it appears that for any p, one can replace α_p in the above theorem by $1/5$.*

Since Theorem 3.20 is not sharp, and the proof involves some tedious and unenlightening calculations, we do not include its details. We merely mention some analogues of the steps used by [12] to prove the case $p = 1$. As done there, one can "linearize" the desired inequality to obtain the following equivalent form: if (X, Y) is a symmetric log-concave vector in $\mathbb{R}^2$ with $h_p(X) = h_p(Y)$, then for every $\theta \in [0, 1]$,

$$h_p(\theta X + (1-\theta)Y) \leq h_p(X) + \frac{1}{\alpha_p} \log\left(\theta^{\alpha_p} + (1-\theta)^{\alpha_p}\right).$$

To prove this form of the theorem, it is convenient as in [12] to divide into cases where θ is "small" and "large". For the latter case, the bound

$$e^{h_p(X+Y)} \leq e^{h_\infty(X+Y)+\frac{\log p}{p-1}} = p^{1/(p-1)}\left(e^{h_\infty(X)} + e^{h_\infty(Y)}\right) \leq p^{1/(p-1)}\left(e^{h_p(X)} + e^{h_p(Y)}\right),$$

easily obtained by combining Lemmata 2.2 and 2.4, suffices. The former case is more involved and relies on proving the following extension of [12, Lemma 1]: If $w : \mathbb{R}^2 \to \mathbb{R}_+$ is a symmetric log-concave density, and we define $f(x) := \int w(x, y)dy$ and $\gamma = \int w(0, y)dy / \int w(x, 0)dx$, then

$$\frac{\int\int -f(x)^{p-2}f'(x)yw(x, y)dxdy}{\int f(x)^p dx} \leq \left(2e(e+2) + \frac{e+1}{p}\right)\gamma.$$

Staring at Theorem 3.16 and Conjecture 3.19, and given that one would expect to be able to interpolate between the $p = 1$ and $p = \infty$ cases, it is natural to pose the following conjecture that would subsume all of the results and conjectures discussed in this section.

Conjecture 3.22 *Fix $\kappa \in [-\infty, \frac{1}{d}]$. For a symmetric κ-concave random vector in $\mathbb{R}^2$, with coordinates (U, V), it holds for any $p \in [1, \infty]$ that*

$$N_p^{1/2}(U+V) \leq N_p^{1/2}(U) + N_p^{1/2}(V),$$

whenever all these quantities are finite. Equivalently, when X is a symmetric κ-concave random vector in $\mathbb{R}^d$, then for any given $p \in [1, \infty]$, the function

$$M_p^X(v) = \begin{cases} N_p^{1/2}(\langle v, X\rangle) & v \neq 0 \\ 0 & v = 0 \end{cases}$$

defines a norm on $\mathbb{R}^d$ when it is finite everywhere.

Given the close connection of the $p = \infty$ case with intersection bodies and Busemann's theorem, one wonders if there is a connection between the unit balls of the conjectured norms M_p^X in Conjecture 3.22 on the one hand, and the so-called L_p-intersection bodies that arise in the dual L_p Brunn-Minkowski theory (see, e.g., Haberl [74]) on the other.

After the first version of this survey was released, Jiange Li (personal communication) has verified that Conjecture 3.22 is true when $p = 0$ (with arbitrary κ) and when $p = 2$ (with $\kappa = 0$, i.e., in the log-concave case).

3.4 Reverse EPI via Rényi Entropy Comparisons

The Rogers-Shephard inequality [136] is a classical and influential inequality in Convex Geometry. It states that for any convex body K in $\mathbb{R}^d$,

$$|K - K| \leq \binom{2d}{d} \mathrm{Vol}(K) \tag{43}$$

where $K - K := \{x - y : x, y \in K\}$. Since $\binom{2d}{d} < 4^d$, this implies that $|K - K|^{1/d} < 4|K|^{1/d}$, complementing the fact that $|K - K|^{1/d} \geq 2|K|^{1/d}$ by the BMI. In particular, the Rogers-Shephard inequality may be thought of as a Reverse BMI. In this section, we discuss integral and entropic liftings of the Rogers-Shephard inequality.

An integral lifting of the Rogers-Shephard inequality was developed by Colesanti [46] (see also [2, 5]). For a real non-negative function f defined in $\mathbb{R}^d$, define the difference function Δf of f,

$$\Delta f(z) := \sup\{\sqrt{f(x)f(-y)} : x, y \in \mathbb{R}^d, \frac{1}{2}(x + y) = z\} \tag{44}$$

It is proved in [46] that if $f : \mathbb{R}^d \to [0, \infty)$ is a log-concave function, then

$$\int_{\mathbb{R}^d} \Delta f(z) dz \leq 2^d \int_{\mathbb{R}^d} f(x) dx, \tag{45}$$

where the equality is attained by multi-dimensional exponential distribution.

On the other hand, an entropic lifting of the Rogers-Shephard inequality was developed by [33]. We develop an extension of their argument and result here. In order to state it, we need to recall the notion of relative entropy between two distributions: if X, Y have densities f, g respectively, then

$$D(X\|Y) = D(f\|g) := \int_{\mathbb{R}^d} f(x) \log \frac{f(x)}{g(x)} dx$$

is the relative entropy between X and Y. By Jensen's inequality, $D(X\|Y) \geq 0$, with equality if and only if the two distributions are identical.

Lemma 3.23 *Suppose $(X, Y) \in \mathbb{R}^d \times \mathbb{R}^d$ has a κ-concave distribution, with $\kappa < 0$. If X and Y are independent, then*

$$h(X-Y) \le \min\{h(X) + D(X\|Y), h(Y) + D(Y\|X)\} + \sum_{i=0}^{d-1} \frac{1-\kappa d}{1-\kappa i}.$$

Proof By affine invariance, the distribution of $X - Y$ is κ-concave, so that one can apply Lemma 3.3 to obtain

$$\begin{aligned} h(X-Y) &\le \log \|f\|_\infty^{-1} + \sum_{i=0}^{d-1} \frac{1-\kappa d}{1-\kappa i} \\ &\le \log f(0)^{-1} + \sum_{i=0}^{d-1} \frac{1-\kappa d}{1-\kappa i}. \end{aligned}$$

Denoting the marginal densities of X and Y by f_1 and f_2, respectively, we have $f(0) = \int_{\mathbb{R}^d} f_1(x) f_2(x) dx$, and hence

$$\begin{aligned} h(X-Y) &\le -\log \int_{\mathbb{R}^d} f_1(x) f_2(x) dx + \sum_{i=0}^{d-1} \frac{1-\kappa d}{1-\kappa i} \\ &\le \int_{\mathbb{R}^d} f_1(x)[-\log f_2(x)] dx + \sum_{i=0}^{d-1} \frac{1-\kappa d}{1-\kappa i} \\ &= h(X) + D(X\|Y) + \sum_{i=0}^{d-1} \frac{1-\kappa d}{1-\kappa i}. \end{aligned}$$

Clearly the roles of X and Y here are interchangeable, yielding the desired bound. □

In the case where the marginal distributions are the same, Lemma 3.23 reduces as follows.

Theorem 3.24 *Suppose* $(X, Y) \in \mathbb{R}^d \times \mathbb{R}^d$ *has a* κ*-concave distribution, with* $\kappa < 0$. *If* X *and* Y *are independent and identically distributed, then*

$$N(X-Y) \le C_\kappa N(X),$$

where

$$C_\kappa = \exp\left\{ \frac{2}{d}(1-d\kappa) \sum_{j=0}^{d-1} \frac{1}{1-j\kappa} \right\}.$$

As $\kappa \to 0$, we recover the fact, obtained in [33], that $N(X-Y) \le e^2 N(X)$ for X, Y i.i.d. with log-concave marginals. We believe that this statement can be tightened, even in dimension 1. Indeed, it is conjectured in [111] that for X, Y i.i.d. with log-concave marginals,

$$N(X - Y) \leq 4N(X)$$

is the tight entropic version of Rogers-Shepard in one dimension, with equality for the one-sided exponential distribution.

3.5 *Reverse Rényi EPI via Convex Ordering*

3.5.1 Convex Ordering and Entropy Maximization

In this section, we build on an elegant approach of Y. Yu [168], who obtained inequalities for Rényi entropy of order $p \in (0, 1]$ for i.i.d. log-concave measures under stochastic ordering assumptions. In particular, we achieve extensions to κ-concave measures with $\kappa < 0$ and impose weaker distributional symmetry assumptions, and observe that the resulting inequalities may be interpreted as Reverse EPIs.

Lemma 3.25 *Let $X \sim f$, $Y \sim g$ be random vectors on $\mathbb{R}^d$. In order to prove*

$$h_p(X) \geq h_p(Y),$$

it suffices to prove

$$\mathbb{E}f^{p-1}(X) \geq \mathbb{E}f^{p-1}(Y), \quad \text{if } p \in (0, 1), \tag{46}$$

$$\mathbb{E}f^{p-1}(X) \leq \mathbb{E}f^{p-1}(Y), \quad \text{if } p \in (1, \infty), \tag{47}$$

$$-\mathbb{E}\log f(X) \geq -\mathbb{E}\log f(Y), \quad \text{if } p = 1. \tag{48}$$

Proof Notice that the expressions in the hypothesis for $p \neq 1$ can be re-written as $\mathbb{E}f^{p-1}(X) = \int_{\mathbb{R}^d} f^{p-1}(x)f(x)dx$ and $\mathbb{E}f^{p-1}(Y) = \int_{\mathbb{R}^d} f^{p-1}(x)g(x)dx$. For $p \in (0, 1)$,

$$\begin{aligned}\int f^p dx &= \left(\int f^{p-1}f\right)^p \left(\int f^p\right)^{1-p} \\ &\overset{(a)}{\geq} \left(\int f^{p-1}g\right)^p \left(\int f^p\right)^{1-p} \\ &\overset{(b)}{\geq} \int g^p dx,\end{aligned}$$

where (a) is from applying the hypothesis and (b) is by Hölder's inequality (applied in the probability space $(\mathbb{R}^d, g\,dx)$). Inequality (46) follows from the fact that $(1-p)^{-1}\log x$ is order-preserving for $p \in (0, 1)$.

When $p \in (1, \infty)$,

$$\int f^p dx = \left(\int f^{p-1} f\right)^p \left(\int f^p\right)^{1-p}$$
$$\leq \left(\int f^{p-1} g\right)^p \left(\int f^p\right)^{1-p}$$
$$\overset{(c)}{\leq} \int g^p dx,$$

where Hölder's inequality is reversed for $p \in (1, \infty)$ accounting for (c). Inequality (47) follows since $(1-p)^{-1} \log x$ is order-reversing for such p.

In the case $p = 1$, we use the hypothesis and then Jensen's inequality to obtain

$$\begin{aligned} h(X) &= -\mathbb{E} \log f(X) \\ &\geq -\mathbb{E} \log f(Y) \\ &\geq -\mathbb{E} \log g(Y) \\ &= h(Y), \end{aligned}$$

which yields inequality (48) and completes the proof of the lemma. □

Of the observations in Lemma 3.25, (46) and (48) were used in [168]; we add (47), which is relevant to Reverse EPIs for κ-concave measures with $\kappa > 0$.

We recall the notion of convex ordering for random vectors.

Definition 3.26 *For random variables X, Y taking values in a linear space V, we say that X dominates Y in the convex order, written $X \geq_{cx} Y$, if $\mathbb{E}\varphi(X) \geq \mathbb{E}\varphi(Y)$ for every convex and continuous function $\varphi : V \to \mathbb{R}$.*

We need a basic lemma relating supports of distributions comparable in the convex ordering.

Lemma 3.27 *Given random vectors $X \sim f$ and $Y \sim g$ such that $Y \leq_{cx} X$, if supp(f) is a convex set, then supp(g) $\subset$ supp(f).*

Proof Take ρ to be the Minkowski functional (Definition 3.15) associated to supp(f) and then define

$$\varphi(x) = \max\{\rho(x) - 1, 0\}.$$

As the maximum of two convex functions, φ is convex. Also observe that φ is identically zero on supp(f) while strictly positive on the complement. By the ordering assumption

$$0 \leq \mathbb{E}(\varphi(Y)) \leq \mathbb{E}(\varphi(X)) = 0.$$

Thus $\mathbb{E}(\varphi(Y)) = 0$, which implies the claim. □

We can now use convex ordering as a criterion to obtain a maximum entropy property of convex measures under certain conditions.

Theorem 3.28 *Let X and Y be random vectors in $\mathbb{R}^d$, with X being κ-concave for some $\kappa \in (-\infty, \frac{1}{d}]$. If $X \geq_{cx} Y$, then*

$$h_p(X) \geq h_p(Y)$$

for $0 \leq p \leq \kappa/(1 - d\kappa) + 1$.

Proof Recall that X is κ-concave if and only if it admits a $s_{\kappa,d}$-concave density f on its support, with $s_{\kappa,d} = \kappa/(1 - d\kappa)$. Thus it follows that for $a \leq s_{\kappa,d}$, f^a is a convex function, (resp. concave) for $a < 0$ (resp. $a > 0$). Our hypothesis is simply that that $p - 1 \leq s_{\kappa,d}$.

For $p < 1$ we can apply the convex ordering to necessarily convex function f^{p-1}, as $\mathbb{E}f^{p-1}(X) \geq \mathbb{E}f^{p-1}(Y)$ and apply Lemma 3.25 under the hypothesis (46).

When $p > 1$ the proof is the same as the application of convex ordering to the concave function f^{p-1} will reverse the inequality to attain $\mathbb{E}f^{p-1}(X) \leq \mathbb{E}f^{p-1}(Y)$ and then invoking Lemma 3.25 under hypothesis (47) will yield the result.

To consider $p = 1$, X must be at least log-concave, in which case we can follow [168] exactly. This amounts to applying convex ordering to $-\log f$ and Lemma 3.25 a final time.

After recalling that the support of a κ-concave measure is a convex set, the $p = 0$ case follows from Lemma 3.27. □

Theorem 3.28 extends a result of Yu [168], who shows that for X log-concave, $h_p(X) \geq h_p(Y)$ for $0 < p \leq 1$ when $X \geq_{cx} Y$. Observe that as κ approaches $1/d$, the upper limit of the range of p for which Theorem 3.28 applies approaches ∞.

Some care should be taken to interpret Theorem 3.28 and the entropy inequalities to come. For example, the t-distribution (see Example 4 after Theorem 3.2) does not have finite p-Rényi entropy when $p \leq \frac{d}{\nu+d}$ and hence the theorem only yields non-trivial results on the interval $(\frac{d}{\nu+d}, 1 - \frac{1}{\nu+d}]$. Notice that in the important special case where X is Cauchy, corresponding to $\nu = 1$, this interval is empty; thus Theorem 3.28 fails to give a maximum entropy characterization of the Cauchy distribution (which is of interest from the point of view of entropic limit theorems).

Definition 3.29 *We say that a family of random vectors $\{X_1, \ldots, X_n\}$ is exchangeable when $(X_{\sigma(1)}, \ldots, X_{\sigma(n)})$ and $(X_1, \ldots, X_n)$ are identically distributed for any permutation σ of $\{1, \ldots, n\}$.*

3.5.2 Results Under an Exchangeability Condition

Let us also remind the reader of the notion of majorization for $a, b \in \mathbb{R}^n$. First we recall that a square matrix is doubly stochastic if its row sums and column sums are all equal to 1.

Definition 3.30 *For vectors $a, b \in \mathbb{R}^n$, we will write $b \prec a$ (and say that b is majorized by a) if there exists a doubly stochastic matrix M such that $Ma = b$.*

There are several equivalent formulations of this notion that are well studied (see, e.g., [147]), but we will not have use for them. Note that if $\mathbf{1}$ is the vector with all coordinates equal to 1, then $\mathbf{1}^T(Ma) = (\mathbf{1}^T M)a = \mathbf{1}^T a$, implying that $b \prec a$ can only hold if the sum of coordinates of a equals the sum of coordinates of b.

Lemma 3.31 *Let $X_1, \ldots, X_n$ be exchangeable random variables taking values in a real vector space V, and let $\varphi : V^n \to \mathbb{R}$ be a convex function symmetric in its coordinates. If $b \prec a$,*

$$\mathbb{E}\varphi(a_1X_1, \ldots, a_nX_n) \geq \mathbb{E}\varphi(b_1X_1, \ldots, b_nX_n).$$

Proof Since every doubly stochastic matrix can be written as the convex combination of permutation matrices by the Birkhoff von-Neumann theorem (see, e.g., [147]), we can write $b \prec a$ as $b = (\sum_\sigma \lambda_\sigma P_\sigma)a$ where $\lambda_i \in [0, 1]$ with $\sum_\sigma \lambda_\sigma = 1$ and P_σ is a permutation matrix. We compute

$$\begin{aligned}
\mathbb{E}\varphi(b_1X_1, \ldots, b_nX_n) &= \mathbb{E}\varphi\left((\sum_\sigma \lambda_\sigma P_\sigma a)_1 X_1, \ldots, (\sum_\sigma \lambda_\sigma P_\sigma a)_n X_n\right) \\
&\leq \sum_\sigma \lambda_\sigma \mathbb{E}\varphi(a_{\sigma(1)}X_1, \ldots, a_{\sigma(n)}X_n) \\
&= \sum_\sigma \lambda_\sigma \mathbb{E}\varphi(a_{\sigma(1)}X_{\sigma(1)}, \ldots, a_{\sigma(n)}X_{\sigma(n)}) \\
&= \sum_\sigma \lambda_\sigma \mathbb{E}\varphi(a_1X_1, \ldots, a_nX_n) \\
&= \mathbb{E}\varphi(a_1X_1, \ldots, a_nX_n),
\end{aligned}$$

where the steps are justified– in order– by definition, convexity, exchangeability, coordinate symmetry, and then algebra. □

Theorem 3.32 *Let $X = (X_1, \ldots, X_n)$ be an exchangeable collection of d-dimensional random vectors. Suppose $b \prec a$ and that $a_1X_1 + \cdots + a_nX_n$ has a s-concave density. Then for any $p \in [0, s+1]$,*

$$h_p(b_1X_1 + \cdots + b_nX_n) \leq h_p(a_1X_1 + \cdots + a_nX_n).$$

Proof Let f denote the s-concave density function of $a_1X_1 + \cdots + a_nX_n$. Thus for $p < 1$ (resp. $p > 1$) the function

$$\varphi(x_1, \ldots, x_n) = f^{p-1}(x_1 + \cdots + x_n)$$

is convex (resp. concave) and clearly symmetric in its coordinates, hence by Lemma 3.31

$$\mathbb{E}\varphi(b_1X_1, \ldots, b_nX_n) \leq \mathbb{E}\varphi(a_1X_1, \ldots, a_nX_n),$$
$$(\text{ resp. } \mathbb{E}\varphi(b_1X_1, \ldots, b_nX_n) \geq \mathbb{E}\varphi(a_1X_n, \ldots, a_nX_n)).$$

But this is exactly,

$$\mathbb{E}f^{p-1}(b_1X_1, \ldots, b_nX_n) \leq \mathbb{E}f^{p-1}(a_1X_n, \ldots, a_nX_n),$$
$$\left(\text{ resp. } \mathbb{E}f^{p-1}(b_1X_1, \ldots, b_nX_n) \geq \mathbb{E}f^{p-1}(a_1X_n, \ldots, a_nX_n)\right),$$

and thus by Lemma 3.25,

$$h_p(b_1X_1 + \cdots + b_nX_n) \leq h_p(a_1X_n + \cdots + a_nX_n).$$

The case $p = 1$ is similar by setting

$$\varphi(x_1, \ldots, x_n) = -\log f(x_1 + \cdots + x_n),$$

and applying Lemma 3.31 and Lemma 3.25. □

Definition 3.33 *For $\Omega \subseteq \mathbb{R}^d$, we define a function $\varphi : \Omega \to \mathbb{R}$ to be Schur-convex in the case that for any $x, y \in \Omega$ with $x \prec y$ we have $\varphi(x) \leq \varphi(y)$.*

Corollary 3.34 *Suppose $X = (X_1, \ldots X_n)$ is an exchangeable collection of random vectors in $\mathbb{R}^d$, with X being κ-concave. Let $\Delta_n = \{\theta \in [0,1]^n : \sum_{i=1}^n \theta_i = 1\}$ be the standard simplex, and define the function $\Phi_{X,p} : \Delta_n \to \mathbb{R}$ by*

$$\Phi_{X,p}(\theta) = h_p(\theta_1X_1 + \cdots + \theta_nX_n).$$

For $p \in [0, s_{\kappa,d} + 1]$, $\Phi_{X,p}$ is a Schur-convex function. In particular, $\Phi_{X,p}$ is maximized by the standard basis elements e_i, and minimized by $(\frac{1}{n}, \ldots, \frac{1}{n})$.

Proof If X is κ-concave, then by affine invariance $\theta_1X_1 + \cdots + \theta_nX_n$ is κ-concave, and hence Theorem 3.32 applies. The extremizers are identified by observing that for any θ in the simplex

$$(1/n, , \cdots, 1/n) \prec \theta \prec e_i,$$

and the corollary follows. □

Of course, using the standard simplex is only a matter of normalization; analogous results are easily obtained by setting $\sum_i \theta_i$ to be any positive constant.

When the coordinates of X_i are assumed to be independent, then X is log-concave if and only if each X_i each log-concave. As a consequence we recover in the $\kappa = 0$ and $p \leq 1$ case, the theorem of Yu in [168].

Theorem 3.35 ([168]) *Let $X_1, \cdots, X_n$ be i.i.d. log-concave random vectors in $\mathbb{R}^d$. Then the function $a \mapsto h_p(a_1X_1 + \cdots + a_nX_n)$ is Schur-convex on the simplex for $p \in (0, 1]$.*

3.5.3 Results Under an Assumption of Identical Marginals

We now show that the exchangeability hypothesis can be loosened in Corollary 3.34.

Theorem 3.36 *Let $X = (X_1, \ldots, X_n)$ be a collection of d–dimensional random vectors with X_i identically distributed and κ-concave. For $p \in [0, s_{\kappa,d} + 1]$, the function $\Phi_{X,p}$ defined in Corollary 3.34 satisfies*

$$\Phi_{X,p}(a) \leq \Phi_{X,p}(e_i).$$

Stated explicitly, for $a \in \Delta_n$, we have

$$h_p(a_1X_1 + \cdots + a_nX_n) \leq h_p(X_1).$$

Proof Let f be the density function of X_1 and $a \in \Delta_n$. If $p < 1$, by Lemma 3.25, it suffices to prove that

$$\mathbb{E}f^{p-1}(a_1X_1 + \cdots + a_nX_n) \leq \mathbb{E}f^{p-1}(X_1).$$

Since f is a $s_{\kappa,d}$-concave function and $p - 1 \leq s_{\kappa,d}$, f is also $(p-1)$-concave, which means that f^{p-1} is convex. Consequently, we have

$$\begin{aligned}\mathbb{E}f^{p-1}(a_1X_1 + \cdots + a_nX_n) &\leq a_1\mathbb{E}f^{p-1}(X_1) + \cdots + a_n\mathbb{E}f^{p-1}(X_n)\\ &= \mathbb{E}f^{p-1}(X_1),\end{aligned}$$

where the equality is by the fact that X_i are identically distributed. The cases of $p > 1$ and $p = 1$ follow similarly. □

Corollary 3.37 *Suppose $X_1, X_2, \cdots, X_n$ are identically distributed and κ-concave. If $p \in [0, s_{\kappa,d} + 1]$, we have the triangle inequality*

$$N_p^{1/2}\left(\sum_{i=1}^n X_i\right) \leq \sum_{i=1}^n N_p^{1/2}(X_i).$$

Moreover, for any $p > s_{\kappa,d} + 1$,

$$N_p^{1/2}\left(\sum_{i=1}^n X_i\right) \leq \frac{(s_{\kappa,d} + 1)^{1/s_{\kappa,d}}}{p^{1/(p-1)}} \sum_{i=1}^n N_p^{1/2}(X_i).$$

Proof We have, by Theorem 3.36, for $p \in [0, s_{\kappa,d} + 1]$,

$$N_p^{1/2}\left(\sum_{i=1}^n X_i\right) \leq N_p^{1/2}(nX_1) = \sum_{i=1}^n N_p^{1/2}(X_i).$$

The second inequality can be derived from Lemma 3.3, combined with Theorem 3.36 and the monotonicity of Rényi entropies:

$$\begin{aligned} N_p^{1/2}\left(\sum_{i=1}^n X_i\right) &\leq N_{s_{\kappa,d}+1}^{1/2}\left(\sum_{i=1}^n X_i\right) \\ &\leq N_{s_{\kappa,d}+1}^{1/2}(nX_i) \\ &= \exp\left(h_{s_{\kappa,d}+1}(X_i)/d + \log n\right) \\ &\leq \exp\left(h_p(X_i)/d + \left[\log n + \frac{\log(s_{\kappa,d}+1)}{s_{\kappa,d}} - \frac{\log p}{p-1}\right]\right) \\ &= \frac{(s_{\kappa,d}+1)^{1/s_{\kappa,d}}}{p^{1/(p-1)}} \sum_{i=1}^n N_p^{1/2}(X_i). \end{aligned}$$

□

Observe that Corollary 3.37 is very reminiscent of Conjectures 3.18 and 3.22; the main difference is that here we have the assumption of identical marginals as opposed to central symmetry of the joint distribution.

We state the next corollary as a direct application of Corollary 3.37 for the log-concave case.

Corollary 3.38 *Suppose $X_1, X_2, \cdots, X_n$ are identically distributed log-concave random vectors in $\mathbb{R}^d$. Then*

$$N_p\left(\sum_{i=1}^n X_i\right) \leq n^2 N_p(X_1) \text{ for } p \in [0, 1], \tag{49}$$

$$N_p\left(\sum_{i=1}^n X_i\right) \leq e^2 p^{2/(1-p)} n^2 N_p(X_1) \leq e^2 n^2 N_p(X_1) \text{ for } p \in (1, \infty]. \tag{50}$$

In particular, if X and X' are identically distributed log-concave random vectors, then

$$N_p(X + X') \leq 4N_p(X) \text{ for } p \in [0, 1],$$

$$N_p(X + X') \leq 4e^2 p^{2/(1-p)} N_p(X) \leq 4e^2 N_p(X) \text{ for } p \in (1, \infty].$$

Cover and Zhang [55] proved the remarkable fact that if X and X' (possibly dependent) have the same log-concave distribution on $\mathbb{R}$, then $h(X+X') \leq h(2X)$ (in fact, they also showed a converse of this fact). As observed by [111], their method also works in the multivariate setting, where it implies that $N(X+X') \leq 4N(X)$ for i.i.d. log-concave X, X'. This fact is recovered by the previous corollary.

Let us finally remark that if we are not interested in an explicit constant, then a version of this inequality already follows from the Reverse EPI of [26]. Indeed,

$$N(X+X') \leq CN(X),$$

since the same unit-determinant affine transformation must put both X and X' in M-position. However, the advantage of the methods we have explored is that we can obtain explicit constants.

3.6 Remarks on Special Positions that Yield Reverse EPI's

Let us recall the definition of isotropic bodies and measures in the convex geometric sense.

Definition 3.39 *A convex body K in $\mathbb{R}^d$ is called isotropic if there exists a constant L_K such that*

$$\frac{1}{|K|^{1+\frac{2}{d}}} \int_K \langle x, \theta \rangle^2 dx = L_K^2,$$

for all unit vectors $\theta \in \mathbb{S}^{d-1}$. More generally, a probability measure μ on $\mathbb{R}^d$ is called isotropic if there exists a constant L_K such that

$$\int_{\mathbb{R}^d} \langle x, \theta \rangle^2 \mu(dx) = L_K^2,$$

for all unit vectors $\theta \in \mathbb{S}^{d-1}$.

The notion of M-position (i.e., a position or choice of affine transformation applied to convex bodies for which a reverse Brunn-Minkowski inequality holds) was first introduced by V. Milman [122]. Alternative approaches to proving the existence of such a position were developed in [124, 130, 72]. It was shown by Bobkov [29] that if the standard Gaussian measure conditioned to lie in a convex body K is isotropic, then the body is in M-position and the reverse BMI applies. The notion of M-position was extended from convex bodies to log-concave measures in [24], and further to convex measures in [26]. Using this extension, together with the sufficient condition obtained in [29], one can give an explicit description of a position for which a reverse EPI applies with a universal– but not explicit– constant.

Nonetheless there are other explicit positions for which one can get reverse EPIs with explicit (but not dimension-independent) constants. One instance of such is obtained from an extension to convex measures obtained by Bobkov [28] for Hensley's theorem (which had earlier been extended from convex sets to log-concave functions by Ball [9]).

Theorem 3.40 ([28]) *For a symmetric, convex probability measure μ on $\mathbb{R}^d$ with density f such that the body Λ_f^{d-k} is isotropic, we have for any linear two subspaces H_1, H_2 of codimension k,*

$$\int_{H_1} f dx \leq C_k \int_{H_2} f dx.$$

What is more, $C_k < \left(\frac{1}{2}e^2\pi k\right)^{\frac{k}{2}}$.

As a consequence we have the following reverse ∞-Rényi EPI in the isotropic context.

Corollary 3.41 *Suppose the joint distribution of the random vector $(X, Y) \in \mathbb{R}^d \times \mathbb{R}^d$ is symmetric and convex, with density $f = f(x, y)$. If the body Λ_f^d is isotropic, then*

$$N_\infty(X+Y) \leq \pi e^2 d \min\{N_\infty(X), N_\infty(Y)\}.$$

Proof Define two d-dimensional subspaces of $\mathbb{R}^d$: $H_1 := \{x = 0\}$, $H_2 := \{x + y = 0\}$. Computing directly and applying Theorem 3.40 we have our result as follows.

$$\begin{aligned}
\frac{N_\infty(X+Y)}{N_\infty(X)} &= \left(\frac{\|f_X\|_\infty}{\|f_{X+Y}\|_\infty}\right)^{\frac{2}{d}} \\
&= \left(\frac{\int_{\mathbb{R}^d} f(0,z)dz}{\int_{\mathbb{R}^d} f(z,-z)dz}\right)^{\frac{2}{d}} \\
&= \left(\frac{2^{\frac{d}{2}} \int_{H_1} f}{\int_{H_2} f}\right)^{\frac{2}{d}} \\
&\leq \pi e^2 d.
\end{aligned}$$

□

4 The Relationship Between Functional and Entropic Liftings

In this section, we observe that the integral lifting of an inequality in Convex Geometry may sometimes be seen as a Rényi entropic lifting.

We start by considering integral and entropic liftings of a classical inequality in Convex Geometry, namely the Blaschke-Santaló inequality. For a convex body $K \subset \mathbb{R}^d$ with $0 \in \mathrm{int}(K)$, the polar K° of K is defined as

$$K^\circ = \{y \in \mathbb{R}^d : \langle x, y \rangle \leq 1 \text{ for all } x \in K\},$$

and, more generally, the polar K^z with respect to $z \in \mathrm{int}(K)$ by $(K - z)^\circ$. There is a unique point $s \in \mathrm{int}(K)$, called the Santaló point of K, such that the volume product $|K||K^s|$ is minimal– it turns out that this point is such that the barycenter of K^s is 0. The Blaschke-Santaló inequality states that

$$|K||K^s| \leq |B_2^d|^2,$$

with equality if and only if K is an ellipsoid. In particular, the volume product $|K||K^\circ|$ of a centrally symmetric convex body K is maximized by the Euclidean ball. This inequality was proved by Blaschke [21] in dimensions 2 and 3, and by Santaló [142] in general dimension; the equality conditions were settled by Petty [129]. There have been many subsequent proofs; see [19] for a recent Fourier analytic proof as well as a discussion of the earlier literature.

More generally, if K, L are compact sets in $\mathbb{R}^d$, then

$$|K| \cdot |L| \leq \omega_d^2 \max_{x \in K, y \in L} |\langle x, y \rangle|^d. \tag{51}$$

The inequality (51) implies the Blaschke-Santaló inequality by taking K to be a symmetric convex body, and L to be the polar of K.

Let us now describe an integral lifting of the inequality (51), which was proved by Lehec [89, 90] building on earlier work of Ball [11], Artstein-Klartag-Milman [6], and Fradelizi-Meyer [69].

Let f and g be non-negative Borel functions on $\mathbb{R}^d$ satisfying the duality relation

$$\forall x, y \in \mathbb{R}^d,\ f(x)g(y) \leq e^{-\langle x, y \rangle}.$$

If f (or g) has its barycenter (defined as $(\int f)^{-1} \int x f(x) dx$) at 0, then

$$\int_{\mathbb{R}^d} f(x)dx \int_{\mathbb{R}^d} g(y)dy \leq (2\pi)^d.$$

The inequality (51) also has an entropic lifting. For any two independent random vectors X and Y in $\mathbb{R}^d$, Lutwak-Yang-Zhang [102] showed that

$$N(X) \cdot N(Y) \leq \frac{4\pi^2 e^2}{d} \mathbf{E}\big[|\langle X, Y \rangle|^2\big], \tag{52}$$

with equality achieved for Gaussians. They also have an even more general (and still sharp) statement that bounds $[N_p(X)N_p(Y)]^{p/2}$ in terms of $\mathbf{E}[|\langle X, Y \rangle|^p]$, with extremizers being certain generalized Gaussian distributions. As $p \to \infty$, the expression $(\mathbf{E}[|\langle X, Y\rangle|^p])^{1/p}$ approaches the essential supremum of $|\langle X, Y\rangle|$, which in the case that X and Y are uniformly distributed on convex bodies is just the maximum that appears in the right side of inequality (51). Thus the Blaschke-Santaló inequality appears as the L_∞ instance of the family of inequalities proved by Lutwak-Yang-Zhang [102], whereas the entropic lifting (52) is the L_2 instance of the same family. This perspective of entropy inequalities as being tied to an L_2-analogue of the Brunn-Minkowski theory is greatly developed in a series of papers by Lutwak, Yang, Zhang, sometimes with additional coauthors (see, e.g., [101] and the references therein), but this is beyond the scope of this survey.

For a function $V : \mathbb{R}^d \to \mathbb{R}$, its Legendre transform $\mathcal{L}V$ is defined by

$$\mathcal{L}V(x) = \sup_y [\langle x, y\rangle - V(y)] .$$

For $f = e^{-V}$ log-concave, following Klartag and V. Milman [88], we define its polar by

$$f^\circ = e^{-\mathcal{L}V} .$$

Some basic properties of the polar are collected below.

Lemma 4.1 *Let f be a non-negative function on $\mathbb{R}^d$.*

1. *If f is log-concave, then*

$$(f^\circ)^\circ = f. \tag{53}$$

2. *If g is also a non-negative function on $\mathbb{R}^d$, and the "Asplund product" of f and g is defined by $f \star g(x) = \sup_{x_1+x_2=x} f(x_1)g(x_2)$, then*

$$(f \star g)^\circ = f^\circ g^\circ . \tag{54}$$

3. *For any linear map u: $\mathbb{R}^d \to \mathbb{R}^d$ with full rank, we have the composition identity*

$$f^\circ \circ u = \left(f \circ u^{-T}\right)^\circ , \tag{55}$$

where u^{-T} is the inverse of the adjoint of u.

4. *If $f(x)$ takes its maximum value at $x = 0$, one has*

$$\sup f^\circ = \frac{1}{\sup f} . \tag{56}$$

5. *For any $p > 0$,*

$$(f^\circ)^p(x) = (f^p)^\circ(px). \tag{57}$$

Proof Write $f := e^{-V}$ for a function $V : \mathbb{R}^d \to \mathbb{R}$. The first two properties are left as an exercise for the reader– these are also standard facts about the Legendre transform and its relation to the infimal convolution of convex functions (see, e.g., [135]). For the third, we have

$$\left(f^\circ \circ u\right)(x) = e^{-\sup_y[\langle ux,y\rangle - V(y)]} = e^{-\sup_y\left[\langle x,u^T y\rangle - V(y)\right]}$$
$$= e^{-\sup_y\left[\langle x,y\rangle - V(u^{-T}y)\right]} = \left(f \circ u^{-T}\right)^\circ(x),$$

which proves the property.

For the fourth, observe that we have, for any $x \in \mathbb{R}^d$,

$$\mathcal{L}V(x) = \sup_y\,[\langle x, y\rangle - V(y)] \geq -V(0).$$

On the other hand,

$$\mathcal{L}V(0) = \sup_y\,[-V(y)] = -V(0).$$

Thus we have proved that $\inf \mathcal{L}V = -V(0)$, which is equivalent to the desired property.

The last property is checked by writing $(f^\circ)^p(x) = e^{-\sup_y[\langle px,y\rangle - pV(y)]}$. □

Bourgain and V. Milman [36] proved a *reverse* form of the Blaschke-Santaló inequality, which asserts that there is a universal positive constant c such that

$$|K| \cdot |K^\circ| \geq c^d,$$

for any symmetric convex body K in $\mathbb{R}^d$, for any dimension d. Klartag and V. Milman [88] obtained a functional lifting of this reverse inequality.

Theorem 4.2 ([88]) *There exists a universal constant $c > 0$ such that for any dimension d and for any log-concave function $f : \mathbb{R}^d \to [0, \infty)$ centered at the origin (in the sense that $f(0)$ is the maximum value of f) with $0 < \int_{\mathbb{R}^d} f < \infty$,*

$$c^d < \left(\int_{\mathbb{R}^d} f\right)\left(\int_{\mathbb{R}^d} f^\circ\right) < (2\pi)^d.$$

Note that the upper bound here is just a special case of the integral lifting of the Blaschke-Santaló inequality discussed earlier.

We observe that Theorem 4.2 can be thought of in information-theoretic terms, namely as a type of certainty/uncertainty principle.

Theorem 4.3 *Let $X \sim f$ be a log-concave random vector in $\mathbb{R}^d$, which is centered at the origin in the sense that f is maximized there. Let X° be a random vector in $\mathbb{R}^d$ drawn from the density $f^\circ / \int_{\mathbb{R}^d} f^\circ$. Define the constants*

$$A_{p,d} := \frac{d(\log 2\pi - \log p - p \log c)}{1-p},$$

$$B_{p,d} := \frac{d(\log c - \log p - p \log 2\pi)}{1-p},$$

where the constant c is the same as in Theorem 4.2. Then, for $p > 1$, we have

$$\max\{d \log c, A_{p,d}\} \le h_p(X) + h_p(X^\circ) \le \min\{d(\log 2\pi + 2), B_{p,d}\}, \tag{58}$$

and for $p < 1$, we have

$$\max\{d \log c, B_{p,d}\} \le h_p(X) + h_p(X^\circ) \le \min\left\{ d\left(\frac{2\log p}{p-1} + \log 2\pi\right), A_{p,d}\right\} \tag{59}$$

In particular, if $p = \infty$,

$$d \log c \le h_\infty(X) + h_\infty(X^\circ) \le d \log 2\pi, \tag{60}$$

and for $p = 1$,

$$d \log c \le h(X) + h(X^\circ) \le d(\log 2\pi + 2). \tag{61}$$

Proof We have

$$h_p(X) + h_p(X^\circ) = \frac{\log\left[\int f^p \int (f^\circ)^p\right] - p \log \int f^\circ}{1-p}. \tag{62}$$

By property (57), we have $\int (f^\circ)^p = \frac{1}{p^d} \int (f^p)^\circ$. So by (62):

$$h_p(X) + h_p(X^\circ) = \frac{\log\left[\int f^p \int (f^p)^\circ\right] - d \log p - p \log \int f^\circ}{1-p}.$$

Thus, by applying Theorem 4.2 twice, if $p > 1$:

$$h_p(X) + h_p(X^\circ) \ge \frac{d \log 2\pi - d \log p - p \log \int f^\circ}{1-p} \ge A_{p,d}.$$

On the other hand,

$$h_p(X) + h_p(X^\circ) \le \frac{d \log c - d \log p - p \log \int f^\circ}{1-p} \le B_{p,d}.$$

Therefore we have

$$A_{p,d} \le h_p(X) + h_p(X^\circ) \le B_{p,d}. \tag{63}$$

A similar argument for $p < 1$ gives

$$B_{p,d} \le h_p(X) + h_p(X^\circ) \le A_{p,d}. \tag{64}$$

Letting $p \to \infty$, we have (60). For $p = 1$, by Lemma 2.4 and (60),

$$\begin{aligned} n\log c \le h_\infty(X) + h_\infty(X^\circ) \le h(X) + h(X^\circ) \le h_\infty(X) + h_\infty(X^\circ) + 2n \\ \le n(\log 2\pi + 2), \end{aligned}$$

which provides (61). Thus for $p > 1$, by (60), (61) and Lemma 2.2, we also have

$$n\log c \le h_\infty(X) + h_\infty(X^\circ) \le h_p(X) + h_p(X^\circ) \le h(X) + h(X^\circ) \le n(\log 2\pi + 2).$$

Combining with (63) provides (58), which provides the theorem. For $p < 1$, we have, by (61) and Lemma 2.2, we have

$$d\log c \le h(X) + h(X^\circ) \le h_p(X) + h_p(X^\circ).$$

Combining this with (64) provides the left most inequality of (59). And by applying Lemma 2.4 on $h_p(f) - h_\infty(f)$ and by (60), we have

$$h_p(X) + h_p(X^\circ) \le \frac{2d\log p}{p-1} + h_\infty(X) + h_\infty(X^\circ) \le \frac{2d\log p}{p-1} + d\log 2\pi.$$

Combining this with (64) gives (59). □

Klartag and Milman [88] prove a reverse Prékopa-Leindler inequality (Reverse PLI).

Theorem 4.4 ([88]) *Given $f, g\colon \mathbb{R}^d \to [0,\infty)$ be even log-concave functions with $f(0) = g(0) = 1$, then there exist u_f, u_g in $SL(d)$ such that $\bar{f} = f \circ u_f$, $\bar{g} = g \circ u_g$ satisfy*

$$\left(\int \bar{f} \star \bar{g}\right)^{\frac{1}{d}} \le C\left(\left(\int \bar{f}\right)^{\frac{1}{d}} + \left(\int \bar{g}\right)^{\frac{1}{d}}\right),$$

where $C > 0$ is a universal constant, u_f depends solely on f, and u_g depends solely on g.

We observe that the Reverse PLI can be proved from the Positional Reverse Rényi EPI we proved earlier, modulo the reverse functional Blaschke-Santaló inequality of Klartag-Milman.

Proposition 4.5 *Theorems 3.8 and 4.2 together imply Theorem 4.4.*

Proof Let $f, g \colon \mathbb{R}^d \to [0, \infty)$ be even log-concave functions with $f(0) = g(0) = 1$. Now by property (56), $\|f^\circ\|_\infty = 1$ as well. Now apply reversed ∞-EPI on a pair of independent random vectors X and Y with density functions $f^\circ / \int f^\circ$ and $g^\circ / \int g^\circ$, respectively, there exist linear maps $u_1, u_2 \in SL(d)$ depending solely on f and g, respectively, such that

$$\left(\int \frac{(f^\circ \circ u_1(x)) \cdot (g^\circ \circ u_2(x))}{\int f^\circ \cdot \int g^\circ} dx\right)^{-\frac{2}{d}} = N_\infty(u_1(X) + u_2(Y))$$

$$\lesssim N_\infty(X) + N_\infty(Y) = \left\|\frac{f^\circ}{\int f^\circ}\right\|_\infty^{-\frac{2}{d}} + \left\|\frac{g^\circ}{\int g^\circ}\right\|_\infty^{-\frac{2}{d}} = \left(\int f^\circ\right)^{\frac{2}{d}} + \left(\int g^\circ\right)^{\frac{2}{d}}.$$

Therefore we have

$$\left(\int (f^\circ \circ u_1(x)) \cdot (g^\circ \circ u_2(x))\, dx\right)^{-\frac{2}{d}} \lesssim \left(\int f^\circ\right)^{-\frac{2}{d}} + \left(\int g^\circ\right)^{-\frac{2}{d}}. \quad (65)$$

Thus by Theorem 4.2, we have the right-hand side of (65):

$$\left(\int f^\circ\right)^{-\frac{2}{d}} + \left(\int g^\circ\right)^{-\frac{2}{d}} \lesssim \left(\int f\right)^{\frac{2}{d}} + \left(\int g\right)^{\frac{2}{d}}. \quad (66)$$

On the other hand, by properties (53), (54) and (55), we have the right hand side of (65):

$$\left(\int (f^\circ \circ u_1(x)) \cdot (g^\circ \circ u_2(x))\, dx\right)^{-\frac{2}{d}} \gtrsim \left(\int (f \circ u_1^{-t}) \star (g \circ u_2^{-t})\right)^{\frac{2}{d}}. \quad (67)$$

Denote $u_f := u_1^{-t}$, $u_g := u_2^{-t}$; $\bar{f} := f \circ u_f$, $\bar{g} := g \circ u_g$, and combining (65) (66) and (67) provides Theorem 4.4. □

5 Concluding Remarks

One productive point of view put forward by Lutwak, Yang and Zhang is that the correct analogy is between entropy inequalities and the inequalities of the L^2-Brunn-Minkowski theory rather than the standard Brunn-Minkowski theory. While we did not have space to pursue this direction in our survey apart from a brief discussion in Section 4, we refer to [101] and the references therein for details.

A central question when considering integral or entropic liftings of Convex Geometry is whether there exist integral and entropic analogues of mixed volumes. Recent work of Bobkov-Colesanti-Fragala [31] has shown that an integral lifting of intrinsic volumes does exist, and Milman-Rotem [121, 120] independently showed this as well as an integral lifting of mixed volumes more generally. A fully satisfactory theory of "intrinsic entropies" or "mixed entropies" is yet to emerge, although some promising preliminary results in this vein can be found in [78].

It is also natural to explore nonlinear generalizations, to ambient spaces that are manifolds or groups. Log-concave (and convex) measures can be put into an even broader context by viewing them as instances of curvature in metric measure spaces. Indeed, thanks to path-breaking work of [151, 98], it was realized that one can give meaning (synthetically) to the notion of a lower bound on Ricci curvature for a metric space $(\mathcal{X}, d)$ equipped with a measure μ (thus allowing for geometry beyond the traditional setting of Riemannian manifolds). In particular, they extended the celebrated Curvature-Dimension condition $CD(K, N)$ of Bakry and Émery [8] to metric measure spaces $(\mathcal{X}, d, \mu)$; the simplest case $CD(K, \infty)$ is defined by a "displacement convexity" (or convexity along optimal transport paths) property of the relative entropy functional $D(\cdot\|\mu)$. For Riemannian manifolds, the $CD(K, N)$ condition is satisfied if and only if the manifold has dimension at most N and Ricci curvature at least K, while Euclidean space $\mathbb{R}^d$ equipped with a log-concave measure may be thought of as having non-negative Ricci curvature in the sense that it satisfies $CD(0, d)$. Moreover, $\mathbb{R}^d$ equipped with a convex measure may be interpreted as a $CD(K, N)$ space with effective dimension N being negative (other examples can be found in [119]). In these more general settings (where there may not be a group structure), it is not entirely clear whether there are natural formulations of entropy power inequalities. Even for the case of Lie groups, almost nothing seems to be known.

One may also seek discrete analogs of the phenomena studied in this survey, which are closely related to investigations in additive combinatorics. In discrete settings, additive structure plays a role as or more important than that of convexity. The Cauchy-Davenport inequality is an analog of the Brunn-Minkowski inequality in cyclic groups of prime or infinite order, with arithmetic progressions being the extremal objects (see, e.g., [157]); extensions to the integer lattice are also known [140, 71, 149]. A probabilistic lifting of the Cauchy-Davenport inequality for the integers is presented in [163]. Sharp lower bounds on entropies of sums in terms of those of summands are still not known for most countable groups; partial results in this direction may be found in [156, 75, 77, 165]. Such bounds are also relevant to the study of information-theoretic approaches to discrete limit theorems, such as those that involve distributional convergence to the Poisson or compound Poisson distributions of sums of random variables taking values in the nonnegative integers; we refer the interested reader to [81, 83, 169, 170, 14] for further details. Probabilistic liftings of other "sumset inequalities" from additive combinatorics can be found in [104, 141, 112, 113, 156, 110, 1, 111, 94].

There are other connections between notions of entropy and convex geometry that we have not discussed in this paper (see, e.g., [23, 7, 164, 41, 67, 64, 93]).

Acknowledgements The authors are grateful to Eric Carlen, Bernardo González Merino, Igal Sason, Tomasz Tkocz, Elisabeth Werner, and an anonymous reviewer for useful comments and references.

References

[1] E. Abbe, J. Li, and M. Madiman. Entropies of weighted sums in cyclic groups and an application to polar codes. *Preprint,* `arXiv:1512.00135`, 2015.

[2] D. Alonso-Gutiérrez, B. González Merino, C. H. Jiménez, and R. Villa. Rogers–Shephard inequality for log-concave functions. *J. Funct. Anal.*, 271(11):n, 2016.

[3] S. Artstein, K. M. Ball, F. Barthe, and A. Naor. On the rate of convergence in the entropic central limit theorem. *Probab. Theory Related Fields*, 129(3):381–390, 2004.

[4] S. Artstein, K. M. Ball, F. Barthe, and A. Naor. Solution of Shannon's problem on the monotonicity of entropy. *J. Amer. Math. Soc.*, 17(4):975–982 (electronic), 2004.

[5] S. Artstein-Avidan, K. Einhorn, D. I. Florentin, and Y. Ostrover. On Godbersen's conjecture. *Geom. Dedicata*, 178:337–350, 2015.

[6] S. Artstein-Avidan, B. Klartag, and V. Milman. The Santaló point of a function, and a functional form of the Santaló inequality. *Mathematika*, 51(1-2):33–48 (2005), 2004.

[7] S. Artstein-Avidan, B. Klartag, C. Schütt, and E. Werner. Functional affine-isoperimetry and an inverse logarithmic Sobolev inequality. *J. Funct. Anal.*, 262(9):4181–4204, 2012.

[8] D. Bakry and M. Émery. Diffusions hypercontractives. In *Séminaire de probabilités, XIX, 1983/84*, volume 1123 of *Lecture Notes in Math.*, pages 177–206. Springer, Berlin, 1985.

[9] K. Ball. Cube slicing in $\mathbf{R}^n$. *Proc. Amer. Math. Soc.*, 97(3):465–473, 1986.

[10] K. Ball. *Isometric problems in ℓ^p and sections of convex sets.* PhD thesis, University of Cambridge, UK, 1986.

[11] K. Ball. Logarithmically concave functions and sections of convex sets in $\mathbf{R}^n$. *Studia Math.*, 88(1):69–84, 1988.

[12] K. Ball, P. Nayar, and T. Tkocz. A reverse entropy power inequality for log-concave random vectors. *Preprint,* `arXiv:1509.05926`, 2015.

[13] K. Ball and V. H. Nguyen. Entropy jumps for isotropic log-concave random vectors and spectral gap. *Studia Math.*, 213(1):81–96, 2012.

[14] A. D. Barbour, O. Johnson, I. Kontoyiannis, and M. Madiman. Compound Poisson approximation via information functionals. *Electron. J. Probab.*, 15(42):1344–1368, 2010.

[15] F. Barthe. Optimal Young's inequality and its converse: a simple proof. *Geom. Funct. Anal.*, 8(2):234–242, 1998.

[16] F. Barthe. Restricted Prékopa-Leindler inequality. *Pacific J. Math.*, 189(2):211–222, 1999.

[17] Y. Baryshnikov, R. Ghrist, and M. Wright. Hadwiger's Theorem for definable functions. *Adv. Math.*, 245:573–586, 2013.

[18] W. Beckner. Inequalities in Fourier analysis. *Ann. of Math. (2)*, 102(1):159–182, 1975.

[19] G. Bianchi and M. Kelly. A Fourier analytic proof of the Blaschke-Santaló Inequality. *Proc. Amer. Math. Soc.*, 143(11):4901–4912, 2015.

[20] N.M. Blachman. The convolution inequality for entropy powers. *IEEE Trans. Information Theory*, IT-11:267–271, 1965.

[21] W. Blaschke. Uber affine Geometrie VII: Neue Extremeingenschaften von Ellipse und Ellipsoid. *Ber. Verh. Sächs. Akad. Wiss., Math. Phys. Kl.*, 69:412–420, 1917.

[22] S. Bobkov, M. Fradelizi, J. Li, and M. Madiman. When can one invert Hölder's inequality? (and why one may want to). *Preprint*, 2016.

[23] S. Bobkov and M. Madiman. Concentration of the information in data with log-concave distributions. *Ann. Probab.*, 39(4):1528–1543, 2011.

[24] S. Bobkov and M. Madiman. Dimensional behaviour of entropy and information. *C. R. Acad. Sci. Paris Sér. I Math.*, 349:201–204, Février 2011.

[25] S. Bobkov and M. Madiman. The entropy per coordinate of a random vector is highly constrained under convexity conditions. *IEEE Trans. Inform. Theory*, 57(8):4940–4954, August 2011.

[26] S. Bobkov and M. Madiman. Reverse Brunn-Minkowski and reverse entropy power inequalities for convex measures. *J. Funct. Anal.*, 262:3309–3339, 2012.

[27] S. Bobkov and A. Marsiglietti. Variants of entropy power inequality. *Preprint*, `arXiv:1609.04897`, 2016.

[28] S. G. Bobkov. Convex bodies and norms associated to convex measures. *Probab. Theory Related Fields*, 147(1-2):303–332, 2010.

[29] S. G. Bobkov. On Milman's ellipsoids and M-position of convex bodies. In C. Houdré, M. Ledoux, E. Milman, and M. Milman, editors, *Concentration, Functional Inequalities and Isoperimetry*, volume 545 of *Contemp. Math.*, pages 23–33. Amer. Math. Soc., 2011.

[30] S. G. Bobkov and G. P. Chistyakov. Entropy power inequality for the Rényi entropy. *IEEE Trans. Inform. Theory*, 61(2):708–714, February 2015.

[31] S. G. Bobkov, A. Colesanti, and I. Fragalà. Quermassintegrals of quasi-concave functions and generalized Prékopa-Leindler inequalities. *Manuscripta Math.*, 143(1-2):131–169, 2014.

[32] S. G. Bobkov, M. Madiman, and L. Wang. Fractional generalizations of Young and Brunn-Minkowski inequalities. In C. Houdré, M. Ledoux, E. Milman, and M. Milman, editors, *Concentration, Functional Inequalities and Isoperimetry*, volume 545 of *Contemp. Math.*, pages 35–53. Amer. Math. Soc., 2011.

[33] S. G. Bobkov and M. M. Madiman. On the problem of reversibility of the entropy power inequality. In *Limit theorems in probability, statistics and number theory*, volume 42 of *Springer Proc. Math. Stat.*, pages 61–74. Springer, Heidelberg, 2013. Available online at `arXiv:1111.6807`.

[34] C. Borell. Convex measures on locally convex spaces. *Ark. Mat.*, 12:239–252, 1974.

[35] C. Borell. Convex set functions in d-space. *Period. Math. Hungar.*, 6(2):111–136, 1975.

[36] J. Bourgain and V. D. Milman. New volume ratio properties for convex symmetric bodies in $\mathbf{R}^n$. *Invent. Math.*, 88(2):319–340, 1987.

[37] H. J. Brascamp and E. H. Lieb. Best constants in Young's inequality, its converse, and its generalization to more than three functions. *Advances in Math.*, 20(2):151–173, 1976.

[38] H. J. Brascamp and E. H. Lieb. On extensions of the Brunn-Minkowski and Prékopa-Leindler theorems, including inequalities for log concave functions, and with an application to the diffusion equation. *J. Functional Analysis*, 22(4):366–389, 1976.

[39] A. Burchard. A short course on rearrangement inequalities. *Available online at* http://www.math.utoronto.ca/almut/rearrange.pdf, June 2009.

[40] H. Busemann. A theorem on convex bodies of the Brunn-Minkowski type. *Proc. Nat. Acad. Sci. U. S. A.*, 35:27–31, 1949.

[41] U. Caglar and E. M. Werner. Divergence for s-concave and log concave functions. *Adv. Math.*, 257:219–247, 2014.

[42] E. A. Carlen and A. Soffer. Entropy production by block variable summation and central limit theorems. *Comm. Math. Phys.*, 140, 1991.

[43] M. Christ. Near-extremizers of Young's inequality for $\mathbb{R}^d$. *Preprint*, `arXiv:1112.4875`, 2011.

[44] M. Christ. Near equality in the Brunn-Minkowski inequality. *Preprint*, `arXiv:1207.5062`, 2012.

[45] M. Christ. Near equality in the two-dimensional Brunn-Minkowski inequality. *Preprint*, `arXiv:1206.1965`, 2012.

[46] A. Colesanti. Functional inequalities related to the Rogers-Shephard inequality. *Mathematika*, 53(1):81–101 (2007), 2006.

[47] A. Colesanti. Log concave functions. *Preprint*, 2016.

[48] D. Cordero-Erausquin and M. Ledoux. The geometry of Euclidean convolution inequalities and entropy. *Proc. Amer. Math. Soc.*, 138(8):2755–2769, 2010.

[49] J. Costa, A. Hero, and C. Vignat. On solutions to multivariate maximum alpha-entropy problems. *Lecture Notes in Computer Science*, 2683(EMMCVPR 2003, Lisbon, 7-9 July 2003):211–228, 2003.

[50] M.H.M. Costa. A new entropy power inequality. *IEEE Trans. Inform. Theory*, 31(6): 751–760, 1985.

[51] M.H.M. Costa and T.M. Cover. On the similarity of the entropy power inequality and the Brunn-Minkowski inequality. *IEEE Trans. Inform. Theory*, 30(6):837–839, 1984.
[52] T. Courtade. Strengthening the entropy power inequality. *Preprint,* `arXiv:1602.03033`, 2016.
[53] T. Courtade, M. Fathi, and A. Pananjady. Wasserstein Stability of the Entropy Power Inequality for Log-Concave Densities. *Preprint,* `arXiv:1610.07969`, 2016.
[54] T. M. Cover and J. A. Thomas. *Elements of information theory*. Wiley-Interscience [John Wiley & Sons], Hoboken, NJ, second edition, 2006.
[55] T. M. Cover and Z. Zhang. On the maximum entropy of the sum of two dependent random variables. *IEEE Trans. Inform. Theory*, 40(4):1244–1246, 1994.
[56] A. Dembo, T.M. Cover, and J.A. Thomas. Information-theoretic inequalities. *IEEE Trans. Inform. Theory*, 37(6):1501–1518, 1991.
[57] V. I. Diskant. Stability of the solution of a Minkowski equation. *Sibirsk. Mat. Ž.*, 14:669–673, 696, 1973.
[58] Yu. S. Eliseeva, F. Götze, and A. Yu. Zaitsev. Arak inequalities for concentration functions and the Littlewood–Offord problem. *Preprint,* `arXiv:1506.09034`, 2015.
[59] W. R. Emerson and F. P. Greenleaf. Asymptotic behavior of products $C^p = C + \cdots + C$ in locally compact abelian groups. *Trans. Amer. Math. Soc.*, 145:171–204, 1969.
[60] A. Figalli and D. Jerison. Quantitative stability for sumsets in $\mathbb{R}^n$. *J. Eur. Math. Soc. (JEMS)*, 17(5):1079–1106, 2015.
[61] A. Figalli, F. Maggi, and A. Pratelli. A refined Brunn-Minkowski inequality for convex sets. In *Annales de l'Institut Henri Poincare (C) Non Linear Analysis*. Elsevier, 2009.
[62] A. Figalli, F. Maggi, and A. Pratelli. A mass transportation approach to quantitative isoperimetric inequalities. *Invent. Math.*, 182(1):167–211, 2010.
[63] M. Fradelizi, A. Giannopoulos, and M. Meyer. Some inequalities about mixed volumes. *Israel J. Math.*, 135:157–179, 2003.
[64] M. Fradelizi, J. Li, and M. Madiman. Concentration of information content for convex measures. *Preprint,* `arXiv:1512.01490`, 2015.
[65] M. Fradelizi, M. Madiman, A. Marsiglietti, and A. Zvavitch. Do Minkowski averages get progressively more convex? *C. R. Acad. Sci. Paris Sér. I Math.*, 354(2):185–189, February 2016.
[66] M. Fradelizi, M. Madiman, A. Marsiglietti, and A. Zvavitch. On the monotonicity of Minkowski sums towards convexity. *Preprint*, 2016.
[67] M. Fradelizi, M. Madiman, and L. Wang. Optimal concentration of information content for log-concave densities. In C. Houdré, D. Mason, P. Reynaud-Bouret, and J. Rosinski, editors, *High Dimensional Probability VII: The Cargèse Volume*, Progress in Probability. Birkhäuser, Basel, 2016. Available online at `arXiv:1508.04093`.
[68] M. Fradelizi and A. Marsiglietti. On the analogue of the concavity of entropy power in the Brunn-Minkowski theory. *Adv. in Appl. Math.*, 57:1–20, 2014.
[69] M. Fradelizi and M. Meyer. Some functional forms of Blaschke-Santaló inequality. *Math. Z.*, 256(2):379–395, 2007.
[70] R. J. Gardner. The Brunn-Minkowski inequality. *Bull. Amer. Math. Soc. (N.S.)*, 39(3): 355–405 (electronic), 2002.
[71] R. J. Gardner and P. Gronchi. A Brunn-Minkowski inequality for the integer lattice. *Trans. Amer. Math. Soc.*, 353(10):3995–4024 (electronic), 2001.
[72] A. Giannopoulos, G. Paouris, and B.-H. Vritsiou. The isotropic position and the reverse Santaló inequality. *Israel J. Math.*, 203(1):1–22, 2014.
[73] H. Groemer. On the Brunn-Minkowski theorem. *Geom. Dedicata*, 27(3):357–371, 1988.
[74] C. Haberl. L_p intersection bodies. *Adv. Math.*, 217(6):2599–2624, 2008.
[75] S. Haghighatshoar, E. Abbe, and E. Telatar. A new entropy power inequality for integer-valued random variables. *IEEE Trans. Inform. Th.*, 60(7):3787–3796, July 2014.
[76] D. Hensley. Slicing convex bodies—bounds for slice area in terms of the body's covariance. *Proc. Amer. Math. Soc.*, 79(4):619–625, 1980.

[77] V. Jog and V. Anantharam. The entropy power inequality and Mrs. Gerber's lemma for groups of order 2^n. *IEEE Trans. Inform. Theory*, 60(7):3773–3786, 2014.

[78] V. Jog and V. Anantharam. On the geometry of convex typical sets. In *Proc. IEEE Intl. Symp. Inform. Theory*, Hong Kong, China, June 2015.

[79] O. Johnson. A conditional entropy power inequality for dependent variables. *IEEE Trans. Inform. Theory*, 50(8):1581–1583, 2004.

[80] O. Johnson. An information-theoretic central limit theorem for finitely susceptible FKG systems. *Teor. Veroyatn. Primen.*, 50(2):331–343, 2005.

[81] O. Johnson. Log-concavity and the maximum entropy property of the Poisson distribution. *Stochastic Process. Appl.*, 117(6):791–802, 2007.

[82] O. Johnson and A.R. Barron. Fisher information inequalities and the central limit theorem. *Probab. Theory Related Fields*, 129(3):391–409, 2004.

[83] O. Johnson, I. Kontoyiannis, and M. Madiman. Log-concavity, ultra-log-concavity, and a maximum entropy property of discrete compound Poisson measures. *Discrete Appl. Math.*, 161:1232–1250, 2013. DOI: 10.1016/j.dam.2011.08.025.

[84] O. Johnson and C. Vignat. Some results concerning maximum Rényi entropy distributions. *Ann. Inst. H. Poincaré Probab. Statist.*, 43(3):339–351, 2007.

[85] Oliver Johnson. *Information theory and the central limit theorem*. Imperial College Press, London, 2004.

[86] T. H. Kjeldsen. From measuring tool to geometrical object: Minkowski's development of the concept of convex bodies. *Arch. Hist. Exact Sci.*, 62(1):59–89, 2008.

[87] T. H. Kjeldsen. Egg-forms and measure-bodies: different mathematical practices in the early history of the modern theory of convexity. *Sci. Context*, 22(1):85–113, 2009.

[88] B. Klartag and V. D. Milman. Geometry of log-concave functions and measures. *Geom. Dedicata*, 112:169–182, 2005.

[89] J. Lehec. A direct proof of the functional Santaló inequality. *C. R. Math. Acad. Sci. Paris*, 347(1-2):55–58, 2009.

[90] J. Lehec. Partitions and functional Santaló inequalities. *Arch. Math. (Basel)*, 92(1):89–94, 2009.

[91] L. Leindler. On a certain converse of Hölder's inequality. In *Linear operators and approximation (Proc. Conf., Oberwolfach, 1971)*, pages 182–184. Internat. Ser. Numer. Math., Vol. 20. Birkhäuser, Basel, 1972.

[92] L. Leindler. On a certain converse of Hölder's inequality. II. *Acta Sci. Math. (Szeged)*, 33(3-4):217–223, 1972.

[93] J. Li, M. Fradelizi, and M. Madiman. Information concentration for convex measures. In *Proc. IEEE Intl. Symp. Inform. Theory*, Barcelona, Spain, July 2016.

[94] J. Li and M. Madiman. A combinatorial approach to small ball inequalities for sums and differences. *Preprint*, `arXiv:1601.03927`, 2016.

[95] W. V. Li and Q.-M. Shao. Gaussian processes: inequalities, small ball probabilities and applications. In *Stochastic processes: theory and methods*, volume 19 of *Handbook of Statist.*, pages 533–597. North-Holland, Amsterdam, 2001.

[96] E. H. Lieb. Proof of an entropy conjecture of Wehrl. *Comm. Math. Phys.*, 62(1):35–41, 1978.

[97] G. Livshyts, G. Paouris, and P. Pivovarov. On sharp bounds for marginal densities of product measures. *Preprint*, `arXiv:1507.07949`, 2015.

[98] J. Lott and C. Villani. Ricci curvature for metric-measure spaces via optimal transport. *Ann. of Math. (2)*, 169(3):903–991, 2009.

[99] L. A. Lusternik. Die Brunn-Minkowskische ungleichung fur beliebige messbare mengen. *C. R. (Doklady) Acad. Sci. URSS*, 8:55–58, 1935.

[100] E. Lutwak. The Brunn-Minkowski-Firey theory. I. Mixed volumes and the Minkowski problem. *J. Differential Geom.*, 38(1):131–150, 1993.

[101] E. Lutwak, S. Lv, D. Yang, and G. Zhang. Affine moments of a random vector. *IEEE Trans. Inform. Theory*, 59(9):5592–5599, September 2013.

[102] E. Lutwak, D. Yang, and G. Zhang. Moment-entropy inequalities. *Ann. Probab.*, 32(1B):757–774, 2004.

[103] E. Lutwak, D. Yang, and G. Zhang. Moment-entropy inequalities for a random vector. *IEEE Trans. Inform. Theory*, 53(4):1603–1607, 2007.

[104] M. Madiman. On the entropy of sums. In *Proc. IEEE Inform. Theory Workshop*, pages 303–307. Porto, Portugal, 2008.

[105] M. Madiman. A primer on entropic limit theorems. *Preprint*, 2017.

[106] M. Madiman and A.R. Barron. The monotonicity of information in the central limit theorem and entropy power inequalities. In *Proc. IEEE Intl. Symp. Inform. Theory*, pages 1021–1025. Seattle, July 2006.

[107] M. Madiman and A.R. Barron. Generalized entropy power inequalities and monotonicity properties of information. *IEEE Trans. Inform. Theory*, 53(7):2317–2329, July 2007.

[108] M. Madiman and F. Ghassemi. The entropy power of sums is fractionally superadditive. In *Proc. IEEE Intl. Symp. Inform. Theory*, pages 295–298. Seoul, Korea, 2009.

[109] M. Madiman and F. Ghassemi. Combinatorial entropy power inequalities: A preliminary study of the Stam region. *Preprint*, 2016.

[110] M. Madiman and I. Kontoyiannis. The entropies of the sum and the difference of two IID random variables are not too different. In *Proc. IEEE Intl. Symp. Inform. Theory*, Austin, Texas, June 2010.

[111] M. Madiman and I. Kontoyiannis. Entropy bounds on abelian groups and the Ruzsa divergence. *Preprint*, `arXiv:1508.04089`, 2015.

[112] M. Madiman, A. Marcus, and P. Tetali. Information-theoretic inequalities in additive combinatorics. In *Proc. IEEE Inform. Theory Workshop*, Cairo, Egypt, January 2010.

[113] M. Madiman, A. Marcus, and P. Tetali. Entropy and set cardinality inequalities for partition-determined functions. *Random Struct. Alg.*, 40:399–424, 2012.

[114] M. Madiman, J. Melbourne, and P. Xu. Rogozin's convolution inequality for locally compact groups. *Preprint*, 2016.

[115] M. Madiman, L. Wang, and S. Bobkov. Some applications of the nonasymptotic equipartition property of log-concave distributions. *Preprint*, 2016.

[116] M. Meyer and S. Reisner. Characterizations of affinely-rotation-invariant log-concave measures by section-centroid location. In *Geometric aspects of functional analysis (1989–90)*, volume 1469 of *Lecture Notes in Math.*, pages 145–152. Springer, Berlin, 1991.

[117] M. Meyer and S. Reisner. A geometric property of the boundary of symmetric convex bodies and convexity of flotation surfaces. *Geom. Dedicata*, 37(3):327–337, 1991.

[118] M. Meyer and S. Reisner. The convex intersection body of a convex body. *Glasg. Math. J.*, 53(3):523–534, 2011.

[119] E. Milman. Sharp isoperimetric inequalities and model spaces for the curvature-dimension-diameter condition. *J. Eur. Math. Soc. (JEMS)*, 17(5):1041–1078, 2015.

[120] V. Milman and L. Rotem. α-concave functions and a functional extension of mixed volumes. *Electron. Res. Announc. Math. Sci.*, 20:1–11, 2013.

[121] V. Milman and L. Rotem. Mixed integrals and related inequalities. *J. Funct. Anal.*, 264(2):570–604, 2013.

[122] V. D. Milman. Inégalité de Brunn-Minkowski inverse et applications à la théorie locale des espaces normés. *C. R. Acad. Sci. Paris Sér. I Math.*, 302(1):25–28, 1986.

[123] V. D. Milman. Entropy point of view on some geometric inequalities. *C. R. Acad. Sci. Paris Sér. I Math.*, 306(14):611–615, 1988.

[124] V. D. Milman. Isomorphic symmetrizations and geometric inequalities. In *Geometric aspects of functional analysis (1986/87)*, volume 1317 of *Lecture Notes in Math.*, pages 107–131. Springer, Berlin, 1988.

[125] V. D. Milman. Geometrization of probability. In *Geometry and dynamics of groups and spaces*, volume 265 of *Progr. Math.*, pages 647–667. Birkhäuser, Basel, 2008.

[126] V. D. Milman and A. Pajor. Isotropic position and inertia ellipsoids and zonoids of the unit ball of a normed n-dimensional space. In *Geometric aspects of functional analysis (1987–88)*, volume 1376 of *Lecture Notes in Math.*, pages 64–104. Springer, Berlin, 1989.

[127] H. H. Nguyen and V. H. Vu. Small ball probability, inverse theorems, and applications. In *Erdös centennial*, volume 25 of *Bolyai Soc. Math. Stud.*, pages 409–463. János Bolyai Math. Soc., Budapest, 2013.

[128] M. Payaró and D. P. Palomar. Hessian and concavity of mutual information, differential entropy, and entropy power in linear vector Gaussian channels. *IEEE Trans. Inform. Theory*, 55(8):3613–3628, 2009.

[129] C. M. Petty. Affine isoperimetric problems. In *Discrete geometry and convexity (New York, 1982)*, volume 440 of *Ann. New York Acad. Sci.*, pages 113–127. New York Acad. Sci., New York, 1985.

[130] G. Pisier. *The volume of convex bodies and Banach space geometry*, volume 94 of *Cambridge Tracts in Mathematics*. Cambridge University Press, Cambridge, 1989.

[131] A. Prékopa. Logarithmic concave measures with application to stochastic programming. *Acta Sci. Math. (Szeged)*, 32:301–316, 1971.

[132] A. Prékopa. On logarithmic concave measures and functions. *Acta Sci. Math. (Szeged)*, 34:335–343, 1973.

[133] E. Ram and I. Sason. On Rényi Entropy Power Inequalities. *Preprint*, 2016.

[134] O. Rioul. Information theoretic proofs of entropy power inequalities. *IEEE Trans. Inform. Theory*, 57(1):33–55, 2011.

[135] R.T. Rockafellar. *Convex Analysis*. Princeton University Press, Princeton, NJ, 1997. Reprint of the 1970 original, Princeton Paperbacks.

[136] C. A. Rogers and G. C. Shephard. The difference body of a convex body. *Arch. Math. (Basel)*, 8:220–233, 1957.

[137] B. A. Rogozin. An estimate for the maximum of the convolution of bounded densities. *Teor. Veroyatnost. i Primenen.*, 32(1):53–61, 1987.

[138] M. Rudelson and R. Vershynin. Smallest singular value of a random rectangular matrix. *Comm. Pure Appl. Math.*, 62(12):1707–1739, 2009.

[139] M. Rudelson and R. Vershynin. Non-asymptotic theory of random matrices: extreme singular values. In *Proceedings of the International Congress of Mathematicians. Volume III*, pages 1576–1602. Hindustan Book Agency, New Delhi, 2010.

[140] I. Z. Ruzsa. Generalized arithmetical progressions and sumsets. *Acta Math. Hungar.*, 65(4):379–388, 1994.

[141] I. Z. Ruzsa. Entropy and sumsets. *Random Struct. Alg.*, 34:1–10, 2009.

[142] L. A. Santaló. An affine invariant for convex bodies of n-dimensional space. *Portugaliae Math.*, 8:155–161, 1949.

[143] G. Savaré and G. Toscani. The concavity of Rènyi entropy power. *IEEE Trans. Inform. Theory*, 60(5):2687–2693, May 2014.

[144] A. Segal. Remark on stability of Brunn-Minkowski and isoperimetric inequalities for convex bodies. In *Geometric aspects of functional analysis*, volume 2050 of *Lecture Notes in Math.*, pages 381–391. Springer, Heidelberg, 2012.

[145] C.E. Shannon. A mathematical theory of communication. *Bell System Tech. J.*, 27:379–423, 623–656, 1948.

[146] D. Shlyakhtenko. Shannon's monotonicity problem for free and classical entropy. *Proc. Natl. Acad. Sci. USA*, 104(39):15254–15258 (electronic), 2007. With an appendix by Hanne Schultz.

[147] B. Simon. *Convexity: An analytic viewpoint*, volume 187 of *Cambridge Tracts in Mathematics*. Cambridge University Press, Cambridge, 2011.

[148] A.J. Stam. Some inequalities satisfied by the quantities of information of Fisher and Shannon. *Information and Control*, 2:101–112, 1959.

[149] Y. V. Stanchescu. An upper bound for d-dimensional difference sets. *Combinatorica*, 21(4):591–595, 2001.

[150] R. M. Starr. Quasi-equilibria in markets with non-convex preferences. *Econometrica*, 37(1):25–38, January 1969.

[151] K.-T. Sturm. On the geometry of metric measure spaces. I. *Acta Math.*, 196(1):65–131, 2006.

[152] S. J. Szarek and D. Voiculescu. Volumes of restricted Minkowski sums and the free analogue of the entropy power inequality. *Comm. Math. Phys.*, 178(3):563–570, 1996.

[153] S. J. Szarek and D. Voiculescu. Shannon's entropy power inequality via restricted Minkowski sums. In *Geometric aspects of functional analysis*, volume 1745 of *Lecture Notes in Math.*, pages 257–262. Springer, Berlin, 2000.

[154] S. Takano. The inequalities of Fisher information and entropy power for dependent variables. In *Probability theory and mathematical statistics (Tokyo, 1995)*, pages 460–470. World Sci. Publ., River Edge, NJ, 1996.

[155] S. Takano. Entropy and a limit theorem for some dependent variables. In *Proceedings of Prague Stochastics '98*, volume 2, pages 549–552. Union of Czech Mathematicians and Physicists, 1998.

[156] T. Tao. Sumset and inverse sumset theory for Shannon entropy. *Combin. Probab. Comput.*, 19(4):603–639, 2010.

[157] T. Tao and V. Vu. *Additive combinatorics*, volume 105 of *Cambridge Studies in Advanced Mathematics*. Cambridge University Press, Cambridge, 2006.

[158] T. Tao and V. Vu. From the Littlewood-Offord problem to the circular law: universality of the spectral distribution of random matrices. *Bull. Amer. Math. Soc. (N.S.)*, 46(3):377–396, 2009.

[159] G. Toscani. A Strengthened Entropy Power Inequality for Log-Concave Densities. *IEEE Trans. Inform. Theory*, 61(12):6550–6559, 2015.

[160] A. M. Tulino and S. Verdú. Monotonic decrease of the non-gaussianness of the sum of independent random variables: A simple proof. *IEEE Trans. Inform. Theory*, 52(9):4295–7, September 2006.

[161] C. Villani. A short proof of the "concavity of entropy power". *IEEE Trans. Inform. Theory*, 46(4):1695–1696, 2000.

[162] L. Wang and M. Madiman. Beyond the entropy power inequality, via rearrangements. *IEEE Trans. Inform. Theory*, 60(9):5116–5137, September 2014.

[163] L. Wang, J. O. Woo, and M. Madiman. A lower bound on the Rényi entropy of convolutions in the integers. In *Proc. IEEE Intl. Symp. Inform. Theory*, pages 2829–2833, Honolulu, Hawaii, July 2014.

[164] E. M. Werner. Rényi divergence and L_p-affine surface area for convex bodies. *Adv. Math.*, 230(3):1040–1059, 2012.

[165] J. O. Woo and M. Madiman. A discrete entropy power inequality for uniform distributions. In *Proc. IEEE Intl. Symp. Inform. Theory*, Hong Kong, China, June 2015.

[166] P. Xu, J. Melbourne, and M. Madiman. Reverse entropy power inequalities for s-concave densities. In *Proc. IEEE Intl. Symp. Inform. Theory.*, pages 2284–2288, Barcelona, Spain, July 2016.

[167] W. H. Young. On the multiplication of successions of Fourier constants. *Proc. Roy. Soc. Lond. Series A*, 87:331—339, 1912.

[168] Y. Yu. Letter to the editor: On an inequality of Karlin and Rinott concerning weighted sums of i.i.d. random variables. *Adv. in Appl. Probab.*, 40(4):1223–1226, 2008.

[169] Y. Yu. Monotonic convergence in an information-theoretic law of small numbers. *IEEE Trans. Inform. Theory*, 55(12):5412–5422, 2009.

[170] Y. Yu. On the entropy of compound distributions on nonnegative integers. *IEEE Trans. Inform. Theory*, 55(8):3645–3650, 2009.

[171] R. Zamir and M. Feder. A generalization of the entropy power inequality with applications. *IEEE Trans. Inform. Theory*, 39(5):1723–1728, 1993.

[172] R. Zamir and M. Feder. On the volume of the Minkowski sum of line sets and the entropy-power inequality. *IEEE Trans. Inform. Theory*, 44(7):3039–3063, 1998.

Log-Concave Functions

Andrea Colesanti

Abstract We attempt to provide a description of the geometric theory of log-concave functions. We present the main aspects of this theory: operations between log-concave functions; duality; inequalities including the Prékopa-Leindler inequality and the functional form of Blaschke-Santaló inequality and its converse; functional versions of area measure and mixed volumes; valuations on log-concave functions.

Keywords Log-concave functions • Convex bodies • Functional inequalities • Valuations

2010 *Mathematics Subject Classification*. 26B25, 52A39.

1 Introduction

A function f is log-concave if it is of the form

$$f = e^{-u}$$

where u is convex. This simple structure might suggest that there cannot be anything too deep or interesting behind. Moreover, it is clear that log-concave functions are in one-to-one correspondence with convex functions, for which there exists a satisfactory and consolidated theory. Why to develop yet another theory?

Despite these considerations, which may occur to those who meet these functions for the first time, the theory of log-concave functions is rich, young, and promising. There are two main reasons for that. The first comes from probability theory: many important examples of probability measures on $\mathbb{R}^n$, starting with the Gaussian measure, have a log-concave density. These measures are referred to as log-concave

A. Colesanti (✉)
Dipartimento di Matematica e Informatica "U. Dini", Università degli Studi di Firenze, Firenze, Italy
e-mail: colesant@math.unifi.it

E. Carlen et al. (eds.), *Convexity and Concentration*, The IMA Volumes in Mathematics and its Applications 161, DOI 10.1007/978-1-4939-7005-6_15

probability measures (and thanks to a celebrated results of Borell they admit an equivalent and more direct characterization, see [14, 15]). They have been attracting more and more interest over the last years. Typical results that have been proved for these measures are: Poincaré (or spectral gap) and log-Sobolev inequalities, concentration phenomena, isoperimetric type inequalities, central limit theorems, and so on (see [45] for a survey).

The second motivation comes from convex geometry and gives rise to the geometric theory of log-concave functions, which is the theme of this paper. There is a natural way to embed the set of convex bodies in that of log-concave functions, and there are surprisingly many analogies between the theory of convex bodies and that of log-concave functions. The extension of notions and propositions from the context of convex bodies to the more recent theory of log-concave functions is sometimes called *geometrization of analysis*. The seeds of this process were the Prékopa-Leindler inequality (see [42, 48]), recognized as the functional version of the Brunn-Minkowski inequality, and the discovery of a functional form of the Blaschke-Santaló inequality due to Ball (see [8]). A strong impulse to the development of geometrization of analysis was then given by the innovative ideas of Artstein-Avidan, Klartag and Milman who, through a series of papers (see [3, 5, 6], and [45]), widened the perspectives of the study of log-concave functions and transformed this subject into a more structured theory. In the course of this paper, we will see how many authors have then contributed in recent years to enrich this theory with new results, concepts, and directions for future developments.

Here we try to provide a picture of the current state of the art in this area. We will start from the beginning. In Section 3, we give a precise definition of the space of log-concave functions we work with, denoted by $\mathcal{L}^n$, and we describe basic properties of these functions. Moreover, we define the operations that are commonly used to add such functions and to multiply them by non-negative reals. Once equipped with these operations $\mathcal{L}^n$ is a convex cone of functions, just like the family of convex bodies $\mathcal{K}^n$ with respect to the Minkowski addition and the corresponding multiplication by positive scalars.

Section 4 is entirely devoted to the notion of duality. The most natural way to define the dual of a log-concave function $f = e^{-u}$ is to set

$$f^\circ := e^{-u^*}$$

where u^* is the Fenchel (or Legendre) transform of the convex function u. The effectiveness of this definition will be confirmed by the inequalities reported in the subsequent Section 5. In Section 4, we recall the basic properties of this duality relation and the characterization result due to Artstein-Avidan and Milman, which ensures that the duality mapping which takes f in f° is characterized by two elementary properties only: monotonicity, and idempotence. In the same section, we will also see a different duality relation, due to Artstein-Avidan and Milman as well, which can be applied to the subclass of $\mathcal{L}^n$ formed by geometric log-concave functions.

Inequalities are the salt of the earth, as every analyst knows, and log-concave functions are a very fertile ground by this point of view. In Section 5, we review the two main examples of inequalities in this area: the Prékopa-Leindler inequality and the functional versions of the Blaschke-Santaló inequality together with its converse. Concerning the Prékopa-Leindler inequality, we also explain its connection with the Brunn-Minkowski inequality, and we show how its infinitesimal form leads to a Poincaré inequality due to Brascamp and Lieb. In the same section we also introduce the notion of the difference function of a log-concave function and an inequality which can be interpreted as the functional version of the Rogers-Shephard inequality for the volume of the difference body of a convex body.

The analogy between convex bodies and log-concave functions has its imperfections. Here is a first discrepancy: in convex geometry the important notions of mixed volumes and mixed area measures are originated by the polynomiality of the volume of Minkowski linear combinations of convex bodies. This property fails to be true in the case of log-concave functions, at least if the usual addition (the one introduced in Section 3) is in use. Nevertheless, there have been some attempts to overcome this difficulty. In Section 6, we describe two constructions that lead to the definition of functional versions of area measure and mixed volumes for log-concave functions.

A second aspect in which the geometric theory of log-concave functions, at present, differs from that of convex bodies is given by valuations. The theory of valuations on convex bodies is one of the most active and prolific parts of convex geometry (see, for instance, Chapter 6 of [55] for an updated survey on this subject). Two milestones in this area are the Hadwiger theorem which characterizes continuous and rigid motion invariant valuations, and McMullen's decomposition theorem for continuous and translation invariant valuations. On the other hand, the corresponding theory of valuations on the space of log-concave functions is still moving the first steps, and it is not clear whether neat characterization results will be achieved in the functional setting as well. The situation is depicted in Section 7.

In the appendix of the paper we collected some of the main notions and results from convex geometry, described in a very synthetic way, for the reader's convenience.

2 Notations

We work in the n-dimensional Euclidean space $\mathbb{R}^n$, $n \geq 1$, endowed with the usual scalar product (x, y) and norm $\|x\|$. B_n denotes the unit ball of $\mathbb{R}^n$.

If A is a subset of $\mathbb{R}^n$, we denote by I_A its *indicatrix* function, defined in $\mathbb{R}^n$ as follows:

$$I_A(x) = \begin{cases} 0 & \text{if } x \in A, \\ \infty & \text{if } x \notin A. \end{cases}$$

The characteristic function of A will be denoted by χ_A:

$$\chi_A(x) = \begin{cases} 1 & \text{if } x \in A, \\ 0 & \text{if } x \notin A. \end{cases}$$

The Lebesgue measure of a (measurable) set $A \subset \mathbb{R}^n$ will be denoted by $V_n(A)$ (and sometimes called the volume of A) and

$$\int_A f dx$$

stands for the integral of a function f over A, with respect to the Lebesgue measure.

A *convex body* is a compact, convex subset of $\mathbb{R}^n$; the family of convex bodies will be denoted by $\mathcal{K}^n$. Some notions and constructions regarding convex bodies, directly used in this paper, are recalled in the appendix. For an exhaustive presentation of the theory of convex bodies the reader is referred to [55].

3 The Space $\mathcal{L}^n$

3.1 The Spaces $\mathcal{C}^n$ and $\mathcal{L}^n$

In order to define the space of log-concave functions, which we will be working with, in a precise way, we start by the definition of a specific space of *convex* functions. The typical convex function u that we will consider, is defined on the whole space $\mathbb{R}^n$ and attains, possibly, the value ∞. The domain of u is the set

$$\text{dom}(u) = \{x \in \mathbb{R}^n \,:\, u(x) < \infty\}.$$

By the convexity of u, dom(u) is a convex set. The function u is *proper* if its domain is not empty.

Definition 3.1 *We set*

$$\mathcal{C}^n = \left\{ u \,:\, \mathbb{R} \to \mathbb{R}^n \cup \{\infty\} \,:\, u \text{ convex and s.t.} \lim_{\|x\| \to \infty} u(x) = \infty \right\}$$

and

$$\mathcal{L}^n = e^{-\mathcal{C}^n} = \{f = e^{-u} \,:\, u \in \mathcal{C}^n\}.$$

Clearly in the previous definition we adopt the convention $e^{-\infty} = 0$. $\mathcal{L}^n$ is the space of log-concave functions which we will be working with. Note that the *support* of a function $f = e^{-u} \in \mathcal{L}^n$, i.e. the set

$$\text{sprt}(f) = \{x \in \mathbb{R}^n \,:\, f(x) > 0\}$$

coincides with dom(u).

Remark 3.2 As an alternative to the previous definition (to avoid the use of convex functions), one could proceed as follows. A function $f : \mathbb{R}^n \to [0, \infty)$ is said log-concave if

$$f((1-t)x_0 + tx_1) \geq f(x_0)^{1-t} f(x_1)^t, \quad \forall\, x_0, x_1 \in \mathbb{R}^n, \quad \forall\, t \in [0,1]$$

(with the convention: $0^\alpha = 0$ for every $\alpha \geq 0$). Then $\mathcal{L}^n$ is the set of all log-concave functions f such that

$$\lim_{\|x\| \to \infty} f(x) = 0.$$

There are clearly many examples of functions belonging to $\mathcal{L}^n$. We choose two of them which are particularly meaningful for our purposes.

Example 3.3 Let K be a convex body; then $I_K \in \mathcal{C}^n$. As a consequence the function e^{-I_K}, which is nothing but the characteristic function of K, belongs to $\mathcal{L}^n$.

This simple fact provides a one-to-one correspondence between the family of convex bodies and a subset of log-concave functions. In other words, $\mathcal{K}^n$ can be seen as a subset of $\mathcal{L}^n$. We will see that this embedding is in perfect harmony with the natural algebraic structure of $\mathcal{L}^n$ and $\mathcal{K}^n$.

Example 3.4 Another prototype of log-concave function is the Gaussian function

$$f(x) = e^{-\frac{\|x\|^2}{2}}$$

which clearly belongs to $\mathcal{L}^n$.

Remark 3.5 By convexity and the behavior at infinity, any function $u \in \mathcal{C}^n$ is bounded from below. As a consequence

$$f \in \mathcal{L}^n \;\Rightarrow\; \sup_{\mathbb{R}^n} f < \infty.$$

3.2 *Operations on $\mathcal{L}^n$*

We will now define an addition and a multiplication by non-negative reals on $\mathcal{L}^n$. With these operations $\mathcal{L}^n$ becomes a *cone* (but not a vector space) of functions, just like the family of convex bodies $\mathcal{K}^n$ with respect to the Minkowski addition and dilations, is a cone of sets. The operations that we are going to introduce are widely accepted to be the natural ones for $\mathcal{L}^n$. Their construction is not straightforward; the following stepwise procedure might be of some help for the reader.

Let u and v be in $\mathcal{C}^n$; their *infimal convolution*, denoted by $u \Box v$, is defined as follows:

$$(u\Box v)(x) = \inf_{y\in\mathbb{R}^n}\{u(y) + v(x-y)\}.$$

This operation is thoroughly studied in convex analysis (see, for instance, the monograph [51] by Rockafellar, to which we will refer for its properties). As a first fact, we have that $u\Box v \in \mathcal{C}^n$, i.e. this is an internal operation of $\mathcal{C}^n$ (see, for instance, [25, Prop. 2.6]). The infimal convolution has the following nice geometric interpretation (which can be easily verified): $u\Box v$ is the function whose epigraph is the vector sum of the epigraphs of u and v:

$$\mathrm{epi}(u\Box v) = \{x + y \,:\, x \in \mathrm{epi}(u),\ y \in \mathrm{epi}(v)\} = \mathrm{epi}(u) + \mathrm{epi}(v),$$

where, for $w \in \mathcal{C}^n$

$$\mathrm{epi}(w) = \{(x, y) \in \mathbb{R}^n \times \mathbb{R} \,:\, y \geq w(x)\}.$$

Naturally associated to $\Box$ there is a multiplication by positive reals: for $u \in \mathcal{C}^n$ and $\alpha > 0$ we set

$$(\alpha \times u)(x) = \alpha\, u\left(\frac{x}{\alpha}\right).$$

This definition can be extended to the case $\alpha = 0$ by setting

$$0 \times u = I_{\{0\}};$$

the reason being that $I_{\{0\}}$ acts as the identity element: $I_{\{0\}}\Box u = u$ for every $u \in \mathcal{C}^n$. Note that

$$u\Box u = 2 \times u \quad \forall\, u \in \mathcal{C}^n,$$

as it follows easily from the convexity of u.

We are now ready to define the corresponding operations on $\mathcal{L}^n$.

Definition 3.6 *Let $f = e^{-u}, g = e^{-v} \in \mathcal{L}^n$ and let $\alpha, \beta \geq 0$. We define the function $\alpha \cdot f \oplus \beta \cdot g$ as follows:*

$$(\alpha \cdot f \oplus \beta \cdot g) = e^{-(\alpha\times u\Box\beta\times v)}.$$

According to the previous definitions, when $\alpha, \beta > 0$ we have that[1]

$$(\alpha \cdot f \oplus \beta \cdot g)(x) = \sup_{y\in\mathbb{R}^n} f\left(\frac{x-y}{\alpha}\right)^{\alpha} g\left(\frac{y}{\beta}\right)^{\beta}. \tag{1}$$

[1] For this reason the sum defined here is sometimes referred to as the *Asplund product*, see, for instance, [3].

Example 3.7 As an instructive and remarkable example, let us see how these operations act on characteristic functions of convex bodies. Let $K, L \in \mathcal{K}^n$, and $\alpha, \beta \geq 0$. The *Minkowski linear combination* of K and L with coefficients α and β is

$$\alpha K + \beta L = \{\alpha x + \beta y \, : \, x \in K, \, y \in L\}.$$

The reader may check, as a simple exercise, the following identity

$$\alpha \cdot \chi_K \oplus \beta \cdot \chi_L = \chi_{\alpha K + \beta L}.$$

As $\mathcal{C}^n$ is closed with respect to $\Box$ and $\times$ (see [25, Prop. 2.6]), we have the following result.

Proposition 3.8 *Let $f, g \in \mathcal{L}^n$ and $\alpha, \beta \geq 0$. Then $\alpha \cdot f \oplus \beta \cdot g \in \mathcal{L}^n$.*

3.3 *The Volume Functional*

In the parallelism between convex geometry and the theory of log-concave functions it is important to find the corresponding notion of the volume of a convex body, in the functional setting. The natural candidate is the $L^1(\mathbb{R}^n)$-norm. Given $f \in \mathcal{L}^n$ we set

$$I(f) := \int_{\mathbb{R}^n} f(x)dx.$$

To prove that this integral is always finite we exploit the following lemma (see Lemma 2.5 in [25]).

Lemma 3.9 *Let $u \in \mathcal{C}^n$; then there exists $a > 0$ and $b \in \mathbb{R}$ such that*

$$u(x) \geq a\|x\| + b \quad \forall\, x \in \mathbb{R}^n.$$

As a consequence, if $f = e^{-u} \in \mathcal{L}^n$, we have that

$$f(x) \leq Ce^{-a\|x\|} \quad \forall\, x \in \mathbb{R}^n$$

for some $a > 0$ and $C > 0$. This implies that

$$I(f) < \infty \quad \forall f \in \mathcal{L}^n,$$

i.e.

$$\mathcal{L}^n \subset L^1(\mathbb{R}^n).$$

We will refer to the quantity $I(f)$ as the integral or the *volume* functional, evaluated at f. Note that if K is a convex body and $f = \chi_K$, then

$$I(f) = I(\chi_K) = \int_K dx = V_n(K).$$

3.4 p-Concave and Quasi-Concave Functions

A one parameter family of sets of functions which includes log-concave functions is that of p-concave functions, as the parameter p ranges in $\mathbb{R} \cup \{\pm\infty\}$. Roughly speaking a function is p-concave if its p-th power is concave in the usual sense, but the precise definition requires some preparation.

Given $p \in \mathbb{R} \cup \{\pm\infty\}$, $a, b \geq 0$ and $t \in [0, 1]$, the p-th mean of a and b, with weights t and $(1-t)$ is

$$M_p(a, b; t) := ((1-t)a^p + tb^p)^{1/p}$$

if $p > 0$. For $p < 0$, we adopt the same definition if $a > 0$ and $b > 0$, while if $ab = 0$ we simply set $M_p(a, b; t) = 0$. For $p = 0$:

$$M_0(a, b; t) := a^{1-t}b^t.$$

Finally, we set

$$M_\infty(a, b; t) := \max\{a, b\}, \quad M_{-\infty}(a, b; t) := \min\{a, b\}.$$

A non-negative function f defined on $\mathbb{R}^n$ is said to be p-concave if

$$f((1-t)x + ty) \geq M_p(f(x), f(y); t) \quad \forall\, x, y \in \mathbb{R}^n,\ \forall\, t \in [0, 1].$$

For $p = 0$, we have the condition of log-concavity; for $p = 1$, this clearly gives back the notion of concave functions; for $p = -\infty$, the above conditions identifies the so-called quasi-concave functions, which can be characterized by the convexity of their super-level sets.

In the course of this paper we will see that some of the results that we present for log-concave functions admit a corresponding form for p-concave functions.

4 Duality

The notion of conjugate, or dual, function of a log-concave function that we introduce here (following, for instance, [3]) is based on the well-known relation of duality in the realm of convex functions, provided by the Fenchel, or Legendre, transform, that we briefly recall. Let u be a convex function in $\mathbb{R}^n$; we set

$$u^*(y) = \sup_{x \in \mathbb{R}^n} (x, y) - u(x), \quad \forall\, y \in \mathbb{R}^n.$$

Remark 4.1 Being the supremum of linear functions, u^* is convex. Moreover, unless $u \equiv \infty$, $u^*(y) > -\infty$ for every y. If we require additionally that $u \in \mathcal{C}^n$ (and $u \not\equiv \infty$), then u^* is proper (see [25, Lemma 2.5]). On the other hand, $u \in \mathcal{C}^n$ does not imply, in general, $u^* \in \mathcal{C}^n$. Indeed, for $u = I_{\{0\}}$ we have $u^* \equiv 0$.

Definition 4.2 *For* $f = e^{-u} \in \mathcal{L}^n$, *we set*

$$f^\circ = e^{-u^*}.$$

A more direct characterization of f° is

$$f^\circ(y) = \inf_{x \in \mathbb{R}^n} \left[\frac{e^{-(x,y)}}{f(x)} \right]$$

(where the involved quotient has to be intended as ∞ when the denominator vanishes). Hence f° is a log-concave function (which does not necessarily belong to $\mathcal{L}^n$).

The idempotence relation (that one would expect)

$$(u^*)^* = u \tag{2}$$

has to be handled with care, as it is not always true in $\mathcal{C}^n$. This depends on the fact that the Fenchel conjugate of a function is always lower semi-continuous (l.s.c., for brevity), while u need not to have this property. On the other hand, this is the only possible obstacle for (2).

Proposition 4.3 *Let* $u \in \mathcal{C}^n$ *be l.s.c., then* (2) *holds.*

Corollary 4.4 *Let* $f \in \mathcal{L}^n$ *be upper semi-continuous (u.s.c.). Then*

$$(f^\circ)^\circ = f. \tag{3}$$

Examples 1. Let K be a convex body and I_K be its indicatrix function. Then we have

$$(I_K)^*(y) = \sup_{x \in K} (x, y) =: h_K(y) \quad \forall\, y \in \mathbb{R}^n.$$

Here, following the standard notations, we denoted by h_K the *support function* of the convex body K (see the appendix).

2. The Gaussian function is the unique element of $\mathcal{L}^n$ which is self-dual:

$$f = e^{-\frac{\|x\|^2}{2}} \Leftrightarrow f^\circ \equiv f.$$

Remark 4.5 The Fenchel transform gives another interpretation of the inf-convolution operation and, consequently, of the addition that we have defined on $\mathcal{L}^n$. Indeed, if u and v are in $\mathcal{C}^n$ and $\alpha, \beta \geq 0$, then:

$$(\alpha \times u \,\square\, \beta \times v)^* = \alpha u^* + \beta v^* \;\Rightarrow\; \alpha \times u \,\square\, \beta \times v = (\alpha u^* + \beta v^*)^*,$$

if the function on the left-hand side of the last equality is l.s.c. (see [25, Prop. 2.1]). Hence, given $f = e^{-u}, g = e^{-v} \in \mathcal{L}^n$ (such that $\alpha \cdot f \oplus \beta \cdot g$ is u.s.c.) we have

$$\alpha \cdot f \oplus \beta \cdot g = e^{-(\alpha u^* + \beta v^*)^*}. \tag{4}$$

In other words, the algebraic structure that we have set on $\mathcal{L}^n$ coincides with the usual addition of functions and multiplication by non-negative reals, applied to the conjugates of the exponents (with sign changed).

4.1 Characterization of Duality

In the papers [5] and [6], Arstein-Avidan and Milman established several powerful characterizations of duality relations in the class of convex and log-concave functions (as well as in other classes of functions). The space of convex functions in which they work is slightly different from ours. They denote by $Cvx(\mathbb{R}^n)$ the space of functions $u : \mathbb{R}^n \to \mathbb{R} \cup \{\pm\infty\}$, which are convex and l.s.c. One of their results is the following characterizations of the Fenchel conjugate, proved in [6].

Theorem 4.6 (Artstein-Avidan, Milman) *Let $\mathcal{T} : Cvx(\mathbb{R}^n) \to Cvx(\mathbb{R}^n)$ be such that:*

(1) *$\mathcal{T}\mathcal{T}u = u$ for every $u \in Cvx(\mathbb{R}^n)$;*
(2) *$u \leq v$ in $\mathbb{R}^n$ implies $\mathcal{T}(u) \geq \mathcal{T}(v)$ in $\mathbb{R}^n$.*

Then $\mathcal{T}$ coincides essentially with the Fenchel conjugate: there exist $C_0 \in \mathbb{R}$, $v_0 \in \mathbb{R}^n$ and an invertible symmetric linear transformation B of $\mathbb{R}^n$ such that for every $u \in Cvx(\mathbb{R}^n)$,

$$\mathcal{T}(u)(y) = u^*(By + v_0) + (x, v_0) + C_0, \quad \forall y \in \mathbb{R}^n.$$

A direct consequence of the previous result, is a characterization of the conjugate that we have introduced before for log-concave functions. Following the notation of [5] and [6] we set

$$LC(\mathbb{R}^n) = \{f = e^{-u} \, : \, u \in Cvx(\mathbb{R}^n)\}.$$

Theorem 4.7 (Artstein-Avidan, Milman) *Let $\mathcal{T} : LC(\mathbb{R}^n) \to LC(\mathbb{R}^n)$ be such that:*

(1) *$\mathcal{T}\mathcal{T}f = f$ for every $f \in LC(\mathbb{R}^n)$;*
(2) *$f \leq g$ in $\mathbb{R}^n$ implies $\mathcal{T}(f) \geq \mathcal{T}(g)$ in $\mathbb{R}^n$.*

Then there exist $C_0 \in \mathbb{R}$, $v_0 \in \mathbb{R}^n$ and an invertible symmetric linear transformation B of $\mathbb{R}^n$ such that for every $f \in LC(\mathbb{R}^n)$,

$$\mathcal{T}(f)(y) = C_0 e^{-(v_0,x)} f^\circ(Bx + v_0) \quad \forall y \in \mathbb{R}^n.$$

4.2 Geometric Log-Concave Functions and a Related Duality Transform

In the paper [7] the authors introduce a special subclass of $Cvx(\mathbb{R}^n)$, called the class of geometric convex functions, and denoted by $Cvx_0(\mathbb{R}^n)$. A function $u \in Cvx(\mathbb{R}^n)$ belongs to $Cvx_0(\mathbb{R}^n)$ if

$$\inf_{\mathbb{R}^n} u = \min_{\mathbb{R}^n} u = u(0) = 0.$$

Correspondingly, they define the class of geometric log-concave functions as follows:

$$LC_g(\mathbb{R}^n) = \{f = e^{-u} \, : \, u \in Cvx_0(\mathbb{R}^n)\}.$$

Note in particular that if $f \in LC_g(\mathbb{R}^n)$, then

$$0 \leq f(x) \leq 1 = f(0) = \max_{\mathbb{R}^n} f \quad \forall x \in \mathbb{R}^n.$$

For $u \in Cvx_0(\mathbb{R}^n)$ the set

$$u^{-1}(0) = \{x \, : \, u(x) = 0\}$$

is closed (by semicontinuity), convex and it contains the origin, even if not necessarily as an interior point. As an extension of the notion of polar set of a convex body having the origin in its interior (see the appendix), we set

$$(u^{-1}(0))^\circ = \{x \in \mathbb{R}^n \, : \, (x, y) \leq 1 \ \forall \, y \in u^{-1}(0)\}.$$

The new duality transform introduced in [7], denoted by $\mathcal{A}$, is defined, for $u \in Cvx_0(\mathbb{R}^n)$, by

$$(\mathcal{A}u)(x) = \begin{cases} \sup_{\{y : u(y) > 0\}} \dfrac{(x, y) - 1}{u(y)} & \text{if } x \in (u^{-1}(0))^\circ, \\ \infty & \text{otherwise.} \end{cases}$$

Many interesting properties of this transform are proved in [7]; among them, we mention that $\mathcal{A}$ is order reversing and it is an involution, i.e.

$$\mathcal{A}(\mathcal{A}u) = u \quad \forall\, u \in Cvx_0(\mathbb{R}^n). \tag{5}$$

As in the case of Fenchel transform, these features can be used to characterize this operator, together with the Fenchel transform itself.

Theorem 4.8 (Artstein-Avidan, Milman) *Let $n \geq 2$ and $\mathcal{T} : Cvx_0(\mathbb{R}^n) \to Cvx_0(\mathbb{R}^n)$ be a transform which is order reversing and is an involution. Then either*

$$\mathcal{T}u = (u^*) \circ B \quad \forall\, u \in Cvx_0(\mathbb{R}^n),$$

or

$$\mathcal{T}u = C_0(\mathcal{A}u) \circ B \quad \forall\, u \in Cvx_0(\mathbb{R}^n),$$

where B is an invertible linear transformation of $\mathbb{R}^n$, $C_0 \in \mathbb{R}$.

As an application, a corresponding characterization result can be derived for the case of geometric log-concave functions.

5 Inequalities

5.1 The Prékopa-Leindler Inequality

Let f, g, h be non-negative measurable functions defined in $\mathbb{R}^n$, and let t be a parameter which ranges in $[0, 1]$. Assume that the following condition holds:

$$f((1-t)x_0 + tx_1) \geq g(x_0)^{1-t} h(x_1)^t \quad \forall\, x_0, x_1 \in \mathbb{R}^n. \tag{6}$$

In other words, f which is evaluated at the convex linear combination of any two points is greater than the geometric mean of g and h at those points. Then the integral of f is greater than the geometric mean of the integrals of g and h:

$$\int_{\mathbb{R}^n} f dx \geq \left(\int_{\mathbb{R}^n} g dx\right)^{1-t} \left(\int_{\mathbb{R}^n} h dx\right)^t. \tag{7}$$

Inequality (7) is the general form of the Prékopa-Leindler inequality; it was proved in [42, 48], and [49].

Though the inequality (7) in itself is rather simple, the condition behind it, i.e. (6), is unusual as it is not a point-wise condition but involves the values of f, g, and h at different points. It will become clearer once it is written using the operations that we have introduced for log-concave functions. In fact, our next aim is to discover how Prékopa-Leindler is naturally connected to log-concavity. As a first step in this direction, we observe that, given g and h, one could rewrite inequality (7) replacing f by the smallest function which verifies (6). Namely, let

$$\bar{f}(z) = \sup_{(1-t)x+ty=z} g^{1-t}(x)h^t(y). \tag{8}$$

Then, if $\bar{f}$ is measurable[2], (7) holds for the triple $\bar{f}, g, h$. In view of (1), if $g, h \in \mathcal{L}^n$ then

$$\bar{f} = (1-t) \cdot g \oplus t \cdot g \in \mathcal{L}^n.$$

The second observation concerns equality conditions in (7), in which log-concave functions intervene directly. Note first that if $f = g = h$ (for which we trivially have equality in (7)), then (6) is equivalent to say that these functions are log-concave. Moreover, the converse of this claim is basically true, due to the following result proved by Dubuc (see [28, Theorem 12]). Assume that f, g, and h are such that (6) is verified and equality holds in (7); then there exist a log-concave function F, a vector x_0, and constants $c_1, c_2, \alpha, \beta \geq 0$ such that:

$$f(x) = F(x) \quad \text{a.e. in } \mathbb{R}^n,$$

$$g(x) = c_1 F(\alpha x + x_0) \quad \text{a.e. in } \mathbb{R}^n,$$

$$h(x) = c_2 F(\beta x + x_0) \quad \text{a.e. in } \mathbb{R}^n.$$

In view of what we have seen so far, we may rephrase (7) in the realm of log-concave functions in the following way.

Theorem 5.1 *Let $g, h \in \mathcal{L}^n$ and let $t \in [0,1]$. Then*

$$\int_{\mathbb{R}^n} [(1-t) \cdot g \oplus t \cdot h] dx \geq \left(\int_{\mathbb{R}^n} g dx \right)^{1-t} \left(\int_{\mathbb{R}^n} h dx \right)^t, \tag{9}$$

i.e.

$$I((1-t) \cdot g \oplus t \cdot h) \geq I(g)^{t-1} I(h)^t.$$

Moreover, equality holds if and only if g coincide with a multiple of h up to a translation and a dilation of the coordinates.

[2] In general the measurability of g and h does not imply that of f. See [34] for more information on this point.

Written in this form, the Překopa-Leindler inequality is clearly equivalent to the following statement: *the volume functional I is log-concave in the space* $\mathcal{L}^n$. This point of view will be important to derive the infinitesimal form of this inequality. In the sequel we will refer to the Prékopa-Leindler inequality in the form (9) as to (PL).

We note here an important consequence of (PL), which was emphasized and exploited in various ways in [17].

Theorem 5.2 *Let* $F = F(x, y)$ *be defined in* $\mathbb{R}^n \times \mathbb{R}^m$, *and assume that* F *is log-concave. Then the function* $f : \mathbb{R}^n \to \mathbb{R}$ *defined by*

$$f(x) = \int_{\mathbb{R}^m} F(x, y)dy$$

is log-concave.

The proof is a simple application of (PL).

We conclude this part with some further remarks on the Prékopa-Leindler inequality.

Remark 5.3 One way to look at (PL) is as a reverse form of the Hölder inequality. Indeed, an equivalent formulation of Hölder inequality is the following: if g and h are non-negative measurable functions defined on $\mathbb{R}^n$, and $t \in [0, 1]$,

$$\int_{\mathbb{R}^n} g^{1-t} h^t dx \le \left(\int_{\mathbb{R}^n} g dx \right)^{1-t} \left(\int_{\mathbb{R}^n} h dx \right)^t .$$

Prékopa-Leindler inequality asserts that the previous inequality is reversed if the geometric mean of g and h is replaced by the supremum of their geometric means, in the sense of (8).

Remark 5.4 A more general form of (PL) is the Borell-Brascamp-Lieb inequality (see, for instance, Section 10 of [34]). This inequality asserts that if f, g, h are non-negative measurable functions defined on $\mathbb{R}^n$ such that for some $p > -\frac{1}{n}$ and $t \in [0, 1]$

$$f((1-t)x + ty) \ge M_p(g(x), h(y); t) \quad \forall\, x, y \in \mathbb{R}^n,$$

then

$$\int_{\mathbb{R}^n} f dx \ge M_{\frac{p}{np+1}} \left(\int_{\mathbb{R}^n} g dx, \int_{\mathbb{R}^n} h dx; t \right).$$

Here we have used the definition of p-mean introduced in subsection 3.4.

In the same way as (PL) has a special meaning for log-concave functions, Borell-Brascamp-Lieb inequality is suited to p-concave functions.

Remark 5.5 Prékopa-Leindler inequality can also be seen as a special case of a very general class of inequalities proved by Barthe in [9]. One way (even if limiting) of looking at Barthe's inequalities is as a multifunctional version of (PL). Barthe's inequalities are in turn the reverse form of Brascamp-Lieb inequalities, which have as a simple special case the Hölder inequality. A neat presentation of these inequalities can be found in [34], Section 15.

5.2 *Proof of the Prékopa-Leindler Inequality*

For completeness we supply a proof of the Prékopa-Leindler inequality in its formulation (9), i.e. restricted to log-concave functions (omitting the characterization of equality conditions).

As preliminary steps, note that if one of the functions g and h is identically zero then the inequality is trivial. Hence we assume that $g \not\equiv 0$ and $h \not\equiv 0$. Moreover, as it is easy to check, it is not restrictive to assume

$$\sup_{\mathbb{R}^n} g = \sup_{\mathbb{R}^n} h = 1 \tag{10}$$

(see also Remark 3.5).

The rest of the proof proceeds by induction on the dimension n. For simplicity we will set

$$f = (1-t) \cdot g \oplus t \cdot h$$

throughout. For convenience of notations we will in general denote by x, y, and z the variable of f, g, and h, respectively.

The case $n = 1$. Fix $s \in [0, 1]$; by the definition of the operations $\cdot$ and $\oplus$, we have the following set inclusion

$$\{x : f(x) \geq s\} \supset (1-t)\{y : g(y) \geq s\} + t\{z : h(z) \geq s\}.$$

As f, g, and h are log-concave, their super-level sets are intervals, and, by the behavior of these functions at infinity, they are bounded. Note that if if I and J are bounded interval of the real line we have

$$V_1(I + J) = V_1(I) + V_1(J)$$

(which is the one-dimensional version of the Brunn-Minkowski inequality, in the case of "convex sets"). Hence

$$V_1(\{x : f(x) \geq s\}) \geq (1-t)V_1(\{y : g(y) \geq s\}) + tV_1(\{z : h(z) \geq s\}).$$

Now we integrate between 0 and 1 and use the layer cake principle

$$\int_{\mathbb{R}} f\,dx \geq (1-t)\int_{\mathbb{R}} g\,dy + t\int_{\mathbb{R}} h\,dz \geq \left(\int_{\mathbb{R}} g\,dy\right)^{1-t} \left(\int_{\mathbb{R}} h\,dz\right)^{t}$$

where we have used the arithmetic-geometric mean inequality. This concludes the proof in dimensional one. Note that in this case one obtains (under the assumption (10)) a stronger inequality, namely the integral of f is greater than the arithmetic mean of those of g and h.

The case $n \geq 1$. Assume that the inequality is true up to dimension $(n-1)$. Fix $\bar{y}_n$ and $\bar{z}_n$ in $\mathbb{R}$, and let $\bar{x}_n = (1-t)\bar{y}_n + t\bar{z}_n$. Moreover let $\bar{f}, \bar{g}, \bar{h} : \mathbb{R}^{n-1} \to \mathbb{R}$ be defined by

$$\bar{f}(x_1, \dots, x_{n-1}) = f(x_1, \dots, x_{n-1}, \bar{x}_n), \ \ \bar{g}(y_1, \dots, y_{n-1}) = g(y_1, \dots, y_{n-1}, \bar{y}_n),$$
$$\bar{h}(z_1, \dots, z_{n-1}) = h(z_1, \dots, z_{n-1}, \bar{z}_n).$$

As x_n is the convex linear combination of y_n and z_n, and as $f = (1-t) \cdot g \oplus t \cdot h$, we have that $\bar{f}$, $\bar{g}$, and $\bar{h}$ verify the assumption of (PL), so that, by induction,

$$\int_{\mathbb{R}^{n-1}} \bar{f} dx \geq \left(\int_{\mathbb{R}^{n-1}} \bar{g} dy \right)^{1-t} \left(\int_{\mathbb{R}^{n-1}} \bar{h} dz \right)^{t}. \tag{11}$$

Next define $F, G, H : \mathbb{R} \to \mathbb{R}$ as

$$F(x) = \int_{\mathbb{R}^{n-1}} f(x_1, \dots, x_{n-1}, x) dx_1 \dots dx_{n-1},$$
$$G(y) = \int_{\mathbb{R}^{n-1}} g(y_1, \dots, y_{n-1}, y) dy_1 \dots dy_{n-1},$$
$$H(z) = \int_{\mathbb{R}^{n-1}} h(z_1, \dots, z_{n-1}, z) dz_1 \dots dz_{n-1}.$$

By Theorem 5.2 these are log-concave functions; moreover, (11) is exactly condition (6) for them. Hence, by induction,

$$\int_{\mathbb{R}} F dx \geq \left(\int_{\mathbb{R}} G dy \right)^{1-t} \left(\int_{\mathbb{R}} H dz \right)^{t},$$

and this is nothing but the required inequality for f, g, h. □

5.3 *Prékopa-Leindler and Brunn-Minkowski Inequality*

One way to understand the importance of Prékopa-Leindler inequality is to set it in relation to the Brunn-Minkowski inequality, one of the most important results in convex geometry.

Theorem 5.6 (Brunn-Minkowski inequality) *Let K and L be convex bodies and* $t \in [0,1]$. *Then*

$$[V_n((1-t)K+tL)]^{1/n} \geq (1-t)[V_n(K)]^{1/n} + t[V_n(L)]^{1/n}. \tag{12}$$

In case both K and L have non-empty interior, equality holds if and only if they are homothetic, i.e. they coincide up to a translation and a rotation.

The article [34] by Gardner contains an exhaustive survey on this result. Here we only mention that Brunn-Minkowski inequality ((BM) for brevity) is a special case of the family of Aleksandrov-Fenchel inequalities (see [55]), and that a simple argument leads in few lines from this inequality to the isoperimetric inequalitiy (restricted to convex bodies):

$$V_n(K) \leq c\,\left[\mathcal{H}^{n-1}(\partial K)\right]^{n/(n-1)} \quad \forall\, K \in \mathcal{K}^n\,:\, \mathrm{int}(K) \neq \emptyset. \tag{13}$$

Here c is a dimensional constant and $\mathcal{H}^{n-1}$ is the $(n-1)$-dimensional Hausdorff measure. Moreover equality holds if and only if K is a ball. The argument to deduce (13) from (12) is rather known and can be found, for instance, in [34].

In what follows we show that (BM) can be easily proved through (PL).

Proof of the Brunn-Minkowsi inequality. Let K, L, and t be as in Theorem 5.6. Let

$$g = \chi_K, \quad h = \chi_L, \quad f = (1-t)\cdot g \oplus t \cdot h.$$

As we saw in example 3.7,

$$f = \chi_{(1-t)K+tL}.$$

By (PL) we get

$$V_n((1-t)K+tL) \geq V_n(K)^{1-t}\,V_n(L)^t. \tag{14}$$

This is usually referred to as the multplicative form of the Brunn-Minkowski inequality. From that, by exploiting the homogeneity of volume, (BM) in its standard form can be deduced as follows. Given K, L, and t as above, assume that the volumes of K and L are strictly positive (the general case can be obtained by approximation). Let

$$\bar{K} = \frac{1}{V_n(K)^{1/n}}\,K, \quad \bar{L} = \frac{1}{V_n(L)^{1/n}}\,L,$$

so that

$$V_n(\bar{K}) = V_m(\bar{L}) = 1.$$

We set also

$$\bar{t} = \frac{V_n(L)^{1/n}}{(1-t)V_n(K)^{1/n} + tV_n(L)^{1/n}}\, t.$$

Applying (14) to $\bar{K}$, $\bar{L}$, and $\bar{t}$ leads to

$$1 \leq V_n((1-\bar{t})\bar{K} + \bar{t}\bar{L}) = V_n\left(\frac{1}{(1-t)V_n(K)^{1/n} + tV_n(L)^{1/n}}(1-t)K + tL\right).$$

□

5.4 The Infinitesimal Form of (PL)

Both Prékopa-Leindler and Brunn-Minkowski inequalities are concavity inequalities. More precisely, (BM) asserts that the volume functional to the power $1/n$ is concave on the family of convex bodies $\mathcal{K}^n$, while, according to (PL), the logarithm of the integral functional I is concave on $\mathcal{L}^n$. The concavity of a functional F can be expressed by the usual inequality:

$$F((1-t)x_0 + tx_1) \geq (1-t)F(x_0) + tF(x_1) \quad \forall\, x_0, x_1;\ \forall\, t \in [0,1],$$

or by its infinitesimal version

$$D^2F(x) \leq 0 \quad \forall\, x, \tag{15}$$

where $D^2F(x)$ denotes the *second variation* of F at x (if it exists, and whatever its meaning can be). The infinitesimal form of the Brunn-Minkowski inequality has been investigated in [24], where it is shown that (15) provides a class of Poincaré type inequalities on the unit sphere of $\mathbb{R}^n$. Here we will show that correspondingly, the infinitesimal form of (PL) is equivalent to a class of (known) inequalities, also of Poincaré type, on $\mathbb{R}^n$, with respect to log-concave probability measures. These inequalities have been proved by Brascamp and Lieb in [17].

Theorem 5.7 (Brascamp-Lieb) *Let $f = e^{-u} \in \mathcal{L}^n$ and assume that $u \in C^2(\mathbb{R}^n)$ and $D^2u(x) > 0$ for every $x \in \mathbb{R}^n$. Then for every $\phi \in C^1(\mathbb{R}^n)$ such that*

$$\int_{\mathbb{R}^n} \phi f dx = 0,$$

the following inequality holds:

$$\int_{\mathbb{R}^n} \phi^2 f dx \leq \int_{\mathbb{R}^n} ((D^2u)^{-1}\nabla\phi, \nabla\phi) f dx. \tag{16}$$

Remark 5.8 When

$$u(x) = \frac{\|x\|^2}{2},$$

i.e. f is the Gaussian function, (16) becomes the usual Poincaré inequality in Gauss space:

$$\int_{\mathbb{R}^n} \phi^2 d\gamma_n(x) \leq \int_{\mathbb{R}^n} \|\nabla\phi\|^2 d\gamma_n(x) \tag{17}$$

for every $\phi \in C^1(\mathbb{R}^n)$ such that

$$\int_{\mathbb{R}^n} \phi d\gamma_n(x) = 0, \tag{18}$$

where γ_n is the standard Gaussian probability measure. Note that (17) is sharp, indeed it becomes an equality when ϕ is a linear function. In general, the left-hand side of (16) is a weighted $L^2(\mathbb{R}^n, \mu)$-norm of $\nabla\phi$ (squared), where μ is the measure with density f. Note, however, that (16) admits extremal functions (i.e., for which equality holds) for every choice of f; this will be clear from the proof that we present in the sequel.

Proof of Theorem 5.7. We will consider a special type of log-concave functions. Let $u \in C^2(\mathbb{R}^n) \cap \mathcal{C}^n$ be such that

$$cI_n \leq D^2u(x) \quad \forall\, x \in \mathbb{R}^n, \tag{19}$$

where I_n is the $n \times n$ identity matrix and $c > 0$. We denote by $\mathcal{C}_s^n$ the space formed by these functions and set

$$\mathcal{L}_s^n := e^{-\mathcal{C}_s^n} \subset \mathcal{L}^n.$$

We set

$$(\mathcal{C}_s^n)^* = \{u^* \,:\, u \in \mathcal{C}_s^n\}.$$

By standard facts from convex analysis (see, for instance, [51]), if $u \in \mathcal{C}_s^n$ then $u^* \in C^2(\mathbb{R}^n)$; moreover ∇u is a diffeomorphism between $\mathbb{R}^n$ and itself and

$$\nabla u^* = (\nabla u)^{-1}; \tag{20}$$

$$u^*(y) = ((\nabla u)^{-1}(y), y) - u((\nabla u)^{-1}y) \quad \forall\, y \in \mathbb{R}^n; \tag{21}$$

$$D^2u^*(y) = (D^2u((\nabla u)^{-1}(y)))^{-1} \quad \forall\, y \in \mathbb{R}^n. \tag{22}$$

Let $f = e^{-u} \in \mathcal{L}_s^n$; the functional I is defined by

$$I(f) = \int_{\mathbb{R}^n} f(x)dx.$$

By the change of variable $y = \nabla u(x)$ and by the previous relations we get

$$I(f) = \int_{\mathbb{R}^n} e^{u^*(y)-(y,\nabla u^*(y))} \det(D^2u^*(y))dy.$$

In other words, $I(e^{-u})$ can be expressed as an integral functional depending on u^*. Given $v \in (\mathcal{C}_s^n)^*$ set

$$J(v) = \int_{\mathbb{R}^n} e^{v(y)-(y,\nabla v(y))} \det(D^2v(y))dy.$$

By Remark 4.5, Prékopa-Leindler inequality in its form (9), restricted to $\mathcal{L}_s^n$, is equivalent to say that

$$J : (\mathcal{C}_s^n)^* \to \mathbb{R} \quad \text{is } log - concave \tag{23}$$

where now log-concavity is with respect to the usual addition of functions in $(\mathcal{C}_s^n)^*$. The previous relation is the key step of the proof. We will now determine the second variation of $\ln(J)$ at $v \in (C^n)^*$. Let $\psi \in C_c^\infty(\mathbb{R}^n)$ (i.e. $\psi \in C^\infty(\mathbb{R}^n)$ and it has compact support). There exists $\epsilon > 0$ such that

$$v_s = v + s\psi \quad \text{is convex for every } s \in [-\epsilon, \epsilon].$$

Set

$$g(s) = J(v_s).$$

Then $\ln(g(s))$ is concave in $[-\epsilon, \epsilon]$, so that

$$g(0)g''(0) - g'^2(0) \leq 0. \tag{24}$$

After computing $g'(0)$ and $g''(0)$ and returning to the variable x, inequality (24) will turn out to be nothing but the Poincaré inequality of Brascamp and Lieb.

For simplicity, from now on we will restrict ourselves to the one-dimensional case, but the same computation can be done for general dimension (at the price of some additional technical difficulties, consisting in suitable integration by parts formulas), as shown in [24] for the case of the Brunn-Minkowski inequality.

So now v and ψ are functions of one real variable; we denote by v', v'', ψ', ψ'' their first and second derivatives, respectively. The function $g(s)$ takes the form

$$g(s) = \int_{\mathbb{R}} e^{v_s(y) - yv_s'(y)} v_s''(y)dy.$$

Then

$$g'(s) = \int_{\mathbb{R}} e^{v_s(y) - yv_s'(y)} [(\psi(y) - y\psi'(y))v_s''(y) + \psi''(y)]dy.$$

Note that

$$\int_{\mathbb{R}} e^{v_s(y) - yv_s'(y)} \psi''(y)dy = \int_{\mathbb{R}} y\psi'(y)v_s''(y)dy$$

after an integration by parts (no boundary term appears as ψ has bounded support). Then

$$g'(s) = \int_{\mathbb{R}} e^{v_s(y) - yv_s'(y)} \psi(y)v_s''(y)dy.$$

Differentiating again (this time at $s = 0$) we get

$$\begin{aligned} g''(0) &= \int_{\mathbb{R}} e^{v(y) - yv'(y)} \psi(y)[(\psi(y) - y\psi'(y))v''(y) + \psi''(y)]dy \\ &= \int_{\mathbb{R}} e^{v(y) - yv'(y)} [\psi(y)v''(y) - (\psi'(y))^2]dy, \end{aligned}$$

where we have integrated by parts again in the second equality. Now set

$$\phi(x) = \psi(u'(x)).$$

Note that $\phi \in C_c^\infty(\mathbb{R})$; moreover, any $\phi \in C_c^\infty(\mathbb{R})$ can be written in the previous form for a suitable ψ. We have:

$$g'(0) = \int_{\mathbb{R}} \phi(x)f(x)dx,$$

and

$$g''(0) = \int_{\mathbb{R}} \phi^2(x)f(x) - \int_{\mathbb{R}} \frac{(\phi'(x))^2}{u''(x)}dx. \tag{25}$$

Hence (18) is equivalent to $g'(0) = 0$. If we now replace (25) in (24) we obtain the desired inequality. $\square$

Remark 5.9 There are several other examples of the argument used to derive "differential" inequalities (i.e., involving the gradient, or derivatives in general) like Poincaré, Sobolev, and log-Sobolev inequalities, starting from Prékopa-Leindler or Brunn-Minkowski inequality; see, for instance, [11, 12], and the more recent paper [13].

5.5 Functional Blaschke-Santaló Inequality and Its Converse

One of the most fascinating open problems in convex geometry is the Mahler conjecture, concerning the optimal lower bound for the so-called *volume product* of a convex body. If $K \in \mathcal{K}^n$ and the origin is an interior point of K, the polar body (with respect to 0) of K is the set

$$K^\circ = \{x \in \mathbb{R}^n \, : \, (x, y) \leq 1 \; \forall \, y \in K\}.$$

K° is also a convex body. More generally, if K has non-empty interior and z is an interior point of K, the polar body of K with respect to z is

$$K^z := (K - z)^\circ.$$

It can be proved that there exists an interior point of K, the *Santaló point*, for which $V_n(K^z)$ is minimum (see [55]).

Roughly speaking, the polar body of a large set is small and vice versa; this suggests to consider the following quantity:

$$\mathcal{P}(K) = V_n(K)V_n(K^z),$$

where z is the Santaló point of K, called the volume product of K. $\mathcal{P}$ is invariant under affine transformations of $\mathbb{R}^n$ and in particular it does not change if K is dilated (or shrunk). It is relatively easy to see that it admits a maximum and a minimum as K ranges in $\mathcal{K}^n$. Then it becomes interesting to find such extremal values and the corresponding extremizers.

The Blashcke-Santaló inequality asserts that

$$\mathcal{P}(K) \leq \mathcal{P}(B_n) \quad \forall K \in \mathcal{K}^n,$$

(we recall that B_n is the unit ball) and equality holds if and only if K is an ellipsoid (see, for instance, [55]). On the other hand, the problem of finding the minimum of $\mathcal{P}$ is still open, in dimension $n \geq 3$. The Mahler conjecture asserts that

$$\mathcal{P}(K) \geq \mathcal{P}(\Delta) \quad \forall \, K \in \mathcal{K}^n$$

where Δ is a simplex. Correspondingly, in the case of symmetric convex bodies it is conjectured that

$$\mathcal{P}(K) \geq \mathcal{P}(Q) \quad \forall K \in \mathcal{K}^n, \text{ symmetric}$$

where Q is a cube. The validity of these conjectures has been established in the plane by Mahler himself, and, in higher dimension, for some special classes of convex bodies; among them we mention zonoids and unconditional convex bodies. Anyway it would be impossible to give even a synthetic account of all the contributions and results that appeared in the last decades in this area. A recent and updated account can be found in [54]. We mention, as this result has a specific counterpart for log-concave functions, that the best known lower bound for the volume product of symmetric convex bodies (asymptotically optimal with respect n as n tends to ∞) has been established by Bourgain and Milman (see [16]):

$$\mathcal{P}(K) \geq c^n \mathcal{P}(Q), \quad \forall K \in \mathcal{K}^n, \text{ symmetric,} \tag{26}$$

where c is a constant independent of n. For a recent improvement of the constant c as well as for different proofs of (26), we again refer the reader to [54] (see in particular Section 8).

Within the framework that we have been describing so far, where results from convex geometry are systematically transferred to the space of log-concave functions, it is natural to expect a functional counterpart of the volume product of convex bodies, and related upper and lower bounds. Given a log concave function $f = e^{-u} \in \mathcal{L}^n$, we have seen that we can define

$$f^\circ = e^{-u^*}$$

where u^* is the Fenchel conjugate of u (see Section 4). Hence we are led to introduce the following quantity

$$\mathcal{P}(f) := \int_{\mathbb{R}^n} f dx \int_{\mathbb{R}^n} f^\circ dx = I(f)\, I(f^\circ)$$

as a counterpart of the volume product of a convex body. On the other hand, as suggested by the case of convex bodies, it could be important to introduce also a parameter $z \in \mathbb{R}^n$, as the center of polarity. Hence, given $f \in \mathcal{L}^n$ and $z \in \mathbb{R}^n$, we set

$$f_z(x) = f(x - z) \quad \forall\, x \in \mathbb{R}^n,$$

and more generally we consider

$$\mathcal{P}(f_z) := \int_{\mathbb{R}^n} f_z dx \int_{\mathbb{R}^n} (f_z)^\circ dx.$$

The functional Blaschke-Santaló inequality, i.e. an optimal upper bound for $\mathcal{P}(f_z)$, was established in [3] where the authors prove that for every $f \in \mathcal{L}^n$ (with positive integral), if we set

$$z_0 = \frac{1}{I(f)} \int_{\mathbb{R}^n} x f(x) dx$$

then

$$\mathcal{P}(f_{z_0}) \leq (2\pi)^n \tag{27}$$

and equality holds if and only if f is (up to a translation of the coordinate system) a Gaussian function, i.e. is of the form

$$f(x) = e^{-(Ax,x)}$$

where A is a positive definite matrix. In the special case of even functions, for which we have $z_0 = 0$, this result was achieved by Ball in [8]. A different proof (which in particular does not exploit its geometric counterpart) of the result by Artstein, Klartag, and Milman was given by Lehec in [40]. We also mention that an interesting extension of (27) was given in [30] (see also [41]).

In a similar way, the reverse Blaschke-Santaló inequality (26) has been extended to the functional case. In [38] the authors proved that there exists an absolute constant $c > 0$ (i.e., c does not depend on the dimension n) such that

$$\mathcal{P}(f_0) \geq c^n$$

for every $f \in \mathcal{L}^n$ even. This result has been improved in various ways in the papers [32] and [33] (see also [29]). We also mention that in [31] a sharp lower bound for the functional $\mathcal{P}(f)$ has been given for *unconditional* log-concave functions f (i.e., even with respect to each coordinate). This corresponds to the solution of the Mahler conjecture in the case of unconditional convex bodies.

5.6 Functional Rogers-Shephard Inequality

Given a convex body K in $\mathbb{R}^n$, its *difference body* DK is defined by

$$DK = K + (-K) = \{x + y : x \in K, -y \in K\}.$$

DK is a centrally symmetric convex body, and, in a sense, any measurement of how far is K from DK could serve as a measure of asymmetry of K. The discrepancy between K and DK can be identified via the volume ratio:

$$\frac{V_n(K)}{V_n(DK)}.$$

If we apply the Brunn-Minkowski inequality to K and $-K$, we immediately get

$$V(DK) \geq 2^n V_n(K).$$

The celebrated Rogers-Shephard inequality (see [50]) provides a corresponding upper bound:

$$V_n(DK) \leq \binom{2n}{n} V_n(K). \tag{28}$$

Equality holds in the previous inequality if and only if K is a simplex.

It is natural to wonder whether these facts may find any correspondence for log-concave functions. This question was studied in [23]. The first step is to define a notion of difference function of a log-concave function. Let $f \in \mathcal{L}^n$; we first set

$$\bar{f}(x) = f(-x) \quad \forall\, x \in \mathbb{R}^n$$

(clearly $\bar{f} \in \mathcal{L}^n$). Then we define

$$\Delta f = \frac{1}{2} \cdot f \oplus \frac{1}{2} \cdot \bar{f}.$$

In more explicit terms:

$$\Delta f(x) = \sup \left\{ \sqrt{f(y) f(-z)} \;:\; x = \frac{y+z}{2} \right\}$$

(in fact Δf corresponds to the difference body rescaled by the factor $\frac{1}{2}$).

To get a lower bound for the integral of the difference function we may use the Prékopa-Leindler inequality and obtain:

$$I(\Delta f) = \int_{\mathbb{R}^n} \Delta f dx \geq \int_{\mathbb{R}^n} f dx = I(f).$$

In [23] the following inequality was proved:

$$\int_{\mathbb{R}^n} \Delta f dx \leq 2^n \int_{\mathbb{R}^n} f dx \quad \forall f \in \mathcal{L}^n. \tag{29}$$

The previous inequality is sharp. One extremizer is the function f defined by

$$f(x) = f(x_1, \ldots, x_n) = \begin{cases} e^{-\sum_{i=1}^n x_i} & \text{if } x_i \geq 0 \text{ for every } i = 1 \ldots, n, \\ 0 & \text{otherwise.} \end{cases}$$

All other extremizers can be obtained by the previous function by an affine change of variable and the multiplication by a positive constant (see [23]).

The results of [23] have been recently extended and complemented in the papers [1] and [2], where the authors obtain considerable new developments. To describe an example of their results, given f and g in $\mathcal{L}^n$ one may consider

$$\Delta(f, g) = \frac{1}{2} \cdot f \oplus \frac{1}{2} \cdot \bar{g}.$$

In the above-mentioned papers, among other results the authors establish optimal upper bounds for the integral of $\Delta(f, g)$, which in the case $f = g$ returns the inequality (29).

5.7 The Functional Affine Isoperimetric Inequality

We conclude this section by mentioning yet another inequality for log-concave functions. As we recalled in the introduction, among the main results that can be proved for log-concave probability measures there are log-Sobolev type inequalities (we refer the reader to [39] for this type of inequalities). In the paper [4] the authors prove a reverse form of the standard log-Sobolev inequality (in the case of the Lebesgue measure). The proof of this inequality is based on an important geometric inequality in convex geometry; the *affine isoperimetric inequality*, involving the affine surface area. We refer the reader to [55] for this notion.

The research started in [4] is continued in the papers [19, 20, 21], and [22]. In particular, in these papers several possible functional extensions of the notion of affine surface area are proposed, along with functional versions of the affine isoperimetric inequality.

6 Area Measures and Mixed Volumes

6.1 The First Variation of the Total Mass Functional

Given two convex bodies K and L, for $\epsilon > 0$ consider the following perturbation of K: $K_\epsilon := K + \epsilon L$. The volume of K_ϵ, as a function of ϵ, is a polynomial and hence admits right derivative at $\epsilon = 0$:

$$\lim_{\epsilon \to 0+} \frac{V_n(K + \epsilon L) - V_n(K)}{\epsilon} =: V(\underbrace{K, \dots, K}_{(n-1)\text{-times}}, L) = V(K, \dots, K, L). \tag{30}$$

Here we used the standard notations for *mixed volumes* of convex bodies (see the appendix). The mixed volumes $V(K, \dots, K, L)$, when K is fixed and L ranges in $\mathcal{K}^n$, can be computed using the *area measure* of K. Indeed, there exists a unique non-negative Radon measure on $\mathbb{S}^{n-1}$, called the area measure of K and denoted by $S_{n-1}(K, \cdot)$, such that

$$V(K, \dots, K, L) = \frac{1}{n} \int_{\mathbb{S}^{n-1}} h_L(x) dS_{n-1}(K, x) \quad \forall L \in \mathcal{K}^n. \tag{31}$$

According to (30) we may say that $V(K, \dots, K, L)$ is the directional derivative of the volume functional at K along the direction L. Moreover, as support function behaves linearly with respect to Minkowski addition (see the appendix), (31) tells us that the first variation of the volume at K is precisely the area measure of K. Note also that if we choose L to be the unit ball B_n of $\mathbb{R}^n$, then we have (under the assumption that K has non-empty interior) that the derivative in (30) is the perimeter of K:

$$V(K, \dots, K, B_n) = \lim_{\epsilon \to 0+} \frac{V_n(K + \epsilon L) - V_n(K)}{\epsilon} = \mathcal{H}^{n-1}(\partial K)$$

where $\mathcal{H}^{n-1}$ stands for the $(n-1)$-dimensional Hausdorff measure.

One could try to follow a similar path to define a notion of area measure of a log-concave function f, replacing the volume functional by the integral of $f \in \mathcal{L}^n$

$$I(f) = \int_{\mathbb{R}^n} f(x) dx.$$

Then the idea is to compute the first variation of I and deduce as a consequence a surrogate of the area measure. More precisely, in view of (30) and (31), the problem of computing the following limit arises:

$$\delta I(f, g) := \lim_{\epsilon \to 0+} \frac{I(f \oplus \epsilon \cdot g) - I(f)}{\epsilon} \tag{32}$$

where $f,\ g \in \mathcal{L}^n$. Here a first striking difference between the geometric and the functional setting appears. While the volume of the linear combination of convex bodies is always polynomial in the coefficients, this is not the case for functions. Indeed (see, for instance, [25]) there are examples in which $\delta I(f, g) = \infty$.

The idea to compute the limit (32) appeared for the first time in the papers [38, 52] and [53], for a specific choice of the function f (the density of the Gaussian measure), in order to define a notion of mean width (one of the intrinsic volumes) of log-concave functions. The computation of the same limit for general f and g was then considered in [25].

Even if the limit (32) exists (finite of infinite) under the sole assumption $I(f) > 0$ (see [25] and [38]), explicit formulas for it (e.g., similar to (31)) have been found only under quite restrictive assumptions. To give an example of such formulas we rephrase Theorem 4.5 in [25]. This result needs some preparation. First of all we denote by $C^2_+(\mathbb{R}^n)$ the set of functions u from $C^2(\mathbb{R}^n)$ such that $D^2u > 0$ in $\mathbb{R}^n$. Next we define

$$\mathcal{C}^n_s = \left\{ u \in \mathcal{C}^n \,:\, u < \infty \text{ in } \mathbb{R}^n,\ u \in C^2_+(\mathbb{R}^n),\ \lim_{|x|\to\infty} \frac{u(x)}{|x|} = +\infty \right\}$$

and

$$\mathcal{L}^n_s = e^{-\mathcal{C}^n_s} = \{e^{-u} \,:\, u \in \mathcal{C}^n_s\} \subset \mathcal{L}^n.$$

Given $f = e^{-u}$ and $g = e^{-v} \in \mathcal{L}^n_s$, we say that g is an *admissible perturbation of* f if there exists a constant $c > 0$ such that

$$u^* - cv^* \quad \text{is convex in } \mathbb{R}^n.$$

This condition can be viewed as the fact the convexity of u^* *controls* that of v^*.

Theorem 6.1 *Let* $f = e^{-u}$, $g = e^{-v} \in \mathcal{L}^n_s$ *and assume that* g *is an admissible perturbation of* f. *Then* $\delta I(f,g)$ *exists, is finite, and is given by*

$$\delta I(f,g) = \int_{\mathbb{R}^n} v^*(\nabla u(x)) f(x)\, dx. \tag{33}$$

Using a different point of view, we may consider the measure $\tilde{\mu}_f$ on $\mathbb{R}^n$, with density f with respect to the Lebesgue measure. Then we define μ_f as the push-forward of $\tilde{\mu}$ through the gradient map ∇u. At this regard note that, as $f = e^{-u} \in \mathcal{L}^n_s$, ∇u is a diffeomorphism between $\mathbb{R}^n$ and itself. Then (33) is equivalent to

$$\delta I(f,g) = \int_{\mathbb{R}^n} v^*(y)\, d\mu_f(y) = \int_{\mathbb{R}^n} (-\ln(g))^*(y)\, d\mu_f(y). \tag{34}$$

Roughly speaking, as the linear structure on $\mathcal{L}^n$ is the usual addition and multiplication by scalars, transferred to the conjugates of the exponents (with minus sign), (34) says that the measure μ_f is the first variation of the functional I at the function f; for this reason this measure could be interpreted as the area measure of f. Note that this fact cannot be considered to be too general: if we change the assumptions on f (i.e., the fact that $f \in \mathcal{L}^n_s$) then the expression of $\delta I(f,g)$ may change significantly (see, for instance, Theorem 4.6 in [25]) .

It is interesting to note that the measure μ_f was studied also by Cordero-Erausquin and Klartag in [27], with a different perspective.

6.2 Mixed Volumes of Log-Concave Functions

As we saw in the previous section, if we endow $\mathcal{L}^n$ with the addition defined in Section 3.2, the total mass functional of linear combinations of log-concave functions is in general not a polynomial in the coefficients. This is a clear indication that, within the frame of this linear structure, it is not possible to define mixed volumes of generic log-concave functions. On the other hand, there exists a choice of the operations on $\mathcal{L}^n$ which permits to define mixed volumes. These facts were established mainly in the papers [46] and [47] (see also [10] for related results), and here we briefly describe the main points of this construction.

As we said, we have to abandon for a moment the addition previously defined on $\mathcal{L}^n$ and introduce a new one. Given $f, g \in \mathcal{L}^n$ we set

$$(f\tilde{+}g)(z) = \sup\{\min\{f(x), g(y)\} \, : \, x + y = z\}. \tag{35}$$

This apparently intricate definition has in fact a simple geometric interpretation:

$$\{z \in \mathbb{R}^n \, : \, (f\tilde{+}g)(z) \geq t\} = \{x \in \mathbb{R}^n \, : f(x) \geq t\} + \{y \in \mathbb{R}^n \, : \, g(y) \geq t\}$$

for every $t > 0$ such that each of the two sets on the right-hand side is non-empty. In other words, the super-level sets of $f\tilde{+}g$ are the Minkowski addition of the corresponding super-level sets of f and g.

The addition (35) preserves log-concavity (see, for instance, [46]), and then it is an internal operation of $\mathcal{L}^n$ (but it is in fact also natural for quasi-concave functions; see [10, 46]).

A notion of multiplication by non-negative scalars is naturally associated to the previous addition: for $f \in \mathcal{L}^n$ and $\lambda > 0$ we define $\lambda\tilde{\cdot}f$ by

$$(\lambda\tilde{\cdot}f)(x) = f\left(\frac{x}{\lambda}\right).$$

In this new frame, the functional I evaluated at linear combinations of log-concave functions admits a polynomial expansion. More precisely, the following theorem, proved in [46], provides the definition of mixed volumes of log-concave functions.

Theorem 6.2 *There exists a function $V \, : \, (\mathcal{L}^n)^n \to \mathbb{R}$ such that, for every $m \in \mathbb{N}$, $f_1, \ldots, f_m \in \mathcal{L}^n$ and $\lambda_1, \ldots, \lambda_m > 0$,*

$$I(\lambda_1\tilde{\cdot}f_1\tilde{+}\ldots\tilde{+}\lambda_m\tilde{\cdot}f_m) = \sum_{i_1,\ldots,i_n=1}^{m} \lambda_{i_1}\cdots\lambda_{i_n} V(f_{i_1},\ldots,f_{i_n}).$$

In [46] the authors prove several inequalities for mixed volumes of log-concave (and, more generally, quasi-concave) functions, including versions of the Bunn-Minkowski and Alexandrov-Fenchel inequalities.

As in the case of convex bodies, several interesting special cases of mixed volumes can be enucleated. For instance, if we fix $f \in \mathcal{L}^n$ and consider, for $i \in \{0, \ldots, n\}$,

$$V_i(f) := V(\underbrace{f, \ldots, f}_{i\text{-times}}, \underbrace{I_{B_n}, \ldots, I_{B_n}}_{(n-i)\text{-times}}),$$

we have a notion which can be regarded as the i-intrinsic volume of f. These quantities have been studied in [10] and [46].

7 Valuations on $\mathcal{L}^n$

We start by valuations on convex bodies. A (real-valued) valuation on $\mathcal{K}^n$ is a mapping $\sigma : \mathcal{K}^n \to \mathbb{R}$ such that

$$\sigma(K \cup L) + \sigma(K \cap L) = \sigma(K) + \sigma(L) \quad \forall\, K, L \in \mathcal{K}^n \text{ s.t. } K \cup L \in \mathcal{K}^n. \tag{36}$$

The previous relation establishes a finite additivity property of σ. A typical example of valuation is the volume (i.e., the Lebesgue measure), which, as a measure, is countably additive and then fulfills (36). Another, simple, example is provided by the Euler characteristic, which is constantly 1 on $\mathcal{K}^n$ and then it obviously verifies (36). Note that both volume and Euler characteristic are also continuous with respect to Hausdorff metric, and invariant under rigid motions of $\mathbb{R}^n$. Surprisingly, there are other examples of this type; namely each intrinsic volume V_i, $i = 0, \ldots, n$, (see the appendix for a brief presentation) is a rigid motion invariant and continuous valuation on $\mathcal{K}^n$.

The celebrated Hadwiger theorem (see [35, 36, 37]), asserts that, conversely, every rigid motion invariant and continuous valuation can be written as the linear combination of intrinsic volumes; in particular the vector space of such valuations has finite dimension n and $\{V_0, \ldots, V_n\}$ is a basis of this space. If rigid motion invariance is replaced by the weaker assumption of translation invariance, still the relevant space of valuations preserves a rather strong algebraic structure. It was proved by McMullen (see [44]) that any translation invariant and continuous valuation σ on $\mathcal{K}^n$ can be written as

$$\sigma = \sum_{i=0}^{n} \sigma_i$$

where σ_i has the same property of σ and it is i-homogeneous with respect to dilations.

The results that we have mentioned are two of the milestones in this area and stimulated a great development of the theory of valuations on convex bodies, which now counts many ramifications. The reader may find an updated survey on this subject in [55, chapter 6].

The richness of this part of convex geometry recently motivated the start of a parallel theory of valuations on spaces of functions. Coherently with the theme of this article, we restrict ourselves to valuations on $\mathcal{L}^n$; the reader may find a survey of the existing literature on this field of research in [18, 26] and [43].

A mapping $\mu : \mathcal{L}^n \to \mathbb{R}$ is called a (real-valued) valuation if

$$\mu(f \vee g) + \mu(f \wedge g) = \mu(f) + \mu(g), \quad \forall f, g \in \mathcal{L}^n \text{ s.t. } u \vee g \in \mathcal{L}^n,$$

where "$\vee$" and "$\wedge$" denote the point-wise maximum and minimum, respectively (note that the minimum of two functions in $\mathcal{L}^n$ is still in $\mathcal{L}^n$). In other words, sets are replaced by functions and union and intersection are replaced by maximum and minimum. One reason for this definition is that, when restricted to characteristic functions, it gives back the ordinary notion of valuation on relevant sets.

Having the picture of valuations on $\mathcal{K}^n$ in mind, it becomes interesting to consider valuations μ on $\mathcal{L}^n$ which are:

- invariant with respect to some group G of transformations of $\mathbb{R}^n$:

$$\mu(f \circ T) = \mu(f) \quad \forall f \in \mathcal{L}^n, \forall T \in G;$$

- continuous with respect to some topology τ in $\mathcal{L}^n$:

$$f_i \to f \quad \text{as } i \to \infty \text{ w.r.t. } \tau \quad \Rightarrow \quad \mu(f_i) \to \mu(f).$$

The investigation in this area is still at the beginning, and satisfactory characterizations of valuations with the previous properties are not known. At this regard we report a result which can be deduced from [18], preceded by some preparatory material.

Let μ be a valuation defined on $\mathcal{L}^n$. For G we chose the group of rigid motions of $\mathbb{R}^n$; hence we assume that μ is rigid motion invariant.

Next we want to define a continuity property for μ. Note that, while in $\mathcal{K}^n$ the choice of the topology induced by the Hausdforff metric is natural and effective, the situation in $\mathcal{L}^n$ is rather different. For a discussion on this topic we refer the reader to [26, section 4.1]. Here we consider the following notion of continuous valuation on $\mathcal{L}^n$. A sequence f_i, $i \in \mathbb{N}$, contained in $\mathcal{L}^n$, is said to converge to $f \in \mathcal{L}^n$ if:

- f_i is increasing with respect to i;
- $f_i \leq f$ in $\mathbb{R}^n$ for every i;
- f_i converges to f point-wise in the *relative interior* of the support of f (see the definition in the appendix).

Given this definition, we say that μ is continuous if

$$\lim_{i \to \infty} \mu(f_i) = \mu(f),$$

whenever a sequence f_i converges to f in the way specified above. To be able to characterize μ we need two additional properties: μ is increasing, i.e.

$$f_1 \leq f_2 \quad \text{in } \mathbb{R}^n \quad \Rightarrow \quad \mu(f_1) \leq \mu(f_2);$$

and μ is simple, i.e.

$$f \equiv 0 \quad \text{a.e. in } \mathbb{R}^n \quad \Rightarrow \quad \mu(f) = 0.$$

Theorem 7.1 *μ is a rigid motion invariant, continuous, increasing and simple valuation on $\mathcal{L}^n$ if and only if there exists a function $F : \mathbb{R}_+ \to \mathbb{R}_+$ such that*

$$\mu(f) = \int_{\mathbb{R}^n} F(f(x))dx, \tag{37}$$

and, moreover, F is continuous, increasing, vanishes at 0 and verifies the following integrability condition:

$$\int_0^1 \frac{(-\ln(t))^{n-1}}{t} F(t)dt < +\infty. \tag{38}$$

The proof is a direct application of the results proved in [18] for valuations on the space of convex functions $\mathcal{C}^n$. Indeed, we set $\bar{\mu} : \mathcal{C}^n \to \mathbb{R}$ defined by

$$\bar{\mu}(u) = \mu(e^{-u}) \quad \forall\, u \in \mathcal{C}^n$$

$\bar{\mu}$ inherits the features of μ. The valuation property follows immediately from the monotonicity of the exponential function. Rigid motion invariance and monotonicity are straightforward (note that $\bar{\mu}$ is decreasing). As for continuity, the reader may check that the convergence that we have introduced in $\mathcal{L}^n$ induces precisely the one defined in [18]. The property of being simple for μ implies that $\bar{\mu}(u) = 0$ for every $u \in \mathcal{C}^n$ such that $u \equiv \infty$ a.e. in $\mathbb{R}^n$. Hence we may apply Theorem 1.3 in [18], and deduce the integral representation (37). The integrability condition (38) follows from (1.5) in [18].

Other type of valuations on $\mathcal{L}^n$ can be generated by taking weighed means of intrinsic volumes of super-level sets. More precisely, let $f \in \mathcal{L}^n$. For every $t > 0$ the set

$$L_f(t) = \mathrm{cl}(\{x \in \mathbb{R}^n : f(x) \geq t\})$$

(where "cl" denotes the closure) is (either empty or) a compact convex set, i.e. a convex body, by the properties of f. Note that, for every $f, g \in \mathcal{L}^n$,

$$L_{f \vee g} = L_f(t) \cap L_g(t), \quad L_{f \wedge g} = L_f(t) \cup L_g(t).$$

Using these relations and the valuation property of intrinsic volumes (see (41)) we easily get, for an arbitrary $i \in \{0, \ldots, n\}$,

$$V_i(L_{f\vee g}(t)) + V_i(L_{f\wedge g}(t)) = V_i(L_f(t)) + V_i(L_g(t)).$$

In other words, the map $\mathcal{L}^n \to \mathbb{R}$:

$$f \longrightarrow V_i(L_f(t))$$

is a valuation on $\mathcal{L}^n$. More generally we may multiply this function by a non-negative number depending on t and sum over different values of t (keeping i fixed). The result will be again a valuation. The most general way to do it is to consider a continuous sum, that is an integral. In other words, we may take the application:

$$\mu(f) = \int_0^{\infty} V_i(L_f(t))d\nu(t) \tag{39}$$

where ν is a Radon measure. These type of valuations have been considered in [18] for convex functions. In particular, it follows from condition (1.11) in [18] that $\mu(f)$ is finite for every $f \in \mathcal{L}^n$ if and only if ν verifies the integrability condition

$$\int_0^1 \frac{(-\ln(t))^i}{t} d\nu(t) < +\infty. \tag{40}$$

Moreover, μ is homogeneous with respect to dilations of $\mathbb{R}^n$. More precisely, given $f \in \mathcal{L}^n$ and $\lambda > 0$, define the function f_λ as

$$f_\lambda(x) = f\left(\frac{x}{\lambda}\right) \quad \forall\, x \in \mathbb{R}^n.$$

Then

$$\mu(f_\lambda) = \lambda^i \mu(f) \quad \forall f \in \mathcal{L}^n.$$

By theorem 1.4 in [18] and an argument similar to that used in the proof of theorem 7.1, we obtain the following result.

Theorem 7.2 *A mapping $\mu : \mathcal{L}^n \to \mathbb{R}$ is a rigid motion invariant, continuous, monotone valuation, which is in addition homogeneous of some order α, if and only if $\alpha = i \in \{0, \ldots, n\}$, and μ can be written in the form* (39), *for some measure ν verifying condition* (40).

Remark 7.3 In the case $i = n$ formulas (37) and (39) are the same via the layer cake principle.

It would be very interesting to remove part of the assumptions (e.g., monotonicity or homogeneity) in theorems 7.1 and 7.2 and deduce corresponding characterization results.

Appendix A. Basic Notions of Convex Geometry

This part of the paper contains some notions and constructions of convex geometry that are directly invoked throughout this paper. Our main reference text on the theory of convex bodies is the monograph [55].

A.1 Convex Bodies and Their Dimension

We denote by $\mathcal{K}^n$ the class of convex bodies, i.e. compact convex subsets of $\mathbb{R}^n$.

Given a convex body K its dimension is the largest integer $k \in \{0, \dots, n\}$ such that there exists a k-dimensional hyperplane of $\mathbb{R}^n$ containing K. In particular, if K has non-empty interior, then its dimension is n. The *relative interior* of K is the set of points $x \in K$ such that there exists a k-dimensional ball centered at x included in K, where k is the dimension of K. If the dimension of K is n, then the relative interior coincides with usual interior.

A.2 Minkowski Addition

The *Minkowski linear combination* of $K, L \in \mathcal{K}^n$ with coefficients $\alpha, \beta \geq 0$ is

$$\alpha K + \beta L = \{\alpha x + \beta y \, : \, x \in K, \, y \in L\}.$$

It is easy to check that this is still a convex body.

A.3 Support Function

The *support function* of a convex body K is defined as:

$$h_K \, : \, \mathbb{R}^n \to \mathbb{R}, \quad h_K(x) = \sup_{y \in K} (x, y).$$

This is a 1-homogeneous convex function in $\mathbb{R}^n$. Vice versa, to each 1-homogeneous convex function h we may assign a unique convex body K such that $h = h_K$. Support functions and Minkowski additions interact in a very simple way; indeed, for every K and L in $\mathcal{K}^n$ and $\alpha, \beta \geq 0$ we have

$$h_{\alpha K + \beta L} = \alpha h_K + \beta h_L.$$

A.4 Hausdorff Metric

$\mathcal{K}^n$ can be naturally equipped with a metric: the Hausdorff metric d_H. One way to define d_H is as the $L^\infty(\mathbb{S}^{n-1})$ distance of support functions, restricted to the unit sphere:

$$d_H(K, L) = \|h_K - h_L\|_{L^\infty(\mathbb{S}^{n-1})} = \max\{|h_K(x) - h_L(x)| \, : \, x \in \mathbb{S}^{n-1}\}.$$

Hausdorff metric has many useful properties; in particular, we note that $\mathcal{K}^n$ is a locally compact space with respect to d_H.

A.5 Intrinsic Volumes

An easy way to define intrinsic volumes of convex bodies is through the Steiner formula. Let K be a convex body and let B_n denote the closed unit ball of $\mathbb{R}^n$. For $\epsilon > 0$ the set

$$K + \epsilon B_n = \{x + \epsilon y \, : \, x \in K, \, y \in B\} = \{y \in \mathbb{R}^n \, : \, \mathrm{dist}(x, K) \leq \epsilon\}$$

is called the parallel set of K and denoted by K_ϵ. The Steiner formula asserts that the volume of K_ϵ is a polynomial in ϵ. The coefficients of this polynomial are, up to dimensional constants, the intrinsic volumes $V_0(K), \ldots, V_n(K)$ of K:

$$V_n(K_\epsilon) = \sum_{i=0}^{n} V_i(K)\epsilon^{n-i}\kappa_{n-i}.$$

Here κ_j denotes the j-dimensional volume of the unit ball in $\mathbb{R}^j$, for every $j \in \mathbb{N}$. Among the very basic properties of intrinsic volumes, we mention that: V_0 is constantly 1 for every K; V_n is the volume; V_{n-1} is $(n-1)$-dimensional Hausdorff measure of the boundary (only for those bodies with non-empty interior). Moreover, intrinsic volumes are continuous with respect to Hausdorff metric, rigid motion invariant, monotone, and homogeneous with respect to dilations (V_i is i-homogeneous). Finally, each intrinsic volume is a valuation

$$V_i(K \cup L) + V_i(K \cap L) = V_i(K) + V_i(L) \tag{41}$$

for every K and L in $\mathcal{K}^n$, such that $K \cup L \in \mathcal{K}^n$. Hadwiger's theorem claims that every rigid motion invariant and continuous valuation can be written as the linear combination of intrinsic volumes.

A.6 Mixed Volumes

The Steiner formula is just an example of the polynomiality of the volume of linear combinations of convex bodies. A more general version of it leads to the notions of mixed volumes. Let $m \in \mathbb{N}$ and $K_1, \dots, K_m$ be convex bodies; given $\lambda_1, \dots, \lambda_m \geq 0$, the volume of the convex body $\lambda_1 K_1 + \cdots + \lambda_m K_m$ is a homogeneous polynomial of degree n in the variables λ_i's, and its coefficients are the mixed volumes of the involved bodies. The following more precise statement is a part of Theorem 5.16 in [55]. There exists a function $V : (\mathcal{K}^n)^n \to \mathbb{R}_+$, the *mixed volume*, such that

$$V_n(\lambda_1 K_1 + \cdots + \lambda_m K_m) = \sum_{i_1,\dots,i_n=1}^{m} \lambda_{i_1} \cdots \lambda_{i_n} V(K_{i_1}, \dots, K_{i_n})$$

for every $K_1, \dots, K_m \in K^n$ and $\lambda_1, \dots, \lambda_m \geq 0$. Hence a mixed volume is a function of n convex bodies. Mixed volumes have a number of interesting properties. In particular they are non-negative, symmetric, and continuous; moreover, they are linear and monotone with respect to each entry.

A.7 The Polar Body

The *polar* of a convex body K, having the origin as an interior point, is the set

$$K^\circ = \{y : (x, y) \leq 1 \quad \forall x \in K\}.$$

This is again a convex body, with the origin in its interior, and $(K^\circ)^\circ = K$.

Acknowledgements The author would like to thank the anonymous referee for the careful reading of the paper and his/her corrections and valuable suggestions.

This research was partially supported by G.N.A.M.P.A (INdAM) and by the FIR project 2013: *Geometrical and qualitative aspects of PDE's*.

References

[1] D. Alonso-Gutierrez, B. González, C. H. Jiménez, R. Villa, Rogers-Shephard inequality for log-concave functions, preprint (2015).
[2] S. Artstein-Avidan, K. Einhorn, D.Y. Florentin, Y. Ostrover, On Godbersen's conjecture, Geom. Dedicata 178 (2015), 337–350.
[3] S. Artstein-Avidan, B. Klartag, V. Milman, The Santaló point of a function, and a functional form of the Santaló inequality, Mathematika 51 (2004), 33–48.
[4] S. Artstein-Avidan, B. Klartag, C. Shütt, E. Werner, Functional affine-isoperimetry and an inverse logarithmic Sobolev inequality, J. Funct. Anal. 262 (2012), no. 9, 4181–4204.
[5] S. Artstein-Avidan, V. Milman, A characterization of the concept of duality, Electron. Res. Announc. Math. Sci. 14 (2007), 42–59.

[6] S. Artstein-Avidan, V. Milman, The concept of duality in convex analysis, and the characterization of the Legendre transform, Ann. of Math. 169 (2009), 661–674.
[7] S. Artstein-Avidan, V. Milman, Hidden structures in the class of convex functions and a new duality transform, J. Eur. Math. Soc. 13 (2011), 975–1004.
[8] K. Ball, PhD Thesis, University of Cambridge (1988).
[9] F. Barthe, On a reverse form of the Brascamp-Lieb inequality, Invent. Math. 134 (1998), 335–361.
[10] S. Bobkov, A. Colesanti, I. Fragalá, Quermassintegrals of quasi-concave functions and generalized Prékopa-Leindler inequalities, Manuscripta Math. (2013), 1–39 (electronic version).
[11] S. Bobkov, M. Ledoux, From Brunn-Minkowski to sharp Sobolev inequalities, Ann. Mat. Pura Appl. 187 (2008), 369–384.
[12] S. Bobkov, M. Ledoux, From Brunn-Minkowski to Brascamp-Lieb and to logarithmic Sobolev inequalities, Geom. Funct. Anal. 10 (2000), 1028–1052.
[13] F. Bolley, I. Gentil, A. Guillin, Dimensional improvements of the logarithmic Sobolev, Talagrand and Brascamp-Lieb inequalities, preprint.
[14] C. Borell, Convex measures on locally convex spaces, Ark. Mat. 12 (1974), 239–252.
[15] C. Borell, Convex set functions in d-space, Period. Math. Hungar. 6 (1975), 111–136.
[16] J. Bourgain, V. Milman, New volume ratio properties for convex symmetric bodies in $\mathbb{R}^n$, Invent. Math. 88 (1987), 319–340.
[17] H. J. Brascamp and E. H. Lieb, On extensions of the Brunn-Minkowski and Prékopa-Leindler theorems, including inequalities for log-concave functions and with an application to the diffusion equation, J. Functional Analysis, 22 (4) (1976), 366–389.
[18] L. Cavallina, A. Colesanti, Monotone valuations on the space of convex functions, Anal. Geom. Metr. Spaces 3 (2015), 167–211.
[19] U. Caglar, M. Fradelizi, O. Guédon, J. Lehec, C. Schütt, E. Werner, Functional versions of the L_p-affine surface area and entropy inequalities, to appear in Int. Math. Res. Not.
[20] U. Caglar, E. Werner, Divergence for s-concave and log-concave functions, Adv. Math. 257 (2014), 219–247.
[21] U. Caglar, E. Werner, Mixed f-divergence and inequalities for log-concave functions, to appear on Proc. London Math. Soc.
[22] U. Caglar, D. Ye, Orlicz affine isoperimetric inequalities for functions, preprint.
[23] A. Colesanti, Functional inequalities related to the Rogers-Shephard inequality, Mathematika 53 (2006), 81–101.
[24] A. Colesanti, From the Brunn-Minkowski inequality to a class of Poincaré-type inequalities, Commun. Contemp. Math. 10 (2008), 765–772.
[25] A. Colesanti and I. Fragalà, The first variation of the total mass of log-concave functions and related inequalities, Adv. Math. 244 (2013), 708–749.
[26] A. Colesanti, N. Lombardi, Valuations on the space of quasi-concave functions, in B. Klartag, E. Milman (Eds.), Geometric Aspects of Functional Analysis, Lecture Notes in Mathematics 2169, Springer, Berlin, 2017.
[27] D. Cordero-Erausquin, B. Klartag, Moment measures, J. Functional Analysis 268 (2015), 3834–3866.
[28] S. Dubuc, Critères de convexité et inégalités intégrales, Ann. Inst. Fourier (Grenoble) 27 (1977), 135–165.
[29] M. Fradelizi, Y. Gordon, M. Meyer, S. Reisner, On the case of equality for an inverse Santaló functional inequality, Adv. Geom. 10 (2010), 621–630.
[30] M. Fradelizi, M. Meyer, Some functional forms of Blaschke-Santaló inequality. Math. Z. 256 (2007), 379–395.
[31] M. Fradelizi, M. Meyer, Increasing functions and inverse Santaló inequality for unconditional functions, Positivity 12 (2008), 407–420.
[32] M. Fradelizi, M. Meyer, Some functional inverse Santaló inequalities. Adv. Math. 218 (2008), 1430–1452.
[33] M. Fradelizi, M. Meyer, Functional inequalities related to Mahler's conjecture, Monatsh. Math. 159 (2010), 13–25.

[34] R. J. Gardner, The Brunn-Minkowski inequality, Bull. Amer. Math. Soc. 39 (2002), 355–405.
[35] H. Hadwiger, *Vorlesungen über Inhalt, Oberfläche und Isoperimetrie*, Springer-Verlag, Berlin-Göttingen-Heidelberg, 1957.
[36] D. Klain, A short proof of Hadwiger's characterization theorem, Mathematika 42 (1995), 329–339.
[37] D. Klain, G. Rota, *Introduction to geometric probability*, Cambridge University Press, New York, 1997.
[38] B. Klartag and V. D. Milman. Geometry of log-concave functions and measures, Geom. Dedicata 112 (2005), 169–182.
[39] M. Ledoux, *The concentration of measure phenomenon*, Math. Surveys Monogr., vol. 89, Amer. Math. Soc., 2005.
[40] J. Lehec, A direct proof of the functional Santaló inequality, C. R. Acad. Sci. Paris, Ser. 347 (2009), 55–58.
[41] J. Lehec, Partitions and functional Santaló inequalities, Arch. Math. (Basel) 92 (2009), 89–94.
[42] L. Leindler, On certain converse of Hölder's inequality II, Acta Math. Sci. (Szeged) 33 (1972), 217–223.
[43] M. Ludwig, Valuations on function spaces, Adv. Geom. 11 (2011), 745–756.
[44] P. McMullen, Valuations and Euler type relations on certain classes of convex polytopes, Proc. London Math. Soc. 35 (1977), 113–135.
[45] V. Milman, Geometrization of probability. In: Geometry and Dynamics of Groups and Spaces, Progr. Math. 265, M. Kapranov et al. (eds.), Birkhäuser, 647–667 (2008).
[46] V. Milman, L. Rotem, Mixed integrals and related inequalities, J. Funct. Analysis 264 (2013), 570–604.
[47] V. Milman, L. Rotem, α-concave functions and a functional extension of mixed volumes, Electr. Res. Announcements Math. Sci. 20 (2013), 1–11.
[48] A. Prékopa, Logarithmic concave measures with application to stochastic programming, Acta Sci. Math. (Szeged) 32 (1971), 301–315.
[49] A. Prékopa, On logarithmic concave measures and functions, Acta Sci. Math. (Szeged) 34 (1975), 335–343.
[50] C. A. Rogers, G. C. Shephard, The difference body of a convex body, Arch. Math. 8 (1957), 220–233.
[51] R. T. Rockafellar, *Convex Analysis*. Princeton University Press, Princeton, New Jersey, 1970.
[52] L. Rotem, PhD Thesis, University of Tel Aviv (2011).
[53] L. Rotem, On the mean width of log-concave functions, in: B. Klartag, S. Mendelson, V. Milman (Eds.), Geometric Aspects of Functional Analysis, Vol. 2050, Springer, Berlin, 2012, 355–372.
[54] D. Ryabogin, A. Zvavitch, Analytic methods in convex geometry, IMPAN lecture notes, 2 (2014), 89–183.
[55] R. Schneider, *Convex Bodies: The Brunn-Minkowski Theory*. Second expanded edition. Cambridge University Press, Cambridge, 2014.

On Some Problems Concerning Log-Concave Random Vectors

Rafał Latała

1 Introduction

A Radon measure μ on a locally convex linear space F is called *logarithmically concave* (*log-concave* in short) if for any compact nonempty sets $K, L \subset F$ and $\lambda \in [0,1]$, $\mu(\lambda K + (1-\lambda)L) \geq \mu(K)^{\lambda}\mu(L)^{1-\lambda}$. A random vector with values in F is called *log-concave* if its distribution is logarithmically concave.

The class of log-concave measures is closed under affine transformations, convolutions, and weak limits. By the result of Borell [4] an n-dimensional vector with a full dimensional support is log-concave iff it has a log-concave density, i.e. a density of the form e^{-h}, where h is a convex function with values in $(-\infty, \infty]$. A typical example of a log-concave vector is a vector uniformly distributed over a convex body. It may be shown that the class of log-concave distributions on $\mathbb{R}^n$ is the smallest class that contains uniform distributions on convex bodies and is closed under affine transformations and weak limits.

Every full-dimensional logarithmically concave probability measure on $\mathbb{R}^n$ may be affinely transformed into an *isotropic* distribution, i.e. a distribution with mean zero and identity covariance matrix.

In recent years the study of log-concave vectors attracted attention of many researchers, cf. monographs [2] and [5]. There are reasons to believe that logarithmically concave isotropic distributions have similar properties as product distributions. The most important results confirming this belief are the central limit theorem of Klartag [9] and Paouris' large deviation for Euclidean norms [21]. However, many important questions concerning log-concave measures are still open – in this note we present and discuss some of them.

R. Latała (✉)
Institute of Mathematics, University of Warsaw, Banacha 2, 02-097 Warszawa, Poland
e-mail: rlatala@mimuw.edu.pl

E. Carlen et al. (eds.), *Convexity and Concentration*, The IMA Volumes in Mathematics and its Applications 161, DOI 10.1007/978-1-4939-7005-6_16

Notation. By $\langle \cdot, \cdot \rangle$ we denote the standard scalar product on $\mathbb{R}^n$. For $x \in \mathbb{R}^n$ we put $\|x\|_p = (\sum_{i=1}^n |x_i|^p)^{1/p}$ for $1 \le p < \infty$ and $\|x\|_\infty = \max_i |x_i|$, we also use $|x|$ for $\|x\|_2$. We set B_p^n for a unit ball in l_p^n, i.e.. $B_p^n = \{x \in \mathbb{R}^n : \|x\|_p \le 1\}$. $\mathcal{B}(\mathbb{R}^n)$ stands for the family of Borel sets on $\mathbb{R}^n$.

By a letter C we denote absolute constants, value of C may differ at each occurence. Whenever we want to fix a value of an absolute constant we use letters $C_1, C_2, \ldots$.

2 Optimal Concentration

Let ν be a symmetric exponential measure with parameter 1, i.e. the measure on the real line with the density $\frac{1}{2}e^{-|x|}$. Talagrand [23] (see also [17] for a simpler proof based on a functional inequality) showed that the product measure ν^n satisfies the following two-sided concentration inequality

$$\forall_{A \in \mathcal{B}(\mathbb{R}^n)}\ \forall_{t>0}\ \nu^n(A) \ge \frac{1}{2} \ \Rightarrow\ 1 - \nu^n(A + C\sqrt{t}B_2^d + CtB_1^d) \le e^{-t}(1 - \nu^n(A)).$$

This is a very strong result – a simple transportation of measure argument shows that it yields the Gaussian concentration inequality

$$\forall_{A \in \mathcal{B}(\mathbb{R}^n)}\ \forall_{t>0}\ \gamma_n(A) \ge \frac{1}{2} \ \Rightarrow\ 1 - \gamma_n(A + C\sqrt{t}B_2^n) \le e^{-t}(1 - \gamma_n(A)),$$

where γ_n is the canonical Gaussian measure on $\mathbb{R}^n$, i.e. the measure with the density $(2\pi)^{-d/2}\exp(-|x|^2/2)$.

It is natural to ask if similar inequalities may be derived for other measures. To answer this question we should first find a right way to enlarge sets.

Definition 1 Let μ be a probability measure on $\mathbb{R}^n$, for $p \ge 1$ we define the following sets

$$\mathcal{M}_p(\mu) := \Big\{v \in \mathbb{R}^n : \int |\langle v, x\rangle|^p d\mu(x) \le 1\Big\},$$

and

$$\mathcal{Z}_p(\mu) := (\mathcal{M}_p(\mu))^\circ = \Big\{x \in \mathbb{R}^n : |\langle v, x\rangle|^p \le \int |\langle v, y\rangle|^p d\mu(y) \text{ for all } v \in \mathbb{R}^n\Big\}.$$

Sets $\mathcal{Z}_p(\mu_K)$ for $p \ge 1$, when μ_K is the uniform distribution on the convex body K are called *L_p-centroid bodies of K*. They were introduced (under a different normalization) in [16], their properties were also investigated in [21]. Observe that for isotropic measures $\mathcal{M}_2(\mu) = \mathcal{Z}_2(\mu) = B_2^n$.

Obviously $\mathcal{M}_p(\mu) \subset \mathcal{M}_q(\mu)$ and $\mathcal{Z}_p(\mu) \supset \mathcal{Z}_q(\mu)$ for $p \geq q$. Next definition allows to reverse these inclusions.

Definition 2 We say that moments of a probability measure μ on $\mathbb{R}^n$ grow α-regularly for some $\alpha \in [1, \infty)$ if for any $p \geq q \geq 2$ and $v \in \mathbb{R}^n$,

$$\left(\int |\langle v, x\rangle|^p d\mu(x)\right)^{1/p} \leq \alpha \frac{p}{q} \left(\int |\langle v, x\rangle|^q d\mu(x)\right)^{1/q}.$$

It is easy to see that for measures with α-regular growth of moments and $p \geq q \geq 2$ we have $\alpha\frac{p}{q}\mathcal{M}_p(\mu) \supset \mathcal{M}_q(\mu)$ and $\mathcal{Z}_p(\mu) \subset \alpha\frac{p}{q}\mathcal{Z}_q(\mu)$.

Moments of log-concave measures grow 3-regularly (1-regularly for symmetric measures and 2-regularly for centered measures). The following easy observation was noted in [15].

Proposition 3 *Suppose that μ is a symmetric probability measure on $\mathbb{R}^n$ with α-regular growth of moments. Let K be a convex set such that for any halfspace A,*

$$\mu(A) \geq \frac{1}{2} \Rightarrow 1 - \mu(A + K) \leq \frac{1}{2}e^{-p}.$$

Then $K \supset c(\alpha)\mathcal{Z}_p$ if $p \geq p(\alpha)$, where $c(\alpha)$ and $p(\alpha)$ depend only on α.

The above motivates the following definition.

Definition 4 We say that a measure μ satisfies the *optimal concentration inequality with constant β* (CI(β) in short) if

$$\forall_{p\geq 2}\, \forall_{A\in\mathcal{B}(\mathbb{R}^n)}\, \mu(A) \geq \frac{1}{2} \Rightarrow 1 - \mu(A + \beta\mathcal{Z}_p(\mu)) \leq e^{-p}(1 - \mu(A)).$$

By the result of Gluskin and Kwapien [6], $\mathcal{M}_p(\nu^n) \sim p^{-1}B_\infty^n \cap p^{-1/2}B_2^n$, so $\mathcal{Z}_p(\nu^n) \sim pB_1^n + p^{1/2}B_2^n$. Therefore Talagrand's two-sided concentration inequality states that ν^n satisfy CI(β) with $\beta \leq C$.

Remark 5 By Proposition 2.7 in [15] CI(β) may be equivalently stated as

$$\forall_{p\geq 2}\, \forall_{A\in\mathcal{B}(\mathbb{R}^n)}\, \mu(A + \beta\mathcal{Z}_p(\mu)) \geq \min\left\{\frac{1}{2}, e^p\mu(A)\right\}. \tag{1}$$

In [15] a very strong conjecture was posed that every symmetric log-concave measure on $\mathbb{R}^n$ satisfy CI(β) with a uniform constant β. Unfortunately there are very few examples supporting this conjecture.

Theorem 6 *The following probability measures satisfy the optimal concentration inequality with an absolute constant β:*

i) symmetric product log-concave measures;
ii) uniform distributions on B_p^n-balls, $1 \leq p \leq \infty$;
iii) rotationally invariant logconcave measures.

Parts i) ii) were showed in [15], iii) may be showed using a radial transportation and the Gaussian concentration inequality.

Property CI(β) is invariant under linear transformations, so it is enough to study it for isotropic measures. For isotropic log-concave measures and $p \geq 2$ we have $\mathcal{Z}_p(\mu) \subset p\mathcal{Z}_2(\mu) = pB_2^n$, so CI($\beta$) implies the *exponential concentration*:

$$\forall_{A\in\mathcal{B}(\mathbb{R}^n)}\ \mu(A) \geq \frac{1}{2} \ \Rightarrow\ 1-\mu(A+\beta pB_2^n) \leq e^{-p} \text{ for } p \geq 2.$$

By the result of E. Milman [20] the exponential concentration for log-concave measures is equivalent to *Cheeger's inequality*:

$$\forall_{A\in\mathcal{B}(\mathbb{R}^n)}\ \mu^+(A) := \lim_{t\to 0+} \frac{\mu(A+tB_2^n)-\mu(A)}{t} \geq \frac{1}{\beta'}\min\{\mu(A), 1-\mu(A)\},$$

and constants β, β' are comparable up to universal multiplicative factors. The long-standing open conjecture of Kannan, Lovasz, and Simonovits [8] states that isotropic log-concave probability measures satisfy Cheeger's inequality with a uniform constant.

The best known bound for the exponential concentration constant for isotropic log-concave measures $\beta \leq Cn^{1/3}\sqrt{\log n}$ is due to Eldan [7]. We will show a weaker estimate for the CI constant.

Proposition 7 *Every centered log-concave probability measure on $\mathbb{R}^n$ satisfies the optimal concentration inequality with constant $\beta \leq C\sqrt{n}$.*

Our proof is based on the following two simple lemmas.

Lemma 8 *Let μ be a probabilistic measure on $\mathbb{R}^n$. Then*

$$\mu(10\lambda\mathcal{Z}_p(\mu)) \geq 1-\lambda^{-p} \quad \textit{for } p \geq n,\ \lambda \geq 1.$$

Proof Let $T = \{u_1, \ldots, u_N\}$ be a 1/2-net in $\mathcal{M}_p(\mu)$ of cardinality $N \leq 5^n$, i.e. such set $T \subset \mathcal{M}_p(\mu)$ that $\mathcal{M}_p(\mu) \subset T + \frac{1}{2}\mathcal{M}_p(\mu)$. Then the condition $x \notin \mathcal{Z}_p(\mu)$ implies $\langle u_j, x\rangle > 1/2$ for some $j \leq N$. Hence

$$\begin{aligned} 1-\mu(10\lambda\mathcal{Z}_p(\mu)) = \mu(\mathbb{R}^n \setminus 10\lambda\mathcal{Z}_p(\mu)) &\leq \sum_{j=1}^{N} \mu\{x\in\mathbb{R}^n\colon \langle u_j, x\rangle > 5\lambda\} \\ &\leq N(5\lambda)^{-p} \leq \lambda^{-p}, \end{aligned}$$

where the second inequality follows by Chebyshev's inequality. □

Lemma 9 *Let μ be a log-concave probability measure on $\mathbb{R}^n$ and K be a symmetric convex set such that $\mu(K) \geq 1-e^{-p}$ for some $p \geq 2$. Then for any Borel set A in $\mathbb{R}^n$,*

$$\mu(A+9K) \geq \min\left\{\frac{1}{2}, e^p\mu(A)\right\}.$$

Proof By Borell's lemma [4] we have for $t \geq 1$,

$$1-\mu(tK) \leq \mu(K)\left(\frac{1-\mu(K)}{\mu(K)}\right)^{\frac{t+1}{2}} \leq e^{-\frac{t+1}{3}p}.$$

Let $\mu(A) = e^{-u}$ for some $u \geq 0$. Set

$$\tilde{u} := \max\{u, 2p\} \quad \text{and} \quad \tilde{A} := A \cap 4\frac{\tilde{u}}{p}K.$$

We have

$$\mu(\tilde{A}) \geq \mu(A) - \left(1-\mu\left(4\frac{\tilde{u}}{p}K\right)\right) \geq e^{-u} - e^{-\frac{p}{3}}e^{-\frac{4\tilde{u}}{3}} \geq \frac{1}{2}e^{-\tilde{u}}.$$

Observe that if $x \in \tilde{A}$ then $\frac{2p}{\tilde{u}}x \in 8K$, therefore $(1-\frac{2p}{\tilde{u}})\tilde{A} \subset \tilde{A} + 8K$ and

$$\mu(A+9K) \geq \mu(\tilde{A}+8K+K) \geq \mu\left(\left(1-\frac{2p}{\tilde{u}}\right)\tilde{A} + \frac{2p}{\tilde{u}}K\right) \geq \mu(\tilde{A})^{1-\frac{2p}{\tilde{u}}}\mu(K)^{\frac{2p}{\tilde{u}}}$$
$$\geq \left(\frac{1}{2}e^{-\tilde{u}}\right)^{1-\frac{2p}{\tilde{u}}}\left(\frac{1}{2}\right)^{\frac{2p}{\tilde{u}}} = \frac{1}{2}e^{2p-\tilde{u}} \geq \min\left\{\frac{1}{2}, e^{p}\mu(A)\right\}.$$

□

Proof of Proposition 7. By the linear invariance we may and will assume that μ is isotropic.

Applying Lemma 8 with $\lambda = e$ and Lemma 9 with $K = 10e\mathcal{Z}_p(\mu)$ we see that (1) holds with $\beta = 90e$ for $p \geq n$. For $p \geq \sqrt{n}$ we have $2\sqrt{n}\mathcal{Z}_p(\mu) \supset \mathcal{Z}_{p\sqrt{n}}(\mu)$ and we get (1) with $\beta = 180e\sqrt{n}$ in this case.

The Paouris inequality (4) gives

$$1-\mu(C_1 t\sqrt{n}B_2^n) \leq e^{-t\sqrt{n}} \quad \text{for } t \geq 1.$$

Together with Lemma 9 this yields for any Borel set A and $t \geq 1$,

$$\mu(A + 9C_1 t\sqrt{n}B_2^n) \geq \min\left\{\frac{1}{2}, e^{t\sqrt{n}}\mu(A)\right\}.$$

Using the above bound for $t = 1$ and the inclusion $\mathcal{Z}_p(\mu) \supset \mathcal{Z}_2(\mu) = B_2^n$ we obtain (1) with $\beta = 9C_1\sqrt{n}$ for $2 \leq p \leq \sqrt{n}$. □

It would be of interest to improve the estimate from Proposition 7 to $\beta \leq Cn^{1/2-\varepsilon}$ for some $\varepsilon > 0$. Suppose that we are able to show that

$$\mu\left(C_2\sqrt{\frac{n}{p}}\mathcal{Z}_p(\mu)\right) \geq 1 - e^{-p} \quad \text{for } 2 \leq p \leq n. \tag{2}$$

Then (assuming again that μ is isotropic)
i) if $p \leq p_0 := n^{1/9}(\log n)^{-1/3}$, we obtain by the Eldan's bound on Cheeger's constant

$$\mu(A + Cn^{4/9}(\log n)^{1/6}\mathcal{Z}_p(\mu)) \geq \mu(A + Cp_0 n^{1/3}\sqrt{\log n}B_2^n) \geq \min\left\{\frac{1}{2}, e^{p_0}\mu(A)\right\}$$

$$\geq \min\left\{\frac{1}{2}, e^p\mu(A)\right\}.$$

ii) if $p_0 \leq p \leq n$, then by (2) and Lemma 9,

$$\mu(A + 9C_2 n^{4/9}(\log n)^{1/6}\mathcal{Z}_p(\mu)) \geq \mu\left(A + 9C_2\sqrt{\frac{n}{p}}\mathcal{Z}_p(\mu)\right) \geq \min\left\{\frac{1}{2}, e^p\mu(A)\right\}.$$

So (2) would yield CI(β) for μ with $\beta \leq Cn^{4/9}(\log n)^{1/3}$. Unfortunately we do not know whether (2) holds for symmetric log-concave measures (we are able to show it in the unconditional case).

A measure μ on $\mathbb{R}^n$ is called *unconditional* if it is invariant under symmetries with respect to coordinate axes. If μ is a log-concave, isotropic, and unconditional measure on $\mathbb{R}^n$, then the result of Bobkov and Nazarov [3] yields $\mathcal{Z}_p(\mu) \subset C\mathcal{Z}_p(\nu^n)$. Therefore property CI($\beta$) yields two-level concentration inequality for such measures

$$\forall_{A\in\mathcal{B}(\mathbb{R}^n)}\ \forall_{t>0}\ \mu(A) \geq \frac{1}{2} \Rightarrow 1-\mu(A+C\beta(\sqrt{t}B_2^d+tB_1^d)) \leq e^{-t}(1-\mu(A)). \quad (3)$$

Klartag [10] showed that unconditional isotropic log-concave measures satisfy exponential concentration inequality with a constant $\beta \leq C\log n$. We do not know if similar bound for β holds for the optimal concentration inequality or its weaker form (3).

3 Weak and Strong Moments

One of the fundamental properties of log-concave vectors is the Paouris inequality [21] (see also [1] for a shorter proof).

Theorem 10 *For any log-concave vector X in $\mathbb{R}^n$,*

$$(\mathbb{E}|X|^p)^{1/p} \leq C(\mathbb{E}|X| + \sigma_X(p)) \quad \textit{for}\ p \geq 1,$$

where

$$\sigma_X(p) := \sup_{|v|\leq 1}(\mathbb{E}|\langle v, X\rangle|^p)^{1/p}.$$

Equivalently, in terms of tails we have

$$\mathbb{P}(|X| \geq Ct\mathbb{E}|X|) \leq \exp\left(-\sigma_X^{-1}(t\mathbb{E}|X|)\right) \quad \text{for } t \geq 1.$$

Observe that if X is additionally isotropic then $\sigma_X(p) \leq p\sigma_X(2) = p$ for $p \geq 2$ and $\mathbb{E}|X| \leq (\mathbb{E}|X|^2)^{1/2} = \sqrt{n}$, so we get

$$\mathbb{P}(|X| \geq Ct\sqrt{n}) \leq e^{-t\sqrt{n}} \quad \text{for } t > 1 \text{ and isotropic log-concave vector } X. \tag{4}$$

It would be very valuable to have a reasonable characterization of random vectors which satisfy the Paouris inequality. The following example shows that the regular growth of moments is not enough.

Example 11 Let $Y = \sqrt{n}gU$, where U has a uniform distribution on S^{n-1} and g is the standard normal $\mathcal{N}(0, 1)$ r.v., independent of U. Then it is easy to see that Y is isotropic, rotationally invariant and for any seminorm on $\mathbb{R}^n$

$$(\mathbb{E}\|Y\|^p)^{1/p} = \sqrt{n}(\mathbb{E}|g|^p)^{1/p}(\mathbb{E}\|U\|^p)^{1/p} \sim \sqrt{pn}(\mathbb{E}\|U\|^p)^{1/p} \quad \text{for } p \geq 1.$$

In particular this implies that for any $v \in \mathbb{R}^n$,

$$(\mathbb{E}|\langle v, Y\rangle|^p)^{1/p} \leq C\frac{p}{q}(\mathbb{E}|\langle v, Y\rangle|^q)^{1/q} \quad \text{for } p \geq q \geq 2.$$

So moments of Y grow C regularly. Moreover

$$(\mathbb{E}|Y|^p)^{1/p} \sim \sqrt{pn}, \quad (\mathbb{E}|Y|^2)^{1/2} = \sqrt{n}, \quad \sigma_Y(p) \leq Cp,$$

thus for $1 \ll p \ll n$, $(\mathbb{E}|Y|^p)^{1/p} \gg (\mathbb{E}|Y|^2)^{1/2} + \sigma_Y(p)$.

It is natural to ask whether Theorem 10 may be generalized to non-Euclidean norms. In [11] the following conjecture was formulated and discussed.

Conjecture 12 *There exists a universal constant C such that for any n-dimensional log-concave vector X and any norm* $\|\ \|$ *on* $\mathbb{R}^n$,

$$(\mathbb{E}\|X\|^p)^{1/p} \leq C\left(\mathbb{E}\|X\| + \sup_{\|v\|_*\leq 1}(\mathbb{E}|\langle v, X\rangle|^p)^{1/p}\right) \quad \textit{for } p \geq 1,$$

where $\|v\|_* = \sup\{|\langle v, x\rangle|\colon \|x\| \leq 1\}$ *denotes the dual norm on* $\mathbb{R}^n$.

Note that obviously for any random vector X and $p \geq 1$,

$$(\mathbb{E}\|X\|^p)^{1/p} \geq \max\left\{\mathbb{E}\|X\|, \sup_{\|v\|_*\leq 1}(\mathbb{E}|\langle v, X\rangle|^p)^{1/p}\right\}.$$

The following simple observation from [15] shows that the optimal concentration yields comparison of weak and strong moments.

Proposition 13 *Suppose that the law of an n-dimensional random vector X is α-regular and satisfies the optimal concentration inequality with constant β. Then for any norm $\| \ \|$ on $\mathbb{R}^n$,*

$$(\mathbb{E}|\|X\| - \mathbb{E}\|X\||^p)^{1/p} \le C\alpha\beta \sup_{\|v\|_* \le 1} (\mathbb{E}|\langle v, X\rangle|^p)^{1/p} \quad \textit{for } p \ge 1.$$

Recall that log-concave measures are 3-regular. Therefore if the law of X is of one of three types listed in Theorem 6 then for any norm $\| \ \|$,

$$(\mathbb{E}\|X\|^p)^{1/p} \le \mathbb{E}\|X\| + C \sup_{\|v\|_* \le 1} (\mathbb{E}|\langle v, X\rangle|^p)^{1/p} \quad \text{for } p \ge 1.$$

We do not know if such inequality is satisfied for Euclidean norms and arbitrary log-concave vectors, i.e. whether Paouris inequality holds with the constant 1 in front of $\mathbb{E}|X|$. This question is related to the so-called variance conjecture, discussed in [2].

The following extension of the Paouris inequality was shown in [13].

Theorem 14 *Let X be a log-concave vector with values in a normed space $(F, \| \ \|)$ which may be isometrically embedded in ℓ_r for some $r \in [1, \infty)$. Then for $p \ge 1$,*

$$(\mathbb{E}\|X\|^p)^{1/p} \le Cr \left(\mathbb{E}\|X\| + \sup_{\varphi \in F^*, \|\varphi\|_* \le 1} (\mathbb{E}|\varphi(X)|^p)^{1/p} \right).$$

Remark 15 Let X and F be as above. Then by Chebyshev's inequality we obtain large deviation estimate for $\|X\|$:

$$\mathbb{P}(\|X\| \ge Crt\mathbb{E}\|X\|) \le \exp\left(-\sigma_{X,F}^{-1}(t\mathbb{E}\|X\|)\right) \quad \text{for } t \ge 1,$$

where

$$\sigma_{X,F}(p) := \sup_{\varphi \in F^*, \|\varphi\|_* \le 1} (\mathbb{E}\varphi(X)^p)^{1/p} \quad \text{for } p \ge 1$$

denotes the weak p-th moment of $\|X\|$.

Remark 16 If $i\colon F \to \ell_r$ is a nonisometric embedding and $\lambda = \|i\|_{F \to \ell_r} \|i^{-1}\|_{i(F) \to F}$, then we may define another norm on F by $\|x\|' := \|i(x)\| / \|i\|_{F \to \ell_r}$. Obviously $(F, \| \ \|')$ isometrically embeds in ℓ_r, moreover $\|x\|' \le \|x\| \le \lambda \|x\|'$ for $x \in F$. Hence Theorem 14 gives

$$(\mathbb{E}\|X\|^p)^{1/p} \le \lambda(\mathbb{E}(\|X\|')^p)^{1/p} \le C_2 r\lambda \left(\mathbb{E}\|X\|' + \sup_{\varphi\in F^*, \|\varphi\|'_* \le 1} (\mathbb{E}|\varphi(X)|^p)^{1/p} \right)$$

$$\le C_2 r\lambda \left(\mathbb{E}\|X\| + \sup_{\varphi\in F^*, \|\varphi\|_* \le 1} (\mathbb{E}|\varphi(X)|^p)^{1/p} \right).$$

Since log-concavity is preserved under linear transformations and, by the Hahn-Banach theorem, any linear functional on a subspace of l_r is a restriction of a functional on the whole l_r with the same norm, it is enough to prove Theorem 14 for $F = l_r$. An easy approximation argument shows that we may consider finite dimensional spaces l_r^n. This way Theorem 14 reduces to the following finite dimensional statement.

Theorem 17 *Let X be a log-concave vector in $\mathbb{R}^n$ and $r \in [1,\infty)$. Then*

$$(\mathbb{E}\|X\|_r^p)^{1/p} \le Cr\,(\mathbb{E}\|X\|_r + \sigma_{r,X}(p)) \quad \textit{for } p \ge 1,$$

where

$$\sigma_{r,X}(p) := \sigma_{X,l_r^n}(p) = \sup_{\|v\|_{r'} \le 1} (\mathbb{E}|\langle v, X\rangle|^p)^{1/p}$$

and r' denotes the Hölder's dual of r, i.e. $r' = \frac{r}{r-1}$ for $r > 1$ and $r' = \infty$ for $r = 1$.

Any finite dimensional space embeds isometrically in ℓ_∞, so to show Conjecture 12 it is enough to establish Theorem 17 (with a universal constant in place of Cr) for $r = \infty$. Such a result was shown for isotropic log-concave vectors.

Theorem 18 ([12]) *Let X be an isotropic log-concave vector in $\mathbb{R}^n$. Then for any $a_1, \ldots, a_n$ and $p \ge 1$,*

$$(\mathbb{E}\max_i |a_i X_i|^p)^{1/p} \le C \left(\mathbb{E}\max_i |a_i X_i| + \max_i (\mathbb{E}|a_i X_i|^p)^{1/p} \right) \quad \textit{for } p \ge 1.$$

However a linear image of an isotropic vector does not have to be isotropic, so to establish the conjecture we need to consider either isotropic vectors and an arbitrary norm or vectors with a general covariance structure and the standard ℓ_∞-norm.

In the case of unconditional vectors slightly more is known.

Theorem 19 ([11]) *Let X be an n-dimensional isotropic, unconditional, log-concave vector and $Y = (Y_1, \ldots, Y_n)$, where Y_i are independent symmetric exponential r.v's with variance 1 (i.e., with the density $2^{-1/2}\exp(-\sqrt{2}|x|)$). Then for any norm $\|\ \|$ on $\mathbb{R}^n$ and $p \ge 1$,*

$$(\mathbb{E}\|X\|^p)^{1/p} \le C \left(\mathbb{E}\|Y\| + \sup_{\|v\|_* \le 1} (\mathbb{E}|\langle v, X\rangle|^p)^{1/p} \right).$$

Proof is based on the Talagrand generic-chaining type two-sided estimate of $\mathbb{E}\|Y\|$ [24] and the Bobkov-Nazarov [3] bound for the joint d.f. of X, which implies $(\mathbb{E}|\langle v, X\rangle|^p)^{1/p} \lesssim C(\mathbb{E}|\langle v, Y\rangle|^p)^{1/p}$ for $p \gtrsim 1$ and $v \subset \mathbb{R}^n$.

Using the easy estimate $\mathbb{E}\|Y\| \leq C \log n\, \mathbb{E}\|X\|$ we get the following.

Corollary 20 *For any n-dimensional unconditional, log-concave vector X, any norm* $\| \ \|$ *on* $\mathbb{R}^n$ *and* $p \geq 1$ *one has*

$$(\mathbb{E}\|X\|^p)^{1/p} \leq C\left(\log n\, \mathbb{E}\|X\| + \sup_{\|v\|_* \leq 1} (\mathbb{E}|\langle v, X\rangle|^p)^{1/p}\right).$$

The Maurey-Pisier result [18] implies $\mathbb{E}\|Y\| \leq C\mathbb{E}\|X\|$ in spaces with nontrivial cotype.

Corollary 21 *Let* $2 \leq q < \infty$ *and* $F = (\mathbb{R}^n, \| \ \|)$ *has a q-cotype constant bounded by* $\beta < \infty$. *Then for any n-dimensional unconditional, log-concave vector X and* $p \geq 1$,

$$(\mathbb{E}\|X\|^p)^{1/p} \leq C(q, \beta)\left(\mathbb{E}\|X\| + \sup_{\|v\|_* \leq 1} (\mathbb{E}|\langle v, X\rangle|^p)^{1/p}\right).$$

where $C(q, \beta)$ *is a constant that depends only on q and* β.

For a class of invariant measures Conjecture 12 was established in [12].

Proposition 22 *Let X be an n-dimensional random vector with the density of the form* $e^{-\varphi(\|x\|_r)}$, *where* $1 \leq r \leq \infty$ *and* $\varphi\colon [0, \infty) \to (-\infty, \infty]$ *is nondecreasing and convex. Then for any norm* $\| \ \|$ *on* $\mathbb{R}^n$ *and any* $p \geq 1$,

$$(\mathbb{E}\|X\|^p)^{1/p} \leq C(r)\mathbb{E}\|X\| + C \sup_{\|v\|_* \leq 1} (\mathbb{E}|\langle v, X\rangle|^p)^{1/p}.$$

4 Sudakov Minoration

For any norm $\| \ \|$ on $\mathbb{R}^n$ we have

$$\|x\| = \sup_{\|v\|_* \leq 1} \langle v, x\rangle = \frac{1}{2} \sup_{\|v\|_*, \|w\|_* \leq 1} \langle v - w, x\rangle \quad \text{for } x \in \mathbb{R}^n.$$

Thus to estimate the mean of a norm of a random vector X one needs to investigate $\mathbb{E}\sup_{v,w \in V} \langle v - w, X\rangle$ for bounded subsets V in $\mathbb{R}^n$.

There are numerous powerful methods to estimate suprema of stochastic processes (cf. the monograph [25]), let us however present only a very easy upper bound. Namely for any $p \geq 1$,

$$\mathbb{E}\sup_{v,w\in V}\langle v-w,X\rangle \le \Big(\mathbb{E}\sup_{v,w\in V}|\langle v-w,X\rangle|^p\Big)^{1/p} \le \Big(\mathbb{E}\sum_{v,w\in V}|\langle v-w,X\rangle|^p\Big)^{1/p}$$
$$\le |V|^{2/p}\sup_{v,w\in V}(\mathbb{E}|\langle v-w,X\rangle|^p)^{1/p}.$$

In particular,

$$\mathbb{E}\sup_{v,w\in V}\langle v-w,X\rangle \le e^2\sup_{v,w\in V}(\mathbb{E}|\langle v-w,X\rangle|^p)^{1/p} \quad \text{if } |V|\le e^p.$$

It is natural to ask when the above estimate may be reversed. Namely, when is it true that if the set $V\subset\mathbb{R}^n$ has large cardinality (say at least e^p) and variables $(\langle v,X\rangle)_{v\in V}$ are A-separated with respect to the L_p-distance then $\mathbb{E}\sup_{v,w\in V}\langle v-w,X\rangle$ is at least of the order of A? The following definition gives a more precise formulation of such property.

Definition 23 Let X be a random n-dimensional vector. We say that X *satisfies the L_p-Sudakov minoration principle with a constant* $\kappa>0$ (SMP$_p(\kappa)$ in short) if for any nonempty set $V\subset\mathbb{R}^n$ with $|V|\ge e^p$ such that

$$d_{X,p}(v,w) := (\mathbb{E}|\langle v-w,X\rangle|^p)^{1/p}\ge A \quad \text{for all } v,w\in V,\ v\ne w, \tag{6}$$

we have

$$\mathbb{E}\sup_{v,w\in V}\langle v-w,X\rangle\ge\kappa A. \tag{7}$$

A random vector X *satisfies the Sudakov minoration principle with a constant* κ (SMP(κ) in short) if it satisfies SMP$_p(\kappa)$ for any $p\ge 1$.

Example 24 If X has the canonical n-dimensional Gaussian distribution $\mathcal{N}(0,I_n)$ then $(\mathbb{E}|\langle v,X\rangle|^p)^{1/p}=\gamma_p|v|$, where $\gamma_p=(\mathbb{E}|\mathcal{N}(0,1)|^p)^{1/p}\sim\sqrt{p}$ for $p\ge 1$. Hence condition (6) is equivalent to $|v-w|\ge A/\gamma_p$ for distinct vectors $v,w\in V$ and the classical Sudakov minoration principle for Gaussian processes [22] then yields

$$\mathbb{E}\sup_{v,w\in V}\langle v-w,X\rangle = 2\mathbb{E}\sup_{v\in V}\langle v,X\rangle \ge \frac{A}{C\gamma_p}\sqrt{\log|V|}\ge\frac{A}{C}$$

provided that $|V|\ge e^p$. Therefore X satisfies the Sudakov minoration principle with a universal constant. In fact it is not hard to see that for centered Gaussian vectors the Sudakov minoration principle in the sense of Definition 23 is formally equivalent to the minoration property established by Sudakov.

The Sudakov minoration principle for vectors X with independent coordinates was investigated in detail in [14]. It was shown there that for SMP in such a case the sufficient (and necessary if coordinates of X have identical distribution) condition is the regular growth of moments of coordinates of X, i.e. the existence of $\alpha < \infty$ such that $(\mathbb{E}|X_i|^p)^{1/p} \leq \alpha\frac{p}{q}(\mathbb{E}|X_i|^q)^{1/q}$ for all i and $p \geq q \geq 1$. In particular log-concave vectors X with independent coordinates satisfy SMP with a universal constant κ.

In the sequel we will discuss the following conjecture.

Conjecture 25 *Every n-dimensional log-concave random vector satisfies the Sudakov-minoration principle with a universal constant.*

Remark 26 Suppose that X is log-concave and (6) is satisfied, but $|V| = e^q$ with $1 \leq q \leq p$. Since $d_{X,q}(v,w) \geq \frac{q}{3p}d_{X,p}(v,w)$, the Sudakov minoration principle for a log-concave vector X implies the following formally stronger statement – for any nonempty $V \subset \mathbb{R}^n$ and $A > 0$,

$$\mathbb{E}\sup_{v,w\in V}\langle v-w, X\rangle \geq \frac{\kappa}{C}\sup_{p\geq 1}\min\left\{\frac{A}{p}\log N(V, d_{X,p}, A), A\right\},$$

where $N(V, d, \varepsilon)$ denotes the minimal number of balls in metric d of radius ε that cover V.

The Sudakov minoration principle and Conjecture 25 were posed independently by Shahar Mendelson, Emanuel Milman, and Grigoris Paouris (unpublished) and by the author in [12]. In [19] there is discussed approach to the Sudakov minoration and its dual version based on variants of the Johnson-Lindenstrauss dimension reduction lemma. The results presented below were proven in [12].

It is easy to see that the Sudakov minoration property is affinely invariant, so it is enough to investigate it only for isotropic random vectors. Using the fact that isotropic log-concave vectors satisfy exponential concentration with constant Cn^γ with $\gamma < 1/2$ one may show that the lower bound (6) holds for special classes of sets.

Proposition 27 *Suppose that X is an n-dimensional log-concave random vector, $p \geq 2$, $V \subset \mathbb{R}^n$ satisfies* (6) *and* $\mathrm{Cov}(\langle v, X\rangle, \langle w, X\rangle) = 0$ *for $v, w \in V$ with $v \neq w$. Then* (7) *holds with a universal constant κ provided that $|V| \geq e^p$.*

In the case of general sets we know at the moment only the following much weaker form of the Sudakov minoration principle.

Theorem 28 *Let X be a log-concave vector, $p \geq 1$ and $V \subset \mathbb{R}^n$ be such that $|V| \geq e^{e^p}$ and* (6) *holds. Then*

$$\mathbb{E}\sup_{v,w\in V}\langle v-w, X\rangle \geq \frac{1}{C}A.$$

Stronger bounds may be derived in the unconditional case. Comparing unconditional log-concave vectors with vectors with independent symmetric exponential coordinates one gets the following bound on κ.

Proposition 29 *Suppose that X is an n-dimensional log-concave unconditional vector. Then X satisfies* $\mathrm{SMP}(1/C\log(n+1))$.

The next result presents a bound on κ independent of dimension but under a stronger assumptions on the cardinality of V than in the definition of SMP.

Theorem 30 *Let X be a log-concave unconditional vector in $\mathbb{R}^n$, $p \geq 1$ and $V \subset \mathbb{R}^n$ be such that $|V| \geq e^{p^2}$ and* (6) *holds. Then*

$$\mathbb{E}\sup_{v,w\in V}\langle v-w,X\rangle = 2\mathbb{E}\sup_{v\in V}\langle v,X\rangle \geq \frac{1}{C}A.$$

Remark 31 Theorems 28 and 30 may be rephrased in terms of entropy numbers as in Remark 26. Namely, for any nonempty set $V \subset \mathbb{R}^n$ and log-concave vector X,

$$\mathbb{E}\sup_{v,w\in V}\langle v-w,X\rangle \geq \frac{1}{C}\sup_{p\geq 1,A>0}\min\left\{\frac{A}{p}\log\log N(V,d_{X,p},A),A\right\}.$$

If X is unconditional and log-concave, then

$$\mathbb{E}\sup_{v,w\in V}\langle v-w,X\rangle \geq \frac{1}{C}\sup_{p\geq 1,A>0}\min\left\{\frac{A}{p}\sqrt{\log N(V,d_{X,p},A)},A\right\}.$$

We know that a class of invariant log-concave vectors satisfy $\mathrm{SMP}(\kappa)$ with uniform κ.

Theorem 32 *All n-dimensional random vectors with densities of the form* $\exp(-\varphi(\|x\|_p))$, *where $1 \leq p \leq \infty$ and $\varphi\colon[0,\infty)\to(-\infty,\infty]$ is nondecreasing and convex satisfy the Sudakov minoration principle with a universal constant. In particular all rotationally invariant log-concave random vectors satisfy the Sudakov minoration principle with a universal constant.*

One of the important consequences of the SMP-property is the following comparison-type result for random vectors.

Proposition 33 *Suppose that a random vector X in $\mathbb{R}^n$ satisfies* $\mathrm{SMP}(\kappa)$. *Let Y be a random n-dimensional vector such that $\mathbb{E}|\langle v,Y\rangle|^p \leq \mathbb{E}|\langle v,X\rangle|^p$ for all $p\geq 1$, $v\in\mathbb{R}^n$. Then for any norm $\|\ \|$ on $\mathbb{R}^n$ and $p \geq 1$,*

$$\begin{aligned}(\mathbb{E}\|Y\|^p)^{1/p} &\leq C\Big(\frac{1}{\kappa}\log_+\Big(\frac{en}{p}\Big)\mathbb{E}\|X\| + \sup_{\|v\|_*\leq 1}(\mathbb{E}|\langle v,Y\rangle|^p)^{1/p}\Big)\\ &\leq C\Big(\frac{1}{\kappa}\log_+\Big(\frac{en}{p}\Big)+1\Big)(\mathbb{E}\|X\|^p)^{1/p}.\end{aligned} \tag{8}$$

As a consequence we know that for random vectors which satisfy Sudakov minoration principle weak and strong moments are comparable up to a logarithmic factor.

Corollary 34 *Suppose that X is an n-dimensional random vector, which satisfies* SMP(κ). *Then for any norm $\| \ \|$ on $\mathbb{R}^n$ and any $p \geq 1$,*

$$\left(\mathbb{E}\|X\|^p\right)^{1/p} \leq C\Big(\frac{1}{\kappa}\log_+\Big(\frac{en}{p}\Big)\mathbb{E}\|X\| + \sup_{\|v\|_* \leq 1}\left(\mathbb{E}|\langle v, X\rangle|^p\right)^{1/p}\Big).$$

Acknowledgements This research was supported by the National Science Centre, Poland grant 2015/18/A/ST1/00553.

References

[1] R. Adamczak, R. Latała, A. E. Litvak, K. Oleszkiewicz, A. Pajor and N. Tomczak-Jaegermann, *A short proof of Paouris' inequality*, Canad. Math. Bull. **57** (2014), 3–8.
[2] D. Alonso-Gutiérrez and J. Bastero, *Approaching the Kannan-Lovász-Simonovits and variance conjectures*, Lecture Notes in Mathematics **2131**, Springer, Cham, 2015
[3] S. Bobkov and F. L. Nazarov, *On convex bodies and log-concave probability measures with unconditional basis*, in: Geometric aspects of functional analysis, 53–69, Lecture Notes in Math. **1807**, Springer, Berlin, 2003
[4] C. Borell, *Convex measures on locally convex spaces*, Ark. Math. **12** (1974), 239–252.
[5] S. Brazitikos, A. Giannopoulos, P. Valettas and B. H. Vritsiou, *Geometry of isotropic convex bodies*, Mathematical Surveys and Monographs **196**, American Mathematical Society, Providence, RI, 2014.
[6] E .D. Gluskin and S. Kwapień *Tail and moment estimates for sums of independent random variables with logarithmically concave tails*, Studia Math. **114** (1995) 303–309.
[7] R. Eldan, *Thin shell implies spectral gap up to polylog via a stochastic localization scheme*, Geom. Funct. Anal. **23** (2013), 532–569.
[8] R. Kannan, L. Lovász and M. Simonovits, *Isoperimetric problems for convex bodies and a localization lemma*, Discrete Comput. Geom. **13** (1995), 541–559.
[9] B. Klartag, *A central limit theorem for convex sets*, Invent. Math. **168** (2007), 91–131.
[10] B. Klartag, *A Berry-Esseen type inequality for convex bodies with an unconditional basis*, Probab. Theory Related Fields, **145** (2009), 1–33.
[11] R. Latała, *Weak and strong moments of random vectors*, in: Marcinkiewicz centenary volume, 115–121, Banach Center Publ. **95**, Polish Acad. Sci. Inst. Math., Warsaw, 2011.
[12] R. Latała, *Sudakov-type minoration for log-concave vectors*, Studia Math. **223** (2014), 251–274.
[13] R. Latała and M. Strzelecka, *Weak and strong moments of l_r-norms of log-concave vectors*, Proc. Amer. Math. Soc. **144** (2016), 3597–3608.
[14] R. Latała and T. Tkocz, *A note on suprema of canonical processes based on random variables with regular moments*, Electron. J. Probab. **20** (2015), no. 36, 1–17.
[15] R. Latała and J. O. Wojtaszczyk, *On the infimum convolution inequality*, Studia Math. **189** (2008), 147–187.
[16] E. Lutvak and G. Zhang, *Blaschke-Santaló inequalities*, J. Differential Geom. **47** (1997), 1–16.
[17] B. Maurey, *Some deviation inequalities*, Geom. Funct. Anal. **1** (1991), 188–197.
[18] B. Maurey and G. Pisier, *Séries de variables aléatoires vectorielles indépendantes et propriétés géométriques des espaces de Banach*, Studia Math. **58** (1976), 45–90.
[19] S. Mendelson, E. Milman and G. Paouris, *Generalized Sudakov via dimension reduction - a program*, arXiv:1610.09287.

[20] E. Milman, *On the role of convexity in isoperimetry, spectral gap and concentration*, Invent. Math. **177** (2009), 1–43.

[21] G. Paouris, *Concentration of mass on convex bodies*, Geom. Funct. Anal. **16** (2006), 1021–1049.

[22] V. N. Sudakov, *Gaussian measures, Cauchy measures and ε-entropy*, Soviet Math. Dokl. **10** (1969), 310–313.

[23] M. Talagrand, *A new isoperimetric inequality and the concentration of measure phenomenon*, in: Israel Seminar (GAFA), Lecture Notes in Math. **1469**, 94–124, Springer, Berlin 1991

[24] M. Talagrand, *The supremum of some canonical processes* Amer. J. Math. **116** (1994), 283–325.

[25] M. Talagrand, *Upper and lower bounds for stochastic processes. Modern methods and classical problems*, Ergebnisse der Mathematik und ihrer Grenzgebiete. 3. Folge. A Series of Modern Surveys in Mathematics **60**, Springer, Heidelberg, 2014.

Stability Results for Some Geometric Inequalities and Their Functional Versions

Umut Caglar and Elisabeth M. Werner

Abstract The Blaschke Santaló inequality and the L_p affine isoperimetric inequalities are major inequalities in convex geometry and they have a wide range of applications. Functional versions of the Blaschke Santaló inequality have been established over the years through many contributions. More recently and ongoing, such functional versions have been established for the L_p affine isoperimetric inequalities as well. These functional versions involve notions from information theory, like entropy and divergence.

We list stability versions for the geometric inequalities as well as for their functional counterparts. Both are known for the Blaschke Santaló inequality. Stability versions for the L_p affine isoperimetric inequalities in the case of convex bodies have only been known in all dimensions for $p = 1$ and for $p > 1$ only for convex bodies in the plane. Here, we prove almost optimal stability results for the L_p affine isoperimetric inequalities, for all p, for all convex bodies, for all dimensions. Moreover, we give stability versions for the corresponding functional versions of the L_p affine isoperimetric inequalities, namely the reverse log Sobolev inequality, the L_p affine isoperimetric inequalities for log concave functions, and certain divergence inequalities.

Keywords Entropy • Divergence • Affine isoperimetric inequalities • Log Sobolev inequalities

1991 *Mathematics Subject Classification.* 46B, 52A20, 60B, 35J

U. Caglar (✉)
Department of Mathematics and Statistics, Florida International University, Miami, FL 33199, USA
e-mail: ucaglar@fiu.edu

E.M. Werner
Department of Mathematics, Case Western Reserve University, Cleveland, OH 44106, USA

Université de Lille 1, UFR de Mathématique, 59655 Villeneuve d'Ascq, France
e-mail: elisabeth.werner@case.edu

E. Carlen et al. (eds.), *Convexity and Concentration*, The IMA Volumes in Mathematics and its Applications 161, DOI 10.1007/978-1-4939-7005-6_17

1 Introduction and Background

We present stability results for several geometric and functional inequalities. Our main focus will be on geometric inequalities coming from affine convex geometry, namely the Blaschke Santaló inequality, e.g., [24, 55], and the L_p affine-isoperimetric and related inequalities [12, 21, 45, 51, 66] and also their functional counterparts, which includes the functional Blaschke Santaló inequality [5, 7, 22, 35] and the recently established divergence and entropy inequalities [6, 17, 20]. These inequalities are fundamental in convex geometry and geometric analysis, e.g., [10, 29, 30, 45, 46, 48, 49, 60, 64, 66, 67], and they have applications throughout mathematics. We only quote: approximation theory of convex bodies by polytopes [11, 27, 37, 54, 57, 61], affine curvature flows [3, 4, 62, 63], information theory [6, 17, 18, 20, 51, 65], valuation theory [2, 28, 29, 38, 40, 41, 42, 52, 56], and partial differential equations [43]. Therefore, it is important to know stability results of those inequalities.

Stability results answer the following question: Is the inequality that we consider sensitive to small perturbations? In other words, if a function almost attains the equality in a given inequality, is it possible to say that then this function is close to the minimizers of the inequality? For the Blaschke Santaló inequality and the functional Blaschke Santaló inequality such stability results have been established in [8] and [9], respectively. Stability results for the L_p-affine isoperimetric inequalities for convex bodies were proved in [13] for $p = 1$ and dimension $n \geq 3$. In [32, 33], stability results for the L_p-affine isoperimetric inequality were proved in dimension 2 and for $p \geq 1$.

We present here stability results for the L_p-affine isoperimetric inequalities for all p and in all dimensions. Stability results for the corresponding functional versions of these inequalities are also given.

Throughout, we will assume that K is a convex body in $\mathbb{R}^n$, i.e., a convex compact subset of $\mathbb{R}^n$ with non-empty interior $\text{int}(K)$. We denote by ∂K the boundary of K and by $\text{vol}(K)$ or $|K|$ its n-dimensional volume. B_2^n is the Euclidean unit ball centered at 0 and $S^{n-1} = \partial B_2^n$ its boundary. The standard inner product on $\mathbb{R}^n$ is $\langle , \rangle$. It induces the Euclidean norm, denoted by $\| \cdot \|_2$. We will use the Banach-Mazur distance $d_{BM}(K, L)$ to measure the distance between the convex bodies K and L,

$$d_{BM}(K, L) = \min\{\alpha \geq 1 : K - x \subset T(L - y) \subset \alpha(K - x),\\ \text{for } T \in GL(n), x, y \in \mathbb{R}^n\}.$$

In the case when K and L are 0-symmetric, x and y can be taken to be 0,

$$d_{BM}(K, L) = \min\{\alpha \geq 1 : K \subset T(L) \subset \alpha\, K, \ \text{for } T \in GL(n)\}.$$

2 Stability in Inequalities for Convex Bodies

2.1 The Blaschke Santaló Inequality

Let K be a convex body in $\mathbb{R}^n$ such that $0 \in \text{int}(K)$. The polar K° of K is defined as

$$K^\circ = \{y \in \mathbb{R}^n \ : \ \langle x, y \rangle \leq 1 \ \text{ for all } \ x \in K\}$$

and, more generally, the polar K^z with respect to $z \in \text{int}(K)$ by $(K - z)^\circ$. The classical Blaschke Santaló inequality (see, e.g., [55]) states that there is a unique point $s \in \text{int}(K)$, the Santaló point of K, such that the volume product $|K||K^s|$ is minimal and that

$$|K|\,|K^s| \leq |B_2^n|^2$$

with equality if and only if K is an ellipsoid.

Ball and Böröczky [8] proved the following stability version of the Blaschke Santaló inequality. It will be one of the tools to prove stability versions for the L_p-affine isoperimetric inequalities.

Theorem 1 ([8]) *Let K be a convex body in $\mathbb{R}^n$, $n \geq 3$, with Santaló point at 0. If $|K||K^\circ| > (1 - \varepsilon)|B_2^n|^2$, for $\varepsilon \in (0, \frac{1}{2})$, then for some $\gamma > 0$, depending only on n, we have*

$$d_{BM}(K, B_2^n) < 1 + \gamma \varepsilon^{\frac{1}{3(n+1)}} |\log \varepsilon|^{\frac{4}{3(n+1)}}.$$

Remark It was noted in [8] that if K is 0-symmetric, then the exponent $\frac{1}{3(n+1)}$ occurring in Theorem 1 can be replaced by $\frac{2}{3(n+1)}$. Moreover, it was also noted in [8] that taking K to be the convex body resulting from B_2^n by cutting off two opposite caps of volume ε, shows that the exponent $\frac{1}{(3(n+1)}$ cannot be replaced by anything larger than $\frac{2}{n+1}$, even for 0-symmetric convex bodies with axial rotational symmetry. Therefore the exponent of ε is of the correct order.

2.2 L_p-Affine Isoperimetric Inequalities

Now we turn to stability results for the L_p-affine isoperimetric inequalities for convex bodies. These inequalities involve the L_p-affine surface areas which are a central part of the rapidly developing L_p and Orlicz Brunn Minkowski theory and are the focus of intensive investigations (see, e.g., [19, 23, 25, 26, 30, 39, 40, 41, 42, 43, 44, 45, 46, 47, 56, 57, 58, 59, 60, 61, 62, 63, 64, 65, 66]).

The L_p-affine surface area $as_p(K)$ of a convex body K in $\mathbb{R}^n$ was introduced by Lutwak for all $p > 1$ in his seminal paper [45] and for all other p by Schütt and Werner [60](see also [31]). The case $p = 1$ is the classical affine surface area introduced by Blaschke in dimensions 2 and 3 [12] (see also [36, 59]).

Let $p \in \mathbb{R}$, $p \neq -n$ and assume that K is a convex body with centroid or Santaló point at the origin. Then

$$as_p(K) = \int_{\partial K} \frac{\kappa(x)^{\frac{p}{n+p}}}{\langle x, N(x)\rangle^{\frac{n(p-1)}{n+p}}} d\mu_K(x), \tag{1}$$

where $N(x)$ is the unit outer normal in $x \in \partial K$, the boundary of K, $\kappa(x)$ is the (generalized) Gaussian curvature in x and μ_K is the surface area measure on ∂K. In particular, for $p = 0$

$$as_0(K) = \int_{\partial K} \langle x, N_K(x)\rangle \, d\mu_K(x) = n|K|.$$

For $p = 1$,

$$as_1(K) = \int_{\partial K} \kappa_K(x)^{\frac{1}{n+1}} \, d\mu_K(x)$$

is the classical affine surface area which is independent of the position of K in space. Note also that $as_p(B_2^n) = \mathrm{vol}_{n-1}(\partial B_2^n) = n|B_2^n|$ for all $p \neq -n$. If the boundary of K is sufficiently smooth, (1) can be written as an integral over the boundary S^{n-1} of the Euclidean unit ball B_2^n,

$$as_p(K) = \int_{S^{n-1}} \frac{f_K(u)^{\frac{n}{n+p}}}{h_K(u)^{\frac{n(p-1)}{n+p}}} d\sigma(u).$$

Here, σ is the usual surface area measure on S^{n-1}, $h_K(u) = \max_{x \in K}\langle x, u\rangle$ is the support function of K in direction $u \in S^{n-1}$, and $f_K(u)$ is the curvature function, i.e. the reciprocal of the Gaussian curvature $\kappa_K(x)$ at this point $x \in \partial K$ that has u as outer normal. In particular, for $p = \pm\infty$,

$$as_{\pm\infty}(K) = \int_{S^{n-1}} \frac{1}{h_K(u)^n} d\sigma(u) = n|K^\circ|. \tag{2}$$

The L_p-affine surface area is invariant under linear transformations T with determinant 1. More precisely (see, e.g., [60]), if $T : \mathbb{R}^n \to \mathbb{R}^n$ is a linear, invertible map, then

$$as_p(T(K)) = |\mathrm{det} T|^{\frac{n-p}{n+p}} as_p(K). \tag{3}$$

The L_p-affine surface area is a valuation [40, 42, 58], i.e., for convex bodies K and L such that $K \cup L$ is convex,

$$as_p(K \cup L) + as_p(K \cap L) = as_p(K) + as_p(L).$$

Valuations have become a major topic in convex geometry in recent years. We refer to, e.g., [2, 28, 29, 38, 40, 41, 42, 52, 56].

We now state the L_p-affine isoperimetric inequalities for the quantities $as_p(K)$. They were proved by Lutwak for $p > 1$ [45] and for all other p by Werner and Ye [66]. The case $p = 1$ is the classical affine isoperimetric inequality [12, 21].

Theorem 2 ($p = 1$ [12, 21], $p > 1$ [45], all other p [66]) *Let K be a convex body with centroid at the origin.*

(i) If $p > 0$, then

$$\frac{as_p(K)}{as_p(B_2^n)} \leq \left(\frac{|K|}{|B_2^n|}\right)^{\frac{n-p}{n+p}},$$

with equality if and only if K is an ellipsoid. For $p = 0$, equality holds trivially for all K.

(ii) If $-n < p < 0$, then

$$\frac{as_p(K)}{as_p(B_2^n)} \geq \left(\frac{|K|}{|B_2^n|}\right)^{\frac{n-p}{n+p}},$$

with equality if and only if K is an ellipsoid.

(iii) If K is in addition in C_+^2 and if $p < -n$, then

$$c^{\frac{np}{n+p}} \left(\frac{|K|}{|B_2^n|}\right)^{\frac{n-p}{n+p}} \leq \frac{as_p(K)}{as_p(B_2^n)}.$$

The constant c in (iii) is the constant from the inverse Blaschke Santaló inequality due to Bourgain and Milman [15]. This constant has recently been improved by Kuperberg [34] (see also [50] for a different proof).

2.3 Stability for the L_p-Affine Isoperimetric Inequality for Convex Bodies

Stability results for the L_p-affine isoperimetric inequalities for convex bodies were proved by Böröczky [13] for $p = 1$ and dimension $n \geq 3$. Ivaki [32, 33] gave stability results for the L_p-affine isoperimetric inequality in dimension 2 and $p \geq 1$. We present here stability results for the L_p-affine isoperimetric inequalities for all p and in all dimensions. Before we do so, we first quote the results by Böröczky [13] and Ivaki [33].

Theorem 3 ([13]) *If K is a convex body in $\mathbb{R}^n$, $n \geq 3$, and*

$$\left(\frac{as_1(K)}{as_1(B_2^n)}\right)^{n+1} > (1-\epsilon)\left(\frac{|K|}{|B_2^n|}\right)^{n-1} \quad \text{for } \epsilon \in \left(0, \frac{1}{2}\right), \tag{4}$$

then for some $\gamma > 0$, depending only on n, we have

$$d_{BM}(K, B_2^n) < 1 + \gamma \varepsilon^{\frac{1}{6n}} |\log \varepsilon|^{\frac{1}{6}}.$$

Later, in [8], the above approximation was improved to

$$d_{BM}(K, B_2^n) < 1 + \gamma \varepsilon^{\frac{1}{3(n+1)}} |\log \varepsilon|^{\frac{4}{3(n+1)}}.$$

Ivaki [33] gave a stability version for the Blaschke Santaló inequality from which the following stability result for the L_p-affine isoperimetric inequality in dimension 2 and $p \geq 1$ follows easily.

Theorem 4 ([33]) *Let K be an origin symmetric convex body in $\mathbb{R}^2$, and $p \geq 1$. There exists an $\epsilon_p > 0$, depending on p, such that the following holds. If for an ϵ, $0 < \epsilon < \epsilon_p$,*

$$\left(\frac{as_p(K)}{2\pi}\right)^{p+2} > (1-\epsilon)^p \left(\frac{area(K)}{\pi}\right)^{2-p}$$

then for some $\gamma > 0$, we have

$$d_{BM}(K, B_2^2) < 1 + \gamma \varepsilon^{\frac{1}{2}}. \tag{5}$$

The same author also considered the case when K is a not necessarily origin symmetric convex body in $\mathbb{R}^2$ [33]. Then the order of approximation becomes $\frac{1}{4}$ instead of $\frac{1}{2}$. Note also that there are results in dimension $n = 2$ by Böröczky and Makai [14] on stability of the Blaschke Santaló inequality, from which a stability result of the form (5) for the L_p-affine isoperimetric inequality in dimension 2 follows easily. But the order of approximation in the origin-symmetric case is $1/3$ and in the general case $1/6$.

We now present almost optimal stability results for the L_p-affine isoperimetric inequalities, for all p, for all convex bodies, for all dimensions. To do so, we use the above stability version of the Blaschke Santaló inequality by Ball and Böröczky [8], together with inequalities proved in [66].

Theorem 5 *Let K be a convex body in $\mathbb{R}^n$, $n \geq 3$, with Santaló point or centroid at 0.*

(i) *Let $p > 0$. If $\left(\frac{as_p(K)}{as_p(B_2^n)}\right)^{n+p} > (1-\varepsilon)^p \left(\frac{|K|}{|B_2^n|}\right)^{n-p}$, then for some $\gamma > 0$, depending only on n, we have*

$$d_{BM}(K, B_2^n) < 1 + \gamma \varepsilon^{\frac{1}{3(n+1)}} |\log \varepsilon|^{\frac{4}{3(n+1)}}.$$

(ii) Let $-n < p < 0$. *If* $\left(\frac{as_p(K)}{as_p(B_2^n)}\right)^{n+p} < (1-\varepsilon)^p \left(\frac{|K|}{|B_2^n|}\right)^{n-p}$, *then for some* $\gamma > 0$, *depending only on n, we have*

$$d_{BM}(K, B_2^n) < 1 + \gamma \varepsilon^{\frac{1}{3(n+1)}} |\log \varepsilon|^{\frac{4}{3(n+1)}}.$$

Remarks (i) If K is 0-symmetric, then $\varepsilon^{\frac{1}{3(n+1)}}$ can be replaced by $\varepsilon^{\frac{2}{3(n+1)}}$. This follows from [8]. See also the Remark after Theorem 1.

(ii) The example in [8] already quoted in the Remark after Theorem 1 shows that $\varepsilon^{\frac{1}{3(n+1)}}$ cannot be replaced by anything smaller than $\varepsilon^{\frac{2}{n-1}}$, even for 0-symmetric convex bodies with axial rotational symmetry. Indeed, let K be the convex body obtained from B_2^n by removing two opposite caps of volume ε each. Then

$$\left(\frac{as_p(K)}{as_p(B_2^n)}\right)^{n+p} > (1 - k\varepsilon^{\frac{n-1}{n+1}})^p \left(\frac{|K|}{|B_2^n|}\right)^{n-p} = (1-\delta)^p \left(\frac{|K|}{|B_2^n|}\right)^{n-p},$$

where we have put $\delta = k\varepsilon^{\frac{n-1}{n+1}}$ and where k is a constant that depends on n only, except for $0 < p < n$, where it also depends on p. And $d_{BM}(K, B_2^n) = 1 + \gamma\delta^{\frac{2}{n-1}}$.

Proof of Theorem 5. (i) As $as_p(B_2^n) = n|B_2^n|$, we observe that the inequality

$$\left(\frac{as_p(K)}{as_p(B_2^n)}\right)^{n+p} > (1-\varepsilon)^p \left(\frac{|K|}{|B_2^n|}\right)^{n-p}$$

is equivalent to the inequality

$$as_p(K)^{n+p} > (1-\varepsilon)^p n^{n+p} |K|^{n-p} |B_2^n|^{2p}. \tag{6}$$

It was proved in [66] that for all $p > 0$,

$$as_p(K)^{n+p} \le n^{n+p} |K|^n |K^\circ|^p.$$

Hence we get from the assumption that

$$n^{n+p} |K|^n |K^\circ|^p > (1-\varepsilon)^p n^{n+p} |K|^{n-p} |B_2^n|^{2p},$$

or equivalently, that

$$|K||K^\circ| > (1-\varepsilon)\, |B_2^n|^2,$$

and we conclude with the Ball and Böröczky stability result in Theorem 1.

(ii) The proof of (ii) is done similarly. We use the inequality

$$as_p(K)^{n+p} \ge n^{n+p} |K|^n |K^\circ|^p,$$

which holds for $-n < p < 0$ and which was also proved in [66]. □

Another stability result for the L_p-affine isoperimetric inequalities for convex bodies is obtained as a corollary to Proposition 17 below. We list it now, as we want to compare the two. Let K be a convex body in $\mathbb{R}^n$ with 0 in its interior and let the function ψ of Proposition 17 be $\psi(x) = \|x\|_K^2/2$, where $\|\cdot\|_K$ is the gauge function of the convex body K,

$$\|x\|_K = \min\{\alpha \geq 0 : \; x \in \alpha K\} = \max_{y \in K^\circ} \langle x, y \rangle.$$

Let

$$as_\lambda(\psi) = \int_{\mathbb{R}^n} e^{(2\lambda-1)\psi(x) - \lambda\langle \nabla\psi, x\rangle} \left(\det\left(\nabla^2\psi(x)\right)\right)^\lambda dx \tag{7}$$

be the L_λ-affine surface area of the function ψ. This quantity is discussed in detail in Section 3.3. Differentiating $\psi(x) = \|x\|_K^2/2$, we get $\langle x, \nabla\psi(x)\rangle = 2\psi(x)$. Thus, for $\psi(x) = \|x\|_K^2/2$, the expression (7) simplifies to

$$as_\lambda(\psi) = \int_{\mathbb{R}^n} \left(\det \nabla^2\psi(x)\right)^\lambda e^{-\psi(x)} dx. \tag{8}$$

Note that for the Euclidean norm $\|.\|_2$, $as_\lambda\left(\frac{\|\cdot\|_2^2}{2}\right) = (2\pi)^{\frac{n}{2}}$ and it was proved in [20] that

$$\frac{as_\lambda\left(\frac{\|\cdot\|_K^2}{2}\right)}{as_\lambda\left(\frac{\|\cdot\|_2^2}{2}\right)} = \frac{as_p(K)}{as_p(B_2^n)}, \tag{9}$$

where λ and p are related by $\lambda = \frac{p}{n+p}$. Together with Proposition 17, this immediately implies another stability result for the L_p-affine isoperimetric inequalities for convex bodies.

Corollary 6 *Let K be a convex body in $\mathbb{R}^n$ with the centroid or the Santaló point at the origin.*
(i) Let $0 < p \leq \infty$ and suppose that for some $\varepsilon \in (0, \varepsilon_0)$,

$$\frac{as_p(K)}{as_p(B_2^n)} > (1-\varepsilon)^{\frac{p}{n+p}} \left(\frac{|K|}{|B_2^n|}\right)^{\frac{n-p}{n+p}}.$$

(i) Let $-n < p < 0$ and suppose that for some $\varepsilon \in (0, \varepsilon_0)$,

$$\frac{as_p(K)}{as_p(B_2^n)} < (1-\varepsilon)^{\frac{p}{n+p}} \left(\frac{|K|}{|B_2^n|}\right)^{\frac{n-p}{n+p}}.$$

Then, in both cases (i) and (ii), there exist $c > 0$ and a positive definite matrix A such that

$$\int_{R(\varepsilon)B_2^n} \left| \|Ax\|_K^2 - \|x\|_2^2 - c \right| dx < \eta \varepsilon^{\frac{1}{129n^2}},$$

where $R(\varepsilon) = \frac{|\log \varepsilon|^{\frac{1}{2}}}{8n}$ *and* ε_0, η *depend on* n.

Proof It is easy to see (e.g., [20]) that

$$|K| = \frac{1}{2^{\frac{n}{2}} \Gamma\left(1 + \frac{n}{2}\right)} \int e^{-\frac{\|x\|_K^2}{2}} dx.$$

As $|B_2^n| = \frac{\pi^{\frac{n}{2}}}{\Gamma(1+\frac{n}{2})}$, we get, with $\psi(x) = \frac{\|x\|_K^2}{2}$, by (9) and the assumptions of the theorem, that for $0 < p \leq \infty$,

$$as_\lambda(\psi) > (1-\varepsilon)^\lambda (2\pi)^{n\lambda} \left(\int e^{-\psi(x)} dx \right)^{1-2\lambda}.$$

We have also used that $\lambda = \frac{p}{n+p}$. The result for $0 < p \leq \infty$ then follows immediately from Proposition 17. The case $-n < p < 0$ is treated similarly. □

Remarks In general, one cannot deduce Theorem 5 from Corollary 6. However, it follows from Theorem 5 that there exists $T \in GL(n)$ and $x_0, y_0 \in \mathbb{R}^n$ such that

$$K - x_0 \subset T(B_2^n - y_0) \subset \left(1 + \gamma \varepsilon^{\frac{1}{3(n+1)}} |\log \varepsilon|^{\frac{4}{3(n+1)}}\right) (K - x_0).$$

For simplicity, assume that $x_0 = y_0 = 0$, which corresponds to the case that K is 0-symmetric. Then this means that for all $x \in \mathbb{R}^n$,

$$\left| \|x\|_K - \|T(x)\|_2 \right| \leq \|T\| \left(\gamma \varepsilon^{\frac{1}{3(n+1)}} |\log \varepsilon|^{\frac{4}{3(n+1)}}\right) \|x\|_2$$

and thus

$$\int_{R(\varepsilon)B_2^n} \left| \|x\|_K^2 - \|T(x)\|_2^2 \right| dx$$

$$\leq \left(1 + \gamma \varepsilon^{\frac{1}{3(n+1)}} |\log \varepsilon|^{\frac{4}{3(n+1)}}\right) |B_2^n| \, \|T\|^2 R^{n+2}(\varepsilon) \left(\gamma \varepsilon^{\frac{1}{3(n+1)}} |\log \varepsilon|^{\frac{4}{3(n+1)}}\right)$$

$$= \left(1 + \gamma \varepsilon^{\frac{1}{3(n+1)}} |\log \varepsilon|^{\frac{4}{3(n+1)}}\right) \frac{|B_2^n|}{(8n)^{n+2}} \|T\|^2 \left(\gamma \varepsilon^{\frac{1}{3(n+1)}} |\log \varepsilon|^{\frac{4}{3(n+1)} + \frac{n+2}{2}}\right).$$

Hence, allowing general T, the exponent of ε can be improved.

2.4 Stability Result for the Entropy Power Ω_K

An affine invariant quantity that is closely related to the L_p-affine surface areas is the entropy power Ω_K. It was introduced in [51] as the limit of L_p-affine surface areas,

$$\Omega_K = \lim_{p\to\infty}\left(\frac{as_p(K)}{n|K^\circ|}\right)^{n+p}. \tag{10}$$

The quantity Ω_K is related to the relative entropy of the cone measures of K and K°. We refer to [51] for the details and only mention an affine isoperimetric inequality for Ω_K proved in [51].

Theorem 7 ([51]) *If K is a convex body of volume 1, then*

$$\Omega_{K^\circ} \leq \Omega_{\left(\frac{B_2^n}{|B_2^n|^{\frac{1}{n}}}\right)^\circ}. \tag{11}$$

Equality holds if and only if K is a normalized ellipsoid.

We now use the previous theorems to prove stability results for inequality (11). Using the invariant property (3) and the fact that $as_p(B_2^n) = n|B_2^n|$, this inequality can be written as

$$\Omega_{K^\circ} \leq |B_2^n|^{2n}.$$

Theorem 8 *Let K be a convex body in $\mathbb{R}^n$, $n \geq 3$, of volume 1 and such that the Santaló point or the centroid is at 0. Suppose that for some $\varepsilon \in (0, \frac{1}{2})$,*

$$\Omega_{K^\circ} > (1-\varepsilon)|B_2^n|^{2n}. \tag{12}$$

Then for some $\gamma > 0$, depending only on n, we have

$$d_{BM}(K^\circ, B_2^n) < 1 + \gamma\left(\frac{2\varepsilon}{n}\right)^{\frac{1}{3(n+1)}}\left|\log\frac{2\varepsilon}{n}\right|^{\frac{4}{3(n+1)}}.$$

Remarks similar to the ones after Theorem 5 hold.

Proof It was shown in [66] that $\left(\frac{as_p(K^\circ)}{n|K|}\right)^{n+p}$ is decreasing in $p \in (0,\infty)$. By definition (7), $\lim_{p\to\infty}\left(\frac{as_p(K^\circ)}{n|K|}\right)^{n+p} = \Omega_{K^\circ}$. Therefore we get with assumption (12) that for all $p > 0$

$$\left(\frac{as_p(K^\circ)}{n|K|}\right)^{n+p} > (1-\varepsilon)|B_2^n|^{2n}.$$

Or, equivalently, as $|K| = 1$,

$$as_p(K^\circ)^{n+p} > (1-\varepsilon)n^{n+p}|K|^{n+p}|B_2^n|^{2n} = (1-\varepsilon)n^{n+p}|B_2^n|^{2p}\,|B_2^n|^{2(n-p)}$$
$$\geq (1-\varepsilon)n^{n+p}|K^\circ|^{n-p}|B_2^n|^{2p}.$$

In the last inequality we have used the Blaschke Santaló inequality $|K|\,|K^\circ| \leq |B_2^n|^2$, which we can apply as long as $n-p \geq 0$. Note that for all $\varepsilon \in (0, \frac{1}{2})$ and $p > 0$

$$1-\varepsilon > \left(1 - \frac{2\varepsilon}{p}\right)^p.$$

Hence, using the elementary inequality above, we get for all $0 < p \leq n$ that

$$as_p(K^\circ)^{n+p} > \left(1 - \frac{2\varepsilon}{p}\right)^p n^{n+p}|K^\circ|^{n-p}|B_2^n|^{2p}.$$

Inequality (6) and the arguments used after it imply that for all $0 < p \leq n$,

$$d_{BM}(K^\circ, B_2^n) < 1 + \gamma\left(\frac{2\varepsilon}{p}\right)^{\frac{1}{3(n+1)}} \left|\log\frac{2\varepsilon}{p}\right|^{\frac{4}{3(n+1)}}.$$

Since the right-hand side of above equation is decreasing in p, minimizing over p in the interval $(0, n]$ gives the result. □

The second stability result and the corresponding comparisons (see the Remark after Corollary 6) are obtained accordingly. We skip the proof.

Theorem 9 *Let K be a convex body in $\mathbb{R}^n$, $n \geq 3$, of volume 1 and with Santaló point or centroid at 0, such that $\Omega_{K^\circ} > (1-\varepsilon)|B_2^n|^{2n}$. Then there exists $c > 0$ and a positive definite matrix A such that*

$$\int_{R(\varepsilon)B_2^n} \left|\|Ax\|_K^2 - |x|_2^2 - c\right| dx < \eta\varepsilon^{\frac{1}{129n^2}},$$

$R(\varepsilon) = \frac{|\log\varepsilon|^{\frac{1}{2}}}{8n}$ *and ε_0, η depend on n.*

3 Stability Results for Functional Inequalities

3.1 Stability for the Functional Blaschke Santaló Inequality

We will first state a functional version of the Blaschke Santaló inequality. To do so, we recall that the Legendre transform of a function $\psi : \mathbb{R}^n \to \mathbb{R} \cup \{+\infty\}$ at $z \in \mathbb{R}^n$ is defined by

$$\mathcal{L}_z\psi(y) = \sup_{x\in\mathbb{R}^n} \left(\langle x - z, y\rangle - \psi(x)\right), \quad \text{for } y \in \mathbb{R}^n. \tag{13}$$

The function $L_z\psi : \mathbb{R}^n \to \mathbb{R} \cup \{+\infty\}$ is always convex and lower semicontinuous. If ψ is convex, lower semicontinuous and $\psi < +\infty$, then $L_zL_z\psi = \psi$. When $z = 0$, we write

$$\psi^*(y) = \mathcal{L}_0\psi(y) = \sup_x \left(\langle x, y\rangle - \psi(x)\right). \tag{14}$$

Work by K.M. Ball [7], S. Artstein-Avidan, B. Klartag, V.D.Milman [5], M. Fradelizi, M. Meyer [22], and J. Lehec [35] led to the functional version of the Blaschke Santaló inequality which we now state.

Theorem 10 ([5, 7, 22, 35]) *Let $\rho : \mathbb{R} \to \mathbb{R}_+$ be a log-concave non-increasing function and $\psi : \mathbb{R}^n \to \mathbb{R} \cup \{+\infty\}$ be measurable. Then*

$$\inf_{z\in\mathbb{R}^n} \int_{\mathbb{R}^n} \rho(\psi(x))dx \int_{\mathbb{R}^n} \rho(\mathcal{L}_z\psi(x))dx \le \left(\int_{\mathbb{R}^n} \rho\left(\frac{\|x\|_2^2}{2}\right)dx\right)^2.$$

If ρ is decreasing, there is equality if and only if there exist a, b, c in $\mathbb{R}$, $a < 0$, $z \in \mathbb{R}^n$ and a positive definite matrix $A : \mathbb{R}^n \to \mathbb{R}^n$ such that

$$\psi(x) = \frac{\|A(x+z)\|_2^2}{2} + c, \quad \textit{for } x \in \mathbb{R}^n$$

and moreover either $c = 0$, or $\rho(t) = e^{at+b}$, for $t > -|c|$.

Remark If $\rho(t) = e^{-t}$ and if $\varphi = e^{-\psi}$ has centroid at 0, i.e., $\int_{\mathbb{R}^n} xe^{-\psi}dx = 0$, then the inequality of the above theorem simplifies to

$$\begin{aligned}\int_{\mathbb{R}^n} \rho(\psi(x))dx \int_{\mathbb{R}^n} \rho(\mathcal{L}_z\psi(x))dx &= \left(\int_{\mathbb{R}^n} e^{-\psi(x)}dx\right)\left(\int_{\mathbb{R}^n} e^{-\psi^*(x)}dx\right)\\ &\le \left(\int_{\mathbb{R}^n} e^{-\frac{\|x\|_2^2}{2}}dx\right)^2.\end{aligned} \tag{15}$$

Barthe, Böröczky, and Fradelizi [9] established the following stability theorem for the functional Blaschke Santaló inequality.

Theorem 11 ([9]) *Let $\rho : \mathbb{R} \to \mathbb{R}_+$ be a log-concave and decreasing function with $\int_{\mathbb{R}_+} \rho < \infty$. Let $\psi : \mathbb{R}^n \to \mathbb{R}$ be a convex, measurable function. Assume that for some $\varepsilon \in (0, \varepsilon_0)$ and all $z \in \mathbb{R}^n$ the following inequality holds*

$$\int_{\mathbb{R}^n} \rho(\psi(x))dx \int_{\mathbb{R}^n} \rho(\mathcal{L}_z\psi(x))dx > (1-\varepsilon)\left(\int_{\mathbb{R}^n} \rho\left(\frac{\|x\|_2^2}{2}\right)dx\right)^2.$$

Then there exists some $z \in \mathbb{R}^n$, $c \in \mathbb{R}$ and a positive definite $n \times n$ matrix A such that

$$\int_{R(\varepsilon)B_2^n} \left| \frac{\|x\|_2^2}{2} + c - \psi(Ax + z) \right| dx < \eta \varepsilon^{\frac{1}{129n^2}},$$

where $\lim_{\varepsilon \to 0} R(\varepsilon) = \infty$ *and* $\varepsilon_0, \eta, R(\varepsilon)$ *depend on n and ρ.*

3.2 Stability for Divergence Inequalities

A function $\varphi : \mathbb{R}^n \to [0, \infty)$ is log concave, if it is of the form $\varphi(x) = e^{-\psi(x)}$, where $\psi : \mathbb{R}^n \to \mathbb{R}$ is a convex function. Recall that we say that $\varphi = e^{-\psi}$ has centroid at 0, respectively, the Santaló point, at 0 if,

$$\int x\varphi(x)dx = \int xe^{-\psi(x)}dx = 0, \text{ respectively } \int xe^{-\psi^*(x)}dx = 0.$$

The following entropy inequality for log concave functions was established in [17], Corollary 13.

Theorem 12 ([17]) *Let $\varphi : \mathbb{R}^n \to [0, \infty)$ be a log-concave function that has centroid or Santaló point at* 0. *Let $f : (0, \infty) \to \mathbb{R}$ be a convex, decreasing function. Then*

$$\int_{supp(\varphi)} \varphi f\left(e^{\langle \frac{\nabla\varphi}{\varphi}, x\rangle} \varphi^{-2} \left(\det\left(\nabla^2\left(-\log\varphi\right)\right)\right)\right) \geq f\left(\frac{(2\pi)^n}{\left(\int \varphi dx\right)^2}\right) \left(\int_{supp(\varphi)} \varphi dx\right). \quad (16)$$

If f is a concave, increasing function, the inequality is reversed.
Equality holds in both cases if and only if $\varphi(x) = ce^{-\langle Ax,x\rangle}$, where c is a positive constant and A is an $n \times n$ positive definite matrix.

Theorem 12 was proved under the assumptions that the convex or concave functions f and the log concave functions φ have enough smoothness and integrability properties so that the expressions considered in the above statement make sense. Thus, in this section, we will make the same assumptions on f and φ, i.e., we will assume that $\varphi^\circ \in L^1(\mathrm{supp}(\varphi), dx)$, the Lebesgue integrable functions on the support of φ, that

$$\varphi \in C^2(\mathrm{supp}(\varphi)) \cap L^1(\mathbb{R}^n, dx), \quad (17)$$

where $C^2(\mathrm{supp}(\varphi))$ denotes the twice continuously differentiable functions on their support, and that

$$\varphi f\left(\frac{e^{\frac{\langle \nabla\varphi, x\rangle}{\varphi}}}{\varphi^2} \det\left(\nabla^2\left(-\log\varphi\right)\right)\right) \in L^1(\mathrm{supp}(\varphi), dx). \quad (18)$$

Recall that $\varphi(x) = e^{-\psi(x)}$ and put $d\mu = e^{-\psi}dx$. Then the left-hand side of inequality (16) can be written as

$$\int_{\mathbb{R}^n} f\left(e^{2\psi-\langle\nabla\psi,x\rangle}\det\left(\nabla^2\psi\right)\right)d\mu.$$

It was shown in [17] that the left-hand side of the inequality (16) is the natural definition of f-divergence $D_f(\varphi)$ for a log concave function φ, so that (16) can be rewritten as

$$D_f(\varphi) \geq f\left(\frac{(2\pi)^n}{\left(\int\varphi dx\right)^2}\right)\left(\int_{\text{supp}(\varphi)}\varphi dx\right). \tag{19}$$

In information theory, probability theory, and statistics, an f-divergence is a function that measures the difference between two (probability) distributions. We refer to, e.g., [17] for details and references about f-divergence.

Theorem 13 *Let $f : (0,\infty) \to \mathbb{R}$ be a concave, strictly increasing function. Let $\psi : \mathbb{R}^n \to \mathbb{R}$ be a convex function such that $e^{-\psi} \in C^2(\mathbb{R}^n)$ and such that $\int_{\mathbb{R}^n} xe^{-\psi(x)}dx = 0$ or $\int_{\mathbb{R}^n} xe^{-\psi^*(x)}dx = 0$. Suppose that for some $\varepsilon \in (0,\varepsilon_0)$,*

$$\int_{\mathbb{R}^n} f\left(e^{2\psi-\langle\nabla\psi,x\rangle}\det\left(\nabla^2\psi\right)\right)d\mu >$$
$$f\left(\frac{(2\pi)^n}{\left(\int_{\mathbb{R}^n}d\mu\right)^2}\right)\left(\int_{\mathbb{R}^n}d\mu\right) - \varepsilon f'\left(\frac{(2\pi)^n}{\left(\int_{\mathbb{R}^n}d\mu\right)^2}\right)\left(\int_{\mathbb{R}^n}d\mu\right)^{-1}.$$

Then there exist $c > 0$ and a positive definite matrix A such that

$$\int_{R(\varepsilon)B_2^n}\left|\frac{\|x\|_2^2}{2} + c - \psi(Ax)\right|dx < \eta\varepsilon^{\frac{1}{129n^2}},$$

where $\lim_{\varepsilon\to 0} R(\varepsilon) = \infty$ and $\varepsilon_0, \eta, R(\varepsilon)$ depend on n.

The analogue stability result holds, if f is convex and strictly decreasing.

Proof We treat the case when f is concave and strictly increasing. The case when f is convex and strictly decreasing is done similarly. We set $d\nu = \frac{e^{-\psi}dx}{\int e^{-\psi}dx} = \frac{\mu}{\int d\mu}$. Then ν is a probability measure and by Jensen's inequality and a change of variable,

$$\left(\int d\mu\right)\int_{\mathbb{R}^n} f\left(e^{(2\psi(x)-\langle\nabla\psi,x\rangle)}\left(\det\left(\nabla^2\psi(x)\right)\right)\right)d\nu \leq$$
$$\left(\int d\mu\right) f\left(\int_{\mathbb{R}^n} e^{(2\psi(x)-\langle\nabla\psi,x\rangle)}\left(\det\left(\nabla^2\psi(x)\right)\right)d\nu\right)$$

$$= f\left(\frac{1}{\int d\mu}\int_{\mathbb{R}^n} e^{-\psi^*(x)}dx\right)\left(\int d\mu\right).$$

Thus, by the assumption of the theorem, we get

$$\begin{aligned} f&\left(\frac{1}{\int d\mu}\int_{\mathbb{R}^n} e^{-\psi^*(x)}dx\right)\left(\int d\mu\right) \\ &> \left(\int d\mu\right) f\left(\frac{(2\pi)^n}{(\int d\mu)^2}\right) - \frac{\varepsilon}{\int d\mu} f'\left(\frac{(2\pi)^n}{(\int d\mu)^2}\right) \\ &\geq \left(\int d\mu\right) f\left(\frac{(2\pi)^n - \varepsilon}{(\int d\mu)^2}\right). \end{aligned}$$

The last inequality holds as by Taylor's theorem and the assumptions on f (i.e., $f'' \leq 0$), for ε small enough, there is a real number τ such that

$$\begin{aligned} f\left(\frac{(2\pi)^n - \varepsilon}{(\int d\mu)^2}\right) &= f\left(\frac{(2\pi)^n}{(\int d\mu)^2}\right) - \frac{\varepsilon}{(\int d\mu)^2} f'\left(\frac{(2\pi)^n}{(\int d\mu)^2}\right) + \frac{\varepsilon^2}{2\left(\int d\mu\right)^4} f''(\tau) \\ &\leq f\left(\frac{(2\pi)^n}{(\int d\mu)^2}\right) - \frac{\varepsilon}{(\int d\mu)^2} f'\left(\frac{(2\pi)^n}{(\int d\mu)^2}\right). \end{aligned}$$

Therefore we arrive at

$$f\left(\frac{1}{\int d\mu}\int_{\mathbb{R}^n} e^{-\psi^*(x)}dx\right) > f\left(\frac{(2\pi)^n - \varepsilon}{(\int d\mu)^2}\right).$$

Since f is strictly increasing we conclude that

$$\frac{1}{\int d\mu}\int_{\mathbb{R}^n} e^{-\psi^*(x)}dx > \frac{(2\pi)^n - \varepsilon}{(\int d\mu)^2},$$

which is equivalent to

$$\left(\int_{\mathbb{R}^n} e^{-\psi(x)}dx\right)\left(\int_{\mathbb{R}^n} e^{-\psi^*(x)}dx\right) > (2\pi)^n - \varepsilon.$$

From that we get

$$\left(\int_{\mathbb{R}^n} e^{-\psi(x)}dx\right)\left(\int_{\mathbb{R}^n} e^{-\psi^*(x)}dx\right) > (1-\varepsilon)\,(2\pi)^n.$$

As μ has its centroid at 0, we have by (15) that

$$\inf_{z\in\mathbb{R}^n}\left(\int_{\mathbb{R}^n} e^{-\psi(x)}dx\right)\left(\int_{\mathbb{R}^n} e^{-\mathcal{L}_z\psi(y)}\,dy\right)=\left(\int_{\mathbb{R}^n} e^{-\psi}\,dx\right)\left(\int_{\mathbb{R}^n} e^{-\psi^*(y)}\,dy\right)$$

and the theorem follows from the result by Barthe, Böröczky and Fradelizi [9], Theorem 11, with $\rho(t)=e^{-t}$. □

3.3 Stability for the Reverse Log Sobolev Inequality

We now prove a stability result for the reverse log Sobolev inequality. This inequality was first proved by Artstein-Avidan, Klartag, Schütt, and Werner [6] under strong smoothness assumptions. Those were subsequently removed in [20] and there, also equality characterization was achieved.

We first recall the reverse log Sobolev inequality. Let γ_n be the standard Gaussian measure on $\mathbb{R}^n$. For a log-concave probability measure μ on $\mathbb{R}^n$ with density $e^{-\psi}$, i.e., $\psi=-\log(d\mu/dx)$, let

$$S(\mu)=\int_{\mathbb{R}^n}\psi\,d\mu$$

be the Shannon entropy of μ.

Theorem 14 ([6, 20]) *Let μ be a log-concave probability measure on $\mathbb{R}^n$ with density $e^{-\psi}$ with respect to the Lebesgue measure. Then*

$$\int_{\mathbb{R}^n}\log\left(\det(\nabla^2\psi)\right)\,d\mu\le 2\ \left(S(\gamma_n)-S(\mu)\right). \tag{20}$$

Equality holds if and only if μ is Gaussian (with arbitrary mean and positive definite covariance matrix).

Inequality (20) is a reverse log Sobolev inequality as it can be shown that the log Sobolev inequality is equivalent to

$$2\left(S(\gamma_n)-S(\mu)\right)\le n\log\left(\frac{\int_{\mathbb{R}^n}\Delta\psi\,d\mu}{n}\right),$$

where Δ is the Laplacian. We refer to, e.g., [6, 20] for the details.

Note that inequality (20) follows from inequality (16) with $f(t)=\log t$. However, because of the assumptions on φ in Theorem 13, the result would only hold under those assumptions and not in the full generality stated in Theorem 14. Similarly, a stability result for Theorem 14 follows from Theorem 13 with $f(t)=\log t$. But again, because of the assumptions of Theorem 13, the result would only hold for those ψ such that $e^{-\psi}$ is in $C^2(\mathbb{R}^n)$ and has centroid at 0. We can prove a stability result for Theorem 14 without these assumptions. The proof is similar to the one of Theorem 13. We include it for completeness. But first we need to recall various items.

For a convex function $\psi : \mathbb{R}^n \to \mathbb{R} \cup \{+\infty\}$, we define D_ψ to be the convex domain of ψ, $D_\psi = \{x \in \mathbb{R}^n, \psi(x) < +\infty\}$. We always consider convex functions ψ such that $\text{int}\left(D_\psi\right) \neq \emptyset$. In the general case, when ψ is neither smooth nor strictly convex, the gradient of ψ, denoted by $\nabla\psi$, exists almost everywhere by Rademacher's theorem (e.g., [53]), and a theorem of Alexandrov [1], Busemann and Feller [16], guarantees the existence of its Hessian $\nabla^2\psi$ almost everywhere in $\text{int}\left(D_\psi\right)$. We let X_ψ be the set of points of $\text{int}\left(D_\psi\right)$ at which its Hessian $\nabla^2\psi$ in the sense of Alexandrov, Busemann, and Feller exists and is invertible. Then, by definition of the Legendre transform, for a convex function $\psi : \mathbb{R}^n \to \mathbb{R} \cup \{+\infty\}$ we have

$$\psi(x) + \psi^*(y) \geq \langle x, y \rangle$$

for every $x, y \in \mathbb{R}^n$, and with equality if and only if $x \in D_\psi$ and $y = \nabla\psi(x)$, i.e.,

$$\psi^*(\nabla\psi(x)) = \langle x, \nabla\psi(x) \rangle - \psi(x), \quad \text{a.e. in } \mathrm{D}_{\llcorner}. \tag{21}$$

Theorem 15 *Let $\psi : \mathbb{R}^n \to \mathbb{R} \cup \{+\infty\}$ be a convex function and let μ be a log-concave probability measure on $\mathbb{R}^n$ with density $e^{-\psi}$ with respect to Lebesgue measure. Suppose that for some $\varepsilon \in (0, \varepsilon_0)$,*

$$\int_{\mathbb{R}^n} \log\left(\det(\nabla^2\psi)\right) d\mu > 2\left(S(\gamma_n) - S(\mu)\right) - \varepsilon.$$

Then there exist $c > 0$ and a positive definite matrix A such that

$$\int_{R(\varepsilon)B_2^n} \left| \frac{\|x\|_2^2}{2} + c - \psi(Ax) \right| dx < \eta\varepsilon^{\frac{1}{129n^2}},$$

where $\lim_{\varepsilon\to 0} R(\varepsilon) = \infty$ and $\varepsilon_0, \eta, R(\varepsilon)$ depend on n.

Proof Both terms of the inequality are invariant under translations of the measure μ, so we can assume that μ has its centroid at 0.

Put $\varepsilon = \log\beta > 0$. Since $S(\gamma_n) = \frac{n}{2}\log(2\pi e)$, the inequality of the theorem turns into

$$\int_{D_\psi} \log\left(\beta \det(\nabla^2\psi)\right) d\mu + 2\int_{D_\psi} \psi \, d\mu > \log(2\pi e)^n,$$

which in turn is equivalent to

$$\int_{D_\psi} \log\big(\beta \det(\nabla^2\psi)\big)\, d\mu + \int_{D_\psi} \log\left(e^{2\psi}\right)\, d\mu - n > \log(2\pi)^n. \tag{22}$$

We now use the divergence theorem and get

$$\int_{D_\psi} \langle x, \nabla\psi(x)\rangle\, d\mu = \int_{\mathrm{int}(D_\psi)} \mathrm{div}(x)\, d\mu - \int_{\partial D_\psi} \langle x, N_{D_\psi}(x)\rangle e^{-\psi(x)} d\sigma_{D_\psi},$$

where $N_{D_\psi}(x)$ is an exterior normal to the convex set D_ψ at the point x and σ_{D_ψ} is the surface area measure on ∂D_ψ. Since D_ψ is convex, the centroid 0 of μ is in D_ψ. Thus $\langle x, N_{D_\psi}(x)\rangle \geq 0$ for every $x \in \partial D_\psi$ and $\mathrm{div}(x) = n$ hence

$$-n \leq -\int_{D_\psi} \langle x, \nabla\psi(x)\rangle\, d\mu = \int_{D_\psi} \log\left(e^{-\langle x, \nabla\psi(x)\rangle}\right) d\mu$$

Thus we get from inequality (22),

$$\int_{D_\psi} \log\big(\beta \det(\nabla^2\psi)\, e^{2\psi(x)-\langle x, \nabla\psi(x)\rangle}\big)\, d\mu > \log(2\pi)^n.$$

With Jensen's inequality, and as $d\mu = e^{-\psi} dx$,

$$\beta \int_{D_\psi} \det(\nabla^2\psi)\, e^{\psi(x)-\langle x, \nabla\psi(x)\rangle} dx > (2\pi)^n. \tag{23}$$

By (21),

$$\int_{D_\psi} \det(\nabla^2\psi)\, e^{\psi(x)-\langle x, \nabla\psi(x)\rangle} dx = \int_{D_\psi} \det(\nabla^2\psi)\, e^{-\psi^*(\nabla\psi(x))} dx.$$

The change of variable $y = \nabla\psi(x)$ gives

$$\int_{D_\psi} e^{-\psi^*(\nabla\psi(x))} \det(\nabla^2\psi(x))\, dx = \int_{D_{\psi^*}} e^{-\psi^*(y)}\, dy, \tag{24}$$

and inequality (23) becomes

$$\int_{D_{\psi^*}} e^{-\psi^*(y)}\, dy > \frac{1}{\beta}(2\pi)^n.$$

As $\int_{D_\psi} e^{-\psi} dx = 1$ and $\beta^{-1} = e^{-\varepsilon} \geq 1 - \varepsilon$, we therefore get that

$$\left(\int_{\mathbb{R}^n} e^{-\psi} dx\right)\left(\int_{\mathbb{R}^n} e^{-\psi^*(y)} dy\right) \geq \left(\int_{D_\psi} e^{-\psi} dx\right)\left(\int_{D_{\psi^*}} e^{-\psi^*(y)} dy\right)$$
$$> (1-\varepsilon)(2\pi)^n.$$

As μ has its centroid at 0, we have by (15) that

$$\inf_{z\in\mathbb{R}^n}\left(\int_{\mathbb{R}^n} e^{-\psi(x)} dx\right)\left(\int_{\mathbb{R}^n} e^{-\mathcal{L}_z\psi(y)} dy\right) = \left(\int_{\mathbb{R}^n} e^{-\psi} dx\right)\left(\int_{\mathbb{R}^n} e^{-\psi^*(y)} dy\right).$$

The theorem now follows from Theorem 11, the stability result for the functional Blaschke Santaló inequality, due to Barthe, Böröczky, and Fradelizi [9]. □

3.4 Stability for the L_λ-Affine Isoperimetric Inequality for Log Concave Functions

The following divergence inequalities were proved in [17]. In fact, inequalities (25), (26) and consequently (16) are special cases of a more general divergence inequality proved in [17].

For $0 < \lambda < 1$, it says

$$\int\left(e^{2\psi-\langle\nabla\psi,x\rangle} \det\left(\nabla^2\psi\right)\right)^\lambda d\mu \leq \left(\frac{\int_{\mathbb{R}^n} e^{-\psi^*} dx}{\int_{\mathbb{R}^n} d\mu}\right)^\lambda \left(\int_{\mathbb{R}^n} d\mu\right) \tag{25}$$

and for $\lambda \notin [0,1]$,

$$\int\left(e^{2\psi-\langle\nabla\psi,x\rangle} \det\left(\nabla^2\psi\right)\right)^\lambda d\mu \geq \left(\frac{\int_{\mathbb{R}^n} e^{-\psi^*} dx}{\int_{\mathbb{R}^n} d\mu}\right)^\lambda \left(\int_{\mathbb{R}^n} d\mu\right). \tag{26}$$

The left-hand sides of the above inequalities are the L_λ-affine surface areas $as_\lambda(\psi)$. For a general log concave function $\varphi = e^{-\psi}$ (and not just a log concave function in $C^2(\mathbb{R}^n)$) they were introduced in [20],

$$as_\lambda(\psi) = \int_{X_\psi} e^{(2\lambda-1)\psi(x)-\lambda\langle\nabla\psi,x\rangle}\left(\det\left(\nabla^2\psi(x)\right)\right)^\lambda dx. \tag{27}$$

Since $\det\left(\nabla^2\psi(x)\right) = 0$ outside X_ψ, the integral may be taken on D_ψ for $\lambda > 0$. In particular,

$$as_0(\psi) = \int_{X_\psi} e^{-\psi(x)} dx \quad \text{and} \quad as_1(\psi) = \int_{X_{\psi^*}} e^{-\psi^*(x)} dx.$$

Assume now that $\int xe^{-\psi(x)}dx = 0$ or $\int xe^{-\psi^*(x)}dx = 0$. Then we can apply the functional Blaschke Santaló inequality (15) and get from (25) that for $\lambda \in [0, 1]$,

$$as_\lambda(\psi) \leq (2\pi)^{n\lambda} \left(\int_{\mathbb{R}^n} e^{-\psi(x)} dx \right)^{1-2\lambda}.$$

Similarly, for $\lambda \leq 0$, we get from (26)

$$as_\lambda(\psi) \geq (2\pi)^{n\lambda} \left(\int_{\mathbb{R}^n} e^{-\psi(x)} dx \right)^{1-2\lambda},$$

provided that $\varphi \in C^2(\mathbb{R}^n)$, which is the assumption on φ in inequality (16). However, these inequalities hold without such a strong smoothness assumption. This, together with characterization of equality, was proved in [20].

Theorem 16 ([20]) *Let $\psi : \mathbb{R}^n \to \mathbb{R} \cup \{\infty\}$ be a convex function. For $\lambda \in [0, 1]$,*

$$as_\lambda(\psi) \leq (2\pi)^{n\lambda} \left(\int_{X_\psi} e^{-\psi(x)} dx \right)^{1-2\lambda} \tag{28}$$

and for $\lambda \leq 0$,

$$as_\lambda(\psi) \geq (2\pi)^{n\lambda} \left(\int_{X_\psi} e^{-\psi(x)} dx \right)^{1-2\lambda}. \tag{29}$$

For $\lambda = 0$ equality holds trivially in these inequalities. Moreover, for $0 < \lambda \leq 1$, or $\lambda < 0$, equality holds in above inequalities if and only if $\psi(x) = \frac{1}{2}\langle Ax, x\rangle + c$, where A is a positive definite $n \times n$ matrix and c is a constant.

A stability result for these inequalities is again an immediate consequence of Theorem 13. But again, we would then get the stability result for log concave functions $\varphi \in C^2(\mathbb{R}^n)$ only, so we include the proof for general functions.

Proposition 17 *Let $\psi : \mathbb{R}^n \to \mathbb{R} \cup \{+\infty\}$ be a convex function such that $\int xe^{-\psi(x)}dx = 0$ or $\int xe^{-\psi^*(x)}dx = 0$.*

(i) Let $0 < \lambda \leq 1$ and suppose that for some $\varepsilon \in (0, \varepsilon_0)$,

$$as_\lambda(\psi) > (1-\varepsilon)^\lambda (2\pi)^{n\lambda} \left(\int_{X_\psi} e^{-\psi(x)} dx \right)^{1-2\lambda}.$$

(ii) Let $\lambda < 0$ and suppose that for some $\varepsilon \in (0, \varepsilon_0)$,

$$as_\lambda(\psi) < (1-\varepsilon)^\lambda (2\pi)^{n\lambda} \left(\int_{X_\psi} e^{-\psi(x)} dx \right)^{1-2\lambda}.$$

Then, in both cases (i) and (ii), there exists $c > 0$ and a positive definite matrix A such that

$$\int_{R(\varepsilon)B_2^n} \left| \frac{\|x\|_2^2}{2} + c - \psi(Ax) \right| dx < \eta \varepsilon^{\frac{1}{129n^2}},$$

where $\lim_{\varepsilon \to 0} R(\varepsilon) = \infty$ *and* $\varepsilon_0, \eta, R(\varepsilon)$ *depend on* n.

Proof (i) The case $\lambda = 1$ is the stability case for the functional Blaschke Santaló inequality of Theorem 11. Therefore we can assume that $0 < \lambda < 1$. We put $d\mu = e^{-\psi} dx$. By Hölder's inequality with $p = 1/\lambda$ and $q = 1/(1-\lambda)$,

$$\begin{aligned}
as_\lambda(\psi) &= \int_{X_\psi} e^{\lambda(2\psi(x) - \langle \nabla\psi, x\rangle)} \left(\det\left(\nabla^2 \psi(x)\right)\right)^\lambda d\mu \\
&\le \left(\int_{X_\psi} e^{2\psi(x) - \langle \nabla\psi, x\rangle} \det\left(\nabla^2 \psi(x)\right) d\mu \right)^\lambda \left(\int_{X_\psi} d\mu \right)^{1-\lambda} \\
&= \left(\int_{D_\psi} e^{\psi(x) - \langle \nabla\psi, x\rangle} \det\left(\nabla^2 \psi(x)\right) dx \right)^\lambda \left(\int_{X_\psi} e^{-\psi(x)} dx \right)^{1-\lambda} \\
&\le \left(\int_{\mathbb{R}^n} e^{-\psi^*(x)} dx \right)^\lambda \left(\int_{X_\psi} e^{-\psi(x)} dx \right)^{1-\lambda},
\end{aligned}$$

where, in the last equality, we have used (21) and (24). Therefore, by the assumption (i) of the proposition

$$\left(\int_{\mathbb{R}^n} e^{-\psi^*(x)} dx \right)^\lambda \left(\int_{X_\psi} e^{-\psi(x)} dx \right)^{1-\lambda} > (1-\varepsilon)^\lambda (2\pi)^{n\lambda} \left(\int_{X_\psi} e^{-\psi(x)} dx \right)^{1-2\lambda},$$

which is equivalent to

$$\begin{aligned}
\left(\int_{\mathbb{R}^n} e^{-\psi^*(x)} dx \right) \left(\int_{\mathbb{R}^n} e^{-\psi(x)} dx \right) &> \left(\int_{\mathbb{R}^n} e^{-\psi^*(x)} dx \right) \left(\int_{X_\psi} e^{-\psi(x)} dx \right) \\
&> (1-\varepsilon)(2\pi)^n,
\end{aligned}$$

and the result is again a consequence of Theorem 11 by Barthe, Böröczky, and Fradelizi [9].

Similarly, in the case (ii) the proposition follows by applying the reverse Hölder inequality. □

The following Blaschke Santaló type inequality follows directly from inequality (28). It was also proved, together with its equality characterization in [20].

Corollary 18 ([20]) *Let $\lambda \in [0, \frac{1}{2}]$ and let $\psi : \mathbb{R}^n \to \mathbb{R} \cup \{+\infty\}$ be a convex function such that $\int xe^{-\psi(x)}dx = 0$ or $\int xe^{-\psi^*(x)}dx = 0$. Then*

$$as_\lambda(\psi)\, as_\lambda((\psi^*)) \leq (2\pi)^n.$$

Equality holds if and only if there exist $a \in \mathbb{R}$ and a positive definite matrix A such that $\psi(x) = \frac{1}{2}\langle Ax, x\rangle + a$, for every $x \in \mathbb{R}^n$.

We have the following stability result as a direct consequence of Theorem 11.

Proposition 19 *Let $\psi : \mathbb{R}^n \to \mathbb{R} \cup \{+\infty\}$ be a convex function such that $\int xe^{-\psi(x)}dx = 0$ or $\int xe^{-\psi^*(x)}dx = 0$. Let $0 \leq \lambda \leq \frac{1}{2}$ and suppose that for some $\varepsilon \in (0, \varepsilon_0)$,*

$$as_\lambda(\psi)\, as_\lambda((\psi^*)) \geq (1 - \epsilon)(2\pi)^n.$$

Then, there exist $c > 0$ and a positive definite matrix A such that

$$\int_{R(\varepsilon)B_2^n} \left| \frac{\|x\|_2^2}{2} + c - \psi(Ax) \right| dx < \eta \varepsilon^{\frac{1}{129n^2}},$$

where $\lim_{\varepsilon \to 0} R(\varepsilon) = \infty$ and $\varepsilon_0, \eta, R(\varepsilon)$ depend on n.

Acknowledgements Elisabeth M. Werner is partially supported by an NSF grant.

References

[1] A.D. ALEKSANDROV *Almost everywhere existence of the second differential of a convex function and some properties of convex surfaces connected with it*, Leningrad State Univ. Annals [Uchenye Zapiski] Math. Ser. 6 (1939), 3–35.

[2] S. ALESKER, *Continuous rotation invariant valuations on convex sets*, Ann. of Math. 149, (1999), 977–1005.

[3] B. ANDREWS, *Gauss curvature flow: the fate of the rolling stones*, Invent. Math. 138, (1999), 151–161.

[4] B. ANDREWS, *The affine curve-lengthening flow*, J. Reine Angew. Math. 506, (1999), 43–83.

[5] S. ARTSTEIN-AVIDAN, B. KLARTAG, V. MILMAN, *The Santaló point of a function, and a functional form of Santaló inequality*, Mathematika 51 (2004), 33–48.

[6] S. ARTSTEIN-AVIDAN, B. KLARTAG, C. SCHÜTT AND E. M. WERNER, *Functional affine-isoperimetry and an inverse logarithmic Sobolev inequality*, Journal of Functional Analysis 262, (2012), 4181–4204.

[7] K. BALL, *Isometric problems in l_p and sections of convex sets*, PhD dissertation, University of Cambridge (1986).

[8] K. BALL AND K. BÖRÖCZKY, *Stability of some versions of the Prékopa Leindler inequality*, Monatshefte für Mathematik 163(1), (2011), 1–14.

[9] F. BARTHE, K. BÖRÖCZKY AND M. FRADELIZI, *Stability of the functional form of the Blaschke-Santaló inequality*, Monatshefte für Mathematik, (2013) DOI: 0.1007/s00605-013-0499-9.

[10] A. BERNIG, J. H. G. FU, AND G. SOLANES, *Integral geometry of complex space forms*, Geom. Funct. Anal. 24, (2014), 403–492.
[11] K. BÖRÖCZKY, *Polytopal approximation bounding the number of k-faces*, Journal of Approximation Theory 102, (2000), 263–285.
[12] W. BLASCHKE, *Vorlesungen über Differentialgeometrie II, Affine Differentialgeometrie*. Springer Verlag, Berlin, (1923).
[13] K. BÖRÖCZKY, *Stability of the Blaschke-Santaló and the affine isoperimetric inequalities*, Advances in Math. 225, (2010), 1914–1928.
[14] K. BÖRÖCZKY AND MAKAI, *Volume product in the plane II, upper estimates: the polygonal case and stability*, in preparation.
[15] J. BOURGAIN AND V.D. MILMAN, *New volume ratio properties for convex symmetric bodies in* $\mathbb{R}^n$, Invent. Math. 88, (1987), 319–340.
[16] H. BUSEMANN AND W. FELLER, *Krümmungseigenschaften konvexer Flächen*, Acta Math. 66 (1935), 1–47.
[17] U. CAGLAR AND E. M. WERNER, *Divergence for s-concave and log concave functions*, Advances in Math. 257 (2014), 219–247.
[18] U. CAGLAR AND E. WERNER, *Mixed f-divergence and inequalities for log concave functions*, Proceedings London Math. Soc. 110, (2015), 271–290.
[19] U. CAGLAR AND D. YE, *Orlicz affine isoperimetric inequalities for functions*, arXiv:1506.02974 (2015).
[20] U. CAGLAR, M. FRADELIZI, O. GUÉDON, J. LEHEC, C. SCHÜTT AND E. M. WERNER, *Functional versions of L_p-affine surface area and entropy inequalities*, International Mathematics Research Notices 2015; doi: 10.1093/imrn/rnv151.
[21] A. DEICKE, *Über die Finsler-Räume mit $A_i = 0$*, Archiv Math. 4 (1953), 45–51.
[22] M. FRADELIZI AND M. MEYER, *Some functional forms of Blaschke-Santaló inequality*, Math. Z. 256, No. 2, (2007), 379–395.
[23] R. J. GARDNER AND G. ZHANG, *Affine inequalities and radial mean bodies,* Amer. J. Math. 120, No.3, (1998), 505–528.
[24] R. GARDNER, *Geometric tomography*, Encyclopedia of Mathematics and its Applications vol. 58, Cambridge Univ. Press, Cambridge, 2006.
[25] R. J. GARDNER, *The dual Brunn-Minkowski theory for bounded Borel sets: Dual affine quermassintegrals and inequalities*, Advances in Math. 216, (2007), 358–386.
[26] R. GARDNER, D. HUG AND W. WEIL, *The Orlicz-Brunn-Minkowski theory: a general framework, additions, and inequalities*, J. Differential Geometry, 97, No. 3, (2014), 427–476.
[27] P. M. GRUBER, *Aspects of approximation of convex bodies*, Handbook of Convex Geometry, vol.A, North Holland, (1993), 321–345.
[28] C. HABERL, *Blaschke valuations*, Amer. J. Math. 133, (2011), 717–751.
[29] C. HABERL, *Minkowski valuations intertwining with the special linear group*, J. Eur. Math. Soc. (JEMS) 14, (2012), 1565–159.
[30] C. HABERL AND F. SCHUSTER, *General L_p affine isoperimetric inequalities*, J. Differential Geometry 83, (2009), 1–26.
[31] D. HUG, *Curvature Relations and Affine Surface Area for a General Convex Body and its Polar*, Results in Mathematics 29, (1996), 233–248.
[32] M. N. IVAKI, *On stability of the p-affine isoperimetric inequality*, The Journal of Geometric Analysis 24 (2014), 1898–1911.
[33] M. N. IVAKI, *Stability of the Blaschke-Santaló inequality in the plane*, Monatshefte für Mathematik 177 (2015), 451–459.
[34] G. KUPERBERG, *From the Mahler conjecture to Gauss linking integrals*, Geom. Funct. Anal. 18, No. 3, (2008), 870–892.
[35] J. LEHEC, *A simple proof of the functional Santaló inequality*, C. R. Acad. Sci. Paris. Sér.I 347 (2009), 55–58.
[36] K. LEICHTWEISS, *Affine Geometry of Convex bodies*, Johann Ambrosius Barth Verlag, Heidelberg, (1998).
[37] M. LUDWIG, *Asymptotic approximation of smooth convex bodies by general polytopes*, Mathematika 46, (1999), 103–125.

[38] M. LUDWIG, *Ellipsoids and matrix valued valuations*, Duke Math. J. 119, (2003), 159–188.
[39] M. LUDWIG, *Intersection bodies and valuations*, Amer. J. Math. 128, (2006), 1409–1428.
[40] M. LUDWIG, *General affine surface areas*, Advances in Math. 224, (2010), 2346–2360.
[41] M. LUDWIG, *Minkowski areas and valuations*, J. Differential Geometry, 86, (2010), 133–162.
[42] M. LUDWIG AND M. REITZNER, *A classification of $SL(n)$ invariant valuations,* Ann. of Math. 172, (2010), 1223–1271.
[43] E. LUTWAK AND V. OLIKER, *On the regularity of solutions to a generalization of the Minkowski problem*, J. Differential Geometry 41, (1995), 227–246.
[44] E. LUTWAK, *Extended affine surface area,* Advances in Math. 85, (1991), 39–68.
[45] E. LUTWAK, *The Brunn-Minkowski-Firey theory II : Affine and geominimal surface areas*, Advances in Math. 118, (1996), 244–294.
[46] E. LUTWAK, D. YANG AND G. ZHANG, *Sharp Affine L_p Sobolev inequalities*, J. Differential Geometry 62, (2002), 17–38.
[47] E. LUTWAK, D. YANG AND G. ZHANG, *Volume inequalities for subspaces of L_p*, J. Differential Geometry 68, (2004), 159–184.
[48] M. MEYER AND E. M. WERNER, *The Santaló-regions of a convex body,* Transactions of the AMS 350, (1998), 4569–4591.
[49] M. MEYER AND E. WERNER, *On the p-affine surface area*, Advances in Math. 152, (2000), 288–313.
[50] F. NAZAROV, *The Hörmander proof of the Bourgain-Milman theorem*, Geometric Aspects of Functional Analysis 2050, (2012), 335–343.
[51] G. PAOURIS AND E. M. WERNER, *Relative entropy of cone measures and Lp centroid bodies*, Proceedings London Math. Soc. 104, (2012), 253–286.
[52] L. PARAPATITS AND T. WANNERER, *On the Inverse Klain Map*, Duke Math. J. 162, (2013), 1895–1922.
[53] J. M. BORWEIN AND J.D. VANDERWERFF, *Convex Functions: Constructions, Characterizations and Counterexamples*, Cambridge Univ. Press, 2010.
[54] M. REITZNER, *Random points on the boundary of smooth convex bodies*, Trans. Amer. Math. Soc. 354, (2002), 2243–2278.
[55] R. SCHNEIDER, *Convex Bodies: The Brunn-Minkowski theory,* Encyclopedia of Mathematics and its Applications vol. 151, Cambridge Univ. Press, Cambridge, 2014.
[56] F. SCHUSTER, *Crofton measures and Minkowski valuations*, Duke Math. J. 154, (2010), 1–30.
[57] C. SCHÜTT, *The convex floating body and polyhedral approximation*, Israel J. Math. 73, (1991), 65–77.
[58] C. SCHÜTT, *On the affine surface area*, Proc. Amer. Math. Soc. 118, (1993), 1213–1218.
[59] C. SCHÜTT AND E. M. WERNER, *The convex floating body*, Math. Scand. 66, (1990), 275–290.
[60] C. SCHÜTT AND E. WERNER, *Surface bodies and p-affine surface area*, Advances in Math. 187, (2004), 98–145.
[61] C. SCHÜTT AND E. M. WERNER, *Random polytopes of points chosen from the boundary of a convex body,* Geometric aspects of functional analysis, vol. 1807 of Lecture Notes in Math., Springer-Verlag, (2002), 241–422.
[62] A. STANCU, *The Discrete Planar L_0-Minkowski Problem*, Advances in Math. 167, (2002), 160 174.
[63] A. STANCU, *On the number of solutions to the discrete two-dimensional L_0-Minkowski problem*, Advances in Math. 180, (2003), 290–323.
[64] E. M. WERNER, *On L_p-affine surface areas*, Indiana Univ. Math. J. 56, No. 5, (2007), 2305–2324.
[65] E. M. WERNER, *Rényi Divergence and L_p-affine surface area for convex bodies*, Advances in Math. 230, (2012), 1040–1059.
[66] E. M. WERNER AND D. YE, *New L_p affine isoperimetric inequalities*, Advances in Math. 218, (2008), 762–780.
[67] E. WERNER AND D.YE, *Inequalities for mixed p affine surface area*, Math. Annalen 347, (2010), 703–737.

Measures of Sections of Convex Bodies

Alexander Koldobsky

Abstract This article is a survey of recent results on slicing inequalities for convex bodies. The focus is on the setting of arbitrary measures in place of volume.

1 Introduction

The study of volume of sections of convex bodies is a classical direction in convex geometry. It is well developed and has numerous applications; see [G3, K4]. The question of what happens if volume is replaced by an arbitrary measure on a convex body has not been considered until very recently, mostly because it is hard to believe that difficult geometric results can hold in such generality. However, in 2005 Zvavitch [Zv] proved that the solution to the Busemann-Petty problem, one of the signature problems in convex geometry, remains exactly the same if volume is replaced by an arbitrary measure with continuous density. It has recently been shown [K6, KM, K8, K9, K10, K11, KP] that several partial results on the slicing problem, a major open question in the area, can also be extended to arbitrary measures. For example, it was proved in [K11] that the slicing problem for sections of proportional dimensions has an affirmative answer which can be extended to the setting of arbitrary measures. It is not clear yet whether these results are representative of something bigger, or it is just an isolated event. We let the reader make the judgement.

2 The Slicing Problem for Measures

The slicing problem [Bo1, Bo2, Ba5, MP] asks whether there exists an absolute constant C so that for every origin-symmetric convex body K in $\mathbb{R}^n$ of volume 1 there is a hyperplane section of K whose $(n-1)$-dimensional volume is greater

A. Koldobsky (✉)
Department of Mathematics, University of Missouri, Columbia, MO 65211, USA
e-mail: koldobskiya@missouri.edu

E. Carlen et al. (eds.), *Convexity and Concentration*, The IMA Volumes in Mathematics and its Applications 161, DOI 10.1007/978-1-4939-7005-6_18

than $1/C$. In other words, does there exist a constant C so that for any $n \in \mathbb{N}$ and any origin-symmetric convex body K in $\mathbb{R}^n$

$$|K|^{\frac{n-1}{n}} \leq C \max_{\xi \in S^{n-1}} |K \cap \xi^\perp|, \tag{1}$$

where $\xi^\perp$ is the central hyperplane in $\mathbb{R}^n$ perpendicular to ξ, and $|K|$ stands for volume of proper dimension? The best current result $C \leq O(n^{1/4})$ is due to Klartag [Kla2], who slightly improved an earlier estimate of Bourgain [Bo3]. The answer is known to be affirmative for some special classes of convex bodies, including unconditional convex bodies (as initially observed by Bourgain; see also [MP, J2, BN]), unit balls of subspaces of L_p [Ba4, J1, M1], intersection bodies [G3, Th.9.4.11], zonoids, duals of bodies with bounded volume ratio [MP], the Schatten classes [KMP], k-intersection bodies [KPY, K10]. Other partial results on the problem include [Ba3, BKM, DP, Da, GPV, Kla1, KlaK, Pa, EK, BaN]; see the book [BGVV] for details.

Iterating (1) one gets the lower dimensional slicing problem asking whether the inequality

$$|K|^{\frac{n-k}{n}} \leq C^k \max_{H \in Gr_{n-k}} |K \cap H| \tag{2}$$

holds with an absolute constant C, where $1 \leq k \leq n-1$ and Gr_{n-k} is the Grassmanian of $(n-k)$-dimensional subspaces of $\mathbb{R}^n$.

Inequality (2) was proved in [K11] in the case where $k \geq \lambda n$, $0 < \lambda < 1$, with the constant $C = C(\lambda)$ dependent only on λ. Moreover, this was proved in [K11] for arbitrary measures in place of volume. We consider the following generalization of the slicing problem to arbitrary measures and to sections of arbitrary codimension.

Problem 1 *Does there exist an absolute constant C so that for every $n \in \mathbb{N}$, every integer $1 \leq k < n$, every origin-symmetric convex body K in $\mathbb{R}^n$, and every measure μ with non-negative even continuous density f in $\mathbb{R}^n$,*

$$\mu(K) \leq C^k \max_{H \in Gr_{n-k}} \mu(K \cap H)\, |K|^{k/n}. \tag{3}$$

Here $\mu(B) = \int_B f$ for every compact set B in $\mathbb{R}^n$, and $\mu(B \cap H) = \int_{B \cap H} f$ is the result of integration of the restriction of f to H with respect to Lebesgue measure in H. The case of volume corresponds to $f \equiv 1$.

In some cases we will write (3) in an equivalent form

$$\mu(K) \leq C^k \frac{n}{n-k} c_{n,k} \max_{H \in Gr_{n-k}} \mu(K \cap H)\, |K|^{k/n}, \tag{4}$$

where $c_{n,k} = |B_2^n|^{\frac{n-k}{n}}/|B_2^{n-k}|$, and B_2^n is the unit Euclidean ball in $\mathbb{R}^n$. It is easy to see that $c_{n,k} \in (e^{-k/2}, 1)$, and $\frac{n}{n-k} \in (1, e^k)$, so these constants can be incorporated in the constant C.

Surprisingly, many partial results on the original slicing problem can be extended to the setting of arbitrary measures. Inequality (3) holds true in the following cases:

- for arbitrary n, K, μ and $k \geq \lambda n$, where $\lambda \in (0, 1)$, with the constant C dependent only on λ, [K11];
- for all n, K, μ, k, with $C \leq O(\sqrt{n})$, [K8, K9]. The symmetry condition was later removed in [CGL];
- for intersection bodies K (see definition below), with an absolute constant C, [K6] for $k = 1$, [KM] for all k;
- for the unit balls of finite dimensional subspaces of L_p, $p > 2$, with C depending only on p, and $C \leq O(\sqrt{p})$, [KP];
- for the unit balls of n-dimensional normed spaces that embed in L_p, $p \in (-n, 2]$, with C depending only on p, [K10];
- for unconditional convex bodies, with an absolute constant C, [K11];
- for duals of convex bodies with bounded volume ratio, with an absolute constant C, [K11];
- for $k = 1$ and log-concave measures μ, with $C \leq O(n^{1/4})$, [KZ];
- a discrete version of inequality (3) was established in [AHZ] with the constant depending only on the dimension.

The proofs of these results are based on stability in volume comparison problems introduced in [K5] and developed in [K6, KM, K7, K8, K9, K10, K13]. Stability reduces Problem 1 to estimating the outer volume ratio distance from a convex body to the classes of generalized intersection bodies. The concept of an intersection body was introduced by Lutwak [Lu] in connection with the Busemann-Petty problem.

A closed bounded set K in $\mathbb{R}^n$ is called a star body if every straight line passing through the origin crosses the boundary of K at exactly two points different from the origin, the origin is an interior point of K, and the boundary of K is continuous.

For $1 \leq k \leq n-1$, the classes $\mathcal{BP}_k^n$ of generalized k-intersection bodies in $\mathbb{R}^n$ were introduced by Zhang [Z3]. The case $k = 1$ represents the original class of intersection bodies $\mathcal{I}_n = \mathcal{BP}_1^n$ of Lutwak [Lu]. We define $\mathcal{BP}_k^n$ as the closure in the radial metric of radial k-sums of finite collections of origin-symmetric ellipsoids (the equivalence of this definition to the original definitions of Lutwak and Zhang was established by Goodey and Weil [GW] for $k = 1$ and by Grinberg and Zhang [GrZ] for arbitrary k.) Recall that the radial k-sum of star bodies K and L in $\mathbb{R}^n$ is a new star body $K +_k L$ whose radius in every direction $\xi \in S^{n-1}$ is given by

$$r_{K+_kL}^k(\xi) = r_K^k(\xi) + r_L^k(\xi).$$

The radial metric in the class of origin-symmetric star bodies is defined by

$$\rho(K,L) = \sup_{\xi\in S^{n-1}} |r_K(\xi) - r_L(\xi)|.$$

The following stability theorem was proved in [K11] (see [K6, KM] for slightly different versions).

Theorem 1 ([K11]) *Suppose that $1 \le k \le n-1$, K is a generalized k-intersection body in $\mathbb{R}^n$, f is an even continuous non-negative function on K, and $\varepsilon > 0$. If*

$$\int_{K\cap H} f \le \varepsilon, \qquad \forall H \in Gr_{n-k},$$

then

$$\int_K f \le \frac{n}{n-k}\, c_{n,k}\, |K|^{k/n}\varepsilon.$$

The constant is the best possible. Recall that $c_{n,k} \in (e^{-k/2}, 1)$.

Define the outer volume ratio distance from an origin-symmetric star body K in $\mathbb{R}^n$ to the class $\mathcal{BP}_k^n$ of generalized k-intersection bodies by

$$\text{o.v.r.}(K, \mathcal{BP}_k^n) = \inf\left\{ \left(\frac{|D|}{|K|}\right)^{1/n} : K \subset D,\ D \in \mathcal{BP}_k^n \right\}.$$

Theorem 1 immediately implies a slicing inequality for arbitrary measures and origin-symmetric star bodies.

Corollary 1 *Let K be an origin-symmetric star body in $\mathbb{R}^n$. Then for any measure μ with even continuous density on K we have*

$$\mu(K) \le \left(\text{o.v.r.}(K, \mathcal{BP}_k^n)\right)^k \frac{n}{n-k}\, c_{n,k} \max_{H\in Gr_{n-k}} \mu(K\cap H)\, |K|^{k/n}.$$

Thus, stability reduces Problem 1 to estimating the outer volume ratio distance from K to the class of generalized k-intersection bodies. The results on Problem 1 mentioned above were all obtained by estimating this distance by means of various techniques from the local theory of Banach spaces. For example, the solution to the slicing problem for sections of proportional dimensions follows from an estimate obtained in [KPZ]: for any origin-symmetric convex body K in $\mathbb{R}^n$ and any $1 \le k \le n-1$,

$$\text{o.v.r.}(K, \mathcal{BP}_k^n) \le C_0 \sqrt{\frac{n}{k}} \left(\log\left(\frac{en}{k}\right)\right)^{3/2}, \tag{5}$$

where C_0 is an absolute constant. The proof of this estimate in [KPZ] is quite involved. It uses covering numbers, Pisier's generalization of Milman's reverse Brunn-Minkowski inequality, properties of intersection bodies. Combining this with Corollary 1, one gets

Theorem 2 ([K11]) *If the codimension of sections k satisfies $\lambda n \leq k$ for some $\lambda \in (0, 1)$, then for every origin-symmetric convex body K in $\mathbb{R}^n$ and every measure μ with continuous non-negative density in $\mathbb{R}^n$,*

$$\mu(K) \leq C^k \left(\sqrt{\frac{(1 - \log \lambda)^3}{\lambda}} \right)^k \max_{H \in Gr_{n-k}} \mu(K \cap H)\, |K|^{k/n},$$

where C is an absolute constant.

For arbitrary K, μ and k the best result so far is the following $\sqrt{n}$ estimate; see [K8, K9]. By John's theorem, for any origin-symmetric convex body K there exists an ellipsoid $\mathcal{E}$ so that $\frac{1}{\sqrt{n}}\mathcal{E} \subset K \subset \mathcal{E}$. Since every ellipsoid is a generalized k-intersection body for every k, we get that

$$\text{o.v.r.}(K, \mathcal{BP}_k^n) \leq \sqrt{n}.$$

By Corollary 1,

$$\mu(K) \leq n^{k/2} \frac{n}{n-k} c_{n,k} \max_{H \in Gr_{n-k}} \mu(K \cap H)\, |K|^{k/n}.$$

3 The Isomorphic Busemann-Petty Problem

In 1956, Busemann and Petty [BP] asked the following question. Let K, L be origin-symmetric convex bodies in $\mathbb{R}^n$ such that

$$\left|K \cap \xi^\perp\right| \leq \left|L \cap \xi^\perp\right|, \qquad \forall \xi \in S^{n-1}. \tag{6}$$

Does it necessarily follow that $|K| \leq |L|$? The problem was solved at the end of the 1990s in a sequence of papers [LR, Ba1, Gi, Bo4, Lu, P, G1, G2, Z1, K1, K2, Z2, GKS]; see [K4, p.3] or [G3, p.343] for the solution and its history. The answer is affirmative if $n \leq 4$, and it is negative if $n \geq 5$.

The lower dimensional Busemann-Petty problem asks the same question for sections of lower dimensions. Suppose that $1 \leq k \leq n-1$, and K, L are origin-symmetric convex bodies in $\mathbb{R}^n$ such that

$$|K \cap H| \leq |L \cap H|, \qquad \forall H \in Gr_{n-k}. \tag{7}$$

Does it follow that $|K| \le |L|$? It was proved in [BZ] (see also [K3, K4, RZ, M2] for different proofs) that the answer is negative if the dimension of sections $n - k > 3$. The problem is still open for two- and three-dimensional sections ($n - k = 2, 3,\ n \ge 5$).

Since the answer to the Busemann-Petty problem is negative in most dimensions, it makes sense to ask whether the inequality for volumes holds up to an absolute constant, namely, does there exist an absolute constant C such that inequalities (6) imply $|K| \le C\,|L|$? This question is known as the isomorphic Busemann-Petty problem, and in the hyperplane case it is equivalent to the slicing problem; see [MP]. A version of this problem for sections of proportional dimensions was proved in [K12]. For an alternative proof, see [CGL].

Theorem 3 ([K12]) *Suppose that* $0 < \lambda < 1,\ k > \lambda n$, *and* K, L *are origin-symmetric convex bodies in* $\mathbb{R}^n$ *satisfying the inequalities*

$$|K \cap H| \le |L \cap H|, \qquad \forall H \in Gr_{n-k}.$$

Then

$$|K|^{\frac{n-k}{n}} \le (C(\lambda))^k\, |L|^{\frac{n-k}{n}},$$

where $C(\lambda)$ *depends only on* λ.

This result implies Theorem 2 in the case of volume. It is not clear, however, whether Theorem 2 can be directly used to prove Theorem 3.

Zvavitch [Zv] has found a remarkable generalization of the Busemann-Petty problem to arbitrary measures in place of volume. Suppose that $1 \le k < n$, μ is a measure with even continuous density f in $\mathbb{R}^n$, and K and L are origin-symmetric convex bodies in $\mathbb{R}^n$ so that

$$\mu(K \cap \xi^\perp) \le \mu(L \cap \xi^\perp), \qquad \forall \xi \in S^{n-1}. \tag{8}$$

Does it necessarily follow that $\mu(K) \le \mu(L)$? The answer is the same as for volume—affirmative if $n \le 4$ and negative if $n \ge 5$. An isomorphic version was recently proved in [KZ], namely, for every dimension n inequalities (8) imply $\mu(K) \le \sqrt{n}\,\mu(L)$. It is not known whether the constant $\sqrt{n}$ is optimal for arbitrary measures.

The proof of the $\sqrt{n}$ estimate in the isomorphic Busemann-Petty problem for arbitrary measures in [KZ] is based on the following argument. Denote by

$$d_{BM}(K, L) = \inf\{d > 0 : \exists T \in GL(n) : K \subset TL \subset dK\}$$

the Banach-Mazur distance between two origin-symmetric star bodies L and K in $\mathbb{R}^n$, and let

$$d_I(K) = \min\{d_{BM}(K, D) : D \in \mathcal{I}_n\}$$

denote the Banach-Mazur distance from K to the class $\mathcal{I}_n = \mathcal{BP}_1^n$ of intersection bodies.

Theorem 4 *For any measure μ with continuous, non-negative even density f on $\mathbb{R}^n$ and any two origin-symmetric star bodies $K, L \subset \mathbb{R}^n$ such that*

$$\mu(K \cap \xi^\perp) \leq \mu(L \cap \xi^\perp), \qquad \forall \xi \in S^{n-1}, \tag{9}$$

we have

$$\mu(K) \leq d_I(K)\mu(L).$$

If K is convex, John's theorem implies that $d_I(K) \leq \sqrt{n}$, so the $\sqrt{n}$ estimate follows. It is not clear whether the Banach-Mazur distance in Theorem 4 can be replaced by the outer volume ratio distance from K to the class of intersection bodies. Also there is no known direct connection between the isomorphic Busemann-Petty problem for arbitrary measures and Problem 1.

In the case where the measure μ is log concave, the constant $\sqrt{n}$ can be improved to $cn^{1/4}$; see [KZ, Th. 4] or [CGL, Th.1.2].

4 Projections of Convex Bodies

The projection analog of the Busemann-Petty problem is known as Shephard's problem, posed in 1964 in [Sh]. Denote by $K|\xi^\perp$ the orthogonal projection of K to $\xi^\perp$. Suppose that K and L are origin-symmetric convex bodies in $\mathbb{R}^n$ so that $|K|\xi^\perp| \leq |L|\xi^\perp|$ for every $\xi \in S^{n-1}$. Does it follow that $|K| \leq |L|$? The problem was solved by Petty [Pe] and Schneider [Sch], independently, and the answer is affirmative only in dimension 2.

Both solutions use the fact that the answer to Shephard's problem is affirmative in every dimension under the additional assumption that L is a projection body. An origin symmetric convex body L in $\mathbb{R}^n$ is called a projection body if there exists another convex body K so that the support function of L in every direction is equal to the volume of the hyperplane projection of K to this direction: for every $\xi \in S^{n-1}$,

$$h_L(\xi) = |K|\xi^\perp|.$$

The support function $h_L(\xi) = \max_{x \in L} |(\xi, x)|$ is equal to the dual norm $\|\xi\|_{L^*}$, where L^* denotes the polar body of L.

Separation in Shephard's problem was proved in [K5].

Theorem 5 ([K5]) *Suppose that $\varepsilon > 0$, K and L are origin-symmetric convex bodies in $\mathbb{R}^n$, and L is a projection body. If $|K|\xi^\perp| \leq |L|\xi^\perp| - \varepsilon$ for every $\xi \in S^{n-1}$, then $|K|^{\frac{n-1}{n}} \leq |L|^{\frac{n-1}{n}} - c_{n,1}\varepsilon$, where $c_{n,1}$ is the same constant as in Theorem 1; recall that $c_{n,1} > 1/\sqrt{e}$.*

Stability in Shephard's problem turned out to be more difficult, and it was proved in [K13] only up to a logarithmic term and under an additional assumption that the body L is isotropic. Recall that a convex body D in $\mathbb{R}^n$ is isotropic if $|D| = 1$ and $\int_D (x, \xi)^2 dx$ is a constant function of $\xi \in S^{n-1}$. Every convex body has a linear image that is isotropic; see [BGVV].

Theorem 6 ([K13]) *Suppose that $\varepsilon > 0$, K and L are origin-symmetric convex bodies in $\mathbb{R}^n$, and L is a projection body which is a dilate of an isotropic body. If $|K|\xi^\perp| \le |L|\xi^\perp| + \varepsilon$ for every $\xi \in S^{n-1}$, then $|K|^{\frac{n-1}{n}} \le |L|^{\frac{n-1}{n}} + C\varepsilon \log^2 n$, where C is an absolute constant.*

The proof is based on an estimate for the mean width of a convex body obtained by E.Milman [M3].

The projection analog of the slicing problem reads as

$$|K|^{\frac{n-1}{n}} \ge c \min_{\xi \in S^{n-1}} |K|\xi^\perp|,$$

and it was solved by Ball [Ba2], who proved that c may and has to be of the order $1/\sqrt{n}$.

The possibility of extension of Shephard's problem and related stability and separation results to arbitrary measures is an open question. Also, the lower dimensional Shephard problem was solved by Goodey and Zhang [GZ], but stability and separation for the lower dimensional case have not been established.

Stability and separation for projections have an interesting application to surface area. If L is a projection body, so is $L + \varepsilon B_2^n$ for every $\varepsilon > 0$. Applying separation in Shephard's problem to this pair of bodies, dividing by ε and sending ε to zero, one gets a hyperplane inequality for surface area (see [K7]): if L is a projection body, then

$$S(L) \ge c \min_{\xi \in S^{n-1}} S(L|\xi^\perp)\, |L|^{\frac{1}{n}}. \tag{10}$$

On the other hand, applying stability to any projection body L which is a dilate of a body in isotropic position (see [K13])

$$S(L) \le C \log^2 n \max_{\xi \in S^{n-1}} S(L|\xi^\perp)\, |L|^{\frac{1}{n}}. \tag{11}$$

Here c and C are absolute constants, and $S(L)$ is surface area. Versions of these inequalities for arbitrary convex bodies were recently established in [GKV].

Acknowledgements This work was supported in part by the US National Science Foundation grant DMS-1265155.

References

[AHZ] M. ALEXANDER, M. HENK AND A. ZVAVITCH, A discrete version of Koldobsky's slicing inequality, arXiv:1511.02702.

[Ba1] K. BALL, Some remarks on the geometry of convex sets, Geometric aspects of functional analysis (1986/87), Lecture Notes in Math. **1317**, Springer-Verlag, Berlin-Heidelberg-New York, 1988, 224–231.

[Ba2] K. BALL, Shadows of convex bodies, Trans. Amer. Math. Soc. **327** (1991), 891–901.

[Ba3] K. BALL, Logarithmically concave functions and sections of convex sets in $\mathbb{R}^n$, Studia Math. **88** (1988), 69–84.

[Ba4] K. BALL, Normed spaces with a weak-Gordon-Lewis property, Functional analysis (Austin, TX, 1987/1989), 36–47, Lecture Notes in Math., **1470**, Springer, Berlin, 1991.

[Ba5] K. BALL, Isometric problems in ℓ_p and sections of convex sets, Ph.D. dissertation, Trinity College, Cambridge (1986).

[BaN] K. BALL AND V. H. NGUYEN, Entropy jumps for isotropic log-concave random vectors and spectral gap, Studia Math. **213** (2012), 81–96.

[BN] S. BOBKOV AND F. NAZAROV, On convex bodies and log-concave probability measures with unconditional basis, Geometric aspects of functional analysis (Milman-Schechtman, eds), Lecture Notes in Math. **1807** (2003), 53–69.

[Bo1] J. BOURGAIN, On high-dimensional maximal functions associated to convex bodies, Amer. J. Math. **108** (1986), 1467–1476.

[Bo2] J. BOURGAIN, Geometry of Banach spaces and harmonic analysis, Proceedings of the International Congress of Mathematicians (Berkeley, CA, 1986), Amer. Math. Soc., Providence, RI, 1987, 871–878.

[Bo3] J. BOURGAIN, On the distribution of polynomials on high-dimensional convex sets, Lecture Notes in Math. **1469** (1991), 127–137.

[Bo4] J. BOURGAIN, On the Busemann-Petty problem for perturbations of the ball, Geom. Funct. Anal. **1** (1991), 1–13.

[BKM] J. BOURGAIN, B. KLARTAG, V. MILMAN, Symmetrization and isotropic constants of convex bodies, Geometric Aspects of Functional Analysis, Lecture Notes in Math. **1850** (2004), 101–116.

[BZ] J. BOURGAIN, GAOYONG ZHANG, On a generalization of the Busemann-Petty problem, Convex geometric analysis (Berkeley, CA, 1996), 65–76, Math. Sci. Res. Inst. Publ., 34, Cambridge Univ. Press, Cambridge, 1999.

[BGVV] S. BRAZITIKOS, A. GIANNOPOULOS, P. VALETTAS AND B. VRITSIOU, Geometry of isotropic log-concave measures, Amer. Math. Soc., Providence, RI, 2014.

[BP] H. BUSEMANN AND C. M. PETTY, Problems on convex bodies, Math. Scand. **4** (1956), 88–94.

[CGL] G. CHASAPIS, A. GIANNOPOULOS AND D. LIAKOPOULOS, Estimates for measures of lower dimensional sections of convex bodies, preprint; arXiv:1512.08393.

[DP] N. DAFNIS AND G. PAOURIS, Small ball probability estimates, ψ_2-behavior and the hyperplane conjecture, J. Funct. Anal. **258** (2010), 1933–1964.

[Da] S. DAR, Remarks on Bourgain's problem on slicing of convex bodies, Operator theory, Advances and Applications **77** (1995), 61–66.

[EK] R. ELDAN AND B. KLARTAG, Approximately gaussian marginals and the hyperplane conjecture, Proceedings of the Workshop on "Concentration, Functional Inequalities and Isoperimetry", Contemp. Math. **545** (2011), 55–68.

[G1] R. J. GARDNER, Intersection bodies and the Busemann-Petty problem, Trans. Amer. Math. Soc. **342** (1994), 435–445.

[G2] R. J. GARDNER, A positive answer to the Busemann-Petty problem in three dimensions, Annals of Math. **140** (1994), 435–447.

[G3] R. J. GARDNER, Geometric tomography, Second edition, Cambridge University Press, Cambridge, 2006, 492 p.

[GKS] R.J. Gardner, A. Koldobsky, Th. Schlumprecht, An analytic solution to the Busemann-Petty problem on sections of convex bodies, Annals of Math. **149** (1999), 691–703.

[Gi] A. Giannopoulos, A note on a problem of H. Busemann and C. M. Petty concerning sections of symmetric convex bodies, Mathematika **37** (1990), 239–244.

[GKV] A. Giannopoulos, A. Koldobsky and P. Valettas, Inequalities for the surface area of projections of convex bodies, prerpint; arXiv:1601.05600.

[GPV] A. Giannopoulos, G. Paouris and B. Vritsiou, A remark on the slicing problem, J. Funct. Anal. **262** (2012), 1062–1086.

[GW] P. Goodey and W. Weil, Intersection bodies and ellipsoids, Mathematika **42** (1958), 295–304.

[GZ] P. Goodey and Gaoyong Zhang, Inequalities between projection functions of convex bodies, Amer. J. Math. **120** (1998) 345–367.

[GrZ] E. Grinberg and Gaoyong Zhang, Convolutions, transforms, and convex bodies, Proc. London Math. Soc. **78** (1999), 77–115.

[J1] M. Junge, On the hyperplane conjecture for quotient spaces of L_p, Forum Math. **6** (1994), 617–635.

[J2] M. Junge, Proportional subspaces of spaces with unconditional basis have good volume properties, Geometric aspects of functional analysis (Israel Seminar, 1992–1994), 121–129, Oper. Theory Adv. Appl., 77, Birkhauser, Basel, 1995.

[Kla1] B. Klartag, An isomorphic version of the slicing problem, J. Funct. Anal. **218** (2005), 372–394.

[Kla2] B. Klartag, On convex perturbations with a bounded isotropic constant, Geom. Funct. Anal. **16** (2006), 1274–1290.

[KlaK] B. Klartag and G. Kozma, On the hyperplane conjecture for random convex sets, Israel J. Math. **170** (2009), 253–268.

[K1] A. Koldobsky, Intersection bodies, positive definite distributions and the Busemann-Petty problem, Amer. J. Math. **120** (1998), 827–840.

[K2] A. Koldobsky, Intersection bodies in $\mathbb{R}^4$, Adv. Math. **136** (1998), 1–14.

[K3] A. Koldobsky, A functional analytic approach to intersection bodies, Geom. Funct. Anal. **10** (2000), 1507–1526.

[K4] A. Koldobsky, Fourier analysis in convex geometry, Amer. Math. Soc., Providence RI, 2005, 170 p.

[K5] A. Koldobsky Stability in the Busemann-Petty and Shephard problems, Adv. Math. **228** (2011), 2145–2161.

[K6] A. Koldobsky, A hyperplane inequality for measures of convex bodies in $\mathbb{R}^n$, $n \leq 4$, Discrete Comput. Geom. **47** (2012), 538–547.

[K7] A. Koldobsky, Stability and separation in volume comparison problems, Math. Model. Nat. Phenom. **8** (2012), 159–169.

[K8] A. Koldobsky, A $\sqrt{n}$ estimate for measures of hyperplane sections of convex bodies, Adv. Math. **254** (2014), 33–40.

[K9] A. Koldobsky, Estimates for measures of sections of convex bodies, GAFA Seminar Volume, B.Klartag and E.Milman, editors, Lect. Notes in Math. **2116** (2014), 261–271.

[K10] A. Koldobsky, Slicing inequalities for subspaces of L_p, Proc. Amer. Math. Soc. **144** (2016), 787–795.

[K11] A. Koldobsky, Slicing inequalities for measures of convex bodies, Adv. Math. **283** (2015), 473–488.

[K12] A. Koldobsky, Isomorphic Busemann-Petty problem for sections of proportional dimensions, Adv. in Appl. Math. **71** (2015), 138–145.

[K13] A. Koldobsky, Stability inequalities for projections of convex bodies, Discrete Comput. Geom. **57** (2017), 152–163.

[KM] A. Koldobsky and Dan Ma, Stability and slicing inequalities for intersection bodies, Geom. Dedicata **162** (2013), 325–335.

[KP] A. Koldobsky and A. Pajor, A remark on measures of sections of L_p-balls, in: Geometric Aspects of Functional Analysis, Israel Seminar (2014–2016), ed. by B.Klartag and E.Milman. Lecture Notes in Math, to appear.

[KPY] A. KOLDOBSKY, A. PAJOR AND V. YASKIN, Inequalities of the Kahane-Khinchin type and sections of L_p-balls, Studia Math. **184** (2008), 217–231.

[KPZ] A. KOLDOBSKY, G. PAOURIS AND M. ZYMONOPOULOU, Isomorphic properties of intersection bodies, J. Funct. Anal. **261** (2011), 2697–2716.

[KZ] A. KOLDOBSKY AND A. ZVAVITCH, An isomorphic version of the Busemann-Petty problem for arbitrary measures, Geom. Dedicata **174** (2015), 261–277.

[KMP] H. KÖNIG, M. MEYER, A. PAJOR, The isotropy constants of Schatten classes are bounded, Math. Ann. **312** (1998), 773–783.

[LR] D. G. LARMAN AND C. A. ROGERS, The existence of a centrally symmetric convex body with central sections that are unexpectedly small, Mathematika **22** (1975), 164–175.

[Lu] E. LUTWAK, Intersection bodies and dual mixed volumes, Advances in Math. **71** (1988), 232–261.

[M1] E. MILMAN, Dual mixed volumes and the slicing problem, Adv. Math. **207** (2006), 566–598.

[M2] E. MILMAN, Generalized intersection bodies, J. Funct. Anal. **240** (2006), 530–567.

[M3] E. MILMAN, On the mean-width of isotropic convex bodies and their associated L_p-centroid bodies, preprint, arXiv:1402.0209.

[MP] V. D. MILMAN AND A. PAJOR, Isotropic position and inertia ellipsoids and zonoids of the unit ball of a normed n-dimensional space, in: Geometric Aspects of Functional Analysis, ed. by J. Lindenstrauss and V. D. Milman, Lecture Notes in Mathematics **1376**, Springer, Heidelberg, 1989, pp. 64–104.

[Pa] G. PAOURIS, On the isotropic constant of non-symmetric convex bodies, Geom. Aspects of Funct. Analysis. Israel Seminar 1996–2000, Lect. Notes in Math., **1745** (2000), 239–244.

[P] M. PAPADIMITRAKIS, On the Busemann-Petty problem about convex, centrally symmetric bodies in $\mathbb{R}^n$, Mathematika **39** (1992), 258–266

[Pe] C. M. PETTY, Projection bodies, Proc. Coll. Convexity (Copenhagen 1965), Kobenhavns Univ. Mat. Inst., 234–241.

[RZ] B. RUBIN AND GAOYONG ZHANG, Generalizations of the Busemann-Petty problem for sections of convex bodies, J. Funct. Anal. **213** (2004), 473–501.

[Sch] R. SCHNEIDER, Zu einem problem von Shephard über die projektionen konvexer Körper, Math. Z. **101** (1967), 71–82.

[Sh] G. C. SHEPHARD, Shadow systems of convex bodies, Israel J. Math. **2** (1964), 229–306.

[Z1] GAOYONG ZHANG, Intersection bodies and Busemann-Petty inequalities in $\mathbb{R}^4$, Annals of Math. **140** (1994), 331–346.

[Z2] GAOYONG ZHANG, A positive answer to the Busemann-Petty problem in four dimensions, Annals of Math. **149** (1999), 535–543.

[Z3] GAOYONG ZHANG, Sections of convex bodies, Amer. J. Math. **118** (1996), 319–340.

[Zv] A. ZVAVITCH, The Busemann-Petty problem for arbitrary measures, Math. Ann. **331** (2005), 867–887.

On Isoperimetric Functions of Probability Measures Having Log-Concave Densities with Respect to the Standard Normal Law

Sergey G. Bobkov

Abstract Isoperimetric inequalities are discussed for one-dimensional probability distributions having log-concave densities with respect to the standard Gaussian measure.

Suppose that a probability measure μ on $\mathbf{R}^n$ has a log-concave density f with respect to the standard n-dimensional Gaussian measure γ_n, that is,

$$f(x) = e^{-\frac{1}{2}|x|^2 - V(x)}, \qquad x \in \mathbf{R}^n,$$

for some convex function $V : \mathbf{R}^n \to (-\infty, \infty]$. One may also say that μ is log-concave with respect to γ_n. In this case, an important theorem due to D. Bakry and M. Ledoux [B-L] asserts that μ satisfies a Gaussian-type isoperimetric inequality

$$\mu^+(A) \geq \varphi\Big(\Phi^{-1}(\mu(A))\Big), \tag{1}$$

relating the "size" $\mu(A)$ of an arbitrary Borel subset $A \subset \mathbf{R}^n$ to its μ-perimeter

$$\mu^+(A) = \liminf_{\varepsilon \downarrow 0} \frac{\mu(A_\varepsilon) - \mu(A)}{\varepsilon}$$

(where A_ε stands for the Euclidean ε-neighborhood of A). Here, Φ^{-1} denotes the inverse to the normal distribution function $\Phi(x) = \gamma_1((-\infty, x])$ with density $\varphi(x) = \frac{1}{\sqrt{2\pi}} e^{-\frac{1}{2}x^2}$ $(x \in \mathbf{R})$. In other words, the isoperimetric function of μ,

$$I_\mu(p) = \inf_{\mu(A)=p} \mu^+(A), \qquad 0 < p < 1$$

(called also an isoperimetric profile) dominates the isoperimetric function $I(p) = \varphi(\Phi^{-1}(p))$ of the measure γ_n, i.e., one has

$$I_\mu(p) \geq I(p) \tag{2}$$

S.G. Bobkov (✉)
University of Minnesota, Minneapolis, MN 55455, USA
e-mail: bobkov@math.umn.edu

E. Carlen et al. (eds.), *Convexity and Concentration*, The IMA Volumes in Mathematics and its Applications 161, DOI 10.1007/978-1-4939-7005-6_19

for all p. The original proof of (1)–(2) given in [B-L] is based on semi-group arguments and a functional form proposed in [B2]. As was shown by L. A. Caffarelli [C], all μ's under consideration represent contractions of γ_n, so the proof of (1)–(2) may be reduced to the purely Gaussian case. An alternative localization approach to the Bakry-Ledoux theorem was later proposed in [B3]; cf. also [B4] for an extension of (1) to a larger class of probability measures. Another approach unifying a number of analytic and isoperimetric inequalities of Gaussian type has been recently developed by P. Ivanisvili and A. Volberg [I-V].

Recently, Raphaël Bouyrie raised the question of whether or not the inequality (2) is strict, even in dimension one, assuming that μ is symmetric and non-Gaussian. Although we do not know the original motivation, this question seems to be rather interesting in itself and not so elementary. Here we give an affirmative answer, involving some arguments from [B3] which were used to prove (1)–(2) in dimension one. Thus, we have:

Theorem 3 *Let μ be a symmetric probability measure on* **R** *which is log-concave with respect to the standard Gaussian measure γ_1. If μ is not Gaussian, then its isoperimetric function satisfies*

$$I_\mu(p) > I(p) \quad \textit{for all } p \in (0,1).$$

Equivalently, the coincidence $I_\mu(p_0) = I(p_0)$ for some p_0 causes μ to be Gaussian. Of course, this is not true at all without the log-concavity hypothesis (with respect to γ_1). For example, consider the class of symmetric probability measures μ on **R** having log-concave densities f with respect to the linear Lebesgue measure (the class of log-concave measures). In this case, the isoperimetric functions have the form

$$J(p) = I_\mu(p) = f(F^{-1}(p)), \tag{3}$$

where F^{-1} is the inverse to the distribution function

$$F(x) = \mu((-\infty, x]) = \int_{-\infty}^{x} f(y)\, dy$$

restricted to the support interval (cf. [B1]). Here, J may be an arbitrary positive concave function on $(0,1)$, symmetric about the point $1/2$. Hence, in this class it may easily happen that $J(p) \geq I(p)$ on $(0,1)$ with equality only at two points p_0 and $1 - p_0$ (or even for one point $p_0 = 1/2$, only). Let us also mention that the property $J \geq I$ is another way to say that μ represents a Lipschitz transform of γ_1.

Assuming that V is of class C^2 in the representation (1), we find from (3) that

$$\begin{aligned} V''(x) = -1 - \left(\frac{f'(x)}{f(x)}\right)' &= -1 - (J'(F(x))' \\ = -1 - J''(F(x))f(x) &= -1 - J''(p)J(p), \qquad p = F(x). \end{aligned}$$

Hence, in terms of the isoperimetric function, the log-concavity with respect to γ_1 is equivalent to the relation

$$J''(p)J(p) \leq -1.$$

For such functions (that are also symmetric about $1/2$), Theorem 3 may be stated as follows: If $J''(p)J(p) = -1$ for some $p \in (0, 1)$, then this equality holds true for all p (in which case, necessarily $J = I$). It might be natural to try to prove Theorem 3 using this formulation. However, we prefer to choose a different route, which allows one to avoid the C^2-assumption on the density f, and which also suggests a possible way to quantify the assertion of this theorem. To be more precise, we have:

Theorem 4 *Let μ be a probability measure supported on the interval $(-a, a) \subset \mathbf{R}$ with density $e^{-V(x)}\varphi(x)$, where V is an even, convex function, which is differentiable and increasing on $(0, a)$. Then the isoperimetric function of μ satisfies*

$$I_\mu(p) \geq \frac{1}{2\,\Phi(V'(x))}\, e^{-\frac{1}{2}V'(x)^2 - V'(x)y}\, \varphi(y), \tag{4}$$

where $p = \mu((-\infty, x])$, $x \in (-a, 0)$, and

$$y = -V'(x) + \Phi^{-1}\left(2p\,\Phi(V'(x))\right).$$

A similar bound also holds for $p > 1/2$, by using $I_\mu(1-p) = I_\mu(p)$.

One can check that equality in (4) is attained for the family of probability measures $\mu = \mu_\lambda$ with densities

$$\varphi_\lambda(x) = \frac{1}{Z}\, e^{-\lambda|x|}\, \varphi(x), \qquad x \in \mathbf{R}, \tag{5}$$

where λ is an arbitrary positive parameter and $Z = Z(\lambda)$ is a normalizing constant.

We now turn to the proofs. As a first step, let us verify Theorem 3 in the particular case of measures μ_λ described in (5).

Lemma 3 *Given $\lambda > 0$, we have $I_{\mu_\lambda}(p) > I(p)$ for all $p \in (0, 1)$.*

Proof According to (3), the isoperimetric function of μ_λ is given by

$$I_{\mu_\lambda}(p) = \varphi_\lambda\Big(\Phi_\lambda^{-1}(p)\Big),$$

where Φ_λ denotes the distribution function of μ_λ. Therefore, we need to show that $\Phi_\lambda(y) = \Phi(x) \Rightarrow \varphi_\lambda(y) > \varphi(x)$ for all $x, y \in \mathbf{R}$, where one may additionally assume that $x \leq 0$ (using the symmetry).

The increasing map $T(x) = \Phi_\lambda^{-1}(\Phi(x))$ pushes forward γ_1 to μ_λ, so that $\Phi_\lambda(T(x)) = \Phi(x)$. After differentiation we have

$$\varphi_\lambda(T(x))T'(x) = \varphi(x).$$

Hence, it is sufficient to see that $T'(x) < 1$ for all $x < 0$. To this aim, first note that

$$\int_{-\infty}^{x} e^{\lambda y}\,\varphi(y)\,dy = e^{\lambda^2/2}\,\Phi(x-\lambda), \qquad Z = 2\int_{-\infty}^{0} e^{\lambda y}\,\varphi(y)\,dy = 2e^{\lambda^2/2}\,\Phi(-\lambda).$$

Hence, the distribution function of μ_λ is described as

$$\Phi_\lambda(x) = \mu_\lambda((-\infty,x]) = \frac{\Phi(x-\lambda)}{2\Phi(-\lambda)}, \qquad x \leq 0,$$

and, by the symmetry, $\Phi_\lambda(x) = 1 - \Phi_\lambda(-x)$ for $x \geq 0$. It follows that

$$T(x) = \Phi^{-1}(\alpha\Phi(x)) + \lambda \qquad (\alpha = 2\Phi(-\lambda),\ x \leq 0),$$

so, putting $x = \Phi^{-1}(p)$, we get

$$T'(x) = \frac{\alpha\varphi(x)}{I(\alpha\Phi(x))} = \frac{\alpha I(p)}{I(\alpha p)}.$$

But the last ratio is smaller than 1, since $\alpha < 1$ and since $I(p)/p$ is a decreasing function. The latter property is true for any positive, strictly concave function I on $(0,1)$, which follows from the representation

$$\frac{I(p)}{p} = \frac{I(0+)}{p} + \int_0^1 I'(ps)\,ds. \tag{6}$$

This proves the lemma.

Lemma 4 *Let μ be a symmetric probability measure, which is log-concave with respect to γ_1 with density $f = e^{-V}\varphi$. Suppose that V is monotone in some neighborhood of a point $x \in \mathbf{R}$, and let $p = \mu((-\infty,x])$. Then*

$$I_\mu(p) \geq I_{\mu_\lambda}(p) \quad \textit{for some } \lambda > 0.$$

Proof We prove a stronger statement: Let a positive finite measure μ have density $f(y) = e^{-V(y)}\varphi(y)$ for some convex even function $V : \mathbf{R} \to (-\infty,\infty]$, finite on the interval $(-a,a)$. If a point $x \in (-a,0)$ is such that

$$\mu((-a,x]) \geq p, \qquad \mu((x,0]) \geq \frac{1}{2} - p \qquad \Big(0 < p < \frac{1}{2}\Big). \tag{7}$$

and if V is monotone in some neighborhood of x, then $f(x) \geq I_{\mu_\lambda}(p)$ for some $\lambda > 0$.

To simplify this assertion, let $l(y) = c - \lambda y$ be an affine function which is tangent to $V(y)$ at x, with necessarily $\lambda > 0$ in view of the monotonicity assumption on V. We extend l from the negative half-axis $(-\infty,0)$ to $(0,\infty)$ to get an even function, and as a result we obtain a new positive measure μ_0 with density

$$f_0(y) = Ce^{-\lambda|y|}\varphi(y).$$

Since $l(x) = V(x)$ and $l \le V$ everywhere on $(-a, a)$, we have $f \le f_0$, so that $\mu_0((-a,x]) \ge p$ and $\mu_0((x,0]) \ge \frac{1}{2} - p$. Therefore, in our stronger statement we are reduced to the class of densities of type $f = C\varphi_\lambda$, where C is an arbitrary positive parameter.

For such densities, we have

$$\mu((-\infty,x]) = C\Phi_\lambda(x), \quad \mu((x,0]) = C\Big(\frac{1}{2} - \Phi_\lambda(x)\Big), \qquad f(x) = C\varphi_\lambda(x),$$

and involving the assumption (7), we get a constraint on C, namely,

$$C \ge C_0 = \max\left\{\frac{p}{\Phi_\lambda(x)}, \frac{\frac{1}{2} - p}{\frac{1}{2} - \Phi_\lambda(x)}\right\}. \tag{8}$$

Since $C = C_0$ is the worst situation in our conclusion, it remains to show that

$$C_0\,\varphi_\lambda(x) \ge \varphi_\lambda(\Phi_\lambda^{-1}(p)) \equiv I_{\mu_\lambda}(p)$$

with C_0 defined in (8). Putting $q = \Phi_\lambda(x)$, this is the same as

$$\max\left\{\frac{p}{q}, \frac{\frac{1}{2} - p}{\frac{1}{2} - q}\right\} I_{\mu_\lambda}(q) \ge I_{\mu_\lambda}(p).$$

If $p \ge q$, it holds true, since $\frac{p}{q} I_{\mu_\lambda}(q) \ge I_{\mu_\lambda}(p)$, which in turn follows from the fact that the function I_{μ_λ} is strictly concave (so that $I_{\mu_\lambda}(p)/p$ is strictly decreasing). In case $p \le q$, we use

$$\frac{\frac{1}{2} - p}{\frac{1}{2} - q} I_{\mu_\lambda}(q) \ge I_{\mu_\lambda}(p),$$

or equivalently, after the change $p' = \frac{1}{2} - p$, $q' = \frac{1}{2} - q$,

$$\frac{p'}{q'} I_{\mu_\lambda}\Big(\frac{1}{2} - q'\Big) \ge I_{\mu_\lambda}\Big(\frac{1}{2} - p'\Big).$$

Here again $p' \ge q'$ and we deal with the concave function $\tilde{I}(p') = I_{\mu_\lambda}(\frac{1}{2} - p')$ on the interval $(0, 1/2)$. Hence, $\tilde{I}(p')/p'$ is strictly decreasing, which is seen from the general identity (6).

Lemma 4 is proved.

Proof of Theorem 3 If a probability measure μ on the line is log-concave with respect to γ_1, it has a density

$$f(x) = e^{-V(x)}\varphi(x),$$

for some convex even function V on the interval $(-a, a)$, finite or not, and one may put $V = \infty$ outside that interval. Since V attains its minimum at zero, necessarily $V(0) < 0$, as long as μ is non-Gaussian. In particular, in this case

$$I_\mu(1/2) = f(0) > \varphi(0) = I(1/2).$$

Moreover, let $[-x_0, x_0]$ be the longest interval, where V is constant, so that $V(x_0-) = V(0)$. Then similarly

$$I_\mu(p) = I_\mu(1/2) > I(1/2)$$

for all $p \in [p_0, 1 - p_0]$, where $p_0 = \mu((-a, x_0])$.

In case $0 < p < p_0$, $p = \mu((-a, x])$, the point x necessarily belongs to the interval $(-a, x_0)$, where V is strictly decreasing. Therefore, one may apply Lemma 4 and combine it with Lemma 3, which then leads to the required assertion $I_\mu(p) \geq I_{\mu_\lambda}(p) > I(p)$.

Theorem 3 is thus proved.

Proof of Theorem 4 If an even, convex function V in the representation $f = e^{-V}\varphi$ for the density of μ is differentiable and is increasing on $(0, a)$, the assumption of Lemma 4 is fulfilled for all points $x \neq 0$ from the supporting interval of the measure μ. In this case, since the tangent affine function in the proof of Lemma 4 is given by $l(y) = V(x) + V'(x)(y-x)$, necessarily $\lambda = \lambda(x) = -V'(x)$ $(-a < x < 0)$. Hence, we obtain that

$$I_\mu(p) \geq I_{\mu_{\lambda(x)}}(p), \qquad p = \mu((-a, x]). \tag{9}$$

The expression $I_{\mu_{\lambda(x)}}(p)$ may be written in a more explicit form. Recall that, for $0 < p < 1/2$,

$$y \equiv \Phi_\lambda^{-1}(p) = \Phi^{-1}(\alpha p) + \lambda, \qquad Z = 2e^{\lambda^2/2}\,\Phi(-\lambda),$$

where $\alpha = 2\Phi(-\lambda)$, so that

$$I_{\mu_\lambda}(p) = \frac{1}{Z}\,e^{-\lambda\,|y|}\,\varphi(y) = \frac{1}{2\Phi(-\lambda)}\,e^{-\lambda^2/2+\lambda y}\,\varphi(y).$$

Hence, (9) turns into (4), thus proving Theorem 4.

Acknowledgements This research was partially supported by the Alexander von Humboldt Foundation and NSF grant DMS-1612961.

References

[B-L] D. Bakry, M. Ledoux. Lévy-Gromov's isoperimetric inequality for an infinite-dimensional diffusion generator. Invent. Math. 123 (1996), no. 2, 259–281.

[B1] S. G. Bobkov. Extremal properties of half-spaces for log-concave distributions. Ann. Probab. 24 (1996), no. 1, 35–48.

[B2] S. G. Bobkov. An isoperimetric inequality on the discrete cube, and an elementary proof of the isoperimetric inequality in Gauss space. Ann. Probab. 25 (1997), no. 1, 206–214.

[B3] S. G. Bobkov. A localized proof of the isoperimetric Bakry-Ledoux inequality and some applications [in Russian]. Teor. Veroyatnost. Primenen. 47 (2002), no. 2, 340–346; English translation: Theory Probab. Appl. 47 (2003), no. 2, 308–314.

[B4] S. G. Bobkov. Perturbations in the Gaussian isoperimetric inequality. J. Math. Sciences (New York), vol. 166 (2010), no. 3, 225–238. Translated from: Problems in Math. Analysis, 45 (2010), 3–14.

[C] L. A. Caffarelli. Monotonicity properties of optimal transportation and the FKG and related inequalities. Commun. Math. Phys. 214 (2000), 547–563.

[I-V] P. Ivanisvili, A. Volberg. Isoperimetric functional inequalities via the maximum principle: the exterior differential systems approach. Preprint (2015), arXiv:1511.06895.

Counting Integer Points in Higher-Dimensional Polytopes

Alexander Barvinok

Abstract We survey some computationally efficient formulas to estimate the number of integer or 0-1 points in polytopes. In many interesting cases, the formulas are asymptotically exact when the dimension of the polytopes grows. The polytopes are defined as the intersection of the non-negative orthant or the unit cube with an affine subspace, while the main ingredient of the formulas comes from solving a convex optimization problem on the polytope.

1 Introduction

Computationally efficient counting of integer points in polyhedra is a much-studied topic with applications in combinatorics, algebra, representation theory, mathematical programming, statistics, compiler optimization, and social choice theory, see, for example, survey [13] and [28], [12], [32], [26] for more recent developments.

Here we are given a convex polytope $P \subset \mathbf{R}^d$, the standard integer lattice $\mathbf{Z}^d \subset \mathbf{R}^d$ and we want to compute exactly or approximately the number $|P \cap \mathbf{Z}^d|$ of integer points in P. The polytope P can be given as the convex hull of the set of its vertices (which we assume to be points with rational coordinates) or as the set of solutions to a finite system of linear inequalities (which we assume to have integer coefficients), or somehow indirectly, for example, by an oracle, which, given a point $x \in \mathbf{R}^d$, reports whether x lies in P.

If the dimension d of the ambient space is fixed in advance and not a part of the input, then all the above ways to define P are more or less computationally equivalent and there is a polynomial time algorithm, which, given such a $P \subset \mathbf{R}^d$, computes $|P \cap \mathbf{Z}^d|$ exactly, see [1] for the underlying theory and [15] and [37] for implementations, respectively, `Latte` and `barvinok`.

The situation changes dramatically if the ambient dimension d is not fixed in advance and can be a part of the input. To start with, it begins to matter how the

A. Barvinok (✉)
Department of Mathematics, University of Michigan, 48109-1043 Ann Arbor, MI, USA
e-mail: barvinok@umich.edu

E. Carlen et al. (eds.), *Convexity and Concentration*, The IMA Volumes
in Mathematics and its Applications 161, DOI 10.1007/978-1-4939-7005-6_20

polytope P is defined, as there are polytopes with many vertices and relatively few facets and vice versa. More importantly, even testing whether the polytope contains at least one integer point becomes a hard (NP-complete) problem. For some classes of polytopes, however, the problem is still manageable.

As is well known, even when P is a simplex defined in $\mathbf{R}^d$ by a system

$$a_1x_1 + \ldots + a_dx_d = b \quad \text{and} \quad x_1, \ldots, x_d \geq 0,$$

where $a_1, \ldots, a_d$ and b are positive integers, testing whether $P \cap \mathbf{Z}^d = \emptyset$ is an NP-complete problem, known as the *knapsack problem*. However, one can count integer points approximately (there is a fully polynomial time approximation scheme) in the polytope defined by the system

$$\sum_{j=1}^{d} a_{ij}x_j \leq b_i \quad \text{for} \quad i = 1, \ldots, m \qquad \text{and} \qquad x_1, \ldots, x_d \geq 0,$$

where a_{ij} and b_i are positive integers, provided the number m of inequalities is fixed in advance [18] (note that $x_1 = \ldots = x_d = 0$ is always a solution). The algorithm is based on dynamic programming.

One can try to approximate the number of integer points in P by the volume of P, which is a much easier problem computationally, see [36]. In general, $\operatorname{vol}(P)$ is a poor approximation for $|P \cap \mathbf{Z}^d|$, as the example of the unit cube shows: the volume of the cube defined by the inequalities $0 \leq x_k \leq 1$ for $k = 1, \ldots, d$ is 1 while the number of integer points is 2^d. However, it is shown in [24] that if a polytope $P \subset \mathbf{R}^d$ with m facets contains a ball of radius $d\sqrt{\log m}$, then the volume of P provides a close approximation to the number of integer points in P.

In this paper, we discuss the method introduced in [5] and its ramifications. The main goal of the method is to provide a quick way to get an estimate of the number of integer points, which, in particular, will be computationally feasible even in very high dimensions where other approaches may fail. The centerpiece of the approach is the following family of heuristic formulas.

1.1 The Gaussian Formulas

In what follows, the following function plays an important role:

$$g(x) = (x+1)\ln(x+1) - x\ln x \quad \text{for} \quad x \geq 0. \tag{1}$$

As is easy to see, $g(x)$ is strictly increasing and concave. We extend it to a function on the non-negative orthant $\mathbf{R}^n_+$ by

$$g(x) = \sum_{i=1}^{n} g(x_i) \quad \text{where} \quad x = (x_1, \ldots, x_n). \tag{2}$$

We suppose that the polytope $P \subset \mathbf{R}^n$ is defined as the intersection of an affine subspace with the non-negative orhant $\mathbf{R}^n_+$. More precisely, P is defined by the system

$$P = \left\{ x = (x_1, \ldots, x_n) : \ Ax = b, \ x_1, \ldots, x_n \geq 0 \right\}, \tag{3}$$

where A is an $r \times n$ integer matrix of $\operatorname{rank}(A) = r < n$ and b is an integer r-vector. We also assume that P is bounded (and hence is a polytope) and has a non-empty interior, that is, contains a point $x = (x_1, \ldots, x_n)$ where $x_k > 0$ for $k = 1, \ldots, n$, in which case

$$d = \dim P = n - r.$$

For the purpose of counting integer points, any d-dimensional polytope P with n facets can be represented in this form by a change of variables.

We consider the following optimization problem

$$\text{Find} \qquad \max g(x) \tag{4}$$

$$\text{Subject to} \qquad x \in P \tag{5}$$

Since the function g of (1)–(2) is strictly concave and the polytope P of (3) is bounded, the maximum in (4)–(5) is attained at a unique point

$$z = (z_1, \ldots, z_n),$$

and computing this point z is computationally easy, both in theory and in practice, cf. [31]. Moreover, it is not hard to show that since P has a non-empty interior and the right derivative $\partial g / \partial x_k$ at $x = 0$ is $+\infty$, we have $z_k > 0$ for $k = 1, \ldots, n$. We define an $r \times r$ matrix $B = \left(b_{ij}\right)$ by

$$b_{ij} = \sum_{k=1}^{n} a_{ik} a_{jk} \left(z_k + z_k^2\right).$$

Let $\Lambda \subset \mathbf{Z}^r$ be the lattice generated by the columns of A. Since $\operatorname{rank}(A) = r$, we have $\operatorname{rank}(\Lambda) = r$ and we define $\det \Lambda$ as the volume of the fundamental domain of Λ. Computing $\det \Lambda$ is also computationally easy, see, for example, [22]. Now we are ready to state our first main formula:

$$|P \cap \mathbf{Z}^n| \approx \frac{e^{g(z)} \det \Lambda}{(2\pi)^{r/2} \sqrt{\det B}}. \tag{6}$$

We also consider the following modification of the problem. Suppose that a polytope P is defined as the intersection of an affine subspace with the unit cube

$$[0, 1]^n = \left\{ (x_1, \ldots, x_n) : \ 0 \leq x_i \leq 1 \quad \text{for} \quad i = 1, \ldots, n \right\},$$

that is,

$$P = \left\{ x = (x_1, \ldots, x_n) : \ Ax = b, \ 0 \leq x_1, \ldots, x_n \leq 1 \right\},$$

where A is an $r \times n$ integer matrix of $\operatorname{rank}(A) = r < n$ and b is an integer r-vector. We also assume that the P has a non-empty interior, that is, contains a point $x = (x_1, \ldots, x_n)$ such that $0 < x_k < 1$ for $k = 1, \ldots, n$, in which case

$$d = \dim P = n - r.$$

We want to estimate the number $|P \cap \{0, 1\}^n|$ of 0-1 vectors in P. We define

$$h(x) = x \ln \frac{1}{x} + (1 - x) \ln \frac{1}{1 - x} \quad \text{for} \quad 0 \leq x \leq 1, \tag{7}$$

extend it to the unit cube $[0, 1]^n$ by

$$h(x) = \sum_{k=1}^{n} h(x_k) \quad \text{where} \quad x = (x_1, \ldots, x_n) \tag{8}$$

and solve the optimization problem

$$\text{Find} \qquad \max h(x) \tag{9}$$

$$\text{Subject to} \qquad x \in P. \tag{10}$$

As before, since the function h strictly concave, the maximum of h is attained at a unique point $z = (z_1, \ldots, z_n)$, which can be computed efficiently. Similarly, since the polytope P has a non-empty interior and since the right, respectively left, derivative $\partial h / \partial x_k$ at $x_k = 0$, respectively at $x_k = 1$, is equal to $+\infty$, respectively to $-\infty$, it is not hard to show that $0 < z_k < 1$ for $k = 1, \ldots, n$. We define an $r \times r$ matrix $b = \left(b_{ij}\right)$ by

$$b_{ij} = \sum_{k=1}^{n} a_{ik} a_{jk} \left(z_k - z_k^2\right)$$

and state that

$$|P \cap \{0, 1\}^n| \approx \frac{e^{h(z)} \det \Lambda}{(2\pi)^{r/2} \sqrt{\det B}} \tag{11}$$

One legitimate question is what do signs "≈" in the formulas (6) and (11) mean exactly? For example, if P has no integer points at all, in what sense 0 in the left-hand side is approximately equal to a positive number in the right-hand side? Another legitimate concern is that while the left-hand sides of (6) and (11) depend on the polytope P, the right-hand sides depend also on the matrix A, which, given P, can be chosen in a variety of different ways. We will address these concerns in due time; at this point, we consider (6) and (11) as heuristic formulas. Before we give any justification for them, we demonstrate that at least in some cases, they compute reasonable approximations. The following three examples are taken from [14].

1.2 Example: 2-Way Transportation Polytopes and 2-Way Contingency Tables

Here we want to compute the number of $m \times n$ matrices with non-negative integer entries x_{ij} and with prescribed row and column sums

$$\sum_{j=1}^{n} x_{ij} = r_i \quad \text{for} \quad i = 1, \ldots, m \quad \text{and} \quad \sum_{i=1}^{m} x_{ij} = c_j \quad \text{for} \quad j = 1, \ldots, n. \qquad (12)$$

We assume that $r_i, c_j > 0$ for all i and j. The corresponding polytope P of $m \times n$ non-negative matrices with row sums $R = (r_1, \ldots, r_m)$ and column sums $C = (c_1, \ldots, c_n)$ is known as a *2-way transportation polytope*, see [16]. The polytope P is non-empty if and only if the balance condition

$$r_1 + \ldots + r_m = c_1 + \ldots + c_n$$

is satisfied, in which case P is the intersection of the non-negative orthant $\mathbf{R}_+^{m \times n}$ with an affine subspace of codimension $m + n - 1$. In statistics, integer points in P are known as *2-way contingency tables* whereas row and column sums are referred to as *margins*. For example, the exact number of 4×4 non-negative integer matrices with row sums $R = (220, 215, 93, 64)$ and column sums $C = (108, 286, 71, 127)$ is $1,225,914,276,768,514 \approx 1.2 \times 10^{15}$. Contingency tables with those margins were considered, in particular, in [17] in connection with a study of the correlation between the eye color and hair color. It turns out that the formula (6) approximates the true number of contingency tables within a relative error of 0.06. We note that in this case, the polytope is defined by $4+4-1 = 7$ independent linear equations and 16 non-negativity constraints, so to obtain the matrix A of $\text{rank}(A) = 7$ in (3) we remove one of the row/column sum constraints (12) (it does not matter which one).

1.3 Example: 3-Way Transportation Polytopes and 3-Way Contingency Tables

Here we want to compute the number of $m \times n \times s$ arrays with non-negative integer entries x_{ijk} and prescribed sums over coordinate affine hyperplanes,

$$a_i = \sum_{j,k} x_{ijk}, \quad b_j = \sum_{i,k} x_{ijk} \quad \text{and} \quad c_k = \sum_{i,j} x_{ijk}. \tag{13}$$

The sums $A = (a_1, \ldots, a_m)$, $B = (b_1, \ldots, b_n)$ and $C = (c_1, \ldots, c_s)$ are called *margins*, or, sometimes, *1-margins*. We assume that $a_i, b_j, c_k > 0$ for all i, j and k. The corresponding polytope P of $m \times n \times s$ non-negative arrays with margins A, B, and C is known as a *3-way axial transportation polytope*, see [16]. The polytope is non-empty if and only if the following balance conditions

$$a_1 + \ldots + a_m = b_1 + \ldots + b_n = c_1 + \ldots + c_s$$

are satisfied, in which case P is the intersection of the non-negative orthant $\mathbf{R}_+^{m\times n\times s}$ with an affine subspace of codimension $m + n + s - 2$. In statistics, integer points in P are referred to as *3-way contingency tables* whereas sums A, B, and C are called *margins* or *1-margins*. For example, the exact number of $3 \times 3 \times 3$ non-negative integer arrays with margins $A = (31, 22, 87)$, $B = (50, 13, 77)$ and $C = (42, 87, 11)$ is $8,846,838,772,161,591 \approx 8.8 \times 10^{15}$ (computed in [14] using `LattE`). It turns out that the formula (6) approximates the true number within a relative error of 0.002. We note that in this case, the polytope is defined by $3 + 3 + 3 - 2 = 7$ independent linear equations and 27 non-negativity constraints. To apply the formula, we remove one constraint from two of the three A, B, or C balance conditions in (13) (again, it does not matter which).

1.4 Example: Graphs with a Given Degree Sequence

Given a vector $D = (d_1, \ldots, d_n)$ of positive integers, we are interested in the number of graphs (undirected, without loops or multiple edges) on the set $\{1, \ldots, n\}$ such that the degree of vertex k is d_k. Geometrically, we want to count 0-1 points in the polytope $P(D) \subset \mathbf{R}^{n(n-1)/2}$ of $n \times n$ symmetric matrices with entries $0 \leq x_{ij} \leq 1$ satisfying

$$x_{ii} = 0 \quad \text{and} \quad \sum_{j=1}^{n} x_{ij} = d_i \quad \text{for} \quad i = 1, \ldots, n.$$

Without loss of generality, we assume that

$$d_1 \geq d_2 \geq \ldots \geq d_n.$$

The Erdős-Gallai Theorem, see, for example, Theorem 6.3.6 of [10], states that the necessary and sufficient condition for the polytope $P(D)$ to be non-empty are the inequalities

$$\sum_{i=1}^{k} d_i \leq k(k-1) + \sum_{i=k+1}^{n} \min\{k, d_i\} \quad \text{for} \quad k = 1, \ldots, n. \tag{14}$$

and for 0-1 point to exist in this polytope is the congruence

$$d_1 + \ldots + d_n \equiv 0 \mod 2. \tag{15}$$

The polytope has a non-empty interior if the inequalities in (14) are strict. This is the case, for example, when the graph is regular, that is, $d_1 = \ldots = d_n$. For example, if $n = 17$ and $d_1 = \ldots = d_{17} = 4$ (so that (15) is also satisfied) the exact number of graphs is $28,797,220,460,586,826,422,720 \approx 2.9 \times 10^{22}$, whereas the formula (11) approximates the true number within a relative error of 0.028.

Examples 1.2–1.4 indicate that the formulas (6) and (11) must be doing something right, at least in some cases. In the following section, we discuss the intuition underlying (6) and (11).

2 Two Lemmas and a Rationale for the Gaussian Formulas

First, we recall some probability.

Let us fix $p, q > 0$ such that $p + q = 1$. A random variable X taking non-negative integer values k is called *geometric* if

$$\mathbf{P}(X = k) = pq^k \quad \text{for} \quad k = 0, 1, \ldots, .$$

Then we have

$$\mathbf{E}(X) = \frac{q}{p} \quad \text{and} \quad \mathbf{var}(X) = \frac{q}{p^2}.$$

If we denote

$$\mathbf{E}(X) = z$$

for some $z > 0$, then $p = 1/(1 + z)$, $q = z/(1 + z)$ and

$$\mathbf{var}(X) = z + z^2.$$

A random variable X is called *Bernoulli* if it takes values 0 and 1 and

$$\mathbf{P}(X = 0) = q \quad \text{and} \quad \mathbf{P}(X = 1) = p.$$

Hence we have

$$\mathbf{E}(X) = p \quad \text{and} \quad \mathbf{var}(X) = pq.$$

If we denote

$$\mathbf{E}(X) = z$$

for some $0 < z < 1$, then $p = z$, $q = 1 - z$ and

$$\mathbf{var}(X) = z - z^2.$$

Formulas (6) and (11) are built on the following simple Lemma 1 and Lemma 2, respectively.

Lemma 1 *Let $\mathscr{A} \subset \mathbf{R}^n$ be an affine subspace such that the intersection*

$$P = \mathscr{A} \cap \mathbf{R}^n_+$$

is bounded and contains a point $x = (x_1, \ldots, x_n)$ where $x_k > 0$ for $k = 1, \ldots, n$.

Then the function

$$g(x) = \sum_{k=1}^{n} \Big((x_k + 1) \ln (x_k + 1) - x_k \ln x_k \Big) \quad for \quad x \in \mathbf{R}^n_+, \ x = (x_1, \ldots, x_n),$$

attains its maximum on P at a unique point $z = (z_1, \ldots, z_n)$ where $z_k > 0$ for $k = 1, \ldots, n$.

If $X = (X_1, \ldots, X_n)$ is a vector of independent geometric random variables X_k such that

$$\mathbf{E}\,(X_k) = z_k \quad for \quad k = 1, \ldots, n,$$

then

$$\mathbf{P}(X = m) = e^{-g(z)} \quad for\ all \quad m \in P \cap \mathbf{Z}^n.$$

In other words, Lemma 1 asserts that the vector of independent geometric random variables $X = (X_1, \ldots, X_n)$ whose expectations are found from solving the optimization problem (4)–(5) hits every integer point $m \in P$ with the same probability equal to $e^{-g(z)}$.

Lemma 2 *Let $\mathscr{A} \subset \mathbf{R}^n$ be an affine subspace such that the intersection*

$$P = \mathscr{A} \cap [0, 1]^n$$

contains a point $x = (x_1, \ldots, x_n)$ where $0 < x_k < 1$ for $k = 1, \ldots, n$.

Then the function

$$h(x) = \sum_{k=1}^{n} \left(x_k \ln \frac{1}{x_k} + (1 - x_k) \ln \frac{1}{1 - x_k} \right) \quad for \quad x \in [0, 1]^n, \ x = (x_1, \ldots, x_n),$$

attains its maximum on P at a unique point $z = (z_1, \ldots, z_n)$ where $0 < z_k < 1$ for $k = 1, \ldots, n$.

If $X = (X_1, \ldots, X_n)$ *is a vector of independent Bernoulli random variables* X_k *such that*

$$\mathbf{E}(X_k) = z_k \quad \textit{for} \quad k = 1, \ldots, n,$$

then

$$\mathbf{P}(X = m) = e^{-h(z)} \quad \textit{for all} \quad m \in P \cap \{0, 1\}^n.$$

In other words, Lemma 2 asserts that the vector of independent Bernoulli random variables $X = (X_1, \ldots, X_n)$ whose expectations are found from solving the optimization problem (9)–(10) hits every 0-1 point $m \in P$ with the same probability equal to $e^{-h(z)}$.

The proof of Lemma 1 and Lemma 2 is a simple exercise in Lagrange multipliers [5]. We discuss below the general context of the results explaining where the functions $g(x)$ and $h(x)$ come from.

2.1 *The Maximum Entropy Principle*

Let X be a discrete random variable taking values in $\mathbf{R}^n$. Assuming that X takes values from a discrete set $S \subset \mathbf{R}^n$, the *entropy* of X is defined as

$$\mathbf{ent}(X) = \sum_{s \in S} \mathbf{P}(X = s) \ln \frac{1}{\mathbf{P}(X = s)},$$

see, for example, [25]. In our case, we have $S = \mathbf{Z}^n_+$, the set of non-negative integer n-vectors or $S = \{0, 1\}^n$, the set of n-vectors with 0-1 coordinates.

Let us fix a set $S \subset \mathbf{R}^n$ of values of X. Then $\mathbf{ent}(X)$ is a non-negative strictly concave function of the probabilities $\mathbf{P}(X = s)$. If S is finite, then $\mathbf{ent}(X)$ attains its maximum value $\ln |S|$ if X is uniform on S so that $\mathbf{P}(X = s) = 1/|S|$ for all $s \in S$. If S is infinite, then the entropy of a random variable X with values in S can be arbitrarily large.

Let $\mathscr{A} \subset \mathbf{R}^n$ be an affine subspace such that the intersection $\mathscr{A} \cap S$ is non-empty and finite and assume that there exists a random variable X that has the maximum entropy among all random variables with values in S and expectation in $\mathscr{A}$. The restriction of such an X onto $\mathscr{A} \cap S$ is necessarily uniform, since otherwise by reapportioning the probabilities $\mathbf{P}(X = s)$ for $s \in \mathscr{A} \cap S$ we can increase the entropy of X while keeping $\mathbf{E}(X) \in \mathscr{A}$ (here we use that $\mathscr{A}$ is an affine subspace, so that $\alpha_1 s_1 + \ldots + \alpha_k s_k$ lies in $\mathscr{A}$ if $s_1, \ldots, s_k$ lie in $\mathscr{A}$ and $\alpha_1 + \ldots + \alpha_k = 1$). Assuming that $\mathscr{A}$ is defined in $\mathbf{R}^n$ by a finite system of linear equations

$$\langle a_i, x \rangle = \beta_i, \quad i \in I,$$

where $a_i \in \mathbf{R}^n$, $\beta_i \in \mathbf{R}$ and $\langle \cdot, \cdot \rangle$ is the standard inner product in $\mathbf{R}^n$, from the Lagrange optimality condition we conclude that

$$\mathbf{P}(X = s) = \exp\left\{-\lambda - \sum_{i \in I} \lambda_i \langle a_i, s\rangle\right\} \quad \text{for all} \quad s \in S$$

and some real λ and λ_i, $i \in I$. Then

$$\mathbf{P}(X = s) = \exp\left\{-\lambda - \sum_{i \in I} \lambda_i \beta_i\right\} \quad \text{for all} \quad s \in \mathscr{A} \cap S$$

and

$$\mathbf{ent}(X) = \lambda \sum_{s \in S} \mathbf{P}(X = s) + \sum_{i \in I} \lambda_i \langle a_i, \mathbf{E}(X)\rangle = \lambda + \sum_{i \in I} \lambda_i \beta_i,$$

so we conclude that

$$\mathbf{P}(X = s) = e^{-\mathbf{ent}(X)} \quad \text{for all} \quad s \in \mathscr{A} \cap S.$$

Suppose that, in addition, S is the direct product $S = S_1 \times \ldots \times S_n$ where $S_k \subset \mathbf{R}$ for $k = 1, \ldots, n$ and $\mathscr{A} \subset \mathbf{R}^n$ is an affine subspace as above. Let $X = (X_1, \ldots, X_n)$ be a random variable of the maximum entropy (if exists) in the class of random variables with values in S and expectation in $\mathscr{A}$. Then we must have $X_1, \ldots, X_n$ independent since otherwise by choosing $X' = \left(X_1', \ldots, X_n'\right)$ where $X_1', \ldots, X_n'$ are independent copies of $X_1, \ldots, X_n$ we obtain a random variable X' from the same class with $\mathbf{ent}(X') > \mathbf{ent}(X)$, cf. [25]. This is generally true if $\mathscr{A} \subset \mathbf{R}^n$ is a subset that can be defined by equations of the type $f_1(x_1) + \ldots + f_n(x_n) = 0$ for some univariate functions $f_1, \ldots, f_n$, in particular if $\mathscr{A}$ is an affine subspace. Since $X_1, \ldots, X_n$ are independent, we have $\mathbf{ent}(X) = \mathbf{ent}(X_1) + \ldots + \mathbf{ent}(X_n)$.

In Lemma 1, we have $S = \mathbf{Z}_+^n$, so that the random variable $X = (X_1, \ldots, X_n)$ of the largest entropy is necessarily a vector of independent random variables X_k with values in $\mathbf{Z}_+$. It turns out that in the class of all random variables with values in $\mathbf{Z}_+$ and a given expectation x, the geometric random variable has the largest entropy equal to the function $g(x)$ of (1). Hence the random vector X in Lemma 1 is the random vector of the largest entropy among all random vectors with values in $\mathbf{Z}_+^n$ and expectation in $\mathscr{A}$ (it is also not hard to deduce from the above that such a maximum entropy vector indeed exists). The meaning of the function $g : \mathbf{R}_+^n \longrightarrow \mathbf{R}_+$ in (2) and Lemma 1 is that $g(x)$ is the maximum entropy of a random vector X with values in $\mathbf{Z}_+^n$ and expectation $x \in \mathbf{R}_+^n$.

In Lemma 2, we have $S = \{0, 1\}^n$, so that the random variable $X = (X_1, \ldots, X_n)$ of the largest entropy (which exists by a compactness argument) is necessarily a vector of independent Bernoulli random variables X_k. In this case, the function $h(x)$ of (7) is the entropy of the Bernoulli random variable with expectation x. Hence the random vector X in Lemma 2 is the random vector of the largest entropy among all random vectors with values in $\{0, 1\}^n$ and expectation in $\mathscr{A}$. The meaning of the function $h : [0, 1]^n \longrightarrow \mathbf{R}_+$ in (8) and Lemma 2 is that $h(x)$ is the maximum entropy of a random vector with values in $\{0, 1\}^n$ and expectation $x \in [0, 1]^n$.

This approach fits a more general framework of statistics [19] and physics [23]. Suppose we know that a random variable X satisfies certain constraints but don't know anything else. We want to know what distribution X is likely to have. From the perspective of a mathematician, the question does not make much sense as X may have any distribution satisfying given constraints. A statistician, however, may rephrase the wording asking instead what should the "null hypothesis" regarding the distribution of X be. The "maximum entropy principle" states that the null hypothesis should be that X has the maximum entropy distribution in the class of all distributions satisfying given constraints [19]. From the point of view of a physicist, the maximum entropy principle may sound attractive since if we don't know any other information about X then maybe we are not supposed to know and then it looks natural to choose the distribution in the given class of the maximum uncertainty, that is, of the largest entropy. For example, if X is the velocity of a random molecule of oxygen in the room, the temperature of the air in the room tells us the average squared speed $\mathbf{E}\|X\|^2$. As we don't know anything else, the maximum entropy principle tells us to assume that X is Gaussian, as among all random variables with given variance, the Gaussian random variable has the largest entropy. Thus we have arrived to the Maxwell-Boltzmann law, see [23] for more examples.

Given a polytope P that is the intersection of the non-negative orthant with an affine subspace $\mathscr{A}$, let X be a random integer, respectively, 0-1 point in P, sampled from the uniform distribution. Tautologically, X has the largest entropy among all random vectors with values in $\mathbf{Z}_+^n$, respectively $\{0, 1\}^n$, which, additionally, land in $\mathscr{A}$. We approximate X by a more manageable random vector that has the largest entropy among all random vectors with values in $\mathbf{Z}_+^n$, respectively $\{0, 1\}^n$, and with expectation in $\mathscr{A}$. In the next several sections, we present some evidence that this approximation is often accurate.

2.2 A Rationale for the Gaussian Formula

Now we can provide a rationale for formulas (6) and (11). Suppose that $P = \mathscr{A} \cap \mathbf{R}_+^n$, where the affine subspaces $\mathscr{A}$ is defined by a system of linear equations $Ax = b$, where A is an $r \times n$ integer matrix of $\operatorname{rank}(A) = r < n$ and b is an integer r-vector as in (3). We assume that P is bounded and has a non-empty interior. Let $X = (X_1, \ldots, X_n)^T$ be the random vector of Lemma 1, which we interpret as a column n-vector. From Lemma 1, we can write

$$|P \cap \mathbf{Z}^n| = e^{g(z)}\mathbf{P}(AX = b). \tag{16}$$

Let $a_1, \ldots, a_n$ be the columns of A so that $A = [a_1, \ldots, a_n]$. From Lemma 1 we have

$$\mathbf{E}(AX) = \sum_{k=1}^{n} \mathbf{E}(X_k) a_k = \sum_{k=1}^{n} z_k a_k = Az = b,$$

while the covariance matrix $B = (b_{ij})$ of AX is computed as

$$b_{ij} = \mathbf{cov}\left(\sum_{k=1}^{n} a_{ik} X_k, \sum_{k=1}^{n} a_{jk} X_k\right) = \sum_{1 \leq k_1, k_2 \leq n} a_{ik_1} a_{jk_2} \mathbf{cov}\left(X_{k_1}, X_{k_2}\right)$$
$$= \sum_{k=1}^{n} a_{ik} a_{jk} \mathbf{var}(X_k) = \sum_{k=1}^{n} a_{ik} a_{jk} \left(z_k + z_k^2\right).$$

The random variable $AX = X_1 a_1 + \ldots + X_n a_n$ is a linear combination of vectors $a_1, \ldots, a_n$ with independent random coefficients $X_1, \ldots, X_n$, so it is not unreasonable to suspect that if $n \gg r$ then the distribution of AX in the vicinity of its expectation b can be close to Gaussian with covariance matrix B. Then in the spirit of the Local Central Limit Theorem we approximate in (16)

$$\mathbf{P}(AX = b) \approx \frac{\det \Lambda}{(2\pi)^{r/2} \sqrt{\det B}}, \tag{17}$$

where $d = n - r$ and Λ is the lattice in $\mathbf{R}^r$ generated by the columns of A. The right hand side of (17) is an estimate of the probability that a Gaussian random vector with expectation b and covariance matrix B lands in the Voronoi region of b consisting of the points in $\mathbf{R}^r$ that are closer to b than to any other point of Λ. The volume of the Voronoi region of b is exactly $\det \Lambda$, and we assume that the Gaussian density does not change much across the Voronoi region from its maximum value at b. Hence we obtain (6) from (17).

We note that in Example 1.2 we approximate a sum of 16 independent random 7-vectors by a Gaussian random vector while in Example 1.3 we approximate a sum of 27 independent random 7-vectors by a Gaussian random vector. Not surprisingly, in the latter case we get a better precision. In Example 1.4, we approximate a sum of 136 independent random 17-vectors by a Gaussian random vector.

Similarly, let $X = (X_1, \ldots, X_n)^T$ be the random vector of Lemma 2, which we interpret as a column n-vector. From Lemma 2, we can write

$$|P \cap \{0, 1\}^n| = e^{h(z)} \mathbf{P}(AX = b). \tag{18}$$

As above, we have $AX = b$ while the covariance matrix $B = (b_{ij})$ of AX is defined by

$$b_{ij} = \sum_{k=1}^{n} a_{ik} a_{jk} \mathbf{var}(X_k) = \sum_{k=1}^{n} a_{ik} a_{jk} \left(z_k - z_k^2\right)$$

and we obtain (11) as above.

Before we discuss the validity of the Gaussian formulas (6) and (11), we discuss some bounds on the number of integer and 0-1 points that follow from (16) and (18).

3 Concentration and Anti-Concentration Bounds for the Number of Integer Points

One immediate corollary of formulas (16) and (18) are the bounds

$$|P \cap \mathbf{Z}^n| \leq e^{g(z)} \quad \text{and} \quad |P \cap \{0,1\}^n| \leq e^{h(z)}, \tag{19}$$

where z is the solution of the optimization problems (4)–(5) and (9)–(10), respectively. One can show that the bounds (19) capture the logarithmic order of $|P \cap \mathbf{Z}^n|$, respectively $|P \cap \{0,1\}^n|$ asymptotically, as the dimension grows, for the polytopes of non-negative integer [2], respectively, 0-1 matrices [3] with prescribed row and column sums, except in sparse cases, see also Section 5.4.

It was noticed in [34] that the bounds (19) can be improved if one uses anti-concentration inequalities. The following result is obtained in [34].

Theorem 1 *Let $P \subset \mathbf{R}^n$ be a polytope that is the intersection of the non-negative orthant $\mathbf{R}^n_+$ and an affine subspace defined by a system of linear equations $Ax = b$, where A is an $r \times n$ matrix of rank r and b is an r-vector. Suppose that P contains a point $x = (x_1, \ldots, x_n)$ where $x_k > 0$ for $k = 1, \ldots, n$ and let $z = (z_1, \ldots, z_n)$ be the unique point at which the strictly concave function*

$$g(x) = \sum_{k=1}^{n} \Big((x_k + 1) \ln (x_k + 1) - x_k \ln x_k \Big) \quad \text{for} \quad x \in \mathbf{R}^n_+, \quad x = (x_1, \ldots, x_n),$$

attains its maximum on P. Then

$$|P \cap \mathbf{Z}^n| \leq e^{g(z)} \min_{k_1, \ldots, k_r} \frac{1}{(1 + z_{k_1}) \cdots (1 + z_{k_r})},$$

where the minimum is taken over all collections $1 \leq k_1, \ldots, k_r \leq n$ of indices of linearly independent sets of columns of the matrix A.

For example, for the polytope P in Example 1.2, the bound of Theorem 1 is off by a factor of about 5,800 from the true value. The proof of Theorem 1 follows from Lemma 1 and some simple anti-concentration bounds.

Definition 1 Let X be a discrete random vector taking values in $\mathbf{R}^n$. We define the *concentration constant* of X as

$$\gamma(X) = \max_{u \in \mathbf{R}^n} \mathbf{P}(X = u).$$

In words, $\gamma(X)$ is the largest probability of a value of X.

Lemma 3 *Let X and Y be independent discrete random variables with values in* $\mathbf{R}^n$. *Then*

$$\gamma(X+Y) \leq \min\{\gamma(X), \gamma(Y)\}.$$

Proof Let w be a value of $X+Y$. Then

$$\mathbf{P}(X+Y=w) = \sum_{u,v:\, u+v=w} \mathbf{P}(X=u)\mathbf{P}(Y=v) \leq \gamma(X) \sum_{v} \mathbf{P}(Y=v) = \gamma(X)$$

and hence $\gamma(X+Y) \leq \gamma(X)$. Similarly, $\gamma(X+Y) \leq \gamma(Y)$ and the proof follows. □

Now we are ready to prove Theorem 1.

Proof (of Theorem 1) We use Lemma 1. From (16), we have

$$|P \cap \mathbf{Z}^n| = e^{g(z)}\mathbf{P}(AX=b) \leq e^{g(z)}\gamma(AX).$$

Here $AX = X_1 a_1 + \ldots + X_n a_n$, where $a_1, \ldots, a_n$ are the columns of A and $X_1, \ldots, X_n$ are independent geometric random variables with $\mathbf{E}(X_k) = z_k$ for $k = 1, \ldots, n$. In particular,

$$\gamma(X_k) = \mathbf{P}(X_k = 0) = \frac{1}{1+z_k}.$$

Suppose that columns $a_{k_1}, \ldots, a_{k_r}$ are linearly independent. Then

$$\gamma\left(X_{k_1} a_{k_1} + \ldots + X_{k_r} a_{k_r}\right) \leq \gamma\left(X_{k_1}\right) \cdots \gamma\left(X_{k_r}\right) = \frac{1}{(1+z_{k_1}) \cdots (1+z_{k_r})}$$

and the proof follows by Lemma 3. □

Similarly, one obtains an upper bound for the number of 0-1 points in a polytope.

Theorem 2 *Let* $P \subset \mathbf{R}^n$ *be a polytope that is the intersection of the unit cube* $[0,1]^n$ *and the affine subspace defined by a system of linear equations* $Ax = b$, *where* A *is an* $r \times n$ *matrix of rank* r *and* b *is an* r*-vector. Suppose that* P *contains a point* $x = (x_1, \ldots, x_n)$ *where* $x_k > 0$ *for* $k = 1, \ldots, n$ *and let* $z = (z_1, \ldots, z_n)$ *be the unique point at which the strictly concave function*

$$h(x) = \sum_{k=1}^{n} \left(x_k \ln \frac{1}{x_k} + (1-x_k) \ln \frac{1}{1-x_k} \right) \quad \text{for} \quad x \in [0,1]^n, \quad x = (x_1, \ldots, x_n),$$

attains its maximum on P. *Then*

$$|P \cap \{0,1\}^n| \leq e^{h(z)} \min_{k_1, \ldots, k_r} \max\left\{z_{k_1}, 1-z_{k_1}\right\} \cdots \max\left\{z_{k_r}, 1-z_{k_r}\right\},$$

where the outer minimum is taken over all collections $1 \leq k_1, \ldots, k_r \leq n$ *of indices of linearly independent sets of columns of the matrix A.*

We note that having $z = (z_1, \ldots, z_n)$ computed, the minima in Theorem 1 and Theorem 2 can be found efficiently in $O(rn)$ time by a greedy algorithm of finding a minimum weight basis of the matroid represented by the matrix A. Namely, first we find a non-zero column a_{k_1} of the minimum weight $1/(1 + z_{k_1})$ in Theorem 1, respectively, weight $\max\{z_{k_1}, 1 - z_{k_1}\}$ in Theorem 2, and at each step we select a column a_{k_s} of the minimum weight among all columns that are linearly independent of the previously selected columns, cf., for example, Section 12.4 of [33].

The bounds of Theorem 1 and Theorem 2 are related to the Littlewood-Offord problem, see, for example, Chapter 7 of [35].

In general, one cannot hope to obtain non-trivial lower bounds for the number of integer or 0-1 points in P that would depend smoothly on the matrix A and vector b defining the affine span of P, since P may fail to contain any integer point because of some arithmetic issues (for example, if the integer points in P enumerate 17-regular graphs with 1001 vertex, cf. (15)). However, if we are willing to "fatten" the polytope a bit, we can obtain some lower bounds via concentration inequalities. The case of 0-1 points is particularly simple.

In what follows, $\|\cdot\|_\infty$ is the usual ℓ^∞ norm in $\mathbf{R}^r$.

Theorem 3 *Let the polytope P, matrix* $A = (a_{ij})$ *of size* $r \times n$*, vector b and point* $z = (z_1, \ldots, z_n)$ *be as in Theorem 2. Let*

$$\alpha(A) = \max_{i=1,\ldots,r} \sqrt{\sum_{j=1}^{n} a_{ij}^2} \quad \textit{and} \quad \omega(z) = \sqrt{\sum_{k=1}^{n} \ln^2\left(\frac{z_k}{1 - z_k}\right)}.$$

Let

$$\widehat{P} = \left\{x \in [0, 1]^n : \quad \|Ax - b\|_\infty \leq \alpha(A)\sqrt{1 + \ln r}\right\}.$$

Then

$$\left|\widehat{P} \cap \{0, 1\}^n\right| \geq \frac{1}{3} e^{h(z) - \omega(z)}.$$

Proof Let $b = (\beta_1, \ldots, \beta_r)$ and let $X_1, \ldots, X_n$ be the random variables of Lemma 2. Applying the Hoeffding inequality, see, for example, Theorem 5.2 of [29], we obtain

$$\mathbf{P}\left(\left|\sum_{k=1}^{n} a_{ik} X_k - \beta_i\right| \geq \alpha(A)\sqrt{1 + \ln r}\right) \leq \frac{1}{3r^2} \quad \text{for all} \quad i = 1, \ldots, r. \qquad (20)$$

Similarly, from the Hoeffding inequality it follows that

$$\mathbf{P}\left(\sum_{k=1}^{n}\left(X_k \ln z_k + (1 - X_k)\ln(1 - z_k)\right) \geq -h(z) + \omega(z)\right) \leq \frac{1}{3}. \tag{21}$$

Combining (20)–(21) we conclude that

$$\mathbf{P}\Bigg(\max_{i=1,\dots,r}\left|\sum_{k=1}^{n} a_{ik}X_k - \beta_i\right| \leq \alpha(A)\sqrt{1+\ln r} \quad \text{and}$$
$$\sum_{k=1}^{n}\left(X_k \ln z_k + (1 - X_k)\ln(1 - z_k)\right) \leq -h(z) + \omega(z)\Bigg) \geq \frac{1}{3}.$$

Since

$$\mathbf{P}\left(X_1 = x_1, \dots, X_n = x_n\right) = \prod_{k=1}^{n} z_k^{x_k}(1 - z_k)^{1-x_k} \quad \text{for all} \quad (x_1, \dots, x_n) \in \{0, 1\}^n,$$

the proof follows. □

If the coordinates z_k remain separated from 0 and 1 and are of about the same order, the asymptotic of $\ln|\widehat{P} \cap \{0,1\}^n|$ as n grows is captured by $h(z)$ and the error $\alpha(A)\sqrt{1+\ln r}$ is of a smaller order than the entries of the vector b.

A similar to Theorem 3 result can be obtained for integer points, but since we are dealing with unbounded random variables the bounds are not so good as in the 0-1 case.

Theorem 4 *Let the polytope P, matrix $A = \left(a_{ij}\right)$ of size $r \times n$, vector b and point $z = (z_1, \dots, z_n)$ be as in Theorem 1. Suppose that $|a_{ij}| \leq 1$ for all i, j and that every row of A contains not more than $m \geq 1$ non-zero entries. Suppose further that*

$$\delta \leq z_k \leq \zeta \quad \textit{for} \quad k = 1, \dots, n$$

for some $\delta \leq 1$ and $\zeta \geq 1$. Let

$$\tau(A, z) = \zeta \ln(3 + \ln r)\sqrt{m} \quad \textit{and} \quad \omega(z) = \frac{3\zeta}{\delta}\sqrt{n}$$

and let

$$\widehat{P} = \left\{x \in \mathbf{R}^n_+ : \quad \|Ax - b\|_\infty \leq \tau(A, z)\right\}.$$

Then

$$\left|\widehat{P} \cap \mathbf{Z}^n\right| \geq \frac{1}{3}e^{g(z)-\omega(z)}.$$

Proof Let $b = (\beta_1, \ldots, \beta_r)$ and let $X_1, \ldots, X_n$ be the random variables of Lemma 1.

We need some concentration inequalities for geometric random variables, see for example, [4]. If X is geometric random variable with $\mathbf{E}(X) = z$, then

$$\mathbf{E}\left(e^{-\lambda X}\right) \leq \exp\left\{-\lambda z + \frac{\lambda^2}{2}\left(z+z^2\right)\right\} \quad \text{for} \quad 0 \leq \lambda \leq 2 \qquad \text{and}$$

$$\mathbf{E}\left(e^{\lambda X}\right) \leq \exp\left\{\lambda z + 2\lambda^2\left(z+z^2\right)\right\} \quad \text{for} \quad 0 \leq \lambda \leq \min\{1/3, 1/2z\}.$$

It follows that

$$\mathbf{P}\left(\left|\sum_{j=1}^{n} a_{ij}X_j - \beta_i\right| \geq \tau\right) \ \leq \ 2\exp\left\{-\lambda\tau + 2\lambda^2\left(z+z^2\right)m\right\}$$

for any $0 \leq \lambda \leq \min\{1/3, 1/2\zeta\}$, any $\tau \geq 0$ and all $i = 1, \ldots, r$. Choosing

$$\lambda = \frac{1}{3\zeta\sqrt{m}},$$

we conclude that

$$\mathbf{P}\left(\left|\sum_{j=1}^{n} a_{ij}X_j\right| \ \geq \ \tau(A, z)\right) \ \leq \ \frac{1}{5r} \quad \text{for} \quad i = 1, \ldots, r. \tag{22}$$

Similarly we obtain

$$\mathbf{P}\left(\sum_{k=1}^{n}\left(\ln\frac{1}{1+z_k} + X_k \ln\frac{z_k}{z_k+1}\right) \ \geq \ -g(z) + \omega(z)\right) \ \leq \ \frac{1}{3}. \tag{23}$$

Combining (22)–(23), we conclude that

$$\mathbf{P}\Bigg(\max_{i=1,\ldots,r}\left|\sum_{j=1}^{n} a_{ij}X_j\right| \ \leq \ \tau(A, z) \quad \text{and}$$

$$\sum_{k=1}^{n}\left(\ln\frac{1}{1+z_k} + X_k \ln\frac{z_k}{z_k+1}\right) \ \leq \ -g(z) + \omega(z)\Bigg) \ \geq \ \frac{1}{3}.$$

Since

$$\mathbf{P}\left(X_1 = x_1, \ldots, X_n = x_n\right) = \prod_{k=1}^{n}\left(\frac{1}{z_k+1}\left(\frac{z_k}{z_k+1}\right)^{X_k}\right)$$

$$\text{for all} \quad (x_1, \ldots, x_n) \in \mathbf{Z}_+^n,$$

the proof follows. □

Again, as long as z_k remain separated from 0 and are of about the same order, the asymptotic of $\ln |\widehat{P} \cap \mathbf{Z}^n|$ as n grows is captured by $g(z)$ while the error $\tau(A, z)$ is of a smaller order than the coordinates of b.

4 The Fourier analysis for the Gaussian Formulas

To establish the validity of the Gaussian formulas (6) and (11), we take a closer look at the probabilities $\mathbf{P}(AX = b)$ in the formulas (16) and (18). Lemma 4 and Lemma 5 below provide some identities that express the probabilities as integrals of the characteristic function of X [5].

Lemma 4 *Let the polytope $P \subset \mathbf{R}^n$, point $z = (z_1, \ldots, z_n)$ and vector $X = (X_1, \ldots, X_n)$ of independent geometric random variables be as in Lemma 1.*

Suppose that $P = \mathscr{A} \cap \mathbf{R}^n_+$, where $\mathscr{A}$ is the affine subspace defined by a system of linear equations $Ax = b$, where A is an $r \times n$ integer matrix of $\operatorname{rank}(A) = r$ and b is an integer r-vector. Let $a_1, \ldots, a_n$ be the columns of A, interpreted as integer r-vectors.

Let $\Pi \subset \mathbf{R}^r$ be the parallelepiped of the points $t = (t_1, \ldots, t_r)$ such that

$$-\pi \leq t_k \leq \pi \quad \text{for} \quad k = 1, \ldots, r.$$

Then

$$\mathbf{P}(AX = b) = \frac{1}{(2\pi)^r} \int_\Pi e^{-i\langle t, b\rangle} \prod_{j=1}^n \frac{1}{1 + z_j - z_j e^{i\langle a_j, t\rangle}} \, dt, \tag{24}$$

where $\langle \cdot, \cdot \rangle$ is the inner product in $\mathbf{R}^r$, dt is the Lebesgue measure on Π and $i^2 = -1$.

Proof We have $\mathbf{P}\left(X_j = k\right) = p_j q_j^k$ for $k = 0, 1, \ldots$, and some $p_j, q_j > 0$ such that $p_j + q_j = 1$ and $z_j = q_j / p_j$. Hence

$$\prod_{j=1}^n \frac{1}{1 + z_j - z_j e^{i\langle a_j, t\rangle}} = \prod_{j=1}^n \frac{p_j}{1 - q_j e^{i\langle a_j, t\rangle}}$$

$$= p_1 \cdots p_n \sum_{k_1, \ldots, k_n \geq 0} q_1^{k_1} \cdots q_n^{k_n} \exp\left\{ i \sum_{j=1}^n k_j \langle a_j, t\rangle \right\}.$$

Since for an integer c, we have

$$\frac{1}{2\pi} \int_{-\pi}^{\pi} e^{ict} \, dt = \begin{cases} 1 & \text{if } c = 0, \\ 0 & \text{if } c \neq 0, \end{cases}$$

we have

$$\frac{1}{(2\pi)^r}\int_{\Pi} e^{-i\langle t,b\rangle}\exp\left\{i\sum_{j=1}^{n} k_j\langle a_j,t\rangle\right\}\,dt=\begin{cases}1 & \text{if } \sum_{j=1}^{n} k_j a_j=b,\\ 0 & \text{if } \sum_{j=1}^{n} k_j a_j\neq b,\end{cases}$$

and

$$\frac{1}{(2\pi)^r}\int_{\Pi} e^{-i\langle t,b\rangle}\prod_{j=1}^{n}\frac{1}{1+z_j-z_je^{i\langle a_j,t\rangle}}\,dt$$
$$=p_1\cdots p_n\sum_{\substack{k_1,\ldots,k_n\geq 0\\ k_1a_1+\ldots+k_na_n=b}} q_1^{k_1}\cdots q_n^{k_n}=\mathbf{P}(AX=b).$$

□

Similarly, in the case of 0-1 points, we obtain the following result.

Lemma 5 *Let the polytope $P\subset\mathbf{R}^n$, point $z=(z_1,\ldots,z_n)$ and vector $X=(X_1,\ldots,X_n)$ of independent Bernoulli random variables be as in Lemma 2.*

Suppose that $P=\mathscr{A}\cap[0,1]^n$, where $\mathscr{A}$ is an affine subspace in $\mathbf{R}^n$ defined as in Lemma 4 and let $\Pi\subset\mathbf{R}^r$ be the parallelepiped as in Lemma 4.
Then

$$\mathbf{P}(AX=b)=\frac{1}{(2\pi)^r}\int_{\Pi} e^{-i\langle t,b\rangle}\prod_{j=1}^{n}\left(1-z_j+z_je^{i\langle a_j,t\rangle}\right)\,dt. \tag{25}$$

We obtain the Gaussian formulas (6) and (11) if we show that the bulk of the contribution of the integrals (24), respectively (25) come from an approximation of the integrand in a small neighborhood of $t=0$ in Π.

4.1 Integrals in a Neighborhood of $t=0$

For $t\approx 0$, using that $\sum_{j=1}^{n} z_ja_j=b$, we obtain from the Taylor series expansion

$$e^{-i\langle t,b\rangle}\prod_{j=1}^{n}\frac{1}{1+z_j-z_je^{i\langle a_j,t\rangle}}=\exp\left\{-q(t)-if(t)+h(t)+O(t^5)\right\},$$

where

$$q(t)=\frac{1}{2}\sum_{j=1}^{n}\left(z_j+z_j^2\right)\langle a_j,t\rangle^2, \tag{26}$$

$$f(t) = \frac{1}{6} \sum_{j=1}^{n} z_j(z_j+1)(2z_j+1)\langle a_j, t\rangle^3 \quad \text{and} \tag{27}$$

$$h(t) = \frac{1}{24} \sum_{j=1}^{n} z_j(z_j+1)(6z_j^2+6z_j+1)\langle a_j, t\rangle^4. \tag{28}$$

Similarly,

$$e^{-i\langle t,b\rangle} \prod_{j=1}^{n} \left(1 - z_j + z_j e^{i\langle a_j, t\rangle}\right) = \exp\left\{-q(t) - if(t) + h(t) + O(t^5)\right\}$$

where

$$q(t) = \frac{1}{2} \sum_{j=1}^{n} \left(z_j - z_j^2\right) \langle a_j, t\rangle^2, \tag{29}$$

$$f(t) = \frac{1}{6} \sum_{j=1}^{n} z_j(1-z_j)(1-2z_j)\langle a_j, t\rangle^3 \quad \text{and} \tag{30}$$

$$h(t) = \frac{1}{24} \sum_{j=1}^{n} z_j(1-z_j)(6z_j^2-6z_j+1)\langle a_j, t\rangle^4. \tag{31}$$

When $\det \Lambda = 1$, the Gaussian formulas (6) and (11) state that

$$\mathbf{P}(AX = b) \approx \frac{1}{(2\pi)^r} \int_{\mathbf{R}^r} e^{-q(t)}\, dt$$

for the quadratic forms (26), respectively (29). To prove that $q(t)$ dominates higher order terms in a neighborhood of $t = 0$, it usually suffices to show that the eigenvalues of $q(t)$ are sufficiently large, and that, in turn, depends on the metric properties of the polytope P.

To prove that the integrals (24) and (25) are negligible outside of a neighborhood of $t = 0$, we need to use arithmetic properties of P. The following result from [5] comes in handy.

Lemma 6 *Let A be an $r \times n$ integer matrix with columns $a_1, \ldots, a_n$ and let $u_1, \ldots, u_r$ be the standard basis of $\mathbf{R}^r$. For $k = 1, \ldots, r$, let $Y_k \subset \mathbf{Z}^r$ be a non-empty finite set such that $Ay = u_k$ for all $y \in Y_k$. Let ρ_k be the largest eigenvalue of the quadratic form $\psi_k : \mathbf{R}^n \longrightarrow \mathbf{R}$,*

$$\psi_k(x) = \frac{1}{|Y_k|} \sum_{y \in Y_k} \langle y, x\rangle^2.$$

Let $t = (t_1, \ldots, t_k)$ *where* $-\pi \le t_k \le \pi$ *for* $k = 1, \ldots, r$.

- *Suppose that* $z_j - z_j^2 \ge \alpha$ *for some* $\alpha > 0$ *and* $j = 1, \ldots, n$. *Then*

$$\left| \prod_{j=1}^{n} \left(1 - z_j + z_j e^{i\langle a_j, t\rangle}\right) \right| \le \exp\left\{-\frac{\alpha t_k^2}{5\rho_k}\right\}.$$

- *Suppose that* $z_j + z_j^2 \ge \alpha$ *for some* $\alpha > 0$ *and* $j = 1, \ldots, n$. *Then*

$$\left| \prod_{j=1}^{n} \frac{1}{1 + z_j - z_j e^{i\langle a_j, t\rangle}} \right| \le \left(1 + \frac{2}{5}\alpha\pi^2\right)^{-\lfloor t_k^2/\rho_k\pi^2 \rfloor}.$$

Proof We reproduce the proof of the first bound only, since the second bound is proven similarly. Let us denote

$$F(t) = \prod_{j=1}^{n} \left(1 - z_j + z_j e^{i\langle a_j, t\rangle}\right),$$

so that

$$|F(t)|^2 = \prod_{j=1}^{n} \left((1 - z_j)^2 + 2z_j(1 - z_j)\cos\langle a_j, t\rangle + z_j^2\right).$$

For real numbers ξ and η we write $\xi \equiv \eta \bmod 2\pi$ if $\xi - \eta$ is an integer multiple of 2π. Let

$$-\pi \le \gamma_j \le \pi \quad \text{for} \quad j = 1, \ldots, n$$

be reals such that

$$\langle a_j, t\rangle \equiv \gamma_j \mod 2\pi \quad \text{for} \quad j = 1, \ldots, n.$$

Using that

$$\cos\gamma \le 1 - \frac{\gamma^2}{5} \quad \text{for} \quad -\pi \le \gamma \le \pi,$$

we obtain

$$|F(t)|^2 = \prod_{j=1}^{n} \left((1 - z_j)^2 + 2z_j(1 - z_j)\cos\gamma_j + z_j^2\right) = \prod_{j=1}^{n} \left(1 - \frac{2z_j(1 - z_j)}{5}\gamma_j^2\right)$$
$$\le \exp\left\{-\frac{2\alpha}{5}\sum_{j=1}^{n}\gamma_j^2\right\}.$$

Let

$$c = (\gamma_1, \ldots, \gamma_n).$$

Then, for every $y \in Y_k$ we have

$$t_k = \langle u_k, t \rangle = \langle Ay, t \rangle = \langle y, A^T t \rangle \equiv \langle y, c \rangle \mod 2\pi.$$

Since $|t_k| \leq \pi$, we must have $|\langle y, c \rangle| \geq |t_k|$ and, therefore,

$$\sum_{j=1}^{n} \gamma_j^2 = \|c\|^2 \geq \frac{1}{\rho_k} \psi_k(c) = \frac{1}{\rho_k |Y_k|} \sum_{y \in Y_k} \langle y, c \rangle^2 \geq \frac{t_k^2}{\rho_k}$$

and the proof follows. □

Lemma 6 is applicable only if the columns of A generate the whole lattice $\mathbf{Z}^r$, so $\det \Lambda = 1$ in (6) and (11). Moreover, to get ρ_k sufficiently small, $\rho_k \ll 1$, we have to present sufficiently short vectors y distributed sufficiently uniformly in $\mathbf{Z}^n$ such that $Ay = u_k$.

As is shown in [5], Lemma 6 and simple eigenvalue estimates of the quadratic forms (26) and (29) suffice to show that the Gaussian formulas (6) and (11) for the number of multi-way contingency tables, cf. Example 1.3, are asymptotically exact, provided the number of "ways" is at least 5 (hence 3-way tables of Example 1.3 are not covered). It took a substantially more refined analysis of what happens in the neighborhood of 0, to show that the Gaussian formula is asymptotically exact also for 3- and 4- way contingency tables [8]. We state the result of [8] in the case of 3-way tables of Example 1.3, omitting technical bounds on the rate of convergence.

To make the formula (6) applicable, we drop two of the constraints as described in Example 1.3.

Theorem 5 *Let us fix a* $0 < \delta < 1$ *and let* P *be a 3-axial transportation polytope of* $m \times n \times s$ *arrays with positive integer margins* $A = (a_1, \ldots, a_m)$, $B = (b_1, \ldots, b_n)$ *and* $C = (c_1, \ldots, c_s)$ *such that*

$$a_1 + \ldots + a_m = b_1 + \ldots + b_n = c_1 + \ldots + c_s.$$

Suppose that

$$\delta w \leq m, n, s \leq w$$

for some positive integer w. *Suppose further that the solution* $z = (z_{ijk})$ *of the optimization problem (4)–(5) satisfies*

$$\delta \leq z_{ijk} \leq \frac{1}{\delta} \quad \text{for all} \quad i, j, k.$$

Then the Gaussian formula (6) approximates the number $|P \cap \mathbf{Z}^{m \times n \times s}|$ *of integer points in P within a relative error of* $\gamma(\delta)w^{-0.5}$ *for some constant* $\gamma(\delta) > 0$.

Similarly, it is shown in [8] that the Gaussian formula (11) is asymptotically exact for 0-1 points in multi-way transportation polytopes of 3- and more "ways."

4.2 Corrections to the Gaussian Formulas

Curiously, the Gaussian formula (6) is *not* asymptotically exact in the case of 2-way contingency tables of Example 1.2 and the Gaussian formula (11) is *not* asymptotically exact in the case of polytopes of graphs of a given degree sequence of Example 1.4. The formulas need a correction based on the 3rd and 4th moments, also known as the Edgeworth correction, see, for example, [27].

We calculate the correction term as follows. Let q be the quadratic form of (26), respectively (29). Let us consider the Gaussian probability measure on $\mathbf{R}^r$ with density proportional to e^{-q}. We define

$$\mu = \mathbf{E} f^2 \quad \text{and} \quad \nu = \mathbf{E}\, h,$$

for functions $f, h : \mathbf{R}^r \longrightarrow \mathbf{R}$ defined by (27)–(28), respectively, (30)–(31). We define the *corrected Gaussian formula* by

$$|P \cap \mathbf{Z}|^n \approx \frac{e^{g(z)} \det \Lambda}{(2\pi)^{r/2}\sqrt{\det B}} \exp\left\{-\frac{\mu}{2} + \nu\right\} \tag{32}$$

respectively by

$$|P \cap \{0,1\}|^n \approx \frac{e^{h(z)} \det \Lambda}{(2\pi)^{r/2}\sqrt{\det B}} \exp\left\{-\frac{\mu}{2} + \nu\right\}. \tag{33}$$

We note that it is easy to compute μ and ν, as it is easy to integrate polynomials of a low degree over the Gaussian measure. The following result for the number of 2-way contingency tables was obtained in [6]. Again, to make the formula (6) applicable, we drop one row or column sum constraint, as is described in Example 1.2.

Theorem 6 *Let us fix* $0 < \delta < 1$ *and let P be the polytope of* $m \times n$ *non-negative integer matrices with positive integer row sums* $R = (r_1, \ldots, r_m)$ *and column sums* $C = (c_1, \ldots, c_n)$ *such that*

$$r_1 + \ldots + r_m = c_1 + \ldots + c_n.$$

Suppose that $\delta m \leq n$ *and* $\delta n \leq m$ *and that the solution* $z = (z_{ij})$ *to the optimization problem (4)–(5) satisfies*

$$\delta s \leq z_{ij} \leq s \quad \textit{for all} \quad i, j$$

and some $s \geq 1$. Then the corrected Gaussian formula (32) approximates the number $|P \cap \mathbf{Z}^{m\times n}|$ of integer points in P within a relative error of $(m+n)^{-\gamma(\delta)}$ for some $\gamma(\delta) > 0$.

The proof of Theorem 6 requires a more careful analysis of the integral (24) outside of $t = 0$ than that provided by Lemma 6 although the main action is in the analysis of the integral in a neighborhood of $t = 0$. It turns out that the function $f(t)$ defined by (27) behaves roughly as a Gaussian random variable on the probability space of $\mathbf{R}^r$ endowed with probability density proportional to e^{-q} for the quadratic form (26). This happens because many of the linear functions $\langle a_j, t\rangle$ entering (27) are weakly correlated. Hence we have

$$\mathbf{E}\, e^{-if} \approx \exp\left\{-\frac{1}{2}\mathbf{E} f^2\right\} = e^{-\mu/2}.$$

Similarly, the function $h(t)$ defined by (28) is almost constant on that probability space, so we have

$$\mathbf{E}\, e^{h} \approx \exp\left\{\mathbf{E}\, h\right\} = e^{\nu}.$$

Earlier a similar to the Theorem 6 result was obtained, in a different language but by using the asymptotic analysis of a similar kind, in [11] in the particular case when all row sums are (almost) equal and all column sums are (almost) equal: $r_1 = \ldots = r_m$ and $c_1 = \ldots = c_n$.

The following result for the number of graphs with a given degree sequence, see Example 1.4, is obtained in [7].

Theorem 7 *Let us fix $0 < \delta < 1/2$ and let P be the polytope of $n \times n$ symmetric matrices with zero diagonal and positive integer row (column) sums $d_1, \ldots, d_n$ satisfying $d_1 + \ldots + d_n \equiv 0 \bmod 2$. Suppose that the solution $z = (z_{ij})$ to the optimization problem (9)–(10) satisfies*

$$\delta \leq z_{ij} \leq 1-\delta \quad \textit{for all} \quad 1 \leq i \neq j \leq n.$$

Then the corrected Gaussian formula (33) approximates the number $|P \cap \{0,1\}^{m\times n}|$ of 0-1 points in P within a relative error of $(m+n)^{-\gamma(\delta)}$ for some $\gamma(\delta) > 0$.

Curiously, in this case points the outside of a neighborhood the origin (more precisely, the corners of the parallelepiped Π) contribute to the integral (25), see Section 4.1. If $d_1 + \ldots + d_n \equiv 0 \bmod 2$, the contribution from the outside doubles the contribution of the origin while if $d_1 + \ldots + d_n \equiv 1 \bmod 2$, the contribution from the outside cancels the contribution from the origin, making, predictably, the integral 0.

A similar to the Theorem 7 result in a different language but by using the asymptotic analysis of a similar kind was obtained earlier in [30] in the particular case when degrees $d_1, \ldots, d_n$ are (almost) equal.

5 Concluding Remarks and Open Questions

5.1 *Why Are the Formulas (Sometimes) So Precise?*

The Gaussian formulas (6) and (11) provide amazingly close estimates in Examples 1.2–1.4 and in many other cases, see [14]. Although we can prove that in Example 1.3 the Gaussian formula is asymptotically exact while in Examples 1.2 and 1.4 the corrected Gaussian formulas are asymptotically exact, the error bounds we are able to prove are nowhere close to the actual errors. It would be nice to develop new methods that would allow us to bound errors more accurately.

5.2 *Is the Uniformity Condition for z_k Really Necessary?*

To obtain asymptotic estimates in Theorems 4, 5, 6 and 7, we require the coordinates z_k in the optimal solution of the optimization problems (4)–(5) and (9)–(10) to be within a constant factor of each other, and only in Theorem 3 we allow ourselves more flexibility. It is not clear to what extent that uniformity condition is really necessary. In the case of graphs with a given degree sequence, see Example 1.4 and Theorem 7, the conditions are satisfied if the degree sequence $D = (d_1, \ldots, d_n)$ lies sufficiently deep inside the polytope defined by the Erdős–Gallai conditions (14). This can be related to the fact that the number of graphs with degree sequence D begins to oscillate rapidly when D approaches the boundary of the conditions (14) and hence there is no chance to approximate that number by an analytic formula such as the Gaussian formula (11), see [7] for a discussion.

The case of 2-way contingency tables of Example 1.2 and Theorem 6 is more mysterious. It is shown in [4] that for the margins $R = (3n, n, \ldots, n)$ and $C = (3n, n, \ldots, n)$ of an $n \times n$ contingency table, in the solution $Z = (z_{ij})$ of the optimization problem (4)–(5), which is naturally arranged into an $n \times n$ matrix, the entries z_{ij} are *not* uniform: z_{11} grows linearly with n while all other entries remain bounded by a constant. Curiously, for the margins $R' = (2n, n, \ldots, n)$ and $C' = (2n, n, \ldots, n)$ of an $n \times n$ table, all entries z_{ij} stay bounded by a constant as n grows. It is not clear whether there is indeed a sharp phase transition in the number of contingency tables as the margins evolve from (R', C') to (R, C) which would preclude the existence of an analytic formula approximating the number of tables across the transition.

5.3 *What a Random Integer or 0-1 Point of a Polytope Looks Like?*

Reasoning along similar lines as in the proofs of Theorem 3 and Theorem 4, one can show that if $\ln |P \cap \mathbf{Z}^n| \approx g(z)$ then a random integer point sampled from the uniform distribution on $P \cap \mathbf{Z}^n$ looks roughly like a vector of geometric random variables in Lemma 1 as far as various statistics averaging over several coordinates are concerned. In [4] it is shown that a random 2-way contingency table with given margins indeed behaves like that under some technical conditions. Similarly, if $\ln |P \cap \{0, 1\}^n| \approx h(z)$ then a random 0-1 point sampled from the uniform distribution in $P \cap \{0, 1\}^n$ looks roughly like a vector of independent Bernoulli random variables of Lemma 2 with respect to averaging statistics. In [7] it is shown that a random graph with a given degree sequence behaves like that under some technical conditions (for bipartite graphs, one obtains similar, but much more robust results [3]).

What is not clear is the true extent at which the distributions match. For example, is it true that an individual coordinates of a random point in $P \cap \mathbf{Z}^n$ behaves roughly as a geometric random variable? In relation to the example of Section 5.2, is it true that the expectation of the $(1, 1)$ entry of a random non-negative integer $n \times n$ matrix with row sums $(3n, n, \ldots, n)$ and column sums $(3n, n, \ldots, n)$ grows linearly in n? By definition, the k-th coordinate of a random point in $P \cap \{0, 1\}^n$ is a Bernoulli random variable, but one can ask whether its expectation is close to z_k, where $z = (z_1, \ldots, z_n)$ is the solution in the optimization problem (9)–(10).

Finally, it would be interesting to find out whether knowing optimal solutions z in problems (4)–(5) and (9)–(10) can help in solving problems of discrete (integer, respectively Boolean) optimization. If a random point in $P \cap \mathbf{Z}^n$ (a random point in $P \cap \{0, 1\}^n$) behaves as a vector of independent geometric random variables (as a vector of independent Bernoulli random variables) with expectation z, then z is roughly the average of integer points in P (0-1 points in P) and if we are interested in finding at least one integer (0-1) point in P, it may make sense to look for such a point near z. Recall that z is the solution of a convex optimization problem and hence can be found easily.

5.4 *What If the Gaussian Model is Not Applicable?*

As follows from Section 2.2, the Gaussian model may work only for reasonably "dense" instances, where the optimal solution $z = (z_1, \ldots, z_n)$ in problems (4)–(5) and (9)–(10) has entries bounded below away from 0. For example, if we try to apply the Gaussian formula (11) to estimate the number of $n \times n$ matrices with 0-1 entries and row and column sums equal 1, we get a rather poor approximation (there are $n!$ such matrices). In this case, for the Bernoulli random variables X_{ij} of Lemma 2 we have $\mathbf{E}\left(X_{ij}\right) = 1/n$ and hence the random vector AX in Lemma 2 is much closer to the vector of independent Poisson random variables than to a vector of Gaussian random variables. It would be interesting to develop the Poisson model in the general sparse case.

We note that there are many results on the enumeration of 0-1 and integer matrices with prescribed margins and graphs with a given degree sequence in the sparse case, that is, where the margins and degrees are relatively small, see [21], [20], [9] and the references therein.

Acknowledgements This work was partially supported by NSF Grant DMS 1361541.

References

1. Barvinok, A.: Integer Points in Polyhedra. Zurich Lectures in Advanced Mathematics. European Mathematical Society (EMS), Zürich (2008)
2. Barvinok, A.: Asymptotic estimates for the number of contingency tables, integer flows, and volumes of transportation polytopes. International Mathematics Research Notices. IMRN **2009**, no. 2, 348–385 (2009)
3. Barvinok, A.: On the number of matrices and a random matrix with prescribed row and column sums and 0-1 entries. Advances in Mathematics **224**, no. 1, 316–339 (2010)
4. Barvinok, A.: What does a random contingency table look like? Combinatorics, Probability and Computing **19**, no. 4, 517–539 (2010)
5. Barvinok, A., Hartigan, J. A.: Maximum entropy Gaussian approximations for the number of integer points and volumes of polytopes. Advances in Applied Mathematics **45**, no. 2, 252–289 (2010)
6. Barvinok, A., Hartigan, J. A.: An asymptotic formula for the number of non-negative integer matrices with prescribed row and column sums. Transactions of the American Mathematical Society **364**, no. 8, 4323–4368 (2012)
7. Barvinok, A., Hartigan, J. A.: The number of graphs and a random graph with a given degree sequence. Random Structures & Algorithms **42**, no. 3, 301–348 (2013)
8. Benson-Putnins, D.: Counting integer points in multi-index transportation polytopes. Preprint `arXiv:1402.4715` (2014)
9. Blinovsky, V., Greenhill, C.: Asymptotic enumeration of sparse uniform hypergraphs with given degrees. European Journal of Combinatorics **51**, 287–296 (2016)
10. Brualdi, R. A., Ryser, H. J.: Combinatorial Matrix Theory. Encyclopedia of Mathematics and its Applications, **39**. Cambridge University Press, Cambridge (1991)
11. Canfield, E.R., McKay B.D.: Asymptotic enumeration of integer matrices with large equal row and column sums. Combinatorica **30**, no. 6, 655–680 (2010)
12. Christandl, M., Doran, B., Walter, M. : Computing multiplicities of Lie group representations. In: 2012 IEEE 53rd Annual Symposium on Foundations of Computer Science – FOCS 2012 , pp. 639–648, IEEE Computer Society, Los Alamitos, CA, (2012)
13. De Loera, J. A.: The many aspects of counting lattice points in polytopes. Mathematische Semesterberichte **52**, no. 2, 175–195 (2005)
14. De Loera, J.A.: Appendix: details on experiments (counting and estimating lattice points). Optima **81**, 17–22 (2009)
15. De Loera, J.A., Hemmecke, R., Tauzer, J., Yoshida, R.: Effective lattice point counting in rational convex polytopes. Journal of Symbolic Computation **38**, no. 4, 1273–1302 (2004)
16. De Loera, J.A., Kim, E.D.: Combinatorics and geometry of transportation polytopes: an update. In: Discrete Geometry and Algebraic Combinatorics, Contemporary Mathematics **625**, pp. 37–76, American Mathematical Society, Providence, R.I. (2014)
17. Diaconis, P., Efron, B.: Testing for independence in a two-way table: new interpretations of the chi-square statistic. With discussions and with a reply by the authors. The Annals of Statistics **13**, no. 3, 845–913 (1985)
18. Dyer, M.: Approximate counting by dynamic programming. In: Proceedings of the Thirty-Fifth Annual ACM Symposium on Theory of Computing, pp. 693–699, ACM, New York (2003)
19. Good, I. J.: Maximum entropy for hypothesis formulation, especially for multidimensional contingency tables. Annals of Mathematical Statistics **34**, 911–934 (1963)
20. Greenhill, C., McKay, B. D.: Asymptotic enumeration of sparse nonnegative integer matrices with specified row and column sums. Advance in Applied Mathematics **41**, no. 4, 459–481 (2008)

21. Greenhill, C., McKay, B. D., Wang, X.: Asymptotic enumeration of sparse 0-1 matrices with irregular row and column sums. Journal of Combinatorial Theory. Series A **113**, no. 2, 291–324 (2006)
22. Grötschel, M., Lovász, L., Schrijver, A. Geometric Algorithms and Combinatorial Optimization. Second edition. Algorithms and Combinatorics, **2**, Springer-Verlag, Berlin (1993)
23. Jaynes, E. T.: Information theory and statistical mechanics. Physical Review (2) **106**, 620–630 (1957)
24. Kannan, R., Vempala, S.: Sampling lattice points. In: Proceedings of the Twenty-Ninth Annual ACM Symposium on the Theory of Computing, pp. 696–700, ACM, New York (1999)
25. Khinchin, A. I.: Mathematical Foundations of Information Theory. Translated by R. A. Silverman and M. D. Friedman, Dover Publications, Inc., New York, N. Y. (1957)
26. Klebanov, V.: Precise quantitative information flow analysis – a symbolic approach. Theoretical Computer Science **538**, 124–139 (2014)
27. Kolassa, J. E.: Series Approximation Methods in Statistics. Second edition. Lecture Notes in Statistics, **88**, Springer-Verlag, New York (1997)
28. Lepelley, D., Louichi, A., Smaoui, H.: On Ehrhart polynomials and probability calculations in voting theory. Social Choice and Welfare **30**, issue 3, 363–383 (2008)
29. McDiarmid, C.: On the method of bounded differences. In: Surveys in combinatorics, 1989 (Norwich, 1989), pp. 148–188, London Mathematical Society Lecture Note Series **141**, Cambridge Univ. Press, Cambridge (1989)
30. McKay, B. D., Wormald, N. C.: Asymptotic enumeration by degree sequence of graphs of high degree. European Journal of Combinatorics **11**, no. 6, 565–580 (1990)
31. Nesterov, Yu., Nemirovskii, A.: Interior-Point Polynomial Algorithms in Convex Programming. SIAM Studies in Applied Mathematics, **13**, Society for Industrial and Applied Mathematics (SIAM), Philadelphia, PA (1994)
32. Pak, I., Panova, G.: On the complexity of computing Kronecker coefficients. Computational Complexity, posted online 2 September 2015, DOI 10.1007/s00037-015-0109-4 (2015)
33. Papadimitriou, C. H., Steiglitz, K.: Combinatorial Optimization: Algorithms and Complexity. Corrected reprint of the 1982 original, Dover Publications, Inc., Mineola, NY (1998)
34. Shapiro, A.: Bounds on the number of integer points in a polytope via concentration estimates. Preprint `arXiv:1011.6252` (2010)
35. Tao, T., Vu, V.: Additive Combinatorics, Cambridge Studies in Advanced Mathematics, **105**, Cambridge University Press, Cambridge (2006)
36. Vempala, S.S.: Recent progress and open problems in algorithmic convex geometry. In: 30th International Conference on Foundations of Software Technology and Theoretical Computer Science, pp. 42–64, LIPIcs. Leibniz International Proceedings in Informatics, **8**, Schloss Dagstuhl. Leibniz-Zent. Inform., Wadern (2010)
37. Verdoolaege, S., Seghir, R., Beyls, K., Loechner, V., Bruynooghe, M.: Counting integer points in parametric polytopes using Barvinok's rational functions. Algorithmica **48**, no. 1, 37–66 (2007)

The Chain Rule Operator Equation for Polynomials and Entire Functions

Hermann König and Vitali Milman

Abstract After a short survey on operator functional equations we study the solutions of the chain rule operator equation

$$T(f \circ g) = (Tf) \circ g \cdot Tg$$

for maps T on spaces of polynomials and entire functions. In comparison with operators on C^k-spaces with $k \in \mathbb{N} \cup \{\infty\}$, which were studied before, there are entirely different solutions. However, these yield discontinuous maps T. Under a mild continuity assumption on T we determine all solutions of the chain rule equation on spaces of real or complex polynomials or analytic functions. The normalization condition $T(-2\,\mathrm{Id}) = -2$ yields $Tf = f'$ as the only solution.

1 Introduction and Results

Various fundamental operations in analysis and geometry such as derivatives, multiplicative maps, the Fourier transform or the duality of convex bodies essentially may be characterized by very simple properties or by operator equations on classical function spaces. We concentrate in this paper on characterizations of the derivative by the chain rule. To put things into perspective, we first mention a few recent results characterizing other classical operations in analysis. Let us start with multiplicative maps. On the real line, we have the standard

Lemma 1 *Assume that* $K : \mathbb{R} \to \mathbb{R}$ *is measurable, not identically zero and* multiplicative *in the sense that* $K(uv) = K(u)K(v)$ *for all* $u, v \in \mathbb{R}$. *Then there exists some* $p \in \mathbb{R}$ *such that either* $K(u) = |u|^p$ *or* $K(u) = |u|^p \mathrm{sgn}\,(u)$ *for all* $u \in \mathbb{R}$. *If* K *is continuous at* 0, *in the first case* $p \geq 0$, *in the second one* $p > 0$.

H. König (✉)
Mathematisches Seminar, Universität Kiel, 24098 Kiel, Germany
e-mail: hkoenig@math.uni-kiel.de

V. Milman
School of Mathematical Sciences, Tel Aviv University, Ramat Aviv, Tel Aviv 69978, Israel
e-mail: milman@post.tau.ac.il

E. Carlen et al. (eds.), *Convexity and Concentration*, The IMA Volumes
in Mathematics and its Applications 161, DOI 10.1007/978-1-4939-7005-6_21

The *additive* version of this lemma, i.e. that measurable additive functions $L : \mathbb{R} \to \mathbb{R}$, $L(u + v) = L(u) + L(v)$ are linear, is due to Banach [B] and Sierpinski [S]; the reduction to this case using logarithms is straightforward, cf. Aczél [A] . A beautiful proof of the result of [B] and [S] was given by Alexiewicz and Orlicz [AO]. A classical result of Milgram [M] extends Lemma 1 to bijective transformations on spaces of continuous functions:

Theorem 2 *Let M be a real topological manifold and $C(M)$ denote the space of real valued continuous functions on M . Assume that $T : C(M) \to C(M)$ is bijective and* multiplicative, *i.e.*

$$T(f \cdot g) = Tf \cdot Tg \quad ; \quad f, g \in C(M) .$$

Then there is a continuous function $p : M \to \mathbb{R}_+$ and a homeomorphism $v : M \to M$ such that

$$(Tf)(x) = |f(v(x))|^{p(x)} \operatorname{sgn}(f(v(x))) \; ; \; f \in C(M), \; x \in M .$$

Hence, up to homeomorphic change of variables in M, T is of power type. The analogue of Theorem 2 for C^k-functions when $k \in \mathbb{N}$ is much more recent and due to Mrčun and Šemrl [MS]. An alternative proof which works also for $k = \infty$ was given by Artstein-Avidan, Faifman and Milman [AFM] who also extended the result to complex-valued functions. They used ideas of Mrčun [Mr]. In this case $p = 1$ and T turns out to be linear and continuous:

Theorem 3 *Let $k \in \mathbb{N} \cup \{\infty\}$ and M be a C^k-manifold. Assume that $T : C^k(M) \to C^k(M)$ is a bijective and multiplicative operator,*

$$T(f \cdot g) = Tf \cdot Tg \; ; \; f, g \in C^k(M) .$$

Then there is a C^k-diffeomorphism $v : M \to M$ such that

$$(Tf)(x) = f(v(x)) \; ; \; f \in C^k(M) \; , \; x \in M .$$

On the Schwartz space $\mathcal{S}(\mathbb{R}^n)$ of rapidly decreasing functions $f : \mathbb{R}^n \to \mathbb{C}$, there are two natural multiplications: the pointwise multiplication and the convolution. The Fourier transform $\mathcal{F}$ is bijective on $\mathcal{S}(\mathbb{R}^n)$ and exchanges these multiplications with one another. It was shown in [AFM], following work by Alesker, Artstein-Avidan, Faifman and Milman [AAFM], that it is the only operator with these properties, up to linear maps:

Theorem 4 *Assume that $T : \mathcal{S}(\mathbb{R}^n) \to \mathcal{S}(\mathbb{R}^n)$ is bijective and exchanges the usual multiplication with the convolution, in the sense that*

$$T(f \cdot g) = Tf * Tg \quad ; f, g \in \mathcal{S}(\mathbb{R}^n)$$

and

$$T(f * g) = Tf \cdot Tg \quad ; f, g \in \mathcal{S}(\mathbb{R}^n) \ .$$

Then there is a linear map $v : \mathbb{R}^n \to \mathbb{R}^n$ *with* $|\det v| = 1$ *such that either* $Tf = \mathcal{F}(f \circ v)$ *or* $Tf = \overline{\mathcal{F}(f \circ v)}$, *where* $\mathcal{F}$ *denotes the Fourier transform.*

Theorem 4 is a direct consequence of the analogue of Theorem 3 for $\mathcal{S}(\mathbb{R}^n)$ instead of $C^\infty(\mathbb{R}^n)$: Both $\mathcal{F} \circ T$ and $T \circ \mathcal{F}$ will be multiplicative with respect to pointwise multiplication so that by Theorem 3 for a suitable C^∞-homeomorphism v which in this case turns out to be a linear map $v : \mathbb{R}^n \to \mathbb{R}^n$ with $|\det v| = 1$, $(\mathcal{F} \circ T)(f)(x) = f(v(x))$ or $\overline{f(v(x))}$, the latter being a second possibility for complex-valued functions.

We now turn to characterizations of the derivative. Its action on products of real-valued C^1-functions obviously is given by the Leibniz rule. We have the following characterization of the solutions of the Leibniz rule:

Theorem 5 *Let* $k \in \mathbb{N} \cup \{0\}$. *Assume that* $T : C^k(\mathbb{R}) \to C(\mathbb{R})$ *is a map satisfying the Leibniz rule*

$$T(f \cdot g) = Tf \cdot g + f \cdot Tg \quad ; f, g \in C^k(\mathbb{R}) \ .$$

Then there are continuous functions $a, b \in C(\mathbb{R})$ *such that, if* $k \in \mathbb{N}$,

$$Tf = a \cdot f' + b \cdot f \ \ln |f| \quad ; f \in C^k(\mathbb{R}) \ .$$

If $k = 0$, *the only possibility is* $Tf = b \cdot f \ln |f|$.

The case $k = 0$ is due to Goldmann-Šemrl [GS], the case $k \in \mathbb{N}$ was shown in König-Milman [KM]. Suitable initial conditions like $T(c) = 0$ for a suitable constant function with value $c \notin \{0, 1\}$ imply that $b = 0$, and hence that $Tf = a \cdot f'$ is essentially the derivative, up to multiplication by a continuous function.

The main topic of the present paper is to what extent the chain rule characterizes the derivative on various classical spaces of analysis. To formulate the results precisely, we introduce some notations. Let $X = C^k(\mathbb{R})$, $Y = C(\mathbb{R})$, $k \in \mathbb{N} \cup \{\infty\}$ and $T : X \to Y$ satisfy the chain rule operator equation

$$T(f \circ g) = (Tf) \circ g \cdot Tg \quad ; \quad f, g \in X \ . \tag{1}$$

In [AKM] we found all solutions of (1) under a mild condition of non-degeneration of T. They are of the form

$$Tf = \frac{H \circ f}{H} \, |f'|^p \, \{\operatorname{sgn} f'\} \ , \tag{2}$$

where $p \geq 0$, $H \in C(\mathbb{R})$, $H > 0$ and the term $\{\operatorname{sgn} f'\}$ might appear or might not appear. No continuity was assumed in [AKM]; however, it is a consequence of formula (2). The normalization $T(-2\,\mathrm{Id}) = -2$ leads to $Tf = f'$ being the only solution of (1).

The method of proof in [AKM] does not work for maps T on polynomial spaces. Also, we could not determine the solutions in the complex case for maps T on spaces of entire functions. We now consider equation (1) in these cases: For $\mathbb{K} \in \{\mathbb{R}, \mathbb{C}\}$, denote by $\mathcal{P} = \mathcal{P}(\mathbb{K})$ the space of polynomials with coefficients in $\mathbb{K}$, by $\mathcal{P}_n = \mathcal{P}_n(\mathbb{K})$ the subspace of polynomials of degree at most n, $n \in \mathbb{N}$, by $\mathcal{C} = \mathcal{C}(\mathbb{K})$ the space of continuous functions from $\mathbb{K}$ to $\mathbb{K}$ and by $\mathcal{E} = \mathcal{E}(\mathbb{K})$ the space of real-analytic functions ($\mathbb{K} = \mathbb{R}$) or entire functions ($\mathbb{K} = \mathbb{C}$). Let $X \in \{\mathcal{P}, \mathcal{P}_n, \mathcal{E}\}$ and $Y \in \{\mathcal{C}, \mathcal{P}, \mathcal{E}\}$ and $T : X \to Y$ be an operator satisfying the chain rule operator equation (1). For $X = \mathcal{P}$ and $Y = \mathcal{C}$ there are solutions of (1) which are very different from the solutions (2) for $X = C^k(\mathbb{R})$:

Example For each prime number $p \in \mathbb{N}$, choose an arbitrary value $c_p \geq 0$ with $c_q > 0$ for at least one prime number q and define $T : \mathcal{P} \to \mathcal{C}$ by

$$Tf = \prod_{j=1}^{r} c_{p_j}$$

(constant function) if $f \in \mathcal{P}$ is a polynomial of degree $\deg f = \prod_{j=1}^{r} p_j$ (possible repetition of primes). Since $\deg(f \circ g) = \deg f \cdot \deg g$ for $f, g \in \mathcal{P}$, T satisfies (1). This solution clearly is not continuous in any reasonable sense since polynomials of degree n may converge, e.g., uniformly on compact sets to lower degree polynomials. We may multiply these solutions by those of (2)

$$Tf = \prod_{j=1}^{r} c_{p_j} \, \frac{H \circ f}{H} \, |f'|^p \, (\operatorname{sgn} f')^m$$

to get further solutions of the chain rule equation (1). We have not been able to determine whether this yields the general solution of (1) for maps $T : \mathcal{P} \to \mathcal{C}$. This leads to number theoretic questions.

However, we found the general solution of (1) under a mild continuity assumption on $T : X \to Y$ for $X \in \{\mathcal{P}, \mathcal{P}_n, \mathcal{E}\}$ and $Y \in \{\mathcal{C}, \mathcal{P}, \mathcal{E}\}$. We call the operator T *pointwise continuous* if for any sequence of functions $f_n \in X$ converging uniformly on compact subsets of $\mathbb{K}$ to a function $f \in X$, we have pointwise convergence at $0 \in \mathbb{K}$, i.e. $\lim_{n\to\infty}(Tf_n)(0) = (Tf)(0)$ in $\mathbb{K}$. Under this assumption, the non-zero solutions of (1) are essentially of the same form as (2), where -depending on Y- possibly $p \in \mathbb{N}_0$ or restrictions on H apply. In this and the following we use the notation $\operatorname{sgn} \xi = \frac{\xi}{|\xi|}$ for $\xi \in \mathbb{K} \setminus \{0\}$.

Theorem 6 *Let $\mathbb{K} \in \{\mathbb{R}, \mathbb{C}\}$, $X \in \{\mathcal{P}, \mathcal{P}_n\}$ and $Y \in \{\mathcal{C}, \mathcal{P}\}$ and suppose that $T : X \to Y$, $T \neq 0$ satisfies the chain rule operator equation*

$$T(f \circ g) = (Tf) \circ g \cdot Tg \quad ; \quad f, g, f \circ g \in X \tag{1}$$

and is pointwise continuous. Then there is a continuous nowhere vanishing function $H \in \mathcal{C}(\mathbb{K})$ and there are $p \in \mathbb{K}$ with $Re(p) \geq 0$ and $m \in \mathbb{Z}$ such that

$$Tf = \frac{H \circ f}{H} \, |f'|^p \, (\operatorname{sgn} f')^m \, . \tag{3}$$

For $\mathbb{K} = \mathbb{R}$, $m \in \{0, 1\}$ suffices and $H > 0$. For $p = 0$, only $m = 0$ yields a solution T with range in $\mathcal{C}(\mathbb{K})$. If $Y = \mathcal{P}$, H is constant and $p = m \in \mathbb{N}_0$ is a non-negative integer, with T being of the form $Tf = f'^m$.

Theorem 7 $\mathbb{K} \in \{\mathbb{R}, \mathbb{C}\}$ *and $X = Y = \mathcal{E}(\mathbb{K})$. Suppose that $T : X \to Y$, $T \neq 0$ satisfies the chain rule operator equation*

$$T(f \circ g) = (Tf) \circ g \cdot Tg \quad ; \quad f, g \in X \tag{1}$$

and is pointwise continuous. Then there is a real-analytic ($\mathbb{K} = \mathbb{R}$) or entire function ($\mathbb{K} = \mathbb{C}$) $h \in \mathcal{E}(\mathbb{K})$ and there is $m \in \mathbb{N}_0$ such that

$$Tf = \exp(h \circ f - h) \, f'^m \, . \tag{4}$$

The functions H and h of Theorems 6 and 7 are related by $H = \exp(h)$. In both theorems, $Tf = f'$ is the only solution of (1) satisfying the additional condition $T(-2\,\mathrm{Id}) = -2$. It is not clear to us whether Theorem 7 holds without the continuity assumption on T.

2 Proof of Theorems 6 and 7

In the proofs of Theorem 6 and 7 we need

Lemma 8 *Let $\mathbb{K} \in \{\mathbb{R}, \mathbb{C}\}$ and $\phi : \mathbb{K} \to \mathbb{K}$ be a non-zero continuous multiplicative function*

$$\phi(\xi\eta) = \phi(\xi)\phi(\eta) \quad ; \quad \xi, \eta \in \mathbb{K} \, .$$

Then $\phi(0) = 0$ and there are $p \in \mathbb{K}$ with $Re(p) \geq 0$ and $m \in \mathbb{Z}$ such that

$$\phi(\xi) = |\xi|^p \, (\operatorname{sgn} \xi)^m \quad ; \quad \operatorname{sgn} \xi = \frac{\xi}{|\xi|} \; (\xi \neq 0) \, .$$

For $Re(p) = 0$, only $m = 0$ is allowed to guarantee the continuity of ϕ. In the real case $m \in \{0, 1\}$.

Remark $Re(p) \geq 0$ is needed since ϕ is bounded near zero.

Proof In the real case this is well known, cf. Aczél [A], Theorem 3 of 2.1.2 or [AKM], Lemma 13. Consider $\mathbb{K} = \mathbb{C}$ and define a continuous function $f : \mathbb{C} \to \mathbb{C}$ by $f(z) := \phi(e^z)$, $z \in \mathbb{C}$. Then $f(z+w) = f(z)f(w)$ for all $z, w \in \mathbb{C}$. By Theorem 3 of Section 5.1.1 of Aczél [A] there are $\alpha, \beta \in \mathbb{C}$ such that $f(z) = e^{\alpha z + \beta \bar{z}}$ which implies $\phi(\xi) = \xi^\alpha \bar{\xi}^\beta$. Writing $\xi = |\xi| \operatorname{sgn} \xi$, we get

$$\phi(\xi) = |\xi|^{\alpha+\beta} (\operatorname{sgn} \xi)^{\alpha-\beta} = |\xi|^p (\operatorname{sgn} \xi)^q$$

with $p = \alpha + \beta, q = \alpha - \beta \in \mathbb{C}$. Writing $\operatorname{sgn} \xi = e^{i\theta}$ with $\theta \in [0, 2\pi]$, the continuity of ϕ requires $e^0 = e^{i2\pi q}$, i.e. $q = m \in \mathbb{Z}$. The boundedness of ϕ near 0 implies $Re(p) \geq 0$. Clearly, for $\mathbb{K} = \mathbb{R}$ only $m \in \{0, 1\}$ are needed, giving the two solutions $|\xi|^p$ and $|\xi|^p (\operatorname{sgn} \xi)$ for $p > 0$ (and $\mathbb{1}$ for $p = 0$). □

Proof of Theorem 6. (a) For $n \in \mathbb{N}$, let $T : \mathcal{P}_n \to \mathcal{C}$ be a non-zero map satisfying the chain rule operator equation (1). Since $T \neq 0$, there is $g \in \mathcal{P}_n$ and $x_1 \in \mathbb{K}$ such that $Tg(x_1) \neq 0$. Given any $x_0 \in \mathbb{K}$, consider the shift map $S(x) := x + x_1 - x_0$, $S \in \mathcal{P}_1 \subset \mathcal{P}_n$ and let $f := g \circ S$. Then by (1), $0 \neq Tg(x_1) = T(f \circ S^{-1})(x_1) = Tf(x_0)T(S^{-1})(x_1)$. Hence $Tf(x_0) \neq 0$. Again by (1), $Tf = T(f \circ \mathrm{Id}) = (Tf) \circ \mathrm{Id} \cdot T(\mathrm{Id}) = Tf \cdot T(\mathrm{Id}), f \in \mathcal{P}_n$. Therefore $T(\mathrm{Id})(x_0) = 1$ and $T(\mathrm{Id}) = \mathbb{1}$ is the function identically equal to 1. For $x \in \mathbb{K}$ consider again a shift map $S_x \in \mathcal{P}_1 \subset \mathcal{P}_n$ given by $S_x(y) = x + y$, $y \in \mathbb{K}$. Then by (1)

$$1 = T(\mathrm{Id}) = T(S_{-x} \circ S_x) = T(S_{-x}) \circ S_x \cdot T(S_x) .$$

Hence for any $y \in \mathbb{K}$, $T(S_x)(y) \neq 0$. In particular, $T(S_x)(0) \neq 0$ for all $x \in \mathbb{K}$. For any $f \in \mathcal{P}_n$, again by (1),

$$T(f \circ S_x) = (Tf) \circ S_x \cdot T(S_x) \quad , \quad T(f \circ S_x)(0) = (Tf)(x) \cdot T(S_x)(0) .$$

We conclude that for any $x \in \mathbb{K}$

$$Tf(x) = \frac{T(f \circ S_x)(0)}{T(S_x)(0)} . \tag{5}$$

Equation (5) and the pointwise continuity of T imply for a sequence $f_n \in X$ converging uniformly on compacta to $f \in X$, we have $\lim_{n\to\infty}(Tf_n)(x) - (Tf)(x)$ for any $x \in \mathbb{K}$, and not only for $x = 0$. By (5) it suffices to determine the values of Tf at 0 for any $f \in \mathcal{P}_n$. Since for any $f \in \mathcal{P}_n, f(x) = \sum_{j=0}^n \frac{f^{(j)}(0)}{j!} x^j$ is determined by the function and derivative values $(f^{(j)}(0))_{0 \leq j \leq n}$, $(Tf)(0)$ is a function of these values. Hence there is $F_n : \mathbb{K}^{n+1} \to \mathbb{K}$ such that

$$(Tf)(0) = F_n(f(0), f'(0), \cdots, f^{(n)}(0)) . \tag{6}$$

Since $(f \circ S_x)^{(j)} = f^{(j)} \circ S_x$ in view of $S_x' = \mathbb{1}$, (5) and (6) imply that

$$(Tf)(x) = \frac{F_n(f(x), f'(x), \cdots, f^{(n)}(x))}{F_n(x, 1, 0, \cdots, 0)} \tag{7}$$

with $F_n(x, 1, 0, \cdots, 0) = T(S_x)(0) \neq 0$ for any $x \in \mathbb{K}$.

(b) Fix $x \in \mathbb{K}$ and define $\tilde{F}_{n,x} : \mathbb{K}^n \to \mathbb{K}$ by

$$\tilde{F}_{n,x}(\xi_1, \cdots, \xi_n) := \frac{F_n(x, \xi_1, \cdots, \xi_n)}{F_n(x, 1, \cdots, 0)} . \tag{8}$$

For any $(\eta_1, \cdots, \eta_n) \in \mathbb{K}^n$ there is a polynomial $g \in \mathcal{P}_n$ with $g(x) = x$ and $g^{(j)}(x) = \eta_j$ for all $j \in \{1, \cdots, n\}$. For any $\xi_1 \in \mathbb{K}$, define $f \in \mathcal{P}_1 \subset \mathcal{P}_n$ by $f(y) := \xi_1(y-x)+x$. Then $(f \circ g)^{(n)}(x) = \xi_1 \eta_n$ and $(g \circ f)^{(n)}(x) = \xi_1^n \eta_n$. Thus by (7) and (8)

$$\begin{aligned}
\tilde{F}_{n,x}(\xi_1\eta_1, \xi_1\eta_2, \cdots, \xi_1\eta_n) &= \tilde{F}_{n,x}((f \circ g)'(x), \cdots, (f \circ g)^{(n)}(x)) \\
&= T(f \circ g)(x) = (Tf)(x) \cdot (Tg)(x) = (Tg)(x) \cdot (Tf)(x) \\
&= T(g \circ f)(x) = \tilde{F}_{n,x}(\xi_1\eta_1, \xi_1^2\eta_2, \cdots, \xi_1^n\eta_n) . \qquad (9)
\end{aligned}$$

For any $(t_1, \cdots, t_n) \in \mathbb{K}^n$ and $\alpha \in \mathbb{K}$, $\alpha \neq 0$, we may choose $(\eta_1, \cdots, \eta_n) \in \mathbb{K}^n$ with $\eta_i = t_i/\alpha$, $i \in \{1, \cdots, n\}$. Then by (9) with $\xi_1 = \alpha$

$$\tilde{F}_{n,x}(t_1, \cdots, t_n) = \tilde{F}_{n,x}(t_1, \alpha t_2, \alpha^2 t_3, \cdots, \alpha^{n-1} t_n) . \tag{10}$$

Fixing $t_1 \in \mathbb{K}$ we show that $G : \mathbb{K}^{n-1} \to \mathbb{K}$ defined by $G(t_2, \cdots, t_n) := \tilde{F}_{n,x}(t_1, t_2, t_3, \cdots, t_n)$ is continuous at zero: If $t^{(m)} = (t_2^{(m)}, \cdots, t_n^{(m)}) \in \mathbb{K}^{n-1}$, $(m \in \mathbb{N})$, is a sequence converging to $0 \in \mathbb{K}^{n-1}$, choose polynomials $f_m \in \mathcal{P}_n$ with $f_m(x) = x$, $f_m'(x) = t_1$ and $f_m^{(j)}(x) = t_j^{(m)}$ for $j \in \{2, \cdots, n\}$, $m \in \mathbb{N}$. Clearly f_m converges uniformly on compact sets to f where $f(y) = t_1(y - x) + x$. By the assumption of pointwise continuity of T, (7) and (8)

$$\begin{aligned}
G(t_2^{(m)}, \cdots, t_n^{(m)}) &= \tilde{F}_{n,x}(t_1, t_2^{(m)}, \cdots, t_n^{(m)}) = (Tf_m)(x) \\
&\to (Tf)(x) = \tilde{F}_{n,x}(t_1, 0, \cdots, 0) = G(0, \cdots, 0) ,
\end{aligned}$$

i.e., G is continuous at $0 \in \mathbb{K}^{n-1}$. Thus, letting $\alpha \to 0$ in (10), we find

$$\tilde{F}_{n,x}(t_1, t_2, \cdots, t_n) = \lim_{\alpha \to 0} \tilde{F}_{n,x}(t_1, \alpha t_2, \cdots, \alpha^{n-1} t_n) = \tilde{F}_{n,x}(t_1, 0, \cdots, 0)$$

This shows for arbitrary $(t_1, \cdots, t_n) \in \mathbb{K}^n$ that

$$\tilde{F}_{n,x}(t_1, t_2, \cdots, t_n) = \tilde{F}_{n,x}(t_1, 0, \cdots, 0) , \tag{11}$$

i.e., $\tilde{F}_{n,x}$ does not depend on the variables $(t_2, \cdots, t_n) \in \mathbb{K}^{n-1}$.

(c) For any $f \in \mathcal{P}_n$ with $f(x) = x, f'(x) = \xi_1$ we have by (7), (8), and (11)

$$Tf(x) = \tilde{F}_{n,x}(\xi_1, 0, \cdots, 0) = \frac{F_n(x, \xi_1, 0, \cdots, 0)}{F_n(x, 1, 0, \cdots, 0)} =: \phi(x, \xi_1) . \tag{12}$$

If $g \in \mathcal{P}_n$ satisfies $g(x) = x$, $g'(x) = \eta_1$, we have by (1)

$$\phi(x, \xi_1 \eta_1) = T(f \circ g)(x) = (Tf)(x) \cdot (Tg)(x) = \phi(x, \xi_1)\phi(x, \eta_1) .$$

Therefore $\phi(x, \cdot)$ is multiplicative on $\mathbb{K}$ for every fixed x. It is also continuous: for $\xi_1^{(m)} \to \xi_1$ in $\mathbb{K}$ put $f_m(y) := \xi_1^{(m)}(y - x) + x, f(y) := \xi_1(y - x) + x$. Then $f_m \to f$ converges uniformly on compact sets and

$$\phi(x, \xi_1^{(m)}) = (Tf_m)(x) \to (Tf)(x) = \phi(x, \xi_1)$$

by the pointwise continuity of T. By Lemma 8, $\phi(x, 0) = 0$ and there are $m(x) \in \mathbb{Z}$ and $p(x) \in \mathbb{K}$ with $Re(p(x)) \geq 0$ such that

$$\phi(x, \xi_1) = |\xi_1|^{p(x)} (\operatorname{sgn} \xi_1)^{m(x)} , \ \operatorname{sgn} \xi_1 = \frac{\xi_1}{|\xi_1|} , \ \xi_1 \neq 0 \tag{13}$$

with $m(x) = 0$ if $Re(p(x)) = 0$ and $m(x) \in \{0, 1\}$ if $\mathbb{K} = \mathbb{R}$.

Let $H(x) := T(S_x)(0) = F_n(x, 1, 0, \cdots, 0)$. H never vanishes and is continuous since T is pointwise continuous. For any $f \in \mathcal{P}_n$, by (7), (11), (12) and (13)

$$\begin{aligned} Tf(x) &= \frac{F_n(f(x), f'(x), 0, \cdots, 0)}{F_n(x, 1, 0, \cdots, 0)} = \frac{\phi(f(x), f'(x))H(f(x))}{H(x)} \\ &= \frac{H \circ f(x)}{H(x)} |f'(x)|^{p(f(x))} (\operatorname{sgn} f'(x))^{m(f(x))} . \end{aligned} \tag{14}$$

Choosing $f(x) = 2x$, we find that p is a continuous function since Tf and H are continuous. Actually, we show that p is a constant function: Choose arbitrary $x, y, z \in \mathbb{K}$ and any functions $f, g \in \mathcal{P}_n$ with $g(x) = y, f(y) = z$. Then by (14) and (1)

$$|f'(y)g'(x)|^{p(yz)} (\operatorname{sgn} f'(y)g'(x))^{m(yz)} = |f'(y)|^{p(z)} (\operatorname{sgn} f'(y))^{m(z)} |g'(x)|^{p(y)} (\operatorname{sgn} g'(x))^{m(y)} .$$

Applying this first to polynomials with $f'(y) > 0$, $g'(x) > 0$ but otherwise arbitrary, we find that $p(yz) = p(z) = p(y) =: p$ for all $y, z \in \mathbb{K}$, i.e. p is constant. Using $(\operatorname{sgn} f'(y))^{m(yz)} = (\operatorname{sgn} f'(y))^{m(z)}$ for $f \in \mathcal{P}_n$ with arbitrary y, z and $\operatorname{sgn} f'(y)$, we also find that $m(yz) = m(z) = m(y) =: m \in \mathbb{Z}$ is constant. With $p = p(f(x))$ and $m = m(f(x))$, (14) gives the general solution for $X = \mathcal{P}_n$, $Y = \mathcal{C}$, both for $\mathbb{K} = \mathbb{R}$ and $\mathbb{K} = \mathbb{C}$.

(d) Since (14) is independent of $n \in \mathbb{N}$, this is also the general solution for $X = \mathcal{P}$ and $Y = \mathcal{C}$. In the case that all functions in the range of T are polynomials, $Y = \mathcal{P}$, all functions

$$Tf = \frac{H \circ f}{H} \, |f'|^p \, (\text{sgn} f')^m \ ; \ m \in \mathbb{Z}, p \in \mathbb{K}, Re(p) \geq 0, f \in \mathcal{P}$$

have to be polynomials. For $f(x) = \frac{1}{2}x^2$ with $f'(x) = x$ this means that

$Tf(x) = \frac{H(\frac{1}{2}x^2)}{H(x)} \, |x|^p \, (\text{sgn}\, x)^m$, $Tf \in \mathcal{P}$. For $p = 0$ also $m = 0$ and $Tf = \frac{H \circ f}{H}$. For $p > 0$, Tf has a zero of order p in $x_0 = 0$. Since Tf is a polynomial, it follows that $p \in \mathbb{N}$ is a positive integer and $Tf(x) = x^p g(x)$ with $g \in \mathcal{P}$, $g(0) \neq 0$. This implies that $m \in \mathbb{Z}$ has to be such that $x^p = |x|^p \, (\text{sgn}\, x)^m$, i.e. that $Tf = \frac{H \circ f}{H} f'^p \in \mathcal{P}$ for all $f \in \mathcal{P}$ with $p \in \mathbb{N}_0$.

Applying this to linear functions $f(x) = ax + b, f(y) = \frac{1}{a}y - \frac{b}{a} = x$, we find that $p(x) := \frac{H(ax+b)}{H(x)}$ and $\frac{H(x)}{H(ax+b)} = \frac{1}{p(x)}$ are polynomials in x. Therefore $\frac{H(ax+b)}{H(x)} =: c_{a,b}$ is constant in x for any fixed values $a, b \in \mathbb{K}$. In particular, $\frac{H(2x)}{H(x)} = \frac{H(2*0)}{H(0)} = c_{2,0} = 1$ and $\frac{H(x+b)}{H(x)} = c_{1,b}$ for any fixed $b \in \mathbb{K}$ and all $x \in \mathbb{K}$. We find

$$H(2x + 2b) = H(x + b) = c_{1,b}H(x) = c_{1,b}H(2x) = H(2x + b) \ ; \ x, b \in \mathbb{K} \, .$$

Therefore $H(y + d) = H(y)$ for all $y, b \in \mathbb{K}$. Hence H is constant and $\frac{H \circ f}{H} = 1$ for all $f \in \mathcal{P}$. We conclude that $Tf = (f')^p, p \in \mathbb{N}_0$. □

Proof of Theorem 7. Since $\mathcal{P}(\mathbb{K}) \subset X = Y = \mathcal{E}(\mathbb{K}) \subset C(\mathbb{K})$, Theorem 6 yields that T, restricted to $\mathcal{P}(\mathbb{K})$, has the form

$$Tf = \frac{H \circ f}{H} \, |f'|^p \, (\text{sgn} f')^m \quad ; \quad f \in \mathcal{P}(\mathbb{K}) \, , \tag{15}$$

with $m \in \mathbb{Z}$, $p \in \mathbb{K}$, $Re(p) \geq 0$. We also know that $x \to H(x) = T(S_x)(0)$ is continuous by the pointwise continuity of T. Let $c \in \mathbb{K}$, $c \neq 0$ be arbitrary. Applying (15) to $f(z) = cz$ and using that $Tf \in \mathcal{E}(\mathbb{K})$, we get that $z \to \frac{H(cz)}{H(z)}$ is in $\mathcal{E}(\mathbb{K})$, i.e. analytic on $\mathbb{K}$. Since H is nowhere zero, there exists an analytic function $k(c, \cdot) \in \mathcal{E}(\mathbb{K})$ such that $\frac{H(cz)}{H(z)} = \exp(k(c, z))$, with $k(c, 0) = 0$. For $c, d \in \mathbb{K}$ we find

$$\exp(k(cd, z)) = \frac{H(cdz)}{H(z)} = \frac{H(cdz)}{H(dz)} \frac{H(dz)}{H(z)} = \exp(k(c, dz) + k(d, z)) \, ,$$

hence $k(cd, z) = k(c, dz) + k(d, z)$. In particular, for $z = 1$, $k(c, d) = k(cd, 1) - k(d, 1)$. Let $h(d) := k(d, 1)$ for $d \neq 0$. Then $k(c, d) = h(cd) - h(d)$, and with d replaced by z, $k(c, z) = h(cz) - h(z)$. Since H is continuous, k is continuous as a function of both variables. Therefore, $\lim_{c \to 0} k(c, z) = \lim_{c \to 0} h(cz) - h(z) =: h(0) - h(z)$ exists z-uniformly on compact sets. Thus $k(c, \cdot) \in \mathcal{E}(\mathbb{K})$ for all $c \in \mathbb{K}$ implies that $h \in \mathcal{E}(\mathbb{K})$. For $w, z \in \mathbb{K} \setminus \{0\}$, define $c \in \mathbb{K}$ by $w = cz$. Then

$$\frac{H(w)}{H(z)} = \exp(k(c, z)) = \exp(h(w) - h(z)) \, .$$

This extends by continuity to $w = 0$ or $z = 0$. Hence we have for all $f \in \mathcal{P}(\mathbb{K})$ that $\frac{H \circ f}{H} = \exp(h \circ f - h)$. Since Tf, $\frac{H}{H \circ f}$ are in $\mathcal{E}(\mathbb{K})$, also $|f'|^p \, (\mathrm{sgn} f')^m$ has to be real-analytic ($\mathbb{K} = \mathbb{R}$) or analytic ($\mathbb{K} = \mathbb{C}$) for all polynomials f, requiring that $p = m \in \mathbb{N}_0$, taking into account also that $Re(p) \geq 0$ and $m \in \mathbb{Z}$. Therefore

$$Tf = \exp(h \circ f - h) \, f'^m \quad , \quad f \in \mathcal{P}(\mathbb{K}) \tag{16}$$

with $m \in \mathbb{N}_0$. Given any function $f \in \mathcal{E}(\mathbb{K})$, its n-th order Taylor polynomials $p_n(f) \in \mathcal{P}(\mathbb{K})$ converge uniformly on compacta to f. By the assumption of pointwise continuity of T, we have $\lim_{n \to \infty} T(p_n(f))(z) = Tf(z)$ for any $z \in \mathbb{K}$. Further $\lim_{n \to \infty} h \circ p_n(f)(z) = h \circ f(z)$ and $\lim_{n \to \infty} p_n(f)'(z) = f'(z)$. Therefore (16) holds for all $f \in \mathcal{E}(\mathbb{K})$.

It is clear in both Theorem 6 and Theorem 7 that the condition $T(-2 \, \mathrm{Id}) = -2$ implies $p = m = 1$ and that H and h are constant, i.e. $Tf = f'$. □

Acknowledgements We would like to express our thanks to the referee whose suggestions and comments made the paper more readable and, in particular, shortened the proof of Theorem 6. Further, we would like to thank Pawel Domanski for discussions concerning the chain rule equation for entire functions.

Hermann König was supported in part by Minerva.

Vitali Milman was supported in part by the Alexander von Humboldt Foundation, by Minerva, by ISF grant 826/13 and by BSF grant 0361-4561.

References

[A] J. Aczél; Lectures on functional equations and their applications, Academic Press, 1966.

[AAFM] S. Alesker, S. Artstein-Avidan, D. Faifman, V. Milman; A characterization of product preserving maps with applications to a characterization of the Fourier transform, Illinois J. Math. 54 (2010), 1115–1132.

[AFM] S. Artstein-Avidan, D. Faifman, V. Milman; On multiplicative maps of continuous and smooth functions, Geometric aspects of functional analysis, Lecture Notes in Math. 2050, 35–59, Springer, Heidelberg, 2012.

[AKM] S. Artstein-Avidan, H. König, V. Milman; The chain rule as a functional equation, J. Funct. Anal. 259 (2010), 2999–3024.

[AO] A. Alexiewicz, W. Orlicz; Remarque sur l'équation fonctionelle $f(x+y) = f(x) + f(y)$, Fund. Math. 33 (1945), 314–315.

[B] S. Banach; Sur l'équation fonctionelle $f(x + y) = f(x) + f(y)$, Fund. Math. 1 (1920), 123–124.

[GS] H. Goldmann, P. Šemrl; Multiplicative derivations on $C(X)$, Monatsh. Math. 121 (1996), 189–197.

[KM] H. König, V. Milman; Characterizing the derivative and the entropy function by the Leibniz rule, with an appendix by D. Faifman, J. Funct. Anal. 261 (2011), 1325–1344.

[M] A.N. Milgram; Multiplicative semigroups of continuous functions, Duke Math. J. 16 (1940), 377–383.

[Mr] J. Mrčun; On isomorphisms of algebras of smooth functions, Proc. Amer. Math. Soc. 133 (2005), 3109–3113.
[MS] J. Mrčun, P. Šemrl; Multplicative bijections between algebras of differentiable functions, Ann. Acad. Sci. Fenn. Math. 32 (2007), 471–480.
[S] W. Sierpinski; Sur l'équation fonctionelle $f(x+y) = f(x) + f(y)$, Fund. Math. 1 (1920), 116–122.

ISBN 978-1493970049 -9

9 781493 970049

Printed in Japan

H0050283